Introduction to Numerical Ordinary and Partial Differential Equations Using MATLAB®

Contents

---

[1] An asterisk that precedes a section indicates that the section may be skipped without a significant loss of continuity to the main development of the text.

PREFACE

MATLAB® is an abbreviation for MATrix LABoratory and it is ideally suited for computations involving matrices. Since all of the sciences routinely collect data in the form of (spreadsheet) matrices, MATLAB turns out to be particularly suitable for the analysis of mathematical problems in an assortment of fields. MATLAB is very easy to learn how to use and has tremendous graphical capabilities. Many schools have site licenses and student editions of the software are available at special affordable rates. MATLAB is perhaps the most commonly used mathematical software in the general scientific fields (from biology, physics, and engineering to fields like business and finance) and is used by numerous university in mathematics departments.

MATERIAL

The book is an undergraduate-level textbook giving a thorough introduction to the various aspects of numerically solving problems involving differential equations, both partial (PDEs) and ordinary (ODEs). It is largely self-contained with the prerequisite of a basic course in single-variable calculus and it covers all of the needed topics from numerical analysis. For the material on partial differential equations, apart from the basic concept of a partial derivative, only certain portions rely on facts from multivariable calculus and these are not essential to the main development with the only exception being in the final chapter on the finite element method. The book is made up of the following three parts:

Part I: Introduction to MATLAB and Numerical Preliminaries (Chapters 1–7). This part introduces the reader to the MATLAB software and its graphical capabilities, and shows how to write programs with it. The needed numerical analysis preparation is also done here and there is a chapter on floating point arithmetic. The basic element in MATLAB is a matrix and MATLAB is very good at manipulating and working with them. As numerous methods for differential equations problems amount to a discretization into a matrix problem, MATLAB is an ideal tool for the subject. An extensive chapter is given on matrices and linear systems which integrates theory and applications with MATLAB's prowess.

Part II: Ordinary Differential Equations (Chapters 8–10). Chapter 8 gives an applications-based introduction to ordinary differential equations, and progressively introduces a plethora of numerical methods for solving initial value problems involving a single first order ODE. Applications include population dynamics and numerous problems in physics. The various numerical methods are compared and error analysis is done. Chapter 9 adapts the methods of the previous chapter for initial value problems of higher order and systems of ODEs. Applications that are extensively investigated include predator-prey problems, epidemiology models, chaos, and numerous physical problems. The geometric theory on topics such as phase-plane analysis, stability, and the Poincaré-Bendixson theorem is presented and corroborated with numerical experiments. Chapter 10 covers two-point boundary value problems for second-order ODEs. The very successful (linear and nonlinear) shooting methods are presented and advocated as the methods of choice for such problems. The chapter also includes sections on finite difference methods and Rayleigh-Ritz methods. These two methods are the one-dimensional analogues of the main methods that will be used for solving boundary value problems for PDE in Part III.

Part III: Partial Differential Equations (Chapters 11–13). After a brief section on the three-dimensional graphical capabilities of MATLAB, Chapter 11 introduces partial differential equations based on the model problem of heat flow and steady-state distribution. This model allows us to introduce many concepts of elliptic and parabolic PDEs. The remainder of this chapter focuses on finite difference methods for solving elliptic boundary value problems. Although the schemes for hyperbolic and parabolic problems are usually simpler to write down and use, elliptic problems are much more stable and so attention to stability issues can be deferred. All sorts of boundary conditions are considered and much theory (both mathematical and numerical) is presented and investigated. Chapter 12 begins with a discussion on hyperbolic PDE and the model wave equation. The remaining sections show to how use finite difference methods to solve well-posed problems involving both hyperbolic and parabolic PDEs. Finally, Chapter 13 gives an introduction to the finite element method (FEM). This method is much more versatile in dealing with irregular-shaped domains and various boundary conditions than are the finite difference methods, whose use is most often restricted to rectangular domains. The FEM is based on breaking the domain up into smaller pieces that can be of any shape. We mostly use triangular elements, since MATLAB has some nice tools to help us effectively triangulate a domain once we decide on a deployment of nodes. The techniques presented in this chapter will enable the reader to numerically solve any elliptic boundary value problem of the form:

$$\begin{cases} \text{(PDE)} & -\nabla\cdot(p\nabla u)+qu=f & \text{on } \Omega \\ \text{(BCs)} & u=g & \text{on } \Gamma_1 , \\ & \bar{n}\cdot\nabla u+ru=h & \text{on } \Gamma_2 \end{cases}$$

for which a solution exists. Here Ω is any domain in the plane whose boundary is made up of pieces determined by graphs of functions (simply or multiply connected), and Γ_1 and Γ_2 partition its boundary. Existence and uniqueness theorems are given that help to determine when such problems are well-posed. This is quite a general class of problems that has numerous applications.

INTENDED AUDIENCE AND STYLE OF THIS BOOK

The text easily includes enough material for a one-year course, but several one-semester/quarter courses can be taught out of it. One useful feature is the large number of exercises that span from routine computations to help solidify newly learned skills to more advanced conceptual and theoretical questions and new applications. Some sections are marked with an asterisk to indicate that they should be considered as optional; their deletion would cause no major disruption to the main themes of the text. Some of these optional sections are more theoretical than the others (e.g., Section 10.5: Rayleigh-Ritz methods), while others present applications in a particular related area (e.g., Section 7.2: Introduction to Computer Graphics). To facilitate readability of the text, we employ the following font conventions: Regular text is printed in the (current) Times New Roman font, MATLAB inputs and commands appear in `Courier New` font, whereas MATLAB output is printed in Ariel font. Essential vocabulary words are set in **bold type**, while less essential vocabulary is set in *italics*.

Over the past six years I have been teaching numerous courses in numerical analysis, discrete mathematics, and mathematical modeling at the University of Guam. Prior to this, at the University of Hawaii, I had been teaching more theoretically based courses in an assortment of mathematical subjects. In my education at the University of Michigan and the University of Maryland, apart from being given much good solid training in both pure and applied areas of mathematics, I was also imparted with a tremendous appreciation for the interesting and rich history of mathematics. This book brings together a conceptual and rigorous approach to many different areas of numerical differential equations, along with a practical approach for making the most out of the MATLAB computing environment to solve problems and gain further understanding. It also includes numerous historical comments (and portraits) on key mathematicians who have made contributions to the various areas under investigation. It teaches how to make the most of mathematical theory and computational efficiency. At the University of Guam, I have been able to pick and choose many of the topics that I would cover in such classes. Throughout these courses I was using the MATLAB computing environment as an integral component, and most portions of the text have been classroom tested.

I was motivated to write this book precisely because I could not find single books that were suitable to use for several courses that I was teaching. Often I would find that I would need to put several books on reserve at the library since no single textbook would cover all of the needs of these courses and it would be unreasonable to require the students to purchase a large number of textbooks. A major problem was coming up with suitable homework problems to assign that involved interesting applications and that forced the student to combine conceptual thinking along with experiments on the computer. I started off by writing out my own homework assignments and as these problems and my lecture notes began to reach a sizeable volume, I decided it was time to expand them into a book. There are many decent books on how to use MATLAB, there are other books on programming, and still others on theory and modeling with differential equations. There does not seem to exist, however, a comprehensive treatment of all of these topics in the market. This book is designed primarily to fill this important gap in the textbook market. It encourages students to make the most out of both the heavy computational machinery of MATLAB through efficiently designed programs and their own conceptual thinking. It emphasizes using computer experiments to motivate mathematical theory and discovery. Sports legend Yogi Berra once said, "In theory there is no difference between theory and practice. In practice there is." This quote arguably rings more true for differential equations than for any other branch of mathematics. Much can be learned about differential equations ‚by doing computer experiments and this practice is continually encouraged and emphasized throughout the text.

There are four intended uses of this book:

1. *A standalone textbook for courses in numerical differential equations.* It could be used for a one-semester course allowing for a flexible coverage of topics in ordinary and/or partial differential equations. It could also be used for a two-semester course in numerical differential equations. The coverage of Part I topics could vary, of course, depending on the level of preparedness of the students.

2. *A textbook for a course in numerical analysis.* Apart from the extensive coverage of differential equations, the text includes designated coverage of many of the standard topics in numerical analysis such as rootfinding (Chapter 6), floating point arithmetic (Chapter 5), solving linear systems (direct and iterative methods), and numerical linear algebra (Chapter 7). Other numerical analysis topics such as interpolation, numerical differentiation, and integration are covered as they are needed.

3. *An accompanying text for a more traditional course in ordinary and/or partial differential equations* that could be used to introduce and use (as time and interest permits) the very important numerical tools of the subject. The ftp site for this book includes all of the programs (M-files) developed in the text and they can be copied into the user's computers and used to obtain numerical solutions of a great variety of problems in differential equations. For such usage, the amount of time spent learning about programming these codes can be variable, depending on the interests and time constraints of the particular class.

4. *A book for self-study* by any science student or practitioner who uses differential equations and would like to learn more about the subject and/or about MATLAB.

The programs and codes in the book have all been developed to work with the latest versions of MATLAB (Student Versions or Professional Versions).[1] All of the M-files developed in the text and the exercises for the reader can be downloaded from book's ftp site:

`ftp://ftp.wiley.com/public/sci_tech_med/numerical_differential/`

Although it is essentially optional throughout the book, when convenient we occasionally use MATLAB's Symbolic Toolbox that comes with the Student

---

[1] The codes and M-files in this book have been tested on MATLAB versions 5, 6, and 7. The (very) rare instances where a version-specific issue arises are carefully explained. One added feature of later versions is the extended menu options that make many tasks easier than they used to be. A good example of this is the improvements in the MATLAB graphics window. Many features of a graph can be easily modified directly using (user-friendly) menu options. In older versions, such editing had to be done by entering the correct "handle graphics" commands into the MATLAB command window.

Version (but is optional with the Professional Version). Each chapter has many detailed worked-out examples for all of the material that is introduced. Additionally, the text is punctuated with numerous "Exercises for the Reader" that reinforce the reader's active participation. Detailed solutions to all of these are given in an appendix at the back of the book.

ACKNOWLEDGMENTS

Many individuals and groups have assisted me in various ways that have led to the development of this book and I would like to take this space to express my appreciation to some of them. I would like to thank my students who have taken my courses (very often as electives) and who have read through preliminary versions of parts of the book and offered useful feedback that has improved the pedagogy of this text. The people at MathWorks (the company that develops MATLAB), in particular, Courtney Esposito, have been very supportive in providing me with software and high-quality technical support, whenever I needed it.

During my preparation of the material, I was in constant need of getting hold of journal articles and books in the various subject areas. Despite the limited collection and the budget constraints of the University of Guam library, librarian Moses Francisco deserves special mention. He has always been able to do an outstanding job in getting the materials that I needed in a timely fashion. His conscientiousness, efficiency, and friendly demeanor have been an enlightening experience and the book has benefited greatly from his assistance. I would also like to mention acquisitions manager Roque Iriarte, who has been very helpful in obtaining important new books for our collection.

Feedback from reviewers of this book has been very helpful. These reviewers include: Chris Gardiner (Eastern Michigan University), Mark Gockenbach (Michigan Tech), Murli Gupta (George Washington University), Jenny Switkes (Cal Poly Pomona), Robin Young (University of Massachusetts), and Richard Zalik (Auburn University). Among these, I owe special thanks to Drs. Gockenbach and Zalik; each carefully read through major portions of the text (Gockenbach read through the entire manuscript) and have provided extensive suggestions, scholarly remarks, and corrections. I would like to thank Robert Krasny (University of Michigan) for several useful discussions on numerical linear algebra.

The historical accounts throughout the text have benefited from the extensive MacTutor website. The book includes several photographs of mathematicians who have made contributions to the areas under investigation. I thank Benoit Mandelbrot for permitting the inclusion of his photograph. I thank Dan May and MetLife archives for providing me with and allowing me to include a company photo of Alfred Lotka. I am very grateful to George Phillips for extending permission to me to include his photographs of John Crank and Phyllis Nicolson. Peter Lax has kindly contacted the son of Richard Courant on my behalf to obtain permission for me to include a photograph of Courant. Two very interesting air foil mesh graphics that appear in Chapter 13 were created by Tim Barth of NASA's Jet Propulsion Laboratory; I am grateful to him for allowing their inclusion.

I have had many wonderful teachers throughout my years and I would like to express my appreciation to all of them. I would like to make special mention of some of them. First, back in middle school, I spent a year in a parochial school with a teacher, Sister Jarlaeth, who had a tremendous impact in kindling my interest in mathematics; my experience with her led me to develop a newfound respect for education. Although Sister Jarlaeth has passed, her kindness and caring for students and the learning process will live on with me forever. It was her example that made me decide to become a mathematics professor as well as a teacher who cares. Several years later when I arrived in Ann Arbor, Michigan for the mathematics PhD program, I had intended to complete my PhD in an area of abstract algebra, an area in which I was very well prepared and interested. During my first year, however, I was so enormously impressed and enlightened by the analysis courses that I needed to take, that I soon decided to change my area of focus to analysis. I would particularly like to thank my analysis professors Peter Duren, Fred Gehring, M. S. ("Ram") Ramanujan, and the late Allen Shields. Their cordial, rigorous, and elegant lectures replete with many historical asides were a most delightful experience.

I thank my colleagues at the University of Guam for their support and encouragement of my teaching many MATLAB-based mathematics courses. Portions of this book were completed while I was spending semesters at the National University of Ireland and (as a visiting professor) at the University of Missouri at Columbia. I would like to thank my hosts and the mathematics departments at these institutions for their hospitality and for providing such stimulating atmospheres in which to work.

Last, but certainly not least, I have two more individuals to thank. My mother, Christa Stanoyevitch, has encouraged me throughout the project and has done a superb job proofreading the entire book. Her extreme conscientiousness and ample corrections and suggestions have significantly improved the readability of this book. I would like to also thank my good friend Sandra Su-Chin Wu for assistance whenever I needed it with the many technical aspects of getting this book into a professional form. Near the end of this project, she provided essential help in getting this book into its final form. Inevitably, there will remain some typos and perhaps more serious mistakes. I take full responsibility for these and would be grateful to any readers who could direct my attention to any such oversights.

Chapter 1: MATLAB Basics

1.1: WHAT IS MATLAB?

As a student who has already taken courses at least up through calculus, you most likely have seen the power of graphing calculators and perhaps those with symbolic capabilities. MATLAB adds a whole new exciting set of capabilities as a powerful computing tool. Here are a few of the advantages you will enjoy when using MATLAB, as compared to a graphing calculator:

1. It is easy to learn and use. You will be entering commands on your big, familiar computer keyboard rather than on a tiny little keypad where sometimes each key has four different symbols attached.
2. The graphics that MATLAB produces are of very high resolution. They can be easily copied to other documents (with simple clicks of your mouse) and printed out in black/white or color format. The same is true of any numerical and algebraic MATLAB inputs and outputs.
3. MATLAB is an abbreviation for MATrix LABoratory. It is ideally suited for calculations and manipulations involving matrices. This is particularly useful for computer users since the spreadsheet (the basic element for recording data on a computer) is just a matrix.
4. MATLAB has many built-in programs and you can interactively use them to create new programs to perform your desired tasks. It enables you to take advantage of the full computing power of your computer, which has much more memory and speed than a graphing calculator.
5. MATLAB's language is based on the C-family of computer languages. People experienced with such languages will find the transition to MATLAB natural and people who learn MATLAB without much computer background will, as a fringe benefit, be learning skills that will be useful in the future if they need to learn more computer languages.
6. MATLAB is heavily used by mathematicians, scientists, and engineers and there is a tremendous amount of interesting programs and information available on the Internet (much of it is free). It is a powerful computing environment that continues to evolve.

We wish here and now to present a disclaimer. MATLAB is a spectacularly vast computing environment and our plan is not to discuss all of its capabilities, but rather to give a decent survey of enough of them so as to provide the reader with a powerful new arsenal of uses of MATLAB for solving a variety of problems in mathematics and other sciences. Several good books have been written just on using MATLAB; see, for example, references [HiHi-00], [HuLiRo-01], [PSMI-98], and [HaLi-00].[1]

1.2: STARTING AND ENDING A MATLAB SESSION

We assume that MATLAB has been installed on the system that you are using.[2] Instructions for starting MATLAB are similar to those for starting any installed software on your system. For example, on most windows-based systems, you should be able to simply double click on MATLAB's icon. Once MATLAB is started, a command window should pop up with a **prompt**: >> (or EDU>> if you are using the Student Version). In what follows, if we tell you to enter something like >> 2+2 (on the command window), you enter 2+2 only at the prompt—which is already there waiting for you to type something. Before we begin our first brief tutorial, we point out that there is a way to create a file containing all interactions with a particular MATLAB session. The command diary will do this. Here is how it works: Say you want to save the session we are about to start to your floppy disk, which you have inserted in the a:/-drive. After the prompt type:

```
>> diary a:/tutor1.txt
```

NOTE: If you are running MATLAB in a computer laboratory or on someone else's machine, you should always save things to your portable storage device or personal account. This will be considerate to the limitations of hard drive space on the machines you are using and will give you better assurance that the files still will be available when you need them.

This causes MATLAB to create a text file called tutor1.txt in your a:/- drive called tutor1.txt, which, until you end the current MATLAB session, will be a carbon copy of your entire session on the command window. You can later open it up to edit, print, copy, etc. It is perhaps a good idea to try this out once to see how it works and how you like it (and we will do this in the next section), but in practice, most MATLAB users will often just copy the important parts of their MATLAB session and paste them appropriately in an open word processing window of their choice.

On most platforms, you can end a MATLAB session by clicking down your left mouse button after you have moved the cursor to the "File" menu (located on the upper-left corner of the MATLAB command window). This will cause a menu of commands to appear that you can choose from. With the mouse button still held down, slide the cursor down to the "Exit MATLAB" option and release it. This

---

[1] Citations in square brackets refer to items in the References section in the back of this book.
[2] MATLAB is available on numerous computing platforms including PC Windows, Linux, MAC, Solaris, Unix, HP-UX. The functionality and use is essentially platform independent although some external interface tasks may vary.

will end the session. Another way to accomplish the same would be to simply click (and release) the left mouse button after you have slid it on top of the "X" button at the upper-right corner of the command window. Yet another way is to simply enter the command:

```
>> quit
```

Any diary file you created in the session will now be accessible.

1.3: A FIRST MATLAB TUTORIAL

As with all tutorials we present, this is intended to be worked by the reader on a computer with MATLAB installed. Begin by starting a MATLAB session as described earlier. If you like, you may begin a diary as shown in the previous section on which to document this session. MATLAB will not respond to or execute any command until you press the "enter key," and you can edit a command (say, if you made a typo) and press enter no matter where the cursor is located in a given command line. Let us start with some basic calculations: First enter the command:

```
>> 5+3
→ ans = 8
```

The arrow (→) notation indicates that MATLAB has responded by giving us ans = 8. As a general rule we will print MATLAB input in a different font (Courier New) than the main font of the text (Times New Roman). It does not matter to MATLAB whether you leave spaces around the + sign.[3] (This is usually just done to make the printout more legible.) Instead of adding, if we wanted to divide 5 by 3, we would enter (the operation ÷ is represented by the keyboard symbol / in MATLAB)

```
>> 5/3
→ ans =1.6667
```

The output "1.6667" is a four-decimal approximation to the unending decimal approximation. The exact decimal answer here is 1.666666666666... (where the 6's go on forever). The four-decimal display mode is the default format in which MATLAB displays decimal answers. The previous example demonstrates that if the inputs and outputs are integers (no decimals), MATLAB will display them as such. MATLAB does its calculations using about 16 digits—we shall discuss this in greater detail in Chapters 2 and 5. There are several ways of changing how your outputs are displayed. For example, if we enter:

```
>> format long
```

---

[3] The format of actual output that MATLAB gives can vary slightly depending on the platform and version being used. In general it will take up more lines and have more blank spaces than as we have printed it We adopt this convention throughout the book in order to save space.

```
>> 5/3
→ ans =1.66666666666667
```

we will see the previous answer displayed with 15 digits. All subsequent calculations will be displayed in this format until you change it again. To change back to the default format, enter `>> format short`. Other popular formats are `>> format bank` (displays two decimal places, useful for applications to finance) and `>> format rat` (approximates all answers as fractions of small integers and displays them as such). It is not such a good idea to work in `format rat` unless you know for sure the numbers you are working with are fractions as opposed to irrational numbers, like $\pi = 3.14159265...$, whose decimals go on forever without repetition and are impossible to express via fractions.

In MATLAB, a single equals sign (=) stands for "is assigned the value." For example, after switching back to the default format, let us store the following constants into MATLAB's workspace memory:

```
>> format short
>> a = 2.5
→ a = 2.5000
>> b = 64
→ b=64
```

Notice that after each of these commands, MATLAB will produce an output of simply what you have inputted and assigned. You can always suppress the output on any given MATLAB command by tacking on a semicolon (;) at the end of the command (before you press enter). Also, you can put multiple MATLAB commands on a single line by separating them with commas, but these are not necessary after a semicolon. For example, we can introduce two new constants aa and bb without having any output using the single line:

```
>> aa = 11; bb = 4;
```

Once variables have been assigned in a MATLAB session, computations involving them can be done using any of MATLAB's built-in functions. For example, to evaluate $aa + a\sqrt{b}$, we could enter

```
>> aa + a*sqrt(b)
→ ans=31
```

Note that aa stands for the single variable that we introduced above rather than a^2, so the output should be 31. MATLAB has many built-in functions, many of which are listed in the MATLAB Command Index at the end of this book.

MATLAB treats all numerical objects as matrices, which are simply rectangular arrays of numbers. Later we will see how easy and flexible MATLAB is in manipulating such arrays. Suppose we would like to store in MATLAB the following two matrices:

$$A = \begin{bmatrix} 2 & 4 \\ -1 & 6 \end{bmatrix}, \quad B = \begin{bmatrix} 2 & 5 & -3 \\ 1 & 0 & -1 \\ 8 & 4 & 0 \end{bmatrix}.$$

We do so using the following syntax:

```
>> A = [2 4 ; -1 6]
→ A= 2    4
    -1    6
```

```
>> B = [2 5 -3; 1 0 -1; 8 4 0]
→ B= 2   5  -3
     1   0  -1
     8   4   0
```

(note that the rows of a matrix are entered in order and separated by semicolons; also, adjacent entries within a row are given at least one space between). You can see from the outputs that MATLAB displays these matrices pretty much in their mathematical form (but without the brackets).

In MATLAB it is extremely simple to edit a previous command into a new one. Let's say in the matrix B above, we wish to change the bottom-left entry from eight to three. Since the creation of matrix B was the last command we entered, we simply need to press the up-arrow key (↑) once and magically the whole last command appears at the cursor (do this!). If you continue to press this up-arrow key, the preceding commands will continue to appear in order. Try this now! Next press the down arrow key (↓) several times to bring you back down again to the most recent command you entered (i.e., where we defined the matrix B). Now simply use the mouse and/or left- and right-arrow keys to move the cursor to the 8 and change it to 3, then press enter. You have now overwritten your original matrix for B with this modified version. Very nice indeed! But there is more. If on the command line you type a sequence of characters and then press the up-arrow key, MATLAB will then retrieve only those input lines (in order of most recent occurrence) that begin with the sequence of characters typed. Thus for example, if you type a and then up-arrow twice, you would get the line of input where we set aa = 11.

A few more words about "variables" are in order. Variable names can use up to 19 characters, and must begin with a letter, but after this you can use digits and underscores as well. For example, two valid variable names are diffusion22time and Shock_wave_index; however, Final$Amount would not be an acceptable variable name because of the symbol $. Any time that you would like to check on the current status of your variables, just enter the command who:

```
>> who
```

→Your variables are:
A B a aa ans b bb

For more detailed information about all of the variables in the workspace (including the size of all of the matrices) use the command whos:

```
>> whos
```

→	Name	Size	Bytes	Class
	A	2x2	32 double	array
	B	3x3	72 double	array
	a	1x1	8 double	array
	aa	1x1	8 double	array
	ans	1x1	8 double	array
	b	1x1	8 double	array
	bb	1x1	8 double	array

You will notice that MATLAB retains both the number a and the matrix A. <u>MATLAB is case-sensitive</u>. You will also notice that there is the variable ans in the workspace. Whenever you perform an evaluation/calculation in MATLAB, an automatic assignment of the variable ans is made to the most recent result (as the output shows). To clear any variable, say aa, use the command

```
>>clear aa
```

Do this and check with who that aa is no longer in the workspace. If you just enter clear, all variables are erased from the workspace. More importantly, suppose that you have worked hard on a MATLAB session and would like to retain all of your workspace variables for a future session. To save (just) the workspace variables, say to your floppy a:\ drive, make sure you have your disk inserted and enter:

```
>> save a:/tutvars
```

This will create a file on your floppy called tutvars.mat (you could have called it by any other name) with all of your variables. To see how well this system works, go ahead and quit this MATLAB session and start a new one. If you type who you will get no output since we have not yet created any variables in this new session. Now (making sure that the floppy with tutvars is still inserted) enter the command:

```
>> load a:/tutvars
```

If you enter who once again you will notice that all of those old variables are now in our new workspace. You have just made it through your first MATLAB tutorial. End the session now and examine the diary file if you have created one.

If you want more detailed information about any particular MATLAB command, say who, you would simply enter:

```
>> help who
```

and MATLAB would respond with some usage information and related commands.

1.4: VECTORS AND AN INTRODUCTION TO MATLAB GRAPHICS

On any line of input of a MATLAB session, if you enter the percent symbol (%), anything you type after this is ignored by MATLAB's processor and is treated as a comment.[4] This is useful, in particular, when you are writing a complicated program and would like to enhance it with some comments to make it more understandable (both to yourself at a later reading and to others who will read it). Let us now begin a new MATLAB session.

A **vector** is a special kind of matrix with only one row or one column. Here are examples of vectors of each type:

$$x = [1 \quad 2 \quad 3] \qquad y = \begin{bmatrix} 2 \\ -3 \\ 5 \end{bmatrix}.$$

```
>> % We create the above two vectors and one more as variables in our
>> %  MATLAB session.
>>   x = [1 2 3], y = [2 ; -3 ; 5], z = [4   -5   6]
→ x = 1   2   3       y = 2                z = 4  -5   6
                         -3
                          5
```

```
>> %  Next we perform some simple array operations.
>> a = x + z
→a = 5  -3   9
>> b = x + y  %MATLAB needs arrays to be the same size to add/subtract
→??? Error using ==> +
Matrix dimensions must agree.
>> c=x.*z    %term by term multiplication, notice the dot before the *
→ c = 4  -10   18
```

The **transpose** of any matrix A, denoted as A^T or A', consists of the matrix whose rows are (in order) the columns of A and vice versa. For example the transpose of the 2×3 matrix

$$A = \begin{bmatrix} 2 & 4 & 9 \\ 1 & -2 & 5 \end{bmatrix}$$

is the 3×2 matrix

$$A' = \begin{bmatrix} 2 & 1 \\ 4 & -2 \\ 9 & 5 \end{bmatrix}.$$

---

[4] MATLAB's windows usually conform to certain color standards to make codes easier to look through. For example, when a comment is initiated with %, the symbol and everything appearing after it will be shown in green. Also, warning/error messages (as we will soon experience on the next page) appear in red. The default color for input and output is black.

In particular, the transpose of a row vector is a column vector and vice versa.

```
>> y'    %MATLAB uses the  prime ' for the transpose operation
→ ans = 2  -3   5
>> b=x+y'    %cf. with the result for x + | y
→ b = 3  -1   8
>> % We next give some other useful ways to create vectors.
>> % To create a (row) vector having 5 elements linearly spaced
>> % between 0 and 10 you could enter
>> linspace(0,10,5)    %Do this!
→ ans = 0  2.5000  5.0000  7.5000  10.0000
```

We indicate the general syntax of `linspace` as well as another useful way to create vectors (especially big ones!):

v=linspace(F,L,N) →	If F and L are real numbers and N is a positive integer, this command creates a row vector v with: first entry = F, last entry = L, and having N equally spaced entries.
v = F:G:L →	If F and L are real numbers and G is a nonzero real number, this command creates a vector v with: first entry = F, last (possible) entry = L, and gap between entries = G. G is optional with default value 1.

To see an example, enter

```
>> x = 1:.25:2.5  %will overwrite previously stored value of x
→ x = 1.0000  1.2500  1.5000  1.7500  2.0000 2.2500  2.5000
>> y = -2:.5:3
→ y = -2 0000  -1.5000  -1.0000  -0.5000      0   0 5000  1.0000  1.5000  2.0000
2.5000   3.0000
```

EXERCISE FOR THE READER 1.1: Use the `linspace` command above to recreate the vector y that we just built.

The basic way to plot a graph in MATLAB is to give it the *x*-coordinates (as a vector a) and the corresponding *y*-coordinates (as a vector b of the same length) and then use the `plot` command.

plot(a,b) →	If a and b are vectors of the same length, this command will create a plot of the line segments connecting (in order) the points in the *xy*-plane having *x*-coordinates listed in the vector a and corresponding *y*-coordinates in the vector b.

To demonstrate how this works, we begin with some simple vector plots and work our way up to some more involved function plots. The following commands will produce the plot shown in Figure 1.1.

```
>> x = [1 2 3 4]; y = [1 -3 3 0];
>> plot(x,y)
```

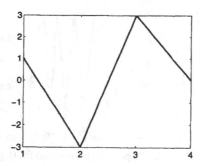

FIGURE 1.1: A simple plot resulting from the command `plot(x,y)` using the vector $x = [1\ 2\ 3\ 4]$ for x-coordinates and the vector $y = [1\ -3\ 3\ 0]$ for corresponding y-coordinates. [5]

Next, we use the same vector approach to graph the function $y = \cos(x^2)$ on $[0,5]$. The finer the grid determined by the vectors you use, the greater the resolution. To see this first hand, enter:

```
>> x = linspace(0,5,5);    % I will be supressing a lot of output, you
>>                         % can drop the ';' to see it
>> y = cos(x.^2);
```

Note the dot (.) before the power operator (^). The dot before an operator changes the default matrix operation to a **component-wise operation**. Thus `x.^2` will create a new vector of the same size as x where each of the entries is just the square of the corresponding entry of x. This is what we want. The command `x^2` would ask MATLAB to multiply the matrix (or row vector) x by itself, which (as we will explain later) is not possible and so would produce an error message.

```
>> plot(x,y)   % produces our first very rough plot of the function
>>             % with only 5 plotting points
```

See Figure 1.2(a) for the resulting plot. Next we do the same plot but using 25 points and then 300 points. The editing techniques of Section 1.2 will be of use as you enter the following commands.

```
>> x = linspace(0,5,25);
>> y = cos(x.^2);
>> plot(x,y)    % a better plot with 25 points.
>> x = linspace(0,5,300);
>> y = cos(x.^2);
>> plot(x,y)    % the plot is starting to look good with 300 points.
```

---

[5] Numerous attributes of a MATLAB plot or other graphic can be modified using the various (very user-friendly) menu options available on the MATLAB graphics window. These include font sizes, line styles, colors, and thicknesses, axis label and tick locations, and numerous other items. To improve readability of this book we will use such features without explicit mention (mostly to make the fonts more readable to accommodate reduced figure sizes).

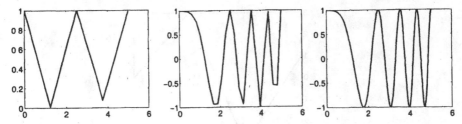

FIGURE 1.2: Plots of the function $y = \cos(x^2)$ on $[0,5]$ with increasing resolution: (a) (left) 5 plotting points, (b) (middle) 25 plotting points, and (c) (right) 300 plotting points.

If you want to add more graphs to an existing plot, enter the command:

```
>> hold on     %do this!
```

All future graphs will be added to the existing one until you enter hold off. To see how this works, let's go ahead and add the graphs of $y = \cos(2x)$ and $y = \cos^2 x$ to our existing plot of $y = \cos(x^2)$ on $[0,5]$. To distinguish these plots, we might want to draw the curves in different styles and perhaps even different colors. Table 1.1 is a summary of the codes you can use in a MATLAB plot command to get different plot styles and colors:

TABLE 1.1: MATLAB codes for plot colors and styles.

Color/Code		Plot Style/Code	
black / k	red / r	solid / –	stars / *
blue / b	white / w	dashed / – –	x-marks / x
cyan / c	yellow / y	dotted / :	circles / o
green / g		dash-dot / - .	plus-marks / +
magenta / m		points / .	pentacles/ p

Suppose that we want to produce a dashed cyan graph of $y = \cos(2x)$ and a dotted red graph of $y = \cos^2 x$ (to be put in with the original graph). We would enter the following:

```
>> y1 = cos (2*x);
>> plot(x,y1,'c--')        %will plot with cyan dashed curve
>> y2 = cos(x).^2;         % cos(x)^2 would produce an error
>> plot(x,y2,'r:')         %will plot in dotted red style
>> hold off                %puts an end to the current graph
```

You should experiment now with a few other options. Note that the last four of the plot styles will put the given object (stars, x-marks, etc.) around each point that is actually plotted. Since we have so many points (300) such plots would look like very thick curves. Thus these last four styles are more appropriate when the density of plot points is small. You can see the colors on your screen, but unless you have a color printer you should make use of the plot styles to distinguish between multiple graphs on printed plots.

Many features can be added to a plot. For example, the steps below show how to label the axes and give your plot a title.

```
>> xlabel('x')
>> ylabel('cos(x.^2), cos(2*x), cos(x).^2')
>> title('Plot created by yourname')
```

Notice at each command how your plot changes; see Figure 1.3 for the final result.

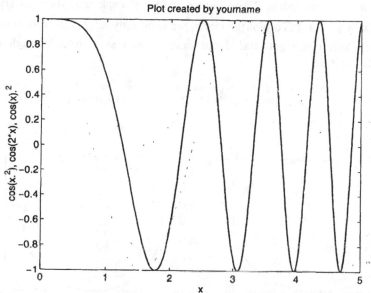

FIGURE 1.3: Plot of three different trigonometric functions done using different colors and styles.

In a MATLAB plot, the points and connecting line segments need not define the graph of a function. For example, to get MATLAB to draw the unit square with vertices $(0,0)$, $(1,0)$, $(1,1)$, $(0,1)$, we could key in the x- and y-coordinates (in an appropriate order so the connecting segments form a square) of these vertices as row vectors. We need to repeat the first vertex at the end so the square gets closed off. Enter:

```
>> x=[0 1 1 0 0]; y=[0 0 1 1 0];
>> plot(x,y)
```

Often in mathematics, the variables x and y are given in terms of an auxiliary variable, say t (thought of as time), rather than y simply being given in terms of (i.e., a function of) x. Such equations are called **parametric equations**, and are easily graphed with MATLAB. Thus parametric equations (in the plane) will look

like: $\begin{cases} x = x(t) \\ y = y(t) \end{cases}$.

These can be used to represent any kind of curve and are thus much more versatile than functions $y = f(x)$, whose graphs must satisfy the vertical line test. MATLAB's plotting format makes plotting parametric equations a simple task. For example, the following parametric equations

$$\begin{cases} x = 2\cos(t) \\ y = 2\sin(t) \end{cases}$$

represent a circle of radius 2 and center (0,0). (Check that they satisfy the equation $x^2 + y^2 = 4$.) To plot the circle, we need only let t run from 0 to 2π (since the whole circle gets traced out exactly once as t runs through these values). Enter:

```
>> t = 0:.01:2*pi;   % a lot of points for decent resolution, as you
>>                   % guessed, 'pi' is how MATLAB denotes π
>> x = 2*cos(t);
>> y = 2*sin(t);
>> plot(x,y)
```

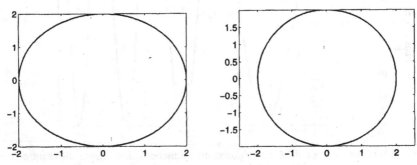

FIGURE 1.4: Parametric plots of the circle $x^2 + y^2 = 4$, (a) (left) first using MATLAB's default rectangular axis setting, and then (b) (right) after the command `axis('equal')` to put the axes into proper perspective.

You will see an ellipse in the figure window (Figure 1.4(a)). This is because MATLAB uses different scales on the x- and y-axes, unless told otherwise. If you enter: `>>axis('equal')`, MATLAB will use the same scale on both axes so the circle appears as it should (Figure 1.4(b)). Do this!

EXERCISE FOR THE READER 1.2: In the same fashion use MATLAB to create a plot of the more complicated parametric equations:

$$\begin{cases} x(t) = 5\cos(t/5) + \cos(2t) \\ y(t) = 5\sin(t/5) + \sin(3t) \end{cases} \quad \text{for} \ \ 0 \le t \le 10\pi.$$

Caution: Do not attempt to plot this one by hand!

If you use the `axis('equal')` command in Exercise for the Reader 1.2, you should be getting the plot pictured in Figure 1.5.

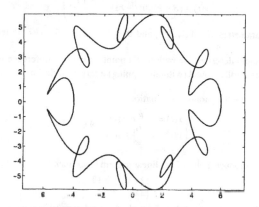

FIGURE 1.5: A complicated MATLAB parametric plot.

EXERCISES 1.4:

1.   Use MATLAB to plot the graph of $y = \sin(x^4)$ for $0 \le x \le 2\pi$, (a) using 200 plotting points, and (b) using 5000 plotting points.

2.   Use MATLAB to plot the graph of $y = e^{-1/x^2}$ for $-3 \le x \le 3$, (a) using 50 plotting points, and (b) using 10,000 plotting points.

NOTE: When MATLAB does any plot, it automatically tries to choose the axes to exactly accommodate all of the plot points. For functions with vertical asymptotes (like the ones in the next two exercises), you will see that this results in rather peculiar-looking plots. To improve the appearance of the plots, you can rescale the axes. This is done by using the following command:

	Resets the axis range for plots to be:
`axis([xmin xmax ymin ymax])` →	$\text{xmin} \le x \le \text{xmax}$ $\text{ymin} \le y \le \text{ymax}$
	Here, the four vector entries can be any real numbers with xmin < xmax, and ymin < ymax.

3.   Use MATLAB to produce a nice plot of the graph of $y = \dfrac{2 - x^2}{x^2 + x - 6}$ on the interval $[-5, 5]$. Experiment a bit with the `axis` command as explained in the above note.

4.   Use MATLAB to plot the graph of $y = \dfrac{x^4 - 16}{x^3 + 2x^2 - 6}$ on the interval $[-1, 5]$. Adjust the axes, as explained in the note preceding Exercise 3, so as to get an attractive plot.

5.   Use MATLAB to plot the circle of radius 3 and center $(-2,1)$.

6.   Use MATLAB to obtain a plot of the *epicycloids* that are given by the following parametric

equations:

$$\begin{cases} x(t) = (R+r)\cos t - r\cos\left(\dfrac{R+r}{r}t\right) \\ y(t) = (R+r)\sin t - r\sin\left(\dfrac{R+r}{r}t\right) \end{cases}, \quad 0 \le t \le 2\pi$$

using first the parameters $R = 4$, $r = 1$, and then $R = 12$, $r = 5$. Use no less than 1000 plotting points.

Note: An epicycloid describes the path that a point on the circumference of a smaller circle (of radius r) makes as it rolls around (without slipping) a larger circle (of radius R).

7.   Use MATLAB to plot the parametric equations:

$$\begin{cases} x(t) = e^{-\sqrt{t}}\cos(t) \\ y(t) = e^{-\sqrt{2t}}\sin(t) \end{cases}, \quad 0 \le t \le 100.$$

8.   Use MATLAB to produce a plot of the linear system (two lines):

$$\begin{cases} 2x + 3y = 13 \\ 2x - y = 1 \end{cases}.$$

Include a label for each line as well as a label of the solution (that you can easily find by hand), all produced by MATLAB.

Hints: You will need the `hold` on command to include so many things in the same graph. To insert the labels, you can use either of the commands below to produce the string of text label at the coordinates (x, y).

`text(x,y,'label')` →	Inserts the text string `label` in the current graphic window at the location of the specified point (x, y).
`gtext('label')` →	Inserts the text string `label` in the current graphic window at the location of exactly where you click your mouse.

9.   Use MATLAB to draw a regular octagon (stop-sign shape). This means that all sides have the same length and all interior angles are equal. Scale the axes accordingly.

10.   By using the `plot` command (repeatedly and appropriately), get MATLAB to produce a circle inscribed in a triangle that is in turn inscribed in another circle, as shown in Figure 1.6.

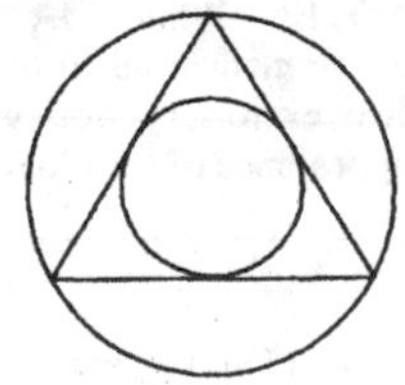

FIGURE 1.6: Illustration for Exercise 10.

11.   By using the `plot` command (repeatedly and appropriately), get MATLAB to produce something as close as possible to the familiar figure on the right. Do not worry about the line/curve thickness for now, but try to get it so that the eyes (dots) are reasonably visible.

1.5: A TUTORIAL INTRODUCTION TO RECURSION ON MATLAB

Getting a calculator or computer to perform a single task is interesting, but what really makes computers such powerful tools is their ability to perform a long series of related tasks. Such multiple tasks often require a program to tell the computer what to do. We will get more into this later, but it is helpful to have a basic idea at this point of how this works. We will now work on a rather elementary problem from finance that will actually bring to light many important concepts. There are several programming commands in MATLAB, but this tutorial will focus on just one of them (while) that is actually quite versatile.

PROBLEM: To pay off a $100,000.00 loan, Beverly pays $1,000.00 at the end of each month after having taken out the loan. The loan charges 8% annual interest (= 8/12% monthly interest) compounded monthly on the unpaid balance. Thus, at the end of the first month, the balance on Beverly's account will be (rounded to two decimals): $100,000 (prev. balance) + $666.27 (interest rounded to two decimals) − $1,000 (payment) = $99,666.67. This continues until Beverly pays off the balance; her last payment might be less than $1,000 (since it will need to cover only the final remaining balance and the last month's interest).

(a) Use MATLAB to draw a plot of Beverly's account balances (on the y-axis) as a function of the number of months (on the x-axis) until the balance is paid off.

(b) Use MATLAB to draw a plot of the accrued interest (on the y-axis) that Beverly has paid as a function of the number of months (on the x-axis).

(c) How many years + months will it take for Beverly to completely pay off her loan? What will her final payment be? How much interest will she have paid off throughout the course of the loan?

(d) Use MATLAB to produce a table of values, with one column being Beverly's outstanding balance given in yearly (12 month) increments, and the second column being her total interest paid, also given in yearly increments. Paste the data you get into your word processor to produce a cleaner table of this data.

(e) Redo part (c) if Beverly were to increase her monthly payments to $1,500.

Our strategy will be as follows: We will get MATLAB to create two vectors B and TI that will stand for Beverly's account balances (after each month) and the total interest accrued. We will set it up so that the last entry in B is zero, corresponding to Beverly's account finally being paid off.

There is another way to construct vectors in MATLAB that will suit us well here. We can simply assign the entries of the vector one by one. Let's first try it with the simple example of the vector $x = [1 \quad 5 \quad -2]$. Start a new MATLAB session and enter:

```
>>x(1) = 1 %specifies the first entry of the vector x, at this point
>>          %x will only have one entry
>>x(2) = 5 %you will see from the output x now has the first two of
>>          %its three components
>>x(3) = -2
```

The trick will be to use **recursion formulas** to automate such a construction of B and TI. This is possible since a single formula shows how to get the next entry of

B or TI if we know the present entry. Such formulas are called recursion formulas and here is what they look like in this case:

$$B(i+1) = B(i) + (.08/12)B(i) - 1000$$
$$TI(i+1) = TI(i) + (.08/12)B(i)$$

In words: The next month's account balance ($B(i+1)$) is the current month's balance ($B(i)$) plus the month's interest on the unpaid balance (($.08/12)B(i)$)) less Beverly's monthly payment. Similarly, the total interest accrued for the next month equals that of the current month plus the current month's interest.

Since these formulas allow us to use the information from any month to get that for the next month, all we really need are the initial values $B(1)$ and $TI(1)$, which are the initial account balance (after zero months) and total interest accrued after zero months. These are of course $100,000.00 and $0.00, respectively.

Caution: It is tempting to call these initial values $B(0)$ and $TI(0)$, respectively. However this cannot be done since they are, in MATLAB, vectors (remember, as far as numerical data is concerned: Everything in MATLAB is a matrix [or a vector]!) rather than functions of time, and indices of matrices and vectors must be positive integers ($i = 1, 2, ...$). This takes some getting used to since i, the index of a vector, often gets mixed up with t, an independent variable, especially by novice MATLAB users.

We begin by initializing the two vectors B and TI as well as the index i.

```
>> B(1)=100000;  TI(1)=0; i=1;
```

Next, making use of the recursion formulas, we wish to get MATLAB to figure out all of the other entries of these vectors. This will require a very useful device called a "while loop". We want the while loop to keep using the recursion formulas until the account balance reaches zero. Of course, if we did not stop using the recursion formulas, the balance would keep getting more and more negative and we would get stuck in what is called an **infinite loop**. The format for a while loop is as follows:

```
>>while    <condition>
. MATLAB commands...
>>end
```

The way it works is that if the <condition> is met, as soon as you enter end, the "...MATLAB commands..." within the loop are executed, one by one, just as if you were typing them in on the command window. After this the <condition> is reevaluated. If it is still met, the "...MATLAB commands..." are again executed in order. If the <condition> is not met, nothing more is done (this is called *exiting the loop*). The process continues. Either it eventually terminates (exits the loop) or it goes on forever (an infinite loop—a bad program). Let's do a simple example before returning to our problem. Before you enter the following commands, try to guess, based on how we just explained while loops, exactly what MATLAB's output will be. Then check your answer with MATLAB's actual output on your screen. If you get it right you are starting to understand the concept of while loops.

```
>> a=1;
>> while a^2 < 5*a
   a=a+2, a^2
   end
```

EXERCISE FOR THE READER 1.3: Analyze and explain each iteration of the above while loop. Note the equation a=a+2 in mathematics makes no sense at all. But remember, in MATLAB the single equal sign means "assignment." So for example, initially $a = 1$. The first run through the while loop the condition is met ($1 = a^2 < 5a = 5$) so a gets reassigned to be $1 + 2 = 3$, and in the same line a^2 is also called to be computed (and listed as output).

Now back to the solution of the problem. We want to continue using the above recursion formulas as long as the balance $B(i)$ remains positive. Since we have already initialized $B(1)$ and $TI(1)$, one possible MATLAB code for creating the rest of these vectors would look like:

```
>> while B(i) > 0
      B(i+1)=B(i)+ 8/12/100*B(1)-1000;  % This and the next are just
         %our recursion formulas.
      TI(i+1)=TI(i)+ 8/12/100*B(i);
      i=i+1;   % this bumps the vector index up by one at each
            % iteration.
   end
```

Notice that MATLAB does nothing, and the prompt does not reappear again, until the while loop is closed off with an end (and you press enter). Although we have suppressed all output, MATLAB has done quite a lot; it has created the vectors B and TI. Observe also that the final balance of zero should be added on as a final component. There is one subtle point that you should be aware of: The value of i after the termination of the while loop is precisely one more than the number of entries of the vectors B and TI thus far created. Try to convince yourself why this is true! Thus we can add on the final entry of B to be zero as follows:

```
>> n=i; B(n)=0;   %We could have just typed 'B(i)=0' but we wanted to
>>                % call 'n' the length of the vector B.
```

Another subtle point is that B (n) was already assigned by the while loop (in the final iteration) but was not a positive balance. This is what caused the while loop to end. So what actually will happen at this stage is that Beverly's last monthly payment will be reduced to cover exactly the outstanding balance plus the last month's interest. Also in the final iteration, the total interest was correctly given by the while loop. To do the required plots, we must first create the time vector. Since time is in months, this is almost just the vector formed by the indices of B (and TI), i.e., it is almost the vector $[1 \ 2 \ 3 \ \cdots \ n]$. But remember there is one slight subtle twist. Time starts off at zero, but the vector index must start off at 1. Thus the time vector will be $[0 \ 1 \ 2 \ \cdots \ n-1]$. We can easily construct it in MATLAB by

```
>> t=0:n-1;    %this is shorthand for 't=0:1:n-1', by default the
>>             %gap size is one.
```

Since we now have constructed all of the vectors, plotting the needed graphs is a simple matter.

```
>> plot(t,B)
>> xlabel('time in months'), ylabel('unpaid balance in dollars')
>> %we add on some descriptive labels on the horizontal and vertical
>> %axis. Before we go on we copy this figure or save it (it is
>> %displayed below)
>> plot(t, TI)
>> xlabel('time in months'), ylabel('total interest paid in dollars')
```

See Figure 1.7 for the MATLAB graphics outputs.

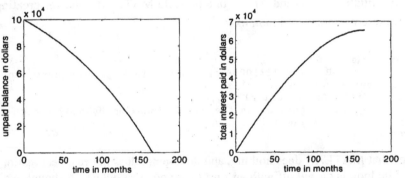

FIGURE 1.7: (a) (top) Graph of the unpaid balance in dollars, as a function of elapsed months in the loan of $100,000 that is being analyzed. (b) (bottom) Graph of the total interest paid in dollars, as a function of elapsed months in the loan of $100,000 that is being analyzed.

We have now taken care of parts (a) and (b). The answer to part (c) is now well within reach. We just have to report the correct components of the appropriate vectors. The time it takes Beverly to pay off her loan is given by the last value of the time vector, i.e.,

```
>> n-1
```
→166.00 =13 years + 10 months (<u>time of loan period</u>).

Her final payment is just the second-to-last component of B, with the final month's interest added to it (that's what Beverly will need to pay to totally clear her account balance to zero):

```
>> format bank   % this puts our dollar answers to the nearest cent.
>> B(n-1)*(1+8/12/100)
→$341.29 (last payment),
```

The total interest paid is just:

```
>> TI(n)
→$65,341.29 (total interest paid)
```

Part (d): Here we simply need to display parts of the two vectors, corresponding to the ends of the first 13 years of the loan and finally the last month (the 10th month after the 13th year). To get MATLAB to generate these two vectors, we could use a while loop as follows:[6]

```
>> k=1; i=1;   %we will use two indices, k will be for the original
>>%              vectors, i will be for the new ones.
>> while k<167
   YB(i)=B(k);   YTI(i)=TI(k);   %we   create   the   two   new   "yearly"
   vectors.
   k=k+12; i=i+1;   %at each iteration, the index of the original
                    %vectors gets bumped up by 12, but that for
                    %the new vectors gets bumped up only by one.
end
```

We next have to add the final component onto each vector (it does not correspond to a year's end). To do this we need to know how many components the yearly vectors already have. If you think about it you will see it is 14, but you could have just as easily asked MATLAB to tell you:

```
>> size(YB) %this command gives the size of any matrix or vector
>>          % (# of rows, # of columns).
→ans = 1.00      14.00
```

```
>>   YB(15)=B(167);  YTI(15)=TI(167);
>>   YB=YB'; YTI=YTI';   %this command reassigns both yearly vectors to
>>                       %be column vectors
```

Before we print them, we would like to print along the left side the column vector of the corresponding years' end. This vector in column form can be created as follows:

```
>>years   = 0:14; years = years' %first we create it as a row vector
>>                               %and then transpose it.
```

We now print out the three columns:

---

[6] A slicker way to enter these vectors would be to use MATLAB's special vector-creating construct that we mentioned earlier as follows: `YB = B(1:12:167)`, and similarly for `YTI`.

```
>> years, YB, YTI    % or better [years, YB, YTI]
years =              YB =             YTI =
```

years	YB	YTI
0	100000.00	0
1.00	95850.02	7850.02
2.00	91355.60	15355.60
3.00	86488.15	22488.15
4.00	81216.69	29216.69
5.00	75507.71	35507.71
6.00	69324.89	41324.89
7.00	62628.90	46628.90
8.00	55377.14	51377.14
9.00	47523.49	55523.49
10.00	39017.99	59017.99
11.00	29806.54	61806.54
12.00	19830.54	63830.54
13.00	9026.54	65026.54
14.00	0.00	65341.29

Finally, by making use of any decent word processing software, we can embellish this rather raw data display into a more elegant form such as Table 1.2.

TABLE 1.2: Summary of annual data for the $100,000 loan that was analyzed in this section.

Years Elapsed:	Account Balance:	Total Interest Paid:
0	$100000.00	$ 0
1	95850.02	7850.02
2	91355.60	15355.60
3	86488.15	22488.15
4	81216.69	29216.69
5	75507.71	35507.71
6	69324.89	41324.89
7	62628.90	46628.90
8	55377.14	51377.14
9	47523.49	55523.49
10	39017.99	59017.99
11	29806.54	61806.54
12	19830.54	63830.54
13	9026.54	65026.54
13 + 10 months	0.00	65341.29

Part (e): We can run the same program but we need only modify the line with the recursion formula for the vector B: It now becomes: B(i+1) = B(i)+I(i+1)-1500;. With this done, we arrive at the following data:

```
>> i-1,    B(i-1)*(1+8/12/100), TI(i)
→ 89 (7 years + 5 months), $693.59(last pmt) , $32693.59 (total interest paid).
```

EXERCISES 1.5:

1. Use a while loop to add all of the odd numbers up: $1 + 3 + 5 + 7 + \cdots$ until the sum exceeds 5 million. What is the actual sum? How many odd numbers were added?

2. Redo Exercise 1 by adding up the even numbers rather than the odd ones.

3. (*Insects from Hell*) An insect population starts off with one pair at year zero. The insects are immortal (i.e., they never die!) and after having lived for two years each pair reproduces another pair (assumed male and female). This continues on indefinitely. So at the end of one year the insect population is still 1 pair (= 2 insects); after two years it is $1 + 1 = 2$ pairs (= 4 insects), since the original pair of insects has reproduced. At the end of the third year it is $1 + 2 = 3$ pairs (the new generation has been alive for only one year, so has not yet reproduced), and after 4 years the population becomes $2 + 3 = 5$ pairs. (a) Find out the insect population (in pairs) at the end of each year from year 1 through year 10. (b) What will the insect population be at the end of 64 years?

HISTORICAL ASIDE: The sequence of populations in this problem: $1, 1, 1 + 1 = 2, 1 + 2 = 3, 2 + 3 = 5, 3 + 5 = 8, \ldots$ was first introduced in the middle ages by the Italian mathematician Leonardo of Pisa (ca. 1180–1250), who is better known by his nickname: Fibonacci (Italian, meaning *son of Bonaccio*). This sequence has numerous applications and has made Fibonacci quite famous. It comes up, for example, in hereditary effects in incest, growth of pineapple cells, and electrical engineering. There is even a mathematical journal named in Fibonacci's honor (the *Fibonacci Quarterly*).

4. Continuing Exercise 3, (a) produce a chart of the insect populations at the end of each 10th year until the end of year 100. (b) Use a while loop to find out how many years it takes for the insect population (in pairs) to exceed 1,000,000,000 pairs.

5. (*Another Insect Population Problem*) In this problem, we also start off with a pair of insects, this time mosquitoes. We still assume that after having lived for two years, each pair reproduces another pair. But now, at the end of three years of life, each pair of mosquitoes reproduces one more pair and then immediately dies. (a) Find out the insect population (in pairs) for each year up through year 10. (b) What will the insect population be at the end of 64 years?

6. Continuing Exercise 5, (a) plot the mosquito (pair) population from the beginning through the end of year 500, as a function of time. (b) How many years does it take for the mosquito population (in pairs) to exceed 1,000,000,000 pairs?

7. When their daughter was born, Mr. and Mrs. de la Hoya began saving for her college education by investing \$5,000 in an annuity account paying 10% interest per year. Each year on their daughter's birthday they invest \$2,000 more in the account. (a) Let A_n denote the amount in the account on their daughter's nth birthday. Show that A_n satisfies the following recursion formulas:

$$A_0 = 5000$$
$$A_n = (1.1)A_{n-1} + 2000.$$

(b) Find the amount that will be in the account when the daughter turns 18.

(c) Print (and nicely label) a table containing the values of n and A_n as n runs from 0 to 18.

8. Louise starts an annuity plan at her work that pays 9% annual interest compounded monthly. She deposits \$200 each month starting on her 25th birthday. Thus at the end of the first month her account balance is exactly \$200. At the end of the second month, she puts in another \$200, but her first deposit has earned her one month's worth of interest. The 9% interest per year means she gets 9%/12 = 0.75% interest per month. Thus the interest she earns in going from the first to second month is .75% of \$200 or \$1.50, and so her balance at the end of the second month is 401.50. This continues, so at the end of the 3rd month, her balance is \$401 50 (old

balance) + .75% of this (interest) + $200 (new deposit) = $604.51. Louise continues to do this throughout her working career until she retires at age 65.

(a) Figure out the balances in Louise's account at her birthdays: 26th, 27th, ..., up through her 65th birthday. Tabulate them neatly in a table (either cut and paste by hand or use your word processor—do not just give the raw MATLAB output, but rather put it in a form so that your average citizen could make sense of it).

(b) At exactly what age (to the nearest month) is Louise when the balance exceeds $100,000? Note that throughout these 40 years Louise will have deposited a total of $200/month × 12 months/yr. × 40 years = $96,000.

9. In saving for retirement, Joe, a government worker, starts an annuity that pays 12% annual interest compounded monthly. He deposits $200.00 into it at the end of each month. He starts this when he is 25 years old. (a) How long will it take for Joe's annuity to reach a value of $1 million? (b) Plot Joe's account balance as a function of time.

10. The **dot product** of two vectors of the same length is defined as follows:
    If

$$x = [x(1) \ x(2) \ \cdots \ x(n)], \quad y = [y(1) \ y(2) \ \cdots \ y(n)]$$

then

$$x \cdot y = \sum_{i=1}^{n} x(i) y(i).$$

The dot product appears and is useful in many areas of math and physics. As an example, check that the dot product of the vectors [2 0 6] and [1 −1 4] is 26. In MATLAB, if x and y are stored as row vectors, then you can get the dot product by typing $x * y'$ (the prime stands for transpose, as in the previous section; Chapter 7 will explain matrix operations in greater detail). Let x and y be the vectors each with 100 components having the forms:

$$x = [1, -1, \ 1, -1, \ \ 1, \ -1, \ \cdots],$$
$$y = [1, \ \ 4, 9, 16, \ 25, \ 36, \ \cdots].$$

Use a while loop in MATLAB to create and store these vectors and then compute their dot product.

Chapter 2: Basic Concepts of Numerical Analysis with Taylor's Theorem

2.1: WHAT IS NUMERICAL ANALYSIS?

Outside the realm of pure mathematics, most practicing scientists and engineers are not concerned with finding exact answers to problems. Indeed, living in a finite universe, we have no way of exactly measuring physical quantities and even if we did, the exact answer would not be of much use. Just a single irrational number, such as

$\pi = 3.14159265358979323846264338327950288419716939937510582097749...,$

where the digits keep going on forever without repetition or any known pattern, has more information in its digits than all the computers in the world could possibly ever store. To help motivate some terminology, we bring forth a couple of examples.

Suppose that Los Angeles County is interested in finding out the amount of water contained in one of its reserve drinking water reservoirs. It hires a contracting firm to measure this amount. The firm begins with a large-scale pumping device to take care of most of the water, leaving only a few gallons. After this, they bring out a more precise device to measure the remainder and come out with a volume of 12,564,832.42 gallons. To get a second opinion, the county hires a more sophisticated engineering firm (that charges 10 times as much) and that uses more advanced measuring devices. Suppose this latter firm came up with the figure 12,564,832.3182. Was the first estimate incorrect? Maybe not, perhaps some evaporation or spilling took place—so there cannot really be an exact answer to this problem. Was it really worth the extra cost to get this more accurate estimate? Most likely not—even an estimate to the nearest gallon would have served equally well for just about any practical purposes.

Suppose next that the Boeing Corporation, in the design and construction of a new 767 model jumbo jet, needs some wing rivets. The engineers have determined the rivets should have a diameter of 2.75 mm with a tolerance (for error) of .000025 mm. Boeing owns a precise machine that will cut such rivets to be of diameter $2.75 \pm .000006$ mm. But they can purchase a much more expensive machine that will produce rivets of diameters $2.75 \pm .0000001$ mm (60 times as accurate). Is it worth it for Boeing to purchase and use this more expensive machine? The aeronautical engineers have determined that such an improvement in rivets would not result in any significant difference in the safety and reliability of the wing and plane; however, if the error exceeds the given tolerance, the wings may become unstable and a safety hazard.

In mathematics, there are many problems and equations (algebraic, differential, and partial differential) whose exact solutions are known to exist but are difficult, very time consuming, or impossible to solve exactly. But for many practical purposes, as evidenced by the previous examples, an estimate to the exact answer will do just fine, provided that we have a guarantee that the error is not too large. So, here is the basic problem in numerical analysis: We are interested in a solution x (= **exact answer**) to a problem or equation. We would like to find an estimate x^* (= **approximation**) so that $|x - x^*|$ (= the **actual error**) is no more than the maximum **tolerance** $(= \varepsilon)$, i.e., $|x - x^*| \le \varepsilon$. The maximum tolerated error will be specified ahead of time and will depend on the particular problem at hand. What makes this approximation problem very often extremely difficult is that we usually do not know x and thus, even after we get x^*, we will have no way of knowing the actual error. But regardless of this, we still need to be able to guarantee that it is less than ε. Often more useful than the actual error is the **relative error**, which measures the error as a ratio in terms of the magnitude of the actual quantity; i.e., it is defined by

$$\text{relative error} = \frac{|x - x^*|}{|x|},$$

provided, of course, that $x \ne 0$.

EXAMPLE 2.1: In the Los Angeles reservoir measurement problem given earlier, suppose we took the exact answer to be the engineering firm's estimate, $x = 12,564,832.3182$ gallons, and the contractor's estimate as the approximation $x^* = 12,564,832.42$. Then the error of this approximation is $|x - x^*| = 0.1018$ gallons, but the relative error (divide this answer by x) is only 8.102×10^{-9}.

EXAMPLE 2.2: In the Boeing Corporation's rivet problem above, the maximum tolerated error is .000025 mm, which translates to a maximum relative error of (divide by $x = 2.75$) 0.000009. The machine they currently have would yield a maximum relative error of 0.000006/2.75 = 0.000002 and the more expensive machine they were considering would guarantee a maximum relative error of no more than 0.0000001/2.75 = 3.6364×10^{-8}.

For the following reasons, we have chosen Taylor's theorem as a means to launch the reader into the realm of numerical analysis. First, Taylor's theorem is at the foundation of most numerical methods for differential equations, the subject of this book. Second, it covers one of those rare situations in numerical analysis where quality error estimates are readily available and thus errors can be controlled and estimated quite effectively. Finally, most readers should have some familiarity with Taylor's theorem from their calculus courses.

Most mathematical functions are very difficult to compute by just using the basic mathematical operations: $+, -, \times, \div$. How, for example, would we compute

$\cos(27°)$ just using these operations? One type of function that is possible to compute in this way is a polynomial. A **polynomial** in the variable x is a function of the form:

$$p(x) = a_n x^n + \cdots + a_2 x^2 + a_1 x + a_0,$$

where $a_n, \cdots a_2, a_1, a_0$ are any real numbers. If $a_n \neq 0$, then we say that the **degree** of $p(x)$ equals n. Taylor's theorem from calculus shows how to use polynomials to approximate a great many mathematical functions to any degree of accuracy. In Section 2.2, we will introduce the special kind of polynomial (called Taylor polynomials) that gets used in this theorem and in Section 2.3 we discuss the theorem and its uses.

EXERCISES 2.1:

1.   If $x = 2$ is approximated by $x^* = 1.96$, find the actual error and the relative error.

2.   If $\pi (= x)$ is approximated by $x^* = 3\frac{1}{8}$ (as was done by the ancient Babylonians, c. 2000 BC), find the actual error and the relative error.

3.   If $x = 10000$ is approximated by $x^* = 9999.96$, find the actual error and the relative error.

4.   If $x = 5280$ feet (one mile) is approximated by $x^* = 5281$ feet, find the actual and relative errors.

5.   If $x = 0.76$ inches and the relative error of an approximation is known to be 0.05, find the possible values for x^*.

6.   If $x = 186.4$ and the relative error of an approximation is known to be 0.001, find the possible values for x^*.

7.   A civil engineering firm wishes to order thick steel cables for the construction of a span bridge. The cables need to measure 2640 feet in length with a maximum tolerated relative error of 0.005. Translate this relative error into an actual tolerated maximum discrepancy from the ideal 2640-foot length.

2.2: TAYLOR POLYNOMIALS

Suppose that we have a mathematical function $f(x)$ that we wish to approximate near $x = a$. The **Taylor polynomial of order** n, $p_n(x)$, for this function **centered at** (or **about**) $x = a$ is that polynomial of degree of at most n that has the same values as $f(x)$ and its first n derivatives at $x = a$. The definition requires that $f(x)$ possess n derivatives at $x = a$. Since derivatives measure rates of change of functions, the Taylor polynomials are designed to mimic the behavior of the function near $x = a$. The following example will demonstrate this property.

EXAMPLE 2.3: Find formulas for, and interpret, the order-zero and order-one Taylor polynomials $p_0(x)$ and $p_1(x)$ of a function $f(x)$ (differentiable) at $x = a$.

SOLUTION: The zero-order polynomial $p_0(x)$ has degree at most zero, and so must be a constant function. But by its definition, we must have $p_0(a) = f(a)$. Since $p_0(x)$ is constant this means that $p_0(x) = f(a)$ (a horizontal line function). The first-order polynomial $p_1(x)$ must satisfy two conditions:

$$p_1(a) = f(a) \text{ and } p_1'(a) = f'(a). \tag{1}$$

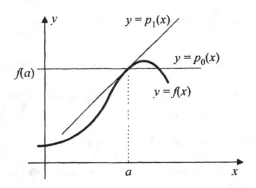

FIGURE 2.1: Illustration of a graph of a function $y = f(x)$ (heavy black curve) together with its zero-order Taylor polynomial $p_0(x)$ (horizontal line) and first-order Taylor polynomial $p_1(x)$ (slanted tangent line).

Since $p_1(x)$ has degree at most one, we can write $p_1(x) = mx + b$, i.e., $p_1(x)$ is just a line with slope m and y-intercept b. If we differentiate this equation and use the second equation in (1), we get that $m = f'(a)$. We now substitute this in for m, put $x = a$ and use the first equation in (1) to find that $f(a) = p_1(a) = f'(a)a + b$. Solving for b gives $b = f(a) - f'(a)a$. So putting this all together yields that $p_1(x) = mx + b = f'(a)x + f(a) - f'(a)a = f(a) + f'(a)(x - a)$. This is just the tangent line to the graph of $y = f(x)$ at $x = a$. These two polynomials are illustrated in Figure 2.1.

In general, it can be shown that the Taylor polynomial of order n is given by the following formula:

$$p_n(x) = f(a) + f'(a)(x-a) + \frac{1}{2}f''(a)(x-a)^2 + \frac{1}{3!}f'''(a)(x-a)^3 +$$
$$\cdots + \frac{1}{n!}f^{(n)}(a)(x-a)^n, \tag{2}$$

where we recall that the factorial of a positive integer k is given by:

$$k! = \begin{cases} 1, & \text{if } k = 0 \\ 1 \cdot 2 \cdot 3 \cdots (k-1) \cdot k, & \text{if } k = 1, 2, 3, \cdots. \end{cases}$$

Since $0! = 1! = 1$, we can use Sigma-notation to rewrite this more simply as:

$$p_n(x) = \sum_{k=0}^{n} \frac{1}{k!} f^{(k)}(a)(x-a)^k. \tag{3}$$

We turn now to some specific examples:

EXAMPLE 2.4: (a) For the function $f(x) = \cos(x)$, compute the following Taylor polynomials at $x = 0$: $p_1(x)$, $p_2(x)$, $p_3(x)$, and $p_8(x)$.
(b) Use MATLAB to find how each of these approximates $\cos(27°)$ and then find the actual error of each of these approximations.
(c) Find a general formula for $p_n(x)$.
(d) Using an appropriate MATLAB graph, estimate the length of the largest interval $[-a, a] = \{|x| \le a\}$ about $x = 0$ that $p_8(x)$ can be used to approximate $f(x)$ with an error always less than or equal to 0.2. What if we want the error to be less than or equal to 0.001?

SOLUTION: Part (a): We see from formula (2) or (3) that each Taylor polynomial is part of any higher-order Taylor polynomial. Since $a = 0$ in this example, formula (2) reduces to:

$$p_n(x) = f(0) + f'(0)x + \frac{1}{2} f''(0)x^2 + \frac{1}{3!} f'''(0)x^3 + \cdots$$

$$+ \frac{1}{n!} f^{(n)}(0)x^n = \sum_{k=0}^{n} \frac{1}{k!} f^{(k)}(0)x^k. \tag{4}$$

A systematic way to calculate these polynomials is by constructing a table for the derivatives:

n	$f^{(n)}(x)$	$f^{(n)}(0)$
0	$\cos(x)$	1
1	$-\sin(x)$	0
2	$-\cos(x)$	-1
3	$\sin(x)$	0
4	$\cos(x)$	1
5	$-\sin(x)$	0

We could continue on, but if one notices the repetitive pattern (when $n = 4$, the derivatives go back to where they began), this can save some time and help with finding a formula for the general Taylor polynomial $p_n(x)$. Using formula (4) in conjunction with the table (and indicated repetition pattern), we conclude that:

$$p_1(x) = 1, \ p_2(x) = 1 - \frac{x^2}{2} = p_3(x), \text{ and } p_8(x) = 1 - \frac{x^2}{2} + \frac{x^4}{4!} - \frac{x^6}{6!} + \frac{x^8}{8!}.$$

Part (b): To use these polynomials to approximate $\cos(27°)$, we of course need to take x to be in radians, i.e., $x = 27° \left(\frac{\pi}{180°} \right) = .4712389....$. Since two of these Taylor polynomials coincide, there are three different approximations at hand for $\cos(27°)$. In order to use MATLAB to make the desired calculations, we introduce a relevant MATLAB function:

> To compute n! in MATLAB, use either: `factorial(n)`, or `gamma(n+1)`

Thus for example, to get $5! = 120$, we could either type `>>factorial(5)` or `>>gamma(6)`. Now we calculate:

```
>> x=27*pi/180;
>> format long
>> p1=1; p2=1-x^2/2
→ 0.88896695048774
>> p8=p2+x^4/gamma(5)-x^6/gamma(7)+x^8/gamma(9)
→ 0.89100652433693
>> abs(p1-cos(x))    %abs, as you guessed, stands for absolute value
→ 0 10899347581163
>> abs(p2-cos(x))    → 0.00203957370062
>> abs(p8-cos(x))    → 1.485654932409375e-010
```

Transcribing these into usual mathematics notation, we have the approximations for $\cos(27°)$:

$$p_1(27°) = 1, \ p_2(27°) = p_3(27°) = .888967..., \ p_8(27°) = .89100694..., $$

which have the corresponding errors:

$$|p_1(27°) - \cos(27°)| = 0.1089...,$$
$$|p_2(27°) - \cos(27°)| = |p_3(27°) - \cos(27°)| = 0.002039..., \text{ and }$$
$$|p_8(27°) - \cos(27°)| = 1.4856 \times 10^{-10}.$$

This demonstrates quite clearly how nicely these Taylor polynomials serve as approximating tools. As expected, the higher degree Taylor polynomials do a better job approximating but take more work to compute.

Part (c): Finding the general formula for the nth-order Taylor polynomial $p_n(x)$ can be a daunting task, if it is even possible. It will be possible if some pattern can be discovered with the derivatives of the corresponding function at $x = a$. In this case, we have already discovered the pattern in part (b), which is quite simple: We just need a nice way to write it down, as a formula in n. It is best to separate into cases where n is even or odd. If n is odd we see $f^{(n)}(0) = 0$, end of story. When n is even $f^{(n)}(0)$ alternates between $+1$ and -1. To get a formula, we write an even n as $2k$, where k will alternate between even and odd integers. The trick is to use either $(-1)^k$ or $(-1)^{k+1}$ for $f^{(2k)}(0)$, which both also alternate between $+1$ and -1. To see which of the two to use, we need only check the starting values at $k = 0$ (corresponding to $n = 0$). Since $f^{(0)}(0) = 1$, we must use $(-1)^k$. Since any odd integer can be written as $n = 2k+1$, in summary we have arrived at the following formula for $f^{(n)}(0)$:

$$f^{(n)}(0) = \begin{cases} (-1)^k, & \text{if } n = 2k \text{ is even,} \\ 0, & \text{if } n = 2k+1 \text{ is odd} \end{cases}$$

Plugging this into equation (4) yields the formulas:

$$p_n(x) = 1 - \frac{x^2}{2!} + \frac{x^4}{4!} - \cdots + (-1)^k \frac{x^{2k}}{(2k)!} = \sum_{j=0}^{k} (-1)^j \frac{x^{2j}}{(2j)!}$$
$$(\text{for } n = 2k \text{ or } n = 2k+1).$$

Part (d): In order to get a rough idea of how well $p_8(x)$ approximates $\cos(x)$, we will first need to try out a few plots. Let us first plot these two functions together on the domain: $-10 \le x \le 10$. This can be done as follows:

```
>> x=-10:.0001:10;
>> y=cos(x);
>> p8=1-x.^2/2+x.^4/gamma(5)-x.^6/gamma(7)+x.^8/gamma(9);
>> plot(x,y,x,p8,'r-.')
```

Notice that we were able to produce both plots (after having constructed the needed vectors) by a single line, without the hold on/hold off method. We have instructed MATLAB to plot the original function $y = \cos(x)$ in the default color and style (blue, solid line) and the approximating function $y = p_8(x)$ as a red plot with the dash/dot style. The resulting plot is the first one shown in Figure 2.2.

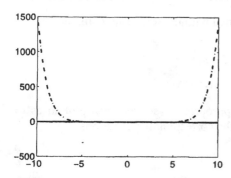 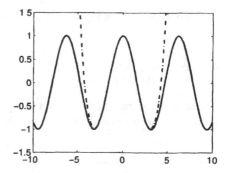

FIGURE 2.2: Graphs of $y = \cos(x)$ (solid) together with the eighth-order Taylor approximating polynomial $y = p_8(x)$ (dash-dot) shown with two different y-ranges.

To answer (even just) the first question, this plot is not going to help much, owing to the fact that the scale on the y-axis is so large (increments are in 200 units and we need our error to be < 0.2). MATLAB always will choose the scales to accommodate all of the points in any given plot. The eighth-degree polynomial $y = p_8(x)$ gets so large at $x = \pm 10$ that the original function $y = \cos(x)$ is dwarfed in comparison so its graph will appear as a flat line (the x-axis). We could redo the plot trying out different ranges for the x-values and eventually arrive at a more satisfactory illustration.[1] Alternatively and more simply, we can work with the existing plot and get MATLAB to manually change the range of the x- and/or y-axes that appear in the plot. The way to do this is with the command:

`axis([xmin xmax ymin ymax])` $\rightarrow$	Changes the range of a plot to : $x\min \le x \le x\max,$ and $y\min \le y \le y\max.$

Thus to keep the x-range the same $[-10, 10]$, but to change the y-range to be $[-1.5, 1.5]$, we would enter the command to create the second plot of Figure 2.2.

```
>> axis([-10 10 -1.5 1.5])
```

We can see now from the second plot above that (certainly) for $-3 \le x \le 3$ we have $|\cos(x) - p_8(x)| \le 0.2$. This graph is, however, unsatisfactory in regards to the second question of the determination of an x-interval for which $|\cos(x) - p_8(x)| \le 0.001$. To answer this latter question and also to get a more satisfactory answer for the first question, we need only look at plots of the actual error $y = |\cos(x) - p_8(x)|$. We do this for two different y-ranges. There is a nice way to get MATLAB to partition its plot window into several (in fact, a matrix of) smaller subwindows.

`subplot(m,n,i)` $\rightarrow$	Causes the plot window to partition into an $m \times n$ matrix of proportionally smaller subwindows, with the next plot going into the ith subwindow (listed in the usual "reading order"—left to right, then top to bottom).

The two error plots in Figure 2.3 were obtained with the following commands:

```
>> subplot(2,1,1)
>> plot(x,abs(y-p8)), axis([-10 10 -.1 .3])
>> subplot(2,1,2)
>> plot(x,abs(y-p8)), axis([-5 5 -.0005 .0015])
```

Notice that the ranges for the axes were appropriately set for each plot so as to make each more suitable to answer each of the corresponding questions.

---

[1] The zoom button $\oplus$ on the graphics window can save some time here. To use it, simply left click on this button with your mouse, then move the mouse to the desired center of the plot at which to zoom and left click (repeatedly). The zoom-out key $\ominus$ works in the analogous fashion.

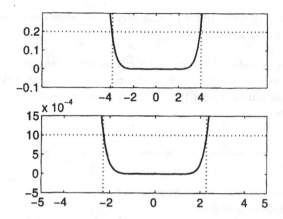

FIGURE 2.3: Plots of the error $y = |\cos(x) - p_8(x)|$ on two different y-ranges. Reference lines were added to help answer the question in part (d) of Example 2.4.

From Figure 2.3, we can deduce that if we want to guarantee an error of at most 0.2, then we can use $p_8(x)$ to approximate $\cos(x)$ anywhere on the interval [−3.8, 3.8], while if we would like the maximum error to be only 0.001, we must shrink the interval of approximation to about [−2.2, 2.2]. In Figure 2.4 we give a MATLAB-generated plot of the function $y = \cos(x)$ along with several of its Taylor polynomials.

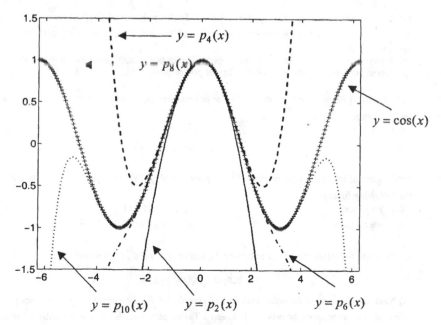

FIGURE 2.4: Some Taylor polynomials for $y = \cos(x)$.

EXERCISE FOR THE READER 2.1: Use MATLAB to produce the plot in Figure 2.4 (without the arrow labels).

It is a rare situation indeed in numerical analysis where we can actually compute the exact errors explicitly. In the next section, we will give Taylor's theorem, which gives us usable estimates for the error even in cases where it cannot be explicitly computed.

EXERCISES 2.2:

1. Find the second- and third-order Taylor polynomials $p_2(x)$ and $p_3(x)$, centered at $x = a$, for the each of the following functions.
    (a) $f(x) = \sin(x)$, $a = 0$ (b) $f(x) = \tan(x)$, $a = 0$
    (c) $f(x) = e^x$, $a = 1$ (d) $f(x) = x^{1/3}$, $a = 8$

2. Repeat Exercise 1 for each of the following:
    (a) $f(x) = \cos(x)$, $a = \pi/2$ (b) $f(x) = \arctan(x)$, $a = 0$
    (c) $f(x) = \ln x$, $a = 1$ (d) $f(x) = \cos(x^2)$, $a = 0$

3. (a) Approximate $\sqrt{65}$ by using the first-order Taylor polynomial of $f(x) = \sqrt{x}$ centered at $x = 64$ (this is tangent line approximation discussed in first-semester calculus) and find the error and the relative error of this approximation.
    (b) Repeat part (a) using instead the second-order Taylor polynomial to do the approximation.
    (c) Repeat part (a) once again, now using the fourth-order Taylor polynomial.

4. (a) Approximate $\sin(92°)$ by using the first-order Taylor polynomial of $f(x) = \sin(x)$ centered at $x = \pi/2$ (tangent line approximation) and find the error and the relative error of this approximation.
    (b) Repeat part (a) using instead the second-order Taylor polynomial to do the approximation.
    (c) Repeat part (a) using the fourth-order Taylor polynomial.

5. Find a general formula for the order n Taylor polynomial $p_n(x)$ centered at $x = 0$ for each of the following functions:
    (a) $y = \sin(x)$ (b) $y = \ln(1 + x)$
    (c) $y = e^x$ (d) $y = \sqrt{x + 1}$

6. Find a general formula for the order n Taylor polynomial $p_n(x)$ centered at $x = 0$ for each of the following functions:
    (a) $y = \tan(x)$ (b) $y = 1/(1 + x)$
    (c) $y = \arctan(x)$ (d) $y = x\sin(x)$

7. (a) Compute the following Taylor polynomials, centered at $x = 0$, of $y = \cos(x^2)$:

$$p_1(x),\ p_2(x),\ p_6(x),\ p_{10}(x).$$

    (b) Next, use the general formula obtained in Example 2.4 for the general Taylor polynomials of $y = \cos(x)$ to write down the order 0, 1, 3, and 5 Taylor polynomials. Replace x with x^2 in each of these polynomials. Compare these with the Taylor polynomials in part (a).

8. Consider the function $f(x) = \sin(3x)$. All of the plots in this problem are to be done with MATLAB on the interval $[-3, 3]$. The Taylor polynomials refer to those of $f(x)$ centered at $x = 0$. Each graph of $f(x)$ should be done with the usual plot settings, while each graph of a Taylor polynomial should be done with the dot style.
   (a) Use the subplot command to create a graphic with 3×2 entries as follows:

The simultaneous graphs of $f(x)$ along with the 1st-order Taylor polynomial (= tangent line).	A graph of the error $\lvert f(x) - p_1(x) \rvert$
The simultaneous graphs of $f(x)$ along with the 3rd-order Taylor polynomial.	A graph of the error $\lvert f(x) - p_3(x) \rvert$
The simultaneous graphs of $f(x)$ along with the 9th-order Taylor polynomial.	A graph of the error $\lvert f(x) - p_9(x) \rvert$

   (b) By looking at your graphs in part (a), estimate on how large an interval $[-a, a]$ about $x = 0$ that the first-order Taylor polynomial would provide an approximation to $f(x)$ with error < 0.25. Answer the same question for p_3 and p_9.

9. (a) Let $f(x) = \ln(1 + x^2)$. Find formulas for the following Taylor polynomials of $f(x)$ centered at $x = 0$: $p_2(x)$, $p_3(x)$, $p_6(x)$. Next, using the subplot command, create a graphic window split in two sides (left and right). On the left, plot (together) the four functions $f(x)$, $p_2(x)$, $p_3(x)$, $p_6(x)$. In the right-side subwindow, plot (together) the corresponding graphs of the three errors: $\lvert f(x) - p_2(x) \rvert$, $\lvert f(x) - p_3(x) \rvert$, and $\lvert f(x) - p_6(x) \rvert$. For the error plot adjust the y-range so as to make it simple to answer the question in part (b). Use different styles/colors to code different functions in a given plot.
   (b) By looking at your graphs in part (a), estimate how large an interval $[-a, a]$ about $x = 0$ on which the second-order Taylor polynomial would provide an approximation to $f(x)$ with error < 0.25. Answer the same question for p_3 and p_6.

10. (a) Let $f(x) = x^2 \sin(x)$. Find formulas for the following Taylor polynomials of $f(x)$ centered at $x = 0$: $p_1(x)$, $p_4(x)$, $p_9(x)$. Next, using the subplot command, get MATLAB to create a graphic window split in two sides (left, and right). On the left, plot (together) the four functions $f(x)$, $p_1(x)$, $p_4(x)$, $p_9(x)$. In the right-side subwindow, plot (together) the corresponding graphs of the three errors: $\lvert f(x) - p_1(x) \rvert$, $\lvert f(x) - p_4(x) \rvert$, and $\lvert f(x) - p_9(x) \rvert$. For the error plot adjust the y-range so as to make it simple to answer the question in part (b). Use different styles/colors to code different functions in a given plot.
   (b) By looking at your graphs in part (a), estimate on how large an interval $[-a, a]$ about $x = 0$ the first-order Taylor polynomial would provide an approximation to $f(x)$ with error < 0.05. Answer the same question for p_4 and p_9.

11. In Example 2.3, we derived the general formula (2) for the zero- and first-order Taylor polynomial.
   (a) Do the same for the second-order Taylor polynomial, i.e., use the definition of the Taylor polynomial $p_2(x)$ to show that (2) is valid when $n = 2$.
   (b) Prove that formula (4) for the Taylor polynomials centered at $x = 0$ is valid for any n.
   (c) Prove that formula (2) is valid for all n.
   Suggestion: For part (c), consider the function $g(x) = f(x + a)$, and apply the result of part

(b) to this function. How are the Taylor polynomials of $g(x)$ at $x = 0$ related to those of $f(x)$ at $x = a$?

12. (*Another Kind of Polynomial Interpolation*) In this problem we compare the fourth-order Taylor polynomial $p_3(x)$ of $y = \cos(x)$ at $x = 0$ (which is actually $p_4(x)$) with the third-order polynomial $p(x) = a_0 + a_1 x + a_2 x^2 + a_3 x^3$, which has the same values and derivative as $\cos(x)$ at the points $x = 0$ and $x = \pi$. This means that $p(x)$ satisfies these four conditions:

$$p(0) = 1 \qquad p'(0) = 0 \qquad p(\pi) = -1 \qquad p'(\pi) = 0.$$

Find the coefficients: $a_0, a_1, a_2,$ and a_3 of $p(x)$, and then plot all three functions together. Discuss the errors of the two different approximating polynomials.

2.3: TAYLOR'S THEOREM

In the examples and problems of previous section we introduced Taylor poly-

nomials $p_n(x)$ of a function $y = f(x)$ (appropriately differentiable) at $x = a$, and we saw that they appear to often serve as great tools for approximating the function near $x = a$. We also have seen that as the order n of the Taylor polynomial increases, so does its effectiveness in approximating $f(x)$. This of course needs to be reconciled with the fact that for larger values of n it is more work to form and compute $p_n(x)$. Additionally, the approximations seemed to improve, in general, when x gets closer to a. This latter observation seems plausible since $p_n(x)$ was constructed using only information about $f(x)$ at

FIGURE 2.5: Brook Taylor[2] (1685–1731), English mathematician.

$x = a$. Taylor's theorem provides precise quantitative estimates for the error

---

[2] Taylor was born in Middlesex, England, and his parents were quite well-rounded people of society. His father was rather strict but instilled in Taylor a love for music and painting. His parents had him educated at home by private tutors until he entered St. John's College in Cambridge when he was 18. He graduated in 1709 after having written his first important mathematical paper a year earlier (this paper was published in 1714). He was elected rather early in his life (1712) to the Royal Society, the election being based more on his potential and perceived mathematical powers rather than on his published works, and two years later he was appointed to the prestigious post of Secretary of the Royal Society. In this same year he was appointed to an important committee that was to settle the issue of "who invented calculus" since both Newton and Leibniz claimed to be the founders. Between 1712 and 1724 Taylor published 13 important mathematical papers on a wide range of subjects including magnetism, logarithms, and capillary action.

Taylor suffered some tragic personal events. His father objected to his marriage (claiming the bride's family was not a "good" one) and after the marriage Taylor and his father cut off all contact until 1723, when his wife died giving birth to what would have been Taylor's first child. Two years later he remarried (this time with his father's blessings) but the following year his new wife also died during childbirth, although this time his daughter survived.

$|f(x) - p_n(x)|$, which can be very useful in choosing an appropriate order n so that $p_n(x)$ will give an approximation within the desired error bounds. We now present Taylor's theorem. For its proof we refer the reader to any decent calculus textbook.

THEOREM 2.1: (*Taylor's Theorem*) Suppose that for a positive integer n, a function $f(x)$ has the property that its $(n+1)$st derivative is continuous on some interval I on the x-axis that contains the value $x = a$. Then the **nth-order remainder** $R_n(x) \equiv f(x) - p_n(x)$ resulting from approximating $f(x)$ by $p_n(x)$ is given by

$$R_n(x) = \frac{f^{(n+1)}(c)}{(n+1)!}(x-a)^{n+1} \quad (x \in I), \tag{5}$$

for some number c, lying between a and x (inclusive); see Figure 2.6.

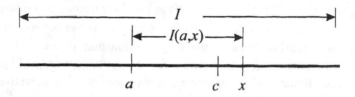

FIGURE 2.6: One possible arrangement of the three points relevant to Taylor's theorem.

REMARKS: (1) Note that like many such theorems in calculus, Taylor's theorem asserts the existence of the number c, but it does not tell us what it is.
(2) By its definition, (the absolute value of) $R_n(x)$ is just the actual error of the approximation of $f(x)$ by its nth-order Taylor polynomial $p_n(x)$, i.e.,

$$\text{error} = |f(x) - p_n(x)| = |R_n(x)|.$$

Since we do not know the exact value of c, we will not be able to calculate the error precisely; however, since we know that c must lie somewhere between a and x on I, call this interval $I(a,x)$, we can estimate that the unknown quantity $|f^{(n+1)}(c)|$ that appears in $R_n(x)$, can be no more than the maximum value of this $(n+1)$st derivative function $|f^{(n+1)}(z)|$ as z runs through the interval $I(a,x)$. In mathematical symbols, this is expressed as:

$$|f^{(n+1)}(c)| \leq \max\{|f^{(n+1)}(z)| : z \in I(a,x)\}. \tag{6}$$

EXAMPLE 2.5: Suppose that we wish to use Taylor polynomials (at $x = 0$) to approximate $e^{0.7}$ with an error less than 0.0001.

(a) Apply Taylor's theorem to find out what order n of a Taylor polynomial we could use for the approximation to guarantee the desired accuracy.
(b) Perform this approximation and then check that the actual error is less than the maximum tolerated error.

SOLUTION: Part (a): Here $f(x) = e^x$, so, since $f(x)$ is its own derivative, we have $f^{(n)}(x) = e^x$ for any n, and so $f^{(n)}(0) = e^0 = 1$. From (4) (or (2) or (3) with $a = 0$), we can therefore write the general Taylor polynomial for $f(x)$ centered at $x = 0$ as

$$p_n(x) = 1 + x + \frac{x^2}{2!} + \frac{x^3}{3!} + \cdots + \frac{x^n}{n!} = \sum_{k=0}^{n} \frac{x^k}{k!},$$

and from (5) (again with $a = 0$), $R_n(0.7) = \dfrac{e^c}{(n+1)!}(0.7)^{n+1}$.

How big can e^c be? For $f(x) = e^x$, this is just the question of finding out the right side of (6). In this case the answer is easy: Since c lies between 0 and 0.7, and e^x is an increasing function, the largest value that e^c can be is $e^{0.7}$. To honestly use Taylor's theorem here (since "we do not know" what $e^{0.7}$ is—that's what we are trying to approximate), let's use the conservative upper bound: $e^c \le e^{0.7} \le e^1 = e < 3$.

Now Taylor's theorem tells us that

$$\text{error} = |e^{0.7} - p_n(0.7)| = |R_n(0.7)| = \frac{e^c}{(n+1)!}(0.7)^{n+1}.$$

(Since all numbers on the right side are nonnegative, we are able to drop absolute value signs.) As was seen above, we can replace e^c with 3 in the right side above, to get

$$\text{something larger than the error} = \frac{3}{(n+1)!}(0.7)^{n+1}.$$

The rest of the plan is simple: We find an n large enough to make the "something larger than actual error" to be less than the desired accuracy 0.0001. Then it will certainly follow that the actual error will also be less than 0.0001. We can continue to compute $3(0.7)^{n+1}/(n+1)!$ until it gets smaller than 0.0001. Better yet, let's use a while loop to get MATLAB to do the work for us; this will also provide us with a good occasion to introduce the remaining relational operators that can be used in any while loops (or subsequent programming). (See Table 2.1.)

TABLE 2.1: Dictionary of MATLAB's relational operators.

Mathematical Relation	MATLAB Code
$>, <$	$>$, $<$
$\geq, \leq$	$>=$, $<=$
$=, \neq$	$==$, $\sim=$

We have already used one of the first pair. For the last one, we reiterate again that the single equal sign in MATLAB is reserved for "assignment." Since it gets used much less often (in MATLAB codes), the ordinary equals in mathematics got stuck with the more cumbersome MATLAB notation.

Now, back to our problem. A simple way to figure out that smallest feasible value of n would be to run the following code:

```
>> n=1;
>> while 3*(0.7)^(n+1)/gamma(n+2) >= 0.0001
      n=n+1;
end
```

This code has no output, but what it does is to keep making n bigger, one-by-one, until that "something larger than the actual error" gets less than 0.0001. The magic value of n that will work is now (by the way the while loop was constructed) simply the last stored value of n:

```
>>n   → 6
```

Part (b): The desired approximation is now: $e^{0.7} \approx p_6(0.7) = \sum_{k=0}^{6} \left.\frac{(0.7)^k}{k!}\right|_{x=0.7}$.

We can do the rest on MATLAB:

```
>> x=0.7;
>> n=0;
>> p6=0;   % we initialize the sum for the Taylor polynomial p6
>> while n<=6
 p6=p6+x^n/gamma(n+1);
 n=n+1;
end
>> p6
→ 2.0137 (approximation)
>> abs(p6-exp(0.7)) %we now check the actual error
→ 1.7889e-005 (this is less than 0.0001, as we knew from Taylor's theorem.)
```

EXERCISE FOR THE READER 2.2: If we use Taylor polynomials of $f(x) = \sqrt{x}$ centered at $x = 16$ to approximate $\sqrt{17} = f(16+1)$, what order Taylor polynomial should be used to ensure that the error of the approximation is less than 10^{-10}? Perform this approximation and then look at the actual error.

For any function $f(x)$, which has infinitely many derivatives at $x = a$, we can form the **Taylor series (centered) at** $x = a$:

$$f(a) + f'(a)(x-a) + \frac{f''(a)}{2}(x-a)^2 + \frac{f'''(a)}{3!}(x-a)^3 + \cdots$$
$$+ \frac{f^{(n)}(a)}{n!}(x-a)^n + \cdots = \sum_{k=0}^{\infty} \frac{f^{(k)}(a)}{k!}(x-a)^k. \qquad (7)$$

Comparing this with (2) and (3), the formulas for the nth Taylor polynomial $p_n(x)$ at $x = a$, we see that the Taylor series is just the infinite series whose first n terms are exactly the same as those of $p_n(x)$. The Taylor series may or may not converge, but if Taylor's theorem shows that the errors $|p_n(x) - f(x)|$ go to zero, then it will follow that the Taylor series above converges to $f(x)$. When $a = 0$ (the most common situation), the series is called the **Maclaurin series** (Figure 2.7).

It is useful to have some Maclaurin series for reference. Anytime we are able to figure out a formula for the general Taylor polynomial at $x = 0$, we can write down the corresponding Maclaurin series. The previous examples we have done yield the Maclaurin series for $\cos(x)$ and e^x. We list these in Table 2.2, as well as a few other examples whose derivations will be left to the exercises.

TABLE 2.2: Some Maclaurin series expansions.

Function	Maclaurin Series	
e^x	$1 + x + \dfrac{x^2}{2!} + \dfrac{x^3}{3!} + \cdots + \dfrac{x^k}{k!} + \cdots$	(8)
$\cos(x)$	$1 - \dfrac{x^2}{2!} + \dfrac{x^4}{4!} + \cdots + \dfrac{(-1)^k x^{2k}}{(2k)!} + \cdots$	(9)
$\sin(x)$	$x - \dfrac{x^3}{3!} + \dfrac{x^5}{5!} \cdots + \dfrac{(-1)^k x^{2k+1}}{(2k+1)!} + \cdots$	(10)
$\arctan(x)$	$x - \dfrac{x^3}{3} + \dfrac{x^5}{5} - \cdots + \dfrac{(-1)^{2k+1} x^{2k+1}}{2k+1} + \cdots$	(11)
$\dfrac{1}{1-x}$	$1 + x + x^2 + x^3 + \cdots + x^k + \cdots$	(12)

One very useful aspect of Maclaurin and Taylor series is that they can be formally combined to obtain new Maclaurin series by all of the algebraic operations (addition, subtraction, multiplication, and division) as well as with substitutions, derivatives, and integrations. These informally obtained expansions are proved to be legitimate in calculus books. The word "formal" here means that all of the above operations on an infinite series should be done as if it were a finite sum. This method is illustrated in the next example.

FIGURE 2.7: Colin Maclaurin[3] (1698–1746), Scottish mathematician.

EXAMPLE 2.6: Using formal manipulations, obtain the Maclaurin series of the functions (a) $x\sin(x^2)$ and (b) $\ln(1+x^2)$.

SOLUTION: Part (a): In (10) simply replace x with x^2 and formally multiply by x (we use the symbol $\sim$ to mean "has the Maclaurin series"):

$$x\sin(x^2) \sim$$
$$x\left(x^2 - \frac{(x^2)^3}{3!} + \frac{(x^2)^5}{5!} \cdots + \frac{(-1)^k (x^2)^{2k+1}}{(2k+1)!} + \cdots \right)$$
$$= x^3 - \frac{x^7}{3!} + \frac{x^{11}}{5!} \cdots + \frac{(-1)^k x^{4k+3}}{(2k+1)!} + \cdots.$$

NOTE: This would have been a lot more work to do by using the definition and looking for patterns.

Part (b): We first formally integrate (12): $-\ln(1-x)$

$$\sim \int (1 + x + x^2 + x^3 + \cdots + x^n + \cdots)dx = C + x + \frac{x^2}{2} + \frac{x^3}{3} + \cdots + \frac{x^{n+1}}{n+1} + \cdots.$$

---

[3] Maclaurin was born in a small village on the river Ruel in Scotland. His father, who was the village minister, died when Colin was only six weeks old. His mother wanted Colin and his brother John to have good education so she moved the family to Dumbarton, which had reputable schools. Colin's mother died when he was only nine years old and he subsequently was cared for by his uncle, also a minister. Colin began studies at Glasgow University when he was 11 years old (it was more common during these times in Scotland for bright youngsters to begin their university studies early—in fact universities competed for them). He graduated at age 14 when he defended an impressive thesis extending Sir Isaac Newton's theory on gravity. He then went on to divinity school with the intention of becoming a minister, but he soon ended this career path and became a chaired mathematics professor at the University of Aberdeen in 1717 at age 19.

Two years later, Maclaurin met the illustrious Sir Isaac Newton and they became good friends. The latter was instrumental in Maclaurin's appointment in this same year as a Fellow of the Royal Society (the highest honor awarded to English academicians) and subsequently in 1725 being appointed to the faculty of the prestigious University of Edinburgh, where he remained for the rest of his career. Maclaurin wrote several important mathematical works, one of which was a joint work with the very famous Leonhard Euler and Daniel Bernoulli on the theory of tides, which was published in 1740 and won the three a coveted prize from the Académie des Sciences in Paris. Maclaurin was also known as an excellent and caring teacher. He married in 1733 and had seven children. He was also known for his kindness and had many friends, including members of the royalty. He was instrumental in his work at the Royal Society of Edinburgh, having it transformed from a purely medical society to a more comprehensive scientific society. During the 1745 invasion of the Jacobite army, Maclaurin was engaged in hard physical labor in the defense of Edinburgh. This work, coupled with an injury from falling off his horse, weakened him to such an extent that he died the following year.

By making $x = 0$, we get that the integration constant C equals zero. Now negate both sides and substitute x with $-x^2$ to obtain:

$$\ln(1+x^2) \sim x^2 - \frac{(-x^2)^2}{2} - \frac{(-x^2)^3}{3} - \cdots - \frac{(-x^2)^{n+1}}{n+1} - \cdots$$

$$\sim x^2 - \frac{x^4}{2} + \frac{x^6}{3} - \cdots + \frac{(-1)^{k+1} x^{2k}}{k} + \cdots.$$

Our next example involves another approximation using Taylor's theorem. Unlike the preceding approximation examples, this one involves an integral where it is impossible to find the antiderivative.

EXAMPLE 2.7: Use Taylor's theorem to evaluate the integral $\int_0^1 \sin(t^2)\,dt$ with an error $< 10^{-7}$.

SOLUTION:　Let us denote the nth-order Taylor polynomial of $\sin(x)$ centered at $x = 0$ by $p_n(x)$. The formulas for each $p_n(x)$ are easily obtained from the Maclaurin series (10). We will estimate $\int_0^1 \sin(t^2)\,dt$ by $\int_0^1 p_n(t^2)\,dt$ for an appropriately large n. We can easily obtain an upper bound for the error of this approximation:

$$\text{error} = \left| \int_0^1 \sin(t^2)\,dt - \int_0^1 p_n(t^2)\,dt \right| \leq \int_0^1 |\sin(t^2) - p_n(t^2)|\,dt \leq \int_0^1 |R_n(t^2)|\,dt,$$

where $R_n(x)$ denotes Taylor's remainder.　Since any derivative of $f(x) = \sin(x)$ is one of $\pm\sin(x)$ or $\pm\cos(x)$, it follows that for any x, $0 \leq x \leq 1$, we have

$$|R_n(x)| = \left| \frac{f^{(n+1)}(c)x^{n+1}}{(n+1)!} \right| \leq \frac{1}{(n+1)!}.$$

Since in the above integrals, t is running from $t = 0$ to $t = 1$, we can substitute $x = t^2$ in this estimate for $|R_n(x)|$. We can use this and continue with the error estimate for the integral to arrive at:

$$\text{error} = \left| \int_0^1 \sin(t^2)\,dt - \int_0^1 p_n(t^2)\,dt \right| \leq \int_0^1 |R_n(t^2)|\,dt \leq \int_0^1 \frac{dt}{(n+1)!} = \frac{1}{(n+1)!}.$$

As in the previous example, let's now get MATLAB to do the rest of the work for us. We first need to determine how large n must be so that the right side above (and hence the actual error) will be less than 10^{-7}.

```
>> n=1;
>> while 1/gamma(n+2) >= 10^(-7)
       n=n+1;
end
>> n → 10
```

So it will be enough to replace $\sin(t^2)$ 10th-order Taylor polynomial evaluated at t^2, $p_n(t^2)$. The simplest way to see the general form of this polynomial (and its integral) will be to replace x with t^2 in the Maclaurin series (10) and then formally integrate it (this will result in the Maclaurin series for $\int_0^x \sin(t^2)\,dt$):

$$\sin(t^2) \sim t^2 - \frac{t^6}{3!} + \frac{t^{10}}{5!} \cdots + \frac{(-1)^k t^{4k+2}}{(2k+1)!} + \cdots \quad \Rightarrow$$

$$\int_0^x \sin(t^2)\,dt \sim \int_0^x \left(t^2 - \frac{t^6}{3!} + \frac{t^{10}}{5!} + \cdots + \frac{(-1)^k t^{4k+2}}{(2k+1)!} + \cdots \right) dt$$

$$\sim C + \frac{x^3}{3} - \frac{x^7}{7\cdot 3!} + \frac{x^{11}}{11\cdot 5!} + \cdots + \frac{(-1)^k x^{4k+3}}{(4k+3)\cdot(2k+1)!} + \cdots.$$

If we substitute $x = 0$, we see that the constant of integration C equals zero. Thus,

$$\int_0^x \sin(t^2)\,dt \sim \frac{x^3}{3} - \frac{x^7}{7\cdot 3!} + \frac{x^{11}}{11\cdot 5!} + \cdots + \frac{(-1)^k x^{4k+3}}{(4k+3)\cdot(2k+1)!} + \cdots \quad .$$

We point out that the formal manipulation here is not really necessary since we could have obtained from (10) an explicit formula for $p_{10}(t^2)$ and then directly integrated this function. In either case, integrating this function from $t = 0$ to $t = 1$ gives the partial sum of the last Maclaurin expansion (for $\int_0^x \sin(t^2)\,dt$) gotten by going up to the $k = 4$ term, since this corresponds to integrating up through the terms of $p_{10}(t^2)$.

```
>> p10=0;
>> k=0;
>> while k<=4
        p10=p10+(-1)^k/(4*k+3)/gamma(2*k+2);
        k=k+1;
end
>> format long
>> p10
→p10 = 0.31026830280668
```

In summary, we have proved the approximation

$$\int_0^1 \sin(t^2)\,dt \approx 0.31026830280668.$$

Taylor's theorem has guaranteed that this is accurate with an error less than 10^{-7}.

EXERCISE FOR THE READER 2.3: Using formal manipulations, find the 10th-order Taylor polynomial centered at $x = 0$ for each of the following functions: (a) $\sin(x^2) - \cos(x^3)$, (b) $\sin^2(x^2)$.

EXERCISES 2.3:

1. In each part below, we give a function $f(x)$, a center a to use for Taylor polynomials, a value for x, and a positive number ε that will stand for the error. The task will be to (carefully) use Taylor's theorem to find a value of the order n of a Taylor polynomial so that the error of the approximation $f(x) \approx p_n(x)$ is less than ε Afterward, perform this approximation and check that the actual error is really less than what it was desired to be.
    (a) $f(x) = \sin(x)$, $a = 0$, $x = 0\ 2\,\mathrm{rad}$, $\varepsilon = 0.0001$
    (b) $f(x) = \tan(x)$, $a = 0$, $x = 5°$, $\varepsilon = 0.0001$
    (c) $f(x) = e^x$, $a = 0$, $x = -0.4$, $\varepsilon = 0\ 00001$
    (d) $f(x) = x^{1/3}$, $a = 27$, $x = 28$, $\varepsilon = 10^{-6}$

2. Follow the directions in Exercise 1 for the following:
    (a) $f(x) = \cos(x)$, $a = \pi/2$, $x = 88°$, $\varepsilon = 0.0001$
    (b) $f(x) = \arctan(x)$, $a = 0$, $x = 1/239$, $\varepsilon = 10^{-8}$
    (c) $f(x) = \ln x$, $a = 1$, $x = 3$, $\varepsilon = 0.00001$
    (d) $f(x) = \cos(x^2)$, $a = 0$, $x = 2.2$, $\varepsilon = 10^{-6}$

3. Using only the Maclaurin series developed in this section, along with formal manipulations, obtain Maclaurin series for the following functions:
    (a) $x^2 \arctan(x)$ (b) $\ln(1+x)$ (c) $\dfrac{x^2+3x}{1-x}$ (d) $\displaystyle\int_0^x \frac{1}{1-t^3}\,dt$

4. Using only the Maclaurin series developed in this section, along with formal manipulations, obtain Maclaurin series for the following functions:
    (a) $\ln(1+x)$ (b) $1/(1+x^2)$ (c) $\arctan(x^2) - \sin(x)$ (d) $\displaystyle\int_0^x \cos(t^5)\,dt$

5. Find the Maclaurin series for $f(x) = \sqrt{1+x}$.

6. Find the Maclaurin series for $f(x) = (1+x)^{1/3}$.

7. (a) Use Taylor's theorem to approximate the integral $\int_0^1 \cos(t^5)\,dt$ with an error less than 10^{-8}.
    (First find a large enough order n for a Taylor polynomial that can be used from the theorem, then actually perform the approximation.)
    (b) How large would n have to be if we wanted the error to be less than 10^{-30}?

8. The **error function** is given by the formula: $\mathrm{erf}(x) = (2/\sqrt{\pi}) \int_0^x e^{-t^2}\,dt$. It is used extensively in probability theory, but unfortunately the integral cannot be evaluated exactly. Use Taylor's theorem to approximate $\mathrm{erf}(2)$ with an error less than 10^{-6}.

9. Since $\tan(\pi/4) = 1$ we obtain $\pi = 4\arctan(1)$. Using the Taylor series for the arctangent, this gives us a scheme to approximate π.
    (a) Using Taylor polynomials of $\arctan(x)$ centered at $x = 0$, how large an order n Taylor polynomial would we need to use in order for $4p_n(1)$ to approximate π with an error less than 10^{-12}?

(b) Perform this approximation

(c) How large an order n would we need for Taylor's theorem to guarantee that $4p_n(1)$ approximates π with an error less than 10^{-50} ?[4]

(d) There are more efficient ways to compute π. One of these dates back to the early 1700s, when Scottish mathematician John Machin (1680–1751) developed the inverse trig identity:

$$\frac{\pi}{4} = \arctan\left(\frac{1}{5}\right) - \arctan\left(\frac{1}{239}\right). \tag{13}$$

to calculate the first 100 decimal places of π. There were no computers back then, so his work was all done by hand and it was important to do it in way where not so many terms needed to be computed. He did it by using Taylor polynomials to approximate each of the two arctangents on the right side of (13).   What order Taylor polynomials would Machin have needed to use (according to Taylor's theorem) to attain his desired accuracy?

(e) Prove identity (13).

Suggestion: For part (d), use the trig identity:

$$\tan(A \pm B) = \frac{\tan A \pm \tan B}{1 \mp \tan A \tan B}$$

to calculate first $\tan(2\tan^{-1}\frac{1}{5})$, then $\tan(4\tan^{-1}\frac{1}{5})$, and finally $\tan(4\tan^{-1}\frac{1}{5} - \tan^{-1}\frac{1}{239})$.

HISTORICAL ASIDE: Since antiquity, the problem of figuring out π to more and more decimals has challenged the mathematical world, and in more recent times the computer world as well. Such tasks can test the powers of computers as well as the methods used to compute them. Even in the 1970s π had been calculated to over 1 million places, and this breakthrough was accomplished using an identity quite similar to (13). See [Bec-71] for an enlightening account of this very interesting history.

10.   (*Numerical Differentiation*)   (a) Use Taylor's theorem to establish the following *forward difference formula*:

$$f'(a) = \frac{f(a+h) - f(a)}{h} - \frac{h}{2}f''(c),$$

for some number c between a and $a + h$, provided that $f'(x)$ is continuous on $[a, a+h]$. This formula is often used as a means of numerically approximating the derivative $f'(a)$ by the simple difference quotient on the right; in this case the error of the approximation would be $|hf''(c)/2|$ and could be made arbitrarily small if we take h sufficiently small.

(b) .With $f(x) = \sinh(x)$, and $a = 0$, how small would we need to take h for the approximation in part (a) to have error less than 10^{-5} ? Do this by first estimating the error, and then (using your value of h) check the actual error using MATLAB. Repeat with an error goal of 10^{-10}.

(c) Use Taylor's theorem to establish the following *central difference formula*:

$$f''(a) = \frac{f(a+h) - 2f(a) + f(a+h)}{h^2} - \frac{h^2}{12}f^{(4)}(c),$$

for some number c between $a-h$ and $a+h$, provided that $f^{(4)}(x)$ is continuous on $[a-h, a+h]$.   This formula is often used as a means of numerically approximating the derivative $f''(a)$ by the simple difference quotient on the right; in this case the error of the approximation would be $|h^2 f^{(4)}(c)/12|$ and could be made arbitrarily small if we take h sufficiently small.

---

[4] Of course, since MATLAB's compiler keeps track of only about 15 digits, such an accurate approximation could not be done without the help of the Symbolic Toolbox (see Appendix A).

(d) Repeat part (b) for the approximation of part (c). Why do the approximations of part (c) seem more efficient, in that they do not require as small an h to achieve the same accuracy?

(e) Can you derive (and prove using Taylor's theorem) an approximation for $f'(x)$ whose error is proportional to h^2 ?

Chapter 3: Introduction to M-Files

3.1: WHAT ARE M-FILES?

Up to now, all of our interactions with MATLAB have been directly through the command window. As we begin to work with more complicated algorithms, it will be preferable to develop standalone programs that can be separately stored and saved into files that can always be called on (by their name) in any MATLAB session. The vehicle for storing such a program in MATLAB is the so-called **M-file**. M-files are programs that are plain-text (ASCII) files written with any word processing program (e.g., Notepad or MS Word) and are called M-files because they will always be stored with the extension <filename>.m.[1] As you begin to use MATLAB more seriously, you will start to amass your own library of M-files (some of these you will have written and others you may have gotten from other sources such as the Internet) and you will need to store them in various places (e.g., certain folders on your own computer, or also on your portable disk for when you do work on another computer). If you wish to make use of (i.e., "call on") some of your M-files during a particular MATLAB session from the command window, you will need to make sure that MATLAB knows where to look for your M-files. This is done by including all possible directories where you have stored M-files in MATLAB's **path**.[2]

M-files are of two types: **script M-files** and **function M-files**. A script M-file is simply a list of MATLAB commands typed out just as you would do on the command line. The script can have any number of lines in it and once it has been saved (to a directory in the path) it can be invoked simply by typing its name in the command window. When this is done, the script is "run" and the effect will be the same as having actually typed the lines one by one in the command window.

---

[1] It is recommended that you use the default MATLAB M-file editor gotten from the "File" menu (on the top left of the command window) and selecting "New"→ "M-File." This editor is designed precisely for writing M-files and contains many features that are helpful in formatting and debugging. Some popular word processing programs (notably MS Word) will automatically attach a certain extension (e.g., ".doc") at the end of any filename you save a document as and it can be difficult to prevent such things. On a Windows/DOS-based PC, one way to change an M-file that you have created in this way to have the needed ".m" extension is to open the DOS command window, change to the directory you have stored your M-file in, and rename the file using the DOS command ren <filename>.m.doc   <filename>.m   (the format is: ren <oldfilename> .oldextension <newfilename>.newextension).

[2] Upon installation, MATLAB sets up the path to include a folder "Work" in its directory, which is the default location for storing M-files. To add other directories to your path, simply select "Set Path" from the "File Menu" and add on the desired path. If you are using a networked computer, you may need to consult with the system administrator on this.

EXAMPLE 3.1: Here is a simple script which assumes that numbers $x0, y0$, and $r > 0$ have been stored in the workspace (before the script is invoked) and that will graph the circle with center $(x0, y0)$ and radius r.

```
t=0:.001:2*pi;
x=x0+r*cos(t);
y=y0+r*sin(t);
plot(x,y)
axis('equal')
```

If the above lines are simply typed as is into a text file and saved as, say, `circdrw.m` into some directory in the path, then at any time later on, if we wish to get MATLAB to draw a circle of radius 2 and center $(5, -2)$, we could simply enter the following in the command window:

```
>> r=2; x0=5; y0= -2;
>> circdrw
```

and *voilà!* the graphic window pops up with the circle we desired. Please remember that any variables created in a script are **global variables,** i.e., they will enter the current workspace when the script is invoked in the command window. One must be careful of this since the script may have been written a long time ago and when it is run the lines of the script are not displayed (only executed).

Function M-files are stored in the same way as script M-files but are quite different in the way they work. Function M-files accept any number of input variables and can output any number of output variables (or none at all). The variables introduced in a function M-file are **local variables,** meaning that they do not remain in the workspace after a function M-file is called in a MATLAB session. Also, the first line of a function M-file must be in the following format:

```
function [<output variables>] = <function_name>(<input variables>)
```

Another important format issue is that the `<function_name>` (which you are free to choose) should coincide exactly with the filename under which you save the function M-file.

EXAMPLE 3.2: We create a function M-file that will do essentially the same thing as the script in the preceding example. There will be three input variables: We will make the first two be the coordinates of the center $(x0, y0)$ of the circle that we wish MATLAB to draw, and the third be the radius. Since there will be no output variables here (only a graphic), our function M-file will look like this:

```
function [ ] = circdrwf(x0,y0,r)
t=0:.001:2*pi;
x=x0+r*cos(t);
y=y0+r*sin(t);
plot(x,y)
axis('equal')
```

In particular, the word function must be in lowercase. We then save this M-file as circdrwf.m in an appropriate directory in the path. Notice we gave this M-file a different name than the one in Example 3.1 (so they may lead a peaceful coexistence if we save them to the same directory). Once this function M-file has been stored we can call on it in any MATLAB session to draw the circle of center $(5, -2)$ and radius 2 by simply entering

```
>> circdrwf(5, -2, 2)
```

We reiterate that, unlike with the script of the first example, after we use a function M-file, none of the variables created in the file will remain in the workspace. As you gain more experience with MATLAB you will be writing a lot of function M-files (and probably very soon find them more useful than script M-files). They are analogous to "functions" in the C-language, "procedures" in PASCAL, and "programs" or "subroutines" in FORTRAN. The <filenames> of M-files can be up to 19 characters long (older versions of MATLAB accepted only length up to 8 characters), and the first character must be a letter. The remaining characters can be letters, digits, or underscore (_).

EXERCISE FOR THE READER 3.1: Write a MATLAB script, call it listp2, that assumes that a positive integer has been stored as n and that will find and output all powers of 2 that are less than or equal to n. Store this script as an M-file listp2.m somewhere in MATLAB's path and then run the script for each of these values of n: $n = 5$, $n = 264$, and $n = 2917$.

EXERCISE FOR THE READER 3.2: Write a function M-file, call it fact, having one input variable—a nonnegative integer n, and the output will be the factorial of n: $n!$. Write this program from scratch (using a while loop) without using a built-in function like gamma. Store this M-file and then run the following evaluations: fact(4), fact(10), fact(0).

Since MATLAB has numerous built-in functions it is often advisable to check first if a proposed M-file name that you are contemplating is already in use. Let's say you are thinking of naming a function M-file you have just written with the name det.m. To check first with MATLAB to see if the name is already in use you can type:

```
>>exist('det')   %possible outputs are 0, 1, 2, 3, 4, 5, 6, 7, 8
→5
```

The output 5 means (as does any positive integer) det is already in use. Let's try again (with a trick often seen on vanity license plates):

```
>>exist('det1')    → 0
```

The output zero means the filename det1 is not yet spoken for so we can safely assign this filename to our new M-file.

EXERCISES 3.1:

1. (a) Write a MATLAB function M-file, call it rectdrwf(1,w), that has two input variables: 1, the length and w, the width, and no output variables, but will produce a graphic of a rectangle with horizontal length = 1 and vertical width = w. Arrange it so that the rectangle sits well inside the graphic window and so that the axes are equally scaled.
   (b) Store this M-file and then run the function rectdrwf(5,3) and rectdrwf(4, 4.5).

2. (a) Write a function M-file, call it segdrwf(x,y), that has two input vectors $x = [x_1 \ x_2 \cdots x_n]$ and $y = [y_1 \ y_2 \cdots y_n]$ of the same size and no output variables, but will produce the graphic gotten by connecting the points $(x_1, y_1), (x_2, y_2), \cdots, (x_n, y_n)$. You might wish to make use of the MATLAB built-in function size(A) that, for an input matrix A, will output its size.
   (b) Run this program with the inputs x = [1 3 5 7 9 1] and y = [1 4 1 4 8 1].
   (c) Determine two vectors x and y so that segdrwf(x,y) will produce an equilateral triangle.

3. Redo Exercise 1, creating a script M-file called rectdrw rather than a function M-file.

4. Redo Exercise 2, creating a script M-file called segdrw rather than a function M-file.

5. (*Finance*) Write a function M-file, call it compintf(r,P,F), that has three input variables: r, the annual interest rate, P, the principal, and F, the future goal. Here is what the function should do: It assumes that we deposit P dollars (assumed positive) into an interest-bearing account that pays 100*r*% interest per year compounded annually. The investment goal is F dollars (F is assumed larger than P, otherwise the goal is reached automatically as soon as the account is opened). The output will be one variable consisting of the number of years it takes for the account balance to first equal or exceed F. Store this M-file and run the following: comintf(0.06, 1000, 100000), comintf(0.085, 1000, 100000), comintf(0.10, 1000, 1000000), and comintf(0.05, 100, 1000000).

   Note: The formula for the account balance after t years is P dollars is invested at 100*r*% compounded annually is $P(1+r)^t$.

6. (*Finance*) Write a function M-file, call it loanperf(r,L,PMT), that has three input variables: r, the annual interest rate, L, the loan amount, and PMT, the monthly payment. There will be one output variable n, the number of months needed to pay off a loan of L dollars made at an annual interest rate of 100*r*% (on the unpaid balance) where at the end of each month a payment of PMT dollars is made. (Of course L and PMT are assumed positive.) You will need to use a while loop construction as in Example 1.1 (Sec. 1.3). After storing this M-file, run the following: loanperf(.0799, 15000, 500), loanperf(.019, 15000, 500), loan(0.99, 22000, 450). What could cause the program loanerf(r, L, PMT) to go into an infinite loop? In the next chapter we will show, among other things, ways to safeguard programs from getting into infinite loops.

7. Redo Exercise 5 writing a script M-file (which assumes the relevant input variables have been assigned) rather than a function M-file.

8. Redo Exercise 6 writing a script M-file (which assumes the relevant input variables have been assigned) rather than a function M-file.

9. Write a function M-file, call it oddfact(n), that inputs a positive integer n and will output the product of all odd positive integers that are less than or equal to n. So, for example, oddfact(8) will be $1 \cdot 3 \cdot 5 \cdot 7 = 105$. Store it and then run this function for the following values: oddfact(5), oddfact(22), oddfact(29). Use MATLAB to find the first

value of n for which oddfact (n) exceeds or equals 1 million, and then 5 trillion.

10.   Write a function M-file, call it evenfact (n), that inputs a positive integer n and will output the product of all even positive integers that are less than or equal to n. So, for example, evenfact (8) will be $2 \cdot 4 \cdot 6 \cdot 8 = 384$. Store it and then run this function for the following values: evenfact (5), evenfact (22), evenfact (29). Get MATLAB to find the first value of n for which evenfact (n) exceeds or equals 1 million, and then 5 trillion. Can you write this M-file without using a while loop, using instead some of MATLAB's built-in functions?

11   Use the error estimate from Example 2.5 (Sec. 2.3) to write a function M-file called expcal (x, err) that does the following: The input variable x is any real number and the other input variable err is any positive number. The output will be an approximation of e^x by a Taylor polynomial $p_n(x)$ based at $x = 0$, where n is the first nonnegative integer such that the error estimate based on Taylor's theorem that was obtained in Example 2.5 gives a guaranteed error less than err. There should be two output variables, n, the order of the Taylor polynomial used, and $y = p_n(x) =$ the approximation. Run this function with the following input data: (2, 0.001), (–6, 10^{-12}), (15, 0.000001), (–30, 10^{-25}). For each of these y-outputs, check with MATLAB's built-in function exp to see if the actual errors are as desired. Is it possible for this program to ever enter into an infinite loop?

12.   Write a function M-file, called coscal (x, err), that does exactly what the function in Exercise 11 does except now for the function $y = \cos(x)$. You will need to obtain a Taylor's theorem estimate for the (actual) error $|\cos(x) - p_n(x)|$. Run this function with the following input data: (0.5, 0.0000001), (–2, 0.0001), ($20°$, 10^{-9}), and ($360020°$, 10^{-9}) (for the last two you will need to convert the inputs to radians). For each of these y-outputs, check with MATLAB's built-in function $\cos(x)$ to see if the actual errors are as desired. Is it possible for this program to ever enter into an infinite loop? Although cos($360020°$) = cos($20°$), the outputs you get will be different; explain this discrepancy.

3.2: CREATING AN M-FILE FOR A MATHEMATICAL FUNCTION

Function M-files can be easily created to store (complicated) mathematical functions that need to be used repeatedly. Another way to store mathematical functions without formally saving them as M-files is to create them as "in-line objects." Unlike M-files, in-line objects are stored only as variables in the current workspace. The following example will illustrate such an M-file construction; in-line objects will be introduced in Chapter 6.

EXAMPLE 3.3:   Write a function M-file, with filename bumpy.m, that will store the function given by the following formula:

$$y = \frac{1}{4\pi}\left[\frac{1}{(x-2)^2 + 1} + \frac{1}{(x+0.5)^4 + 32} + \frac{1}{(x+1)^2 + 2}\right].$$

Once this is done, call on this newly created M-file to evaluate y at $x = 3$ and to sketch a graph of the function from $x = -3$ to $x = 3$.

SOLUTION: After the first "function definition line," there will be only one other line required: the definition of the function above written in MATLAB's language.   Just like in a command window, anything we type after the percent symbol (%) is considered as a comment and will be ignored by MATLAB's processor.   Comment lines that are put in immediately following the function definition line are, however, somewhat more special.   Once a function has been stored   (somewhere   in   MATLAB's   path),   and   you   type   help <function_name> on the command window, MATLAB displays any comments that you inserted in the M-file after the function definition line.   Here is one possibility of a function M-file for the above function:

```
function y = bumpy(x)
% our first function M-file
% x could be a vector
% created by <yourname> on <date>
y=1/(4*pi)*(1./((x-2).^2+1)+1./((x+.5).^4+32)+1./((x+1).^2+2));
```

Some comments are in order. First, notice that there is only one output variable, y, but we have not enclosed it in square brackets. This is possible when there is only one output variable (it would still have been okay to type function [y] = bumpy(x)). In the last line where we typed the definition of y (the output variable), notice that we put a semicolon at the end. This will suppress any duplicate outputs since a function M-file is automatically set up to print the output variable when evaluated.   Also, please look carefully at the placement of parentheses and (especially) the dots when we wrote down the formula. If x is just a number, the dots are not needed, but often we will need to create plots of functions and x will need to be a vector. The placement of dots was explained in Chapter 1.

   The above function M-file should now be saved with the name bumpy.m with the same filename appearing (without the extension) in the function definition line into some directory contained in MATLAB's path (as explained in the previous section). This being done, we can use it just like any of MATLAB's built-in functions, like cos. We now proceed to perform the indicated tasks.

```
>> bumpy(3)
→ 0.0446
>> y     %Remember all the variables in a MATLAB M-file are local only.
→ Undefined function or variable 'y'.

>> x=-3:.01:3;
>> plot(x,bumpy(x))
```

From the plot we see that the function bumpy(x) has two peaks (local maxima) and one valley (local minimum) on the interval $-3 \leq x \leq 3$. MATLAB has many built-in functions to analyze mathematical functions.   Three very important numerical problems are to integrate a function on a specified interval, to find maximums and minimums on a specified interval, and to find zeros or roots of a function. The next example will illustrate how to perform such tasks with MATLAB.

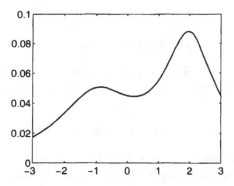

FIGURE 3.1: A graph of the function $y = \text{bumpy}(x)$ of Example 3.3.

EXAMPLE 3.4: For the function bumpy(x) of the previous example, find "good" approximations to the following:

(a) $\int_{-3}^{3} \text{bumpy}(x)\, dx$

(b) The maximum and minimum values of $y = \text{bumpy}(x)$ on the interval $-1.2 \le x \le 1$ (i.e., the height of the left peak and that of the valley) and the corresponding x-coordinates of where these extreme values occur.

(c) A solution of the equation bumpy(x) = 0.08 on the interval $0 \le x \le 2$ (which can be seen to exist by examination of bumpy's graph in Figure 3.1).

SOLUTION: Part (a): The relevant built-in function in MATLAB for doing definite integrals is quad, which is an abbreviation for *quadrature*, a synonym for integration.[3] The syntax is as follows:

quad('function', a, b, tol) ⇥	Approximates the integral $\int_a^b \text{function}(x)\, dx$ with the goal of the error being less than tol.

The function must be stored as an M-file in MATLAB's path or the exact name of a built-in function, and the name must be enclosed in 'single quotes'. [4] If the whole last argument tol is omitted (along with the comma that precedes it), a maximum error goal of 10^{-3} is assumed. If this command is run and you just get

[3] MATLAB has another integrator, quadl, that gives more accurate results for well-behaved integrands. For most purposes, though, quad is quite satisfactory and versatile as a general quadrature tool.

[4] An alternative syntax (that avoids the single quotes) for this and other functions that call on M-files is quad(@function, a, b, tol). Another way to create mathematical functions is to create them as so-called *inline functions* which are stored only in the workspace (as opposed to M-files) and get deleted when the MATLAB session is ended. Inline functions will be introduced in Chapter 6.

an answer (without any additional warnings or error messages), you can safely assume that the approximation is accurate within the sought-after tolerance.

```
>> quad('bumpy',-3,3)     → 0.3061
>> format long
>> ans                     → 0.30608471060690
```

As explained above, this answer should be accurate to at least three decimals. It is actually better than this since if we redo the calculation with a goal of six digits of accuracy, we obtain:

```
>> quad('bumpy',-3,3, 10^(-6))    → 0.30608514875582
```

This latter answer, which is accurate to at least six digits, agrees with the first answer (after rounding) to six digits, so the first answer is already quite accurate. There are limits to how accurate an answer we can get in this way. First of all, MATLAB works with only about 15 or so digits, so we cannot hope for an answer more accurate than this. But roundoff and other errors can occur in large-scale calculations and these can put even further restrictions on the possible accuracy, depending on the problem. We will address this issue in more detail in later chapters. For many practical purposes and applications the quad function and the others we discuss in this example will be perfectly satisfactory and in such cases there will be no need to write new MATLAB M-files to perform such tasks.

Part (b): To (approximately) solve the calculus problem of finding the minimum value of a function on a specified interval, the relevant MATLAB built-in function is fminbnd and the syntax is as follows:

fminbnd('function', a, b, optimset('TolX',tol)) →	Approximates the x-coordinate of the minimum value of function(x) on [a, b] with a goal of the error being <tol.

The usage and syntax comments for quad apply here as well. In particular, if the whole last argument optimset('TolX',tol) is omitted (along with the comma that precedes it), a maximum error goal of 10^{-3} is assumed. Note the syntax for changing the default tolerance goal is a bit different than for quad. This is due to the fact that fminbnd has more options and is capable of doing a lot more than we will have occasion to use it for in this text. For more information, enter help optimset.

```
>> xmin=fminbnd('bumpy',-1.2,1)  %We first find the x-coordinate of
>> %              the valley with a three digit accuracy (at least)
→0.21142776202687
>> xmin=fminbnd('bumpy',-1.2,1, optimset('TolX',1e-6))  %Next let's
>> %                      go for 6 digits of accuracy.
→0.21143721018793 (=x-coordinate of valley)
```

The corresponding y-coordinate (height of the valley) is now gotten by evaluating bumpy(x) at xmin.

```
>> ymin = bumpy(xmin)
```
→ 0.04436776267211 (= <u>y-coordinate of valley</u>)

Since we know that the x-coordinate is accurate to six decimals, a natural question is how accurate is the corresponding y-coordinate that we just obtained? One obvious thing to try would be to estimate this error by plotting bumpy(x) on the interval $x\min - 10^{-6} \le x \le x\min + 10^{-6}$ and then seeing how much the y-coordinates vary on this plot. This maximum variation will be an upper bound for the difference of ymax and the actual value of the y-coordinate for the valley. When we try and plot bumpy(x) on this interval as follows we get the following warning message:

```
>> x=(xmin-10^(-6)):10^(-9):(xmin+10^(-6));
>> plot(x,bumpy(x))
```
→Warning: Requested axes limit range too small; rendering with minimum range allowed by machine precision.

Also, the corresponding plot (which we do not bother reproducing) looks like that of a horizontal line, but the y-tick marks are all marked with the same number (.0444) and similarly for the x-tick marks. This shows that MATLAB's plotting precision works only up to a rather small number of significant digits. Instead we can look at the vector bumpy (x) with x still stored as the vector above, and look at the difference of the maximum less the minimum.

max(v) →	For a vector v, these MATLAB commands will return the maximum entry and
min(v) →	the minimum entry; e.g.: If $v = [2\ 8\ -5\ 0]$ then max (v) →8, and min (v) → −5.

```
>> max(bumpy(x))-min(bumpy(x))
```
→ 1.785377401475330e-014

What this means is that, although xmin was guaranteed only to be accurate to six decimals, the corresponding y-coordinate seems to be accurate to MATLAB precision, which is about 15 digits!

EXERCISE FOR THE READER 3.3: Explain the possibility of such a huge discrepancy between the guaranteed accuracy of xmin (to the actual x-value of where the bottom of the valley occurs) being 10^{-6} and the incredibly smaller value 10^{-14} of the apparent accuracy of the corresponding ymin = bumpy(xmin). Make sure to use some calculus in your explanation.

EXERCISE FOR THE READER 3.4: Explain why the above vector argument does not necessarily guarantee that the error of ymin as an approximation to the actual y-coordinate of the valley is less than 1.8×10^{-14}..

We turn now to the sought-after maximum. Since there is no built-in MATLAB function (analogous to fminbnd) for finding maximums, we must make do with what we have. We can use fminbnd to locate maximums of functions as soon as we make the following observation: The maximum of a function $f(x)$ on an interval I, if it exists, will occur at the same x-value as the minimum value of the negative function $-f(x)$ on I. This is easy to see; just note that the graph of $y = -f(x)$ is obtained by the graph of $y = f(x)$ by turning the latter graph upside-down (more precisely, reflect it over the x-axis), and when a graph is turned upside-down, its peaks become valleys and its valleys become peaks. Let's initially go for six digits of accuracy:

```
>> xmax = fminbnd('-bumpy(x)', -1.2, 1, optimset('TolX',1e-6))
→ -0.86141835836638 (= x-coordinate of left peak)
```

The corresponding y-coordinate is now:

```
>> bumpy(xmax)    → 0.05055706241866 (= y-coordinate of left peak)
```

Part (c): One way to start would be to simultaneously graph $y = \text{bumpy}(x)$ together with the constant function $y = 0.08$ and continue to zoom in on the intersection point. As explained above, though, this graphical approach will limit the attainable accuracy to three or four digits. We must find the root (less than 2) of the equation $\text{bumpy}(x) = 0.08$. This is equivalent to finding a zero of the standard form equation $\text{bumpy}(x) - 0.08 = 0$. The relevant MATLAB function is `fzero` and its usage is as follows:

`fzero('function', a)`	→	Finds a zero of function(x) near the value $x = a$ (if one exists). Goal is machine precision (about 15 digits).

```
>> fzero('bumpy(x)-0.08', 1.5)
→Zero found in the interval: [1.38, 1.62].
→1.61904252091472 (=desired solution)
>> bumpy(ans) %as a check, let's see if this value of x does what we
>>%      want it to do.
→ 0.08000000000000 %Not bad!
```

EXERCISE FOR THE READER 3.5: Write a function M-file with filename `wiggly.m` that will store the following function:

$$y = \sin\left(\exp\left[\frac{1}{(x^2 + .5)^2}\right]\right)\sin(x).$$

(a) Plot this function from $x = -2$ through $x = 2$.
(b) Integrate this function from $x = 0$ to $x = 2$ (use 10^{-5} as your accuracy goal).
(c) Approximate the x-coordinates of both the smallest positive local minimum (valley) and the smallest positive local maximum (peak) from $x = -2$ through $x = 2$.
(d) Approximate the smallest positive solution of $\text{wiggly}(x) = x/2$ (use 10^{-5} as your accuracy goal).

EXERCISES 3.2:

1. (a) Create a MATLAB function M-file for the function $y = f(x) = \exp\left(\sin[\pi/(x+0.001)^2]\right)$
    $+ (x-1)^2$ and then plot this function on the interval $0 \leq x \leq 3$. Do it first using 10 plotting
    points and then using 50 plotting points and finally using 500 points.

    (b) Compute the corresponding integral $\int_1^3 f(x)\,dx$.

    (c) What is the minimum value (y-coordinate) of $f(x)$ on the interval $[1, 10]$? Make sure your
    answer is accurate to within the nearest 1/10,000th.

2. (a) Create a MATLAB function M-file for the function $y = f(x) = \dfrac{1}{x}\sin(x^2) + \dfrac{x^2}{50}$ and then plot
    this function on the interval $0 \leq x \leq 10$. Do it first using 200 plotting points and then using
    5000 plotting points.

    (b) Compute the corresponding integral $\int_1^{10} f(x)\,dx$.

    (c) What is the minimum value (y-coordinate) of $f(x)$ on the interval $[1, 10]$? Make sure your
    answer is accurate to within the nearest 1/10,000th. Find also the corresponding x-coordinate
    with the same accuracy.

3. Evaluate the integral $\int_0^1 \sin(t^2)\,dt$ (with an accuracy goal of 10^{-7}) and compare this with the
    answer obtained in Example 2.7 (Sec. 2.3).

4. (a) Find the smallest positive solution of the equation $\tan(x) = x$ using an accuracy goal of
    10^{-12}. (b) Using calculus, obtain a bound for the actual error.

5. Find all zeros of the polynomial $x^3 + 6x^2 - 14x + 5$.

NOTE: We remind the reader about some facts on polynomials. A polynomial $p(x)$ of degree n can
have at most n roots (that are the x-intercepts of the graph $y = p(x)$). If $x = r$ is a root and if the
derivative $p'(r)$ is not zero, then we say $x = r$ is a root of **multiplicity 1**.
If $p(r) = p'(r) = 0$ but $p''(r) \neq 0$, then we say the root $x = r$ has multiplicity 2. In general we say
$x = r$ is a root of $p(x)$ of multiplicity a, if all of the first $a-1$ derivatives equal zero:

$$p(r) = p'(r) = p''(r) = \cdots p^{(a-1)}(r) = 0 \text{ but } p^{(a)}(r) \neq 0 .$$

Algebraically $x = r$ is a root of multiplicity a means that we can factor $p(x)$ as $(x-r)^a q(x)$ where
$q(x)$ is a polynomial of degree $n-a$. It follows that if we add up all of the multiplicities of all of the
roots of a polynomial, we get the degree of the polynomial. This information is useful in finding all
roots of a polynomial.

6. Find all zeros of the polynomial $2x^4 - 16x^3 - 2x^2 + 25$. For each one, attempt to ascertain its
    multiplicity.

7. Find all zeros of the polynomial

    $$x^6 - \frac{25}{4}x^5 + \frac{4369}{64}x^4 + \frac{8325}{32}x^3 + \frac{13655}{8}x^2 - \frac{325}{32}x + \frac{21125}{8} .$$

    For each one, attempt to ascertain its multiplicity.

8.   Find all zeros of the polynomial

$$x^8 + \frac{136}{5}x^7 + 210x^6 - \frac{165}{5}x^5 - 4094x^4 + \frac{4528}{5}x^3 + 17232x^2 + 320x + 5600 .$$

For each one, attempt to ascertain its multiplicity.

9.   Check that the value $x = 2$ is a zero of both of these polynomials:

$$P(x) = x^8 - 2x^7 + 6x^5 - 12x^4 + 2x^2 - 8$$
$$Q(x) = x^8 - 8x^7 + 28x^6 - 61x^5 + 95x^4 - 112x^3 + 136x^2 - 176x + 112.$$

Next, use `fzero` to seek out this root for each polynomial using $a = 1$ (as a number near the root) and with accuracy goal 10^{-12}.   Compare the outputs and try to explain why the approximation seemed to go better for one of these polynomials than for the other one.

Chapter 4: Programming in MATLAB

4.1: SOME BASIC LOGIC

Computers and their programs are designed to function very logically so that they always proceed by a well-defined set of rules. In order to write effective programs, we must first learn these rules so we can understand what a computer or MATLAB will do in different situations that may arise throughout the course of executing a program. The rules are set forth in the formal science of **logic**. Logic is actually an entire discipline that is considered to be part of both of the larger subjects of philosophy and mathematics. Thus there are whole courses (and even doctoral programs) in logic and any student who wishes to become adept in programming would do well to learn as much as possible about logic. Here in this introduction, we will touch only the surface of this subject, with the hope of supplying enough elements to give the student a working knowledge that will be useful in understanding and writing programs.

The basic element in logic is a **statement**, which is any declarative sentence or mathematical equation, inequality, etc. that has a **truth value** of either **true** or **false**.

EXAMPLE 4.1: For each of the English or mathematical expressions below, indicate which are statements, and for those that are, decide (if possible) the truth value.
(a) Al Gore was Bill Clinton's Vice President.
(b) $3 < 2$
(c) $x + 3 = 5$
(d) If $x = 6$ then $x^2 > 4x$.

SOLUTION: All but (c) are statements. In (c), depending on the value of the variable x, the equation could be either true (if $x = 2$) or false (if $x =$ any other number). The truth values of the other statements are as follows: (a) true, (b) false, and (d) true.

If you enter any mathematical relation (with one of the relational operators from Table 2.1), MATLAB will tell you if the statement is true or false in the following fashion:

Truth Value	MATLAB Code
True	1 (as output)
	Any nonzero number (as input)
False	0 (as input and output)

We shall shortly come to how the input truth values are relevant. For now, let us give some examples of the way MATLAB outputs truth values. In fact, let's use MATLAB to do parts (b) and (d) of the preceding example.

```
>> 3<2
→ 0  (MATLAB is telling us the statement is false.)
>> x=6;  x^2>4*x
→ 1  (MATLAB is telling us the statement is true.)
```

Logical statements can be combined into more complicated compound statements using **logical operators**. We introduce the four basic logical operators in Table 4.1, along with their approximate English translations, MATLAB code symbols, and precise meanings.

TABLE 4.1: The basic logical operators. In the meaning explanation, it is assumed the p and q represent statements whose truth values are known.

Name of Operator	English Approxi -mation	MATLAB Code	Meaning
Negation	not p	~p	~p is true if p is false, and false if p is true.
Conjunction	p and q	p&q	p&q is true if both p and q are true, otherwise it's false.
Disjunction	p or q	p|q	p|q is true in all cases except if p and q are both false, in which case it is also false.
Exclusive Disjunction	p or q (but not both)[1]	xor(p,q)	xor(p,q) is true if exactly one of p or q is true. If p and q are both true or both false then xor(p,q) is false

EXAMPLE 4.2: Determine the truth value of each of the following compound statements.
(a) San Francisco is the capital of California and Egypt is in Africa.
(b) San Francisco is the capital of California or Egypt is in Africa.
(c) San Franciso is not the capital of California.
(d) not $(2 > -4)$
(e) letting $x = 2$, $z = 6$, and $y = -4$: $x^2 + y^2 > z^2/2$ or $zy < x$
(f) letting $x = 2$, $z = 6$, and $y = -4$: $x^2 + y^2 > z^2/2$ or $zy < x$ (but not both)

---

[1] Although most everyone understands the meaning of "and," in spoken English the word "or" is often ambiguous. Sometimes it is intended as the disjunction but other times as the exclusive disjunction For example, if on a long airplane flight the flight attendant asks you, "Would you like chicken or beef?" Certainly here the exclusive disjunction is intended—indeed, if you were hungry and tried to ask for both, you would probably wind up with only one plus an unfriendly flight attendant. On the other hand, if you were to ask a friend about his/her plans for the coming weekend, he/she might reply, "Oh, I might go play some tennis or I may go to Janice's party on Saturday night " In this case the ordinary disjunction is intended You would not be at all surprised if your friend wound up doing both activities. In logic (and mathematics and computer programming) there is no room for such ambiguity, so that is why we have two very precise versions of "or "

SOLUTION: To abbreviate parts (a) through (c) we introduce the symbols:

p = San Francisco is the capital of California.

q = Egypt is in Africa.

From basic geography, Sacremento is California's capital so p is false, and q is certainly true. Statements (a) through (c) can be written as: p and q, p or q, not p, respectively. From what was summarized in Table 4.1 we now see (a) is false, (b) is true, and (c) is true.

For part (d), since $2 > -4$ is true, the negation not $(2 > -4)$ must be false.

For parts (e) and (f), we note that substituting the values of x, y, and z the statements become:

(e) $20 > 18$ or $-24 < 2$ i.e., true or true, so true

(f) $20 > 18$ or $-24 < 2$ (but not both), i.e., true or true (but not both), so false.

MATLAB does not know geography but it could have certainly helped us with the mathematical questions (d) through (f) above. Here is how one could do these on MATLAB:

```
>>  ~(2>-4)
→ 0 (=false)
>> x=2; z=6; y=-4; (x^2+y^2 > z^2/2) | (z*y < x)
→ 1 (=true)
>> x=2; z=6; y=-4;  xor(x^2+y^2 > z^2/2,  z*y < x)
→ 0 (=false)
```

EXERCISES 4.1:

1. For each of the English or mathematical expressions below, indicate which are statements, and for those that are statements, decide (if possible) the truth value.
    (a) Ulysses Grant served as president of the United States.
    (b) Who was Charlie Chaplin?
    (c) With $x = 2$ and $y = 3$ we have $\sqrt{x^2 + y^2} - x + y$.
    (d) What is the population of the United States?

2. For each of the English or mathematical statements below, determine the truth value.
    (a) George Harrison was a member of the Rolling Stones.
    (b) Canada borders the United States or France is in the Pacific Ocean.
    (c) With $x = 2$ and $y = 3$ we have $x^x > y$ or $x^y > y^x$.
    (d) With $x = 2$ and $y = 3$ we have $x^x > y$ or $x^y > y^x$ (but not both).

3. Assume that we are in a MATLAB session in which the following variables have been stored: $x = 6$, $y = 12$, $z = -4$. What outputs would the following MATLAB commands produce? Of course, you should try to figure out these answers on your own and <u>afterward</u> use MATLAB to check your answers.
    (a) >> x + y >= z
    (b) >> xor(z, x-2*y)
    (c) >> (x==2*z)|(x^2>50 & y^2>100)
    (d) >> (x==2*z)|(x^2>50 & y^2>100)

4   The following while loops were separately entered in different MATLAB sessions. What will
     the resulting outputs be? Do this one carefully by hand and then use MATLAB to check your
     answers.

(a)
```
>> i = 1; x=-3;
>> while (i<3) & (x<35)
       x=-x*(i+1)
  end
```

(b)
```
>> i = 1; x=-3;
>> while (i<3) | (x<35)
       x=-x*(i+1)
  end
```

(c)
```
>> i = 1; x=-3;
>> while xor(i<3, x<35)
       x=-x*(i+1)
  end
```

5.   The following while loop was entered in a MATLAB session. What will the resulting output
     be? Do this one carefully by hand and then use MATLAB to check your answers
```
>> i = 1; x=2; y =3;
>> while (i<5) | (x == y)
       x=x*2, y=y+x, i=i+1;
  end
```

4.2: LOGICAL CONTROL FLOW IN MATLAB

Up to this point, the reader has been given a reasonable amount of exposure to
while loops. The while loop is quite universal and is particularly useful in those
situations where it is not initially known how many iterations must be run in the
loop. If we know ahead of time how many iterations we want to run through a
certain recursion, it is more convenient to use a *for loop*. For loops are used to
repeat (iterate) a statement or group of statements a specified number of times.
The format is as follows:

```
>>for n=(start):(gap):(end)
    ...MATLAB commands...
end
```

The **counter** n (which could be any variable of your choice) gets automatically
bumped up at each iteration by the "gap." At each iteration, the "...MATLAB
commands..." are all executed in order (just like they would be if they were to be
entered manually again and again). This continues until the counter meets or
exceeds the "end" number.

EXAMPLE 4.3: To get the feel for how for loops function, we run through some
MATLAB examples. In each case the reader is advised to try to guess the output
of each new command before reading on (or hitting enter in MATLAB) as a
check.

```
>> for n=1:5  % if "gap" is omitted it is assumed to be 1.
   x(n)=n^3;  % we will be creating a vector of cubes of successive
             % integers.
end  %all output has been suppressed, but a vector x has been
>>   %created.
```

```
>>x   %let's display x now
→ x = 1 8 27 64 125
```

Note that since a comma in MATLAB signifies a new line, we could also have written the above for loop in a single line. We do this in the next loop below:

```
>> for k=1:2:10,   x(k)=2; end
>> x  %we display x again.  Try to guess what it now looks like.
→ 2 8 2 64 2 0 2 0 2
```

Observe that there are now nine entries in the vector x. This loop overwrote some of the five entries in the previous vector x (which still remained in MATLAB's workspace). Let us carefully go through each iteration of this loop, explaining exactly what went on at each stage:

$k = 1$ (start) → we redefine $x(1)$ to be 2 (from its original value of 1).

$k = 1 + 2 = 3$ (augment k by gap = 2) → redefine $x(3)$ to be 2 ($x(2)$ was left to its original value of 8).

$k = 3 + 2 = 5$ → redefine $x(5)$ to be 2.

$k = 5 + 2 = 7$ → defines $x(7)$ to be 2 (previously x was a length 5 vector, now it has 7 components), the skipped component $x(6)$ is by default defined to be 0.

$k = 7 + 2 = 9$ → defines $x(9)$ to be 2 and the skipped $x(8)$ to be 0.

$k = 9 + 2 = 11$ (exceeds end = 10 so for loop is exited and thus completed).

The gap in a for loop can even be a negative number, as in the following example that creates a vector in backwards order. The semicolon is omitted to help the reader convince himself or herself how the loop progresses.

```
>> for i=3:-1:1,  y(i)=i, end
→y = 0   0   3
→y = 0   2   3
→y = 1   2   3
```

A very useful tool in programming is the *if-branch*. In its basic form the syntax is as follows:

```
>>if <condition>
   ...MATLAB commands..
end
```

The way such an if-branch works is that if the listed <condition> (which can be any MATLAB statement) is true (i.e., has a nonzero value), then all of the "...MATLAB commands..." listed are executed in order and upon completion the if-branch is then exited. If the <condition> is false then the "...MATLAB commands..." are ignored (i.e., they are bypassed) and the if-branch is immediately exited. As with loops in MATLAB, if-branches may be inserted within loops (or branches) to deal with particular situations that arise. Such loops/branches are said to be **nested**. Sometimes if-branches are used to "raise a flag" if a certain condition arises. The following MATLAB command is often useful for such tasks:

fprintf('<any English /text phrase>') →	Causes MATLAB to print. <any English phrase>

Thus the output of the command `fprintf('Have a nice day!')` will simply be → Have a nice day! This command has a useful feature that allows one to print the values of variables that are currently stored within a text phrase. Here is how such a command would work: We assume that (previously in a MATLAB session) the values $w = 2$ and $h = 9$ have been calculated and stored and we enter:

```
>>fprintf('the width of the rectangle is %d,the length is %d.', w, h)
→the width of the rectangle is 2, the length is 9.»
```

Note that within the "text" each occurrence of %d was replaced, in order, by the (current) values of the variables listed at the end. They were printed as integers (without decimals); if we wanted them printed as floating point numbers, we would use %f in place of %d. Also note that MATLAB unfortunately put the prompt >> at the end of the output rather than on the next line as it usually does. To prevent this, simply add (at the end of your text but before the single right quote) \r—which stands for "carriage return." This carriage return is also useful for splitting up longer groups of text within an `fprintf`.

Sometimes in a nested loop we will want to exit from within an inner loop and at other times we will want exit from the **mother loop** (which is the outermost loop inside of which all the other loops/branches are a part of) and thus halt all operations relating to the mother loop. These two exits can be accomplished using the following useful MATLAB commands:

`break` (anywhere within a loop) →	Causes immediate exit only from the single loop in which `break` was typed.
`return` (anywhere within a nested loop) →	Causes immediate exit from the mother loop, or within a function M-file, immediate exit from M-file (whether or not output has been assigned).

The next example illustrates some of the preceding concepts and commands.

EXAMPLE 4.4: Carefully analyze the following two nested loops, decide what exactly they cause MATLAB to do, and then predict the exact output. After you do this, read on (or use MATLAB) to confirm your predictions.

```
(a)   for n=1:5
          for k=1:3
          a=n+k
          if a>=4, break,   end
          end
      end
```

NOTE: We have inserted tabs to make the nested loops easier to distinguish. Always make certain that each loop/branch is paired with its own `end`.

```
(b)   for n=1:5
          for k=1:3
```

```
        a=n+k
        if a>=4
         fprintf('We stop since a has reached the value %d  \r', a)
          return
        end
      end
    end
```

SOLUTION: Part (a): Both nested loops consist of two loops. The mother loop in each is, as usual, the outermost loop (with counter n). The first loop begins with the mother loop setting the counter n to be 1, then immediately moves to the second loop and sets k to be 1; now in the second loop a is assigned to be the value of $n+k = 1+1 = 2$ and this is printed (since there is no semicolon). Since a = 2 now, the "if-condition" is not satisfied so the if-branch is bypassed and we now iterate the k-loop by bumping up k by 1 (= default gap). Note that the mother loop's n will not get bumped up again until the secondary k-loop runs its course. So now with $k = 2$, a is reassigned as a = $n+k = 1+2 = 3$ and printed, the if-branch is again bypassed and k gets bumped up to 3 (its ending value), a is now reassigned as a = $n+k = 1+3 = 4$ (and printed). The if-branch condition is now met, so the commands within it (in this case only a single "break" command) are run. So we will break out of the k-loop (which is actually redundant at this point since k was at its ending value and the k-loop was about to end anyway). But we are still progressing within the mother n-loop. So now n gets bumped up by 1 to be 2 and we start afresh the k-loop again with $k = 1$. The variable a is now assigned as a = $n+k = 2+1 = 3$ (and printed), the if-branch is bypassed since the condition is not met and k gets bumped up to be 2. Next a gets reassigned as a = $n+k = 2+2 = 4$, and printed. Now the if-branch condition is met so we exit the k-loop (this time prematurely) and n now gets bumped up to be 3. Next entering the k-loop with $k = 1$, a gets set to be $n+k = 3+1 = 4$, and printed, the "if-branch condition" is immediately satisfied, and we exit the k-loop and n now gets bumped up to be 4. As in the last iteration, the k-loop will just reassign a to be 5 and print this, break and n will go to 5 (its final value). In the final stage, a gets assigned as 6, the if-branch breaks us out of the k-loop, and since n is at its ending value, the mother loop exits.

The actual output for part (a) is thus:

→a = 2 a = 3   a = 4   a = 3   a = 4   a = 4   a = 5   a = 6

Part (b): Apart from the fprintf command, the main difference here is the replacement of the break command with the return command. As soon as the if-branch condition is satisfied, the conditions within will be executed and the return will cause the whole nested loop to stop in its tracks. The output will be as follows:

→a =2 a = 3 a = 4   We stop since a has reached the value 4

64 Chapter 4: Programming in MATLAB

EXERCISE FOR THE READER 4.1: Below are two nested loops. Carefully analyze each of them and try to predict resulting outputs and then use MATLAB to verify your predictions. For the second loop the output should be given in the default `format short`.

(a)
```
>>for i=1:5
      i
      if i>2,   fprintf('test'),  end
   end
```

(b)
```
>>for i=1:8, x(i)=0; end  %initialize vector
>>for i=6:-2:2
       for j=1:i
           x(i)=x(i)+1/j;
       end
   end
>>x
```

The basic form of the if-branch as explained above allows one to have MATLAB perform a list of commands in the event that one certain condition is fulfilled. In its more advanced form, if-branches can be set up to perform different sets of commands depending on which situation might arise. In the fullest possible form, the syntax of an if-branch is as follows:

```
>>if <condition_1>
   ...MATLAB commands_1...
elseif <condition_2>
   ...MATLAB commands_2...
   ...
elseif <condition_n>
   ...MATLAB commands_n...
else
   ...MATLAB commands...
end
```

There can be any number of `else if` cases (with subsequent MATLAB commands) and the final `else` is optional. Here is how such an if-branch would function: The first thing to happen is that <condition_1> gets tested. If it tests true (nonzero), then the "...MATLAB commands_1..." are all executed in order, after which the if-branch is exited. If <condition_1> tests false (zero), then the if-branch moves on to the next <condition_2> (associated with the first `elseif`). If this condition tests true, then "...MATLAB commands_2..." are executed and the if-branch is exited, otherwise it moves on to test the next <condition_3>, and so on. If the final `else` is not present, then once the loop goes through testing all of the conditions, and if none were satisfied, the if-branch would exit without performing any tasks. If the `else` is present, then in such a situation the "...MATLAB commands..." after the `else` would be performed as a catch-all to all remaining situations not covered by the conditions listed above.

Our next example will illustrate the use of such an extended if-branch and will also bring forth a rather subtle but important point.

EXAMPLE 4.5: Create a function M-file for the mathematical function defined by the formula:

$$y = \begin{cases} -x^2 - 4x - 2, & \text{if } x < -1 \\ |x|, & \text{if } |x| \le 1, \\ 2 - e^{\sqrt{x-1}}, & \text{if } x > 1 \end{cases}$$

then store this M-file and get MATLAB to plot this function on the interval $-4 \le x \le 4$.

SOLUTION: The M-file can be easily written using an if-branch. If we use the filename $ex4\_5$, here is one possible M-file:

```
function y = ex4_5(x)
if x<-1
       y = -x.^2-4*x-2;
elseif x>1
       y = 2-exp(sqrt(x-1));
else
       y=abs(x);
end
end
```

It is tempting to now obtain the desired plot using the usual command sequence:

```
>>   x=-4:.001:4; y=ex4_5(x); plot(x,y)
```

There is a subtle problem here, though. If we were to enter these commands, we would obtain the graph of (the last function in the formula) $y = |x|$ on the whole interval $[-4, 4]$. Before we explain this and show how to resolve this problem, it would behoove the reader to try to decide what is causing this to happen and to figure out a way to fix this problem.

If we carefully go on to see what went wrong with the last attempt at a plot, we observe that since x is a vector, the first condition $x < -1$ in the if-branch now becomes a bit ambiguous. When asked about such a vector inequality, MATLAB will return a vector of the same size as x that is made up of zeros and ones. In each slot, the vector is 1 if the corresponding component of x satisfies the inequality $(x < -1)$ and 0 if it does not satisfy this inequality. Here is an example:

```
>> [2 -5   3   -2   -1] < -1      %causes MATLAB to test each of the 5
>>%                                inequalities as true (1) or false (0)
 →0   1   0   1   0
```

Here is what MATLAB did in the above attempt to plot our function. Since x is a (large) vector, the first condition $x < -1$ produced another vector as the same size as x made up of (both) 0's and 1's. Since the vector was not all true (1's), the condition as a whole was not satisfied so it moved on to the next condition $x > 1$, which for the same reason was also not satisfied and so bypassed. This left us with the catch-all command y=abs(x) for the whole vector x, which, needless to say, is not what we had intended.

So now how can we fix this problem? One simple fix would be to use a for loop to construct the y-coordinate vector using ex4_5(x) with only scalar values for x. Here is one such way to construct y:

```
>> size(x)    %first we find out
>>%            the size of x
→ 1    8001
>> for n=1:8001
y(n)=ex4_5(x(n));
end
>> plot(x,y)   %now we can
>>%     get the desired plot
```

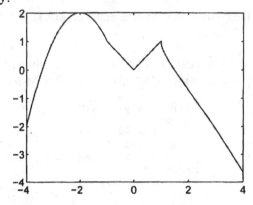

FIGURE 4.1: The plot of the function of Example 4.5.

A more satisfying solution would be to rebuild the M-file in such a way that when vectors are inputted for x, the if-branch testing is done separately for each component. The following program will do the job:

```
function y = ex4_5v2(x)
for i = 1:length(x)
  if x(i)<-1
      y(i) = -x(i).^2-4*x(i)-2;
  elseif x(i)>1
      y(i) = 2-exp(sqrt(x(i)-1));
  else
      y(i)=abs(x(i));
end
```

With this M-file stored in our path the following commands would then indeed produce the desired plot of Figure 4.1:

```
>>  x=-4:.001:4; y=ex4_5v2(x); plot(x,y)
```

In dealing with questions involving integers, MATLAB has several number-theoretic functions available. We mention three here. They will be useful in the following exercise for the reader as well as in the exercises of this section.

floor(x) →	Gives the greatest integer that is $\leq x$ (the **floor** of x).
ceil(x) →	Gives the least integer that is $\geq x$ (the **ceiling** of x).
round(x) →	Gives the nearest integer to x.

For example, floor(2.5) = 2, ceil(2.5) = 3, ceil(−2.5) = −1, and round(−2.2) = −2. Observe that a real number x is an integer exactly when it equals its floor (or ceiling, or round(x) = x).

EXERCISE FOR THE READER 4.2: (a) Write a MATLAB function M-file, call it `sum2sq`, that will take as input a positive integer n and will produce the following output:

(i) In case n cannot be written as a sum of squares (i.e., if it is not possible to write $n = a^2 + b^2$ for some nonnegative integers a and b) then the output should be the statement: "the integer $<n>$ cannot be written as a sum of squares" (where $<n>$ will print as an actual numerical value).

(ii) If n can be written as a sum of squares (i.e., $n = a^2 + b^2$ can be solved for nonnegative integers a and b) then the output should be "the integer $<n>$ can be written as the sum of the squares of $<a>$ and $<b>$" (here again, $<n>$ and also $<a>$ and $<b>$ will print as a actual numerical values) where a and b are actual solutions of the equation.

(b) Run your program with the following inputs: $n = 5$, $n = 25$, $n = 12,233$, $n = 100,000$.

(c) Write a MATLAB program that will determine the largest integer $< 100,000$ that cannot be written as a sum of squares. What is this integer?

(d) Write a MATLAB program that will determine the first integer > 1000 that cannot be written as a sum of squares. What is this integer?

(e) How many integers are there (strictly) between 1000 and 100,000 that cannot be expressed as a sum of the squares of two integers?

A useful MATLAB command syntax for writing interactive script M-files is the following:

`x = input('<Enter input phrase> : ') →`	When a script with this command is run, you will be prompted in command window by the same <Enter input phase> to enter an input for script after which your input will be stored as variable x and the script will be executed.

The command can, of course, also be invoked in a function M-file, or at any time in the MATLAB command window. The next example presents a way to use this command in a nice mathematical experiment.

EXAMPLE 4.6: (*Number Theory: The Collatz Problem*) Suppose we start with any positive integer a_1, and perform the following recursion to define the rest of the sequence $a_1, a_2, a_3, \cdots$:

$$a_{n+1} = \begin{cases} a_n/2, & \text{if } a_n \text{ is even} \\ 3a_n + 1, & \text{if } a_n \text{ is odd} \end{cases}.$$

We note that if a term a_n in this sequence ever reaches 1, then from this point on the sequence will *cycle* through the values 1, 4, 2. For example, if we start with $a_1 = 5$, the recursion formula gives $a_2 = 3 \cdot 5 + 1 = 16$, and then $a_3 = 16/2 = 8$, $a_4 = 8/2 = 4$, $a_5 = 4/2 = 2$, $a_6 = 2/2 = 1$, and so on (4,2,1,4, 2,1,...). Back in 1937, German mathematician Lothar Collatz conjectured that no matter what positive integer we start with for a_1, the above recursively defined sequence will always reach the 1, 4, 2 cycle. Collatz is an example of a mathematician who is more famous for a question he asked than for problems he solved or theorems he proved (although he did significant research in numerical differential equations). The *Collatz conjecture* remains an open problem to this day.[2] Our next example will give a MATLAB script that is useful in examining the Collatz conjecture. Some of the exercises will outline other ways to use MATLAB to run some illuminating experiments on the Collatz conjecture.

EXAMPLE 4.7: We write a script (and save it as `collatz`) that does the following. It will ask for an input for a positive integer to be the initial value $a(1)$ of a Collatz experiment. The program will then run through the Collatz iteration scheme until the sequence reaches the value 1, and so begins to cycle (if ever). The script should output a sentence telling how many iterations were used for this Collatz experiment, and also give the sequence of numbers that were run through until reaching the value of 1.

```
%Collatz script
a(1) = input('Enter a positive integer:   ');
n=1;
while a(n) ~= 1
    if ceil(a(n)/2)==a(n)/2   %tests if a(n) is even
        a(n+1)=a(n)/2;
    else
        a(n+1)=3*a(n)+1;
    end
    n=n+1;
end
fprintf('\r Collatz iteration with initial value a(1)= %d \r', a(1))
fprintf(' took %d iterations before reaching the value 1 and ',n-1)
fprintf(' beginning \r to cycle. The resulting pre-cycling')
fprintf(' sequence is as follows:')
a
clear a %lets us start with a fresh vector a on each run
```

With this script saved as an M-file `collatz`, here is a sample run using $a(1) = 5$:

```
>> collatz
```

[2] The Collatz problem has an interesting history, see, for example [Lag-85] for some details. Many mathematicians have proved interesting results that strongly support the truth of the conjecture. For example, in 1972, the famous Princeton number-theorist J. H. Conway [Con-72] proved that if a Collatz iteration enters into a cycle other than (1,4,2), the cycle must be of length at least 400 (i.e , the cycle itself must consist of at least 400 different numbers). Subsequently, J C. Lagarias (in [Lag-85]) extended Conway's bound from 400 to 275,000! Recent high-speed computer experiments (in 1999, by T. Oliveira e Silvio [OeS-99]) have shown the Collatz conjecture to be true for all initial values of the sequence less than about 2.7×10^{16} . Despite all of these breakthroughs, the problem remains unsolved. P. Erdős, who was undoubtedly one of the most powerful problem-solving mathematicians of the twentieth century, was quoted once as saying "Mathematics is not yet ready for such problems," when talking about the Collatz conjecture In 1996 a prize reward of £1,000 (approx $2,000) was offered for settling the Collatz conjecture Other math problems have (much) higher bounties. For example the *Clay Foundation* (URL: www.claymath org/prizeproblems/statement.htm) has listed seven math problems and offered a prize of $1 million for each one.

Enter a positive integer: 5 (MATLAB gives the first message, we only enter 5, and enter to then get all of the informative output below.)
→Collatz iteration with initial value a(1) = 5 took
5 iterations before reaching the value 1 and beginning
to cycle. The resulting pre-cycling sequence is as follows:
a =5   16   8   4   2   1

EXERCISE FOR THE READER 4.3: (a) Try to understand this script, enter it, and run it with these values: $a(1) = 6, 9, 1, 12, 19, 88, 764$. Explain the purpose of the last command in the above script that cleared the vector a.
(b) Modify the M-file to a new one, called `collctr` (Collatz counter script), which will only give as output the total number of iterations needed for the sequence to reach 1. Make sure to streamline your program so that it does not create the whole vector a (which is not needed here) but rather overwrites new entries for the sequence over the previous ones.

EXERCISES 4.2:

1   Write a MATLAB function M-file, called `sumodsq(n)`, that does the following: The input is a positive integer n. Your function should compute the sum of the squares of all odd integers that do not exceed n:

$$1^2 + 3^2 + 5^2 + \cdots + k^2$$

where k is the largest odd integer that does not exceed n. If this sum is less than 1 million, the output will be the actual sum (a number); if this sum is greater than or equal to 1 million, the output will be the statement "$< n >$ is too big" where $< n >$ will appear as the actual number that was inputted.

2.   Write a function M-file, call it `sevenpow(n)`, that inputs a positive integer n and that figures out how many factors of 7 n has (call this number k) and outputs the statement: "The largest power of 7 which $< n >$ contains as a factor is $< k >$." So for example, if you run sevenpow(98) your output should be the sentence "The largest power of 7 which 98 contains as a factor is 2." Run the commands: `sevenpow(36067)`, `sevenpow(671151153)`, and `sevenpow(3080641535629)`.

3.   (a) Write a function M-file, call it `sumsq(n)`, that will input a positive integer n and output the sum of the squares of all positive integers that are less than or equal to n (sumsq(n) $= 1^2 + 2^2 + 3^2 + \cdots + n^2$). Check and debug this program with the results sumsq(1) = 1, sumsq(3) = 14.
(b) Write a MATLAB loop to determine the largest integer n for which sumsq(n) does not exceed 5,000,000. The output of your code should be this largest integer but nothing else.

4.   (a) Write a MATLAB function M-file, call it `sum2s(n)`, that will take as input a positive integer n and will produce for the output either of the following:
(i) The sum of all of the positive powers of 2 $(2 + 4 + 8 + 16 + ...)$ that do not exceed n, provided this sum is less than 50 million.
(ii) In case the sum in (i) is greater than or equal to 50 million, the output should simply be "overflow."
(b) Run your function with the following inputs: $n = 1$, $n = 10$, $n = 265$, $n = 75,000$, $n = 65,000,000$.

(c) Write a short MATLAB code that will <u>determine the largest integer</u> n for which this program <u>does not</u> produce "overflow"

5   (a) Write a MATLAB function M-file, called bigpro (x), that does the following: The input is a real number x The <u>only</u> output of your function should be the real number formed by the product $x(2x)(3x)(4x)\cdots(nx)$, where n is the first positive integer such that either nx is an integer or $|nx|$ exceeds x^2 (whichever comes first)

(b) Of course, after you write the program you need to debug it. What values should result if we were to use the (correctly created) program to find. bigpro(4), bigpro(2.5), bigpro(12.7)? Run your program for these values as well as for the values $x = -3677/9, x = 233\,6461$, and $x = 125,456.789$.

(c) Find a negative number x that is not an integer such that bigpro(x) is negative.

6.   (*Probability: The Birthday Problem*) This famous problem in probability goes as follows. If there is a room with a party of people and everyone announces his or her birthday, how many people (at least) would there need to be in the room so that there is more than a 50% chance that at least two people have the same birthday?

To solve this problem, we let $P(n) =$ the probability of a common birthday if there are n people in the room. Of course $P(1) = 0$ (no chance of two people having the same birthday if there is only one person in the room), and $P(n) = 1$ when $n > 365$ (there is a 100% chance, i e., guaranteed, two people will have the same birthday if there are more people in the room than days of the year; we ignore leap-year birthdays). We can get an expression for $P(n)$ by calculating the *complementary probability* (i.e., the probability that there will *not* be a common birthday among n different people) This must be

$$\frac{365}{365} \cdot \frac{364}{365} \cdot \frac{363}{365} \cdots \frac{366-n}{365}$$

This can be seen as follows The first person can have any of the 365 possible birthdays, the second person can have only 364 possibilities (since he/she cannot have the same birthday as the first person), the third person is now restricted to only 363 possible birthdays, and so on. We multiply the individual probabilities (fractions) to get the combined probability of no common birthday. Now this is the complementary probability of what we want (i e., it must add to $P(n)$ to give $1 = 100\%$ since it is guaranteed that either there is a common birthday or not). Thus

$$P(n) = 1 - \frac{365}{365} \cdot \frac{364}{365} \cdot \frac{363}{365} \cdots \frac{366-n}{365}$$

(a) Write a MATLAB function M-file for this function $P(n)$. Call the M-file bprob (n) and set it up so that it does the following: If n is a nonnegative integer, the function bprob (n) will output the sentence· "If a room contains $<n>$ people then the probability of a common birthday is $< P(n) >$" where $<n>$ and $< P(n) >$ should be the actual numerical values If n is any other type of number (e.g , a negative number or 2.6) the output should be "Input $<n>$ is not a natural number so the probability is undefined." Save your M-file and then run it for the following values: $n = 3$, $n = 6$, $n = 15$, $n = 90$, $n = 110.5$, and $n = 180$.

(b) Write a MATLAB code that uses your function in part (a) to solve the birthday problem, i e , determine the smallest n for which $P(n) > .5$. More precisely, create a for loop whose <u>only</u> <u>output</u> will be. $n =$ the minimum number needed (for $P(n)$ to be > 5) and the associated probability $P(n)$.

(c) Get MATLAB to draw a neat plot of $P(n)$ vs. n (for all n between 1 and 365), and on the same plot, include the plots of the two horizontal lines with y-intercepts .5 and .9. Interpret the intersections.

7   Write a function M-file, call it pythag (n), that inputs a positive integer n and determines

whether n is the hypotenuse of a right triangle with sides of integer lengths. Thus your program will determine whether the equation $n^2 = a^2 + b^2$ has a solution with a and b both being positive integers.

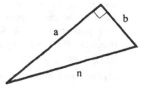

FIGURE 4.2: Pythagorean triples.

Such triples n, a, b are called Pythagorean triples (Figure 4.2). In case there is no solution (as, for example, if $n = 4$), your program should output the statement: "There are no Pythagorean triples with hypotenuse $<n>$." But if there is a solution your output should be a statement that actually gives a specific Pythagorean triple for your value of n. For example, if you type `pythag(5)`, your output should be something like: "There are Pythagorean triples having 5 as the hypotenuse, for example: 3, 4, 5 is one such triple." Run this for several different values of n. Can you find a value of n larger than 1000 that has a Pythagorean triple? Can you find an n that has two different Pythagorean triples associated with it (of course not just by switching a and b)?

HISTORICAL NOTE: Since the ancient times of Pythagoras, mathematicians have tried long and hard to find integer triple solutions of the corresponding equation with exponent 3: $n^3 = a^3 + b^3$. No one has ever succeeded. In the 1700s the amateur French mathematician Pierre Fermat conjectured that no such triples can exist. He claimed to have a truly remarkable proof of this but there was not enough space in the margin of his notes to include it. There has been an incredible amount of research trying to come up with this proof. Just recently, more than 300 years since Fermat stated his conjecture, Princeton mathematician Andrew Wiles came up with a proof. He was subsequently awarded the Fields medal, the most prestigious award in mathematics.

8. (*Plane Geometry*) For an integer n that is at least equal to 3, a *regular n-gon* in the plane is the interior of a set whose boundary consists of n flat edges (sides) each having the same length (and such that the interior angles made by adjacent edges are all equal). When $n = 3$ we get an equilateral triangle, when $n = 4$ we get a square, and when $n = 8$ we get a regular octagon, which is the familiar stop-sign shape. There are regular n-gons for any such value of n; some are pictured in Figure 4.3

FIGURE 4.3: Some regular polygons.

(a) Write a MATLAB function M-file, ngonper1(n,dia), that has two input variables, n = the number of sides of the regular n-gon, and dia = the diameter of the regular n-gons. The *diameter* of an n-gon is the length of the longest possible segment that can be drawn connecting two points on the boundary. When n is even, the diameter segment cuts the n-gons into two congruent (equal) pieces. Assuming that n is an even integer greater than 3 and dia is any positive number, your function should have a single output variable that equals the perimeter of the regular n-gons with diameter = dia. Your solution should include a handwritten mathematical derivation of the formula for this perimeter. This will be the hard part of this exercise, and it should be done, of course, before you write the program. Run your program for the following sets of input data: (i) $n = 4$, dia = $\sqrt{4}$, (ii) $n = 12$, dia = 12, (iii) $n = 1000$, dia = 5000.

(b) Remove the restriction that n is even from your program in part (a). The new function (call it now ngonper (n, dia)) will now do everything that the one you constructed in part (a) did but it will be able to input and deal with any integer n greater than or equal to 3. Again, include with your solution a mathematical derivation of the perimeter formula you are using in your program. Run your program for these sets of values· (i) $n = 3$, dia $= 2$, (ii) $n = 5$, dia $= 4$, (iii) $n = 999$, dia $= 500$

(c) For which values of n (if any) will your function in part (b) continue to give the correct perimeter of an n-gon that is no longer regular? An *irregular n-gon* is the interior of a set in the plane whose boundary consists of n flat edges whose interior angles are not all equal Examples of irregular n-gons include any nonequilateral triangle ($n = 3$), any quadrilateral that is not a square ($n = 4$)   For those n's for which you say things still work, a (handwritten mathematical) proof should be included, and for those n's for which you say things no longer continue to work, a (handwritten) counterexample should be included

9.   (*Plane Geometry*) This exercise consists of doing what is asked for in Exercise 8 (a)(b)(c) but with changing all occurrences of the word "perimeter" to "area " In parts (a) and (b) use the M-file names ngonar1 (n, dia) and ngonarea (n, dia)

10   (*Finance: Compound Interest*) Write a script file called compints that will compute (as output) the future value A in a savings account after prompting the user for the following inputs. the principal P (= amount deposited), the annual interest rate r (as a decimal), the number k of compoundings per year (so quarterly compounding means $k = 4$, monthly means $k = 12$, daily means $k = 365$, etc), and the time t that the money is invested (measured in years).   The relevant formula from finance is $A = P(1 + r/k)^{kt}$.   Run the script using the following sets of inputs· $P = \$10,000$, $r = 8\%$ (08), $k = 4$, and $t = 10$, then changing t to 20, then also changing r to 11%
Suggestion: You probably want to have four separate "input" lines in your script file, the first asking for the principal, etc. Also, to get the printout to look nice, you should switch to format bank inside the script and then (at the very end) switch back to format short.

11.   (*Finance. Compound Interest*) Write a script file called comings that takes the same inputs as in the previous exercise, but instead of producing the output of the future account balance, it should produce a graph of the future value A as a function of time as the time t ranges from zero (day money was invested) until the end of the time period that was entered. Run the script for the three sets of data in the previous problem

12.   (*Finance: Future Value Annuities*) Write a script file called fvanns that will compute (as output) the future value FV in an annuity after prompting the user for the following inputs. the periodic payment PMT (= amount deposited in account per period), the annual interest rate r (as a decimal), the number k of periods per year, that is, the number of compoundings per year (so quarterly compoundings/deposits means $k = 4$, monthly means $k = 12$, bimonthly means $k = 24$, etc.), and the time t that the money is invested (measured in years).   The relevant formula from finance is $FV = PMT((1 + r/k)^{kt} - 1)/(r/k)$. Run the script using the following sets of inputs· $PMT = 200$, $r = 7\%$ (.07), $k = 12$, and $t = 30$, then changing t to 40, then also changing r to 9%. Next change PMT to 400 on each of these three sets of inputs Note, the first set of inputs could correspond to a worker who starts a supplemental retirement plan (say a 401(k)), deposits $200 each month starting at age 35, and continues until he/she plans to retire at age 65 ($t = 30$ years later). The FV will be his/her retirement nest egg at time of retirement. The next set of data could correspond to the same retirement plan but started at age 25 (10 years more time) In each case compare the future value with the total amount of contributions. To encourage such supplemental retirement plans, the federal government allows such contributions (with limits) to be done before taxation.
Suggestion: You probably want to have four separate "input" lines in your script file, the first

asking for the principal, etc.   Also, to get the printout to look nice, you should switch to format bank inside the script and then (at the very end) switch back to format short.

13.   (*Finance: Future Value Annuities*) In this exercise you will be writing a script file that will take the same inputs as in the previous exercise (interactively), but instead of just giving the future value at the end of the time period, this script will produce a graph of the growth of the annuity's value as a function of time.
(a) Base your script on the formula given in the preceding exercise for future value annuities. Call this script fvanngs. Run the script for the same sets of inputs that were given in the previous exercise.
(b) Rewrite the script file, this time constructing the vector of future values using a recursion formula rather than directly (as was asked in part (a)). Call this script fvanng2s. Run the script for the same sets of inputs that were given in the previous exercise.

14.   (*Number Theory: The Collatz Problem*) Write a function M-file, call it collctr, that takes as input, a positive integer an (the first element for a Collatz experiment), and has as output the positive integer n, which equals the number of iterations required for the Collatz iteration to reach the value of 1. What is the first positive integer n for which this number of iterations exceeds 100? 200? 300?

4.3: WRITING GOOD PROGRAMS

Up to this point we have introduced the two ways that programs can be written and stored for MATLAB to use (function M-files and script M-files) and we have also introduced the basic elements of control flow and a few very useful built-in MATLAB functions. To write a good program for a specified task, we will need to put all of our skills together to come up with an M-file that, above all, does what it is supposed to do, is efficient, and is as eloquent as possible. In this section we present some detailed suggestions on how to systematically arrive at such programs. Programming is an art and the reader should not expect to master it easily or in a short time.

STEP 1: **Understand the problem, do some special cases by hand, and draw an outline.** Before you begin to actually type out a program, you should have a firm understanding of what the problem is (that the program will try to solve) and know how to solve it by hand (in theory, at least). Computers are not creative. They can do very well what they are told, but you will need to tell them exactly what to do, so you had better understand how to do what needs to be done. You should do several cases by hand and record the answers. This data will be useful later when you test your program and debug it if necessary. Draw pictures (a flowchart), and write in plain English an explanation of the program, trying to be efficient and avoiding unnecessary tasks that will use up computer time.

STEP 2: **Break up larger programs into smaller module programs.** Larger programs can usually be split up into smaller independent programs. In this way the main program can be considerably reduced in size since it can call on the smaller module programs to perform secondary tasks. Such a strategy has numerous advantages. Smaller programs are easier to write (and debug) than larger ones and they may be used to create other large or improved programs later on down the road.

STEP 3: **Test and debug every program.** This is not an option. You should always test your programs with a variety of inputs (that you have collected output data for in Step 1) to make sure all of the branches and loops function appropriately. Novice and experienced programmers alike are often shocked at how rarely a program works after it is first written. It may take many attempts and changes to finally arrive at a fully functional program, but a lot of valuable experience can be gained in this step. It is one thing to look at a nice program and think one understands it well, but the true test of understanding programming is to be able to create and write good programs. Before saving your program for the first time, always make sure that every "for", "while", or "if" has a matching "end".   One useful scheme when debugging is to temporarily remove all semicolons from the code, perhaps add in some auxiliary output to display, and then run your program on the special cases that you went through by hand in Step 1. You can see first-hand if things are proceeding along the lines that you intended.

STEP 4: **After it finally works, try to make the program as efficient and easy to read as possible.** Look carefully for redundant calculations. Also, try to find ways to perform certain required tasks that use minimal amounts of MATLAB's time. Put in plenty of comments that explain various elements of the program. While writing a complicated program, your mind becomes full of the crucial and delicate details. If you read the same program a few months (or years) later (say, to help you to write a program for a related task), you might find it very difficult to understand without a very time-consuming analysis. Comments you inserted at the time of writing can make such tasks easier and less time consuming. The same applies even more so for other individuals who may need to read and understand your program.

The efficiency mentioned in Step 4 will become a serious issue with certain problems whose programs (even good ones) will push the computer to its limits. We will come up with many examples of such problems this book. We mention here two useful tools in testing efficiency of programs or particular tasks. A **flop** (abbreviation for **floating point operation**) is roughly equivalent to a single addition, subtraction, multiplication, or division of two numbers in full MATLAB precision (rather than a faster addition of two single-digit integers, say). Counting flops is a common way of comparing and evaluating efficiency of various programs and parts thereof. MATLAB has convenient ways of counting flops[3] or elapsed time (`tic/toc`):

---

[3] The `flop` commands in MATLAB are actually no longer available since Version 5 (until further notice). This is due to the fact that, starting with Version 6, the core programs in MATLAB got substantially revised to be much more efficient in performing matrix operations. It was unfortunate that the `flop` counting features could no longer be made to perform in this newer platform (collateral damage) Nonetheless, we will, on occasion, use this function in cases where flop counts will help to

flops(0) ...MATLAB commands... flops	The flops(0) resets the flop counter at zero. The flops tells the number of flops used to execute the "MATLAB commands" in between. (Not available since Version 5, see Footnote 3.)

tic ...MATLAB commands... toc	This tic resets the stopwatch to zero. The toc will tell the elapsed time used to execute the "MATLAB commands."

The results of tic/toc depend not just on the MATLAB program but on the speed of the computer being used, as well as other factors, such as the number of other tasks concurrently being executed on the same computer. Thus the same MATLAB routines will take varying amounts of time on different computers (or even on the same computer under different circumstances). So, unlike flop comparisons, tic/toc comparisons cannot be absolute.

EXAMPLE 4.8: Use the tic/toc commands to compare two different ways of creating the following large vector: $x = [1 \ 2 \ 3 \ \cdots \ 10,000]$. First use the non-loop construction and then use a for loop. The results will be quite shocking, and since we will need to work with such large single vectors quite often, there will be an important lesson to be learned from this example. **When creating large vectors in MATLAB, avoid, if possible, using "loops."**

SOLUTION:

```
>> tic, for n=1:10000, x(n)=n; end, toc
 →elapsed_time =8.9530 (time is measured in seconds)
>> tic, y=1:10000; toc
 →elapsed_time = 0
```

The loop took nearly nine seconds, but the non-loop construction of the same vector went so quickly that the timer did not detect any elapsed time. Let's try to build a bigger vector:

```
>> tic, y=1:100000; toc
 →elapsed_timo = 0.0100
```

This gives some basis for comparison. We see that the non-loop technique built a vector 10 times as large in about 1/1000 of the time that it took the loop construction to build the smaller vector! The flop-counting comparison method would not apply here since no flops were done in these constructions.

Our next example will deal with a concept from linear algebra called the **determinant** of a square matrix, which is a certain important number associated

---

illustrate important points. Readers that do not have access to older versions of MATLAB will not be able to mimic these calculations.

with the matrix. We now give the definition of the determinant of a square $n \times n$ matrix[4]

$$A = \begin{bmatrix} a_{11} & a_{12} & a_{13} & \cdots & a_{1n} \\ a_{21} & a_{22} & a_{23} & \cdots & a_{2n} \\ a_{31} & a_{32} & a_{33} & \cdots & a_{3n} \\ \vdots & \vdots & \vdots & \ddots & \vdots \\ a_{n1} & a_{n2} & a_{n3} & \cdots & a_{nn} \end{bmatrix}.$$

If $n = 1$, so $A = [a_{11}]$, then the determinant of A is simply the number a_{11}. If $n = 2$, so $A = \begin{bmatrix} a_{11} & a_{12} \\ a_{21} & a_{22} \end{bmatrix}$, the determinant of A is defined to be the number $a_{11}a_{22} - a_{12}a_{21}$ that is just the product of the main diagonal entries (top left to bottom right) less the product of the off diagonal entries (top right to bottom left).

For $n = 3$, $A = \begin{bmatrix} a_{11} & a_{12} & a_{13} \\ a_{21} & a_{22} & a_{23} \\ a_{31} & a_{32} & a_{33} \end{bmatrix}$ and the determinant can be defined using the $n = 2$ definition by the so-called cofactor expansion (on the first row). For any entry a_{ij} of the 3×3 matrix A, we define the corresponding **submatrix** A_{ij} to be the 2×2 matrix obtained from A by deleting the row and column of A that contain the entry a_{ij}. Thus, for example,

$$A_{13} = \begin{bmatrix} a_{11} & a_{12} & a_{13} \\ a_{21} & a_{22} & a_{23} \\ a_{31} & a_{32} & a_{33} \end{bmatrix} = \begin{bmatrix} a_{21} & a_{22} \\ a_{31} & a_{32} \end{bmatrix}.$$

Abbreviating the determinant of the matrix A by det(A), the determinant of the 3×3 matrix A is given by the following formula:

$$\det(A) = a_{11} \det(A_{11}) - a_{12} \det(A_{12}) + a_{13} \det(A_{13}).$$

Since we have already shown how to compute the determinant of a 2×2 matrix, the right side can be thus computed. For a general $n \times n$ matrix A, we can compute it with a similar formula in terms of some of its $(n-1) \times (n-1)$ submatrices:

---

[4] The way we define the determinant here is different from what is usually presented as the definition. One can find the formal definition in books on linear algebra such as [HoKu-71].   What we use as our definition is often called *cofactor expansion on the first row*.   See [HoKu-71] for a proof that this is equivalent to the formal definition. The latter is actually more complicated and harder to compute and this is why we chose cofactor expansion

$$\det(A) = a_{11} \det(A_{11}) - a_{12} \det(A_{12}) + a_{13} \det(A_{13}) - \cdots + (-1)^{n+1} a_{1n} \det(A_{1n}).$$

It is proved in linear algebra books (e.g., see [HoKu-71]) that one could instead take the corresponding (cofactor) expansion along any row or column of A, using the following rule to choose the alternating signs: The sign of $\det(A_{ij})$ is $(-1)^{i+j}$.

Below are two MATLAB commands that are used to work with entries and submatrices of a given general matrix A.

A(i,j) →	Represents the entry a_{ij} located in the ith row and the jth column of the matrix A.
A([i1 i2 .. imax], [j1 j2 .jmax]) →	Represents the submatrix of the matrix A formed using the rows $i1, i2, ..., imax$ and columns $j1, j2, ..., jmax$.
A([i1 i2 . imax], :) →	Represents the submatrix of the matrix A formed using the rows $i1, i2, ..., imax$ and all columns.

EXAMPLE 4.9: (a) Write a MATLAB function file, called mydet2(A), that calculates the determinant of a 2×2 matrix A.
(b) Using your function mydet2 of part (a), build a new MATLAB function file called mydet3(A) that computes the determinant of a 3×3 matrix A (by performing cofactor expansion along the first row).
(c) Write a program mydet(A) that will compute the determinant of a square matrix of any size. Test it on the matrices shown below. MATLAB has a built-in function det for computing determinants. Compare the results, flop counts (if available), and times using your function mydet versus MATLAB's program det. Perform this comparison also for a randomly generated 8×8 matrix. Use the following command to generate random matrices:

rand(n,m) → NOTE: rand(n) is equivalent to rand(n,n), and rand to rand(1).	Generates an $n \times m$ matrix whose entries are randomly selected from $0 \le x \le 1$.[5]

$$A = \begin{bmatrix} 2 & 7 & 8 & 10 \\ 0 & -1 & 4 & -9 \\ 0 & 0 & 3 & 6 \\ 0 & 0 & 0 & 5 \end{bmatrix}, \quad A = \begin{bmatrix} 1 & 2 & -1 & -2 & 1 & 2 \\ 0 & 3 & 0 & 2 & 0 & 1 \\ 1 & 0 & 2 & 0 & 3 & 0 \\ 1 & 1 & 1 & 1 & 1 & 1 \\ -2 & -1 & 0 & 1 & 2 & 3 \\ 1 & 2 & 3 & 1 & 2 & 3 \end{bmatrix}$$

---

[5] Actually, the rand function, like any computer algorithm, uses a deterministic program to generate random numbers that is based on a certain seed number (starting value). The numbers generated meet statistical standards for being truly random, but there is a serious drawback that at each fresh start of a MATLAB session, the sequence of numbers generated by successive applications of rand will always result in the same sequence. This problem can be corrected by entering rand('state',sum(100*clock)), which resets the seed number in a somewhat random fashion based on the computer's internal clock. This is useful for creating simulation trials.

SOLUTION: The programs in parts (a) and (b) are quite straightforward:

```
function y = mydet2(A)
y=A(1,1)*A(2,2)-A(1,2)*A(2,1);

function y = mydet3(A)
y=A(1,1)*mydet2(A(2:3,2:3))-A(1,2)*mydet2(A(2:3,[1...
 3]))+A(1,3)*mydet2(A(2:3,1:2));
```

NOTE: The three dots (...) at the end of a line within the second function indicate (in MATLAB) a continuation of the command. This prevents the carriage return from executing a command that did not fit on a single line.

The program for part (c) is not quite so obvious. The reader is strongly urged to try and write one now before reading on.

Without having the mydet program call on itself, the code would have to be an extremely inelegant and long jumble. Since MATLAB allows its functions to (recursively) call on themselves, the program can be elegantly accomplished as follows:

```
function y = mydet(A)
y=0; %initialize y
[n, n] = size(A); %record the size of the square matrix A
if n ==2
    y=mydet2(A);
    return
end
for i=1:n
    y=y+(-1)^(i+1)*A(1,i)*mydet(A(2:n, [1:(i-1) (i+1):n]));
end
```

Let's now run this program side by side with MATLAB's det to compute the requested determinants.

```
>> A=[2  7  8  10; 0 -1  4 -9; 0  0  3  6; 0  0  0  5];
>> A1=[1  2 -1 -2  1  2;0  3  0  2  0  1;1  0  2  0  3  0;1  1  1  1  1  1;...
-2 -1  0  1  2  3; 1  2  3  1  2  3];

>> flops(0), tic,  mydet(A), toc, flops
→ans = -30(=determinant), elapsed_time =  0.0600, ans = 182 (=flop count)
>> flops(0), tic,  mydet(A1), toc, flops
→ans = 324, elapsed_time = 0.1600, ans =5226

>> flops(0), tic,  det(A), toc, flops
→ans =-30, elapsed_time = 0, ans =52

>> flops(0), tic,  det(A1), toc, flops
→ans =324, elapsed_time = 0, ans = 117
```

So far we can see that MATLAB's built-in det works quicker and with a lot less flops than our mydet does. Although mydet still performs reasonably well, check out the flop-count ratios and how they increased as we went from the 4×4 matrix A to the 6×6 matrix $A1$. The ratio of flops `mydet` used to the number that `det` used rose from about a factor of 3.5 to a factor of nearly 50. For larger matrices, the situation quickly gets even more extreme and it becomes no longer practical to use `mydet`. This is evidenced by our next computation with an 8×8 matrix.

```
>> Atest= rand(8); %we suppress output here.
>> flops(0), tic,  det(Atest), toc, flops
→ans = -0.0033, elapsed_time = 0, ans = 326
>> flops(0), tic,  mydet(Atest), toc, flops
→ans = -0.0033, elapsed_time =8.8400, ans = 292178
```

MATLAB's `det` still works with lightning speed (elapsed time was still undetectable) but now `mydet` took a molasses-slow nearly 9 seconds, and the ratio of the flop count went up to nearly 900! If we were to go to a 20×20 matrix, at this pace, our `mydet` would take over 24 years to do! (See Exercise 5 below.) Suprisingly though, MATLAB's `det` can find the determinant of such a matrix in less than 1/100 of a second (on the author's computer) with a flop count of only about 5000. This shows that there are more practical ways of computing (large matrix determinants) than by the definition or by cofactor expansion. In Chapter 7 such a method will be presented.

Each time when the `rand` command is invoked, MATLAB uses a program to generate a random numbers so that in any given MATLAB session, the sequence of "random numbers" generated will always be the same. Random numbers are crucial in the important subject of *simulation*, where trials of certain events that depend on chance (like flipping a coin) need to be tested. In order to assure that the random sequences are different at each start of a MATLAB session, the following command should be issued before starting to use `rand`:

`rand('state',sum(100*clock))` →	This sets the "state" of MATLAB's random number generator in a way that depends in a complicated fashion on the current computer time. It will be different each time MATLAB is started.

Our next exercise for the reader will require the ability to store strings of text into rows of numerical matrices, and then later retrieve them. The following basic example will illustrate how such data transformations can be accomplished: We first create text string T and a numerical vector v:

```
>> T = 'Test', v = [1 2 3 4 5 6]
→T = Test,   v = 1   2   3   4   5   6
```

If we examine how MATLAB has stored each of these two objects, we learn that both are "arrays," but T is a "character array" and v is a "double array" (meaning a matrix of numbers):

```
>> whos T  v  →Name    Size        Bytes  Class
                   T     1x4            8  char array
                   v     1x6           48  double array
```

If we redefine the first four entries of the vector v to be the vector T, we will see that the characters in T get transformed into numbers:

```
>> v(1:4)=T
→v = 84   101   115   116   5   6
```

MATLAB does this with an internal dictionary that translates all letters, numbers, and symbols on the keyboard into integers (between 1 and 256, in a case-sensitive fashion). To transform back to the original characters, we use the char command, as follows:

```
>> U=char(v(1:4))
→U =Test
```

Finally, to call on a stored character (string) array within an fprintf statement, the symbol %s is used as below:

```
>> fprintf('The %s has been performed.', U)
→The Test has been performed.
```

EXERCISE FOR THE READER 4.4: (*Electronic Raffle Drawing Program*)
(a) Create a script M-file, raffledraw, that will randomly choose the winner of a raffle as follows: When run, the first thing the program will do is prompt the user to enter the number of players (this can be any positive integer). Next it will successively ask the user to input the names of the players (in single quotes, as text strings are usually inputted) along with the corresponding weight of each player. The weight of a player can be any positive integer and corresponds to the number of raffle tickets that the player is holding. Then the program will randomly select one of these tickets as the winner and output a phrase indicating the name of the winner.
(b) Run your M-file with the following data on four players: Alfredo has four tickets, Denise has two tickets, Sylvester has two tickets, and Laurie has four tickets. Run it again with the same data.

EXERCISES 4.3:

1. Write a MATLAB function M-file, call it sum3sq(n), that takes as input a positive integer n and as output will do the following. If n can be expressed as a sum of three squares (of positive integers), i.e., if the equation

$$n = a^2 + b^2 + c^2$$

has a solution with a, b, c all positive integers, then the program should output the sentence, "The number $<n>$ can be written as the sum of the squares of the three positive integers $<a>$, $<b>$, and $<c>$" Each of the numbers in brackets must be actual integers that solve the equation. In case the equation has no solution (for a, b, c), the output should be the sentence· "The number $<n>$ cannot be expressed as a sum of the squares of three positive integers" Run your program with the numbers $n = 3$, $n = 7$, $n = 43$, $n = 167$, $n = 994$, $n = 2783$, $n = 25,261$. Do you see a pattern for those integers n for which the equation does/does not have a solution?

2.   Repeat Exercise 1 with "three squares" being replaced by "four squares," so the equation
      becomes:

$$n = a^2 + b^2 + c^2 + d^2.$$

Call your function sum4sq. In each of these problems feel free to run your programs for a
larger set of inputs so as to better understand any patterns that you may perceive.

3.   (*Number Theory: Perfect Numbers*) (a) Write a MATLAB function M-file, call it divsum(n),
      that takes as input a positive integer n and gives as output the sum of all of the proper divisors of
      n. For example, the proper divisors of 10 are 1, 2, and 5 so the output of divsum(10) should
      be 8 (= 1 + 2 + 5). Similarly, divsum(6) should equal 6 since the proper divisors of 6 are 1, 2,
      and 3. Run your program for the following values of n: $n = 10$, $n = 224$, $n = 1410$ (and give
      the outputs).
      (b) In number theory, a **perfect number** is a positive integer n that equals the sum of its proper
      divisors, i.e., $n = $ divsum(n). Thus from above we see that 6 is a perfect number but 10 is not.
      In ancient times perfect numbers were thought to carry special magical properties. Write a
      program that uses your function in part (a) to get MATLAB to find and print all of the perfect
      numbers that are less than 1000. Many questions about perfect numbers still remain perfect
      mysteries even today. For example, it is not known if the list of perfect numbers goes on
      forever.

4.   (*Number Theory: Prime Numbers*) Recall that a positive integer n is called a **prime** number if
      the only positive integers that divide evenly into n are 1 and itself. Thus 4 is not a prime since
      it factors as 2×2. The first few primes are as follows: 2, 3, 5, 7, 11, 13, 17, 19, 23, ... (1 is
      not considered a prime for some technical reasons). There has been a tremendous amount of
      research done on primes, and there still remain many unanswered questions about them that are
      the subject of contemporary research. One of the first questions that comes up about primes is
      whether there are infinitely many of them (i.e., does our list go on forever?). This was answered
      by an ancient Greek mathematician, Euclid, who proved that there are infinitely many primes. It
      is a very time-consuming task to determine if a given (large) number is prime or not (unless it is
      even or ends in 5).
      (a) Write a MATLAB function M-file, call it primeck(n), that will input a positive integer n
      > 1, and will output either the statement: "the number < n > is prime," if indeed, n is prime, or
      the statement "the number < n > is not prime, its smallest prime factor is < k >," if n is not
      prime, and here k will be the actual smallest prime factor of n.
      Test (and debug) your program for effectiveness with the following inputs for n:

$$n = 51, \; n = 53, \; n = 827, \; n = 829.$$

Next test your program for efficiency with the following inputs (depending on how you wrote
your program and also how much memory your computer has, it may take a very long time or
not finish with these tasks)

$$n = 8237, \; n = 38877, \; n = 92173, \; n = 1,875,247, \; n = 2038074747, \; n = 22801763489,$$
$$n = 1689243484681, \; n = 7563374525281.$$

In your solution, make sure to give exactly what the MATLAB printout was; also, next to each
of these larger numbers, write down how much time it took MATLAB to perform the
calculation.
(b) Given enough time (and assuming you are working on a computer that will not run out of
memory) will this MATLAB program always work correctly no matter how large n is? Recall
that MATLAB has an accuracy of about 15 significant digits.

5.   We saw in Example 4.7 that by calculating a determinant by using cofactor expansion, the
      number of flops (additions, subtractions, multiplications, and divisions) increases dramatically.
      For a 2×2 matrix, the number (worst-case scenario, assuming no zero entries) is 3; for a 3×3
      matrix it is 14. What would this number be for a 5×5 matrix?, For a 9×9 matrix? Can you
      determine a general formula for an $n \times n$ matrix?

6   (*Probability and Statistics*) Write a program called cointoss(n) that will have one input
        variable n = a positive integer and will simulate n coin tosses, by (internally) generating a
        sequence of n random numbers (in the range $0 \le x \le 1$) and will count each such number that is
        less than 0 5 as a "HEAD" and each such number that is greater than 0 5 as a "TAIL" If a
        number in the generated sequence turns out to be exactly = 0.5, another simulated coin toss
        should be made (perhaps repeatedly) until a "HEAD" or a "TAIL" comes up There will be only
        one output variable. P = the ratio of the total number of "HEADS" divided by n But the
        program should also cause the following sentence to be printed: "In a trial of < n > coin tosses,
        we had <H> flips resulting in 'HEAD' and <T> flips resulting in 'TAIL,' so 'HEADS' came up
        <100P>% of the time " Here, <H> and <T> are to denote the actual numbers of "HEAD" and
        "TAIL" results. Run your program for the following values of n. 2, 4, 6, 10, 50, 100, 1000,
        5000, 50,000 Is it possible for this program to enter into an infinite loop? Explain!

7.   (*Probability and Statistics*) Write a program similar to the one in the previous exercise except
        that it will not print the sentence, and it will have three output variables: P (as before), H = the
        number of heads, and T = the number of tails Set up a loop to run this program with n = 1000
        fixed for k = 100 times   Collect the outcomes of the variable H as a vector:
        $[h_0, h_1, h_2, \cdots h_{n+1}]$ (with $n+1 = 1001$ entries) where each h_i denotes the number of times that
        the experiment resulted in having exactly h_i heads (so $H = h_i$) and then plot the graph of this
        vector (on the x-axis n runs from 0 to 1001 and on the y-axis we have the h_i -values). Repeat
        this exercise for k = 200 and then k = 500 times.

8.   (*Probability: Random Integers*) Write a MATLAB function M-file, randint(n, k), that has
        two input variables n and k being positive integers. There will be one output variable R, a
        vector with k components $R = [r_1, r_2, \cdots, r_k]$, each of whose entries is a positive integer
        randomly selected from the list {1, 2,   , n }. (Each integer in this list has an equal chance of
        being generated at any time)

9   (*Probability: Random Walks*) Create a MATLAB M-file called ran2walk(n) that simulates
        a random walk in the plane. The input n is the number of steps in the walk The starting point
        of the walk is at the origin (0,0). At each step, random numbers are chosen (with uniform
        distribution) in the interval [−1/2, 1/2] and are added to the present x- and y-coordinates to
        get the next x- and y-coordinates   The MATLAB command rand generates a random
        number in the interval [0, 1], so we must subtract 0 5 from these to get the desired distributions.
        There will be no output variables, but MATLAB will produce a plot of the generated random
        walk
               Run this function for the values n = 8, 25, 75, 250 and (using the subplot option) put them all
        into a single figure.   Repeat once again with the same values. In three dimensions, these
        random walks simulate the chaotic motion of a dust particle that makes many microscopic
        collisions and produces such strange motions. This is because the microscopic particles that
        collide with our particle are also in constant motion. We could easily modify our program by
        adding a third z-coordinate (and using plot3(x,y,z) instead of plot(x,y)) to make a
        program to simulate such three-dimensional random walks Interestingly, each time you run the
        ran2walk function for a fixed value of n, the paths will be different. Try it out a few times.
        Do you notice any sort of qualitative properties about this motion? What are the chances (for a
        fixed n) that the path generated will cross itself? How about in three dimensions? Does the
        motion tend to move the particle away from where it started as n gets large? For these latter
        questions do not worry about proofs, but try to do enough experiments to lead you to make some
        educated hypotheses.

10   (*Probability Estimates by Simulation*) In each part, run a large number of simulations of the
        following experiments and take averages to estimate the indicated quantities.
        (a) Continue to generate random numbers in (0,1) using rand until the accumulated sum

exceeds 1. Let N denote the number of such random numbers that get added up when this sum first exceeds 1. Estimate the *expected value* of N, which can be thought of as the theoretical (long-run) average value of N if the experiment gets repeated indefinitely.

(b) Number a set of cards from 1 to 20, and shuffle them. Turn the cards over one by one and record the number of times K that card number i ($1 \le i \le 20$) occurs at (exactly) the ith draw. Estimate the expected value of K.

Note: Simulation is a very useful tool for obtaining estimates for quantities that can be impossible to estimate analytically; see [Ros-02] for a well-written introduction to this interesting subject. In it the reader can also find a rigorous definition of the expectation of a random variable associated with a (random) experiment. The quantities K and N above are examples of random variables. Their outcomes are numerical quantities associated with the outcomes of (random) experiments. Although the outcomes of random variables are somewhat unpredictable, their long-term averages do exhibit patterns that can be nicely characterized.

For the above two problems, the exact expectations are obtainable using methods of probability; they are $N = e$ and $K = 1$.

The next four exercises will revisit the Collatz conjecture that was introduced in the preceding section.

11. (*Number Theory: The Collatz Problem*) Write a function M-file, call it `collsz`, that takes as input a positive integer `an` (the first element for a Collatz experiment), and has as output a positive integer `size` equaling the size of the largest number in the Collatz iteration sequence before it reaches the value of 1. What is the first positive integer `an` for which this maximum size exceeds the value 100? 1000? 100,000? 1,000,000?

12. (*Number Theory: The Collatz Problem*) Modify the script file, `collatz`, of Example 4.7 in the text to a new one, `collatzg`, that will interactively take the same input and internally construct the same vector a, but instead of producing output on the command window, it should produce a graphic of the vector a's values versus the index of the vector. Arrange the plot to be done using blue pentagrams connected with lines. Run the script using the following inputs: 7, 15, 27, 137, 444, 657.
Note: The syntax for this plot style would be `plot(index, a, bp-)`.

13. (*Number Theory: The Collatz Problem*) If a Collatz experiment is started using a negative integer for $a(1)$, all experiments so far done by researchers have shown that the sequence will eventually cycle. However, in this case, there is more than one possible cycle. Write a script, `collatz2`, that will take an input for a(1) in the same way as script `collatz` in Example 4.7 did, and the script will continue to do the Collatz iteration until it detects a cycle. The output should include the number of iterations done before detecting a cycle as well as the actual cycle vector. Run your script using the following inputs: $-2, -4, -8, -10, -56, -88, -129$.
Suggestion: A cycle can be detected as soon as the same number $a(n)$ has appeared previously in the sequence. So your script will need to store the whole Collatz sequence. For example, each time it has constructed a new sequence element, say $a(20)$, the script should compare with the previous vector elements $a(1)$, $a(20)$, ..., $a(19)$ to see if this new element has previously appeared. If not, the iteration goes on, but if there is a duplication, say, $a(20) = a(15)$, then there will be a cycle and the cycle vector would be $(a(15), a(16), a(17), a(18), a(19))$.

14. (*Number Theory: The Collatz Problem*) Read first the preceding exercise. We consider two cycles as *equivalent* in a Collatz experiment if they contain the same numbers (but not necessarily in the same order). Thus the cycle (1,4,2) has the equivalent forms (4,2,1), and (2,1,4). The program in the previous exercise, if encountering a certain cycle, may output any of the possible equivalent forms, depending on the first duplication encountered. We say that two cycles are *essentially different* if they are not equivalent cycles. In this exercise, you are to use MATLAB to help you figure out the number of essentially different Collatz cycles that

come up from using negative integers for $a(1)$ ranging from -1 to $-20,000$.

Note: The Collatz conjecture can be paraphrased as saying that all Collatz iterations starting with a positive integer must eventually cycle and the resulting cycles are all equivalent to $(4,2,1)$. The essentially different Collatz cycles for negative integer inputs in this problem will cover all that are known to this date. It is also conjectured that there are no more.

Chapter 5: Floating Point Arithmetic and Error Analysis

5.1: FLOATING POINT NUMBERS

We have already mentioned that the data contained in just a single irrational real number such as π has more information in its digits than all the computers in the world could possibly ever store. Then again, it would probably take all the scientific surveyors in the world to look for and not be able to find any scientist who vitally needed, say, the 534th digit of this number. What is usually required in scientific work is to maintain accuracy with a certain number of so-called **significant digits**, which constitutes the portion of a numerical answer to a problem that is trusted to be correct. For example, if we want π to three significant digits, we could use 3.14. A computer can only work with a finite set of numbers; these computer numbers for a given system are usually called floating point numbers. Since there are infinitely many real numbers, what has to happen is that big (infinite) sets of real numbers must get identified with single computer numbers. Floating point numbers are best understood by their relations with numbers in scientific notation, such as 3.14159×10^0, although they need not be written in this form.

A **floating point number system** is determined by a **base** β (any positive integer greater than one), a precision s (any positive integer; this will be the number of significant digits), and two integers m (negative) and M (positive), that determine the exponent range. In such a system, a **floating point number** can always be expressed in the form:

$$\pm .d_1 d_2 \cdots d_s \times \beta^e , \tag{1}$$

where,

$$d_i = 0, 1, 2, \cdots, \text{or } \beta - 1 \text{ but } d_1 \neq 0 \text{ and } m \leq e \leq M .$$

The number zero is represented as $.00 \cdots 0 \times \beta^{-m}$. In a computer, each of the three parts (the sign $\pm$, mantissa $d_1 d_2 \cdots d_s$, and the exponent β) of a floating point number is stored in its own separate fixed-width field. Most contemporary computers and software on the market today (MATLAB included) use **binary arithmetic** ($\beta = 2$). Hand-held calculators use decimal base $\beta = 10$. In the past, other computers have used different bases that were usually powers of two, such as $\beta = 16$ (hexadecimal arithmetic). Of course, such arithmetic (different from base 10) is done in internal calculations only. When the number is displayed, it is converted to decimal form. An important quantity for determining the precision of a given computing system is known as the **unit roundoff** u (or the **machine epsilon**[1]), which is the maximum relative error that can occur when a real number is approximated by a floating point number.[2] For example, one Texas Instruments graphing calculator uses the floating point parameters $\beta = 10$, $s = 12$, $m = -99$, and $M = 99$, which means that this calculator can effectively handle numbers whose absolute values lie (approximately) between 10^{-99} and 10^{99}, and the unit roundoff is $u = 10^{-12}$. MATLAB's arithmetic uses the parameters: $\beta = 2$, $s = 53$, $m = -1074$, and $M = 1023$, which conforms to the **IEEE double precision standard**.[3] This means that MATLAB can effectively handle numbers with absolute values from $2^{-1074} \approx 10^{-324}$ to $2^{1023} \approx 10^{308}$; also the unit roundoff is $u = 2^{-53} \approx 10^{-16}$.

5.2: FLOATING POINT ARITHMETIC: THE BASICS

Many students have gotten accustomed to the reliability and logical precision of exact mathematical arithmetic. When we get the computer to perform calculations for us, we must be aware that floating point arithmetic compounded with roundoff errors can lead to unexpected and undesirable results. Most large-scale numerical algorithms are not exact algorithms, and when such methods are used, attention must be paid to the error estimates. We saw this at a basic level in Chapter 2, and it will reappear again later on several occasions. Here we will talk about different sorts of errors, namely, those that arise and are compounded by the computer's floating point arithmetic. We stress the distinction with the first type of errors. Even if an algorithm is mathematically guaranteed to work, floating point errors may arise and spoil its success. All of our illustrations below will be in base 10 floating point arithmetic, since all of the concepts can be covered and better understood in this familiar setting; changing to a different base is merely a technical issue. To get a feel for the structure of a floating point number system, we begin with an example of a very small system.

---

[1] The rationale for this terminology is that the Greek letter epsilon (ε) is usually used in mathematical analysis to represent a very small number.

[2] There is another characterization of the unit roundoff as the gap between the floating point number 1 and the next floating point number to the right. These two definitions are close, but not quite equivalent; see Example 5.4 and Exercise for the Reader 5.3 for more details on how these two quantities are related, as well as explicit formulas for the unit roundoff.

[3] The IEEE (I-triple-E) is a nonprofit, technical professional association of more than 350,000 individual members in 150 countries. The full name is the Institute of Electrical and Electronics Engineers, Inc. The IEEE single-precision (SP) and double-precision (DP) standards have become the international standard for computers and numerical software. The standards were carefully developed to help avoid some problems and incompatibilities with previous floating point systems. In our notation, the IEEE SP standard specifies $\beta = 2$, $s = 24$, $m = -126$, $= -126$, and $M = 127$ and the IEEE DP standard has $\beta = 2$, $s = 53$, $m = -1022$, and $M = 1023$.

EXAMPLE 5.1: Find all floating point numbers in the system with $\beta = 10$, $s = 1$, $m = -1$, and $M = 1$.

SOLUTION: In this case, it is a simple matter to write down all of the floating point numbers in the system:

$$\pm .1 \times 10^{-1} = \pm .01 \qquad \pm .1 \times 10^{0} = \pm .1 \qquad \pm .1 \times 10^{1} = \pm 1$$
$$\pm .2 \times 10^{-1} = \pm .02 \qquad \pm .2 \times 10^{0} = \pm .2 \qquad \pm .2 \times 10^{1} = \pm 2$$
$$\vdots \qquad\qquad\qquad \vdots \qquad\qquad\qquad \vdots$$
$$\pm .9 \times 10^{-1} = \pm .09 \qquad \pm .9 \times 10^{0} = \pm .9 \qquad \pm .9 \times 10^{1} = \pm 9.$$

Apart from these, there is only $0 = .0 \times 10^{-1}$. Of these 55 numbers, the nonnegative ones are pictured on the number line in Figure 5.1. We stress that the gaps between adjacent floating point numbers are not always the same; in general, these gaps become smaller near zero and more spread out far away from zero (larger numbers).

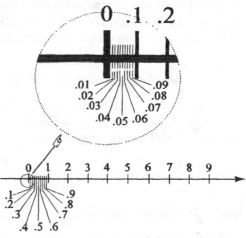

FIGURE 5.1: The nonnegative floating point numbers of Example 5.1. The omitted negative floating point numbers are just (in any floating point number system) the opposites of the positive numbers shown. The situation is typical in that as we approach zero, the density of floating point numbers increases.

Let us now talk about how real numbers get converted to floating point numbers. Any real number x can be expressed in the form

$$x = \pm .d_1 d_2 \cdots d_s d_{s+1} \cdots \times 10^{e} \qquad (2)$$

where there are infinitely many digits (this is the only difference from the floating point representation (1) with $\beta = 10$) and there is no restriction on the exponent's range. The part $.d_1 d_2 \cdots d_s d_{s+1} \cdots$ is called the **mantissa** of x. If x has a finite decimal expansion, we can trail it with an infinite string of zeros to conform with (2). In fact, the representation (2) is unique for any real number (i.e., there is only one way to represent any x in this way) provided that we adopt the convention that an infinite string of 9's not be allowed; such expansions should just be rounded up. (For example, the real number .37999999999... is the same as .38.)

At first glance, it may seem straightforward how to represent a real number x in form (2) by a floating point number of form (1) with $\beta = 10$; simply either chop off or round off any of the digits past the allowed number. But there are serious problems that may arise, stemming from the fact that the exponent e of the real number may be outside the permitted range. Firstly, if $e > M$, this means that x is (probably much) larger than any floating point number and so cannot be represented by one. If such a number were to come up in a computation, the computation is said to have **overflowed**. For example, in the simple setting of Example 5.1, any number $x \geq 10$ would overflow this simple floating point system. Depending on the computer system, overflows will usually result in termination of calculation or a warning. For example, most graphing calculators, when asked to evaluate an expression like e^{5000}, will either produce an error message like "OVERFLOW", and freeze up or perhaps give ∞ as an output. MATLAB behaves similarly to the latter pattern for overflows:

```
>> exp(5000)
→ans = Inf %MATLAB's abbreviation for infinity.
```

Inf (or inf) is MATLAB's way of saying the number is too large to continue to do any more number crunching, except for calculations where the answer will be either Inf or −Inf (a very large negative number). Here are some examples:

```
>> exp(5000)
→ans = Inf % MATLAB tells us we have a very big positive number here

>> 2*exp(5000)
→ans = Inf %No new information

>> exp(5000)/-5
→ans = -Inf %OK now we have a very big negative number.

>> 2*exp(5000)-exp(5000)
→ans = NaN % "NaN" stands for "not a number"
```

The last calculation is more interesting. Obviously, the expression being evaluated is just e^{5000}, which, when evaluated separately, is outputted as Inf. What happens is that MATLAB tries instead to do inf-inf, which is undefined (once large numbers are converted to inf, their relative sizes are lost and it is no longer possible for MATLAB to compare them).[4]

A very different situation occurs if the exponent e of the real number x in (2) is less than m (too small). This means that the real number x has absolute value (usually much) smaller than that of any nonzero floating point number. In this case, a computation is said to have **underflowed**. Most systems will represent an underflow by zero without any warning, and this is what MATLAB does.

---

[4] We mention that the optional "Symbolic Toolbox" for MATLAB allows, among other things, the symbolic manipulation of such expressions. The Symbolic Toolbox does come with the student version of MATLAB. Some of its features are explained in Appendix A.

Underflows, although less serious than overflows, can be a great source of problems in large-scale numerical calculations. Here is a simple example. We know from basic rules of exponents that $e^p e^{-p} = e^{p-p} = e^0 = 1$, but consider the following calculation:

```
>> exp(-5000)
→ans = 0 %this very small number has underflowed to zero

>> exp(5000)*exp(-5000)
→ans = NaN
```

The latter calculation had both underflows and overflows and resulted in $0*\text{Inf}$ ($= 0 \cdot \infty$), which is undefined. We will give another example shortly of some of the tragedies that can occur as the result of underflows. But now we show two simple ways to convert a real number to a floating point number in case there is no overflow or underflow. So we assume the real number x in (2) has exponent e satisfying $m \le e \le M$. The two methods for converting the real number x to its floating point representative $fl(x)$ are as follows:

(i) Chopped (or Truncated) Arithmetic: With this system we simply drop all digits after d_s:

$$fl(x) = fl(\pm .d_1 d_2 \cdots d_s d_{s+1} \cdots \times 10^e) = \pm .d_1 d_2 \cdots d_s \times 10^e.$$

(ii) Rounded Arithmetic: Here we do the usual rounding scheme for the first s significant digits. If $d_{s+1} < 5$, we simply chop as in method (i), but if $d_{s+1} \ge 5$, we need to round up. This may change several digits depending on if there is a string of 9's or not. For example, with $s = 4$, ...2456823... would round to .2457 (one-digit changed), but .2999823 would round to .3000 (four digits changed). So a nice formula as in (i) is not possible. There is, however, an elegant way to describe rounded arithmetic in terms of chopped arithmetic using two steps.

Step 1: Add $5 \times 10^{-(s+1)}$ to the mantissa $.d_1 d_2 \cdots d_s d_{s+1} \cdots$ of x.

Step 2: Now chop as in (i) and retain the sign of x.

EXAMPLE 5.2: The following example parallels some calculations in exact arithmetic with the same calculations in 3-digit floating point arithmetic with $m = -8$ and $M = 8$. The reader is encouraged to go through both sets of calculations, using either MATLAB or a calculator. Note that at each point in a floating point calculation, the numbers need to be chopped accordingly before any math operations can be done.

Exact Arithmetic	Floating Point Arithmetic
$x = \sqrt{3}$	$fl(x) = 1.73 \ (\equiv .173 \times 10^1)$
$x^2 = 3$	$fl(x)^2 = 2.99$

Thus, in floating point arithmetic, we get that $\sqrt{3}^2 = 2.99$. This error is small but understandable.

Exact Arithmetic	Floating Point Arithmetic
$x = \sqrt{1000}$	$fl(x) = 31.6 \; (\equiv .316 \times 10^2)$
$x^2 = 1000$	$fl(x)^2 = 998$

The same calculation with larger numbers, of course, results in a larger error; but relatively it is not much different. A series of small errors can pile up and amount to more catastrophic results, as the next calculations show.

Exact Arithmetic	Floating Point Arithmetic
$x = 1000$	$fl(x) = 1000$
$y = 1/x = .001$	$fl(y) = .001$
$z = 1 + y = 1.001$	$fl(z) = 1$
$w = (z-1) \cdot x^2$	$fl(w) = (1-1) \cdot 1000^2$
$\quad = y \cdot x^2$	$\quad = 0 \cdot 1000^2$
$\quad = \dfrac{1}{x} \cdot x^2$	$\quad = 0$
$\quad = x = 1000$	

The floating point answer of 0 is a ridiculous approximation to the exact answer of 1000! The reason for this tragedy was the conversion of an underflow to zero. By themselves, such conversions are rather innocuous, but when coupled with a sequence of other operations, problematic results can sometimes occur.

When we do not make explicit mention of the exponent range $m \le e \le M$, we assume that the numbers that come up have their exponents in the appropriate range and so there will be no underflows or overflows.

EXERCISE FOR THE READER 5.1: Perform the following calculations in two-digit rounded arithmetic, and compare with the exact answers.

(a) $(.15)^2$

(b) $365,346 \times .4516$

(c) $8001 \div 123$

Our next example should alert the reader that one needs to be cautious on many different fronts when using floating point arithmetic. Many arithmetic rules that we have become accustomed to take for granted sometimes need to be paid careful attention when using floating point arithmetic.

EXAMPLE 5.3: Working in three-digit chopped floating point arithmetic with the exponent e restricted to the range $-8 \le e \le 8$, perform the following tasks:

(a) Compute the infinite series: $\displaystyle\sum_{n=1}^{\infty}\frac{1}{n^2}=1+\frac{1}{4}+\frac{1}{9}+\cdots$

(b) In each part below an equation is given and your task will be to decide how many solutions it will have in this floating point arithmetic. For each part you should give one of these four answers: **NO SOLUTION, EXACTLY ONE SOLUTION, BETWEEN 2 AND 10 SOLUTIONS,** or **MORE THAN 10 SOLUTIONS,** (Work here only with real numbers; take all underflows as zero.)

(i) $3x = 5$

(ii) $x^3 = 0$

SOLUTION: Part (a): Unlike with exact arithmetic, when we sum this infinite series in floating point arithmetic, it is really going to be a finite summation since eventually the terms will be getting too small to have any effect on the accumulated sum. We use the notation $\displaystyle S_N = \sum_{n=1}^{N}\frac{1}{n^2}=1+\frac{1}{2^2}+\frac{1}{3^2}+\cdots+\frac{1}{N^2}$ for the partial sum (a finite sum). To find the infinite sum, we need to calculate (in order) in floating point arithmetic $S_1, S_2, S_3, \cdots$ and continue until these partial sums no longer change. Here are the step-by-step details:

$S_1 = 1$

$S_2 = S_1 + 1/4 = 1 + .25 = 1.25$

$S_3 = S_2 + 1/9 = 1.25 + .111 = 1.36$

$S_4 = S_3 + 1/16 = 1.36 + .0625 = 1.42$

$S_5 = S_4 + 1/25 = 1.42 + .040 = 1.46$

$S_6 = S_5 + 1/36 = 1.46 + .0277 = 1.48$

$S_7 = S_6 + 1/49 = 1.48 + .0204 = 1.50$

$S_8 = S_7 + 1/64 = 1.50 + .0156 = 1.51$

$S_9 = S_8 + 1/81 = 1.51 + .0123 = 1.52$

$S_{10} = S_9 + 1/100 = 1.52 + .010 = 1.53$

$S_{11} = S_{10} + 1/121 = 1.53 + .00826 = 1.53$

We can now stop this infinite process since the terms being added are small enough that when added to the existing partial sum 1.53, their contributions will just get chopped. Thus in the floating point arithmetic of this example, we have computed $\displaystyle\sum_{n=1}^{\infty}\frac{1}{n^2}=1.53$, or more correctly we should write $\displaystyle \text{fl}\left(\sum_{n=1}^{\infty}\frac{1}{n^2}\right)=1.53$.

Compare this result with the result from exact arithmetic $\displaystyle\sum_{n=1}^{\infty}\frac{1}{n^2}=\frac{\pi^2}{6}=1.64....$

Thus in this calculation we were left with only one significant digit of accuracy!

Part (b): (i) The equation $3x = 5$ has, in exact arithmetic, only one solution, $x = 5/3 = 1.666....$ Let's look at the candidates for floating point arithmetic solutions that are in our system. This exact solution has floating point representative 1.66. Checking this in the equation (now working in floating point arithmetic) leads to: $3 \cdot 1.66 = 4.98 \neq 5$. So this will not be a floating point solution. Let's try making the number a bit bigger to 1.67 (this would be the smallest possible jump to the next floating point number in our system). We have (in floating point arithmetic) $3 \cdot 1.67 = 5.01 \neq 5$, so here $3x$ is too large. If these two numbers do not work, no other floating point numbers can (since for other floating point numbers $3x$ would be either less than or equal to 4.98 or greater than or equal to 5.01). Thus we have "NO SOLUTION" to this equation in floating point arithmetic![5]

(ii) As in (i), the equation $x^3 = 0$ has exactly one real number solution, namely $x = 0$. This solution is also a floating point solution. But there are many, many others. The lower bound on the exponent range $-8 \leq e$ is relevant here. Indeed, take any floating point number whose magnitude is less than 10^{-3}, for example, $x = .0006385$. Then $x^3 = (.0006385)^3 = 2.60305... \times 10^{-10} = .260305 \times 10^{-9}$ (in exact arithmetic). In floating point arithmetic, this computation would underflow and hence produce the result $x^3 = 0$. We conclude that in floating point arithmetic, this equation has "MORE THAN 10 SOLUTIONS" (see also Exercise 10 of this section).

EXERCISE FOR THE READER 5.2: Working two-digit rounded floating point arithmetic with the exponent e restricted to the range $-8 \leq e \leq 8$, perform the following tasks:

(a) Compute the infinite series: $\sum_{n=1}^{\infty} \frac{1}{n} = 1 + \frac{1}{2} + \frac{1}{3} + \cdots$

(b) In each part below an equation is given and your task will be to decide how many solutions it will have in this floating point arithmetic. For each part you should give one of these four answers: **NO SOLUTION, EXACTLY ONE SOLUTION, BETWEEN 2 AND 10 SOLUTIONS,** or **MORE THAN 10 SOLUTIONS.** (Work here only with real numbers; take all underflows as zero.)

(i) $x^2 = 100$

(ii) $8x^2 = x^5$

---

[5] We point out that when asked to (numerically) solve this equation in floating point arithmetic, we would simply use the usual (pure) mathematical method but work in floating point arithmetic, i.e., divide both sides by 3. The question of how many solutions there are in floating point arithmetic is a more academic one to help highlight the differences between exact and floating point arithmetic. Indeed, any time one uses a calculator or any floating point arithmetic software to solve any sort of mathematical problem with an exact mathematical method we should be mindful of the fact that the calculation will be done in floating point arithmetic.

EXERCISES 5.2:

NOTE: Unless otherwise specified, assume that all floating point arithmetic in these exercises is done in base 10.

1. In three-digit chopped floating point arithmetic, perform the following operations with these numbers: $a = 10000$, $b = .05$, and $c = 1/3$.
    (a) Write c as a floating point number, i.e., find $fl(c)$.
    (b) Find $a + b$.
    (c) Solve the equation $ax = c$ for x.

2   In three-digit rounded floating point arithmetic, perform the following tasks:
    (a) Find $1.23 + .456$ (b) Find $110,000 - 999$ (c) Find $(.055)^2$

3. In three-digit chopped floating point arithmetic, perform the following tasks:
    (a) Solve the equation $5x + 8 = 0$.
    (b) Use the quadratic formula to solve $1.12x^2 + 88x + 1 = 0$.
    (c) Compute $\displaystyle\sum_{n=1}^{\infty} \frac{1}{n^4} = \frac{1}{1^4} + \frac{1}{2^4} + \frac{1}{3^4} + \cdots$.

4   In three-digit rounded floating point arithmetic, perform the following tasks:
    (a) Solve the equation $5x + 4 = 17$.
    (b) Use the quadratic formula to solve $x^2 - 2.2x + 3 = 0$.
    (c) Compute $\displaystyle\sum_{n=1}^{\infty} \frac{(-1)^n \cdot 10}{n^4 + 2} = \frac{-10}{1^4 + 2} + \frac{10}{2^4 + 2} - \frac{10}{3^4 + 2} + \cdots$.

5. In each part below an equation is given and your task will be to decide how many solutions it will have in 3-digit chopped floating point arithmetic. For each part you should give one of these four answers: NO SOLUTION, EXACTLY ONE SOLUTION, BETWEEN 2 AND 10 SOLUTIONS, or MORE THAN 10 SOLUTIONS. (Work here only with real numbers with exponent e restricted to the range $-8 \le e \le 8$, and take all underflows as zero.)
    (a) $2x + 7 = 16$
    (b) $(x + 5)^2(x + 1/3) = 0$
    (c) $2^x = 20$

6. Repeat the directions of Exercise 5, for the following equations, this time using 3-digit rounded floating point arithmetic with exponent e restricted to the range $-8 \le e \le 8$.
    (a) $2x + 7 = 16$ (b) $x^2 - x = 6$ (c) $\sin(x^2) = 0$

7. Using three-digit chopped floating point arithmetic (in base 10), do the following:
    (a) Compute the sum: $1 + 8 + 27 + 64 + 125 + 216 + 343 + 512 + 729 + 1000 + 1331$, then find the relative error of this floating point answer with the exact arithmetic answer.
    (b) Compute the sum in part (a) in the reverse order, and again find the relative answer of this floating point answer with the exact arithmetic answer.
    (c) If you got different answers in parts (a) and (b), can you explain the discrepancy?

8. Working in two-digit chopped floating point arithmetic, compute the infinite series $\displaystyle\sum_{n=1}^{\infty} \frac{1}{n}$.

9.   Working in two-digit rounded floating point arithmetic, compute the infinite series

$$\sum_{n=2}^{\infty} \frac{1}{n^{3/2} \ln n} .$$

10. In the setting of Example 5 3(b)(ii), exactly how many floating point solutions are there for the equation $x^3 = 0$?

11. (a) Write a MATLAB function M-file `z=rfloatadd(x,y,s)` that has inputs x and y being any two real numbers, a positive integer s, and the output z will be the sum $x + y$ using s-digit rounded floating point arithmetic. The integer s should not be more than 14 so as not to transcend MATLAB's default floating point accuracy.
     (b) Use this program (perhaps in conjunction with loops) to redo Exercise for the Reader 5.2, and Exercise 9.

12. (a) Write a MATLAB function M-file `z=cfloatadd(x,y,s)` that has inputs x and y being any two real numbers, a positive integer s, and the output z will be the sum $x + y$ using s-digit chopped floating point arithmetic. The integer s should not be more than 14 so as not to transcend MATLAB's default floating point accuracy.
     (b) Use this program (perhaps in conjunction with loops) to redo Example 5.3(a), and Exercise 7.

13. (a) How many floating point numbers are there in the system with $\beta = 10$, $s = 2$, $m = -2$, $M = 2$? What is the smallest real number that would cause an overflow in this system?
     (b) How many floating point numbers are there in the system with $\beta = 10$, $s = 3$, $m = -3$, $M = 3$? What is the smallest real number that would cause an overflow in this system?
     (c) Find a formula that depends on s, m, and M that gives the number of floating point numbers in a general base 10 floating point number system ($\beta = 10$). What is the smallest real number that would cause an overflow in this system?

NOTE: In the same fashion as we had with base 10, for any base $\beta > 1$, any nonzero real number x can be expressed in the form:

$$x = \pm .d_1 d_2 \cdots d_s d_{s+1} \cdots \times \beta^e,$$

where there are infinitely many digits $d_i = 0, 1, \cdots, \beta - 1$, and $d_1 \neq 0$. This notation means the following infinite series:

$$x = \pm (d_1 \times \beta^{-1} + d_2 \times \beta^{-2} + \cdots + d_s \times \beta^{-s} + d_{s+1} \times \beta^{-s-1} + \cdots) \times \beta^e.$$

To represent any nonzero real number with its **base** β expansion, we first would determine the exponent e so that the inequality $1/\beta \leq |x|/\beta^e < 1$ is valid. Next we construct the "digits" in order to be as large as possible so that the cumulative sum multiplied by β^e does not exceed $|x|$. As an example, we show here how to get the binary expansions ($\beta = 2$) of each of the numbers $x = 3$ and $x = 1/3$.   For $x = 3$, we get first the exponent $e = 2$, since $1/2 \leq |3|/2^2 < 1$.   Since $(1 \times 2^{-1}) \times 2^2 = 2 < 3$, the first digit d_1 is 1 (in binary arithmetic, the digits can only be zeros or ones). The second digit d_2 is also 1 since the cumulative sum is now

$$(1 \times 2^{-1} + 1 \times 2^{-2}) \times 2^2 = 2 + 1 = 3.$$

Since the cumulative sum has now reached $x = 3$, all remaining digits are zero, and we have the binary expansion of $x = 3$:

$$3 = .1100 \cdots 00 \cdots \times 2^2.$$

Proceeding in the same fashion for $x = 1/3$, we first determine the exponent e to be -1 (since $1/3/2^{-1} = 2/3$ lies in $[1/2, 1)$). We then find the first digit $d_1 = 1$, and cumulative sum is $(1 \times 2^{-1}) \times 2^{-1} = 1/4 < 1/3$. Since $(1 \times 2^{-1} + 1 \times 2^{-2}) \times 2^{-1} = 3/8 > 1/3$, we see that the second digit $d_2 = 0$. Moving along, we get that $d_3 = 1$ and the cumulative sum is

$$(1 \times 2^{-1} + 0 \times 2^{-2} + 1 \times 2^{-3}) \times 2^{-1} = 5/16 < 1/3.$$

Continuing in this fashion, we will find that $d_4 = d_6 = \cdots = d_{2n} = 0$, and $d_5 = d_7 = \cdots = d_{2n+1} = 1$ and so we obtain the binary expansion:

$$1/3 = .101010 \cdots 1010 \cdots \times 2^{-1}.$$

If we require that there be no infinite string of ones (the construction process given above will guarantee this), then these expansions are unique. Exercises 14–19 deal with such representations in nondecimal bases ($\beta \neq 10$).

14. (a) Find the binary expansions of the following real numbers: $x = 1000$, $x = -2$, $x = 2.5$.
    (b) Find the binary expansions of the following real numbers: $x = 5/32$, $x = 2/3$, $x = 1/5$, $x = -0.3$, $x = 1/7$.
    (c) Find the exponent e and the first 5 digits of the binary expansion of π.
    (d) Find the real numbers with the following (terminating) binary expansions: $1010 \cdots 00 \cdots \times 2^8$, $.1110 \cdots 00 \cdots \times 2^{-3}$.

15. (a) Use geometric series to verify the binary expansion of 1/3 that was obtained in the previous note.
    (b) Use geometric series to find the real numbers having the following binary expansions:
    $10010101 \cdots \cdots \times 2^1$, $.11011011 \cdots \cdots \times 2^0$, $.1100011011011 \cdots \cdots \times 2^1$.
    (c) What sort of real numbers will have binary expansions that either end in a sequence of zeros, or repeat, like the one for 1/3 obtained in the note preceding Exercise 14?

16. (a) Write down all floating point numbers in the system with $\beta = 2$, $s = 1$, $m = -1$, $M = 1$. What is the smallest real number that would cause an overflow in this system?
    (b) Write down all floating point numbers in the system with $\beta = 2$, $s = 2$, $m = -1$, $M = 1$. What is the smallest real number that would cause an overflow in this system?
    (c) Write down all floating point numbers in the system with $\beta = 3$, $s = 2$, $m = -1$, $M = 1$. What is the smallest real number that would cause an overflow in this system?

17. (a) How many floating point numbers are there in the system with $\beta = 2$, $s = 3$, $m = -2$, $M = 2$? What is the smallest real number that would cause an overflow in this system?
    (b) How many floating point numbers are there in the system with $\beta = 2$, $s = 2$, $m = -3$, $M = 3$? What is the smallest real number that would cause an overflow in this system?
    (c) Find a formula that depends on s, m, and M that gives the number of floating point numbers in a general binary floating point number system ($\beta = 2$). What is the smallest real number that would cause an overflow in this system?

18. Repeat each part of Exercise 17, this time using base $\beta = 3$.

19. Chopped arithmetic is defined in arbitrary bases exactly the same as was explained for decimal bases in the text. Real numbers must first be converted to their expansion in base β. For rounded floating point arithmetic using s-digits with base β, we simply add $\beta^{-s}/2$ to the mantissa and then chop. Perform the following floating point additions by first converting the numbers to floating point numbers in base $\beta = 2$, doing the operation in two-digit chopped

arithmetic, and then converting back to real numbers. Note that your real numbers may have more digits in them than the number s used in base 2 arithmetic, after conversion

(a) $.2 + 6$ (b) $22 + 7$ (c) $120 + 66$

5.3: FLOATING POINT ARITHMETIC: FURTHER EXAMPLES AND DETAILS

In order to facilitate further discussion on the differences in floating point and exact arithmetic, we introduce the following notation for operations in floating point arithmetic:

$$x \oplus y \equiv \text{fl}(x + y)$$
$$x \ominus y \equiv \text{fl}(x - y)$$
$$x \otimes y \equiv \text{fl}(x \cdot y) \tag{3}$$
$$x \oslash y \equiv \text{fl}(x \div y),$$

(i.e., we put circles around the standard arithmetic operators to represent the corresponding floating point operations). To better illustrate concepts and subtleties of floating point arithmetic without getting into technicalities with different bases, we continue to work only in base $\beta = 10$.

In general, as we have seen, floating point operations can lead to different answers than exact arithmetic operations. In order to track and predict such errors, we first look, in the next example, at the relative error introduced when a real number is approximated by its floating point number representative.

EXAMPLE 5.4: Show that in s-digit chopped floating point arithmetic, the unit roundoff u is 10^{1-s}, and that this number equals the distance from one to the next (larger) floating point number. We recall that the unit roundoff is defined to be the maximum relative error that can occur when a real number is approximated by a floating point number.

SOLUTION: Since $\text{fl}(0)=0$, we may assume that $x \neq 0$. Using the representations (1) and (2) for the floating point and exact numbers, we can estimate the relative error as follows:

$$\left| \frac{x - \text{fl}(x)}{x} \right| = \left| \frac{.d_1 d_2 \cdots d_s d_{s+1} \cdots \times 10^e - .d_1 d_2 \cdots d_s \times 10^e}{.d_1 d_2 \cdots d_s d_{s+1} \cdots \times 10^e} \right| \quad \overset{(s) \text{ slot}}{\underset{(s+1) \text{ slot}}{}}$$

$$= \left| \frac{.00 \cdots 0 d_{s+1} d_{s+2} \cdots \times 10^e}{.d_1 d_2 \cdots d_{s+1} d_{s+2} \cdots \times 10^e} \right| \leq \frac{.00 \cdots 099 \cdots}{.10 \cdots 000 \cdots}$$

$$= \frac{.00 \cdots 100 \cdots}{.10 \cdots 000 \cdots} = \frac{10^{-s}}{10^{-1}} = 10^{1-s}$$

Since equality can occur, this proves that the number on the right side is the unit roundoff. To see that this number u coincides with the gap between the floating point number 1 and the next (larger) floating point number on the right, we write the number 1 in the form (1):

$$1 = .10\cdots00\times10^1,$$

(note there are s digits total on the right, $d_1 = 1$, and $d_2 = d_3 = \cdots = d_s = 0$); we see that the next larger floating point number of this form will be:

$$1 + \text{gap} = .10\cdots01\times10^1.$$

Subtracting gives us the unit roundoff:

$$\text{gap} = .00\cdots01\times10^1 = 10^{-s}\times10^1 = 10^{1-s},$$

as was claimed.

EXERCISE FOR THE READER 5.3: (a) Show that in s-digit rounded floating point arithmetic the unit roundoff is $u = \frac{1}{2}10^{1-s}$, but that the gap from 1 to the next floating point number is still 10^{1-s}.
(b) Show also that in any floating point arithmetic system and for any real number x, we can write

$$\text{fl}(x) = x(1 + \delta), \quad \text{where } |\delta| \leq u. \tag{4}$$

In relation to (4), we also assume that for any single floating point arithmetic operation: $x \odot y$, with "$\odot$" representing any of the floating point arithmetic operations from (3), we can write

$$x \odot y = (x \circ y)(1 + \delta), \quad \text{where } |\delta| \leq u \tag{5}$$

and where "$\circ$" denotes the exact arithmetic operation corresponding to "$\odot$". This assumption turns out to be valid for IEEE (and hence, MATLAB's) arithmetic but for other computing environments may require that the bound on δ be replaced by a small multiple of u. We point out that IEEE standards require that $x \odot y = \text{fl}(x \circ y)$.

In scientific computing, we will often need to do a large number of calculations and the resulting roundoff (or floating point) errors can accumulate. Before we can trust the outputs we get from a computer on an extensive computation, we have to have some confidence of its accuracy to the true answer. There are two major types of errors that can arise: roundoff errors and algorithmic errors. The first results from propagation of floating point errors, and the second arises from mathematical errors in the model used to approximate the true answer to a problem. To decrease mathematical errors, we will need to do more computations, but more computations will increase computer time and roundoff errors. This is a major dilemma of scientific computing! The best strategy and ultimate goal is to try to find efficient algorithms; this point will be applied and reemphasized frequently in the sequel.

To illustrate roundoff errors, we first look at the problem of numerically adding up a set of positive numbers. Our next example will illustrate the following general principle:

A General Principle of Floating Point Arithmetic: When numerically computing a large sum of positive numbers, it is best to start with the smallest number and add in increasing order of magnitude.

Roughly, the reason the principle is valid is that if we start adding the large numbers first, we could build up a rather large cumulative sum. Thus, when we get to adding to this sum some of the smaller numbers, there are much better chances that all or parts of these smaller numbers will have decimals beyond the number of significant digits supported and hence will be lost or corrupted.

EXAMPLE 5.5:   (a) In exact mathematics, addition is associative: $(x+y)+z = x+(y+z)$. Show that in floating point arithmetic, addition is no longer associative.
(b) Show that for adding up a finite sum $S_N = a_1 + a_2 + \cdots + a_N$ of positive numbers (in the order shown), in floating point arithmetic, the error of the floating point answer $\mathrm{fl}(S_N)$ in approximating the exact answer S_N can be estimated as follows:

$$| \mathrm{fl}(S_N) - S_N | \leq u[(N-1)a_1 + (N-1)a_2 + (N-2)a_3 + \tag{6}$$
$$\cdots + 2a_{N-1} + a_N],$$

where u is the unit roundoff.

REMARK: Formula (6), although quite complicated, can be seen to demonstrate the above principle. Remember that u is an extremely small number so that the error on the right will normally be small as long as there are not an inordinate number of terms being added. In any case, the formula makes it clear that the relative contribution of the first term being added is the largest since the error estimate (right side of (6)) is a sum of terms corresponding to each a_i multiplied by the proportionality factor $(N-i)u$. Thus if we are adding $N = 1,000,001$ terms then this proportionality factor is $1,000,000u$ (worst) for a_1 but only u (best) for a_N, and these factors decrease linearly for intermediate terms. Thus it is clear that we should start adding the smaller terms first, and save the larger ones for the end.

SOLUTION: Part (a): We need to find (in some floating point arithmetic system) three numbers x, y, and z such that $(x \oplus y) \oplus z \neq x \oplus (y \oplus z)$. Here is a simple example that also demonstrates the above principle: We use 2-digit chopped arithmetic with $x = 1$, $y = z = .05$. Then,

$$(x \oplus y) \oplus z = (1 \oplus .05) \oplus .05 = 1 \oplus .05 = 1,$$

but

$$x \oplus (y \oplus z) = 1 \oplus (.05 \oplus .05) = 1 \oplus .1 = 1.1 . .$$

This not only provides a counterexample, but since the latter computation (gotten by adding the smaller numbers first) gave the correct answer it also demonstrates the above principle.

Part (b): We continue to use the notation for partial sums that was employed in Example 5.3 (i.e., $S_1 = a_1$, $S_2 = a_1 + a_2$, $S_3 = a_1 + a_2 + a_3$, etc.). By using identity (5) repeatedly, we have:

$$\text{fl}(S_2) = a_1 \oplus a_2 = (a_1 + a_2)(1 + \delta_2) = S_2 + (a_1 + a_2)\delta_2, \quad \text{where } |\delta_2| \le u, \text{ and so}$$

$$\begin{aligned}
\text{fl}(S_3) &= \text{fl}(S_2) \oplus a_3 = (\text{fl}(S_2) + a_3)(1 + \delta_3) \quad \text{where } |\delta_3| \le u \\
&= (S_2 + (a_1 + a_2)\delta_2 + a_3)(1 + \delta_3) \\
&= S_3 + (a_1 + a_2)\delta_2 + (a_1 + a_2 + a_3)\delta_3 + (a_1 + a_2)\delta_2\delta_3 \\
&\approx S_3 + (a_1 + a_2)\delta_2 + (u_1 + a_2 + a_3)\delta_3.
\end{aligned}$$

To get to the last estimate, we ignored the higher-order (last) term of the second-to-last expression. Continuing to estimate in the same fashion leads us to

$$\text{fl}(S_N) \approx S_N + u \begin{cases} a_1(\delta_2 + \delta_3 + \delta_4 + \cdots + \delta_N) \\ + a_2(\delta_2 + \delta_3 + \delta_4 + \cdots + \delta_N) \\ + a_3(\quad\quad \delta_3 + \delta_4 + \cdots + \delta_N) \\ + a_4(\quad\quad\quad\quad \delta_4 + \cdots + \delta_N) \\ \vdots \\ + a_N(\quad\quad\quad\quad\quad\quad \delta_N) \end{cases},$$

where each of the δ_i's arise from application of (5) and thus satisfy $|\delta_i| \le u$. Bounding each of the $|\delta_i|$'s above with u and using the triangle inequality produces the asserted error bound in (6).

EXERCISE FOR THE READER 5.4: From estimate (6) deduce the following (cleaner but weaker) error estimates for the roundoff error in performing a finite summation of positive numbers $S_N = a_1 + a_2 + \cdots + a_N$ in floating point arithmetic:

(a) Error $= |\text{fl}(S_N) - S_N| \le Nu \sum_{n=1}^{N} a_n.$

(b) Relative error $= \left| \dfrac{\text{fl}(S_N) - S_N}{S_N} \right| \le Nu.$

Next, we give some specific examples that will compare these estimates against the actual roundoff errors. Recall from calculus that an (infinite) p-series

$$\sum_{n=1}^{\infty}\frac{1}{n^p} = 1+\frac{1}{2^p}+\frac{1}{3^p}+\cdots$$ converges (i.e., adds up to a finite number) exactly

when $p > 1$, otherwise it diverges (i.e., adds up to infinity). If we ask MATLAB (or any floating point computing system) to add up the terms in any p-series (or any series with positive terms that decrease to zero), eventually the terms will get too small to make any difference when they are added to the accumulated sum, so it will eventually appear that the series has converged (if the computer is given enough time to finish its task).[6] Thus, it is not possible to detect divergence of such a series by asking the computer to perform the summation. Once a series is determined to converge, however, it is possible to get MATLAB to help us estimate the actual sum of the infinite series. The key question in such a problem is to determine how many terms need to be summed in order for the partial sums to approximate the actual sum within the desired tolerance for error. We begin with an example for which the actual sum of the infinite series is known. This will allow us to verify if our accuracy goal is met.

EXAMPLE 5.6: Consider the infinite p-series $\sum_{n=1}^{\infty}\frac{1}{n^2} = 1+\frac{1}{2^2}+\frac{1}{3^2}+\cdots$. Since

$p = 2 > 1$, the series converges to a finite sum S.
(a) How many terms N of this infinite sum would we need to sum up to so that

the corresponding partial sum $\sum_{n=1}^{N}\frac{1}{n^2} = 1+\frac{1}{2^2}+\frac{1}{3^2}+\cdots+\frac{1}{N^2}$ is within an error of

10^{-7} of the actual sum S (i.e., Error $= |S - S_N| \le 10^{-7}$)?
(b) Use MATLAB to perform this summation and compare the result with the exact answer $S = \pi^2/6$ to see if the error goal has been met. Discuss roundoff errors as well.

SOLUTION: Part (a): The mathematical error analysis needed here involves a nice geometric estimation method for the error that is usually taught in calculus courses under the name of the integral test. In estimating the infinite sum

$S = \sum_{n=1}^{\infty}\frac{1}{n^2}$ with the finite partial sum $S_N = \sum_{n=1}^{N}\frac{1}{n^2}$, the error is simply the tail of

the series:

$$\text{Error} = \left|\sum_{n=1}^{\infty}\frac{1}{n^2} - \sum_{n=1}^{N}\frac{1}{n^2}\right| = \sum_{n=N+1}^{\infty}\frac{1}{n^2} = \frac{1}{(N+1)^2}+\frac{1}{(N+2)^2}+\frac{1}{(N+3)^2}+\cdots.$$

---

[6] We point out, however, that many symbolic calculus rules, and in particular, abilities to detect convergence of infinite series, are features available in MATLAB's Symbolic Toolbox (included in the Student Version). See Appendix A.

The problem is to find out how large N must be so that this error is $\leq 10^{-7}$; of course, as is usually the case, we have no way of figuring out this error exactly (if we could, then we could determine S exactly). But we can estimate this "Error" with something larger, let's call it ErrorCap, that we can compute. Each term in the "Error" is represented by an area of a shaded rectangle (with base = 1) in Figure 5.2. Since the totality of the shaded rectangles lies under the graph of $y = 1/x^2$, from $x = N$ to $x = \infty$, we have

$$\text{Error} < \text{Error Cap} \equiv \int_N^\infty \frac{dx}{x^2} = \frac{x^{-1}}{-1}\Bigg]_{x=N}^{x=\infty} = 1/N;$$

and we conclude that our error will be less than 10^{-7}, provided that Error Cap $\leq 10^{-7}$, or $1/N \leq 10^{-7}$, or $N \geq 10^7$.

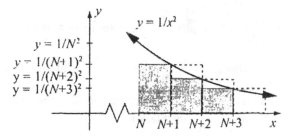

FIGURE 5.2: The areas of the shaded rectangles (that continue on indefinitely to the right) add up to precisely the Error $= \sum_{n=N+1}^{\infty} \frac{1}{n^2}$ of Example 5.6. But since they lie directly under the curve $y = 1/x^2$ from $x = N$ to $x = \infty$, we have Error < Error Cap $= \int_N^\infty \frac{dx}{x^2}$.

Let us now use MATLAB to perform this summation, following the principle of adding up the numbers in order of increasing size.

```
>> Sum=0;  %initialize sum
>> for n=10000000:-1:1
Sum=Sum +1/n^2;
end
>> Sum   →Sum = 1.64493396684823
```

This took about 30 seconds on the author's PC. Since we know that the exact infinite sum is $\pi^2/6$, we can look now at the actual error.

```
>> pi^2/6-Sum
→ans = 9.999999495136080e-008  %This is indeed (just a wee bit) less than the desired
tolerance 10^{-7}.
```

Let us look briefly at (6) (with $a_n = 1/n^2$) to see what kind of a bound on roundoff errors it gives for this computation. At each step, we are adding to the accumulated sum a quotient. Each of these quotients $(1/n^2)$ gives a roundoff error (by (5)) at most $a_n u$. Combining this estimate with (6) gives the bound

$$|fl(S_N) - S_N| \le 2u \left\{ \left[\frac{9999999}{(10000000)^2} \right] + \left[\frac{9999999}{(9999999)^2} \right] + \left[\frac{9999998}{(9999998)^2} \right] + \cdots + \left[\frac{2}{(2)^2} \right] + \left[\frac{1}{(1)^2} \right] \right\}$$

$$\le 2u \left\{ \frac{1}{10000000} + \frac{1}{9999999} + \frac{1}{9999998} + \cdots + \frac{1}{2} + 1 \right\}.$$

The sum in braces is $\sum_{n=1}^{10^7} \frac{1}{n}$. By a picture similar to the one of Figure 5.2, we can estimate this sum as follows: $\sum_{n=1}^{10^7} \frac{1}{n} \le 1 + \int_1^N \ln(x) dx = 1 + \ln(N)$. Since the unit roundoff for MATLAB is 2^{-53}, we can conclude the following upper bound on the roundoff error of the preceding computation:

$$|fl(S_N) - S_N| \le 2u(1 + \ln(10^7)) \approx 3.8 \times 10^{-15}.$$

We thus have confirmation that roundoff error did not play a large role in this computation. Let's now see what happens if we perform the same summation in the opposite order.

```
>> Sum=0;
>> for n=1:10000000
Sum=Sum +1/n^2;
end
>> pi^2/6-Sum          →ans =  1.000009668405966e-007
```

Now the actual error is a bit worse than the desired tolerance of 10^{-7}. The error estimate (6) would also be a lot larger if we reversed the order of summation. Actually for both of these floating point sums the roundoff errors were not very significant; such problems are called *well-conditioned*. The reason for this is that the numbers being added were getting small very fast. In the exercises and later in this book, we will encounter problems that are *ill-conditioned*, in that the roundoff errors can get quite large relative to the amount of arithmetic being done. The main difficulty in the last example was not roundoff error, but the computing time. If we instead wanted our accuracy to be 10^{-10}, the corresponding calculation would take over 8 hours on the same computer! A careful look at the strategy we used, however, will allow us to modify it slightly to get a much more efficient method.

A Better Approach: Referring again to Figure 5.2, we see that by sliding all of the shaded rectangles one unit to the right, the resulting set of rectangles will completely cover the region under the graph of $y = 1/x^2$ from $x = N+1$ to $x = \infty$. This gives the inequality:

$$\text{Error} > \int_{N+1}^{\infty} \frac{dx}{x^2} = \frac{x^{-1}}{-1} \Bigg]_{x=N}^{x=\infty} = 1/(N+1).$$

In conjunction with the previous inequality, we have in summary that:

$$\frac{1}{N+1} < \text{Error} < \frac{1}{N}.$$

If we add to our approximation S_N the average value of these two upper and lower bounds, we will obtain the following much better approximation for S:

$$\tilde{S}_N \equiv S_N + \frac{1}{2}\left[\frac{1}{N} + \frac{1}{N+1}\right].$$

The new error will be at most one-half the length of the interval from $1/(N+1)$ to $1/N$:

$$|S - \tilde{S}_N| \equiv \text{New Error} \le \frac{1}{2}\left[\frac{1}{N} - \frac{1}{N+1}\right] = \frac{1}{2N(N+1)} < \frac{1}{2N^2}.$$

(The elementary last inequality was written so as to make the new error bound easier to use, as we will now see.) With this new scheme much less work will be required to attain the same degree of accuracy. Indeed, if we wanted the error to be less than 10^{-7}, this new scheme would require that the number of terms N needed to sum should satisfy $1/2N^2 \le 10^{-7}$ or $N \ge \sqrt{10^7/2} - 2236.07...$, a far cry less than the 10 million terms needed with the original method! By the same token, to get an error less than 10^{-10}, we would need only take $N = 70,711$! Let us now verify, on MATLAB, this 10-significant-digit approximation:

```
>>   Sum=0; N=70711;
>> for n=N:-1:1
Sum=Sum +1/n^2;
end
>> format long
>> Sum=Sum+(1/N + 1/(N+1))/2        →Sum = 1.64493406684823
>> abs (Sum-pi^2/6)                 →ans =8.881784197001252e-016
```

The actual error here is even better than expected; our approximation is actually as good as machine precision! This is a good example where the (worst-case) error guarantee (New Error) is actually a lot larger than the true error. A careful examination of Figure 5.2 once again should help to make this plausible.

We close this section with an exercise for the reader that deals with the approximation of an alternating series. This one is rather special in that it can be used to approximate the number π and, in particular, we will be able to check the accuracy of our numerical calculations against the theory. We recall from calculus that an alternating series is an infinite series of the form $\sum(-1)^n a_n$, where $a_n > 0$ for each n. Leibniz's Theorem states that if the a_n's decrease, $a_n \geq a_{n+1}$ for each n (sufficiently large), and converge to zero, $a_n \to 0$ as $n \to \infty$, then the infinite series converges to a sum S. It furthermore states that if $S_N = \sum^{N}(-1)^n a_n$ denotes a partial sum (the lower index is left out since it can be any integer), then we have the error estimate: $|S - S_N| \leq a_{N+1}$.

· EXERCISE FOR THE READER 5.5: Use the infinite series expansion:

$$1 - \frac{1}{3} + \frac{1}{5} - \frac{1}{7} + \cdots = \frac{\pi}{4},$$

to estimate π with an error less than 10^{-7}.

The series in the above exercise required the summation of a very large number of terms to get the required degree of accuracy. Such series are said to converge "slowly." Exercise 12 will deal with a better (faster-converging) series to use for approximating π.

EXERCISES 5.3:

NOTE: Unless otherwise specified, assume that all floating point arithmetic in these exercises is done in base 10.

1.   (a) In chopped floating point arithmetic with s digits and exponent range $m \leq e \leq M$, write down (in terms of these parameters s, m, and M) the largest positive floating point number and the smallest positive floating point number.
     (b) Would these answers change if we were to use rounded arithmetic instead?

2.   Recall that for two real numbers x and y, the average value $(x + y)/2$ of x and y lies between the values of x and y.
     (a) Working in a chopped floating point arithmetic system, find an example where $fl((x + y)/2)$ is strictly less than $fl(x)$ and $fl(y)$.
     (b) Repeat part (a) in rounded arithmetic.

3.   (a) In chopped floating point arithmetic with base β, with s digits and exponent range $m \leq e \leq M$, write down (in terms of these parameters β, s, m, and M) the largest positive floating point number and the smallest positive floating point number.
     (b) Would these answers change if we were to use rounded arithmetic instead?

4.   For each of the following arithmetic properties, either explain why the analog will be true in floating point arithmetic, or give an example where it fails. If possible, provide counterexamples that do not involve overflows, but take underflows to be zero.

    (a)   *(Commutativity of Addition)* $x + y = y + x$

    (b)   *(Commutativity of Multiplication)* $x \cdot y = y \cdot x$

    (c)   *(Associativity of Addition)* $x \cdot (y \cdot z) = (x \cdot y) \cdot z$

    (d)   *(Distributive Law)* $x \cdot (y + z) = x \cdot y + x \cdot z$

    (e)   *(Zero Divisors)* $x \cdot y = 0 \Rightarrow x = 0$ or $y = 0$

5.   Consider the infinite series: $1 + \dfrac{1}{8} + \dfrac{1}{27} + \dfrac{1}{64} + \cdots \dfrac{1}{n^3} + \cdots$.

    (a) Does it converge? If it does not, stop here; otherwise continue.

    (b) How many terms would we have to sum to get an approximation to the infinite sum with an absolute error $< 10^{-7}$?

    (c) Obtain such an approximation.

    (d) Using an approach similar to what was done after Example 5.6, add an extra term to the partial sums so as to obtain an improved approximation for the infinite series. How many terms would be required with this improved scheme? Perform this approximation and compare the answer with that obtained in part (c).

6.   Consider the infinite series: $\displaystyle\sum_{n=1}^{\infty} \dfrac{1}{n\sqrt{n}}$.

    (a) Does it converge? If it does not, stop here, otherwise continue.

    (b) How many terms would we have to sum to get an approximation to the infinite sum with an absolute error $1/500$?

    (c) Obtain such an approximation.

    (d) Using an approach similar to what was done after Example 5.6, add an extra term to the partial sums so as to obtain an improved approximation for the infinite series. How many terms would be required with this improved scheme? Perform this approximation and compare the answer with that obtained in part (c).

7.   Consider the infinite series: $\displaystyle\sum_{n=1}^{\infty} \dfrac{(-1)^{n+1}}{n} = 1 - \dfrac{1}{2} + \dfrac{1}{3} - \dfrac{1}{4} + \cdots$.

    (a) Show that this series satisfies the hypothesis of Leibniz's theorem (for n sufficiently large) so that from the theorem, we know the series will converge to a sum S.

    (b) Use Leibniz's theorem to find an integer N so that summing up to the first N terms only will give approximation to the sum with an error less than .0001.

    (c) Obtain such an approximation.

8.   Repeat all parts of Exercise 7 for the series: $\displaystyle\sum_{n=1}^{\infty} (-1)^n \dfrac{\ln n}{n}$.

9.   (a) In an analogous fashion to what was done in Example 5.5, establish the following estimate for floating point multiplications of a set of N positive real numbers, $P_N = a_1 \cdot a_2 \cdots a_N$:

$$| fl(P_N) - P_N | \le Nu. \tag{7}$$

We have, as before, ignored higher-order error terms. Thus, as far as minimizing errors is concerned, unlike for addition, the roundoff errors do not depend substantially on the order of multiplication.

    (b) In forming a product of positive real numbers, is there a good order to multiply so as to minimize the chance of encountering overflows or underflows? Explain your answer with some examples.

10.   (a) Using an argument similar to that employed in Example 5.4, show that in base β chopped

floating point arithmetic the unit roundoff is given by $u = \beta^{1-s}$.

(b) Show that in rounded arithmetic, the unit roundoff is given by $u = \beta^{1-s}/2$.

11.   Compare and contrast a two-digit ($s = 2$) floating point number system in base $\beta = 4$ and a four-digit ($s = 4$) binary ($\beta = 2$) floating point number system.

12.   In Exercise for the Reader 5.5, we used the following infinite series to approximate π:

$$\frac{\pi}{4} = 1 - \frac{1}{3} + \frac{1}{5} - \frac{1}{7} \cdots \implies \pi = 4 - \frac{4}{3} + \frac{4}{5} - \frac{4}{7} \cdots.$$

This alternating series was not a very efficient way to compute π, since it converges very slowly. Even to get an approximation with an accuracy of 10^{-7}, we would need to sum about 20 million terms. In this problem we will give a much more efficient (faster-converging) series for computing π by using Machin's identity ((13) of Chapter 2):

$$\frac{\pi}{4} = 4 \tan^{-1}\left(\frac{1}{5}\right) - \tan^{-1}\left(\frac{1}{239}\right).$$

(a) Use this identity along with the arctangent's MacLaurin series (see equation (11) of Chapter 2) to express π either as a difference of two alternating series, or as a single alternating series. Write your series both in explicit notation (as above) and in sigma notation.

(b) Perform an error analysis to see how many terms you would need to sum in the series (or difference of series) to get an approximation to π with error $< 10^{-7}$. Get MATLAB to perform this summation (in a "good" order) to thus obtain an approximation to π.

(c) How many terms would we need to sum so that the (exact mathematical) error would be less than 10^{-30}?   Of course, MATLAB only uses 16-digit floating point arithmetic so we could not directly use it to get such an approximation to π (Unless we used the symbolic toolbox; see Appendix A).

13.   Here is π, accurate to 30 decimal places:

$$\pi = 3.141592653589793238462643383279\ldots$$

Can you figure out a way to get MATLAB to compute π to 30 decimals of accuracy without using its symbolic capabilities?
Suggestions: What is required here is to build some MATLAB functions that will perform certain mathematical operations with more significant digits (over 30) than what MATLAB usually guarantees (about 15). In order to use the series of Exercise 12, you will need to build functions that will add/subtract, and multiply and divide (at least). Here is an example of a possible syntax for such new function: `z=highaccuracyadd(x,y)` where x and y are vectors containing the mantissa (with over 30 digits) as well as the exponent and sign (can be stored as 1 for plus, 2 for negative in one of the slots). The output z will be another vector of the same size that represents a 30+-significant-digit approximation to the sum $x + y$.   This is actually quite a difficult problem, but it is fun to try.

Chapter 6: Rootfinding

6.1: A BRIEF ACCOUNT OF THE HISTORY OF ROOTFINDING

The mathematical problems and applications of solving equations can be found in the oldest mathematical documents that are known to exist. The *Rhind Mathematical Papyrus*, named after Scotsman A. H. Rhind (1833–1863), who purchased it in a Nile resort town in 1858, was copied in 1650 B.C. from an original that was about 200 years older. It is about 13 inches high and 18 feet long, and it currently rests in a museum in England. This Papyrus contains 84 problems and solutions; many of them linear equations of the form (in modern algebra notation): $ax + b = 0$. It is fortunate that the mild Egyptian climate has so well preserved this document. A typical problem in this papyrus runs as follows: "A heap and its 1/7 part become 19. What is the heap?" In modern notation, this problem amounts to solving the equation $x + (1/7)x = 19$, and is easily solved by basic algebra. The arithmetic during these times was not very well developed and algebra was not yet discovered, so the Egyptians solved this equation with an intricate procedure where they made an initial guess, corrected it and used some complicated arithmetic to arrive at the answer of 16 5/8. The exact origin and dates of the methods in this work are not well documented and it is even possible that many of these methods may have been handed down by Imhotep, who supervised the construction of the pyramids around 3000 B.C.

Algebra derives from the Latin translation of the Arabic word, *al-jabr*, which means "restoring" as it refers to manipulating equations by performing the same operation on both sides. One of the earliest known algebra texts was written by the Islamic mathematician Al-Khwarizmi (c. 780–850), and in this book the quadratic equation $ax^2 + bx + c = 0$ is solved. The Islamic mathematicians did not deal with negative numbers so they had to separate the equation into 6 cases. This important work was translated into Latin, which was the language of scholars and universities in all of the western world during this era. After algebra came into common use in the western world, mathematicians set their sights on solving the next natural equation to look at: the general **cubic** equation $ax^3 + bx^2 + cx + d = 0$. The solution came quite a bit later in the Renaissance era in sixteenth century and the history after this point gets quite interesting.

The Italian mathematician Niccolo Fontana (better known by his nickname Tartaglia; see Figure 6.1) was the first to find the solution of the general cubic equation. It is quite a complicated formula and this is why it is rarely seen in textbooks.

A few years later, Tartaglia's contemporary, Girolamo Cardano[1] (sometimes the English translation "Cardan" is used; see Figure 6.2), had obtained an even more complicated formula (involving radicals of the coefficients) for the solution of the general **quartic** equation $ax^4 + bx^3 + cx^2 + dx + e = 0$. With each extra degree of the polynomial equations solved thus far, the general solution was getting inordinately more complicated, and it became apparent that a general formula for the solution of an nth-degree polynomial equation would be very unlikely and that the best mathematics could hope for was to keep working at obtaining general solutions to higher-order polynomials at one-degree increments.

FIGURE 6.1:
Niccolo Fontana ("Tartaglia") (1491–1557), Italian mathematician.

FIGURE 6.2:
Girolamo Cardano (1501–1576), Italian mathematician.

Three centuries later in 1821, a brilliant, young yet short-lived Norwegian mathematician named Niels Henrik Abel (Figure 6.3) believed he had solved the general **quintic** (5th-degree polynomial) equation, and submitted his work to the Royal Society of Copenhagen for publication. The editor contacted Abel to ask for a numerical example. In his efforts to construct examples, Abel found that his method was flawed, but in doing so he was able to prove that no formula could possibly exist (in terms of radicals and algebraic combinations of the coefficients) for the solution of a general quintic. Such nonexistence results are very deep and this one had ended generations of efforts in this area of rootfinding. [2]

---

[1] When Tartaglia was only 12 years old, he was nearly killed by French soldiers invading his town. He suffered a massive sword cut to his jaw and palate and was left for dead. He managed to survive, but he always wore a beard to hide the disfiguring scar left by his attackers; also his speech was impaired by the sword injury and he developed a very noticeable stutter. (His nickname Tartaglia means stammerer.) He taught mathematics in Venice and became famous in 1535 when he demonstrated publicly his ability of solving cubics, although he did not release his "secret formula." Other Italian mathematicians had publicly stated that such a solution was impossible. The more famous Cardano, located in Milan, was intrigued by Tartaglia's discovery and tried to get the latter to share it with him. At first Tartaglia refused but after Cardano tempted Tartaglia with his connections to the governor, Tartalgia finally acquiesced, but he made Cardano promise never to reveal the formula to anyone and never to even write it down, except in code (so no one could find it after he died). Tartaglia presented his solution to Cardano as a poem, again, so there would be no written record. With his newly acquired knowledge, Cardano was eventually able to solve the general quartic.

[2] Abel lived during a very difficult era in Norway and despite his mathematical wizardry, he was never able to obtain a permanent mathematics professorship. His short life was marked with constant poverty. When he proved his impossibility result for the quintic, he published it immediately on his own but in order to save printing costs, he trimmed down his proof to very bare details and as such it was difficult to read and did not give him the recognition that he was due. He later became close friends with the eminent German mathematician and publisher Leopold Crelle, who recognized Abel's genius and published much of his work. Crelle had even found a suitable professorship for Abel in

At roughly the same time, across the continent in France, another young mathematician, Evariste Galois (Figure 6.4), had worked on the same problems. Galois's life was also tragically cut short, and his brilliant and deep mathematical achievements were not recognized or even published until after his death. Galois invented a whole new area in mathematical group theory and he was able to use his development to show the impossibility of having a general formula (involving radicals and the four basic mathematical operations)

FIGURE 6.4: Evariste Galois[3] (1811–1832), French math-ematician.

FIGURE 6.3: Niels Henrik Abel (1802 –1829), Norwegian mathematician.

for solving the general polynomial equation of degree 5 or more and, furthermore, he obtained results that developed special conditions on polynomial equations under which such formulas could exist.

The work of Abel and Galois had a considerable impact on the development of mathematics and consequences and applications of their theories are still being realized today in the twentyfirst century. Many consequences have evolved from their theories, some resolving the impossibility of several geometric constructions that the Greeks had worked hard at for many years.

Among the first notable consequences of the nonexistence results of Abel and Galois was that pure mathematics would no longer be adequate as a reliable means for rootfinding, and the need for numerical methods became manifest. In the sections that follow we will introduce some **iterative** methods for finding a root of an equation $f(x) = 0$ that is known to exist. In each, a sequence of approximations x_n is constructed that, under appropriate hypotheses, will

---

Berlin, but the good news came too late; Abel had died from tuberculosis shortly before Crelle's letter arrived.

[3] Galois was born to a politically active family in a time of much political unrest in France. His father was mayor in a city near Paris who had committed suicide in 1829 after a local priest had forged the former's name on some libelous public documents. This loss affected Galois considerably and he became quite a political activist. His mathematics teachers from high school through university were astounded by his talent, but he was expelled from his university for publicly criticizing the director of the university for having locked the students inside to prevent them from joining some political riots. He joined a conservative National Guard that was accused of plotting to overthrow the government. His political activities caused him to get sent to prison twice, the second term for a period of six months. While in prison, he apparently fell in love with Stephanie-Felice du Motel, a prison official's daughter. Soon after he got out from prison he was challenged to a duel, the object of which had to do with Stephanie. In this duel he perished. The night before the duel, he wrote out all of his main mathematical discoveries and passed them on to a friend. It was only after this work was posthumously published that Galois's deep achievements were discovered. There is some speculation that Galois's fatal duel was set up to remove him from the political landscape

"converge" to an actual root r. Convergence here simply means that the error $|r - x_n|$ goes to zero as n gets large. The speed of convergence will depend on the particular method being used and possibly also on certain properties of the function $f(x)$.

6.2: THE BISECTION METHOD

This method, illustrated in Figure 6.5, is very easy to understand and write a code for; it has the following basic assumptions:

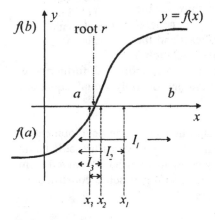

FIGURE 6.5: Illustration of the bisection method. The points x_n are the midpoints of the intervals I_n that get halved in length at each iteration.

ASSUMPTIONS: $f(x)$ is continuous on $[a, b]$, $f(a)$, $f(b)$ have opposite signs, and we are given an error tolerance = tol > 0.

Because of the assumptions, the intermediate value theorem from calculus tells us that $f(x)$ has at least one **root** (meaning a solution of the equation $f(x) = 0$) within the interval $[a, b]$. The method will iteratively construct a sequence x_n that converges to a root r and will stop when it can be guaranteed that $|r - x_n| < $ tol.

The philosophy of the method can be paraphrased as "divide and conquer." In English, the algorithm works as follows: We start with $x_1 = (a + b)/2$ being the midpoint of the interval $[a, b] \equiv [a_1, b_1] \equiv I_1$. We test $f(x_1)$. If it equals zero, we stop since we have found the exact root. If $f(x_1)$ is not zero, then it will have opposite signs either with $f(a)$ or with $f(b)$. In the former case we next look at the interval $[a, x_1] \equiv [a_2, b_2] \equiv I_2$ that now must contain a root of $f(x)$; in the latter case a root will be in $[x_1, b] \equiv [a_2, b_2] \equiv I_2$. The new interval I_2 has length equal to half of that of the original interval. Our next approximation is the new midpoint $x_2 = (a_2 + b_2)/2$. As before, either x_2 will be an exact root or we continue to approximate a root in I_3 that will be in either the left half or right half of I_2. Note that at each iteration, the approximation x_n lies in the interval I_{n+1}, which also contains an actual root. From this it follows that:

$$\text{error} = |x_n - r| \le \text{length}(I_{n+1}) = \frac{\text{length}(I_1)}{2^n} = \frac{b-a}{2^n}. \tag{1}$$

We wish to write a MATLAB M-file that will perform this bisection method for us. We will call our function bisect('function', a, b, tol). This one has four input variables. The first one is an actual mathematical function (with the generic name function) for which a root is sought. The second two variables, a and b, denote the endpoints of an interval at which the function has opposite signs, and the last variable tol denotes the maximum tolerated error. The program should cause the iterations to stop after the error gets below this tolerance (the estimate (1) will be useful here). Before attempting to write the MATLAB M-file, it is always recommended that we work some simple examples by hand. The results will later be used to check our program after we write it.

EXAMPLE 6.1: Consider the function $f(x) = x^5 - 9x^2 - x + 7$.
(a) Show that $f(x)$ has a root on the interval $[1, 2]$.
(b) Use the bisection method to approximate this root with an error < 0.01.
(c) How many iterations would we need to use in the bisection method to guarantee an error < 0.00001?

SOLUTION: Part (a): Since $f(x)$ is a polynomial, it is continuous everywhere. Since $f(1) = 1 - 9 - 1 + 7 = -2 < 0$, and $f(2) = 32 - 36 - 2 + 7 = 1 > 0$, it follows from the intermediate value theorem that $f(x)$ must have a root in the interval $[1, 2]$.

Part (b): Using (1), we can determine the number of iterations required to achieve an error guaranteed to be less than the desired upper bound $0.01 = 1/100$. Since $b - a = 1$, the right side of (1) is $1/2^n$, and clearly $n = 7$ is the first value of n for which this is less than $1/100$. ($1/2^7 = 1/128$.) By the error estimate (1), this means we will need to do 7 iterations. At each step, we will need to evaluate the function $f(x)$ at the new approximation value x_n. If we computed these iterations directly on MATLAB, we would need to enter the formula only once, and make use of MATLAB's editing features. Another way to deal with functions on MATLAB would be to simply store the mathematical function as an M-file. But this latter approach is not so suitable for situations like this where the function gets used only for a particular example. We now show a way to enter a function temporarily into a MATLAB session as a so-called "inline" function:

<fun_name>=inline('<math expression>') →	Causes a mathematical function to be defined (temporarily, only for the current MATLAB session), the name will be <fun_nam>, the formula will be given by <math expression> and the input variables determined by MATLAB when it scans the expression.

`<fun_name>=inline('<math expression>', 'x1', 'x2', ..., 'xn') →`	Works as above but specifies input variables to be x1, x2, ..., xn in the same order.

We enter our function now as an inline function, giving it the convenient and generic name "f."

```
>> f=inline('x^5-9*x^2-x+7')     →f = Inline function:   f(x) = x^5-9*x^2-x+7
```

We may now work with this function in MATLAB just as with other built-in mathematical functions or stored M-file mathematical functions. Its definition will be good for as long as we are in the MATLAB session in which we created this inline function. For example, to evaluate $f(2)$ we can now just type:

```
>> f(2)          → 1
```

For future reference, in writing programs containing mathematical functions as variables, it is better to use the following equivalent (but slightly longer) method:

```
>> feval(f,2)    →1
```

`feval(<funct>, a1,a2,..,an) →`	Returns the value of the stored or inline function `funct(x1,x2, ..., xn)` of n variables at the values `x1=a1, x2=a2, ... , xn=an`.

Let's now use MATLAB to perform the bisection method:

```
>> a1=1; b1=2; x1=(a1+b1)/2, f(x1) %    [a1,b1]=[a,b] and x1 (first
>>% approximation) is the midpoint.  We need to test f(x1).
→x1 = 1.5000 (=first approximation), ans = -7.1563 (value of function at first approximation)

>> a2=x1; b2=b1; x2=(a2+b2)/2, f(x2) %the bisected interval [a2,b2]
>> % is always chosen to be the one where function changes sign.
→x2 =1.7500, ans =-5.8994 (n=2 approximation and value of function)

>> a3=x2; b3=b1; x3=(a3+b3)/2, f(x3)     →x3 =1.8750, ans = -3.3413

>> a4=x3; b4=b3; x4=(a4+b4)/2, f(x4)     →x4 = 1.9375, ans =-1.4198

>> a5=x4; b5=b4; x5=(a5+b5)/2, f(x5)     →x5 = 1.9688, ans = -0.2756

>> a6=x5; b6=b5; x6=(a6+b6)/2, f(x6)
→x6 =1.9844, ans =0.3453 (n=6 approximation and corresponding y-coordinate)

>>a7=a6; b7=x6; x7=(a7+b7)/2, f(x7)      →x7 =1.9766, ans =0.0307
```

The above computations certainly beg to be automated by a loop and this will be done soon in a program.

Part (c): Let's use a MATLAB loop to find out how many iterations are required to guarantee an error < 0.00001:

```
>> n=1; while 1/2^(n)>=0.00001
```

```
n=n+1;
end
>> n    →n =17
```

EXERCISE FOR THE READER 6.1: In the example above we found an approximation $(x7)$ to a root of $f(x)$ that was accurate with an error less than 0.01. For the actual root $x = r$, we of course have $f(r) = 0$, but $f(x7) = 0.0307$. Thus the error of the y-coordinate is over three times as great as that for the x-coordinate. Use calculus to explain this discrepancy.

EXERCISE FOR THE READER 6.2: Consider the function
$$f(x) = \cos(x) - x.$$
(a) Show that $f(x)$ has exactly one root on the interval $[0, \pi/2]$.
(b) Use the bisection method to approximate this root with an error < 0.01.
(c) How many iterations would we need to use in the bisection method to guarantee an error $< 10^{-12}$?

With the experience of the last example behind us, we should now be ready to write our program for the bisection method. In it we will make use of the following built-in MATLAB functions.

sign(x) →	= (the sign of the real number x)=	$\begin{cases} 1, & \text{if } x > 0 \\ 0, & \text{if } x = 0. \\ -1, & \text{if } x < 0 \end{cases}$

Recall that with built-in functions such as quad, some of the input variables were made optional with default values being used if the variables are not specified in calling the function. In order to build such a feature into a function, the following command is useful in writing such an M-file:

nargin (inside the body of a funtion M-file)→	Gives the number of input arguments (that are specified when a function is called).

PROGRAM 6.1: An M-file for the bisection method.

```
function [root, yval] = bisect(varfun, a, b, tol)
% input variables: varfun, a, b, tol
% output variables:  root, yval
% varfun = the string representing a mathematical function (built-in,
% M-file, or inline) of one variable that is assumed to have opposite
% signs at the points x=a,   and x=b.   The program will perform the
% bisection method to approximate a root of varfun in [a,b] with an
% error < tol.   If the tol variable is omitted a default value of
% eps*max(abs(a),abs(b),1)is used.

%we first check to see if there is the needed sign change
ya=feval(varfun,a); yb=feval(varfun,b);
if sign(ya)==sign(yb)
```

```
      error('function has same sign at endpoints')
end

%we assign the default tolerance, if none is specified
if nargin < 4
   tol=eps*max([abs(a) abs(b) 1]);
end

%we now initialize the iteration
an=a; bn=b; n=0;

%finally we set up a loop to perform the bisections
while (b-a)/2^n >= tol
   xn=(an + bn)/2; yn=feval(varfun, xn); n=n+1
   if yn==0
      fprintf('numerically exact root')
      root=xn; yval=yn;
      return
   elseif sign(yn)==sign(ya)
      an=xn; ya=yn;
   else
      bn=xn; yb=yn;
   end
end

root=xn; yval=yn;
```

We will make some more comments on the program as well as the algorithm, but first we show how it would get used in a MATLAB session.

EXAMPLE 6.2: (a) Use the above `bisect` program to perform the indicated approximation of parts (b) and (c) of Example 6.1.
(b) Do the same for the approximation problem of Exercise for the Reader 6.1.

SOLUTION: Part (a): We need only run these commands to get the first approximation (with tol = 0.01):[4]

```
>> f=inline('x^5-9*x^2-x+7', 'x');
>> bisect(f,1,2,.01)
→ans = 1.9766 (This is exactly what we got in Example 6.1(b))
```

By default, a function M-file with more than one output variable will display only the first one (stored into the temporary name "ans"). To display all of the output variables (so in this case also the y-coordinate), use the following syntax:

```
>> [x,y]=bisect(f,1,2,.01)
→x = 1.9844, y =0. 0307
```

To obtain the second approximation, we should use more decimals.

```
>> format long
```

---

[4] We point out that if the function f were instead stored as a function M-file, the syntax for `bisect` would change to `bisect('f',..)` or `bisect(@f,…)`.

```
>> [x,y]=bisect(f,1,2,0.00001)
→x =1.97579193115234, y= 9.717120432028992e-005
```

Part (b): There is nothing very different needed to do this second example:

```
>> g=inline('cos(x)-x')
→g = Inline function: g(x) = cos(x)-x
```

```
>> [x,y]=bisect(g,0,pi/2,0.01)
→x =0.74858262448819, y = -0.01592835281578
```

```
>> [x,y]=bisect(g,0,pi/2,10^(-12))
→x =0 73908513321527, y = -1.888489364887391e-013
```

Some additional comments about our program are now in order. Although it may have seemed a bit more difficult to follow this program than Example 6.1, we have considerably economized by overwriting an or bn at each iteration, as well as xn and yn. There is no need for the function to internally construct vectors of all of the intermediate intervals and approximations if all that we are interested in is the final approximation and perhaps also its y-coordinate. We used the error('message') flag command in the program. If it ever were to come up (i.e., only when the corresponding if-branch's condition is met), then the error 'message' inside would be printed and the function execution would immediately terminate. The general syntax is as follows:

error('message') → (inside the body of a function)	Causes the message to display on the command window and the execution of the function to be immediately terminated.

Notice also that we chose the default tolerance to be $eps \cdot max(|a|, |b|, 1)$, where eps is MATLAB's unit roundoff. Recall that the unit roundoff is the maximum relative error arising from approximating a real number by its floating point representative (see Chapter 5). Although the program would still work if we had just used eps as the default tolerance, in cases where $max(|a|, |b|)$ is much larger than 1, the additional iterations would yield the same floating point approximation as with our chosen default tolerance. In cases where $max(|a|, |b|)$ is much smaller than one, our default tolerance will produce more accurate approximations. As with all function M-files, after having stored bisect, if we were to type help bisect in the command window, MATLAB would display all of the adjacent block of comment lines that immediately follow the function definition line. It is good practice to include comment lines (as we have) that explain various parts of a program.

EXERCISE FOR THE READER 6.3: In some numerical analysis books, the while loop in the above program bisect is rewritten as follows:

```
while (b-a)/2^n >= tol
   xn=(an + bn)/2; yn=feval(varfun, xn); n=n+1;
   if yn*ya > 0
      an=xn; ya=yn;
```

```
  else
       bn=xn; yb=yn;
    end
end
```

The only difference with the corresponding part in our program is with the condition in the if-branch, everything else is identical.

(a) Explain that mathematically, the condition in the if-branch above is equivalent to the one in our program (i.e., both always have the same truth values).

(b) In mathematics there is no smallest positive number. As in Chapter 5, numbers that are too small in MATLAB will underflow to 0. Depending on the version of MATLAB you are using, the smallest positive (distinguishable from 0) number in MATLAB is something like 2.2251e-308. Anything smaller than this will be converted (underflow) to zero. (To see this enter $10^{\wedge}(-400)$). Using these facts, explain why the for loop in our program is better to use than the above modification of it.

(c) Construct a continuous function $f(x)$ with a root at $x = 0$ so that if we apply the bisection program on the interval $[-1, 3]$ with tol = 0.001, the algorithm will work as it is supposed to; however, if we apply the (above) modified program the output will not be within the tolerance 0.001 of $x = 0$.

We close this section with some further comments on the bisection method. It is the oldest of the methods for rootfinding. It is theoretically guaranteed to work as long as the hypotheses are satisfied. Recall the assumptions are that $f(x)$ is continuous on $[a, b]$ and that $f(a)$ and $f(b)$ are of opposite signs. In this case it is said that the interval $[a, b]$ is a **bracket** of the function $f(x)$. The bisection method unfortunately cannot be used to locate zeros of functions that do not possess brackets. For example, the function $y = x^2$ has a zero only at $x = 0$ but otherwise y is always positive so this function has no bracket. Although the bisection method converges rather quickly, other methods that we will introduce will more often work much faster. For a single rootfinding problem, the difference in speed is not much of an issue, but for more complicated or advanced problems that require numerous rootfinding "subproblems," it will be more efficient to use other methods. A big advantage of the bisection method over other methods we will introduce is that the error analysis is so straightforward and we are able to determine the number of necessary iterations quite simply before anything else is done. The **residual** of an approximation x_n to a root $x = r$ of $f(x)$ is the value $f(x_n)$. It is always good practice to examine the residual of approximations to a root. Theoretically the residuals should disintegrate to zero as the approximations get better and better, so it would be a somewhat awkward situation if your approximation to a root had a very large residual. Before beginning any rootfinding problem, it is often most helpful to begin with a (computer-generated) plot of the function.

EXERCISES 6.2:

1. The function $f(x) = \sin(x)$ has a root at $x = \pi$. Find a bracket for this root and use the bisection method with tol $= 10^{-12}$ to obtain an approximation of π that is accurate to 12 decimals. What is the residual?

2. The function $\ln(x) - 1$ has a root at $x = e$. Find a bracket for this root and use the bisection method with tol $= 10^{-12}$ to obtain an approximation of e that is accurate to 12 decimals. What is the residual?

3. Apply the bisection method to find a root of the equation $x^6 + 6x^2 + 2x = 20$ in the interval $[0,2]$ with tolerance 10^{-7}.

4. Apply the bisection method to find a root of the equation $x^9 + 6x^2 + 2x = 3$ in the interval $[-2,-1]$ with tolerance 10^{-7}.

5. Use the bisection method to approximate the smallest positive root of the equation $\tan(x) = x$ with error $< 10^{-10}$.

6. Use the bisection method to approximate the smallest positive root of the equation $e^{2x} = \sin(x) + 1$ with error $< 10^{-10}$.

7. (*Math Finance*) It can be shown[5] that if equal monthly deposits of *PMT* dollars are made into an annuity (interest-bearing account) that pays $100r$% annual interest compounded monthly, then the value $A(t)$ of the account after t years will be given by the formula

    $$A(t) = PMT \frac{(1+r/12)^{12t} - 1}{r/12}.$$

    Suppose Mr. Jones is 30 years old and can afford monthly payments of $350.00 into such an annuity. Mr. Jones would like to plan to be able to retire at age 65 with a $1 million nest egg. Use the bisection method to find the minimum interest rate (= $100r$ %) Mr. Jones will need to shop for in order to reach his retirement goal.

8. (*Math Finance*) It can be shown that to pay off a 30-year house mortgage for an initial loan of *PV* dollars with equal monthly payments of *PMT* dollars and a fixed annual interest rate of $100r$% compounded monthly, the following equation must hold:

    $$PV = PMT \frac{1 - (1+r/12)^{-360}}{r/12}.$$

    (For a 15-year mortgage, change 360 to 180.) Suppose the Bradys wish to buy a house that costs $140,000. They can afford monthly payments of $1,100 to pay off the mortgage. What kind of interest rate $100r$% would they need to be able to afford this house with a 30-year mortgage? How about with a 15-year mortgage? If they went with a 30-year mortgage, how much interest would the Bradys need to pay throughout the course of the loan?

---

[5] See, for example, Chapter 3 and Appendix B of [BaZiBy-02] for detailed explanations and derivations of these and other math finance formulas.

9. Modify the `bisect` program in the text to create a new one, `bisectvv` (stands for: bisection algorithm, vector version), that has the same input variables as `bisect`, but the output variables will now be two vectors x and y that contain all of the successive approximations ($x = [x_1, x_2, \cdots, x_n]$) and the corresponding residuals ($y = [f(x_1), f(x_2), \cdots, f(x_n)]$). Run this program to redo Exercise 3, print only every fourth component, $x_1, y_1, x_5, y_5, x_9, y_9, \cdots$ and also the very last components, x_n, y_n.

10. Modify the `bisect` program in the text to create a new one, `bisectte` (stands for: bisection algorithm, tell everything), that has the same input variables as `bisect`, but this one has no output variables. Instead, it will output at each of the iterations the following phrase: "Iteration n=$< k >$, approximation = $< xn >$, residual = $< yn >$," where the values of k, xn, and yn at each iteration will be the actual numbers. $< k >$ should be an integer and the other two should be floating point numbers. Apply your algorithm to the function $f(x) = 5x^3 - 8x^2 + 2$ with bracket [1, 2] and tolerance = 0.002.

11. Apply the `bisect` program to $f(x) = \tan(x)$ with tol = 0.0001 to each of the following sets of intervals. In each case, is the output (final approximation) within the tolerance of a root? Carefully explain what happens in each case.
    (a) $[a, b] = [5, 7]$, (b) $[a, b] = [4, 7]$, (c) $[a, b] = [4, 5]$.

12. In applying the bisection method to a function $f(x)$ using a bracket $[a, b]$ on which $f(x)$ is known to have exactly one root r, is it possible that x_2 is a better approximation to r than x_5? (This means $|x_2 - r| < |x_5 - r|$.) If no, explain why not; if yes, supply a specific counterexample.

6.3: NEWTON'S METHOD

Under most conditions, when Newton's method works, it converges very quickly, much faster indeed than the bisection method. It is at the foundation of all contemporary state-of-the-art rootfinding programs. The error analysis, however, is quite a bit more awkward than with the bisection method and this will be relegated to Section 6.5. Here we examine various situations in which Newton's method performs outstandingly, where it can fail, and in which it performs poorly.

ASSUMPTIONS: $f(x)$ is a differentiable function that has a root $x = r$ which we wish to accurately approximate. We would like the approximation to be accurate to MATLAB's machine precision of about 15 significant digits.

The idea of the method will be to repeatedly use tangent lines of the function situated at successive approximations to "shoot at" the next approximation. More precisely, the next approximation will equal the x-intercept of the tangent line to the graph of the function taken at the point on the graph corresponding to the current approximation to the root. See Figure 6.6 for an illustration of Newton's method. We begin with an initial approximation x_0 that was perhaps obtained from a plot. It is straightforward to obtain a recursion formula for the next approximation x_{n+1} in terms of the current approximation x_n. The tangent line to the graph of $y = f(x)$ at $x = x_n$ is given by the first-order Taylor polynomial centered at $x = x_n$, which has equation (see equation (3) of Chapter 2): $y = f(x_n) + f'(x_n)(x - x_n)$. The next approximation is the x-intercept of this line and is obtained by setting $y = 0$ and solving for x. Doing this gives us the recursion formula:

$$x_{n+1} = x_n - \frac{f(x_n)}{f'(x_n)}, \tag{2}$$

where it is required that $f'(x_n) \neq 0$.
It is quite a simple task to write a MATLAB program for Newton's method, but following the usual practice, we will begin working through an example "by hand".

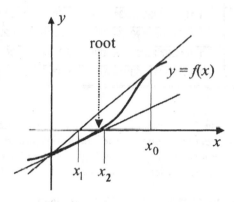

FIGURE 6.6: Illustration of Newton's method. To go from the initial approximation (or guess) x_0 to the next approximation x_1, we simply take the x_1 to be the x-intercept of the tangent line to the graph of $y = f(x)$ at the point $(x_0, f(x_0))$. This procedure gets iterated to obtain successive approximations.

EXAMPLE 6.3: Use Newton's method to approximate $\sqrt[4]{2}$ by performing five iterations on the function $f(x) = x^4 - 2$ using initial guess $x = 1.5$. (Note that [1, 2] is clearly a bracket for the desired root.)

SOLUTION: Since $f'(x) = 4x^3$, the recursion formula (2) becomes:

$$x_{n+1} = x_n - \frac{f(x_n)}{f'(x_n)} = x_n - \frac{x_n^4 - 2}{4x_n^3}.$$

Let's now get MATLAB to find the first five iterations along with the residuals and the errors. For convenient display, we store this data in a 5×3 matrix with the first column containing the approximations, the second the residuals, and the third the errors.

```
>> x(1)=1.5; %initialize, remember zero can't be an index
>> for n=1:5
x(n+1)=x(n)-(x(n)^4-2)/(4*x(n)^3);
A(n, :) = [ x(n+1)    (x(n+1)^4-2)    abs(x(n+1)-2^(1/4))];
end
```

To be able to see how well the approximation went, it is best to use a different format (from the default `format short`) when we display the matrix A.

```
>> format long e
>> A'
```

We display this matrix in Table 6.1.

TABLE 6.1: The successive approximations, residuals, and errors resulting from applying Newton's method to $f(x) = x^4 - 2$ with initial approximation $x_0 = 1.5$.

n	x_n	$f(x_n)$	Error = $|r - x_n|$
1	1.2731481481481e+000	6.2733693232248e-001	8.3941033145427e-002
2	1.1971498203523e+000	5.3969634451728e-002	7.9427053495620e-003
3	1.1892858119092e+000	5.2946012602728e-004	7.8696906514741e-005
4	1.1892071228136e+000	5.2545275686100e-008	7.8109019252537e-009
5	1.1892071150027e+000	1.3322676295502e-015	2.2204460492503e-016

Table 6.1 shows quite clearly just how fast the errors are disintegrating to zero. As we mentioned, if the conditions are right, Newton's method will converge extremely quickly. We will give some clarifications and precise limitations of this comment, but let us first write an M-file for Newton's method.

PROGRAM 6.2: An M-file for Newton's method.[6]

```
function [root, yval] = newton(varfun, dvarfun,  x0, tol, nmax)
% input variables: varfun, dvarfun, x0, tol, nmax
% output variables:  root, yval
% varfun = the string representing a mathematical function (built-in,
% M-file, or inline) and dvarfun = the string representing the
% derivative, x0 = the initial approx.  The program will perform
% Newton's method to approximate a root of varfun near x=x0 until
% either successive approximations differ by less than tol or nmax
% iterations have been completed, whichever comes first.  If the tol
% and nmax variables are omitted, default values of
% eps*max(abs(a),abs(b),1) and 30 are used.
% we assign the default tolerance and maximum number of iterations if
% none are specified
if nargin < 4
   tol=eps*max([abs(a) abs(b) 1]); nmax=30;
end

%we now initialize the iteration
xn=x0;

%finally we set up a loop to perform the approximations
for n=1:nmax
```

---

[6] When one needs to use an apostrophe in a string argument of an `fprintf` statement, the correct syntax is to use a double apostrophe. For example, `fprintf('Newton's')` would produce an error message but `fprintf(Newton''s)` would produce → Newton's.

```
    yn=feval(varfun, xn); ypn=feval(dvarfun, xn);
    if yn == 0
       fprintf('Exact root found\r')
       root = xn; yval = 0;
       return
    end
    if ypn == 0
      error('Zero derivative encountered, Newton''s method failed,
                                   try changing x0')
    end
    xnew=xn-yn/ypn;
    if abs(xnew-xn)<tol
          fprintf('Newton''s method has converged\r')
          root = xnew; yval = feval(varfun, root);
          return
    elseif n==nmax
          fprintf('Maximum number of iterations reached\r')
          root = xnew; yval = feval(varfun, root);
          return
    end
    xn=xnew;
end
```

EXAMPLE 6.4:   (a) Use the above `newton` program to find a root of the equation of Example 6.3 (again using $x_0 = 1.5$).

(b) Next use the program to approximate e by finding a root of the equation $\ln(x) - 1 = 0$. Check the error of this latter approximation.

SOLUTION: Part (a): We temporarily construct some inline functions. Take careful note of the syntax.

```
>> f = inline('x^4-2')        →f =Inline function:  f(x) = x^4-2
>> fp = inline('4*x^3')       →fp =Inline function: fp(x) = 4*x^3
>> format long
>> newton(f, fp, 1.5)
→Newtons method has converged    →ans = 1.18920711500272

>> [x,y]=newton(f,fp,1.5)  %to see also to see the y-value
→Newton's method has converged
→x =1.18920711500272, y = -2.220446049250313e-016
```

Part (b):

```
>> f=inline('log(x)-1'); fp=inline('1/x');
>> [x,y]=newton(f,fp,3)
→Newton's method has converged    →x =2.71828182845905, y = 0
>>abs(exp(1)-x)                    →ans = 4.440892098500626e-016
```

We see that the results of part (a) nicely coincide with the final results of the previous example.

EXERCISE FOR THE READER 6.4: In part (b) of Example 6.4, show using calculus that the y-coordinate corresponding to the approximation of the root e

found is about as far from zero as the x-coordinate is from the root. Explain how floating point arithmetic caused this y-coordinate to be outputted as zero, rather than something like 10^{-17}. (Refer to Chapter 5 for details about floating point arithmetic.)

We next look into some pathologies that can cause Newton's method to fail. Later, in Section 6.5, we will give some theorems that will give some guaranteed error estimates with Newton's method, provided certain hypotheses are satisfied. The first obvious problem (for which we built an error message into our program) is if at any approximation x_n we have $f'(x_n) = 0$. Unless the function at hand is highly oscillatory near the root, such a problem can often be solved simply by trying to reapply Newton's method with a different initial value x_0 (perhaps after examining a more careful plot); see Figure 6.7.

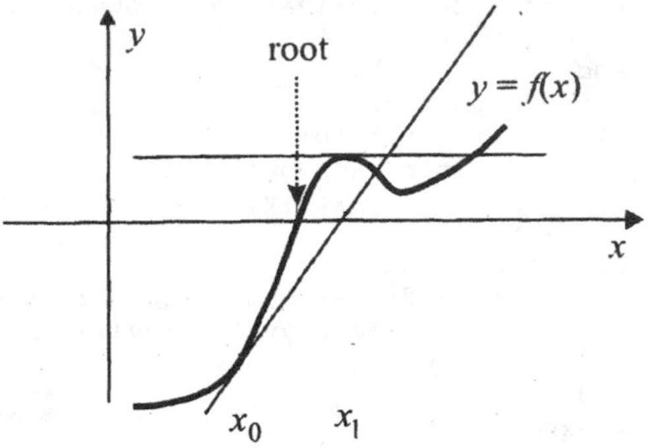

A less obvious problem that can occur in Newton's method is **cycling**. Cycling is said to occur when the sequence of x_ns gets caught in an infinite loop by continuing to run through the same set of fixed values.

We have illustrated the cycling phenomenon with a cycle having just two values. It is possible for such a "Newton cycle" to have any number of values.

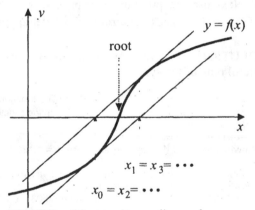

FIGURE 6.7: A zero derivative encountered in Newton's method. Here x_2 is undefined. Possible remedy: Use a different initial value x_0.

FIGURE 6.8: A cycling phenomenon encountered in Newton's method. Possible remedy: Take initial approximation closer to actual root.

EXERCISE FOR THE READER 6.5: (a) Construct explicitly a polynomial $y = p(x)$ with an initial approximation x_0 to a root such that Newton's method will cause cycling.

(b) Draw a picture of a situation where Newton's method enters into a cycle having exactly four values (rather than just two as in Figure 6.8).

Another serious problem with Newton's method is that the approximations can actually sometimes continue to move away from a root.   An illustration is provided by Figure 6.8 if we move x_0 to be a bit farther to the left. This is another reason why it is always recommended to examine residuals when using Newton's method. In the next example we will use a single function to exhibit all three phenomena in Newton's method: convergence, cycling, and divergence.

EXAMPLE 6.5: Consider the function $f(x) = \arctan(x)$, which has a single root at $x = 0$.   The graph is actually similar in appearance (although horizontally shifted) to that of Figure 6.8. Show that there exists a number $a > 0$ such that if we apply Newton's method to find the root $x = 0$ (a purely academic exercise since we know the root) with initial approximation $x_0 > 0$, the following will happen:

(i) If $x_0 < a$, then $x_n \to 0$ (convergence to the root, as desired).

(ii) If $x_0 = a$, then x_n will cycle back and forth between a and $-a$.

(iii) If $x_0 > a$, then $|x_n| \to \infty$ (the approximations actually move farther and farther away from the root).

Next apply the bisection program to approximate this critical value $x = a$ and give some examples of each of (i) and (iii) using the Newton method program.

SOLUTION: Since $f'(x) = \dfrac{1}{1+x^2}$, Newton's recursion formula (2) becomes:

$$x_{n+1} = x_n - \frac{f(x_n)}{f'(x_n)} = x_n - (1 + x_n^2) \arctan(x_n) \equiv g(x_n).$$

(The function $g(x)$ is defined by the above formula.)   Since $g(x)$ is an odd function (i.e., $g(-x) = -g(x)$), we see that we will enter into a cycle (with two numbers) exactly when $(r_1 =)g(x_0) = -x_0$.   (Because then $r_2 = g(x_1) = g(-x_0)$ $= -(x_0) = x_0$, and so on.) Thus, Newton cycles can be found by looking for the positive roots of $g(x) + x = 0$.   Notice that $(g(x) + x)' = 1 - 2x \arctan(x)$ so that the function in parentheses increases (from its initial value of 0 at $x = 0$) until x reaches a certain positive value (the root of $1 - 2x \arctan(x) = 0$) and after this value of x it is strictly decreasing and will eventually become zero (at some value $x = a$) and after this will be negative. Again, since $g(x)$ is an odd function, we can summarize as follows: (i) for $0 < |x| < a$, $|g(x)| < |x|$, (ii) $g(\pm a) = \mp a$, and (iii) for $|x| > a$, $|g(x)| > |x|$.   Once x is situated in any of these three ranges, $g(x)$ will thus be in the same range and by the noted properties of $g(x)$ and of $g(x) + x$ we can conclude the assertions of convergence, cycling, and divergence to infinity as x_0 lies in one of these ranges. We now provide some numerical data that will demonstrate each of these phenomena.

First, since $g(a) = -a$, we may approximate $x = a$ quite quickly by using the bisection method to find the positive root of the function $h(x) = g(x) + x$. We must be careful not to pick up the root $x = 0$ of this function.

```
>> h=inline('2*x-(1+x^2)*atan(x)');
>> h(0.5), h(5) %will show a bracket to the unique positive root
→ans = 0.4204, ans =-25.7084
```

```
>> format long
>> a=bisect(h,.5,5)        → a=1.39174520027073
```

To make things more clear in this example, we use a modification of the newton algorithm, called newtonsh, that works the same as newton, except the output will be a matrix of all of the successive approximations x_n and the corresponding y-values. (Modifying our newton program to get newtonsh is straightforward and is left to the reader.)

```
>> format long e
>> B=newtonsh(f,fp,1)
→Newton's method has converged
(We display the matrix B in Table 6.2.)
```

TABLE 6.2: The result of applying Newton's method to the function $f(x) = \arctan(x)$ with $x_0 = 1 < a$ (critical value). Very fast convergence to the root $x = 0$.

n	x_n	$f(x_n)$
1	-5.707963267948966e-001	-5.186693692550166e-001
2	1.168599039989131e-001	1.163322651138959e-001
3	-1.061022117044716e-003	-1.061021718890093e-003
4	7.963096044106416e-010	7.963096044106416e-010
5	0	0
6	0	0

```
>> B=newtonsh(f,fp,a)        →Maximum number of iterations reached (see Table 6.3)
```

TABLE 6.3: The result of applying Newton's method to the function $f(x) = \arctan(x)$ with $x_0 = a$ (critical value). The approximations cycle.

n	x_n	$f(x_n)$
1	-1.391745200270735e+000	-9.477471335169905e-001
2	1.391745200270735e+000	9.477471335169905e-001
...	...	...
28	1.391745200270735e+000	9.477471335169905e-001
29	-1.391745200270735e+000	-9.477471335169905e-001
30	1.391745200270735e+000	9.477471335169905e-001

```
>> B=newtonsh(f,fp,1.5)
```
→??? Error using ==> newtonsh
→zero derivative encountered, Newton's method failed, try changing x0

We have got our own error flag. We know that the derivative $1/(1+x^2)$ of arctan(x) is never zero. What happened here is that the approximations x_n were getting so large, so quickly (in absolute value) that the derivatives underflowed to zero. To get some output, we redo the above command with a cap on the number of iterations; the results are displayed in Table 6.4.

```
>> B=newtonsh(f,fp,1.5, 0.001, 9)
```
→Maximum number of iterations reached   →B =

TABLE 6.4: The result of applying Newton's method to the function $f(x) = \arctan(x)$ with $x_0 = 1.5 > a$ (critical value). The successive approximations alternate between positive and negative values and their absolute values diverge very quickly to infinity. The corresponding y-values will, of course, alternate between tending to $\pm\pi/2$, the limits of $f(x) = \arctan(x)$ as $x \to \pm\infty$.

n	x_n	$f(x_n)$
1	-1.694079600553820e+000	-1.037546359137891e+000
2	2.321126961438388e+000	1.164002042421975e+000
3	-5.114087836777514e+000	-1.377694528702752e+000
4	3.229568391421002e+001	1.539842326908012e+000
5	-1.575316950821204e+003	-1.570161533990085e+000
6	3.894976007760884e+006	1.570796070053906e+000
7	-2.383028897355213e+013	-1.570796326794855e+000
8	8.920280161123818e+026	1.570796326794897e+000
9	-1.249904599365711e+054	-1.570796326794897e+000

In all of the examples given so far, when Newton's method converged to a root, it did so very quickly. The main reason for this is that each of the roots being approximated was a **simple root**. Geometrically this means that the graph of the differentiable function was not tangent to the x-axis at this root. Thus a root $x = r$ is a simple root of $f(x)$ provided that ($f(r) = 0$ and) $f'(r) \neq 0$. A root r that is not simple is called a **multiple root**

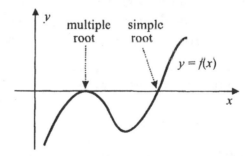

FIGURE 6.9: Illustration of the two types of roots a function can have. Newton's method performs much more effectively in approximating simple roots.

of **order** M ($M > 1$) if $f(r) = f'(r) = f''(r) = \cdots$ $f^{(M-1)}(r) = 0$ but $f^{(M)}(r) \neq 0$ (see Figure 6.9). These definitions were given for polynomials in

Exercises 3.2.   Multiple roots of order 2 are sometimes called **double roots**, order-3 roots are **triple roots**, and so on. If $x = r$ is an order-M root of $f(x)$, it can be shown that $f(x) = (x - r)^M h(x)$ for some continuous function $h(x)$ (see Exercise 13).

EXAMPLE 6.6: How many iterations will it take for Newton's method to approximate the multiple root $x = 0$ of the function $f(x) = x^{21}$ using an initial approximation of $x = 1$ if we want an error < 0.01? How about if we want an error < 0.001?

SOLUTION: We omit the MATLAB commands, but summarize the results. If we first try to run `newton` (or better a variant of it that displays some additional output variables), with a tolerance of 0.01, and a maximum number of iterations = 50, we will get the message that the method converged. It did so after a whopping 33 iterations and the final value of root = 0.184 (which is not within the desired error tolerance) and a (microscopically small) yval = 1.9842e-015. The reason the program gave us the convergence message is because the adjacent root approximations differed by less than the 0.01 tolerance. To get the root to be less than 0.01 we would actually need about .90 iterations!   And to get to an approximation with a 0.001 tolerated error, we would need to go through 135 iterations.   This is a pathetic rate of convergence; even the (usually slower) bisection method would only take 7 iterations for such a small tolerance (why?).

EXERCISES 6.3:

1.   For each of the functions shown below, find Newton's recursion formula (2). Next, using the value of x_0 that is given, find each of x_1, x_2, x_3.

   (a) $f(x) = x^3 - 2x + 5$; $x_0 = -3$ (c) $f(x) = xe^{-x}$; $x_0 = 0.5$

   (b) $f(x) = e^x - 2\cos(x)$; $x_0 = 1$ (d) $f(x) = \ln(x^4) - \cos(x)$; $x_0 = 1$

2.   For each of the functions shown below, find Newton's recursion formula (2). Next, using the value of x_0 that is given, find each of x_1, x_2, x_3.

   (a) $f(x) = x^3 - 15x^2 + 24$; $x_0 = -3$ (c) $f(x) = \ln(x)$; $x_0 = 0.5$

   (b) $f(x) = e^x - 2e^{-x} + 5$; $x_0 = 1$ (d) $f(x) = \sec(x) - 2e^{x^2}$; $x_0 = 1$

3.   Use Newton's method to find the smallest positive root of each equation to 12 digits of accuracy. Indicate the number of iterations used.

   (a) $\tan(x) = x$
   (b) $x\cos(x) = 1$ (c) $4x^2 = e^{x/2} - 2$
                              (d) $(1 + x)\ln(1 + x^2) = \cos(\sqrt{x})$

   Suggestion: You may wish to modify our `newton` program so as to have it display another output variable that gives the number of iterations.

4.   Use Newton's method to find the smallest positive root of each equation to 12 digits of accuracy. Indicate the number of iterations used.

(a) $e^{-x} = x$

(c) $x^4 - 2x^3 + x^2 - 5x + 2 = 0$

(b) $e^x - x^8 = \ln(1 + 2^x)$

(d) $e^x = x^\pi$

5. For the functions given in each of parts (a) through (d) of Exercise 1, use the Newton's method program to find all roots with an error at most 10^{-10}.

6. For the functions given in each of parts (a) through (c) of Exercise 2, use the Newton's method program to find all roots with an error at most 10^{-10}. For part (d) find only the two smallest positive roots.

7. Use Newton's method to find all roots of the polynomial $x^4 - 5x^2 + 2$ with each being accurate to about 15 decimal places (MATLAB's precision limit). What is the multiplicity of each of these roots?

8. Use Newton's method to find all roots of the polynomial $x^6 - 4x^4 - 12x^2 + 2$ with each being accurate to about 15 decimal places (MATLAB's precision limit). What is the multiplicity of each of these roots?

9. (*Finance-Retirement Plans*) Suppose a worker puts in PW dollars at the end of each year into a 401(k) (supplemental retirement plan) annuity and does this for NW years (working years). When the worker retires, he would like to withdraw from the annuity a sum of PR dollars at the end of each year for NR years (retirement years). The annuity pays $100r\%$ annual interest compounded annually on the account balance. If the annuity is set up so that at the end of the NR years, the account balance is zero, then the following equation must hold:

$$PW[(1+r)^{NW} - 1] = PR[1 - (1+r)^{-NR}]$$

(see [BaZiBy-02]). The problem is for the worker to decide on what interest rate is needed to fund this retirement scheme. (Of course, other interesting questions arise that involve solving for different parameters, but any of the other variables can be solved for explicitly.) Use Newton's method to solve the problem for each of the following parameters:

(a) $PW = 2,000$, $PR = 10,000$, $NW = 35$, $NR = 25$

(b) $PW = 5,000$, $PR = 20,000$, $NW = 35$, $NR = 25$

(c) $PW = 5,000$, $PR = 80,000$, $NW = 35$, $NR = 25$

(d) $PW = 5,000$, $PR = 20,000$, $NW = 25$, $NR = 25$

10. For which values of the initial approximation x_0 will Newton's method converge to the root $x = 1$ of $f(x) = x^2 - 1$?

11. For which values of the initial approximation x_0 will Newton's method converge to the root $x = 0$ of $f(x) = \sin^2(x)$?

12. For (approximately) which values of the initial approximation $x_0 > 0$ will Newton's method converge to the root of

(a) $f(x) = \ln(x)$

(c) $f(x) = \sqrt[3]{x}$

(b) $f(x) = x^3$

(d) $f(x) = e^x - 1$

13. The following algorithm for calculating the square root of a number $A > 0$ actually was around for many years before Newton's method:

$$x_{n+1} = \frac{1}{2}\left(x_n + \frac{A}{x_n}\right).$$

(a) Run through five iterations of it to calculate $\sqrt{10}$ starting with $x_0 = 3$. What is the error?

(b) Show that this algorithm can be derived from Newton's method.

14.   Consider each of the following two schemes for approximating π:

SCHEME 1: Apply Newton's method to $f(x) = \cos(x) + 1$ with initial approximation $x_0 = 3$.

SCHEME 2: Apply Newton's method to $f(x) = \sin(x)$ with initial approximation $x_0 = 3$.

Discuss the similarities and differences of each of these two schemes. In particular, explain how accurate a result each scheme could yield (working on MATLAB's precision). Finally use one of these two schemes to approximate π with the greatest possible accuracy (using MATLAB).

15.   Prove that if $f(x)$ has a root $x = r$ of multiplicity M then we can write: $f(x) = (x - r)^M h(x)$ for some continuous function $h(x)$.
Suggestion: Try using L'Hôpital's rule.

6.4: THE SECANT METHOD

When conditions are ripe, Newton's method works very nicely and efficiently. Unlike the bisection method, however, it requires computations of the derivative in addition to the function. Geometrically, the derivative was needed to obtain the tangent line at the current approximation x_n (more precisely at $(x_n, f(x_n))$) whose x-intercept was then taken as the next approximation x_{n+1}. If instead of this tangent line, we use the secant line obtained using the current as well as the previous approximation (i.e., the line passing through the points $(x_n, f(x_n))$ and $(x_{n-1}, f(x_{n-1}))$)

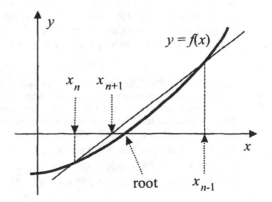

FIGURE 6.10: Illustration of the secant method. Current approximation x_n and previous approximation x_{n-1} are used to obtain a secant line through the graph of $y = f(x)$. The x-intercept of this line will be the next approximation x_{n+1}.

and take the next approximation x_{n+1} to be the x-intercept of this line, we get the so-called **secant method**; see Figure 6.10. Many of the problems that plagued Newton's method can also cause problems for the secant method. In cases where it is inconvenient or expensive to compute derivatives, the secant method is a good replacement for Newton's method. Under certain hypotheses (which were hinted at in the last section) the secant method will converge much faster than the bisection method, although not quite as fast as Newton's method. We will make these comments precise in the next section.

To derive the recursion formula for the secant method we first note that since the three points $(x_{n-1}, f(x_{n-1}))$, $(x_n, f(x_n))$, and $(x_{n+1}, 0)$ all lie on the same (secant) line, we may equate slopes:

$$\frac{0 - f(x_n)}{x_{n+1} - x_n} = \frac{f(x_n) - f(x_{n-1})}{x_n - x_{n-1}}.$$

Solving this equation for x_{n+1} yields the desired recursion formula:

$$x_{n+1} = x_n - f(x_n) \cdot \frac{x_n - x_{n-1}}{f(x_n) - f(x_{n-1})}. \tag{3}$$

Another way to obtain (3) is to replace $f'(x_n)$ in Newton's method (3) with the difference quotient approximation $(f(x_n) - f(x_{n-1}))/(x_n - x_{n-1})$.

EXAMPLE 6.7: We will recompute $\sqrt[4]{2}$ by running through five iterations of the secant method on the function $f(x) = x^4 - 2$ using initial approximations $x_0 = 1.5$ and $x_1 = 1$. Recall in the previous section we had done an analogous computation with Newton's method.

SOLUTION: We can immediately get MATLAB to find the first five iterations along with the residuals and the errors. For convenient display, we store this data in a 5×3 matrix with the first column containing the approximations, the second the residuals, and the third the errors.

```
>> x(1)=1.5; x(2)=1; %initialize, recall zero cannot be an index
>> for n=2:6
x(n+1)=x(n)-f(x(n))*(x(n)-x(n-1))/(f(x(n))-f(x(n-1)));
A(n-1, :) = [ x(n+1)    (x(n+1)^4-2)    abs(x(n+1)-2^(1/4))];
end
>> A
```

Again we display the matrix in tabular form (now in format long).

TABLE 6.5: The successive approximations, residuals, and errors resulting from applying the secant method to $f(x) = x^4 - 2$ with initial approximation $x_0 = 1.5$.

n	x_n	$f(x_n)$	ERROR $= |r - x_n|$
2	1.12307692307692	-0.40911783200868	0.06613019192580
3	1.20829351390874	0.13152179071884	0.01908639890601
4	1.18756281243219	-0.01103858431975	0.00164430257053
5	1.18916801020327	-0.00026305171011	0.00003910479945
6	1.18920719620308	0.00000054624878	0.00000008120036

Compare the results in Table 6.5 with those of Table 6.1, which documented Newton's method for the same problem.

EXERCISE FOR THE READER 6.6: (a) Write a MATLAB M-file called `secant` that has the following input variables: `varfun`, the string representing a mathematical function, `x0` and `x1`, the (different) initial approximations, `tol`, the tolerance, `nmax`, the maximum number of iterations; and output variables: `root`, `yval`, = varfun(root), and `niter`, the number of iterations used. The program should perform the secant method to approximate the root of `varfun(x)` near `x0`, `x1` until either successive approximations differ by less than `tol` or `nmax` iterations have been completed. If the `tol` and `nmax` variables are omitted, default values of `100*eps*max(abs(x0),abs(x1),1)` and 50 are used.
(b) Run your program on the rootfinding problem of Example 6.4. Give the resulting approximation, the residual, the number of iterations used, and the actual (exact) error. Use `x0 = 2` and `x1 = 1.5`.

The secant method uses the current and immediately previous approximations to get a secant line and then shoots the line into the x-axis for the next approximation. A related idea would be to use the current as well as the past two approximations to construct a second-order equation (in general a parabola) and then take as the next approximation the root of this parabola that is closest to the current approximation. This idea is the basis for Muller's method, which is further developed in the exercises of this section. We point out that Muller's method in general will converge a bit quicker than the secant method but not quite as quickly as Newton's method. These facts will be made more precise in the next section.

EXERCISES 6.4:

1. For each of the functions shown below, find the secant method recursion formula (3). Next, using the values of x_0 and x_1 that are given, find each of $x_2, x_3, x_4..$

    (a) $f(x) = x^3 - 2x + 5$; $x_0 = -3, x_1 = -2$

    (b) $f(x) = e^x - 2\cos(x)$; $x_0 = 1, x_1 = 2$

    (c) $f(x) = xe^{-x}$; $x_0 = 0.5, x_1 = 1.0$

    (d) $f(x) = \ln(x^4) - \cos(x)$; $x_0 = 1, x_1 = 1.5$

2. For each of the functions shown below, find the secant method recursion formula (3). Next, using the values of x_0 and x_1 that are given, find each of x_2, x_3, x_4.

    (a) $f(x) = x^3 - 15x^2 + 24$; $x_0 = -3, x_1 = -4$

    (b) $f(x) = e^x - 2e^{-x} + 5$; $x_0 = 1, x_1 = 2$

    (c) $f(x) = \ln(x)$; $x_0 = 0.5, x_1 = 2$

    (d) $f(x) = \sec(x) - 2e^{x^2}$; $x_0 = 1, x_1 = 1.5$

3. Use the secant method to find the smallest positive root of each equation to 12 decimals of accuracy. Indicate the number of iterations used.

   (a) $\tan(x) = x$

   (c) $4x^2 = e^{x/2} - 2$

   (b) $x\cos(x) = 1$

   (d) $(1+x)\ln(1+x^2) = \cos(\sqrt{x})$

   Suggestion: You may wish to modify your secant program so as to have it display another output variable that gives the number of iterations.

4. Use the secant method to find the smallest positive root of each equation to 12 decimals of accuracy. Indicate the number of iterations used.

   (a) $e^{-x} = x$

   (c) $x^4 - 2x^3 + x^2 - 5x + 2 = 0$

   (b) $e^x - x^8 = \ln(1 + 2^x)$

   (d) $e^x = x^\pi$

5. For which values of the initial approximation x_0 will the secant method converge to the root $x = 1$ of $f(x) = x^2 - 1$?

6. For which values of the initial approximation x_0 will the secant method converge to the root $x = 0$ of $f(x) = \sin^2(x)$?

7. For (approximately) which values of the initial approximation $x_0 > 0$ will the secant method converge to the root of

   (a) $f(x) = \ln(x)$

   (c) $f(x) = \sqrt[3]{x}$

   (b) $f(x) = x^3$

   (d) $f(x) = e^x - 1$

NOTE: The next three exercises develop Muller's method, which was briefly described at the end of this section It will be convenient to introduce the following notation for **divided differences** for a continuous function $f(x)$ and distinct points x_1, x_2, x_3 :

$$f[x_1, x_2] \equiv \frac{f(x_2) - f(x_1)}{x_2 - x_1}, \quad f[x_1, x_2, x_3] \equiv \frac{f[x_2, x_3] - f[x_1, x_2]}{x_3 - x_1}.$$

8. Suppose that $f(x)$ is a continuous function and x_0, x_1, x_2 are distinct x-values. Show that the second order polynomial

$$p(x) = f(x_2) + (x - x_2)f[r_2, r_1] + (x - x_2)(x - x_1)f[x_2, x_1, x_0]$$

   is the unique polynomial of degree at most two that passes through the three points: $(x_i, f(x_i))$ for $i = 0, 1, 2$.

   Suggestion: Since $p(x)$ has degree at most two, you need only check that $p(x_i) = f(x_i)$ for $i = 0, 1, 2$. Indeed, if $q(x)$ were another such polynomial then $D(x) \equiv p(x) - q(x)$ would be a polynomial of at most second degree with three roots and this would force $D(x) \equiv 0$ and so $p(x) \equiv q(x)$.

9. (a) Show that the polynomial $p(x)$ of Exercise 8 can be rewritten in the following form:

$$p(x) = f(x_2) + B(x - x_2) + f[x_2, x_1, x_0](x - x_2)^2,$$

   where

$$B = f[x_2, x_1] + (x_2 - x_1)f[x_2, x_1, x_0] = f[x_2, x_1] + f[x_2, x_0] - f[x_0, x_1].$$

   (b) Next, using the quadratic formula show that the roots of $p(x)$ are given by (first thinking of it as a polynomial in the variable $x - x_2$):

$$x - x_2 = \frac{-B \pm \sqrt{B^2 - 4f(x_2)f[x_2, x_1, x_0]}}{2f[x_2, x_1, x_0]}$$

and then show that we can rewrite this as:

$$x = x_2 - \frac{2f(x_2)}{B \pm \sqrt{B^2 - 4f(x_2)f[x_2, x_1, x_0]}} .$$

10. Given the first three approximations for a root of a continuous function $f(x)$: x_0, x_1, x_2, Muller's method will take the next one, x_3, to be that solution in Exercise 9 that is closest to x_2 (the most current approximation). It then continues the process, replacing x_0, x_1, x_2 by x_1, x_2, x_3 to construct the next approximation, x_4.

(a) Show that the latter formula in Exercise 9 is less susceptible to floating point errors than the first one.

(b) Write an M-file, call it `muller`, that will perform Muller's method to find a root. The syntax should be the same as that of the `secant` program in Exercise for the Reader 6.6, except that this one will need three initial approximations in the input rather than 2.

(c) Run your program through six iterations using the function $f(x) = x^4 - 2$ and initial approximations $x_0 = 1, x_1 = 1.5, x_2 = 1.25$ and compare the results and errors with the corresponding ones in Example 6.7 where the secant method was used.

11. Redo Exercise 3 parts (a) through (d), this time using Muller's method as explained in Exercise 10.

12. Redo Exercise 4 parts (a) through (d), this time using Muller's method as explained in Exercise 10.

6.5: ERROR ANALYSIS AND COMPARISON OF ROOTFINDING METHODS

We will shortly show how to accelerate Newton's method in the troublesome cases of multiple roots. This will require us to get into some error analysis. Since some of the details of this section are a bit advanced, we recommend that readers who have not had a course in mathematical analysis simply skim over the section and pass over the technical comments.[7]

The following definition gives us a way to quantify various rates of convergence of rootfinding schemes.

Definition: Suppose that a sequence $\langle x_n \rangle$ converges to a real number r. We say that the **convergence is of order** α (where $\alpha \geq 1$) provided that for some positive number A, the following inequality will eventually be true:

---

[7] Readers who wish to get more comfortable with the notions of mathematical analysis may wish to consult either of the excellent texts [Ros-96] or [Rud-64].

$$|r - x_{n+1}| \le A |r - x_n|^\alpha . \tag{4}$$

For $\alpha = 1$, we need to stipulate $A < 1$ (why?). The word "eventually" here means "for all values of n that are sufficiently large." It is convenient notation to let $e_n = |r - x_n| =$ the error of the nth approximation, which allows us to rewrite (4) in the compact form: $e_{n+1} \le A e_n^\alpha$. For a given sequence with a certain order of convergence, different values of A are certainly possible (indeed, if (4) holds for some number A it will hold also for any bigger number being substituted for A). It is even possible that (4) may hold for all positive numbers A (of course, smaller values of A may require larger starting values of n for validity). In case the greatest lower bound $\hat{A}$ of all such numbers A is positive (i.e., (4) will eventually hold for any number $A > \hat{A}$ but not for numbers $A < \hat{A}$), then we say that $\hat{A}$ is the **asymptotic error constant** of the sequence. In particular, this means there will always be arbitrarily large values of n, for which the error of the $(n+1)$st term is *essentially* proportional to the αth power of that of the nth term and the proportionality constant is approximately $\hat{A}$:

$$|r - x_{n+1}| \approx \hat{A} |r - x_n|^\alpha \quad \text{or} \quad e_{n+1} \approx \hat{A} e_n^\alpha . \tag{5}$$

The word "essentially" here means that we can get the ratios e_{n+1}/e_n^α as close to $\hat{A}$ as desired. In the notation of mathematical analysis, we can paraphrase the definition of $\hat{A}$ by the formula:

$$\hat{A} \equiv \limsup_{n \to \infty} e_{n+1}/e_n^\alpha$$

provided that this limsup is positive. In the remaining case that this greatest lower bound of the numbers A is zero, the asymptotic error constant is undefined (making $\hat{A} = 0$ may seem reasonable but it is not a good idea since it would make (5) fail), but we say that we have **hyperconvergence of order** α.

When $\alpha = 1$, we say there is **linear convergence** and when $\alpha = 2$ we say there is **quadratic convergence**. In general, higher values of α result in speedier convergences and for a given α; smaller values of A result in faster convergences. As an example, suppose $e_n = 0.001 = 1/1000$. In case $\alpha = 1$ and $\hat{A} = 1/2$ we have (approximately and for arbitrarily large indices n) $e_{n+1} \approx 0.0005 = 1/2000$, while if $A = 1/4$, $e_{n+1} \approx 0.00025 = 1/4000$. If $\alpha = 2$ even for $A = 1$ we would have $e_{n+1} \approx (0.001)^2 = 0.000001$ and for $A = 1/4$, $e_{n+1} \approx (1/4)(0.001)^2 = 0.00000025$.

EXERCISE FOR THE READER 6.7: This exercise will give the reader a feel for the various rates of convergence.

(a) Find (if possible) the highest order of convergence of each of the following sequences that have limit equal to zero. For each find also the asymptotic error constant whenever it is defined or whether there is hyper convergence for this order

(i) $e_n = 1/(n+1)$, (ii) $e_n = 2^{-n}$, (iii) $e_n = 10^{-3^n/2^n}$, (iv) $e_n = 10^{-2^n}$, (v) $e_n = 2^{-2^n - n}$

(b) Give an example of a sequence of errors $\langle e_n \rangle$ where the convergence to zero is of order 3.

One point that we want to get across now is that quadratic convergence is extremely fast. We will show that under certain hypotheses, the approximations in Newton's method will converge quadratically to a root whereas those of the bisection method will in general converge only linearly. If we use the secant method the convergence will in general be of order $(1+\sqrt{5})/2 = 1.62....$ We now state these results more precisely in the following theorem.

THEOREM 6.1: (*Convergence Rates for Rootfinding Programs*) Suppose that one of the three methods, bisection, Newton's, or the secant method, is used to produce a sequence $\langle x_n \rangle$ that converges to a root r of a continuous function $f(x)$.

PART A: If the bisection method is used, the convergence is *essentially linear* with constant 1/2. This means that there exist positive numbers $e_n' \ge e_n = |x_n - r|$ that (eventually) satisfy $e_{n+1}' \le (1/2)e_n'$.

PART B: If Newton's method is used and the root r is a simple root and if $f''(x)$ is continuous near the root $x = r$, then the convergence is quadratic with asymptotic error constant $|f''(r)/(2f'(r))|$, except when $f''(r)$ is zero, in which case we have hyperquadratic convergence. But if the root $x = r$ is a multiple root of order M, then the convergence is only linear with asymptotic error constant $A = (M-1)/M$.

PART C: If the secant method is used and if again $f''(x)$ is continuous near the root $x = r$ and the root is simple, then the convergence will be of order $(1+\sqrt{5})/2 = 1.62...$.

Furthermore, the bisection method will always converge as long as the initial approximation x_0 is taken to be within a bracket of $x = r$. Also, under the additional hypothesis that $f''(x)$ is continuous near the root $x = r$, both Newton's and the secant method will always converge to $x = r$, as long as the initial approximation(s) are sufficiently close to $x = r$.

REMARKS: This theorem has a lot of practical use. We essentially already knew what is mentioned about the bisection method. But for Newton's and the secant method, the last statement tells us that as long as we start our approximations "sufficiently close" to a root $x = r$ that we are seeking to approximate, the methods will produce sequences of approximations that converge to $x = r$. The "sufficiently close" requirement is admittedly a bit vague, but at least we can keep retrying initial approximations until we get one that will produce a "good" sequence of approximations. The theorem also tells us that once we get any sequence (from Newton's or the secant method) that converges to a root $x = r$, it will be one that converges at the stated orders. Thus, any initial approximation that produces a sequence converging to a root will produce a great sequence of approximations. A similar analysis of Muller's method will show that if it converges to a simple root, it will do so with order 1.84..., a rate quite halfway between that of the secant and Newton's methods. In general, one cannot say that the bisection method satisfies the definition of linear convergence. The reason for this is that it is possible for some x_n to be coincidentally very close to the root while x_{n+1} is much farther from it (see Exercise 14).

Sketch of Proof of Theorem 6.1: A few details of the proof are quite technical so we will reference out parts of the proof to a more advanced text in numerical analysis. But we will give enough of the proof so as to give the reader a good understanding of what is going on. The main ingredient of the proof is Taylor's theorem. We have already seen the importance of Taylor's theorem as a practical tool for approximations; this proof will demonstrate also its power as a theoretical tool. As mentioned in the remarks above, all statements pertaining to the bisection method easily follow from previous developments. The error analysis estimate (1) makes most of these comments transparent. This estimate implies that $e_n = |r - x_n|$ is at most $(b-a)/2^n$, where $b-a$ is simply the length of the bracket interval $[a, b]$. Sometimes we might get lucky since e_n could conceivably be a lot smaller, but in general we can only guarantee this upper bound for e_n. If we set $e_n' = (b-a)/2^n$, then we easily see that $e_{n+1}' = (1/2)e_n'$ and we obtain the said order of convergence in part A.

The proofs of parts B and C are more difficult. A good approach is to use Taylor's theorem. Let us first deal with the case of a simple root and first with part B (Newton's method).

Since x_n converges to $x = r$, Taylor's theorem allows us to write:

$$f(r) = f(x_n) + f'(x_n)(r - x_n) + \tfrac{1}{2}f''(c_n)(r - x_n)^2,$$

where c_n is a number between $x = r$ and $x = x_n$ (as long as n is large enough for f'' to be continuous between the $x = r$ and $x = x_n$).

The hypotheses imply that $f'(x)$ is nonzero for x near $x = r$. (Reason: $f'(r) \neq 0$ because $x = r$ is a simple root. Since $f'(x)$ (and $f''(x)$) are continuous near $x = r$ we also have $f'(x) \neq 0$ for x close enough to $x = r$.) Thus we can divide both sides of the previous equation by $f'(x_n)$ and since $f(r) = 0$ (remember $x = r$ is a root of $f(x)$), this leads us to:

$$0 = \frac{f(x_n)}{f'(x_n)} + r - x_n + \frac{1}{2} \frac{f''(c_n)}{f'(x_n)} (r - x_n)^2.$$

But from Newton's recursion formula (2) we see that the first term on the right of this equation is just $x_n - x_{n+1}$ and consequently

$$0 = x_n - x_{n+1} + r - x_n + \frac{1}{2} \frac{f''(c_n)}{f'(x_n)} (r - x_n)^2.$$

We cancel the x_n s and then can rewrite the equation as

$$r - x_{n+1} = \frac{-f''(c_n)}{2f'(x_n)} (r - x_n)^2.$$

We take absolute values in this equation to get the desired proportionality relationship of errors:

$$e_{n+1} = \left| \frac{f''(c_n)}{2f'(x_n)} \right| e_n^2. \tag{6}$$

Since the functions f' and f'' are continuous near $x = r$ and x_n (and hence also c_n) converges to r, the statements about the asymptotic error constants or the hyperquadratic convergence now easily follow from (6).

Moving on to part C (still in the case of a simple root), a similar argument to the above leads us to the following corresponding proportionality relationship for errors in the secant method (see Section 2.3 of [Atk-89] for the details):

$$e_{n+1} \approx e_n e_{n-1} \left| \frac{f''(r)}{2f'(r)} \right|. \tag{7}$$

It will be an exact inequality if r is replaced by certain numbers near $x = r$ —as in (6). In (7) we have assumed that $f''(r) \neq 0$. The special case where this second derivative is zero will produce a faster converging sequence, so we deal only with the worse (more general) general case, and the estimates we get here will certainly apply all the more to this special remaining case. From (7) we can actually deduce the precise order $\alpha > 0$ of convergence. To do this we temporarily define the proportionality ratios A_n by the equations $e_{n+1} = A_n e_n^{\alpha}$ (cf., equation (5)).

We can now write:

$$e_{n+1} = A_n e_n^{\alpha} = A_n (A_{n-1} e_{n-1}^{\alpha})^{\alpha} = A_n A_{n-1}^{\alpha} e_{n-1}^{\alpha^2},$$

which gives us that

$$\frac{e_{n+1}}{e_n e_{n-1}} = \frac{A_n A_{n-1}^{\alpha} e_{n-1}^{\alpha^2}}{A_{n-1} e_{n-1}^{\alpha} e_{n-1}} = A_n A_{n-1}^{\alpha-1} e_{n-1}^{\alpha^2-\alpha-1}.$$

Now, as $n \to \infty$, (7) shows the left side of the above equation tends to some positive number. On the right side, however, since $e_n \to 0$, assuming that the A_n's do not get too large (see Section 2.3 of [Atk-89] for a justification of this assumption) this forces the exponent $\alpha^2 - \alpha - 1 = 0$. This equation has only one positive solution, namely $\alpha = (1 + \sqrt{5})/2$.

Next we move on to the case of a multiple root of multiplicity M. We can write $f(x) = (x - r)^M h(x)$ where $h(x)$ is a continuous function with $h(r) \neq 0$. We additionally assume that $h(x)$ is sufficiently differentiable. In particular, we will have $f'(x) = M(x - r)^{M-1} h(x) + (x - r)^M h'(x)$, and so we can rewrite Newton's recursion formula (2) as:

$$x_{n+1} = x_n - \frac{(x_n - r)h(x_n)}{Mh(x_n) + (x_n - r)h'(x_n)} = g(x_n),$$

where

$$g(x) = x - \frac{(x - r)h(x)}{Mh(x) + (x - r)h'(x)}.$$

Since

$$g'(x) = 1 - \frac{h(x)}{Mh(x) + (x - r)h'(x)} - (x - r)\left(\frac{h(x)}{Mh(x) + (x - r)h'(x)}\right)',$$

we have that

$$g'(r) = 1 - \frac{1}{M} = (M - 1)/M > 0$$

(since $M \geq 2$). Taylor's theorem now gives us that

$$x_{n+1} = g(x_n) = g(r) + g'(r)(x_n - r) + g''(c_n)(x_n - r)^2/2$$
$$= r + g'(r)(x_n - r) + g''(c_n)(x_n - r)^2/2$$

(where c_n is a number between r and x_n). We can rewrite this as:

$$e_{n+1} = [(M - 1)/M]e_n + g''(c_n)e_n^2/2.$$

Since $e_n \to 0$ (assuming g'' is continuous at $x = r$), we can divide both sides of this inequality by e_n to get the asserted linear convergence. For details on this latter convergence and also for proofs of the actual convergence guarantee (we have only given sketches of the proofs of the rates of convergence assuming the approximations converge to a root), we refer the interested reader to Chapter 2 of [Atk-89]. QED

From the last part of this proof, it is apparent that in the case we are using Newton's method for a multiple root of order $M > 1$, it would be a better plan to use the modified recursion formula:

$$x_{n+1} = x_n - M \frac{f(x_n)}{f'(x_n)}. \tag{8}$$

Indeed, the proof above shows that with this modification, when applied to an order-M multiple root, Newton's method will again converge quadratically (Exercise 13). This formula, of course, requires knowledge about M.

EXAMPLE 6.8: Modify the Program 6.2, `newton`, into a more general one, call it `newtonmr` that is able to effectively use (8) in cases of multiple roots. Have your program run with the function $f(x) = x^{21}$ again with initial approximation $x = 1$, as was done in Example 6.6 with the ordinary `newton` program.

SOLUTION: We indicate only the changes needed to be made to `newton` to get the new program. We will need one new variable (a sixth one), call it `rootord`, that denotes the order of the root being sought after. In the first if-branch of the program (`if nargin < 4`) we also add the default value `rootord =1`. The only other change needed will be to replace the analogue of (2) in the program with that of (8). If we now run this new program, we will find the exact root $x = 0$. In fact, as you can check with (8) (or by slightly modifying the program to give as output the number of iterations), it takes only a single iteration. Recall from Example 6.6 that if we used the ordinary Newton's method for this function and initial approximation, we would need about 135 iterations to get an approximation (of zero) with error less than 0.001!

In order to effectively use this modified Newton's method for multiple roots, it is necessary to determine the order of a multiple root. One way to do this would be to compare the graphs of the function at hand near the root r in question, together with graphs of successive derivatives of the function, until it is observed that a certain order derivative no longer has a root at r; see also Exercise 15. The order of this derivative will be the order of the root. MATLAB can compute derivatives (and indefinite integrals) provided that you have the Student Version, or the package you have installed includes the Symbolic Math Toolbox; see Appendix A.

For polynomials, MATLAB has a built-in function, `roots`, that will compute all of its roots. Recall that a polynomial of degree n will have exactly n real or complex roots, if we count them according to their multiplicities. Here is the syntax of the `roots` command:

roots([an ... a2 a1 a0]) →	Computes (numerically) all of the n real and complex roots of the polynomial whose coefficients are given by the inputted vector: $$p(x) = a_n x^n + a_{n-1} x^{n-1} + \cdots + a_2 x^2 + a_1 x + a_0.$$

EXAMPLE 6.9: Use MATLAB to find all of the roots of the polynomials

$$p(x) = x^8 - 3x^7 + (9/4)x^6 - 3x^5 + (5/2)x^4 + 3x^3 + (9/4)x^2 + 3x + 1,$$
$$q(x) = x^6 + 2x^5 - 6x^4 - 10x^3 + 13x^2 + 12x - 12.$$

SOLUTION: Let us first store the coefficients of each polynomial as a vector:

```
>>pv=[1 -3  9/4 -3  5/2  3  9/4  3  1]; qv= [1 2  -6  -10 13 12 -12];
>> roots(pv) %this single command will get us all of the roots of
p(x)
→  2.0000 + 0.0000i
   2.0000 - 0.0000i
   0.0000 + 1.0000i
   0.0000 - 1.0000i
  -0.0000 + 1.0000i
  -0.0000 - 1.0000i
  -0.5000 + 0.0000i
  -0.5000 - 0.0000i
```

Since some of these roots are complex, they are all listed as complex numbers. The distinct roots are $x = 2$, $i - i$, and $.5$, each of which are double roots. Since $(x+i)(x-i) = x^2 + 1$ these roots allow us to rewrite $p(x)$ in factored form: $p(x) = (x^2 + 1)^2 (x - 2)^2 (x + 0.5)^2$. The roots of $q(x)$ are similarly obtained·

```
>> roots(qv)
→  1.7321
  -2.0000
  -2.0000
  -1.7321
   1.0000
   1.0000
```

Since the roots of $q(x)$ are all real, they are written as real numbers. We see that $q(x)$ has two double roots, $x = -2$ and $x = 1$, and two simple roots that turn out to be $\pm\sqrt{3}$.

EXERCISE FOR THE READER 6.8: (*Another Approach to Multiple Roots with Newton's Method*). Suppose that $f(x)$ has multiple roots. Show that the function $f(x)/f'(x)$ has the same roots as $f(x)$, but they are all simple. Thus Newton's method could be applied to the latter function with quadratic convergence to determine each of the roots of $f(x)$. What are some problems that could crop up with this approach?

EXERCISES 6.5:

1. Find the highest order of convergence (if defined) of each of the following sequences of errors:

   (a) $e_n = 1/n^5$

   (b) $e_n = e^{-n}$

   (c) $e_n = n^{-n}$

   (d) $e_n = 2^{-n^2}$

2. Find the highest order of convergence (if defined) of each of the following sequences of errors:

   (a) $e_n = 1/\ln(n)^n$

   (b) $e_n = 1/\exp(\exp(n))$

   (c) $e_n = 1/\exp(\exp(\exp(n)))$

   (d) $e_n = 1/n!$

3. For each of the sequences of Exercise 1 that had a well-defined highest order of convergence, determine the asymptotic error constant or indicate if there is hyperconvergence.

4. For each of the sequences of Exercise 2 that had a well-defined highest order of convergence, determine the asymptotic error constant or indicate if there is hyperconvergence.

5. Using just Newton's method or the improvement (8) of it for multiple roots, determine all (real) roots of the polynomial

$$x^8 + 4x^7 - 17x^6 - 84x^5 + 60x^4 + 576x^3 + 252x^2 - 1296x - 1296.$$

Give also the multiplicity of each root and justify these numbers.

6. Using just Newton's method or the improvement (8) of it for multiple roots, determine all (real) roots of the polynomial $x^{10} + x^9 + x^8 - 18x^6 - 18x^5 - 18x^4 + 81x^2 + 81x + 81$. Give also the multiplicity of each root and justify these numbers.

7. (*Fixed Point Iteration*) (a) Assume that $f(x)$ has a root in $[a, b]$, that $g(x) = x - f(x)$ satisfies $a \le g(x) \le b$ for all x in $[a, b]$ and that $|g'(x)| \le \lambda < 1$ for all x in $[a, b]$. Show that the following simple iteration scheme: $x_{n+1} = g(x_n)$, will produce a sequence that converges to a root of $f(x)$ in $[a, b]$.

   (b) Show that $f(x)$ has a unique root in $[a, b]$, provided that all of the hypotheses in part (a) are satisfied.

8. The following algorithm computes the square root of a number $A > 0$:

$$x_{n+1} = \frac{x_n(x_n^2 + 3A)}{3x_n^2 + A}.$$

   (a) Show that it has order of convergence equal to 3 (assuming x_0 has been chosen sufficiently close to $\sqrt{A}$).

   (b) Perform three iterations of it to calculate $\sqrt{10}$ starting with $x_0 = 3$. What is the error?

   (c) Compute the asymptotic error constant.

9. Can you devise a scheme for computing cube roots of positive numbers that, like the one in Exercise 8, has order of convergence equal to 3? If you find one, test it out on $\sqrt[3]{30}$.

10. Prove: If $\beta > \alpha$ and we have a sequence that converges with order β, then the sequence will also converge with order α.

11. Is it possible to have quadratic convergence with asymptotic error constant equal to 3? Either provide an example or explain why not.

12. Prove formula (7) in the proof of Theorem 6.1, in case $f''(r) \neq 0$.

13. Give a careful explanation of how (8) gives quadratic convergence in the case of a root of order $M > 1$, provided that x_0 is sufficiently close to the root.
    Suggestion: Carefully examine the last part of the proof of Theorem 6.1.

14. (*Nonlinear Convergence of the Bisection Method*) (a) Construct a function $f(x)$ that has a root r in an interval $[a, b]$ and that satisfies the requirements of the bisection method but such that x_n does not converge linearly to r.
    (b) Is it possible to have $\limsup_{n \to \infty} e_{n+1} / e_n = \infty$ with the bisection method for a function that satisfies the conditions of part (a)?

15. (a) Explain how Newton's method could be used to detect the order of a root, and then formulate and prove a precise result.
    (b) Use the idea of part (a) to write a MATLAB M-file, newtonorddetect, having a similar syntax to the newton M-file of Program 6.2. Your program should first detect the order of the root, and then use formula (8) (modified Newton's method) to approximate the root. Run your program on several examples involving roots of order 1, 2, 3, and compare the number of iterations used with that of the ordinary Newton's method. In your comparisons, make sure to count the total number of iterations used by newtonorddetect, both in the detection process as well as in the final implementation.
    (c) Run your program of part (b) on the problem of Example 6.8.
    Note: For the last comparisons asked for in part (b), you should modify newton to output the number of iterations used, and include such an output variable in your newtonorddetect program.

Chapter 7: Matrices and Linear Systems

7.1: MATRIX COMPUTATIONS AND MANIPULATIONS WITH MATLAB

I saw my first matrix in my first linear algebra course as an undergraduate, which came after the calculus sequence. A matrix is really just a spreadsheet of numbers, and as computers are having an increasing impact on present-day life and education, the importance of matrices is becoming paramount. Many interesting and important problems can be solved using matrices, and the basic concepts for matrices are quite easy to introduce. Presently, matrices are making their way down into lower levels of mathematics courses and, in some instances, even elementary school curricula. Matrix operations and calculations are simple in principle but in practice they can get quite long. It is often not feasible to perform such calculations by hand except in special situations. Computers, on the other hand, are ideally suited to manipulate matrices and MATLAB has been specially designed to effectively manipulate them. In this section we introduce the basic matrix operations and show how to perform them in MATLAB. We will also present some of the numerous tricks, features, and ideas that can be used in MATLAB to store and edit matrices. In Section 7.2 we present some applications of basic matrix operations to computer graphics and animation. The very brief Section 7.3 introduces concepts related to linear systems and Section 7.4 shows ways to use MATLAB to solve linear systems. Section 7.5 presents an algorithmic and theoretical development of Gaussian elimination and related concepts. In Section 7.6, we introduce norms with the goal of developing some error estimates for numerical solutions of linear systems. Section 7.7 introduces iterative methods for solving linear systems. When conditions are ripe, iterative methods can perform much more quickly and effectively than Gaussian elimination.

We first introduce some definitions and notations from mathematics and then translate them into MATLAB's language. A **matrix** A is a rectangular array of numbers, called *entries*. A generic form for a matrix is shown in Figure 7.1. The **rows** of a matrix are its horizontal strips of entries (labeled from the top) and the **columns** are the vertical strips of entries (labeled from the left). The entry of A that lies in row i and in column j is written as a_{ij} (if either of the indices i or j is greater than a single digit, then the notation inserts a comma: $a_{i,j}$). The matrix A is said to be of size n by m (or an $n \times m$ matrix) if it has n rows and m columns.

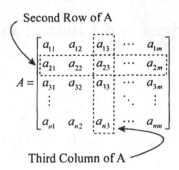

FIGURE 7.1: The anatomy of a matrix A having n rows and m columns. The entry that lies in the second row and the third column is written as a_{23}.

In mathematical notation, the matrix A in Figure 7.1 is sometimes written in the abbreviated form

$$A = [a_{ij}],$$

where its size either is understood from the context or is unimportant. With this notation it is easy to explain how **matrix addition/subtraction** and **scalar multiplication** work. The matrix A can be multiplied by any scalar (real number) α with the obvious definition:

$$\alpha A = [\alpha a_{ij}].$$

Matrices can be added/subtracted only when they have the same size, and the definition is the obvious one: Corresponding entries get added/subtracted, i.e.,

$$[a_{ij}] \pm [b_{ij}] = [a_{ij} \pm b_{ij}].$$

Matrix multiplication, however, is not done in the obvious way. To explain it we first recall the definition of the **dot product** of two vectors of the same size. If $a = [a_1, a_2, \cdots, a_n]$ and $b = [b_1, b_2, \cdots, b_n]$ are two vectors of the same length n (for this definition it does not matter if these vectors are row or column vectors), the dot product of these vectors is defined as follows:

$$a \cdot b = a_1 b_1 + a_2 b_2 + \cdots + a_n b_n = \sum_{k=1}^{n} a_k b_k.$$

Now, if $A = [a_{ij}]$ is an $n \times m$ matrix and $B = [b_{ij}]$ is an $m \times r$ matrix (i.e., the number of columns of A equals the number of rows of B), then the matrix product $C = AB$ is defined and it will be another matrix $C = [c_{ij}]$ of size $n \times r$. To get an entry c_{ij}, we simply take the dot product of the ith row vector of A and the jth column vector of B. Here is a simple example:

EXAMPLE 7.1: Given the matrices:

$$A = \begin{bmatrix} 1 & 0 \\ -3 & 8 \end{bmatrix}, \ B = \begin{bmatrix} 4 & 1 \\ 3 & -6 \end{bmatrix}, \ M = \begin{bmatrix} 4 & -9 \\ 3 & 1 \\ -2 & 5 \end{bmatrix}, \text{ compute the following: } A - 2B,$$

$AB, \ BA, \ AM, \ MA.$

SOLUTION: $A - 2B = \begin{bmatrix} 1 & 0 \\ -3 & 8 \end{bmatrix} - \begin{bmatrix} 8 & 2 \\ 6 & -12 \end{bmatrix} = \begin{bmatrix} -7 & -2 \\ -9 & 20 \end{bmatrix},$

$$AB = \begin{bmatrix} 1 & 0 \\ \boxed{-3 \quad 8} \end{bmatrix} \begin{bmatrix} \boxed{4} & 1 \\ \boxed{3} & -6 \end{bmatrix} = \begin{bmatrix} 4 & 1 \\ \boxed{12} & -51 \end{bmatrix}.$$

(In the product AB, the indicated second row, first column entry was computed by taking the dot product of the corresponding second row of A (shown) with the first column of B (also indicated).) Similarly,

$$BA = \begin{bmatrix} 1 & 8 \\ 21 & -48 \end{bmatrix}, \text{ and } MA = \begin{bmatrix} 31 & -72 \\ 0 & 8 \\ -17 & 40 \end{bmatrix}.$$

The matrix product AM is not defined since the inner dimensions of the sizes of A and M are not the same (i.e., $2 \neq 3$). In particular, these examples show that matrix multiplication is, in general, not commutative; i.e., we cannot say that $AB = BA$ even when both matrix products are possible and have the same sizes.

At first glance, the definition of matrix multiplication might seem too awkward to have much use. But later in this chapter (and in subsequent chapters as well) we will see numerous applications. We digress momentarily to translate these concepts and notations into MATLAB's language. To redo the above example in MATLAB, we would simply enter the matrices (rows separated by semicolons, as shown in Chapter 1) and perform the indicated linear combination and multiplications (since default operations are matrix operations in MATLAB, dots are not used before multiplications and other operators).

MATLAB SOLUTION TO EXAMPLE 7.1:
```
>>   A=[1 0; -3 8]; B=[4 1;3 -6]; M=[4 -9;3 1;-2 5];
>>   A-2*B, A*B, B*A, A*M, M*A
→ans = -7  -2     ans =   4  1     ans =  1   8
       -9  20            12 -51           21 -48
→??? Error using ==> *
Inner matrix dimensions must agree.
→ans = 31  -72
        0   8
      -17  40
```

By hand, matrix multiplication is feasible only for matrices of very small sizes. For example, if we multiply an $n \times n$ (square) matrix by another one of the same size, each entry involves a dot product of two vectors with n components and thus will require n multiplications and $n - 1$ additions (or subtractions). Since there are n^2 entries to compute, this yields a total of $n^2(n+n-1) = 2n^3 - n^2$ floating point operations (flops). For a 5×5 matrix multiplication, this works out to be 225 flops and for a 7×7, this works already to 637 flops, a very tedious hand calculation pushing on the fringes of insanity. As we saw in some experiments in Chapter 4, on a PC with 256 MB of RAM and a 1.6 GHz microprocessor, MATLAB can roughly perform approximately 10 million flops in a few seconds. This means that (replacing the previous bound $2n^3 - n^2$ by the more liberal but easier to deal with bound $2n^3$, setting this equal to 10 million and solving for n) MATLAB can quickly multiply two matrices of size up to about 171×171 (check that a matrix multiplication of two such large matrices works out to about 10 million flops). Actually, not all flop calculations take equal times (this is why the word "flop" has a somewhat vague definition). It turns out that because of MATLAB's specially designed features that are mainly tailored to deal effectively with matrices, MATLAB can actually quickly multiply much larger matrices. On a PC with the aforementioned specs, the following experiment took MATLAB just a few seconds to multiply a pair of 1000×1000 matrices.[1]

```
>> flops(0)
>> A=rand(1000); B=rand(1000); %constructs two 1000 by 1000 random
                               matrices
>> A*B;
>> flops      → 2.0000e + 009
```

This calculation involved 2 billion flops! Later on we will come across applications where such large-scale matrix multiplications come up naturally. Of course, if one (or both) matrices being multiplied have a lot of zero entries, then the computations needed can be greatly reduced. Such matrices (having a very low percentage of nonzero entries) are called **sparse**, and we will later see some special features that MATLAB has to deal with sparse matrices.

We next move on to introduce several of the handy ways that MATLAB offers us to enter and manipulate matrices. For illustration, we assume that the matrices A, B and M of Example 7.1 are still entered in our MATLAB session. The exponential operator (^) in MATLAB works by default as a matrix power. Thus if we enter

```
>> A^2  %matrix squaring
```

we get the square of the matrix A, $A^2 = AA$,

```
→ans = 1    0
       -27  64
```

---

whereas if we precede the operator by a dot, then, as usual, the operator changes to its entry-by-entry analog, and we get the matrix whose entries each equal the square of the corresponding entries of A.

```
>> A.^2  %element squaring
→ans = 1   0
        9  64
```

This works the same way with other operators such as multiplication. Matrix operators such as addition/subtraction, which are the same as element-by-element addition/subtraction, make the dot a nonissue.

To refer to a specific entry a_{ij} in a matrix A, we use the syntax:

A(i,j) →	MATLAB's way of referring to the ith row jth column entry a_{ij} of a matrix A,

which was introduced in Section 4.3. Thus, for example, if we wanted to change the row 1, column 2 entry of our 2×2 matrix A (from 0) to 2, we could enter:

```
>> A(1,2)=2  %without suppressing output, MATLAB shows us the whole
             matrix
→A =  1   2
     -3   8
```

We say that a (square) matrix $D = [d_{ij}]$ is a **diagonal matrix** if all entries, except perhaps those on the main diagonal, are zero (i.e., $d_{ij} = 0$ whenever $i \neq j$). Diagonal matrices (of the same size $n \times n$) are the easiest ones to multiply; indeed, for such a multiplication only n flops are needed:

$$\begin{bmatrix} d_1 & & & \\ & d_2 & & \mathbf{0} \\ & & \ddots & \\ \mathbf{0} & & & d_n \end{bmatrix} \begin{bmatrix} e_1 & & & \\ & e_2 & & \mathbf{0} \\ & & \ddots & \\ \mathbf{0} & & & e_n \end{bmatrix} = \begin{bmatrix} d_1 e_1 & & & \\ & d_2 e_2 & & \mathbf{0} \\ & & \ddots & \\ \mathbf{0} & & & d_n e_n \end{bmatrix}. \tag{1}$$

The large zeros in the above notation are to indicate that all entries in the triangular regions above and below the main diagonal are zeros. There are many matrix-related problems and theorems where things boil down to considerations of diagonal matrices, or minor variations of them.

EXERCISE FOR THE READER 7.1: Prove identity (1).

In MATLAB, we can enter a diagonal matrix using the command `diag` as follows. To create a 5×5 diagonal matrix D with diagonal entries (in order): 1 2 −3 4 5, we could type:

```
>> diag([1 2 -3 4 5])      →ans = 1   0   0   0   0
                                  0   2   0   0   0
```

$$
\begin{array}{ccccc}
0 & 0 & -3 & 0 & 0 \\
0 & 0 & 0 & 4 & 0 \\
0 & 0 & 0 & 0 & 5
\end{array}
$$

One very special diagonal matrix is $n \times n$ (square) **identity matrix** I_n or simply I (if the size is understood or unimportant). It has all the diagonal entries equaling 1 and has the property that whenever it is multiplied (on either side) by a matrix A (so that the multiplication is possible), the product is A, i.e.,

$$
AI = A = IA . \tag{2}
$$

Thus, the identity matrix I plays the role in matrix theory that the number 1 plays in arithmetic; i.e., it is the "(multiplicative) identity." Even easier than with the diag command, we can create identity matrices with the command eye:

```
>> I2=eye(2), I4=eye(4)
→I2 = 1   0       →I4 = 1   0   0   0
       0   1              0   1   0   0
                         0   0   1   0
                         0   0   0   1
```

Let us check identity (2) for our stored 2×2 matrix $A = \begin{bmatrix} 1 & 2 \\ -3 & 8 \end{bmatrix}$:

```
>> I2*A,   A*I2
→ans = 1   2     →ans = 1   2
       -3   8            -3   8
```

To be able to divide one matrix A by another one B, we will actually have to multiply A by the inverse B^{-1} of the matrix B, if the latter exists and the multiplication is possible. It is helpful to think of the analogy with real numbers: To perform a division, say $5 \div 2$, we can recast this as a multiplication $5 \cdot 2^{-1}$, where the inverse of 2 (as with any nonzero real number) is the reciprocal 1/2. The only real number that does not have an inverse is zero; thus we can always divide any real number by any other real number as long as the latter is not zero. Note that the inverse a^{-1} of a real number a has the property that when the two are multiplied the result will be 1 ($a \cdot a^{-1} = 1 = a^{-1} \cdot a$). To translate this concept into matrix theory is simple; since the number 1 translates to the identity matrix, we see that for a matrix B^{-1} to be the **inverse of a matrix** B, we must have

$$
BB^{-1} = I = B^{-1}B. . \tag{3}
$$

In this case we also say that the matrix B is **invertible** (or **nonsingular**). The only way for the equations of (3) to be possible is if B is a square matrix. There are, however, a lot of square matrices that are not invertible. One way to tell whether a square matrix B is invertible is by looking at its determinant $\det(B)$ (which was introduced in Section 4.3), as the following theorem shows:

THEOREM 7.1: (*Invertibility of Square Matrices*)
(1) A square matrix B is invertible exactly when its determinant $\det(B)$ is nonzero.

(2) In case of a 2×2 matrix $B = \begin{bmatrix} a & b \\ c & d \end{bmatrix}$ with determinant $\det(B) \equiv ad - bc \neq 0$, the inverse is given by the formula

$$B^{-1} = \frac{1}{\det(B)} \begin{bmatrix} d & -b \\ -c & a \end{bmatrix}.$$

For a proof of this theorem we refer to any good linear algebra textbook (for example [HoKu-71]). There is an algorithm for computing the inverse of a matrix, which we will briefly discuss later, and there are more complicated formulas for the inverse of a general $n \times n$ matrix, but we will not need to go so far in this direction since MATLAB has some nice built-in functions for finding determinants and inverses. They are as follows:

`inv(A)` →	Numerically computes the inverse of a square matrix A
`det(A)` →	Computes the determinant of a square matrix A

The `inv` command must be used with caution, as the following simple examples show. From the theorem, the matrix $M = \begin{bmatrix} 2 & 3 \\ 1 & 2 \end{bmatrix}$ is easily inverted, and MATLAB confirms the result:

```
>> M=[2 3; 1 2];
>> inv(M)
```

→ ans = 2   -3
          -1   2

However, the matrix $M = \begin{bmatrix} 3 & -6 \\ 2 & -4 \end{bmatrix}$ has $\det(M) = 0$, so from the theorem we know that the inverse does not exist. If we try to get MATLAB to compute this inverse, we get the following:

```
>> M=[3 -6; 2 -4];
>> inv(M)
```
→Warning: Matrix is singular to working precision.
ans = Inf Inf
       Inf Inf

The output does not actually tell us that the matrix is not invertible, but it gives us a meaningless answer (Inf is MATLAB's way of writing ∞) that seems to suggest that there is no inverse. This brings us to a subtle and important point about floating point arithmetic. Since MATLAB, or any computer system, can work only with a finite number of digits, it is not really possible for MATLAB to distinguish between zero and a very small positive or negative number. Furthermore, when doing computations (e.g., in finding an inverse of a (large) matrix,) there are (a lot of) calculations that must be performed and these will introduce roundoff errors. Because of this, something that is actually zero may appear as a nonzero but small number and vice versa (especially after the "noise" of calculations has distorted it). Because of this it is in general impossible to tell if a certain matrix is invertible or not if its determinant is very small. Here is some practical advice on computing inverses. If you get MATLAB to compute an inverse of a square matrix and get a "warning" as above, you should probably reject the output. If you then check the determinant of the matrix, chances are good that it will be very small. Later in this chapter we will introduce the concept of condition numbers for matrices and these will provide a more reliable way to detect so-called **poorly conditioned matrices** that are problematic in linear systems.

Building and storing matrices with MATLAB can be an art. Apart from eye and diag that were already introduced, MATLAB has numerous commands for the construction of special matrices. Two such commonly used commands are ones and zeros.

zeros(n,m) →	Constructs an $n \times m$ matrix whose entries each equal 0.
ones(n,m) →	Constructs an $n \times m$ matrix whose entries each equal 1.

Of course, zeros(n,m) is redundant since we can just use 0*ones(n,m) in its place. But matrices of zeros come up often enough to justify separate mention of this command.

EXAMPLE 7.2: A **tridiagonal** matrix is one whose nonzero entries can lie either on the main diagonal or on the diagonals directly above/below the main diagonal. Consider the 60×60 tridiagonal matrix A shown below:

$$A = \begin{bmatrix} 1 & 1 & 0 & 0 & 0 & 0 & 0 & \cdots & 0 \\ -1 & 1 & 2 & 0 & 0 & 0 & 0 & \cdots & 0 \\ 0 & -1 & 1 & 3 & 0 & 0 & 0 & \cdots & 0 \\ 0 & 0 & -1 & 1 & 1 & 0 & 0 & \cdots & 0 \\ 0 & 0 & 0 & -1 & 1 & 2 & \ddots & \cdots & 0 \\ 0 & 0 & 0 & 0 & -1 & 1 & 3 & \cdots & 0 \\ \vdots & \vdots & \vdots & \vdots & \ddots & \ddots & \ddots & \ddots & \vdots \\ 0 & 0 & 0 & 0 & \cdots & 0 & -1 & 1 \\ 0 & 0 & 0 & 0 & 0 & 0 & 0 & \cdots & 1 \end{bmatrix}.$$

It has 1's straight down the main diagonal, -1's straight down the submain diagonal, the sequence (1,2,3) repeated on the supermain diagonal, and zeros for all other entries.

(a) Store this matrix in MATLAB

(b) Find its determinant and compute and store the inverse matrix as B, if it exists (do not display it). Multiply the determinant of A with that of B.

(c) Print the 6×6 matrix C, which is the submatrix of A whose rows are made up of the rows 30, 32, 34, 36, 38, and 40 of A and whose columns are column numbers 30, 31, 32, 33, 34, and 60 of A.

SOLUTION: Part (a): We start with the 60×60 identity matrix:

```
>> A=eye(60);
```

To put the −1 's along the submain diagonal, we can use the following for loop:

```
>> for i=1:59, A(i+1,i)=-1; end
```

(Note: there are 59 entries in the submain diagonal; they are $A(2, 1)$, $A(3, 2)$, ..., $A(60, 59)$ and each one needs to be changed from 0 to −1.) The supermain diagonal entries can be changed using a similar for loop structure, but since they cycle between the three values 1, 2, 3, we could add in some branching within the for loop to accomplish this cycling. Here is one possible scheme:

```
>> count=1; %initialize counter
>> for i=1:59
if count==1, A(i,i+1)=1;
elseif count==2, A(i,i+1)=2;
else A(i,i+1)=3;
end
count=count+1; %bumps up counter by one
if count==4, count=1; end %cycles counter back to one after it passes
                          %            3
end
```

We can do a brief check of the upper-left 6×6 submatrix to see if A has shaped out the way we want it; we invoke the submatrix features introduced in Section 4.3.

```
>> A(1:6,1:6)    → ans =      1    1    0    0    0    0
                             -1    1    2    0    0    0
                              0   -1    1    3    0    0
                              0    0   -1    1    1    0
                              0    0    0   -1    1    2
```

This looks like what we wanted. Here is another way to construct the supermain diagonal of A. We first construct a vector v that contains the desired supermain diagonal entries:

```
>> vseed=[1 2 3]; v=vseed;
>> for i=1:19
v=[v vseed]; %tacks on "vseed" onto existing v
end
```

Using this vector v, we can reset the supermain diagonal entries of A as we did the submain diagonal entries:

```
>> for i=1:59
A(i,i+1)=v(i);
end
```

Shortly we will give another scheme for building such banded matrices, but it would behoove the reader to understand why the above loop construction does the job.

Part (b):
```
>> det(A)    →ans = 3.6116e + 017
>> B=inv(A); det(A)*det(B)    →ans =1.0000
```

This agrees with a special case of a theorem in linear algebra which states that for any pair of square matrices A and B of the same size, we have:

$$\det(AB) = \det(A) \cdot \det(B). \tag{4}$$

Since it is easy to show that $\det(I) = 1$, from (3) and (4) it follows that $\det(A) \cdot \det(A^{-1}) = 1$.

Part (c): Once again using MATLAB's array features introduced in Section 4.3, we can easily construct the desired submatrix of A as follows:

```
>> C=A(30:2:40, [30:33 59 60])
→ C =      1   3   0   0   0   0
           0  -1   1   2   0   0
           0   0   0  -1   0   0
           0   0   0   0   0   0
           0   0   0   0   0   0
           0   0   0   0   0   0
```

Tridiagonal matrices, like the one in the above example, are special cases of **banded matrices**, which are square matrices with all zero entries except on a certain set of diagonals. Large-sized tridiagonal and banded matrices are good examples of sparse matrices. They arise naturally in the very important finite difference methods that are used to numerically solve ordinary and partial differential equations later in this book. In its full syntax, the `diag` command introduced earlier allows us to put any vector on any diagonal of a matrix.

`diag(v,k)` →	For an integer k and an appropriately sized vector v, this command creates a square matrix with all zero entries, except for the kth diagonal on which will appear the vector v. k = 0 gives the main diagonal, k = 1 gives the supermain diagonal, k = −1 the submain diagonal, etc.

For example, in the construction of the matrix A in the above example, after having constructed the 60×60 identity matrix, we could have put in the −1's in the submain diagonal by entering:

```
>>A=A+ diag(-ones(1,59),-1);
```

and with the vector v constructed as in our solution, we could put the appropriate numbers on the supermain diagonal with the command:

```
>>A=A+ diag(v(1:59),1);
```

EXERCISE FOR THE READER 7.2: (*Random Integer Matrix Generator*) (a) For testing of hypotheses about matrices, it is often more convenient to work with integer-valued matrices rather than (floating point) decimal-valued matrices.

Create a function M-file, called `randint(n,m,k)`, that takes as input three positive integers *n*, *m*, and *k*, and will output an *n×m* matrix whose entries are integers randomly distributed in the set $\{-k, -k+1,..., -1, 0, 1, 2,..., k-1, k\}$.

(b) Use this program to test the formula (4) given in the above example by creating two different 6×6 random integer matrices *A* and *B* and computing $\det(AB) - \det(A)\cdot\det(B)$ to see if it equals zero. Use $k = 9$ so that the matrices have single-digit entries. Repeat this experiment two times (it will produce different random matrices *A* and *B*) to see if it still checks. In each case, give the values of $\det(AB)$.

(c) Keeping $k = 9$, try to repeat the same experiment (again three times) using instead 16×16 sized matrices. Explain what has (probably) happened in terms of MATLAB's default working precision being restricted to about 15 digits.

The matrix arithmetic operations enjoy the following properties (similar to those of real numbers): Here *A* and *B* are any compatible matrices and α is any number.

Commutativity of Addition: $A + B = B + A$. (5)

Associativity: $(A+B)+C = A+(B+C)$, $(AB)C = A(BC)$. (6)

Distributive Laws: $A(B+C) = AB+AC$, $(A+B)C = AC+BC$. (7)

$$\alpha(A+B) = \alpha A + \alpha B, \qquad \alpha(AB) = (\alpha A)B = A(\alpha B).$$ (8)

Each of these matrix identities is true whenever the matrices are of sizes that make all parts defined. Experiments relating to these identities as well as their proofs will be left to the exercises. We point out that matrix multiplication is not commutative; the reader is invited to do an easy counterexample to verify this fact.

We close this section with a few more methods and features for manipulating matrices in MATLAB, which we introduce through a simple MATLAB demonstration.

Let us begin with the following 3×3 matrix:

```
>> A = [1 2 3; 4 5 6; 7 8 9]      → A =    1  2  3
                                           4  5  6
                                           7  8  9
```

We can tack on [10 11 12] as a new bottom row to create a new matrix *B* as follows:

```
>> B=[A;10 11 12]                 → B =    1   2   3
                                           4   5   6
                                           7   8   9
                                          10  11  12
```

If instead we wanted to append this vector as a new column to get a new matrix C, we could add it on as above to the transpose A' of A, and then take transposes again (the transpose operator was introduced in Section 1.3).[2]

```
>> C=[A'; 10 11 12]'          → C =      1  2  3  10
                                          4  5  6  11
                                          7  8  9  12
```

Alternatively, we could start by defining $C = A$ and then introducing the (column) vector as the fourth column of C. The following two commands would give the same result/output.

```
>> C=A;
>> C(:,4)=[10 11 12]'
```

To delete any row/column from a matrix, we can assign it to be the empty vector "[ ]". The following commands will change the matrix A into a new 2×3 matrix obtained from the former by deleting the second row.

```
>>A(2,:)=[]      → A =    1  2  3
                          7  8  9
```

EXERCISES 7.1:

1. In this exercise you will be experimentally verifying the matrix identities (5), (6), (7), and (8) using the following square "test matrices:"

$$A = \begin{bmatrix} 1 & 2 \\ 3 & 4 \end{bmatrix}, \quad B = \begin{bmatrix} 1 & 2 \\ -2 & 2 \end{bmatrix}, \quad C = \begin{bmatrix} 4 & -2 \\ 7 & 6 \end{bmatrix}.$$

For these matrices verify the following:
(a) $A + B = B + A$ (matrix addition is commutative).
(b) $(A + B) + C = A + (B + C)$ (matrix addition is associative).
(c) $(AB)C = A(BC)$ (matrix multiplication is associative).
(d) $A(B + C) = AB + AC$ and $(A + B)C = AC + BC$ (matrix multiplication is distributive).
(e) $\alpha(A + B) = \alpha A + \alpha B$ and $\alpha(AB) = (\alpha A)B = A(\alpha B)$ (for any real number α) (test this last one with $\alpha = 3$).
Note: Such experiments do not constitute proofs. A math experiment can prove only that a mathematical claim is false, however, when a lot of experiments test something to be true, this can give us more of a reason to believe it and then pursue the proof. In the next three exercises, you will be doing some more related experiments. Later exercises in this set will ask for proofs of these identities.

2. Repeat all parts of Exercise 1 using instead the following test matrices:

$$A = \begin{bmatrix} 1 & 2 & 3 \\ 4 & 5 & 6 \\ 7 & 8 & 9 \end{bmatrix}, \quad B = \begin{bmatrix} 1 & 2 & 4 \\ -2 & 2 & 4 \\ -8 & -4 & 8 \end{bmatrix}, \quad C = \begin{bmatrix} 3 & 4 & 7 \\ -2 & -8 & 0 \\ 7 & 3 & 12 \end{bmatrix}.$$

---

[2] Without transposes this could be done directly with a few more keystrokes as follows:
C=[A, [10; 11; 12]]

3.   (a) By making use of the `rand` command, create three 20×20 matrices A, B, and C, each of whose (400) entries are randomly selected numbers in the interval [0, 1]. In this problem you are to check that the identities given in all parts of Exercise 1 continue to test true.

    (b) Repeat this by creating 50×50 sized matrices.

    (c) Repeat again by creating 200×200 sized matrices.

    (d) Finally do it one last time using 1000×1000 sized matrices.

    Suggestion: Of course, here it is not feasible to display all of these matrices and to compare all of the entries of the matrices on both sides by eye. (For part (d) this would entail 1 million comparisons!) The following `max` command will be useful here:

`max(A)` →	If A is a (row or column) vector, this returns the maximum value of the entries of A; if A is an $n \times m$ matrix, it returns the m-length vector whose jth entry equals the maximum value in the jth column of A.
`max(max(A))` →	(From the functionality of the single "max" command) this will return the maximum value of all the entries in A.
`max(max(abs(A)))` →	Since `abs(A)` is the matrix whose entries are the absolute values of the corresponding entries of A, this command will return the maximum absolute value of all the entries in A.

Thus, an easy way to check if two matrices E and F (of the same size) are equal is to check that `max(max(abs(E-F)))` equals zero.

Note: Due to roundoff errors (which should be increasing with the larger-sized matrices), the matrices will not, in general, agree to MATLAB's working precision of about 15 decimals. Your conclusions should take this into consideration.

4.   (a) Use the function `randint` that was constructed in Exercise for the Reader 7.2 to create three 3×3 random integer matrices (use $k = 9$) A, B, C on which you are to test once again each of the parts of Exercise 1.

    (b) Repeat for 20×20 random integer matrices.

    (c) Repeat again for 100×100 random integer matrices.

    (d) If your results checked exactly in parts (b) and (c), explain why things were able to work with such a large-sized matrix in this experiment, whereas in the experiment of checking identity (4) in Exercise for the Reader 7.2, even a moderately sized 16×16 led to problems.

    Suggestion: For parts (b) and (c) you need not print the matrices; refer to the suggestion of the previous exercise.

5.   (a) Build a 20×20 "checkerboard matrix" whose entries alternate from zeros and ones, with a one in the upper-left corner and such that for each entry of this matrix, the immediate upper, lower, left, and right neighbors (if they exist) are different.

    (b) Write a function M-file, call it `checker(n)`, that inputs a positive integer n and outputs an $n \times n$, and run it for $n = 2, 3, 4, 5$ and 6.

6.   Making use of the `randint` M-file of Exercise for the Reader 7.2, perform the following experiments.

    (a) Run through $N = 100$ trials of taking the determinant of a 3×3 matrix with random integer values spread out among –5 through 5. What was the average value of the determinants? What percent of these 100 matrices were not invertible?

    (b) Run through the same experiments with everything the same, except this time let the random integer values be spread out among –10 through 10.

    (c) Repeat part (b) except this time using $N = 500$ trials.

    (d) Repeat part (c) except this time work with 6×6 matrices.

    (e) Repeat part (c) except this time work with 10×10 matrices.

What general patterns have you noticed? Without doing the experiment, what sort of results

would you expect if we were to repeat part (c) with 20×20 matrices?
Note: You need not print all of these matrices or even their determinants in your solution. Just include the relevant MATLAB code and output needed to answer the questions along with the answers.

7. This exercise is similar to the previous one, except this time we will be working with matrices whose entries are real numbers (rather than integers) spread out uniformly randomly in an interval. To generate such an $n \times n$ matrix, for example, if we want the entries to be uniformly randomly spread out over $(-3, 3)$, we could use the command `6*rand(n) - 3*ones(n)`.

(a) Run through $N = 100$ trials of taking the determinant of a 3×3 matrix whose entries are selected uniformly randomly as real numbers in (-3, 3). What was the average value of the determinants? What percent of these 100 matrices were <u>not</u> invertible?

(b) Run through the same experiments with everything the same, except this time let the real number entries be uniformly randomly spread out among -10 through 10.

(c) Repeat part (b) except this time using $N = 500$ trials.

(d) Repeat part (c) except this time work with 6×6 matrices.

(e) Repeat part (c) except this time work with 10×10 matrices.

What general patterns have you noticed? Without doing the experiment, what sort of results would you expect if we were to repeat part (c) with 20×20 matrices?
Note: You need not print all of these matrices or even their determinants in your solution; just include the relevant MATLAB code and output needed to answer the questions along with the answers.

8. Let $M = \begin{bmatrix} 1 & 1 \\ 0 & 1 \end{bmatrix}$, $N = \begin{bmatrix} 1 & 1 \\ 1 & 1 \end{bmatrix}$.

(a) Find $M^2, M^3, M^{26}, N^2, N^3, N^{26}$.

(b) Find general formulas for M^n and N^n where n is any positive integer.

(c) Can you find a 2×2 matrix A of real numbers that satisfies $A^2 = I$ (with $A \neq I$)?

(d) Find a 3×3 matrix $A \neq I$ such that $A^3 = I$. Can you find such a matrix that is not diagonal?
Note: Part (c) may be a bit tricky. If you get stuck, use MATLAB to run some experiments on (what you think might be) possible square roots.

9. Let M and N be the matrices in Exercise 8.

(a) Find (if they exist) the inverses M^{-1} and N^{-1}.

(b) Find square roots of M and N, i.e., find (if possible) two matrices S and T so that:
$S^2 = M$ (i.e., $S = \sqrt{M}$) and $R^2 = N$.

10. (*Discovering Properties and Nonproperties of the Determinant*) For each of the following equations, run through 100 tests with 2×2 matrices whose entries are randomly selected integers within $[-9, 9]$ (using the `randint` M-file of Exercise for the Reader 7.2). For those that test false, record a single counterexample. For those that test true, repeat the same experiment twice more, first using 3×3 and then using 6×6 matrices of the same type. In each case, write your MATLAB code so that if all 100 matrix tests pass as "true" you will have as output (only) something like: "With 100 tests involving 2×2 matrices with random integer entries from -9 to 9, the identity always worked"; while if it fails for a certain matrix, the experiment should stop at this point and output the matrix (or matrices) that give a counterexample. What are your conclusions?

(a) $\det(A') = \det(A)$

(b) $\det(2A) = 4\det(A)$

(c) $\det(-A) = \det(A)$

(d) $\det(A+B) = \det(A) + \det(B)$

(e) If matrix B is obtained from A by replacing one row of A by a number k times the corresponding row of A, then $\det(B) = k \det(A)$.

(f) If one row of A is a constant multiple of another row of A then $\det(A) = 0$.

(g) Two of these identities are not quite correct, in general, but they can be corrected using another of these identities that is correct. Elaborate on this.

Suggestion: For part (f) automate the experiment as follows: After a random integer matrix A is built, randomly select a first row number $r1$ and a different second row number $r2$. Then select randomly an integer k in the range

$[-9, 9]$ Replace the row $r2$ with k times row $r1$. This will be a good possible way to create your test matrices. Use a similar selection process for part (e).

11. (a) Prove that matrix addition is commutative $A + B = B + A$ whenever A and B are two matrices of the same size. (This is identity (5) in the text.)
(b) Prove that matrix addition is associative, $(A+B)+C = A+(B+C)$ whenever A, B, and C are matrices of the same size. (This is the first part of identity (6) in the text.)

12. (a) Prove that the distributive laws for matrices: $A(B+C) = AB + AC$ and $(A+B)C = AC + BC$, whenever the matrices are of appropriate sizes so that a particular identity makes sense. (These are identities (7) in the text.)
(b) Prove that for any real number α, we have that $\alpha(A+B) = \alpha A + \alpha B$, whenever A, B, and C are matrices of the same size and that $\alpha(AB) = (\alpha A)B = A(\alpha B)$ whenever A and B are matrices with AB defined. (These are identities (8) in the text.)

13. Prove that matrix multiplication is associative, $(AB)C = A(BC)$ whenever A, B, and C are matrices so that both sides are defined. (This is the second part of identity (6) in the text.)

14. (*Discovering Facts about Matrices*) As we have seen, many matrix rules closely resemble corresponding rules of arithmetic. But one must be careful since there are some exceptions. One such notable exception we have encountered is that, unlike regular multiplication, matrix multiplication is not commutative; that is, in general we cannot say that $AB = BA$. For each of the statements below about matrices, either give a counterexample, if it is false, or give a proof if it is true. In each identity, assume that the matrices involved can be any matrices for which the expressions are all defined. Also, we use 0 to denote the zero matrix (i.e., all entries are zeros).
(a) $0A = 0$.
(b) If $AB = 0$, then either $A = 0$ or $B = 0$.
(c) If $A^2 = 0$, then $A = 0$.
(d) $(AB)' = B'A'$ (recall A' denotes the transpose of A).
(e) $(AB)^2 = A^2 B^2$.
(f) If A and B are invertible square matrices, then so is AB and $(AB)^{-1} = B^{-1}A^{-1}$.

Suggestion: If you are uncertain of any of these, run some experiments first (as shown in some of the preceding exercises). If your experiments produce a counterexample, you have disproved the assertion. In such a case you merely record the counterexample and move on to the next one.

7.2: INTRODUCTION TO COMPUTER GRAPHICS AND ANIMATION

Computer graphics is the generation and transformation of pictures on the computer. This is a hot topic that has important applications in science and business as well as in Hollywood (computer special effects and animated films). In this section we will show how matrices can be used to perform certain types of geometric operations on "objects." The objects can be either two- or three-dimensional but most of our illustrations will be in the two-dimensional plane. For two-dimensional objects, the rough idea is as follows. We can represent a basic object in the plane as a MATLAB graphic by using the command plot(x,y), where x and y are vectors of the same length. We write x and y as row vectors, stack x on top of y, and we get a $2 \times n$ matrix A where n is the common length of x and y. We can do certain mathematical operations to this matrix to change it into a new matrix $A1$, whose rows are the corresponding vertex vectors $x1$ and $y1$. If we look at plot(x1,y1), we get a transformed version of the original geometrical object.   Many interesting geometric transformations can be realized by simple matrix multiplications, but to make this all work nicely we will need to introduce a new artificial third row of the matrix A, that will simply consist of 1's. If we work instead with these so-called homogeneous coordinates, then all of the common operations of scaling (vertically or horizontally), shifting, rotating, and reflecting can be realized by matrix multiplications of these homogeneous coordinates. We can mix and repeat such transformations to get more complicated geometric transformations; and by putting a series of such plots together we can even make movies.   Another interesting application is the construction of fractal sets.   Fractal sets (or fractals) are beautiful geometric objects that enjoy a certain "self-similarity property," meaning that no matter how closely one magnifies and examines the object, the fine details will always look the same.

Polygons, which we recall are planar regions bounded by a finite set of line segments, are represented by their vertices. If we store the x-coordinates of these vertices and the y-coordinates of these vertices as separate vectors (say the first two rows of a matrix) preserving the order of adjacency, then MATLAB's plot command can easily plot the polygon.

EXAMPLE 7.3: We consider the following "CAT" polygon shown in Figure 7.2. Store the x-coordinates of the vertices of the CAT as the first row vector of a matrix A, and the corresponding y-coordinates as the second row of the same matrix in such a way that MATLAB will be able to reproduce the cat by plotting the second row vector of A versus the first. Afterwards, obtain the plot from MATLAB.

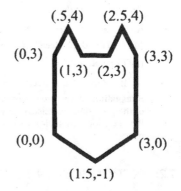

FIGURE 7.2: CAT graphic for Exampe 7.3.

SOLUTION: We can store these nine vertices in a 2×10 matrix A (the first vertex appears twice so the polygon will be closed when we plot it). We may start at any vertex we like, but we must go around the cat in order (either clockwise or counterclockwise). Here is one such matrix that begins at the vertex (0,0) and moves clockwise around the cat.

```
>> A=[0   0   .5   1   2   2.5   3   3   1.5   0; ...
       0   3   4    3   3   4     3   0   -1    0];
```

To reproduce the cat, we plot the second row of A (the y's) versus the first row (the x's):

```
>> plot(A(1,:), A(2,:))
```

   In order to get the cat to fit nicely in the viewing area (recall, MATLAB always sets the view area to just accommodate all the points being plotted), we reset the viewing range to $-2 \le x \le 5$, $-3 \le y \le 6$, and then use the equal setting on the axes so the cat will appear undistorted.

```
>> axis([-2 5 -3 6])
>> axis('equal')
```

The reader should check how each of the last two commands changes the cat graphic; we reproduce only the final plot in Figure 7.3(a). Figure 7.3 actually contains two cats, the original one (white) as well as a gray cat. The gray cat was obtained in the same fashion as the orginal cat, except that the plot command was replaced by the fill command, which works specifically with polygons and whose syntax is as follows:

fill(x,y,color) →	Here x and y are vectors of the x- and y-coordinates of a polygon (preserving adjacency order); color can be either one of the predefined plot colors (as in Table 1.1) in single quotes, (e.g., k would be a black fill) or an RGB-vector [r g b] (with r, g, and b each being numbers in [0,1]) to produce any color; for example, [.5 .5 .5] gives medium gray.

The elements r, g, and b in a color vector determine the amounts of red, green, and blue to use to create a color; any color can be created in this way. For example, $[r\ g\ b] = [1\ 0\ 0]$ would be a pure-red fill; magenta is obtained with the rgb vector [1 0 1], and different tones of gray can be achieved by using equal amounts of red, green, and blue between [0 0 0] (black) and [1 1 1] (white).

For the gray cat in Figure 7.3(b), we used the command fill(A(1,:), A(2,:), [.5 .5 .5]). To get a black cat we could either set the rgb vector to [0 0 0] or replace it with k, which represents the preprogrammed color character for black (see Table 1.1).

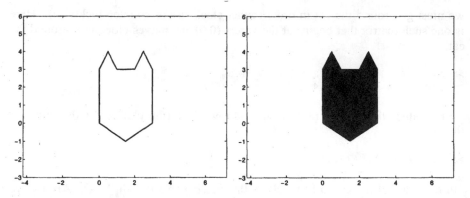

FIGURE 7.3: Two MATLAB versions of the cat polygon: (a) (left) the first white cat was obtained using the `plot` command and (b) (right) the second with the `fill` command.

EXERCISE FOR THE READER 7.3: After experimenting a bit with *rgb* color vectors, get MATLAB to produce an orange cat, a brown cat, and a purple cat. Also, try and find the rgb color vector that best matches the MATLAB built-in color cyan (from Table 1.1, the symbol for cyan is c).

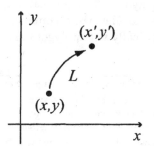

FIGURE 7.4: A linear tranformation *L* in the plane.

A **linear transformation** L on the plane R^2 corresponds to a 2×2 matrix M (Figure 7.4). It transforms any point (x, y) (represented by the column vector $\begin{bmatrix} x \\ y \end{bmatrix}$) to the point $M\begin{bmatrix} x \\ y \end{bmatrix}$ obtained by multiplying it on the left by the matrix M. The reason for the terminology is that a linear transformation preserves the two important linear operations for vectors in the plane: addition and scalar multiplication. That is, letting $P_1 = \begin{bmatrix} x_1 \\ y_1 \end{bmatrix}$, $P_2 = \begin{bmatrix} x_2 \\ y_2 \end{bmatrix}$ be two points in the plane (represented by column vectors), and writing $L(P) = MP$, the linear transformation axioms can be expressed as follows:

$$L(P_1 + P_2) = L(P_1) + L(P_2), \tag{9}$$

$$L(\alpha P_1) = \alpha L(P_1). \tag{10}$$

Both of these are to be valid for any choice of vectors $P_i (i = 1, 2)$ and scalar α. Because $L(P)$ is just MP (the matrix M multiplied by the matrix P), these two identities are consequences of the general properties (7) and (8) for matrices. By the same token, if M is any $n \times n$ matrix, then the transformation $L(P) = MP$

defines a linear transformation (satisfying (9) and (10)) for the space $\mathbf{R}^n$ of n-length vectors. Of course, most of our geomtric applications will deal with the cases $n = 2$ (the plane) or $n = 3$ (3-dimensional (x, y, z) space).

Such tranformations and their generalizations are a basis for what is used in contemporary interactive graphics programs and in the construction of computer videos. If, as in the above example of the CAT, the vertices of a polygon are stored as columns of a matrix A, then, because of the way matrix multiplication works, we can transform each of the vertices at once by multiplying the matrix M of a linear transformation by A. The result will be a new matrix containing the new vertices of the transformed graphic, which can be easily plotted.

We now move on to give some important examples of transformations on the plane $\mathbf{R}^2$.

(1) <u>Scalings:</u> For $a > 0$, $b > 0$ the linear transformation

$$\begin{bmatrix} x' \\ y' \end{bmatrix} = \begin{bmatrix} a & 0 \\ 0 & b \end{bmatrix} \begin{bmatrix} x \\ y \end{bmatrix} = \begin{bmatrix} ax \\ by \end{bmatrix}$$

will scale the horizontal direction with respect to $x = 0$ by a factor of a and the vertical direction with respect to $y = 0$ by a factor of b. If either factor is < 1, there is contraction (shrinkage) toward 0 in the corresponding dircction, while factors > 1 give rise to an expansion (strctching) away from 0 in the corresponding direction. As an example, we use $a = 0.3$ and $b = 1$ to rescale our original CAT (Figure 7.5).

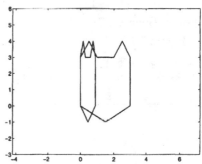

FIGURE 7.5: The scaling of the original cat using factors $a = 0.3$ for horizontal scaling and $b = 1$ (no change) for vertical scaling has produced the narrow-faced cat.

We assume we have left in the graphics window the first (white) cat of Figure 7.3(a).

```
>>M=[.3 0; 0 1];  %store scaling matrix
>>A1=M*A;  %create the vertex matrix of the transformed cat;
>>hold on  %leave the original cat in the window so we can compare
>>plot(A1(1,:), A1(2,:), 'r')  %new cat will be in red
```

Caution: Changes in the axis ranges can also produce scale changes in MATLAB graphics.

(2) <u>Rotations:</u> For a rotation angle θ, the linear tranformation that rotates a point (x, y) an angle θ (counterclockwise) around the origin $(0,0)$ is given by the following linear tranformation:

$$\begin{bmatrix} x' \\ y' \end{bmatrix} = \begin{bmatrix} \cos\theta & -\sin\theta \\ \sin\theta & \cos\theta \end{bmatrix} \begin{bmatrix} x \\ y \end{bmatrix}.$$

(See Exercise 12 for a justification of this.) As an example, we rotate the original cat around the origin using angle $\theta = -\pi/4$ (Figure 7.6). Once again, we assume the graphics window initially contains the original cat of Figure 7.3 before we start to enter the following MATLAB commands:

```
>> M=[cos(-pi/4) -sin(-pi/4); sin(-pi/4) cos(-pi/4)];
>> A1=M*A;, hold on,plot(A1(1,:), A1(2,:), 'r')
```

(3) <u>Reflections:</u> The linear tranformation that reflects points over the x-axis is given by

$$\begin{bmatrix} x' \\ y' \end{bmatrix} = \begin{bmatrix} -1 & 0 \\ 0 & 1 \end{bmatrix}\begin{bmatrix} x \\ y \end{bmatrix} = \begin{bmatrix} -x \\ y \end{bmatrix}.$$

Similary, to reflect points across the y-axis, the linear transformation will use the matrix $M = \begin{bmatrix} 1 & 0 \\ 0 & -1 \end{bmatrix}$. As an example, we reflect our original CAT over the y-axis (Figure 7.7). We assume we have left in the graphics window the first cat of Figure 7.3.

```
>> M=[-1 0; 0 1];
>> A1=M*A; hold on, plot(A1(1,:), A1(2,:), 'r')
```

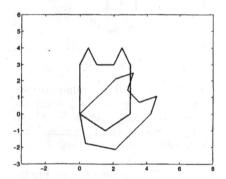
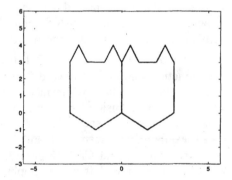

FIGURE 7.6: The rotation (red) of the original CAT (blue) using angle $\theta = -\pi/4$. The point of rotation is the origin (0,0).

FIGURE 7.7: The reflection (left) of the original CAT across the y-axis.

(4) <u>Shifts:</u> Shifts are very simple and important transformations that are not linear transformations. For a fixed (shift) vector $V_0 = (x_0, y_0) \neq (0,0)$ that we identify, when convenient, with the column vector $\begin{bmatrix} x_0 \\ y_0 \end{bmatrix}$, the **shift transformation** T_{V_0} associated with the shift vector V_0 is defined as follows:

$$(x', y') = T_{V_0}(x, y) = (x, y) + V_0 = (x + x_0, y + y_0).$$

What the shift transformation does is simply move all x-coordinates by x_0 units and move all y-coordinates by y_0 units. As an example we show the outcome of applying the shift transformation $T_{(1,1)}$ to our familiar CAT graphic. Rather than a matrix multiplication with the 2×10 CAT vertex matrix, we will need to add the corresponding 2×10 matrix, each of whose 10 columns is the shift column vector $\begin{bmatrix} 1 \\ 1 \end{bmatrix}$ that we are using (Figure 7.8).

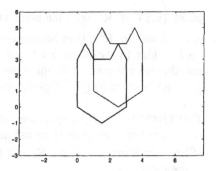

FIGURE 7.8: The shifted CAT (upper right) came from the original CAT using a shift vector (1,1). So the cat was shifted one unit to the right and one unit up.

Once again, we assume the graphics window initially contains the original (white) CAT of Figure 7.3 before we start to enter the following MATLAB commands (and that the CAT vertex matrix A is still in the workspace).

```
>>size(A)  %check size of A   →ans = 2   10
>> V=ones(2,10); A1=A+V; hold on, plot(A1(1,:),A1(2,:), 'r')
```

EXERCISE FOR THE READER 7.4: Explain why the shift transformation is never a linear transformation.

It is unfortunate that the shift transformation cannot be realized as a linear transformation, so that we cannot realize it as using a 2×2 matrix multiplication of our vertex matrix. If we could do this, then all of the important transformations mentioned thus far could be done in the same way and it would make combinations (and in particular making movies) an easier task. Fortunately there is a way around this using so-called homogeneous coordinates. We first point out a more general type of transformation than a linear transformation that includes all linear transformations as well as the shifts. We define it only on the two-dimensional space $\mathbf{R}^2$, but the definition carries over in the obvious way to the three-dimensional space $\mathbf{R}^3$ and higher-dimensional spaces as well. An **affine transformation** on $\mathbf{R}^2$ equals a linear tranformation and/or a shift (applied together). Thus, an affine transformation can be written in the form:

$$\begin{bmatrix} x' \\ y' \end{bmatrix} = M \begin{bmatrix} x \\ y \end{bmatrix} + V_0 = \begin{bmatrix} a & b \\ c & d \end{bmatrix} \begin{bmatrix} x \\ y \end{bmatrix} + \begin{bmatrix} x_0 \\ y_0 \end{bmatrix}. \tag{11}$$

The **homogeneous coordinates** of a point/vector $\begin{bmatrix} x \\ y \end{bmatrix}$ in $\mathbf{R}^2$ is the point/vector

$\begin{bmatrix} x \\ y \\ 1 \end{bmatrix}$ in $\mathbf{R}^3$. Note that the third coordinate of the identified three-dimensional

point is always 1 in homogeneous coordinates. Geometrically, if we identify a point (x, y) of $\mathbf{R}^2$ with the point $(x, y, 0)$ in $\mathbf{R}^3$ (i.e., we identify $\mathbf{R}^2$ with the plane $z = 0$ in $\mathbf{R}^3$), then homogeneous coordinates simply lift all of these points up one unit to the plane $z = 1$. It may seem at first glance that homogeneous coordinates are making things more complicated, but the advantage in computer graphics is given by the following result.

THEOREM 7.2: (*Homogeneous Coordinates*) Any affine transformation on $\mathbf{R}^2$ is a linear transformation if we use homogeneous coordinates. In other words, any affine transformation T on $\mathbf{R}^2$ can be expressed using homogeneous coordinates in the form:

$$\begin{bmatrix} x' \\ y' \\ 1 \end{bmatrix} = T\left(\begin{bmatrix} x \\ y \\ 1 \end{bmatrix}\right) = H \begin{bmatrix} x \\ y \\ 1 \end{bmatrix} \tag{12}$$

(matrix multiplication), where H is some 3×3 matrix.

Proof: The proof of the theorem is both simple and practical; it will show how to form the matrix H in (12) from the parameters in (11) that determine the affine transformation.

Case 1: T is a linear transformation on $\mathbf{R}^2$ with matrix $M = \begin{bmatrix} a & b \\ c & d \end{bmatrix}$, i.e.,

$T\left(\begin{bmatrix} x \\ y \end{bmatrix}\right) = M\begin{bmatrix} x \\ y \end{bmatrix} = \begin{bmatrix} a & b \\ c & d \end{bmatrix}\begin{bmatrix} x \\ y \end{bmatrix}$ (no shift). In this case, the transformation can be

expressed in homogeneous coordinates as:

$$\begin{bmatrix} x' \\ y' \\ 1 \end{bmatrix} = T\left(\begin{bmatrix} x \\ y \\ 1 \end{bmatrix}\right) = \begin{bmatrix} a & b & 0 \\ c & d & 0 \\ 0 & 0 & 1 \end{bmatrix}\begin{bmatrix} x \\ y \\ 1 \end{bmatrix} = H\begin{bmatrix} x \\ y \\ 1 \end{bmatrix}. \tag{13}$$

To check this identity, we simply perform the matrix multiplication:

$$\begin{bmatrix} x' \\ y' \\ 1 \end{bmatrix} = \begin{bmatrix} a & b & 0 \\ c & d & 0 \\ 0 & 0 & 1 \end{bmatrix}\begin{bmatrix} x \\ y \\ 1 \end{bmatrix} = \begin{bmatrix} ax+by+0 \\ cx+dy+0 \\ 0+0+1 \end{bmatrix} \Rightarrow \begin{bmatrix} x' \\ y' \end{bmatrix} = \begin{bmatrix} ax+by \\ cx+dy \end{bmatrix} = M\begin{bmatrix} x \\ y \end{bmatrix},$$

as desired.

Case 2: T is a shift transformation on $\mathbf{R}^2$ with shift vector $V_0 = \begin{bmatrix} x_0 \\ y_0 \end{bmatrix}$, that is,

$T\left(\begin{bmatrix} x \\ y \end{bmatrix}\right) = \begin{bmatrix} x \\ y \end{bmatrix} + \begin{bmatrix} x_0 \\ y_0 \end{bmatrix}$ (so the matrix M in (12) is the identity matrix). In this

case, the transformation can be expressed in homogeneous coordinates as:

$$\begin{bmatrix} x' \\ y' \\ 1 \end{bmatrix} = T\left(\begin{bmatrix} x \\ y \\ 1 \end{bmatrix}\right) = \begin{bmatrix} 1 & 0 & x_0 \\ 0 & 1 & y_0 \\ 0 & 0 & 1 \end{bmatrix}\begin{bmatrix} x \\ y \\ 1 \end{bmatrix} = H\begin{bmatrix} x \\ y \\ 1 \end{bmatrix}. \qquad (14)$$

We leave it to the reader to check, as was done in Case 1, that this homogeneous coordinate linear transformation does indeed represent the shift.

Case 3: The general case (linear transformation plus shift);

$$\begin{bmatrix} x' \\ y' \end{bmatrix} = T\left(\begin{bmatrix} x \\ y \end{bmatrix}\right)\begin{bmatrix} x \\ y \end{bmatrix} = \begin{bmatrix} a & b \\ c & d \end{bmatrix}\begin{bmatrix} x \\ y \end{bmatrix} + \begin{bmatrix} x_0 \\ y_0 \end{bmatrix},$$

can now be realized by putting together the matrices in the preceding two special cases:

$$\begin{bmatrix} x' \\ y' \\ 1 \end{bmatrix} = T\left(\begin{bmatrix} x \\ y \\ 1 \end{bmatrix}\right) = \begin{bmatrix} a & b & x_0 \\ c & d & y_0 \\ 0 & 0 & 1 \end{bmatrix}\begin{bmatrix} x \\ y \\ 1 \end{bmatrix} = H\begin{bmatrix} x \\ y \\ 1 \end{bmatrix}. \qquad (15)$$

We leave it to the reader check this (using the distributive law (7)).

The basic transformations that we have so far mentioned can be combined to greatly expand the mutations that can be performed on graphics. Furthermore, by using homogeneous coordinates, the matrix of such a combination of basic transformations can be obtained by simply multiplying the matrices by the individual basic transformations that are used, in the correct order, of course. The next example illustrates this idea.

EXAMPLE 7.4: Working in homogeneous coordinates, find the transformation that will rotate the CAT about the tip of its chin by an angle of −90°. Express the transformation using the 3×3 matrix M for homogeneous coordinate multiplication, and then get MATLAB to create a plot of the transformed CAT along with the original.

SOLUTION: Since the rotations we have previously introduced will always rotate around the origin $(0,0)$, the way to realize this transformation will be by combining the following three transformations (in order):
(i) First shift coordinates so that the chin gets moved to $(0,0)$. Since the chin has coordinates $(1.5, -1)$, the shift vector should be the opposite so we will use the shift transformation

$$T_{(-1.5,1)} \sim \begin{bmatrix} 1 & 0 & -1.5 \\ 0 & 1 & 1 \\ 0 & 0 & 1 \end{bmatrix} = H_1$$

(the tilde notation is meant to indicate that the shift transformation is represented in homogeneous coordinates by the given 3×3 matrix H_1, as specified by (14)).

(ii) Next rotate (about (0,0)) by $\theta = -90°$. This rotation transformation R has matrix

$$\begin{bmatrix} \cos(-90°) & -\sin(-90°) \\ \sin(-90°) & \cos(-90°) \end{bmatrix} = \begin{bmatrix} 0 & 1 \\ -1 & 0 \end{bmatrix},$$

and so, by (13), in homogeneous coordinates is represented by

$$R \sim \begin{bmatrix} 0 & 1 & 0 \\ -1 & 0 & 0 \\ 0 & 0 & 1 \end{bmatrix} = H_2.$$

(iii) Finally we undo the shift that we started with in (i), using

$$T_{(1\,5,-1)} \sim \begin{bmatrix} 1 & 0 & 1.5 \\ 0 & 1 & -1 \\ 0 & 0 & 1 \end{bmatrix} = H_3.$$

If we multiply each of these matrices (in order) on the left of the original homogeneous coordinates, we obtain the transformed homogeneous coordinates:

$$\begin{bmatrix} x' \\ y' \\ 1 \end{bmatrix} = H_3 H_2 H_1 \begin{bmatrix} x \\ y \\ 1 \end{bmatrix} \equiv M \begin{bmatrix} x \\ y \\ 1 \end{bmatrix},$$

that is, the matrix M of the whole transformation is given by the product $H_3 H_2 H_1$. We now turn things over to MATLAB to compute the matrix M and to plot the before and after plots of the CAT.

```
>> H1=[1 0 -1.5; 0 1 1; 0 0 1]; H2=[0 1 0; -1 0 0; 0 0 1];
>> H3=[1 0 1.5;0 1 -1; 0 0 1];
>> format rat %will give a nicer display of the matrix M
>> M=H3*H2*H1
→M =      0        1      5/2
         -1        0      1/2
          0        0       1
```

We will multiply this matrix M by the matrix AH of homogeneous coordinates corresponding to the matrix A. To form AH, we simply need to tack on a row of ones to the bottom of the matrix A. (See Figure 7.9.)

```
>> AH=A; %start with A
>> size(A) %check the size of A
→ans = 2  10
>> AH(3,:)=ones(1,10); %form the
>> %appropriately sized third row
>> %for AH
>> size(AH)  →ans = 3   10
>> hold on, AH1=M*AH;
>> plot(AH1(1,:), AH1(2,:), 'r')
```

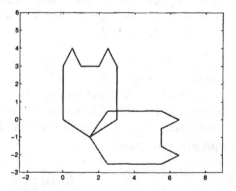

FIGURE 7.9: The red CAT was obtained from the blue cat by rotating −90° about the chin. The plot was obtained using homogeneous coordinates in Example 7.3.

EXERCISE FOR THE READER 7.5:   Working in homogeneous coordinates, (a) find the transformation that will shift the CAT one unit to the right and then horizontally expand it by a factor of 2 (about $x = 0$) to make a "fat CAT". Express the transformation using the 3×3 matrix M for homogeneous coordinate multiplication, and then use MATLAB to create a plot of the transformed fat cat along with the original.

(b) Next, find four transformations each shifting the cat by one of the following shift vectors $(\pm 1, \pm 1)$ (so that all four shift vectors are used) after having rotated the CAT about the central point (1.5, 1.5) by each of the following angles: 30° for the upper-left CAT, −30° for the upper-right CAT, 45° for the lower-left cat, and −45° for the lower-right cat.   Then fill in the four cats with four different (realistic cat) colors, and include the graphic.

We now show how we can put graphics transformations together to create a movie in MATLAB. This can be done in the following two basic steps:

STEPS FOR CREATING A MOVIE IN MATLAB:

Step 1: Construct a sequence of MATLAB graphics that will make up the frames of the movie. After the *j*th frame is constructed, use the command M(:,j) = getframe; to store the frame as the *j*th column of some (movie) matrix *M* .

Step 2: To play the movie, use the command movie(M, rep, fps), where M is the movie matrix constructed in step 1, rep is a positive integer giving the number of times the movie is to be (repeatedly) played, and fps denotes a positive integer giving the speed, in "frames per second," at which the movie is to be played.

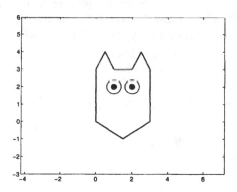

FIGURE 7.10:   The original CAT of Example 7.3 with eyes added, the star of our first cat movie.

Our next example gives a very simple example of a movie. The movie star will of course be the CAT, but this time we will give it eyes (Figure 7.10). For this first example, we do not use matrix transformations, but instead we directly edit (via a loop) the code that generates the graphic. Of course, a textbook cannot play the movie, so the reader is encouraged to rework the example in front of the computer and thus replay the movie.

EXAMPLE 7.5:   Modify the CAT graphic to have a black outline, to have two circular eyes (filled in with yellow), with two smaller black-filled pupils at the center of the eyes. Then make a movie of the cat closing and then reopening its eyes.

SOLUTION: The strategy will be as follows: To create the new CAT with the specified eyes, we use the "hold on" command after having created the basic CAT. Then we `fill` in yellow two circles of radius 0.4 centered at (1, 2) (left eye) and at (2, 2) (right eye); after this we fill in black two smaller circles with radii 0.15 at the same centers (for the pupils). The circles will actually be polygons obtained by parametric equations. To gradually close the eyes, we use a for loop to create CATs with the same outline but whose eyes are shrinking only in the vertical direction.

This could be done with homogeneous coordinate transforms (that would shrink in the y direction each eye but maintain the centers—thus it would have to first shift the eyes down to $y = 0$, shrink and then shift back), or alternatively we could just directly modify the y parametric equations of each eye to put a shrinking scaling factor in front of the sine function to turn the eyes both directly into a shrinking (and later expanding) sequence of ellipses. We proceed with the second approach. Let us first show how to create the CAT with the indicated eyes. We begin with the original CAT (this time with black line color rather than blue), setting the `axis` options as previously, and then enter `hold on`. Assuming this has been done, we can create the eyes as follows:

```
>> t=0:.02:2*pi;    %creates time vector for parametric equations
>> x=1+.4*cos(t); y=2+.4*sin(t);  %creates circle for left eye
>> fill(x,y,'y')  %fills in left eye
>> fill(x+1,y, 'y')  %fills in right eye
>> x=1+.15*cos(t); y=2+.15*sin(t);  %creates circle for left pupil
>> fill(x,y,'k')  %fills in left pupil
>> fill(x+1,y,'k')  %fills in right pupil
```

To make the frames for our movie (and to "get" them), we employ a for loop that goes through the above construction of the "CAT with eyes", except that a factor will be placed in front of the sine term of the y-coordinates of both eyes and pupils. This factor will start at 1, shrink to 0, and then expand back to the value of 1 again. To create such a factor, we need a function with starting value 1 that decreases to zero, then turns around and increases back to 1. One such function that we can use is $(1 + \cos x)/2$ over the interval $[0, 2\pi]$. Below we give one possible implementation of this code:

```
>>t=0:.02:2*pi; counter=1;
>>A=[0  0  .5  1  2  2.5  3  3  1.5  0;... ...
     0  3  4   3  3  4    3  0  -1   0];
>>x=1+.4*cos(t); xp=1+.15*cos(t);
>>for s=0:.2:2*pi
 factor = (cos(s)+1)/2;
 plot(A(1,:), A(2,:), 'k')
 axis([-2 5 -3 6]), axis('equal')
 y=2+.4*factor*sin(t); yp=2+.15*factor*sin(t);
 hold on
 fill(x,y,'y'), fill(x+1,y, 'y'), fill(xp,yp,'k'), fill(xp+1,yp,'k')
 M(:, counter) = getframe;
 hold off, counter=counter+1;
end
```

The movie is now ready for screening. To view it the reader might try one (or both) of the following commands.

```
>> movie(M,4,5)  %slow playing movie, four repeats
>> movie(M,20,75) %much faster play of movie, with 20 repeats
```

EXERCISE FOR THE READER 7.6: (a) Create a MATLAB function M-file, called mkhom(A), that takes a $2 \times m$ matrix of vertices for a graphic (first row has x-coordinates and second row has corresponding y-coordinates) as input and outputs a corresponding $3 \times m$ matrix of homogeneous coordinates for the vertices.
(b) Create a MATLAB function M-file, called rot(Ah,x0,y0, theta) that has inputs, Ah, a matrix of homogeneous coordinates of some graphic, two real numbers, x0, y0 that are the coordinates of

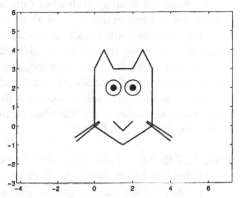

FIGURE 7.11: The more sophisticated cat star of the movie in Exercise for the Reader 7.7 (b).

the center of rotation, and theta , the angle (in radians) of rotation. The output will be the homogeneous coordinate vertex matrix gotten from Ah by rotating the graph an angle theta about the point ($x0$, $y0$).

EXERCISE FOR THE READER 7.7: (a) Recreate the above movie working in homogeneous coordinate transforms on the eyes.
(b) By the same method, create a similar movie that stars a more sophisticated cat, replete with whiskers and a mouth, as shown in Figure 7.11. In this movie, the cat starts off frowning and the pupils will shift first to the left, then to the right, then back to center and finally up, down and back to center again, at which point the cat will wiggle its whiskers up and down twice and change its frown into a smile.

Fractals or **fractal sets** are complicated and interesting sets (in either the plane or three-dimensional space) that have the **self-similarity property** that if one magnifies a certain part of the fractal (any number of times) the details of the structure will look exactly the same.

The computer generation of fractals is also a hot research area and we will look at some of the different methods that are extensively used. Fractals were gradually discovered by mathematicians who were specialists in set theory or function theory, including (among others) the very famous Georg F. L. P. Cantor (1845–1918, German), Waclaw Sierpinski (1882–1969, Polish), Gaston Julia (1893–1978, French) and Giuseppe Peano (1858–1932, Italian) during the late nineteenth and early twentieth centuries. Initially, fractals came up as being pathological objects without any type of unifying themes. Many properties of factals that have shown them to be so useful in an assortment of fields were discovered and popularized by the Polish/French mathematician Benoit Mandelbrot(Figure 7.12).[3] The precise definition of a fractal set takes a lot of preliminaries; we refer to the references, for example, that are cited in the footnote on this page for details. Instead of this, we will jump into some examples. The main point to keep in mind is that all of the examples we give (in the text as well as in the exercises) are actually impossible to print out exactly because of the self-similarity property; the details would require a printer with infinite resolution. Despite this problem, we can use loops or recursion with MATLAB to get some decent renditions of fractals that, as far as the naked eye can tell (your printer's resolution permitting), will be accurate illustrations.

Fractal sets are usually best described by an iterative procedure that runs on forever.

EXAMPLE 7.6: (*The Sierpinski Gasket*) To obtain this fractal set, we begin with an equilateral triangle that we illustrate in gray in Figure 7.13(a); we call this set the *zeroth generation*. By considering the midpoints of each of the sides of this triangle, we can form four (smaller) triangles that are similar to the original. One is upside-down and the other three have the same orientation as the original. We delete this central upside down subtriangle from the zeroth generation to form the *first generation* (Figure 7.13(b)).

FIGURE 7.12: Benoit Mandelbrot (b. 1924) Polish/ French mathematician.

---

[3] Mandelbrot was born in Poland in 1924 and his family moved to France when he was 12 years old. He was introduced to mathematics by his uncle Szolem Mandelbrojt, who was a mathematics professor at the Collège de France. From his early years, though, Mandelbrot showed a strong preference for mathematics that could be applied to other areas rahter than the pure and rather abstruse type of mathematics on which his uncle was working. Since World War II was taking place during his school years, he often was not able to attend school and as a result much of his education was done at home through self-study. He attributes to this informal education the development of his strong geometric intuition. After earning his Ph.D. in France he worked for a short time at Cal Tech and the Institute for Advanced Study (Princeton) for postdoctoral work. He then returned to France to work at the Centre National de la Recherche Scientifique. He stayed at this post for only three years since he was finding it difficult to fully explore his creativity in the formal and traditional mathematics societies that dominated France in the mid-twentieth century (the "Bourbaki School"). He returned to the United States, taking a job as a research fellow with the IBM research laboratories. He found the atmosphere extremely stimulating at IBM and was able to study what he wanted. He discovered numerous applications and properties of fractals; the expanse of applications is well demonstrated by some of the other joint appointments he has held while working at IBM. These include Professor of the Practice of Mathematics at Harvard University, Professor of Engineering at Yale, Professor of Economics at Harvard, and Professor of Physiology at the Einstein College of Medicine.   Many books have been written on fractals and their applications. For a very geometric and accessible treatment (with lots of beautiful pictures of fractals) we cite [Bar-93], along with [Lau-91]; see also [PSJY-92]. More analytic (and mathematically advanced) treatments are nicely done in the books [Fal-85] and [Mat-95].

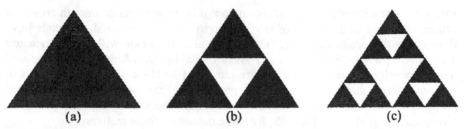

FIGURE 7.13: Generation of the Sierpinski gasket of Example 7.6: (a) the zeroth generation (equilateral triangle), (b) first generation, (c) second generation. The generations continue on forever to form the actual set.

Next, on each of the three (equilateral) triangles that make up this first generation, we again perform the same procedure of deleting the upside-down central subtriangle to obtain the generation-two set (Figure 7.13(c)). This process is to continue on forever and this is how the Sierpinski gasket set is formed. The sixth generation is shown in Figure 7.14.

FIGURE 7.14: Sixth generation of the Sierpinski gasket fractal of Example 7.6.

Notice that higher generations become indistinguishable to the naked eye, and that if we were to focus on one of the three triangles of the first generation, the Sierpinski gasket looks the same in this triangle as does the complete gasket. The same is true if we were to focus on any one of the nine triangles that make up the second generation, and so on.

EXERCISE FOR THE READER 7.8: (a) Show that the nth generation of the Sierpinski triangle is made up of 3^n equilateral triangles. Find the area of each of these nth-generation triangles, assuming that the initial sidelengths are one.
(b) Show that the area of the Sierpinski gasket is zero.
NOTE: It can be shown that the Sierpinski gasket has dimension $\log 4 / \log 3$
$= 1.2618...$, where the *dimension* of a set is a rigorously defined measure of its

true size. For example, any countable union of line segments or smooth arcs is of dimension one and the inside of any polygon is two-dimensional. Fractals have dimensions that are nonintegers. Thus a fractal in the plane will have dimension somewhere (strictly) between 1 and 2 and a fractal in three-dimensional space will have dimension somewhere strictly between 2 and 3. None of the standard sets in two and three dimensions have this property. This noninteger dimensional property is often used as a definition for fractals. The underlying theory is quite advanced; see [Fal-85] or [Mat-95] for more details on these matters.

In order to better understand the self-similarity property of fractals, we first recall from high-school geometry that two triangles are similar if they have the same angles, and consequently their corresponding sides have a fixed ratio. A **similarity transformation** (or **similitude** for short) on $\mathbf{R}^2$ is an affine transformation made up of one or more of the following special transformations: scaling (with both x- and y-factors equal), a reflection, a rotation, and/or a shift. In homogeneous coordinates, it thus follows that a similitude can be expressed in matrix form as follows:

$$\begin{bmatrix} x' \\ y' \\ 1 \end{bmatrix} = T\left(\begin{bmatrix} x \\ y \\ 1 \end{bmatrix} \right) = \begin{bmatrix} s\cos\theta & -s\sin\theta & x_0 \\ \pm s\sin\theta & \pm s\cos\theta & y_0 \\ 0 & 0 & 1 \end{bmatrix} \begin{bmatrix} x \\ y \\ 1 \end{bmatrix} = H \begin{bmatrix} x \\ y \\ 1 \end{bmatrix}, \qquad (16)$$

where s can be any nonzero real number and the signs in the second row of H must be the same. A scaling with both x- and y-factors being equal is customarily called a **dilation**.

EXERCISE FOR THE READER 7.9:   (a) Using Theorem 7.2 (and its proof), justify the correctness of (16).
(b) Show that for any two similar triangles in the plane there is a similitude that transforms one into the other.
(c) Show that if any particular feature (e.g., reflection) is removed from the definition of a similitude, then two similar triangles in the plane can be found, such that one cannot be transformed to the other by this weaker type of transformation.

The self-similarity of a fractal means, roughly, that for the whole fractal (or at least a critical piece of it), a set of similitudes $S_1, S_2, \cdots, S_K$ can be found (the number K of them will depend on the fractal) with the following property: All S_j's have the same scaling factor $s < 1$ so that F can be expressed as the union of the transformed images $F_i = S_i(F)$ and these similar (and smaller) images are **essentially disjoint** in that different ones can have only vertex points or boundary edges in common. Many important methods for the computer generation of fractals will hinge on the discovery of these similitudes $S_1, S_2, \cdots, S_K$. Finding them also has other uses in both the theory and application of fractals. These concepts will be important in Methods 1 and 2 in our solution of the following example.

EXAMPLE 7.7: Write a MATLAB function M-file that will produce graphics for the Sierpinski gasket fractal.

SOLUTION: We deliberately left the precise syntax of the M-file open since we will actually give three different approaches to this problem and produce three different M-files. The first method is a general one that will nicely take advantage of the self-similarity of the Sierpinski gasket and will use homogeneous coordinate transform methods. It was, in fact, used to produce high-quality graphic of Figure 7.14. Our second method will illustrate a different approach, called the **Monte Carlo method**, that will involve an iteration of a random selection process to obtain points on the fractal, and will plot each of the points that get chosen. Because of the randomness of selection, enough iterations produce a reasonably representative sample of points on the fractal and the resulting plot will give a decent depiction of it. Monte Carlo is a city on the French Riviera known for its casinos (it is the European version of Las Vegas). The method gets its name from the random (chance) selection processes it uses. Our third method works similarly to the first but the ideas used to create the M-file are motivated by the special structure of the geometry, in this case of the triangles.

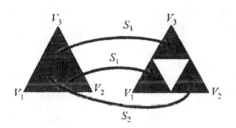

FIGURE 7.15: The three natural similitudes S_1, S_2, S_3 for the Sierpinski gasket with vertices V_1, V_2, V_3 shown on the zeroth and first generations. Since the zeroth generation is an equilateral triangle, so must be the three triangles of the first generation.

Method 1: The Sierpinski gasket has three obvious similitudes, each of which transforms it into one of the three smaller "carbon copies" of it that lie in the three triangles of the first generation (see Figure 7.15). These similitudes have very simple form, involving only a dilation (with factor 0.5) and shifts. The first transformation S_1 involves no shift. Referring to the figure, it is clear that S_2 must shift V_1 to the midpoint of the line segment $V_1 V_2$ that is given by (as a vector) $(V_1 + V_2)/2$.. The shift vector needed to do this, and hence the shift vector for S_2 is $(V_2 - V_1)/2$. (Proof: If we shift V_1 by this vector we get $V_1 + (V_2 - V_1)/2 = (V_2 + V_1)/2$.) Similarly the shift vector for S_3 is $(V_3 - V_1)/2$. It follows that the corresponding matrices for these three similitudes are as given below:

$$S_1 \sim \begin{bmatrix} .5 & 0 & 0 \\ 0 & .5 & 0 \\ 0 & 0 & 1 \end{bmatrix}, S_2 \sim \begin{bmatrix} .5 & 0 & (V_2(1)-V_1(1))/2 \\ 0 & .5 & (V_2(2)-V_1(2))/2 \\ 0 & 0 & 1 \end{bmatrix}, S_3 \sim \begin{bmatrix} .5 & 0 & (V_3(1)-V_1(1))/2 \\ 0 & .5 & (V_3(2)-V_1(2))/2 \\ 0 & 0 & 1 \end{bmatrix}.$$

Program 7.1, sgasket1 (V1,V2,V3,ngen), has four input variables: V1, V2, V3 should be the row vectors representing the vertices $(0,0)$, $(1,\sqrt{3}),(2,0)$ of a particular equilateral triangle, and ngen is the generation number of the Sierpinski gasket to be drawn. The program has no output variables, but will produce a graphic of this generation ngen of the Sierpinski gasket. The idea behind the algorithm is the following. The three triangles making up the generation-one gasket can be obtained by applying each of the three special similitudes S_1,S_2,S_3 to the single generation-zero Gasket. By the same token, each of the nine triangles that comprise the generation-two gasket can be obtained by applying one of the similitudes of S_1,S_2,S_3 to one of the generation-one triangles. In general, the triangles that make up any generation gasket can be obtained as the union of the triangles that result from applying each of the similitudes S_1,S_2,S_3 to each of the previous generation triangles. It works with the equilateral triangle having vertices $(0,0),(1,\sqrt{3}),(2,0)$. The program makes excellent use of recursion.

PROGRAM 7.1: Function M-file for producing a graphic of any generation of the Sierpinski gasket on the special equilateral triangle with vertices $(0,0),(1,\sqrt{3}),(2,0)$ (written with comments in a way to make it easily modified to work for other fractals).

```
function sgasket1(V1,V2,V3,ngen)
%input variables: V1,V2,V3 should be the vertices [0 0], [1,sqrt(3)],
%and [2,0] of a particular equilateral triangle in the plane taken as
%row vectors, ngen is the number of iterations to perform in
%Sierpinski gasket generation.
%The gasket will be drawn in medium gray color.

%first form matrices for similitudes
   S1=[.5 0 0;0 .5 0;0 0 1];
   S2=[.5 0 1; 0 .5 0;0 0 1];
   S3=[.5 0 .5; 0 .5 sqrt(3)/2;0 0 1];

if ngen == 0
   %Fill triangle
   fill([V1(1) V2(1) V3(1) V1(1)], [V1(2) V2(2) V3(2) V1(2)], [.5 .5
.5])
   hold on
else
%recursively invoke the same function on three outer subtriangles
%form homogeneous coordinate matrices for three vertices of triangle
   A=[V1; V2; V3]'; A(3,:)=[1 1 1];
   %next apply the similitudes to this matrix of coordinates
   A1=S1*A; A2=S2*A; A3=S3*A;
%finally, reapply sgasket1 to the corresponding three triangles with
%ngen bumped down by 1. Note, vertex vectors have to be made into
%row vectors using '(transpose).
   sgasket1(A1([1 2],1)', A1([1 2],2)', A1([1 2],3)', ngen-1)
   sgasket1(A2([1 2],1)', A2([1 2],2)', A2([1 2],3)', ngen-1)
   sgasket1(A3([1 2],1)', A3([1 2],2)', A3([1 2],3)', ngen-1)
end
```

To use this program to produce, for example, the generation-one graphic of Figure 7.13(b), one need only enter:

```
>> sgasket1([0 0], [1 sqrt(3)], [2 0], 1)
```

If we wanted to produce a graphic of the more interesting generation-six Sierpinski gasket of Figure 7.15, we would have only to change the last input argument from 1 to 6. Note, however, that this function left the graphics window with a `hold on`. So before doing anything else with the graphics window after having used it, we would need to first enter `hold off`. Alternatively, we could also use the following command:

clf	→	Clears the graphics window.

In addition to recursion, the above program makes good use of MATLAB's elaborate matrix manipulation features. It is important that the reader fully understands how each part of the program works. To this end the following exercise should be useful.

EXERCISE FOR THE READER 7.10: (a) Suppose the above program is invoked with these input variables: $V1 = [0\ 0]$, $V2 = [1\ \sqrt{3}]$, $V3 = [2\ 0]$, ngen = 1. On the first run/iteration, what are the numerical values of each of the following variables:

A, A1, A2, A3, A1([1 2],2), A3([1 2],3)?

(b) Is it possible to modify the above program so that after the graphic is drawn, the screen will be left with `hold off`? If yes, show how to do it; if not, explain.
(c) In the above program, the first three input variables $V1, V2, V3$ seem a bit redundant since we are forced to input them as the vertices of a certain triangle (which gave rise to the special similitudes $S1$, $S2$, and $S3$). Is it possible to rewrite the program so that it has only one input variable ngen? If yes, show how to do it; if not, explain.

Method 2: The Monte Carlo method also will use the special similitudes, but its philosophy is very different from that of the first method. Instead of working on a particular generation of the Sierpinki gasket fractal, it goes all out and tries to produce a decent graphic of the actual fractal. This gets done by plotting a representative set of points on the fractal, a random sample of such. Since so much gets deleted from the original triangle, a good question is What points exactly are left in the Sierpinski gasket? Certainly the vertices of any triangle of any generation will always remain. Such points will be the ones from which the Monte Carlo method samples. Actually there are a lot more points in the fractal than these vertices, although such points are difficult to write down. See one of the books on fractals mentioned earlier for more details.

Here is an outline of how the program will work. We start off with a point we call "Float" that is a vertex of the original (generation-zero) triangle, say $V1$. We then randomly choose one of the similitudes from S_1, S_2, S_3, and apply this to

"Float" to get a new point "New," that will be the corresponding vertex of the generation-one triangle associated with the similitude that was used (lower left for S_1, upper middle for S_3, and lower right for S_2). We plot "New," redefine "Float" = "New," and repeat this process, again randomly selecting one of the similitudes to apply to "Float" to get a new point "New" of the fractal that will be plotted. At the Nth iteration, "New" will be a vertex of one of the Nth-generation triangles (recall there are 3^N such triangles) that will also lie in one of the three generation-one triangles, depending on which of S_1, S_2, S_3 had been randomly chosen. Because of the randomness of choices at each iteration, the points that are plotted usually give a decent rendition of the fractal, as long as a large enough random sample is used (i.e., a large number of iterations).

PROGRAM 7.2: Function M-file for producing a Monte Carlo approximation graphic of Sierpinski gasket, starting with the vertices *V1*, *V2*, and *V3* of any equilateral triangle (written with comments in a way that will make it easily modified to work for other fractals).

```
function [ ] = sgasket2(V1,V2,V3,niter)
%input variables: V1,V2,V3 are vertices of an equilateral triangle in
%the plane taken as row vectors, niter is the number of iterations
%used to obtain points in the fractal.  The output will be a plot of
%all of the points.  If niter is not specified, the default value
%of 5000 is used.
%if only 3 input arguments are given (nargin--3), set niter to
%default
if nargin == 3, niter = 5000; end

%Similitude matrices for Sierpinski gasket.
S1=[.5 0 0;0 .5 0;0 0 1];
S2=[.5 0 (V2(1)-V1(1))/2; 0 .5 (V2(2)-V1(2))/2;0 0 1];
S3=[.5 0 (V3(1)-V1(1))/2; 0 .5 (V3(2)-V1(2))/2;0 0 1];

%Probability vector for Sierpinski gasket has equal probabilities
%(1/3)for choosing one of the three similitudes.
P = [1/3 2/3];

%prepare graphics window for repeated plots of points
clf, axis('equal'); hold on;

%introduce "floating point" (can be any vertex) in homogeneous
%coordinates
Float=[V1(1);V1(2);1];
i = 1; %initialize iteration counter

%Begin iteration for creating new floating points and plotting each
%one that arises.
while i <= niter
   choice = rand;
   if choice < P(1);
      New = S1 * Float;
      plot (New(1), New(2));
   elseif choice < P(2);
      New = S2 * Float;
      plot (New(1), New(2));
   else        New = S3 * Float;
```

```
        plot (New(1), New(2));
    end;
    Float=New;      i = i + 1;
end
hold off
```

Unlike the last program, this one allows us to input the vertices of any equilateral triangle for the generation-zero triangle. The following two commands will invoke the program first with the default 5000 iterations and then with 20,000 (the latter computation took several seconds).

```
>> sgasket2([0 0], [1 sqrt(3)], [2 0])
>> sgasket2([0 0], [1 sqrt(3)], [2 0], 20000)
```

The results are shown in Figure 7.16. The following exercise should help the reader better undertstand how the above algorithm works.

EXERCISE FOR THE READER 7.11: Suppose that we have generated the following random numbers (between zero and one): .5672, .3215, .9543, .4434, .8289, .5661 (written to 4 decimals).
(a) What would be the corresponding sequence of similitudes chosen in the above program from these random numbers?
(b) If we used the vertices [0 0], [1 sqrt(3)], [2 0] in the above program, find the sequence of different "Float" points of the fractal that would arise if the above sequence of random numbers were to come up.
(c) What happens if the vertices entered in the program sgasket2 are those of a nonequilateral triangle? Will the output ever look anything like a Sierpinski gasket? Explain.

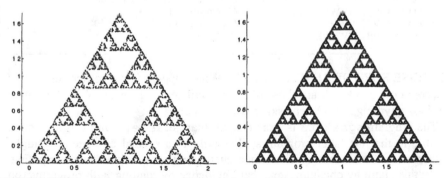

FIGURE 7.16: Monte Carlo renditions of the Sierpinski gasket via the program sgasket2. The left one (a) used 5000 iterations while the right one (b) used 20,000 and took noticeably more time.

Method 3: The last program we write here will actually be the shortest and most versatile of the three. Its drawback is that, unlike the other two, which made use of the specific similitudes associated with the fractal, this program uses the special geometry of the triangle and thus will be more difficult to modify to work for other fractals. The type of geometric/mathematical ideas present in this program, however, are useful in writing other graphics programs. The program sgasket3(V1,V2,V3,ngen) takes as input three vertices V1, V2, V3 of a triangle (written as row vectors), and a positive integer ngen. It will produce a graphic of the ngen-generation Sierpinski gasket, as did the first program. It is again based on the fact that each triangle from a positive generation gasket comes in a very natural way from the triangle of the previous generation in which it lies. Instead of using similitudes and homogeneous coordinates, the program simply uses explicit formulas for the vertices of the $(N+1)$st-generation triangles that lie within a certain Nth-generation triangle. Indeed, for any triangle from any generation of the Sierpinski gasket with vertices V_1, V_2, V_3, three subtriangles of this one form the next generation (see Figure 7.15), each has one vertex from this set, and the other two are the midpoints from this vertex to the other two. For example (again referring to Figure 7.15) the lower-right triangle will have vertices V_2, $(V_1 + V_2)/2$ = the midpoint of V_2V_1, and $(V_2 + V_3)/2$ = the midpoint of V_2V_3. This simple fact, plus recursion, is the idea behind the following program.

PROGRAM 7.3: Function M-file for producing a graphic of any generation of the Sierpinski gasket for an equilateral triangle with vertices *V1*, *V2*, and *V3*.

```
function sgasket3(V1,V2,V3,ngen)
%input variables: V1,V2,V3 are vertices of a triangle in the plane,
%written as row vectors, ngen is the generation of Sierpinski gasket
that will be drawn in medium gray color.
if ngen == 0
%fill triangle
 fill([V1(1) V2(1) V3(1) V1(1)],...
 [V1(2) V2(2) V3(2) V1(2)], [.5 .5 .5])
   hold on
   else
%recursively invoke the same function on three outer subtriangles
   sgasket3(V1, (V1+V2)/2, (V1+V3)/2, ngen-1)
   sgasket3(V2, (V2+V1)/2, (V2+V3)/2, ngen-1)
   sgasket3(V3, (V3+V1)/2, (V3+V2)/2, ngen-1)
end
```

EXERCISE FOR THE READER 7.12: (a) What happens if the vertices entered in the program sgasket3 are those of a nonequilateral triangle? Will the output ever look anything like a Sierpinski gasket? Explain.
(b) The program sgasket3 is more elegant than sgasket1 and it is also more versatile in that the latter program applies only to a special equilateral triangle. Furthermore, it also runs quicker since each iteration involves less computing. Justify this claim by obtaining some hard evidence by running both programs (on the standard equilateral triangle of sgasket1) and comparing tic/toc and flop counts (if available) for each program with the following values for ngen: 1, 3, 6, 8, 10.

Since programs like the one in Method 3 of the above example are usually the most difficult to generalize, we close this section with yet another exercise for the reader that will ask for such a program to draw an interesting and beautiful fractal known as the **von Koch[4] snowflake,** which is illustrated in Figure 7.17. The iteration scheme for this fractal is shown in Figure 7.18.

EXERCISE FOR THE READER 7.13:   Create a MATLAB function, call it snow(n), that will input a positive integer n and will produce the nth generation of the so-called von Koch snowflake fractal.   Note that we start off (generation 0) with an equilateral triangle with sidelength 2.   To get from one generation to the next, we do the following: For each line segment on the boundary, we put up (in the middle of the segment) an equilateral triangle of 1/3 the sidelength.   This construction is illustrated in Figure 7.18, which contains the first few generations of the von Koch snowflake.   Run your program (and include the graphical printout) for the values: $n = 1$, $n = 2$, and $n = 6$.

FIGURE 7.17: The von Koch snowflake fractal. This illustration was produced by the MATLAB program snow(n) of Exercise for the Reader 7.13, with an input value of 6 (generations).

Suggestions: Each generation can be obtained by plotting its set of vertices (using the plot command). You will need to set up a for loop that will be able to produce the next generation's vertices from those of a given generation. It is helpful to think in terms of vectors.

---

[4] The von Koch snowflake was introduced by Swedish mathematician Niels F. H. von Koch (1870–1924) in a 1906 paper *Une méthode géométrique élémentaire pour l'étude de certaines questions de la théorie des courbes planes.* In it he showed that the parametric equations for the curve $(x(t), y(t))$ give an example of functions that are everywhere continuous but nowhere differentiable. Nowhere differentiable, everywhere continuous functions had been first discovered in 1860 by German mathematician Karl Weierstrass (1815–1897), but the constructions known at this time all involved very complicated formulas. Von Koch's example thus gives a curve (of infinite arclength) that is continuous everywhere (no breaks), but that does not have a tangent line at any of its points. The von Koch snowflake has been used in many areas of analysis as a source of examples.

Generation $n = 0$ snowflake:

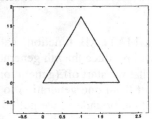

Generation $n = 1$ snowflake:

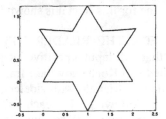

Generation $n = 2$ snowflake:

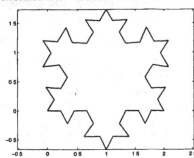

Generation $n = 3$ snowflake:

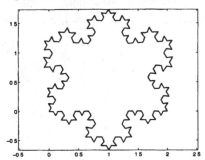

FIGURE 7.18: Some initial generations of the von Koch snowflake. Generation zero is an equilateral triangle (with sidelength 2). To get from any generation to the next, each line segment on the boundary gets replaced by four line segments each having 1/3 of the length of the original segment. The first and fourth segments are at the ends of the original segment and the middle two segments form two sides of an equilateral triangle that protrudes outward.

EXERCISES 7.2:

NOTE: In the problems of this section, the "CAT" refers to the graphic of Example 7.2 (Figure 7.3(a)), the "CAT with eyes" refers to the enhanced version graphic of Example 7.5 (Figure 7.10), and the "full CAT" refers to the further enhanced CAT of Exercise for the Reader 7.7(b) (Figure 7.11). When asked to print a certain transformation of any particular graphic (like the CAT) along with the original, make sure to print the original graphic in one plotting style/color along with the transformed graphic in a different plotting style/color. Also, in printing any graphic, use the `axis(equal)` setting to prevent any distortions and set the axis range to accommodate all of the graphics nicely inside the bounding box

1.   Working in homogeneous coordinates, what is the transformation matrix M that will scale the CAT horizontally by a factor of 2 (to make a "fat CAT") and then shift the cat vertically down a distance 2 and horizontally 1 unit to the left? Create a before and after graphic of the CAT.

2.   Working in homogeneous coordinates, what is the transformation matrix M that will double the size of the horizontal and vertical dimensions of the CAT and then rotate the new CAT by an angle of 45° about the tip of its left ear (the double-sized cat's left ear, that is)? Include a before-and-after graphic of the CAT.

3.   Working in homogeneous coordinates, what is the transformation matrix M that will shift the left eye and pupil of the "CAT with eyes" by 0.5 units and then expand them both by a factor of

2 (away from the centers)? Apply this transformation just to the left eye. Next, perform the analogous transformation to the CAT's right eye and then plot these new eyes along with the outline of the CAT, to get a cat with big eyes.

4. Working in homogeneous coordinates, what is the transformation matrix M that will shrink the "CAT with eyes"'s left eye and left pupil by a factor of 0.5 in the horizontal direction (toward the center of the eye) and then rotate them by an angle of 25° ? Apply this transformation just to the left eye, reflect to get the right eye, and then plot these two along with the outline of the CAT, to get a cat with thinner, slanted eyes.

5. (a) Create a MATLAB function M-file, called reflx(Ah, x0) that has inputs, Ah, a matrix of homogeneous coordinates of some graphic, and a real number x0. The output will be the homogeneous coordinate vertex matrix obtained from Ah by reflecting the graphic over the line $x = x0$. Apply this to the CAT graphic using $x0 = 2$, and give a before-and-after plot.

(b) Create a similar function M-file refly(Ah, y0) for vertical reflections (about the horizontal line $y = y0$) and apply to the CAT using $y0 = 4$ to create a before and after plot.

6. (a) Create a MATLAB function M-file, called shift(Ah,x0,y0) that has as inputs Ah, a matrix of homogeneous coordinates of some graphic, and a pair of real numbers x0, y0. The output will be the homogeneous coordinate vertex matrix obtained from Ah by shifting the graphic using the shift vector (x0, y0). Apply this to the CAT graphic using $x0 = 2$ and $y0 = -1$ and give a before-and-after plot.

(b) Create a MATLAB function M-file, called scale(Ah,a,b,x0,y0) that has inputs Ah, matrix of homogeneous coordinates of some graphic, positive numbers: a and b that represent the horizontal and vertical scaling factors, and a pair of real numbers x0, y0 that represent the coordinates about which the scaling is to be done. The output will be the homogeneous coordinate vertex matrix obtained from Ah by scaling the graphic as indicated. Apply this to the CAT graphic using $a = .25$, $b = 5$ once each with the following sets for $(x0, y0)$: (0,0), (3,0), (0,3), (2.5,4) and create a single plot containing the original CAT along with all four of these smaller, thin cats (use five different colors/plot styles).

7. Working in homogeneous coordinates, what is the transformation matrix M that will reflect an image about the line $y = x$? Create a before-and-after graphic of the CAT.

Suggestion: Rotate first, reflect, and then rotate back again.

8. Working in homogeneous coordinates, what is the transformation matrix M that will shift the left eye and left pupil of the "CAT with eyes" to the left by 0.5 units and then expand them by a factor of 2 (away from the centers)? Apply this transformation just to the left eye, reflect to get the right eye, and then plot these two along with the outline of the "CAT with eyes," to get a cat with big eyes.

9. The **shearing** on $\mathbf{R}^2$ that shears by b in the x-direction and d in the y-direction is the linear transformation whose matrix is $\begin{bmatrix} 1 & b \\ c & 1 \end{bmatrix}$. Apply the shearing to the CAT using several different values of b when $c = 0$, then set $b = 0$ and use several different values of c, and finally apply some shearings using several sets of nonzero values for b and c.

10. (a) Show that the 2×2 matrix $\begin{bmatrix} \cos\theta & -\sin\theta \\ \sin\theta & \cos\theta \end{bmatrix}$, which represents the linear transformation for rotations by angle θ, is invertible, with inverse being the corresponding matrix for rotations by angle $-\theta$.

(b) Does the same relationship hold true for the corresponding 3×3 homogeneous coordinate transform matrices? Justify your answer.

11. (a) Show that the 3×3 matrix $\begin{bmatrix} 1 & 0 & x_0 \\ 0 & 1 & y_0 \\ 0 & 0 & 1 \end{bmatrix}$, which represents the shift with shift vector $\begin{bmatrix} x_0 \\ y_0 \end{bmatrix}$,

is invertible, with its inverse being the corresponding matrix for the shift using the opposite shift vector.

12. Show that the 2×2 matrix $\begin{bmatrix} \cos\theta & -\sin\theta \\ \sin\theta & \cos\theta \end{bmatrix}$ indeed represents the linear transformation for

rotations by angle θ around the origin $(0,0)$.

Suggestion: Let (x,y) have polar coordinates (r,α); then (x',y') has polar coordinates $(r,\alpha+\theta)$. Convert the latter polar coordinates to rectangular coordinates.

13. (*Graphic Art: Rotating Shrinking Squares*) (a) By starting off with a square, and repeatedly shrinking it and rotating it, get MATLAB to create a graphic similar to the one shown in Figure 7.19(a).
(b) Next modify your construction to create a graph similar to the one in Figure 7.19(b) but that uses alternating colors.
Note: This object is not a fractal.

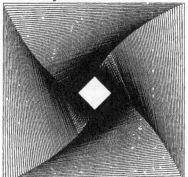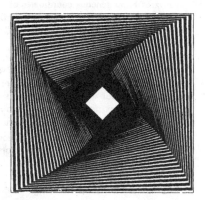

FIGURE 7.19: A rotating and shrinking square of Exercise 13: (a) (left) with no fills; (b) (right) with alternate black-and-white fills.

14. (*Graphic Art: Cat with Eyes Mosaic*) The cat mosaic of Figure 7.20 has been created by taking the original CAT, and creating new pairs of cats (left and right) for each step up. This construction was done with a for loop using 10 iterations (so there are 10 pairs of cats above the original), and could easily have been changed to any number of iterations. Each level upward of cats got scaled to 79% of the preceding level. Also, for symmetry, the left and right cats were shifted upward and to the left and right by the same amounts, but these amounts got smaller (since the cat size did) as we moved upward.
(a) Use MATLAB to create a picture that is similar to that of Figure 7.20, but replace the "CAT with eyes" with the ordinary CAT.
(b) Use MATLAB to create a picture that is similar to that of Figure 7.20.
(c) Use MATLAB to create a picture that is similar to that of Figure 7.20, but replace the "CAT with eyes" with the "full CAT" of Figure 7.11.
Suggestion: You should definitely use a for loop. Experiment a bit with different schemes for horizontal and vertical shifting to get your picture to look like this one.

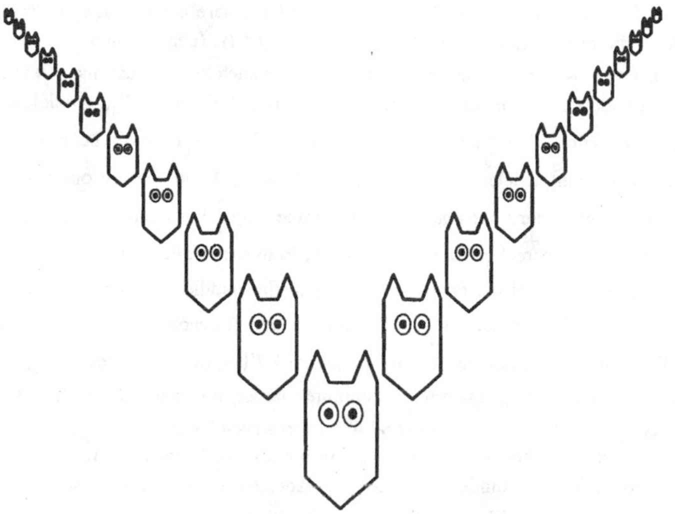

FIGURE 7.20: CAT with eyes mosaic for Exercise 14(b). The original cat (center) has been repeatedly shifted to the left and right, and up, as well as scaled by a factor of 79% each time we go up.

15. (*Movie: "Sudden Impact"*) (a) Create a movie that stars the CAT and proceeds as follows: The cat starts off at the left end of the screen. It then "runs" horizontally towards the right end of the screen. Just as its right side reaches the right side of the screen, it begins to shrink horizontally (but not vertically) until it degenerates into a vertical line segment on the right side of the screen.
(b) Make a movie similar to the one in part (a) except that this one stars the "CAT with eyes" and before it begins to run to the right, its pupils move to the right of the eyes and stay there.
(c) Make a film similar to the one in part (b) except that this one should star the "full CAT" (Figure 7.11) and upon impact with the right wall, the cat's smile changes to a frown.

16. (*Movie: "The Chase"*) (a) Create a movie that stars the "CAT with eyes" and co-stars another smaller version of the same cat (scaled by factors of 0.5 in both the x- and y-directions). The movie starts off with the big cat in the upper left of the screen and the small cat to its right side (very close). Their pupils move directly toward one another to the end of the eyes, and at this point both cats begin moving at constant speed toward the right. When the smaller cat reaches the right side of the screen, it starts moving down while the big cat also starts moving down. Finally, cats stay put in the lower-right corner as their pupils move back to center.
(b) Make the same movie except starring the "full CAT" and costarring a smaller counterpart.

17. (*Movie: "Close Encounter"*) (a) Create a movie that stars the "full CAT" (Figure 7.11) and with the following plot: The cat starts off smiling and then its eyes begin to shift all the way to the lower left. It spots a solid black rock moving horizontally directly toward its mouth level, at constant speed. As the cat spots this rock, its smile changes to a frown. It jumps upward as its pupils move back to center and just misses the rock as it brushes just past the cat's chin. The cat then begins to smile and falls back down to its original position.
(b) Make a film similar to the one in part (a) except that it has the additional feature that the rock is rotating clockwise as it is moving horizontally.
(c) Make a film similar to the one in part (b) except that it has the additional feature that the cat's pupils, after having spotted the rock on the left, slowly roll (along the bottom of the eyes) to the lower-right position, exactly following the rock. Then, after the rock leaves the viewing window, have the cat's pupils move back to center position.

18.   (*Fractal Geometry: The Cantor Square*) The Cantor square is a fractal that starts with the *unit square* in the plane: $C_0 = \{(x,y) : 0 \leq x \leq 1 \text{ and } 0 \leq y \leq 1\}$ (generation zero). To move to the next generation, we delete from this square all points such that at least one of the coordinates is inside the middle 1/3 of the original spread. Thus, to get C_1 from C_0, we delete all the points (x,y) having either $1/3 < x < 2/3$ or $1/3 < y < 2/3$. So C_1 will consist of four smaller squares each having sidelength equal to 1/3 (that of C_0) and sharing one corner vertex with C_0. Future generations are obtained in the same way. For example, to get from C_1 (first generation) to C_2 (second generation) we delete, from each of the four squares of C_1, all points (x,y) that have one of the coordinates lying in the middle 1/3 of the original range (for a certain square of C_1). What will be left is four squares for each of the squares of C_1, leaving a total of 16 squares each having sidelength equal to 1/3 that of the squares of C_1, and thus equal to 1/9. In general, letting this process continue forever, it can be shown by induction that the nth-generation Cantor square consists of 4^n squares each having sidelength $1/3^n$. The Cantor square is the set of points that remains after this process has been continued indefinitely.

        (a) Identify the four similitudes S_1, S_2, S_3, S_4 associated with the Cantor square (an illustration as in Figure 7.16 would be fine) and then, working in homogeneous coordinates, find the matrices of each. Next, following the approach of Method 1 of the solution of Example 7.7, write a function M-file cantorsq1(V1,V2,V3,V4, ngen), that takes as input the vertices V1 = [0 0], V2 = [1 0], V3 = [1 1], and V4 = [0 1] of the unit square and a nonnegative integer ngen and will produce a graphic of the generation ngen Cantor square.

        (b) Write a function M-file cantorsq2(V1,V2,V3,V4, niter) that takes as input the vertices V1, V2, V3 V4 of any square and a positive integer niter and will produce a Monte Carlo generated graphic for the Cantor square as in Method 2 of the solution of Example 7.7. Run your program for the square having sidelength 1 and lower-left vertex $(-1, 2)$ using niter $= 2000$ and niter $= 12,000$.

        (c) Write a function M-file cantorsq3(V1,V2,V3,V4, ngen) that takes as input the vertices V1, V2, V3 V4 of any square and a positive integer ngen and will produce a graphic for the ngen generation Cantor square as did cantorsq1 (but now the square can be any square). Run your program for the square mentioned in part (b) first with ngen = 1 then with ngen = 3. Can this program be written so that it produces a reasonable generalization of the Cantor square when the vertices are those of any rectangle?

19.   (*Fractal Geometry: The Sierpinski Carpet*) The Sierpinski carpet is the fractal that starts with the unit square $\{(x,y) : 0 \leq x \leq 1 \text{ and } 0 \leq y \leq 1\}$ with the central square of 1/3 the sidelength removed (generation zero). To get to the next generation, we punch eight smaller squares out of each of the remaining eight squares of sidelength 1/3 (generation one), as shown in Figure 7.21. Write a function M-file, scarpet2(niter), based on the Monte Carlo method that will take only a single input variable niter and will produce a Monte Carlo approximation of the Sierpinski carpet. You will, of course, need to find the eight similitudes associated with this fractal and get their matrices in homogeneous coordinates. Run your program with inputs niter $= 1000, 2000, 5000,$ and $10,000$.

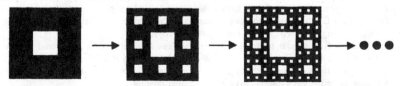

FIGURE 7.21: Illustration of generations zero (left), one (middle), and two (right) of the Sierpinski gasket fractal of Exercises 19, 20, and 21. The fractal consists of the points that remain (shaded) after this process has continued on indefinitely.

20. (*Fractal Geometry: The Sierpinski Carpet*) Read first Exercise 19 (and see Figure 7.21), and if you have not done so yet, identify the eight similitudes $S_1, S_2, \cdots, S_8$ associated with the Sierpinski carpet along with the homogeneous coordinate matrices of each. Next, following the approach of Method 1 of the solution of Example 7.7, write a function M-file scarpet1 (V1, V2, V3, V4, ngen) that takes as input the vertices V1 = [0 0], V2 = [1 0], V3 = [1 1], and V4 = [0 1] of the unit square and a nonnegative integer ngen and will produce a graphic of the generation ngen Cantor square.

    Suggestions: Fill in each outer square in gray, then to get the white central square "punched out," use the hold on and then fill in the smaller square in the color white (rgb vector [1 1 1]). When MATLAB fills a polygon, by default it draws the edges in black. To suppress the edges from being drawn, use the following extra option in the fill commands: fill (xvec, yvec, rgbvec, 'EdgeColor', 'none'). Of course, another nice way to edit a graphic plot from MATLAB is to import the file into a drawing software (such as Adobe Illustrator or Corel Draw) and modify the graphic using the software

21. (*Fractal Geometry: The Sierpinski Carpet*) (a) Write a function M-file called scarpet3 (V1, V2, V3, V4, ngen) that works just like the program scarpet1 of the previous exercise, except that the vertices can be those of any square. Also, base the code not on similitudes, but rather on mathematical formulas for next-generation parameters in terms of present-generation parameters. The approach should be somewhat analogous to that of Method 3 of the solution to Example 7.7.

    (b) Is it possible to modify the sgasket1 program so that it is able to take as input the vertices of any equilateral triangle? If yes, indicate how. If no, explain why not.

22. (*Fractal Geometry: The Fern Leaf*) There are more general ways to construct fractals than those that came up in the text. One generalization of the self similarity approach given in the text allows for transformations that are not invertible (similitudes always are). In this exercise you are to create a function M-file, called fracfern (n), which will input a positive integer n and will produce a graphic for the fern fractal pictured in Figure 7.22, using the Monte Carlo method. For this fractal the four transformations to use are (given by their homogeneous coordinate matrices)

$$S1 = \begin{bmatrix} 0 & 0 & 0 \\ 0 & .16 & 0 \\ 0 & 0 & 1 \end{bmatrix}, \quad S2 = \begin{bmatrix} .85 & .04 & 0 \\ -.04 & .85 & 1.6 \\ 0 & 0 & 0 \end{bmatrix},$$

$$S3 = \begin{bmatrix} .2 & -.26 & 0 \\ .23 & .22 & 1.6 \\ 0 & 0 & 1 \end{bmatrix}, \quad S4 = \begin{bmatrix} -.15 & .28 & 0 \\ .26 & .24 & .44 \\ 0 & 0 & 1 \end{bmatrix},$$

FIGURE 7.22: The fern leaf fractal.

    and the associated probability vector is [.01   .86   .93] (i.e., in the Monte Carlo process, 1% of the time we choose $S1$, 85% of the time we choose $S2$, 7% of the time we choose $S3$, and the remaining 7% of the time we choose $S4$).

    Suggestion: Simply modify the program sgasket2 accordingly.

23. (*Fractal Geometry: The Gosper Island*) (a) Write a function M-file gosper (n) that will input a positive integer n and will produce a graphic of the nth generation of the **Gosper island** fractal, which is defined as follows: Generation zero is a regular hexagon (with, say, unit side lengths). To get from this to generation one, we replace each of the six sides on the boundary of generation zero with three new segments as shown in Figure 7.23. The first few generations of the Gosper island are shown in Figure 7.24.

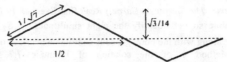

FIGURE 7.23: Iteration scheme for the definition of the Gosper island fractal of Exercise 23. The dotted segment represents a segment of a certain generation of the Gosper island, and the three solid segments represent the corresponding part of the next generation.

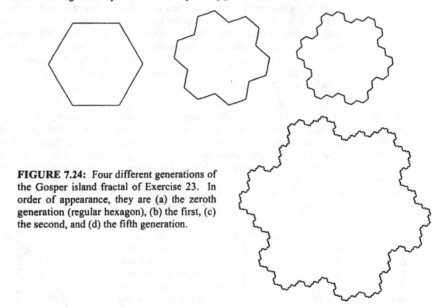

FIGURE 7.24: Four different generations of the Gosper island fractal of Exercise 23. In order of appearance, they are (a) the zeroth generation (regular hexagon), (b) the first, (c) the second, and (d) the fifth generation.

(b) (*Tessellations of the Plane*) It is well known that the only regular polygons that can tessellate (or tile) the plane are the equilateral triangle, the square, and the regular hexagon (honeybees have figured this out). It is an interesting fact that any generation of the Gosper island can also be used to tessellate the plane, as shown in Figure 7.25. Get MATLAB to reproduce each of tessellations that are shown in Figure 7.25.

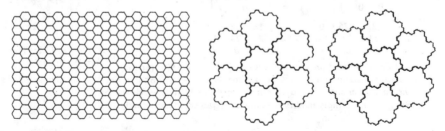

FIGURE 7.25: Tessellations with generations of Gosper islands. The top one (with regular hexagons) is the familiar honeycomb structure.

7.3: NOTATIONS AND CONCEPTS OF LINEAR SYSTEMS

The general linear system in n variables $x_1, x_2, \cdots, x_n$ and n equations can be written as

$$\begin{cases} a_{11}x_1 + a_{12}x_2 + \cdots + a_{1n}x_n = b_1 \\ a_{21}x_1 + a_{22}x_2 + \cdots + a_{2n}x_n = b_2 \\ \quad\cdots \\ a_{n1}x_1 + a_{n2}x_2 + \cdots + a_{nn}x_n = b_n \end{cases} \quad (17)$$

Here, the a_{ij}, and b_j represent given data, and the variables $x_1, x_2, \cdots, x_n$ are the unknowns whose solution is sought. In light of how matrix multiplication is defined, these n equations can be expressed as a single matrix equation:

$$Ax = b, \quad (18)$$

where

$$A = \begin{bmatrix} a_{11} & a_{12} & a_{13} & \cdots & a_{1n} \\ a_{21} & a_{22} & a_{23} & \cdots & a_{2n} \\ \vdots & & & \ddots & \vdots \\ a_{n1} & a_{n2} & a_{n3} & \cdots & a_{nn} \end{bmatrix}, \quad x = \begin{bmatrix} x_1 \\ x_2 \\ \vdots \\ x_n \end{bmatrix}, \quad \text{and} \quad b = \begin{bmatrix} b_1 \\ b_2 \\ \vdots \\ b_n \end{bmatrix}. \quad (19)$$

It is also possible to consider more general linear systems that can contain more or fewer variables than equations, but such systems represent *ill-posed* problems in the sense that they typically do not have unique solutions. Most linear systems that come up in applications will be *well-posed*, meaning that there will exist a unique solution, and thus we will be focusing most of our attention on solving well-posed linear systems. The above linear system is well posed if and only if the coefficient matrix A is invertible, in which case the solution is easily obtained, by left multiplying the matrix equation by A^{-1}: $Ax = b \Rightarrow A^{-1}Ax = A^{-1}b \Rightarrow$

$$x = A^{-1}b. \quad (20)$$

Despite its algebraic simplicity, however, this method of solution, namely computing and then left multiplying by A^{-1}, is, in general, an inefficient way of solving the system. The best general methods for solving linear systems are based on the Gaussian elimination algorithm. Such an algorithm is actually tacitly used by the MATLAB command for matrix division (left divide):

x=A\b →	Solves the matrix equation Ax = b by an elaborate Gaussian elimination procedure.

If the coefficient matrix A is invertible but "close" to being singular and/or if the size of A is large, the system (17) can be difficult to solve numerically. We will make this notion more precise later in this chapter. In case where there are two or three variables, the concepts can be illustrated effectively using geometry.

CASE: n = 2 For convenience of notation, we drop the subscripts and rewrite the system (17) as:

$$\begin{cases} ax + by = e \\ cx + dy = f \end{cases} . \quad (21)$$

The system (21) represents a pair of lines in the plane and the solutions, if any, are the points of intersection. Three possible situations are illustrated in Figure 7.26. We recall (Theorem 7.1) that the coefficient matrix A is nonsingular exactly when $\det(A)$ is nonzero. The singular case thus has the lines being parallel (Figure 7.26(c)). Nearly parallel lines (Figure 7.26(b)) are problematic since they are difficult to distinguish numerically from parallel lines. The determinant alone is not a reliable indicator of near singularity of a system (see Exercise for the Reader 7.14), but the condition number introduced later in this chapter will be.

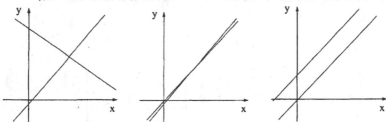

FIGURE 7.26: Three possibilities for the system (21) $\begin{cases} ax + by = e \\ cx + dy = f \end{cases}$: (a) well-conditioned, (b) ill-conditioned (nearly parallel lines), and (c) singular (parallel lines).

EXERCISE FOR THE READER 7.14: Show that for any pair of nearly parallel lines in the plane, it is possible to represent this system by a matrix equation $Ax = b$, which uses a coefficient matrix A with $\det(A) = 1$.

CASE: $n = 3$ A linear equation $ax + by + cz = d$ represents a plane in three-dimensional space $\mathbf{R}^3$. Typically, two planes in $\mathbf{R}^3$ will intersect in a line and a typical intersection of a line with a third plane (and hence the typical intersection of three planes) will be a point. This is the case if the system is nonsingular. There are several ways for such a three-dimensional system to be singular or nearly so. Apart from two of the planes being parallel (making a solution impossible), another way to have a singular system is for one of the planes to be parallel to the line of intersection of the other two. Some of these possibilities are illustrated in Figure 7.27.

For higher-order systems, the geometry is similar although not so easy to visualize since the world we live in has only three (visual) dimensions. For example, in four dimensions, a linear equation $ax + by + cz + dw = e$ will, in general, be a three-dimensional hyperplane in the four-dimensional space $\mathbf{R}^4$. The intersection of two such hyperplanes will typically be a two-dimensional plane in $\mathbf{R}^4$. If we intersect with one more such hyperplane we will (in nonsingular cases) be left with a line in $\mathbf{R}^4$, and finally if we intersect this line with a fourth such hyperplane we will be left with a point, the unique solution, as long as the system is nonsingular.

The variety of singular systems gets extremely complicated for large values of n, as is partially previewed in Figure 7.27 for the case $n = 3$. This makes the determination of near singularity of a linear system a complicated issue, which is why we will need analytical (rather than geometric) ways to detect this.

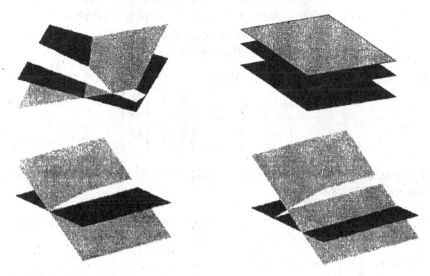

FIGURE 7.27: Four geometric possibilities for a three-dimensional system (18) $Ax = b$. (a) (upper left) Represents a typical nonsingular system, three planes intersect at one common point; (b) (upper right) parallel planes, no solution, a singular system; (c) (lower left) three planes sharing a line, infinitely many solutions, a singular system; (d) (lower right) three different parallel lines arise from intersections of pairs of the three planes, no solution, singular system.[5]

7.4: SOLVING GENERAL LINEAR SYSTEMS WITH MATLAB

The best all-around algorithms for solving linear systems (17) and (18) are based on Gaussian elimination with partial pivoting. MATLAB's default linear system solver is based on this algorithm. In the next section we describe this algorithm; here we will show how to use MATLAB to solve such systems. In the following example we demonstrate three different ways to solve a (nonsingular) linear system in MATLAB by solving the following interpolation problem, and compare flop counts.

EXAMPLE 7.8: *(Polynomial Interpolation)* Find the equation of the polynomial $p(x) = ax^3 + bx^2 + cx + d$ of degree at most 3 that passes through the data points $(-5, 4)$, $(-3, 34)$, $(-1, 16)$, and $(1, 2)$.

---

[5] Note: These graphics were created on MATLAB.

SOLUTION: In general, a set of n data points (with different x-coordinates) can always be interpolated with a polynomial of degree at most $n-1$. Writing out the interpolation equations produces the following four-dimensional linear system:

$$p(-5) = 4 \Rightarrow a(-5)^3 + b(-5)^2 + c(-5) + d = 4$$
$$p(-3) = 34 \Rightarrow a(-3)^3 + b(-3)^2 + c(-3) + d = 34$$
$$p(-1) = 16 \Rightarrow a(-1)^3 + b(-1)^2 + c(-1) + d = 16$$
$$p(1) = 2 \Rightarrow a \cdot 1^3 + b \cdot 1^2 + c \cdot 1 + d = -2.$$

In matrix (18) form this system becomes:

$$\begin{bmatrix} -125 & 25 & -5 & 1 \\ -27 & 9 & -3 & 1 \\ -1 & 1 & -1 & 1 \\ 1 & 1 & 1 & 1 \end{bmatrix} \begin{bmatrix} a \\ b \\ c \\ d \end{bmatrix} = \begin{bmatrix} 4 \\ 34 \\ 16 \\ -2 \end{bmatrix}.$$

We now solve this matrix equation $Ax = b$ in three different ways with MATLAB, and do a flop count[6] for each.

Method 1: Left divide by A. This is the Gaussian elimination method mentioned in the previous section. It is the recommended method.

```
>> format long, A=[-125 25 -5 1;-27 9 -3 1;-1 1 -1 1;1 1 1 1];
   b=[4 34 16 -2]';
>> flops(0);
>> x=A\b
→x =          1.00000000000000
              3.00000000000000
            -10.00000000000000
              4.00000000000000

>> flops     →ans =180
```

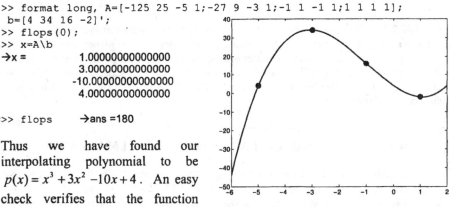

Thus we have found our interpolating polynomial to be $p(x) = x^3 + 3x^2 - 10x + 4$. An easy check verifies that the function indeed interpolates the given data. Its graph along with the data points are shown in Figure 7.28.

FIGURE 7.28: Graph of the interpolating cubic polynomial $p(x) = x^3 + 3x^2 - 10x + 4$ for the four data points that were given in Example 7.8.[7]

---

[6] Flop counts will help us to compare efficiencies of algorithms. Later versions of MATLAB (Version 6 and later) no longer support the flop count feature. We will occasionally tap into the older Version 5 of MATLAB (using flops) to compare flop counts, for purely illustrative purposes.

[7] The MATLAB plot in the figure was created by first plotting the function, as usual, applying hold on, and then plotting each of the points as red dots using the following syntax (e.g., for the data point $(-1, 16)$), plot(-1,16,'ro'). The "EDIT" menu on the graphics window can then be used to enlarge the size of the red dots after they are selected.

Method 2: We compute the inverse of A and multiply this inverse on the left of b.

```
>> flops(0); x=inv(A)*b
→x =          1.00000000000000
              3.00000000000000
            -10.00000000000000
              4.00000000000000
>> flops    →ans = 262
```

We arrived at the same (correct) answer, but with more work. The amount of extra work needed to compute the inverse rather than just solve the system gets worse with larger-sized matrices; moreover, this method is also more prone to errors. We will give more evidence and will substantiate these claims later in this section.

Method 3: This method is more general than the first two since it will work also to solve singular systems that need not have square coefficient matrices. MATLAB has the following useful command:

`rref(Ab) →`	Puts an augmented $n \times (m+1)$ matrix $[A \mid b]$ for the system $Ax = b$ into reduced row echelon form.

The reduced row echelon form of the augmented matrix is a form from which the general solution of the linear system can be easily obtained. In general, a linear system can have (i) no solution, (ii) exactly one solution (nonsingular case), or (iii) infinitely many solutions. We will say more about how to interpret the reduced row echelon form in singular cases later in this section, but in case of a nonsingular (square) coefficient matrix A, the reduced row echelon form of the augmented matrix $[A \mid b]$ will be $[I \mid x]$, where x is the solution vector.

Assuming A and b are still in the workspace, we construct from them the augmented matrix and then use `rref`:

```
>> Ab=A; Ab(:,5)=b;
>> flops(0); rref(Ab)
→ ans =     1   0   0   0   1
            0   1   0   0   3
            0   0   1   0  -10
            0   0   0   1   4
>> flops    →ans = 634
```

Again we obtained the same answer, but with a lot more work (more than triple the flops that were needed in Method 1). Although `rref` is also based in Gaussian elimination, putting things into reduced row echelon form (as is usually taught in linear algebra courses) is a computational overkill of what is needed to solve a nonsingular system. This will be made apparent in the next section.

EXERCISE FOR THE READER 7.15: (a) Find the coefficients a, b, c, and d of the polynomial $p(x) = ax^3 + bx^2 + cx + d$ of degree at most 3 that passes through these data points: $(-2, 4), (1, 3), (2, 5)$, and $(5, -22)$.

(b) Find the equation of the polynomial $p(x) = ax^8 + bx^7 + cx^6 + dx^5 + ex^4 + fx^3 + gx^2 + hx + k$ of degree at most 8 that passes through the following data points: $(-3, -14.5)$, $(-2, -12)$, $(-1, 15.5)$, $(0, 2)$, $(1, -22.5)$, $(2, -112)$, $(3, -224.5)$, $(4, 318)$, $(5, 3729.5)$. Solve the system using each of the three methods shown in Example 7.8 and compare the solutions, computation times (using tic/toc), and flop counts (if available).

In the preceding example, things worked great and the answer MATLAB gave us was the exact one. But the matrix was small and not badly conditioned. In general, solving a large linear system involves a large sequence of arithmetic operations and the floating point errors can add up, sometimes (for poorly conditioned matrices) intolerably fast. The next example will demonstrate such a pathology, and should convince the reader of our recommendation to opt for Method 1 (left divide).

EXAMPLE 7.9: (*The Hilbert Matrix*) A classical example of a matrix that comes up in applications and is very poorly conditioned is

$$H_n = \begin{bmatrix} 1 & 1/2 & 1/3 & \cdots & 1/n \\ 1/2 & 1/3 & 1/4 & \cdots & 1/(n+1) \\ \vdots & & & \ddots & \vdots \\ 1/n & 1/(n+1) & 1/(n+2) & \cdots & 1/(2n-1) \end{bmatrix},$$

the **Hilbert matrix** H_n **of order** n, which is defined above. This matrix can easily be entered into a MATLAB session using a for loop, but MATLAB even has a separate command hilb(n) to create it. In this example we will solve, using each of the three methods from the last example, the equation $Hx = b$, where b is the vector Hx, and where $x = (1\ 1\ 1\ \cdots\ 1)'$ with $n = 12$.

We now proceed with each of the three methods of the solution of the last example to produce three corresponding "solutions," x_meth1, x_meth2, and x_meth3, to the linear system $Hx = b$. Since we know the exact solution to be $x = (1\ 1\ 1\ \cdots\ 1)'$, we will be able to compare both accuracy and speeds of the three methods.

```
>> x=ones(12,1); H=hilb(12); b=H*x;
>> flops(0); x_meth1=H\b10; max(abs(x-x_meth1))      → 0.2385
>> %because each component of the exact solution is 1, this can be
>> % thought of as the maximum relative error in any component.
>> flops        → 995
>> flops(0);
>> x_meth2=inv(H)*b; max(abs(x-x_meth2))→ 0.6976
>> flops        → 1997
>> flops(0);, R=rref([H b]); flops        →ans = 6484
```

If we view row reduced matrix R produced in Method 3 (or just the tenth row by entering R(12,:)), we see that the last row is entirely made up of zeros. Thus, Method 3 leads us to the false conclusion that the linear system is singular.

The results of Method 3 are catastrophic, since the reduced row echelon form produced would imply that there are infinitely many solutions! (We will explain this conclusion shortly.) Methods 1 and 2 both had (unacceptably) large relative errors of about 24% and 70%, respectively. The increasing flop counts in the three methods also shows us that left divide (Method 1) gives us more for less work. Of course, the Hilbert

FIGURE 7.29: David Hilbert[8] (1862–1943), German mathematician.

matrix is quite a pathological one, but such matrices do come up often enough in applications that we must always remember to consider floating point/roundoff error when we solve linear systems. We point out that many problems that arise in applied mathematics involve very well-conditioned linear systems that can be effectively solved (using MATLAB) for dimensions of order 1000! This is the case for many linear systems that arise in the numerical solution of partial differential equations. In Section 7.6, we will examine more closely the concept

---

[8]The Hilbert matrix is but a small morsel of the vast set of mathematical contributions produced in the illustrious career of David Hilbert (Figure 7.29). Hilbert is among the very top echelon of the greatest German mathematicians, and this group contains a lot of competition. Hilbert's ability to transcend the various subfields of mathematics, to delve deeply into difficult problems, and to discover fascinating interrelations was unparalleled. Three years after earning his Ph.D., he submitted a monumental paper containing a whole new and elegant treatment of the so-called *Basis Theorem*, which had been proved by Paul Albert Gordan (1837–1912) in a much more specialized setting using inelegant computational methods. Hilbert submitted his manuscript for publication to the premier German mathematical journal *Mathematische Annalen*, and the paper was sent to Gordan to (anonymously) referee by the editor Felix Christian Klein (1849–1925), also a famous German mathematician and personal friend of Gordan's. It seemed that Gordan, the world-renowned expert in the field of Hilbert's paper, was unable to follow Hilbert's reasoning and rejected the paper for publication on the basis of its incoherence. In response, Hilbert wrote to Klein, " *.. I am not prepared to alter or delete anything, and regarding this paper, I say with all modesty, that this is my last word so long as no definitive and irrefutable objection against my reasoning is raised.*" Hilbert's paper finally appeared in this journal in its original form, and he continued to produce groundbreaking papers and influential books in an assortment of fields spanning all areas of mathematics. He even has a very important branch of analysis named in his honor (Hilbert space theory). In the 1900 International Conference of Mathematics in Paris, Hilbert posed 23 unsolved problems to the mathematical world. These "Hilbert problems" have influenced much mathematical research activity throughout the twentieth century. Several of these problems have been solved, and each such solution marked a major mathematical event. The remaining open problems should continue to influence mathematical thoughts well into the present millennium. In 1895, Hilbert was appointed to a "chair" at the University of Göttingen and although he was tempted with offers from other great universities, he remained at this position for his entire career. Hilbert retired in his birthplace city of Königsberg, which (since he had left it) had become part of Russia after WWII (with the name of the city changed to "Kaliningrad"). Hilbert was made an honorary citizen of this city and in his acceptance speech he gave this now famous quote: "*Wir müssen wissen, wir werden wissen (We must know, we shall know).*"

of the condition number of a matrix and its affect on error bounds for numerical solutions of corresponding linear systems. Just because a matrix is poorly conditioned does not mean that all linear systems involving it will be difficult to solve numerically. The next exercise for the reader gives such an example with the Hilbert matrices.

EXERCISE FOR THE READER 7.16: In this exercise, we will be considering larger analogs of the system studied in Example 7.9. For a positive integer n, we let $c(n)$ be the least common multiple of the integers 1, 2, 3, ..., n, and we define

$b_n = H_n(c(n)e_1')$, where H_n is the nth-order Hilbert matrix and e_1 is the vector $(1,0,0, ..., 0)$ (having n components). We chose $c(n)$ to be as small as possible so that the vector b_n will have all integer components. Note, in Example 7.9, we used $c(10) = 2520$.

(a) For $n = 20$, solve this system using Method 1, time it with tic/toc, and (if available) do a flop count and find the percentage of the largest error in any of the 20 components of x_meth1 to that of the largest component of the exact solution x $(= c(20))$; repeat using Method 2.

(b) Repeat part (a) for $n = 30$.

(c) In parts (a) and (b), you should have found that x_meth1 equaled the exact solution (so the corresponding relative error percentages were zero), but the relative error percentages for x_meth2 grew from 0.00496% in case $n = 10$ (Example 7.9), to about 500% in part (a) and to about 5000% in part (b). Thus the "noise" from the errors has transcended, by far, the values of the exact solution. Continue solving this system using Method 1, for $n = 40$, $n = 50$, and so on, until you start getting errors from the exact solution or the computations start to take too much time, whichever comes first.

Suggestion: MATLAB has a built-in function lcm(a,b) that will find the least common multiple of two positive integers. You can use this, along with a for loop to get MATLAB to easily compute $c(n)$, for any value of n. In each part, you may wish to use max(abs(x-x_meth1)) to detect any errors.

Note: Exercise 31 of Section 7.6 will analyze why things have gone so well with these linear systems.

We close this section by briefly explaining how to interpret the reduced row echelon form to obtain the general solution of a linear system with a singular coefficient matrix that need not be square. Suppose we have a linear system with n equations and m unknowns $x_1, x_2, \cdots, x_m$:

$$\begin{cases} a_{11}x_1 + a_{12}x_2 + \cdots + a_{1m}x_m = b_1 \\ a_{21}x_1 + a_{22}x_2 + \cdots + a_{2m}x_m = b_2 \\ \quad\quad\quad \cdots \\ a_{n1}x_1 + a_{n2}x_2 + \cdots + a_{nm}x_m = b_m \end{cases} . \tag{22}$$

We can write this equation in matrix form $Ax = b$, as before; but in general the coefficient matrix A need not be square. We form the **augmented matrix** of the system by tacking on the vector b as an extra column on the right of A :

$$[A \mid b] = \begin{bmatrix} a_{11} & a_{12} & a_{13} & \cdots & a_{1m} & \vdots & b_1 \\ a_{21} & a_{22} & a_{23} & \cdots & a_{2m} & \vdots & b_2 \\ \vdots & & & \ddots & \vdots & \vdots & \vdots \\ a_{n1} & a_{n2} & a_{n3} & \cdots & a_{nm} & \vdots & b_n \end{bmatrix}.$$

The augmented matrix is said to be in **reduced row echelon form** if the following four conditions are met. Each condition pertains only to the left of the partition line (i.e., the a_{ij} 's):

1. Rows of all zero entries, if any, must be grouped together at the bottom.
2. If a row is not all zeros, the leftmost nonzero entry must equal 1 (such an entry will be called a **leading one** for the row).
3. All entries above and below (in the same column as) a leading one must be zero.
4. If there are more than one leading ones, they must move to the right as we move down to lower rows.

Given an augmented matrix A, the command `rref(Ab)` will output an augmented matrix of the same size (but MATLAB will not show the partition line) that is in reduced row echelon form and that represents an **equivalent** linear system to Ab, meaning that both systems will have the same solution. It is easy to get the solution of any linear system, singular or not, if it is in reduced row echelon form. Since most of our work will be with nonsingular systems (for which `rref` should not be used), we will not say more about how to construct the reduced row echelon form. The algorithm is based on Gaussian elimination, which will be explained in the next section. For more details on the reduced row echelon form, we refer to any textbook on basic linear algebra; see, e.g., [Kol-99], or [Ant-00]. We will only show, in the next example, how to obtain the general solution from the reduced row echelon form.

EXAMPLE 7.10: (a) Which of the following augmented matrices are in reduced row echelon form?

$$M_1 = \begin{bmatrix} 1 & 2 & 0 & \vdots & -2 \\ 0 & 0 & 1 & \vdots & 3 \\ 0 & 0 & 0 & \vdots & 0 \end{bmatrix}, \quad M_2 = \begin{bmatrix} 1 & 2 & 0 & 1 & \vdots & 1 \\ 0 & 1 & 0 & 2 & \vdots & -8 \\ 0 & 0 & 1 & 3 & \vdots & 4 \end{bmatrix}, \quad M_3 = \begin{bmatrix} 1 & 0 & 0 & 1 & \vdots & 1 \\ 0 & 1 & 0 & 2 & \vdots & -8 \\ 0 & 0 & 0 & 0 & \vdots & 4 \end{bmatrix}.$$

(b) For those that are in reduced row echelon form, find the general solution of the corresponding linear system that the matrix represents.

SOLUTION: Part (a): M_1 and M_3 are in reduced row echelon form; M_2 is not. The reader who is inexperienced in this area of linear algebra should carefully verify these claims for each matrix by running through all four of the conditions (i) through (iv).

Part (b): If we put in the variables and equal signs in the three-equation linear system represented by M_1, and then solve for the variables that have leading ones in their places (here x_1 and x_3), we obtain:

$$\begin{cases} x_1 + 2x_2 & = -2 \Rightarrow x_1 = -2 - 2x_2 \\ x_3 = 3 \\ 0 = 0 \end{cases},$$

Thus there are infinitely many solutions: Letting $x_2 = t$, where t is any real number, the general solution can be expressed as

$$\begin{cases} x_1 = -2 - 2t \\ x_2 = t \\ x_3 = 3 \end{cases}, \quad t = \text{any real number}.$$

If we do the same with the augmented matrix M_3, we get $0 = 4$ for the last equation of the system. Since this is impossible, the system has no solution.

Here is a brief summary of how to solve a linear system with MATLAB. For a square matrix A, you should always try x = A\b (left divide). For singular matrices (in particular, nonsquare matrices) use rref on the augmented matrix. There will be either no solution or infinitely many solutions. No solution will always be seen in the reduced row echelon form matrix by a row of zeros before the partition line and a nonzero entry after it (as in the augmented matrix M_3 in the above example). In all other (singular) cases there will be infinitely many solutions; columns without leading ones will correspond to variables that are to be assigned arbitrary real numbers (s, t, u, ...); columns with leading ones correspond to variables that should be solved for in terms of variables of the first type.

EXERCISE FOR THE READER 7.17: Parts (a) and (b): Repeat the instructions of both parts (a) and (b) of the preceding example for the following augmented matrices:

$$M_1 = \begin{bmatrix} 1 & 0 & | & 3 \\ 0 & 1 & | & 2 \\ 0 & 0 & | & 0 \end{bmatrix}, \quad M_2 = \begin{bmatrix} 1 & -2 & 0 & 3 & | & -2 \\ 0 & 0 & 1 & -5 & | & 1 \end{bmatrix}, \quad M_3 = \begin{bmatrix} 1 & 0 & 0 & | & 1 \\ 0 & 0 & 1 & | & 3 \\ 0 & 1 & 0 & | & 4 \end{bmatrix}.$$

Part (c): Using the MATLAB command rref, find the general solutions of the following linear systems:

(i) $\begin{cases} x_1 + 3x_2 + 2x_3 & = 3 \\ 2x_1 + 6x_2 + 2x_3 - 8x_4 & = 4 \end{cases}$

(ii) $\begin{cases} x_1 & - & 2x_2 & + & x_3 & + & x_4 & + & 2x_5 & = & 2 \\ -2x_1 & + & 4x_2 & + & 2x_3 & + & 2x_4 & - & 2x_5 & = & 0 \\ 3x_1 & - & 6x_2 & + & x_3 & + & x_4 & + & 5x_5 & = & 4 \\ -x_1 & + & 2x_2 & + & 3x_3 & + & x_4 & + & x_5 & = & 3 \end{cases}$

EXERCISES 7.4:

1. Use MATLAB to solve the linear system $Ax = b$, with the following choices for A and b. Afterward, check to see that your solution x satisfies $Ax = b$.

(a) $A = \begin{bmatrix} 9 & -5 & 2 \\ 0 & 7 & 5 \\ -1 & -9 & 6 \end{bmatrix}$, $b = \begin{bmatrix} 67 \\ 13 \\ 27 \end{bmatrix}$ (b) $A = \begin{bmatrix} -1 & 1 & 3 & 5 \\ 3 & -4 & -1 & 5 \\ 5 & -1 & 4 & -5 \\ -2 & 3 & -5 & -4 \end{bmatrix}$, $b = \begin{bmatrix} 9 \\ 29 \\ -22 \\ -5 \end{bmatrix}$

(c) $A = \begin{bmatrix} -3 & -3 \\ 1 & -3 \end{bmatrix}$, $b = \begin{bmatrix} 2.928 \\ 3.944 \end{bmatrix}$ (d) $A = \begin{bmatrix} -12 & -20 & 10 \\ -2 & 18 & -1 \\ -3 & 14 & 1 \end{bmatrix}$, $b = \begin{bmatrix} -112.9 \\ 71.21 \\ 45.83 \end{bmatrix}$

2. (*Polynomial Interpolation*) (a) Find the equation of the polynomial of degree at most 2 (parabola) that passes through the data points $(-1, 21)$, $(1, -3)$, $(5, 69)$; then plot the function along with the data points.
    (b) Find the equation of the polynomial of degree at most 3 that passes through the data points: $(-4, -58.8)$, $(2, 9.6)$, $(8, 596.4)$, $(1, 4.2)$; then plot the function along with the data points.
    (c) Find the equation of the polynomial of degree at most 6 that passes through the data points: $(-2, 42)$, $(-1, -29)$, $(-0.5, -16.875)$, $(0, -6)$, $(1, 3)$, $(1.5, 18.375)$, $(2, -110)$;   then plot the function along with the data points.
    (d) Find the equation of the polynomial of degree at most 5 that passes through the data points: $(1, 2)$, $(2, 4)$, $(3, 8)$, $(4, 16)$, $(5, 32)$, $(6, 64)$; then plot the function along with the data points.

3. Find the general solution of each of the following linear systems:
    (a) $\begin{cases} 3x_1 & + & 3x_2 & + & 2x_3 & + & 5x_4 & = & 12 \\ 2x_1 & + & 6x_2 & + & 2x_3 & - & 8x_4 & = & 4 \end{cases}$

    (b) $\begin{cases} x_1 & + & x_2 & + & 3x_3 & & & & x_4 & & 5x_5 & - & 2 \\ x_1 & - & 3x_2 & + & 2x_3 & - & 2x_4 & + & x_5 & = & 2 \\ 2x_1 & - & 7x_2 & + & 3x_3 & - & x_4 & - & 4x_5 & = & 4 \\ -3x_1 & - & 2x_2 & - & 4x_3 & + & 4x_4 & - & x_5 & = & 6 \end{cases}$

    (c) $\begin{cases} 2x_1 & + & 2x_2 & - & 3x_3 & + & x_4 & = & 8 \\ 4x_1 & & & - & 3x_3 & - & 2x_4 & = & 0 \\ -3x_1 & + & 4x_2 & + & x_3 & + & x_4 & = & -1 \end{cases}$

    (d) $\begin{cases} 3x_1 & + & 6x_2 & + & x_3 & - & 2x_4 & + & 12x_5 & & & = & 22 \\ -8x_1 & - & 16x_2 & & & + & 16x_4 & - & 32x_5 & + & x_6 & = & -60 \\ x_1 & + & 2x_2 & & & - & 2x_4 & + & 4x_5 & & & = & 8 \end{cases}$

4. (*Polynomial Interpolation*) In polynomial interpolation problems as in Example 7.8, the coefficient matrix that arises is the so-called **Vandermonde matrix** corresponding to the vector $v = [x_0 \ x_1 \ x_2 \ \cdots \ x_n]$ that is the $(n+1) \times (n+1)$ matrix defined by

$$V = \begin{bmatrix} x_0^n & x_0^{n-1} & \cdots & x_0 & 1 \\ x_1^n & x_1^{n-1} & \cdots & x_1 & 1 \\ \vdots & & & & \vdots \\ x_n^n & x_n^{n-1} & \cdots & x_n & 1 \end{bmatrix}.$$

MATLAB (of course) has a function vander(v), that will create the Vandermonde matrix corresponding to the inputted vector v. Redo parts (a) through (d) of Exercise 2, this time using vander(v).

(e) Write your own version myvander(v), that does what MATLAB's vander(v) does. Check the efficiency of your M-file with that of MATLAB's vander(v) by typing (after you have created and debugged your program) >>type vander to display the code of MATLAB's vander(v).

5.   (*Polynomial Interpolation*)   (a) Write a MATLAB function M-file called pv = polyinterp(x, y) that will input two vectors x and y of the same length (call this length "$n+1$" for now) that correspond to the x- and y-coordinates of $n+1$ data points on which we wish to interpolate with a polynomial of degree at most n. The output will be a vector $pv = [a_n \ a_{n-1} \ \cdots a_2 \ a_1 \ a_0]$ that contains the coefficients of the interpolating polynomial
$p(x) = a_n x^n + a_{n-1} x^{n-1} + \cdots + a_1 x + a_0$.
(b) Use this program to redo part (b) of Exercise 2.
(c) Use this program to redo part (c) of Exercise 2.
(d) Use this program with input vectors $x = \begin{bmatrix} 0 & \pi/2 & \pi & 3\pi/2 & 2\pi & 5\pi/2 & 3\pi & 7\pi/2 & 4\pi \end{bmatrix}$,
and $y = [1 \ 0 \ -1 \ 0 \ 1 \ 0 \ -1 \ 0 \ 1]$. Plot the resulting polynomial along with the data points. Include also the plot (in a different style/color) of a trig function that interpolates this data.

6.   (*Polynomial Interpolation: Asking a Bit More*)   Often in applications, rather than just needing a polynomial (or some other nice interpolating curve) that passes through a set of data points, we also need the interpolating curve to satisfy some additional smoothness requirements.   For   example, consider the design of a railroad transfer segment shown in Figure 7.30.   The curved portion of "interpolating" railroad track needs to do more than just connect the

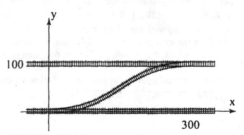

FIGURE 7.30: Exercise 6 asks to find a polynomial function modeling the junction between the two parallel sects of rails.

two parallel tracks; it must do so "smoothly" lest the trains using it would derail. Thus, if we seek a function $y = p(x)$ that models this interpolating track, we see from the figure (and the reference to the xy-coordinate system drawn in) that we would like the center curve of this interpolating track to satisfy the following conditions on the interval $0 \le x \le 300$ feet:

$$p(0) = 0, \ p(300) = 100 \text{ feet}, \ p'(0) = p'(300) = 0.$$

(The last two conditions geometrically will require the graph of $y = p(x)$ to have a horizontal tangent line at the endpoints $x = 0$ and $x = 300$, and thus connect smoothly with the existing tracks.)   If we would like to use a polynomial for this interpolation, since we have four requirements, we should be working with a polynomial of degree at most 3 (that has four parameters): $p(x) = ax^3 + bx^2 + cx + d$.

(a) Set up a linear system for this interpolating polynomial and get MATLAB to solve it.
(b) Next, get MATLAB to graph the rail network (just the rails) including the two sets of parallel

tracks as well as the interpolating rails. Leave a 6-foot vertical distance between each set of adjacent rails.

7.   (*Polynomial Interpolation: Asking a Bit More*)   Three parallel railroad tracks need to be connected by a pair of curved junction segments, as shown in Figure 7.31.

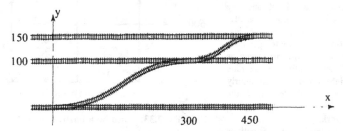

FIGURE 7.31: In Exercise 7, this set of three parallel rails is required to be joined by two sets of smooth junction rails.

(a) If we wish to use a single polynomial function to model (the center curves of) both pairs of junction rails shown in Figure 7.31, what degree polynomials should we use in our model? Set up a linear system to determine the coefficients of this polynomial, then get MATLAB to solve it and determine the polynomial.

(b) Next, get MATLAB to graph the rail network (just the rails) including the three sets of parallel tracks as well as the interpolating rails gotten from the polynomial function you found in part (a). Leave a 6-foot vertical distance between each set of adjacent rails.

(c) Do a separate polynomial interpolation for each of the two junction rails and thus find two different polynomials that model each of the two junctions. Set up the two linear systems, solve them using MATLAB, and then write down the two polynomials.

(d) Next, get MATLAB to graph the rail network (just the rails) including the three sets of parallel tracks as well as the interpolating rails gotten from the polynomial functions you found in part (c). Leave a 6-foot vertical distance between each set of adjacent rails. How does this picture compare with the one in part (b)?

Note: In general it is more efficient to do the piecewise polynomial interpolation that was done in part (d) rather than the single polynomial interpolation in part (b). The advantages become more apparent when there are a lot of data points. This approach is an example of what is called spline interpolation.

8.   (*City Planning: Traffic Logistics*)   The Honolulu street map of Figure 7.32 shows the rush-hour numbers of vehicles per hour that enter or leave the network of four one-way streets.   The variables x_1, x_2, x_3, x_4 represent the traffic flows on the segments shown. For smooth traffic flow, we would like to have equilibrium at each of the four intersections; i.e., the number of incoming cars (per hour) should equal the number of outgoing cars.   For example, at the intersection of Beretania and Piikoi (lower right), we should have $x_1 + 800 = x_2 + 2000$,

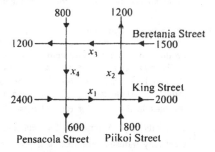

FIGURE 7.32: Rush-hour traffic on some Honolulu streets.

or (after rearranging) $x_1 - x_2 = 1200$.

(a) Obtain a linear system for the smooth traffic flow in the above network by looking at the flows at each of the four intersections.

(b) How many solutions are there? If there are solutions, which, if any, give rise to feasible traffic flow numbers?

(c) Is it possible for one of the four segments in the network to be closed off for construction (a perennial occurrence in Honolulu) so that the network will still be able to support smooth traffic flow? Explain.

9. (*City Planning: Traffic Logistics*) Figure 7.33 shows a busy network of one-way roads in the center of a city. The rush-hour inflows and outflows of vehicles per hour for the network are given as well as a listing of the nine variables that represent hourly flows of vehicles along the one-way segments in the network.

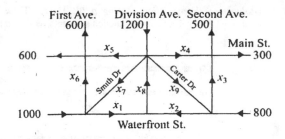

FIGURE 7.33: Rush-hour traffic in a busy city center.

(a) Following the directions of the preceding exercise, use the equilibria at each intersection to obtain a system of linear equations in the nine variables that will govern smooth traffic flow in this network.

(b) Use MATLAB's rref to solve this system. There are going to be infinitely many solutions, but of course not all are going to be feasible answers to the original problem. For example, we cannot have a negative traffic flow on any given street segment, and also the x_j's should be integers. Thus the solutions consist of vectors with eight components where each component is a nonnegative integer.

(c) Considering all of the feasible solutions that were contemplated in part (b), what is the maximum that x_6 can be (in a solution)? What is the minimum?

(d) Repeat part (c) for x_8.

(e) If the city wants to have a parade and close up one of the segments (corresponding to some x_j in the figure) of the town center, is it possible to do this without disrupting the main traffic flow?

(f) If you answered yes to part (e), go further and answer this question. The mayor would like to set up a "Kick the Fat" 5K run through some of the central streets in the city. How many segments (corresponding to some x_j in the figure) could the city cordon off without disrupting the main flow of traffic? Answer the same question if we require, in addition, that the streets which are cordoned off are connected together.

10. (*Economics: Input–Output Analysis*) In any large economy, major industries that are producing essential goods and services need products from other industries in order to meet their own production demands. Such demands need to be accounted for by the other industries in addition to the main consumer demands. The model we will give here is due to Russian ecomomist Wassily Leontief (1906–1999).[9] To present the main ideas, we deal with only three dependent industries: (i) electricity, (ii) steel, and (iii) water. In a certain economy, let us assume that the outside demands for each of these three industries are (annually) d_1 = \$140 million for electricity, d_2 = \$46 million for steel, and d_3 = \$92 million for water. For each dollar that the electricity industry produces each year, assume that it will cost \$0.02 in electricity, \$0.08 in

---

[9] In the 1930s and early 1940s, Leontief did an extensive analysis of the input and output of 500 sectors of the US economy. The calculations were tremendous, and Leontief made use of the first large-scale computer (in 1943) as a necessary tool. He won the Nobel Prize in Economics in 1973 for this research. Educated in Leningrad, Russia (now again St. Petersburg, as it was before 1917), and in Berlin, Leontif subsequently moved to the United States to become a professor of economics at Harvard. His family was quite full of intellectuals: His father was also a professor of economics, his wife (Estelle Marks) was a poet, and his daughter is a professor of art history at the University of California at Berkeley.

steel, and $0.16 in water. Also for each dollar of steel that the steel industry produces each year, assume that it will cost $0.05 in electricity, $0.08 in steel, and $0.10 in water. Assume the corresponding data for producing $1 of water to be $0.12, $0.07, and $0 03. From this data, we can form the so-called **technology matrix**:

$$M \; = \; \begin{array}{c} \text{Electricity demand (per dollar)} \\ \text{Steel demand (per dollar)} \\ \text{Water demand (per dollar)} \end{array} \begin{array}{ccc} \mathbf{E} & \mathbf{S} & \mathbf{W} \\ \begin{bmatrix} .02 & .05 & .12 \\ .08 & .08 & .07 \\ .16 & .10 & .03 \end{bmatrix} \end{array}.$$

(a) Let x_1 = The amount (in dollars) of electricity produced by the electricity industry,

x_2 = The amount of steel produced by the steel industry,

x_3 = The amount of water produced by the water industry, and let

$X = \begin{bmatrix} x_1 \\ x_2 \\ x_3 \end{bmatrix}$. The matrix X is called the **output matrix**. Show/explain why the matrix MX

(called the **internal demand matrix**) gives the total internal costs of electricity, steel, and water that it will collectively cost the three industries to produce the outputs given in X.

(b) For the economy to function, the output of these three major industries must meet both the external demand and the internal demand. The external demand is given by the following *external demand matrix*:

$$D = \begin{bmatrix} d_1 \\ d_2 \\ d_3 \end{bmatrix} = \begin{bmatrix} 140,000,000 \\ 46,000,000 \\ 92,000,000 \end{bmatrix}.$$

(The data was given above.) Thus since the total output X of the industries must meet both the internal MX and external D demands, the matrix X must solve the matrix equation:

$$X = MX + D \Rightarrow (I - M)X = D.$$

It can always be shown that the matrix $I - M$ is nonsingular and thus there is always going to be a unique solution of this problem. Find the solution of this particular input/output problem.

(c) In a particular year, there is a construction boom and the demands go up to these values:

$$d_1 = \$160 \text{ mil.}, d_2 = \$87 \text{ mil.}, d_3 = \$104 \text{ mil}$$

How will the outputs need to change for that year?

(d) In a recessionary year, the external demands drop to the following numbers:

$$d_1 = \$90 \text{ mil.}, d_2 = \$18 \text{ mil.}, d_3 = \$65 \text{ mil}$$

Find the corresponding outputs.

11. (*Economics: Input-Output Analysis*) Suppose, in the economic model of the previous exercise, two additional industries are added: (iv) oil and (v) plastics. In addition to the assumptions of the previous exercise, assume further that it has been determined that for each dollar of electricity produced it will cost $0.18 in oil and $0.03 in plastics, for each dollar of steel produced it will cost $0.07 in oil and $0.01 in plastics, for each dollar of water produced it will cost $0.02 in plastics (but no oil), for each dollar of oil produced it will cost $0.06 in electricity, $0.02 in steel, $0.05 in water, and $0.01 in plastics (but no oil), and finally for each dollar in plastics produced, it will cost $0.02 in electricity, $0.01 in steel, $0.02 in water, $0.22 in oil, and $0.12 in plastics.

(a) Write down the technology matrix.

(b) Assuming the original external demands for the first three industries given in the preceding exercise and external demands of $d_4 = \$188$ mil. for oil and $d_5 = \$35$ mil. for plastics, solve the Leontief model for the resulting output matrix.

(c) Resolve the model using the data in part (c) of Exercise 7, along with $d_4 = \$209$ mil., $d_5 = \$60$ mil.

(d) Resolve the model using the data in part (c) of Exercise 7, along with $d_4 = \$149$ mil., $d_5 = \$16$ mil.

NOTE: (*Combinatorics: Power Sums*) It is often necessary to find the sum of fixed powers of the first several positive integers. Formulas for the sums are well known but it is difficult to remember them all. Here are the first four such *power sums*:

$$\sum_{k=1}^{n} k = 1+2+3+\cdots+n = \frac{1}{2}n(n+1) \tag{23}$$

$$\sum_{k=1}^{n} k^2 = 1+4+9+\cdots+n^2 = \frac{1}{6}n(n+1)(2n+1) \tag{24}$$

$$\sum_{k=1}^{n} k^3 = 1+8+27+\cdots+n^3 = \frac{1}{4}n^2(n+1)^2 \tag{25}$$

$$\sum_{k=1}^{n} k^4 = 1+16+81+\cdots+n^4 = \frac{1}{30}n(n+1)(2n+1)(3n^2+3n-1) \tag{26}$$

Each of these formulas and more general ones can be proved by mathematical induction, but deriving them is more involved. It is a general fact that for any positive integer p, the power sum $\sum_{k=1}^{n} k^p$ can always be expressed as a polynomial $f(n)$ that has degree $p+1$ and has rational coefficients (fractions). See Section 3.54 (p. 199ff) of [Ros-00] for details. The next two exercises will show a way to use linear systems not only to verify, but to derive such formulas.

12.   (*Combinatorics: Power Sums*)   (a) Use the fact (from the general fact mentioned in the preceding note) that $\sum_{k=1}^{n} k$ can be expressed as $f(n)$ where $f(n)$ is a polynomial of degree 2: $f(n) = an^2 + bn + c$, to set up a linear system for a, b, c using $f(1)=1, f(2)=3, f(3)=6$, and use MATLAB (in "format rat") to solve for the coefficients and then verify identity (23).
(b) In a similar fashion, verify identity (24).
(c) In a similar fashion, verify identity (25).
(d) In a similar fashion, verify identity (26).
Note: If you have the Student Version of MATLAB (or have access to the Symbolic Toolbox), the command factor can be used to develop your formulas in factored form—see Appendix A; otherwise leave them in standard polynomial form.

13.   (*Combinatorics: Power Sums*)   (a) By mimicking the approach of the previous exercise, use MATLAB to get a formula for the power sum $\sum_{k=1}^{n} k^5$. Check your formula for the values $n = 5$, and $n = 100$ (using MATLAB, of course). (b) Repeat for the sum $\sum_{k=1}^{n} k^6$.
Note: If you have the Student Version of MATLAB (or have access to the Symbolic Toolbox), the command factor can be used to develop your formulas in factored form—see Appendix A; otherwise leave them in standard polynomial form.

14.   (*Combinatorics: Alternating Power Sums*) For positive integers p and n, the *alternating power sum*:

$$\sum_{k=1}^{n} (-1)^{n-k} k^p = n^p - (n-1)^p + (n-2)^p - \cdots + (-1)^{n-2} 2^p + (-1)^{n-1},$$

like the power sum, can be expressed as $f(n)$ where $f(n)$ is polynomials of degree $n+1$ having rational coefficients. (For details, see [Ros-00], Section 3.17, page 152ff.) As in

Exercise 5, set up linear systems for the coefficients of these polynomial, use MATLAB to solve them, and develop formulas for the alternating power sums for the following values of p. (a) $p = 1$, (b) $p = 2$, (c) $p = 3$, (d) $p = 4$. For each formula you derive, get MATLAB to check it (against the actual power sum) for the following values of n: 10, 250, and 500.

15. (*Linear Algebra: Cramer's Rule*) There is an attractive and explicit formula for solving a nonsingular system $Ax = b$ which expresses the solution of each component of the vector x entirely in terms of determinants. This formula, known as Cramer's rule,[10] is often given in linear algebra courses (because it has nice theoretical applications), but it is a very expensive one to use. Cramer's rule states that the solution $x = [x_1 \ x_2 \ \cdots \ x_n]'$ of the (nonsingular) system $Ax = b$, is given by the following formulas:

$$ x_1 = \frac{\det(A_1)}{\det(A)}, \ x_2 = \frac{\det(A_2)}{\det(A)}, \ \cdots, \ x_n = \frac{\det(A_n)}{\det(A)}, $$

where the $n \times n$ matrix A_i is formed by replacing the ith column of the coefficient matrix A, by the column vector b.

(a) Use Cramer's rule to solve the linear system of Examples 7.7 and 7.8, and compare performance time and (if you have Version 5) flop counts and accuracies with Methods 1, 2, and 3 that were used in the text.

(b) Write a function M-file called x = `cramer(A,b)` that will input a square nonsingular matrix A, a column vector b of the same dimension, and will output the column vector solution of the linear system $Ax = b$ obtained using Cramer's rule. Apply this program to resolve the two systems of part (a).

Note: Of course, you should set up your calculations and programs so that you only ask MATLAB to compute $\det(A)$ once, each time you use Cramer's rule.

7.5: GAUSSIAN ELIMINATION, PIVOTING, AND LU FACTORIZATION

Here is a brief outline of this section. Our goal is a general method for solving the linear system (17) with n variables and n equations, and we will be working with the corresponding augmented matrix $[A \mid b]$ from the resulting matrix equation (18). We will first observe that if a linear system has a triangular matrix A (i.e., all entries below the main diagonal are zero, or all entries above it are zero), then the linear system is very easy to solve. We next introduce the three elementary

---

[10] Gabriel Cramer (1704–1752) was a Swiss mathematician who is credited for introducing his namesake rule in his famous book, *Introduction à l'analyse des lignes courbes algébraique*. Cramer entered his career as a mathematician with impressive speed, earning his PhD at age 18 and being awarded a joint chaired professorship of mathematics at the Académie de Clavin in Geneva. With his shared appointment, he would take turns with his colleague Caldrini teaching for 2 to 3 years. Through this arrangement he was able to do much traveling and made contacts with famous mathematicians throughout Europe; all of the famous mathematicians that met him were most impressed. For example, the prominent Swiss mathematician Johann Bernoulli (1667–1748), insisted that Cramer and only Cramer be allowed to edit the former's collected works. Throughout his career, Cramer was always very productive. He put great energy into his teachings, his researches, and his correspondences with other mathematicians whom he had met. Despite the lack of practicality of Cramer's rule, Cramer actually did a lot of work in applying mathematics to practical areas such as national defense and structural design. Cramer was always in good health until an accidental fall off a carriage. His doctor recommended for him to go to a resort city in the south of France for recuperation, but he passed away on that journey.

row operations that, when performed on an augmented matrix, will lead to another augmented matrix which represents an equivalent linear system that has a triangular coefficient matrix and thus can be easily solved. We then explain the main algorithm, Gaussian elimination with partial pivoting, that will transform any augmented matrix (representing a nonsingular system) into an upper triangular system that is easily solved. This is less work than the usual "Gaussian elimination" taught in linear algebra classes, since the latter brings the augmented matrix all the way into reduced row echelon form. The partial pivoting aspect is mathematically redundant, but is numerically very important since it will help to cut back on floating point arithmetic errors. Gaussian elimination produces a useful factorization of the coefficient matrix,

FIGURE 7.34: Carl Friedrich Gauss (1777–1855), German mathematician.

called the LU factorization, that will considerably cut down the amount of work needed to solve other systems having the same coefficient matrix. We will postpone the error analysis of this procedure to the next section. The main algorithms of this section were invented by the German mathematician Carl F. Gauss.[11]

A square matrix $A = [a_{ij}]$ is called **upper triangular** if all entries below the main diagonal equal zero (i.e., $a_{ij} = 0$ whenever $i > j$). Thus an upper triangular matrix has the following form:

$$\begin{bmatrix} a_{11} & a_{12} & a_{13} & \cdots & a_{1n} \\ & a_{22} & a_{23} & \ldots & a_{2n} \\ & & a_{33} & \cdots & \vdots \\ & \mathbf{0} & & \ddots & \vdots \\ & & & & a_{nn} \end{bmatrix}.$$

---

[11] Carl F. Gauss is acknowledged by many mathematical scholars as the greatest mathematician who ever lived. His potential was discovered early. His first mathematical discovery was the power sum identity (23), and he made this discovery while in the second grade! His teacher, looking to keep young Carl Friedrich occupied for a while asked him to perform the addition of the first 100 integers: $S = 1 + 2 + \ldots + 100$. Two minutes later, Gauss gave the teacher the answer. He did it by rewriting the sum in the reverse order $S = 100 + 99 + \ldots + 1$, adding vertically to the original to get $2S = 101 + 101 + \ldots + 101 = 100 \cdot 101$, so $S = 50 \cdot 101 = 5050$. This idea, of course, yields a general proof of the identity (23). He was noticed by the Duke of Brunswick, who supported Gauss's education and intellectual activity for many years. Gauss's work touched on numerous fields of mathematics, physics, and other sciences. His contributions are too numerous to attempt to do them justice in this short footnote. It is said that very routinely when another mathematician would visit him in his office to present him with a recent mathematical discovery, after hearing about the theorems, Gauss would reach into his file cabinet and pull out some of his own works, which would invariably transcend that of his guest. For many years, until 2001 when the currency in Germany, as well as that of other European countries, changed to the Euro, Germany had honored Gauss by putting his image on the very common 10 Deutsche Mark banknote (value about $5); see Figure 7.34.

Similarly, a square matrix is **lower triangular** if all entries above the main diagonal are zeros. A matrix is triangular if it is of either of these two forms. Many matrix calculations are easy for triangular matrices. The next proposition shows that determinants for triangular matrices are extremely simple to compute.

PROPOSITION 7.3: If $A = [a_{ij}]$ is a triangular matrix, then the determinant of A is the product of the diagonal entries, i.e., $\det(A) = a_{11}a_{22}\cdots a_{nn}$.

The proof can be easily done by mathematical induction and cofactor expansion on the first column or row; we leave it as Exercise 14(a).

From the proposition, it follows that for a triangular matrix to be nonsingular it is equivalent that each of the diagonal entries must be nonzero. For a system $Ax = b$ with an upper triangular coefficient matrix, we can easily solve the system by starting with the last equation, solving for x_n, then using this in the second-to-last equation and solving for x_{n-1}, and continuing to work our way up. Let us make this more explicit. An upper triangular system has the form:

$$
\begin{cases}
a_{11}x_1 + a_{12}x_2 + \cdots \qquad + a_{1n}x_n = b_1 \\
\qquad\quad a_{22}x_2 + \cdots \qquad + a_{2n}x_n = b_2 \\
\qquad\qquad\qquad \ddots \qquad\qquad \vdots \\
\qquad\qquad\quad a_{n-1,n-1}x_{n-1} + a_{n-1,n}x_n = b_{n-1} \\
\qquad\qquad\qquad\qquad\qquad\quad a_{nn}x_n = b_n
\end{cases}
\tag{27}
$$

Assuming A is nonsingular, we have that each diagonal entry a_{ii} is nonzero. Thus, we can start off by solving the last equation of (27):

$$x_n = b_n / a_{nn}.$$

Knowing now the value of x_n, we can then substitute this into the second-to-last equation and solve for x_{n-1}:

$$x_{n-1} = (b_{n-1} - a_{n-1,n}x_n)/a_{n-1,n-1}.$$

Now that we know both x_n and x_{n-1}, we can substitute these into the third-to-last equation and then, similarly, solve for the only remaining unknown in this equation:

$$x_{n-2} = (b_{n-2} - a_{n-2,n-1}x_{n-1} - a_{n-2,n}x_n)/a_{n-2,n-2}.$$

If we continue this process, in general, after having solved for $x_{j+1}, x_{j+2}, \cdots x_n$, we can get x_j by the formula:

$$x_j = \left(b_j - \sum_{k=j+1}^{n} a_{jk} x_k \right) / a_{jj}. \tag{28}$$

This algorithm is called **back substitution**, and is a fast and easy method of solving any upper triangular (nonsingular) linear system. Would it not be nice if all linear systems were so easy to solve? Transforming arbitrary (nonsingular) linear systems into upper triangular form will be the goal of the Gaussian elimination algorithm. For now we record for future reference a simple M-file for the back substitution algorithm.

EXAMPLE 7.11: (a) Create an M-file x=backsubst(U,b) that inputs a nonsingular upper triangular matrix U, and a column vector b of the same dimension and the output will be a column vector x which is the numerical solution of the linear system $Ux = b$ obtained from the back substitution algorithm. (b) Use this algorithm to solve the system $Ux = b$ with

$$U = \begin{bmatrix} 1 & 2 & 3 & 4 \\ 0 & 2 & 3 & 4 \\ 0 & 0 & 3 & 4 \\ 0 & 0 & 0 & 4 \end{bmatrix}, \quad b = \begin{bmatrix} 4 \\ 3 \\ 2 \\ 1 \end{bmatrix}.$$

SOLUTION: Part (a): The M-file is easily written using equation (28).

PROGRAM 7.4: Function M-file solving an upper triangular linear system $Ux = b$.

```
function x=backsubst(U,b)
%solves the upper triangular system Ux=b by back substitution
%inputs:  U - upper triangular matrix,  b - column vector of same
%dimension.
%Output:  x - column vector (solution)
[n m]=size(U);
x(n)=b(n)/U(n,n);
for j=n-1:-1:1
   x(j)=(b(j)-U(j,j+1:n)*x(j+1:n)')/U(j,j);
end
x=x';
```

Notice that MATLAB's matrix multiplication allowed us to replace the sum in (28) with the matrix product shown. Indeed U(j,j+1:n) is a row vector with the same number of entries as the column vector x(j+1:n), so their matrix product will be a real number (1×1 matrix) equaling the sum in (28). The transpose on the x-vector was necessary here to make it a column vector.

Part (b):
```
>> U=[1 2 3 4;0 2 3 4;0 0 3 4;0 0 0 4]; b=[4 3 2 1]';
>> format rat, backsubst(U,b)
→ ans =       1
             1/2
             1/3
             1/4
```

The reader can check that U\b will give the same result.

A lower triangular system $Lx = b$ can be solved with an analogous algorithm called **forward substitution**. Here we start with the first equation to get x_1, then plug this result into the second equation and solve for x_2, and so on.

EXERCISE FOR THE READER 7.18: (a) Write a function M-file called x = fwdsubst (L, b), that will input a lower triangular matrix L, a column vector b of the same dimension, and will output the column vector x solution of the system $Lx = b$ solved by forward substitution. (b) Use this program to solve the system $Lx = b$, where $L = U'$, and U and b are as in Example 7.11.

We now introduce the three **elementary row operations (EROs)** that can be performed on augmented matrices.

(i)   Multiply a row by a <u>nonzero</u> constant.
(ii)   Switch two rows.
(iii) Add a multiple of one row to a different row.

EXAMPLE 7.12: (a) Consider the following augmented matrix:
$Ab = \begin{bmatrix} 1 & 2 & -3 & 5 \\ 2 & 6 & 1 & -1 \\ 0 & 4 & 7 & 8 \end{bmatrix}$. Perform ERO (iii) on this matrix by adding the multiple -2 times row 1 to row 2.

(b) Perform this same ERO on I_3, the 3×3 identity matrix, to obtain a matrix M, and multiply this matrix M on the left of Ab. What do you get?

SOLUTION: Part (a): -2 times row 1 of Ab is $-2[1\ 2\ -3\ 5] = [-2\ -4\ 6\ -10]$. Adding this row vector to row 2 of Ab, produces the new matrix:

$$\begin{bmatrix} 1 & 2 & -3 & 5 \\ 0 & 2 & 7 & -11 \\ 0 & 4 & 7 & 8 \end{bmatrix}.$$

Part (b): Performing this same ERO on I_3 produces the matrix $M = \begin{bmatrix} 1 & 0 & 0 \\ -2 & 1 & 0 \\ 0 & 0 & 1 \end{bmatrix}$,

which when multiplied by Ab gives

$$M(Ab) = \begin{bmatrix} 1 & 0 & 0 \\ -2 & 1 & 0 \\ 0 & 0 & 1 \end{bmatrix} \cdot \begin{bmatrix} 1 & 2 & -3 & 5 \\ 2 & 6 & 1 & -1 \\ 0 & 4 & 7 & 8 \end{bmatrix} = \begin{bmatrix} 1 & 2 & -3 & 5 \\ 0 & 2 & 7 & -11 \\ 0 & 4 & 7 & 8 \end{bmatrix},$$

and we are left with the same matrix that we obtained in part (a). This is no coincidence, as the following theorem shows.

THEOREM 7.4: (*Elementary Matrices*) Let A be any $n \times m$ matrix and $I = I_n$ denote the identity matrix. If B is the matrix obtained from A by performing any particular elementary row operation, and M is the matrix obtained from I by performing this same elementary row operation, then $B = MA$. Also, the matrix M is invertible, and its inverse is the matrix that results from I by performing the inverse elementary row operation on it (i.e., the elementary row operation that will transform M back into I).

Such a matrix M is called an **elementary matrix**. This result is not hard to prove; we refer the reader to any good linear algebra textbook, such as those mentioned in the last section.

It is easy to see that any of these EROs, when applied to the augmented matrix of a linear system, will not alter the solution of the system. Indeed, the first ERO corresponds to simply multiplying the first equation by a nonzero constant (the equation still represents the same line, plane, hyperplane, etc.). The second ERO merely changes the order in which the equations are written; this has no effect on the (joint) solution of the system. To see why the third ERO does not alter solutions of the system is a bit more involved, but not difficult. Indeed, suppose for definiteness that a multiple (say, 2) of the first row is added to the second. This corresponds to a new system where all the equations are the same, except for the second, which equals the old second equation plus twice the first equation. Certainly if we have all of $x_1, x_2, \cdots, x_n$ satisfying the original system, then they will satisfy the new system. Conversely, if all of the equations of the new system are solved by $x_1, x_2, \cdots, x_n$, then this already gives all but the second equation of the original system. But the second equation of the old system is gotten by subtracting twice the first equation of the new system from the second equation (of the new system) and so must also hold.

Each of the EROs is easily programmed into an M-file. We do one of them and leave the other two as exercises.

PROGRAM 7.5: Function M-file for elementary row operation (ii): switching two rows.

```
function B=rowswitch(A,i,j)
   Inputs: a matrix A, and row indices i and j
 % Outputs:  the matrix gotten from A by interchanging row i and row j
[m,n]=size(A);
if i<1|i>m|j<1|j>m
    error('invalid index')
end
B=A;
if i==j
    return
end
B(i,:)=A(j,:);
B(j,:)=A(i,:);
```

It may seem redundant to have included that if-branch for detecting invalid indices, since it would seem that no one in their right mind would use the program with, say, an index equaling 10 with an 8×8 matrix. This program, however, might get used to build more elaborate programs, and in such a program it may not always be crystal clear whether some variable expression is a valid index.

EXERCISE FOR THE READER 7.19: Write similar function M-files B=rowmult(A,i,c) for ERO (i) and B=rowcomb(A,i,j,c) for ERO (iii). The first program will produce the matrix B resulting from A by multiplying the ith row of the latter by c; the second program should replace the ith row of A by c times the jth row plus the ith row.

We next illustrate the Gaussian elimination algorithm, the partial pivoting feature, as well as the *LU* decomposition, by means of a simple example. This will give the reader a feel for the main concepts of this section. Afterward we will develop the general algorithms and comment on some consequences of floating point arithmetic. Remember, the goal of Gaussian elimination is to transform, using only EROs, a (nonsingular) system into an equivalent one that is upper triangular; the latter can then be solved by back substitution.

EXAMPLE 7.13: We will solve the following linear system $Ax = b$ using Gaussian elimination (without partial pivoting):

$$\begin{cases} x_1 & + & 3x_2 & - & x_3 & = & 2 \\ 2x_1 & + & 5x_2 & - & 2x_3 & = & 3 \\ 3x_1 & + & 6x_2 & + & 9x_3 & = & 39 \end{cases}.$$

SOLUTION: For convenience, we will work instead with the corresponding augmented matrix Ab for this system $Ax = b$:

$$Ab = \begin{bmatrix} 1 & 3 & -1 & | & 2 \\ 2 & 5 & -2 & | & 3 \\ 3 & 6 & 9 & | & 39 \end{bmatrix}.$$

For notational convenience, we denote the entries of this or any future augmented matrix as a_{ij}. In computer codes, what is usually done is that at each step the new matrix overwrites the old one to save on memory allocation (not to mention having to invent new variables for unnecessary older matrices). Gaussian elimination starts with the first column, clears out (makes zeros) everything below the main diagonal entry, then proceeds to the second column, and so on.

We begin by zeroing out the entry $a_{21} = 2$: Ab=rowcomb(Ab,1,2,-2).

$$Ab \to M_1(Ab) = \begin{bmatrix} 1 & 0 & 0 \\ -2 & 1 & 0 \\ 0 & 0 & 1 \end{bmatrix} \begin{bmatrix} 1 & 3 & -1 & | & 2 \\ 2 & 5 & -2 & | & 3 \\ 3 & 6 & 9 & | & 39 \end{bmatrix} = \begin{bmatrix} 1 & 3 & -1 & | & 2 \\ 0 & -1 & 0 & | & -1 \\ 3 & 6 & 9 & | & 39 \end{bmatrix},$$

where M_1 is the corresponding elementary matrix as in Theorem 7.4. In the same fashion, we next zero out the next (and last) entry $a_{31} = 3$ of the first column: Ab=rowcomb(Ab,1,3,-3).

$$Ab \to M_2(Ab) = \begin{bmatrix} 1 & 0 & 0 \\ 0 & 1 & 0 \\ -3 & 0 & 1 \end{bmatrix} \begin{bmatrix} 1 & 3 & -1 & | & 2 \\ 0 & -1 & 0 & | & -1 \\ 3 & 6 & 9 & | & 39 \end{bmatrix} = \begin{bmatrix} 1 & 3 & -1 & | & 2 \\ 0 & -1 & 0 & | & -1 \\ 0 & -3 & 12 & | & 33 \end{bmatrix}.$$

We move on to the second column. Here only one entry needs clearing, namely $a_{32} = -3$. We will always use the row with the corresponding diagonal entry to clear out entries below it. Thus, to use the second row to clear out the entry $a_{32} = -3$ in the third row, we should multiply the second row by -3 since $-3 \cdot a_{22} = -3 \cdot (-1) = 3$ added to $a_{32} = -3$ would give zero: Ab=rowcomb(Ab,2,3,-3).

$$Ab \to M_3(Ab) = \begin{bmatrix} 1 & 0 & 0 \\ 0 & 1 & 0 \\ 0 & -3 & 1 \end{bmatrix} \begin{bmatrix} 1 & 3 & -1 & | & 2 \\ 0 & -1 & 0 & | & -1 \\ 0 & -3 & 12 & | & 33 \end{bmatrix} = \begin{bmatrix} 1 & 3 & -1 & | & 2 \\ 0 & -1 & 0 & | & -1 \\ 0 & 0 & 12 & | & 36 \end{bmatrix}.$$

We now have an augmented matrix representing an equivalent upper triangular system. This system (and hence our original system) can now be solved by the back substitution algorithm:

```
>> U=Ab(:,1:3); b=Ab(:,4);
>> x=backsubst(U,b)
→ x =           2
                1
                3
```

NOTE: One could have gone further and similarly cleared out the above diagonal entries and then used the first ERO to scale the diagonal entries to each equal one. This is how one gets to the reduced row echelon form.

To obtain the resulting LU decomposition, we form the product of all the elementary matrices that were used in the above Gaussian elimination: $M = M_3 M_2 M_1$. From what was done, we have that $MA = U$, and hence

$$A = M^{-1}(MA) = M^{-1}U = (M_3 M_2 M_1)^{-1}U = M_1^{-1}M_2^{-1}M_3^{-1}U \equiv LU,$$

where we have defined $L = M_1^{-1}M_2^{-1}M_3^{-1}$. We have used, in the second-to-last equality, the fact that the inverse of a product of invertible matrices is the product of the inverses in the reverse order (see Exercise 7). From Theorem 7.4, each of the inverses M_1^{-1}, M_2^{-1}, and M_3^{-1} is also an elementary matrix corresponding to the inverse elementary row operation of the corresponding original elementary matrix; and furthermore Theorem 7.4 tells us how to multiply such matrices to obtain

$$L = M_1^{-1}M_2^{-1}M_3^{-1} = \begin{bmatrix} 1 & 0 & 0 \\ 2 & 1 & 0 \\ 0 & 0 & 1 \end{bmatrix} \cdot \begin{bmatrix} 1 & 0 & 0 \\ 0 & 1 & 0 \\ 3 & 0 & 1 \end{bmatrix} \cdot \begin{bmatrix} 1 & 0 & 0 \\ 0 & 1 & 0 \\ 0 & 3 & 1 \end{bmatrix} = \begin{bmatrix} 1 & 0 & 0 \\ 2 & 1 & 0 \\ 3 & -3 & 1 \end{bmatrix}.$$

We now have a factorization of the coefficient matrix A as a product LU of a lower triangular and an upper triangular matrix:

$$A = Ab = \begin{bmatrix} 1 & 3 & -1 \\ 2 & 5 & -2 \\ 3 & 6 & 9 \end{bmatrix} = \begin{bmatrix} 1 & 0 & 0 \\ 2 & 1 & 0 \\ 3 & 3 & 1 \end{bmatrix} \cdot \begin{bmatrix} 1 & 3 & -1 \\ 0 & -1 & 0 \\ 0 & 0 & 12 \end{bmatrix} = LU.$$

This factorization, which easily came from the Gaussian elimination algorithm, is a preliminary form of what is known as the LU factorization of A. Once such a factorization is known, any other (nonsingular) system $Ax = c$, having the same coefficient matrix A, can be easily solved in two steps. To see this, rewrite the system as $LUx = c$. First solve $Ly = c$ by forward substitution (works since L is lower triangular), then solve $Ux = y$ by back substitution (works since U is upper triangular). Then x will be the desired solution (Proof: $Ax = (LU)x = L(Ux) = Ly = c$).

We make some observations. Notice that we used only one of the three EROs to perform Gaussian elimination in the above example. Part of the reason for this is that none of the diagonal entries encountered was zero. If this had happened we would have needed to use the rowswitch ERO in order to have nonzero diagonal entries. (This is always possible if the matrix A is nonsingular.) In Gaussian elimination, the diagonal entries that are used to clear out the entries below (using rowcomb) are known as **pivots**. The **partial pivoting** feature, which is often implemented in Gaussian elimination, goes a bit further to assure (by switching the row with the pivot with a lower row, if necessary) that the pivot is as large as possible in absolute value. In exact arithmetic, this partial pivoting has no effect whatsoever, but in floating point arithmetic it can most certainly cut back on errors. The reason for this is that if a pivot turned out to be nonzero, but very small, then its row would need to be multiplied by very large numbers to clear out moderately sized numbers below the pivot. This may cause other numbers in the pivot's row to get multiplied into very large numbers that, when mixed with much smaller numbers, can lead to floating point errors. We will soon give an example to demonstrate this phenomenon.

EXAMPLE 7.14:   Solve the linear system $Ax = b$ of Example 7.13 using Gaussian elimination with partial pivoting.

SOLUTION: The first step would be to switch rows 1 and 3 (to make $|a_{11}|$ as large as possible): Ab=rowswitch(Ab,1,3).

$$Ab \rightarrow P_1(Ab) = \begin{bmatrix} 0 & 0 & 1 \\ 0 & 1 & 0 \\ 1 & 0 & 0 \end{bmatrix} \begin{bmatrix} 1 & 3 & -1 & | & 2 \\ 2 & 5 & -2 & | & 3 \\ 3 & 6 & 9 & | & 39 \end{bmatrix} = \begin{bmatrix} 3 & 6 & 9 & | & 39 \\ 2 & 5 & -2 & | & 3 \\ 1 & 3 & -1 & | & 2 \end{bmatrix}.$$

(We will denote elementary matrices resulting from the rowswitch ERO by P_i's) Next we pivot on the $a_{11} = 3$ entry to clear out the entries below it. To clear out $a_{21} = 2$, we will do Ab=rowcomb(Ab,1,2,-2/3) (i.e., to clear out $a_{21} = 2$, we multiply row 1 by $-a_{21}/a_{11} = -2/3$ and add this to row 2). Similarly, to clear out $a_{31} = 1$, we will do >>Ab=rowcomb(Ab,1,3,-1/3). Combining both of these elementary matrices into a single matrix M_1, we may now write the result of these two EROs as follows:

$$Ab \rightarrow M_1(Ab) = \begin{bmatrix} 1 & 0 & 0 \\ -2/3 & 1 & 0 \\ -1/3 & 0 & 1 \end{bmatrix} \begin{bmatrix} 3 & 6 & 9 & | & 39 \\ 2 & 5 & -2 & | & 3 \\ 1 & 3 & -1 & | & 2 \end{bmatrix} = \begin{bmatrix} 3 & 6 & 9 & | & 39 \\ 0 & 1 & -8 & | & -23 \\ 0 & 1 & -4 & | & -11 \end{bmatrix}.$$

The pivot $a_{22} = 1$ is already as large as possible so we need not switch rows and can clear out the entry $a_{32} = 1$ by doing Ab=rowcomb(Ab,2,3,-1):

$$Ab \rightarrow M_2(Ab) = \begin{bmatrix} 1 & 0 & 0 \\ 0 & 1 & 0 \\ 0 & -1 & 1 \end{bmatrix} \begin{bmatrix} 3 & 6 & 9 & | & 39 \\ 0 & 1 & -8 & | & -23 \\ 0 & 1 & -4 & | & -11 \end{bmatrix} = \begin{bmatrix} 3 & 6 & 9 & | & 39 \\ 0 & 1 & -8 & | & -23 \\ 0 & 0 & 4 & | & 12 \end{bmatrix},$$

and with this the elimination is complete.

Solving this (equivalent) upper triangular system will again yield the above solution. We note that this produces a slightly different factorization: From $M_2 M_1 PA = U$, where U is the left part of the final augmented matrix, we proceed as in the previous example to get $PA = (M_2^{-1}M_1^{-1})U \equiv LU$, i.e.,

$$PA = \begin{bmatrix} 0 & 0 & 1 \\ 0 & 1 & 0 \\ 1 & 0 & 0 \end{bmatrix} \cdot \begin{bmatrix} 1 & 3 & -1 \\ 2 & 5 & -2 \\ 3 & 6 & 9 \end{bmatrix} = \begin{bmatrix} 1 & 0 & 0 \\ 2/3 & 1 & 0 \\ 1/3 & 1 & 1 \end{bmatrix} \cdot \begin{bmatrix} 3 & 6 & 9 \\ 0 & 1 & -8 \\ 0 & 0 & 4 \end{bmatrix} = LU.$$

This is the LU factorization of the matrix A. We now explain the general algorithm of **Gaussian elimination with partial pivoting** and the LU factorization. In terms of EROs and the back substitution, the resulting algorithm is quite compact.

Algorithm for Gaussian Elimination with Partial Pivoting: Given a linear system $Ax = b$ with A an $n \times n$ nonsingular matrix, this algorithm will solve for the solution vector x. The algorithm works on the $n \times (n+1)$ augmented matrix $[A \mid b]$, which we denote by Ab, but whose entries we still denote by a_{ij}.

For $k = 1$ to $n-1$

interchange rows (if necessary) to assure that $|a_{kk}| = \max_{k \leq i \leq n} |a_{ik}|$
---

if $|a_{kk}| = 0$, exit program with message " A is singular".

for $i = k + 1$ to n

$m_{ik} = a_{ik} / a_{kk}$

A = rowcomb($A, k, i, -m_{ik}$)

end i

end k

if $|a_{nn}| = 0$, exit program with message " A is singular".

Apply the back substitution algorithm on the final system (that is now upper triangular) to get solution to the system.

Without the interchanging rows step (unless to avoid a zero pivot), this is **Gaussian elimination** without partial pivoting. From now on, we follow the standard convention of referring to Gaussian elimination with partial pivoting simply as "Gaussian elimination," since it has become the standard algorithm for solving linear systems.

The algorithm can be recast into a matrix factorization algorithm for A. Indeed, at the kth iteration we will, in general, have an elementary matrix P_k corresponding to a row switch or permutation, followed by a matrix M_k that consists of the product of each of the elementary matrices corresponding to the "rowcomb" ERO used to clear out entries below a_{kk}. Letting U denote the upper triangular matrix left at the end of the algorithm, we thus have:

$$M_{n-1}P_{n-1} \cdots M_2 P_2 M_1 P_1 A = U .$$

The *LU* **factorization** (or the *LU* **decomposition**) of A, in general, has the form (see Section 4.4 of [GoVL-83]):

$$PA = LU , \tag{29}$$

where

$$P = P_{n-1}P_{n-2} \cdots P_2 P_1 \quad \text{and} \quad L = P(M_{n-1}P_{n-1} \cdots M_1 P_1)^{-1} , \tag{30}$$

and L is lower triangular.[12] Also, by Theorem 7.4, the matrix PA corresponds to sequentially switching the rows of the matrix A, first corresponding to P_1, next by P_2, and so on. Thus the *LU* factorization A, once known, leads to a quick and practical way to solve any linear system $Ax = b$. First, permute the order of the equations as dictated by the permutation matrix P (do this on the augmented matrix so that b's entries get permuted as well), relabel the system as $Ax = b$, and rewrite it as $LUx = b$. First solve $Ly = c$ by forward substitution (works since L

---

[12] The permutation matrix P in (29) cannot, in general, be dispensed with; see Exercise 6.

is lower triangular), then solve $Ux = y$ by back substitution (works since U is upper triangular). Then x will be the desired solution (Proof:

$$PA = LU \Rightarrow Ax = P^{-1}LUx = P^{-1}L(Ux) = P^{-1}L(y) = P^{-1}Pb = b \,.)$$

This approach is useful if it is needed to solve a lot of linear systems with the same coefficient matrix A. For such situations, we mention that MATLAB has a built-in function lu to compute the LU factorization of a nonsingular matrix A. The syntax is as follows:

[L, U, P]=lu(A) →	For a square singular matrix A, this command will output the lower triangular matrix L, the upper triangular matrix U, and the permutation matrix P of the LU factorization (29) and (30) of the matrix A.

For example, applying this command to the matrix A of Example 7.14 gives:

```
>>A=[1 3 -1; 2 5 -2; 3 6 9]; format rat, [L, U, P]=lu(A)
```

→L =   1   0   0
         2/3   1   0
         1/3   1   1

U =   3   6   9
          0   1   -8
          0   0   4

P =   0   0   1
          0   1   0
          1   0   0

We now wish to translate the above algorithm for Gaussian elimination into a MATLAB program. Before we do this, we make one remark about MATLAB's built-in function max, which we have encountered previously in its default format (the first syntax below):

max(v) →	For a vector v, this command will give the maximum of its components.
[max, index]=max(v) →	With an optional second output variable (that must be declared), max(v) will also give the first index at which this maximum value occurs.

Here, v can be either a row or a column vector. A simple example will illustrate this functionality.

```
>> v=[1 -3 5 -7 9 -11];
>> max(v)                        →ans = 9
>> [max, index] = max(v)         →max = 9, index = 5
>> [max, index] = max(abs(v))    →max = 11, index = 6
```

Another useful tool for programming M-files is the error command:

error('message') →	If this command is encountered with an execution of any M-file, the M-file stops running immediately and displays the message.

PROGRAM 7.6: Function M-file for Gaussian elimination (with partial pivoting) to solve the linear system $Ax = b$, where A is a square nonsingular matrix. This program calls on the previous programs backsubst, rowswitch, and rowcomb.

```
function x=gausselim(A,b)
%Input:   Square matrix A, and column vector b of same dimension
%Output:  Column vector solution x of linear system Ax = b obtained
%by Gaussian elimination with partial pivoting, provided coefficient
%matrix A is nonsingular.
[n,n]=size(A);
Ab=[A';b']'; %form augmented matrix for system

for k=1:n
    [biggest, occured] = max(abs(Ab(k:n,k)));
    if biggest == 0
    error('the coefficient matrix is numerically singular')
    end
    m=k+occured-1;
    Ab=rowswitch(Ab,k, m);
    for j=k+1:n
        Ab=rowcomb(Ab,k,j,-Ab(j,k)/Ab(k,k));
    end
end
% BACK SUBSTITUTION
x=backsubst(Ab(:,1:n),Ab(:,n+1));
```

EXERCISE FOR THE READER 7.20: Use the program gausselim to resolve the Hilbert systems of Example 7.9 and Exercise for the Reader 7.16, and compare to the results of the left divide method that proved most successful in those examples. Apart from additional error messages (regarding condition numbers), how do the results of the above algorithm compare with those of MATLAB's default system solver?

We next give a simple example that will demonstrate the advantages of partial pivoting.

EXAMPLE 7.15: Consider the following linear system: $\begin{bmatrix} 10^{-3} & 1 \\ 1 & 2 \end{bmatrix}\begin{bmatrix} x_1 \\ x_2 \end{bmatrix} = \begin{bmatrix} 1 \\ 3 \end{bmatrix}$,

whose exact solution (starts) to look like $\begin{bmatrix} x_1 \\ x_2 \end{bmatrix} = \begin{bmatrix} 1.0020... \\ .9989... \end{bmatrix}$.

(a) Using floating point arithmetic with three significant digits and chopped arithmetic, solve the system using Gaussian elimination with partial pivoting.
(b) Repeat part (a) in the same arithmetic, except without partial pivoting.

SOLUTION: For each part we show the chain of augmented matrices. Recall, after each individual computation, answers are chopped to three significant digits before any subsequent computations.
Part (a): (with partial pivoting)

$$\begin{bmatrix} .001 & 1 & | & 1 \\ 1 & 2 & | & 3 \end{bmatrix} \xrightarrow{\text{rowswitch}(A,1,2)} \begin{bmatrix} 1 & 2 & | & 3 \\ .001 & 1 & | & 1 \end{bmatrix} \xrightarrow{\text{rowcomb}(A,1,2,-.001)} \begin{bmatrix} 1 & 2 & | & 3 \\ 0 & .998 & | & .997 \end{bmatrix}.$$

Now we use back substitution: $x_2 = .997/.998 = .998$, $x_1 = (3 - 2x_2)/1$ $= (3 - 1.99)/1 = 1.01$. Our computed answer is correct to three decimals in x_2, but has a relative error of about 0.0798% in x_1.

Part (b): (without partial pivoting)

$$\begin{bmatrix} .001 & 1 & | & 1 \\ 1 & 2 & | & 3 \end{bmatrix} \xrightarrow{\text{rowcomb}(A,1,2,-1000)} \begin{bmatrix} 1 & 1 & | & 1 \\ 0 & -998 & | & -997 \end{bmatrix}.$$

Back substitution now gives $x_2 = -997/-998 = .998$, $x_1 = (1 - 1 \cdot x_2)/1$ $= (1 - .998)/1 = .002$. The relative error here is unacceptably large, exceeding 100% in the second component!

EXERCISE FOR THE READER 7.21: Rework the above example using rounded arithmetic rather than chopped, and keeping all else the same.

We close this section with a bit of information on flop counts for the Gaussian elimination algorithm. Note that the partial pivoting adds no flops since permuting rows involves no arithmetic. We will assume the worst-case scenario, in that none of the entries that come up are zeros. In counting flops, we refer to the algorithm above, rather than the MATLAB program. For $k = 1$, we will need to perform $n - 1$ divisions to compute the multipliers m_{i1}, and for each of these multipliers, we will need to do a rowcomb, which will involve n multiplications and n additions/subtractions. (Note: Since the first column entry will be zero and need not be computed, we are only counting columns 2 through $n + 1$ of the augmented matrix.) Thus, associated with the pivot a_{11}, we will have to do $n - 1$ divisions, $(n-1)n$ multiplications, and $(n-1)n$ additions/subtractions. Grouping the divisions and multiplications together, we see that at the $k = 1$ (first) iteration, we will need $(n-1) + (n-1)n = (n-1)(n+1)$ multiplications/divisions and $(n-1)n$ additions/subtractions. In the same fashion, the calculations associated with the pivot a_{22} will involve $n - 2$ divisions plus $(n-2)(n-1)$ multiplications which is $(n-2)n$ multiplications/divisions and $(n-2)(n-1)$ additions/ subtractions. Continuing in this fashion, when we get to the pivot a_{kk}, we will need to do $(n - k)(n - k + 2)$ multiplications/divisions and $(n - k)(n - k + 1)$ additions/ subtractions. Summing from $k = 1$ to $n - 1$ gives the following:

Total multiplications/divisions $\equiv M(n) = \sum_{k=1}^{n-1}(n-k)(n-k+2)$,

Total additions/subtractions $\equiv A(n) = \sum_{k=1}^{n-1}(n-k)(n-k+1)$.

Combining these two sums and regrouping gives:

Grand total flops $\equiv$

$$F(n) = \sum_{k=1}^{n-1} (n-k)(2(n-k)+3) = 2\sum_{k=1}^{n-1} (n-k)^2 + 3\sum_{k=1}^{n-1} (n-k).$$

If we reindex the last two sums, by substituting $j = n - k$, then as k runs from 1 through $n - 1$, so will j (except in the reverse order), so that

$$F(n) = 2\sum_{j=1}^{n-1} j^2 + 3\sum_{j=1}^{n-1} j.$$

We now invoke the power sum identities (23) and (24) to evaluate the above two sums (replace n with $n - 1$ in the identities) and thus rewrite the flop count $F(n)$ as:

$$F(n) = \frac{1}{3}(n-1)n(2n-1) + \frac{3}{2}(n-1)n = \frac{2}{3}n^3 + \text{lower power terms.}$$

The "lower power terms" in the above flop counts can be explicitly computed (simply multiply out the polynomial on the left), but it is the highest order term that grows the fastest and thus is most important for roughly estimating flop counts. The flop count does not include the back substitution algorithm; but a similar analysis shows flop count for the back substitution to be just n^2 (see the Exercise for the Reader 7.22), and we thus can summarize with the following result.

PROPOSITION 7.5: (*Flop Counts for Gaussian Elimination*) In general, the number of flops needed to perform Gaussian elimination to solve a nonsingular system $Ax = b$ with an $n \times n$ coefficient matrix A is

$$\frac{2}{3}n^3 + \text{lower order terms.}$$

EXERCISE FOR THE READER 7.22: Show that for the back substitution algorithm, the number of multiplications/divisions will be $(n^2 + n)/2$, and the number of additions/subtractions will be $(n^2 - n)/2$. Hence, the grand total flops required will be n^2.

By using the natural algorithm for computing the inverse of a matrix, a similar analysis can be used to show the flop count for finding the inverse of a matrix of a nonsingular $n \times n$ matrix to be

$$(8/3)n^3 + \text{lower power terms,}$$

or, in other words, essentially four times that for a single Gaussian elimination (see Exercise 16). Actually, it is possible to modify the algorithm (of Exercise 16) to a more complicated one that can bring this flop count down to

$2n^3$ + lower power terms; but this is still going to be a more expensive and error-prone method than Gaussian elimination, so we reiterate: For solving a single general linear system, Gaussian elimination is the best all-around method.

The next example will give some hard evidence of the rather surprising fact that the computer time required (on MATLAB) to perform an addition/subtraction is about the same as that required to perform a multiplication/division.

EXAMPLE 7.16: In this example, we perform a short experiment to record the time and flops required to add 100 pairs of random floating point numbers. We then do the related experiment involving the same number of divisions.

```
>> A=rand(100); B=rand(100);,
>> tic, for i=1:100, C=A+B; end, toc
→ Elapsed time is 5.778000 seconds.

>> tic, for i=1:100, C=A./B;, end, toc
→ Elapsed time is 5.925000 seconds.
```

The times are roughly of the same magnitude and, indeed, the flop counts are identical and close to the actual number of mathematical operations performed. The reason for the discrepancy in the latter is that, as previously mentioned, a flop is "approximately" equal to one arithmetic operation on the computer; and this is the most useful way to think about a flop.

EXERCISES 7.5:

NOTE: As mentioned in the text, we take "Gaussian elimination" to mean Gaussian elimination with partial pivoting.

1. Solve each of the following linear systems $Ax = b$ using three-digit chopped arithmetic and Gaussian elimination (i) without partial pivoting and then (ii) with partial pivoting. Finally redo the problems (iii) using MATLAB's left divide operator, and then (iv) using exact arithmetic (any method).

    (a) $A = \begin{bmatrix} 2 & -9 \\ 1 & 7.5 \end{bmatrix}$, $b = \begin{bmatrix} 2 \\ -3 \end{bmatrix}$ (b) $A = \begin{bmatrix} .99 & .98 \\ 101 & 100 \end{bmatrix}$, $b = \begin{bmatrix} 1 \\ -1 \end{bmatrix}$

    (c) $A = \begin{bmatrix} 2 & 3 & -1 \\ 4 & 2 & -1 \\ -8 & 2 & 0 \end{bmatrix}$, $b = \begin{bmatrix} 0 \\ 21 \\ -4 \end{bmatrix}$

2. Parts (a) through (c): Repeat all parts of Exercise 1 using two-digit rounded arithmetic.

3. For each square matrix specified, find the LU factorization of the matrix (using Gaussian elimination). Do it first using (i) three-digit chopped arithmetic, then using (ii) exact arithmetic; and finally (iii) compare these with the results using MATLAB's built-in function lu.
    (a) The matrix A in Exercise 1, part (a).
    (b) The matrix A in Exercise 1, part (b).
    (c) The matrix A in Exercise 1, part (c).

4. Parts (a) through (c): Repeat all parts of Exercise 3 using two-digit rounded arithmetic in (i).

5. Consider the following linear system involving the 3×3 Hilbert matrix H_3 as the coefficient matrix:

$$\begin{cases} x_1 + \dfrac{1}{2}x_2 + \dfrac{1}{3}x_3 = 2 \\[2mm] \dfrac{1}{2}x_1 + \dfrac{1}{3}x_2 + \dfrac{1}{4}x_3 = 0 \\[2mm] \dfrac{1}{3}x_1 + \dfrac{1}{4}x_2 + \dfrac{1}{5}x_3 = -1 \end{cases}.$$

(a) Solve the system using two-digit chopped arithmetic and Gaussian elimination without partial pivoting.
(b) Solve the system using two-digit chopped arithmetic and Gaussian elimination.
(c) Solve the system using exact arithmetic (any method).
(d) Find the LU decomposition of the coefficient matrix H_3 by using 2-digit chopped arithmetic and Gaussian elimination.
(e) Find the exact LU decomposition of H_3.

6. (a) Find the LU factorization of the matrix $A = \begin{bmatrix} 0 & 1 \\ 1 & 0 \end{bmatrix}$.

(b) Is it possible to find a lower triangular matrix L and an upper triangular matrix U (not necessarily those in part (a)) such that $A = LU$? Explain why or why not.

7. Suppose that $M_1, M_2, \cdots, M_k$ are invertible matrices of the same size. Prove that their product is invertible with $(M_1 \cdot M_2 \cdots M_k)^{-1} = M_k^{-1} \cdots M_2^{-1} \cdot M_1^{-1}$. In words, "The inverse of the product is the reverse-order product of the inverses."

8. (*Storage and Computational Savings in Solving Tridiagonal Systems*) Just as with any (nonsingular) matrix, we can apply Gaussian elimination to solve tridiagonal systems:

$$\begin{bmatrix} d_1 & a_1 & & & & & \\ b_2 & d_2 & a_2 & & & \text{\Large 0} & \\ & b_3 & d_3 & a_3 & & & \\ & & b_4 & d_4 & a_4 & & \\ & & & \ddots & \ddots & \ddots & \\ \text{\Large 0} & & & & b_{n-1} & d_{n-1} & a_{n-1} \\ & & & & & b_n & d_n \end{bmatrix} \begin{bmatrix} x_1 \\ x_2 \\ x_3 \\ x_4 \\ \vdots \\ x_{n-1} \\ x_n \end{bmatrix} = \begin{bmatrix} r_1 \\ r_2 \\ r_3 \\ r_4 \\ \vdots \\ r_{n-1} \\ r_n \end{bmatrix}. \qquad (31)$$

Here, d's stand for diagonal entries, b's for below-diagonal entries, a's for above-diagonal entries, and r's for right-side entries. We can greatly cut down on storage and unnecessary mathematical operations with zero by making use of the special sparse form of the tridiagonal matrix. The main observation is that at each step of the Gaussian elimination process, we will always be left with a banded matrix with perhaps one additional band above the a's diagonal. (Think about it, and convince yourself. The only way a switch can be done in selecting a pivot is with the row immediately below the diagonal pivot entry.) Thus, we may wish to organize a special algorithm that deals only with the tridiagonal entries of the coefficient matrix.
(a) Show that the Gaussian elimination algorithm, with unnecessary operations involving zeros being omitted, will require no more than $8(n-1)$ flops (multiplications, divisions, additions, subtractions), and the corresponding back substitution will require no more than $5n$ flops. Thus the total number of flops for solving such a system can be reduced to less than $13n$.
(b) Write a program, $x = \texttt{tridiaggauss(d, b,a, r)}$ that inputs the diagonal vector d (of length n) and the above and below diagonal vectors a and b (of length $n-1$) of a nonsingular tridiagonal matrix, the column vector r and will solve the tridiagonal system (31) using the Gaussian elimination algorithm but which overwrites only the four relevant diagonal vectors

(described above; you need to create an $n-2$ length vector for the extra diagonal) and the vector r rather than on the whole matrix. The output should be the solution column vector x.

(c) Test out your algorithm on the system (31) with $n=2$, $n=100$, $n=500$, and $n=1000$ using the following data in the matrices $d_i=4$, $a_i=1$, $b_i=1$, $r=[1\ -1\ 1\ -1\ \cdots]$ and compare results and flop counts with MATLAB's left divide. You should see that your algorithm is much more efficient.

Note: The upper bound $13n$ on flops indicated in part (a) is somewhat liberal; a more careful analysis will show that the coefficient 13 can actually be made a bit smaller (How small can you make it?) But even so, the savings on flops (not to mention storage) are incredible. If we compare $13n$ with the bound $2n^3/3$, for large values of n, we will see that this modified method will allow us to solve extremely large tridiagonal systems that previously would have been out of the question. For example, when $n = 10,000$, this modified method would require storage of $2n + 2(n-1) = 39,998$ entries and less than $13n = 130,000$ flops (this would take a few seconds on MATLAB even on a weak computer); whereas the ordinary Gaussian elimination would require the storage of $n^2 + n = 100,010,000$ entries and approximately $2n^3/3 = 6.66... \times 10^{11}$ flops, an unmanageable task!

9. (*The Thomas Method, an Even Faster Way to Solve Tridiagonal Systems*) By making a few extra assumptions that are usually satisfied in most tridiagonal systems that arise in applications, it is possible to slightly modify the usual Gaussian elimination algorithm to solve the triadiagonal system (31) in just $8n-7$ flops (compared to the upper bound $13n$ of the last problem). The algorithm, known as the **Thomas method,**[13] differs from the usual Gaussian elimation by scaling the diagonal entry to equal 1 at each pivot, and by not doing any row changes (i.e., we forgo partial pivoting, or assume the matrix is of a form that makes it unnecessary; see Exercise 10). This will mean that we will have to keep track only of the above-diagonal entries (the a's vector) and the right-side vector r. The Thomas method algorithm thus proceeds as follows:

Step 1: (Results from rowmult(A, 1, $1/d_1$)): $a_1 = a_1/d_1$, $r_1 = r_1/d_1$.

(We could also add $d_1 = 1$, but since the diagonal entries will always be scaled to equal one, we do not need to explicitly record this change.)

Steps $k = 2$ through $n - 1$: (Results from rowcomb(A, $k-1$,k, $-b_k$) and then rowscale(A, k,

$1/(d_k - b_k a_{k-1})$)):

$$a_k = a_k/(d_k - b_k a_{k-1}), \quad r_k = (r_k - b_k r_{k-1})/(d_k - b_k a_{k-1}).$$

Step n: (Results from same procedure as in steps 2 through $n - 1$, but there is no a_n):

$$r_n = (r_n - b_n r_{n-1})/(d_n - b_n a_{n-1}).$$

This variation of Gaussian elimination has transformed the tridiagonal system into an upper triangular system with the following special form:

$$
\begin{bmatrix}
1 & a_1 & & \\
 & 1 & a_2 & \mathbf{0} \\
 & & \ddots & \ddots \\
\mathbf{0} & & 1 & a_{n-1} \\
 & & & 1
\end{bmatrix}
\begin{bmatrix}
x_1 \\ x_2 \\ \vdots \\ x_{n-1} \\ x_n
\end{bmatrix}
=
\begin{bmatrix}
r_1 \\ r_2 \\ \vdots \\ r_{n-1} \\ r_n
\end{bmatrix},
$$

---

[13] The method is named after the renowned physicist Llewellyn H. Thomas; but it was actually discovered independently by several different individuals working in mathematics and related disciplines. W.F. Ames writes in his book [Ame-77] (p. 52): "The method we describe was discovered independently by many and has been called the Thomas algorithm. Its general description first appeared in widely distributed published form in an article by Bruce et al. [BPRR-53]."

for which the back substitution algorithm takes on the particularly simple form: $x_n = r_n$; then for

$k = n-1, n-2, ..., 2, 1: x_k = r_k - a_k x_{k+1}$.

(a) Write a MATLAB M-file, x=thomas(d, b, a, r) that performs the Thomas method as described above to solve the tridiagonal system (31). The inputs should be the diagonal vector d (of length n) and the above and below diagonal vectors a and b (of length $n-1$) of a nonsingular tridiagonal matrix, and the column vector r. The output should be the computed solution, as a column vector x. Write your program so that it overwrites only the vectors a and r.

(b) Test out your program on the systems of part (c) of Exercise 8, and compare results and flop counts with those for MATLAB's left divide solver. If you have done part (c) of Exercise 8, compare also with the results from the program of the previous exercise.

(c) Do a flop count on the Thomas method to show that the total number of flops needed is $8n - 7$.

NOTE: Looking over the Thomas method, we see that it assumes that $d_1 \neq 0$, and $d_k \neq b_k a_{k-1}$ (for $k = 2$ through n). One might think that to play it safe, it may be better to just use the slightly more expensive modification of Gaussian elimination described in the previous exercise, rather than risk running into problems with the Thomas method. For at most all applications, it turns out that the requirements for the Thomas method indeed are satisfied. Such triadiagonal systems come up naturally in many applications, in particular in finite difference schemes for solving differential equations. One safe approach would be to simply build in a deferral to the previous algorithm in cases where the Thomas algorithm runs into a snag.

10. We say that a square matrix $A = [a_{ij}]$ is **strictly diagonally dominant (by columns)** if for each index k, $1 \le k \le n$, the following condition is met:

$$|a_{kk}| > \sum_{\substack{j=1 \\ j \neq k}}^{n} |a_{kj}|. \tag{32}$$

This condition merely states that each diagonal entry is larger, in absolute value, than the sum of the absolute values of all of the other entries in its column.

(a) Explain why when Gaussian elimination is applied to solve a linear system $Ax = b$ whose coefficient matrix is strictly diagonally dominant by columns, then no row changes will be required.

(b) Explain why the LU factorization of a diagonally dominant by columns matrix A will not have any permutation matrix.

(c) Explain why the requirements for the Thomas method (Exercise 9) will always be met if the coefficient matrix is strictly diagonally dominant by columns.

(d) Which, if any, of the above facts will continue to remain true if the strict diagonal dominance condition (32) is weakened to the following?

$$|a_{kk}| > \sum_{j=k+1}^{n} |a_{kj}|.$$

(That is, we are now only assuming that each diagonal entry is larger, in absolute value, than the sum of the absolute values of the entries that lie in the same column but below it.)

11. Discuss what conditions on the industries must hold in order for the technology matrix M of the Leontief input/output model of Exercise 10 from Section 7.4 to be diagonally dominant by columns (see the preceding exercise).

12. (*Determinants Revisited: Effects of Elementary Row/Column Operations on Determinants*) Prove the following facts about determinants, some of which were previewed in Exercise 10 of Section 7.1.

(a) If the matrix B is obtained from the square matrix A by multiplying one of the latter's rows by a number c (and leaving all other rows the same, i.e., B=rowmult(A,i,c)), then det(B) = $c$det(A).

(b) If the matrix B is obtained from the square matrix A by adding a multiple of the ith row of A to the jth row ($i \neq j$) (i.e., B = rowcomb (A, i, j, c)), then $\det(B) = \det(A)$.

(c) If the matrix B results from the matrix A by switching two rows of the latter (i.e., B=rowswitch(A, i, j)), then $\det(B) = -\det(A)$.

(d) If two rows of a square matrix are the same, then $\det(A) = 0$.

(e) If B is the transpose of A, then $\det(B) = \det(A)$.

Note: In light of the result of part (e), each of the statements in the other parts regarding the effect of a row operation on a determinant has a valid counterpart for the effect of the corresponding column operation on the determinant.

Suggestions: You should make use of identity (20) $\det(AB) = \det(A)\det(B)$, as well as Proposition 7.3 and Theorem 7.4. The results of (a), (b), and (c) can then be proved by calculating determinants of certain elementary matrices. The only difficult thing is for part (c) to show that the determinant of a permutation matrix gotten from the identity matrix by switching two rows equals -1. One way this can be done is by an appropriate (but not at all obvious) matrix factorization. Here is one way to do it for the (only) 2×2 permutation matrix:

$$\begin{bmatrix} 0 & 1 \\ 1 & 0 \end{bmatrix} = \begin{bmatrix} 1 & -1 \\ 0 & 1 \end{bmatrix}\begin{bmatrix} 1 & 0 \\ 1 & 1 \end{bmatrix}\begin{bmatrix} 1 & 0 \\ 0 & -1 \end{bmatrix}\begin{bmatrix} 1 & 1 \\ 0 & 1 \end{bmatrix}.$$

(Check this!) All of the matrix factors on the right are triangular so the determinants of each are easily computed by multiplying diagonal entries (Proposition 7.3), so using (20), we get

$$\det\left(\begin{bmatrix} 0 & 1 \\ 1 & 0 \end{bmatrix}\right) = 1 \cdot 1 \cdot (-1) \cdot 1 = -1.$$

In general, this argument can be made to work for any permutation matrix (obtained by switching two rows of the identity matrix), by carefully generalizing the factorization. For example, here is how the factorization would generalize for a certain 3×3 permutation matrix:

$$\begin{bmatrix} 0 & 0 & 1 \\ 0 & 1 & 0 \\ 1 & 0 & 0 \end{bmatrix} = \begin{bmatrix} 1 & 0 & -1 \\ 0 & 1 & 0 \\ 0 & 0 & 1 \end{bmatrix}\begin{bmatrix} 1 & 0 & 0 \\ 0 & 1 & 0 \\ 1 & 0 & 1 \end{bmatrix}\begin{bmatrix} 1 & 0 & 0 \\ 0 & 1 & 0 \\ 0 & 0 & -1 \end{bmatrix}\begin{bmatrix} 1 & 0 & 1 \\ 0 & 1 & 0 \\ 0 & 0 & 1 \end{bmatrix}.$$

Part (d) can be proved easily from part (c); for part (e) use mathematical induction and cofactor expansion.

13. *(Determinants Revisited: A Better Way to Compute Them)* The Gaussian elimination algorithm provides us with an efficient way to compute determinants. Previously, the only method we gave to compute them was by cofactor expansion, which we introduced in Chapter 4. But we saw that this was an extremely expensive way to compute determinants. A new idea is to use the Gaussian elimination algorithm to transform a square matrix A into an upper triangular matrix. From the previous exercise, each time a rowcomb is done, there will be no effect on the determinant, but each time a rowswitch is done, the determinant is negated. By Proposition 7.3, the determinant of the diagonal matrix is just the product of the diagonal entries. Of course, in the Gaussian elimination algorithm, the column vector b can be removed (if all we are interested in is the determinant). Also, if a singularity is detected, the algorithm should exit and assign $\det(A) = 0$.

(a) Create a function M-file, called y=gaussdet (A), that inputs a square matrix M and outputs the determinant using this algorithm.

(b) Test your program out by computing the determinants of matrices with random integer entries from -9 to 9 of sizes 3×3, 8×8, 20×20, and 80×80 (you need not print the last two matrices) that you can construct using the M-file randint of Exercise for the Reader 7.2. Compare the results, computing times and (if you have Version 5) flop counts with those for MATLAB's built-in det function applied to the same matrices.

(c) Go through an analysis similar to that done at the end of the section to prove a result similar to that of Proposition 7.5 that will give an estimate of the total flop counts for this algorithm, with the highest-order term being accurate.

(d) Obtain a similar flop count for the cofactor expansion method and compare with the answer you got in (c). (The highest-order term will involve factorials rather than powers.)

(e) Use your answer in (c) to obtain a flop count for the amount of flops needed to apply

Cramer's rule to solve a nonsingular linear system $Ax = b$ with A being an $n \times n$ nonsingular matrix.

14. (a) Prove Proposition 7.3.

(b) Prove an analogous formula for the determinant of a square matrix that is upper-left triangular in the sense that all entries above the off-main diagonal are zeros. More precisely, prove that any matrix of the following form,

$$
A = \begin{bmatrix}
a_{11} & a_{12} & a_{13} & \cdots & a_{1n} \\
a_{21} & a_{22} & a_{23} & & \\
\vdots & & & \ddots & \\
a_{n-1,1} & a_{n-1,2} & & & \mathbf{0} \\
a_{n1} & & & &
\end{bmatrix},
$$

has determinant given by $\det(A) = (-1)^k a_{1n} \cdot a_{2,n-1} \cdot a_{3,n-2} \cdots a_{n-1,2} \cdot a_{n1}$, where $n = 2k + i$ $(i = 0,1)$.

Suggestion: Proceed by induction on n, where A is an $n \times n$ matrix. Use cofactor expansion along an appropriate row (or column).

15. (a) Write a function M-file, call it `[L, U, P]= mylu(A)`, that will compute the LU factorization of an inputted nonsingular matrix A.

(b) Apply this function to each of the three coefficient matrices in Exercise 1 as well as the Hilbert matrix H_3, and compare the results (and flop counts) to those with MATLAB's built-in function `lu`. From these comparisons, does your program seem to be as efficient as MATLAB's?

16. (a) Write a function M-file, call it `B=myinv(A)`, that will compute the inverse of an inputted nonsingular matrix A, and otherwise will output the error message: "Matrix detected as numerically singular." Your algorithm should be based on the following fact (which follows from the way that matrix multiplication works). To find an inverse of an $n \times n$ nonsingular matrix A, it is sufficient to solve the following n linear equations:

$$
Ax^1 = \begin{bmatrix} 1 \\ 0 \\ 0 \\ \vdots \\ 0 \end{bmatrix}, \quad Ax^2 = \begin{bmatrix} 0 \\ 1 \\ 0 \\ \vdots \\ 0 \end{bmatrix}, \quad Ax^3 = \begin{bmatrix} 0 \\ 0 \\ 1 \\ \vdots \\ 0 \end{bmatrix}, \quad \cdots Ax^n = \begin{bmatrix} 0 \\ 0 \\ 0 \\ \vdots \\ 1 \end{bmatrix},
$$

where the column vectors on the right sides of these equations are precisely the columns of the $n \times n$ identity matrix. It then would follow that $A\left[x^1 \mid x^2 \mid x^3 \mid \cdots \mid x^n \right] = I$, so that the desired inverse of A is the matrix $A^{-1} = \left[x^1 \mid x^2 \mid x^3 \mid \cdots \mid x^n \right]$. Your algorithm should be based on the LU decomposition, so it gets computed once, rather than doing a complete Gaussian elimination for each of the n equations.

(b) Apply this function to each of the three coefficient matrices in Exercise 1 as well as the Hilbert matrix H_4, and compare the results (and flop counts) to those with MATLAB's built-in function `inv`. From these comparisons, does your program seem to be as efficient as MATLAB's?

(c) Do a flop count similar to the one done for Proposition 7.5 for this algorithm.

Note: For part (a), feel free to use MATLAB's built-in function `lu`; see the comments in the text about how to use the LU factorization to solve linear systems.

7.6: VECTOR AND MATRIX NORMS, ERROR ANALYSIS, AND EIGENDATA

In the last section we introduced the Gaussian elimination (with partial pivoting) algorithm for solving a nonsingular linear system

$$Ax = b, \tag{33}$$

where $A = [a_{ij}]$ is an $n \times n$ coefficient matrix, b is an $n \times 1$ column vector, and x is the $n \times 1$ column vector of variables whose solution is sought. This algorithm is the best all-around general numerical method for solving the linear system (33), but its performance can vary depending on the coefficient matrix A. In this section we will present some practical estimates for the error of the computed solution that will allow us to put some quality control guarantee on the answers that we obtain from (numerical) Gaussian elimination. We need to begin with a practical way to measure the "sizes" of vectors and matrices. We have already used the Euclidean length of a vector v to measure its size, and norms will be a generalization of this concept. We will introduce norms for vectors and matrices in this section, as well as the so-called condition numbers for square matrices. Shortly we will use norms and condition numbers to give precise estimates for the error of the computed solution of (33) (using Gaussian elimination). We will also explain some ideas to try when a system to be solved is poorly conditioned. The theory on modified algorithms that can deal with poorly conditioned systems contains an assortment of algorithms that can perform well if the (poorly conditioned) matrix takes on a special form. If one has the Student Version of MATLAB (or has the Symbolic Toolbox) there is always the option of working in exact arithmetic or with a fixed but greater number of significant digits. The main (and only) disadvantage of working in such arithmetic is that computations move a lot slower, so we will present some concrete criteria that will help us to decide when such a route might be needed. The whole subject of error analysis and refinements for numerically solving linear systems is quite vast and we will not be delving too deeply into it. For more details and additional results, the interested reader is advised to consult one of the following references (listed in order of increasing mathematical sophistication): [Atk-89], [Ort-90], [GoVL-83].

The Euclidean "length" of an n-dimensional vector $x = [x_1 \ x_2 \ \cdots \ x_n]$ is defined by:

$$\text{len}(x) = \sqrt{x_1^2 + x_2^2 + \cdots + x_n^2}. \tag{34}$$

For this definition it is immaterial whether x is a row or column vector. For example, if we are working in two dimensions and if the vector is drawn in the xy-plane from its tail at $(x,y) = (0,0)$ to its tip $(x,y) = (x_1, x_2)$, then $\text{len}(x)$ is (in most cases) the hypotenuse of a right triangle with legs having length $|x_1|$ and $|x_2|$, and so the formula (34) becomes the Pythagorean theorem. In the remaining cases where one of x_1 or x_2 is zero, then $\text{len}(x)$ is simply the absolute value of the other coordinate (in this case also the length of the vector x that will lie on either the x- or y-axis.) From what we know about plane geometry, we can deduce that len(x) has the following properties:

$$\text{len}(x) \geq 0, \text{ and } \text{len}(x) = 0 \text{ if and only if } x = 0(\text{vector}), \tag{35A}$$

$$\text{len}(cx) = |c| \text{ len}(x) \text{ for any scalar } c, \tag{35B}$$

$$\text{len}(x + y) \leq \text{len}(x) + \text{len}(y) \quad (\text{Triangle Inequality}). \tag{35C}$$

Property (35A) is clear (even in n dimensions). Property (35B) corresponds to the geometric fact that when a vector is multiplied by a scalar, the length gets multiplied by the absolute value of the scalar (we learned this early in the chapter). The triangle inequality (35C) corresponds to the geometric fact (in two dimensions) that the length of any side of any triangle can never exceed the sum of the lengths of the other two sides. These properties remain true for general n-dimensional vectors (see Exercise 11 for a more general result).

A **vector norm** for n-dimensional (row or column) vectors $x = [x_1 \ x_2 \ \cdots \ x_n]$ is a way to associate a nonnegative number (default notation: $\|x\|$) with the vector x such that the following three properties hold:

$$\|x\| \geq 0, \|x\| = 0 \text{ if and only if } x = 0(\text{vector}), \tag{36A}$$

$$\|cx\| = |c| \|x\| \text{ for any scalar } c, \tag{36B}$$

$$\|x + y\| \leq \|x\| + \|y\| \quad (\text{Triangle Inequality}). \tag{36C}$$

We have merely transcribed the properties (35) to obtain these three axioms (36) for a norm. It turns out that there is an assortment of useful norms, the aforementioned Euclidean norm being one of them. The one we will use most in this section is the so-called **max norm** (also known as the **infinity norm**) and this is defined as follows:

$$\|x\| = \|x\|_\infty = \max\{|x_i|, 1 \leq i \leq n\} = \max\{|x_1|, |x_2|, \cdots, |x_n|\}. \tag{37}$$

The proper mathematical notation for this vector norm is $\|x\|_\infty$, but since it will be our default vector norm we will often denote it by $\|x\|$ for convenience. The max norm is the simplest of all vector norms, so working with it will allow the complicated general concepts from error analysis to be understood in the simplest possible setting. The price paid for this simplicity will be that some of the resulting error estimates that we obtain using the max norm may be somewhat more liberal than those obtained with other, more complicated norms. Both the max and Euclidean norms are easy to compute on MATLAB (e.g., for the max norm of x we could simply type max(abs(x)), but (of course) MATLAB has built-in functions for both of these vector norms and many others.

norm(x) $\rightarrow$	Computes the length norm len(x) of a (row or column) vector x.

`norm(x, inf)` →	Computes the max norm $\|x\|$ of a (row or column) vector x.

EXAMPLE 7.17: For the two four-dimensional vectors $x = [1, 0, -4, 6]$ and $y = [3, -4, 1, -3]$ find the following:

(a) $\text{len}(x)$, $\text{len}(y)$, $\text{len}(x + y)$

(b) $\|x\|$, $\|y\|$, $\|x + y\|$

SOLUTION: First we do these computations by hand, and then redo them using MATLAB.

Part (a): Using (34) and since $x = [4, -4, -3, 3]$ we get that

$\text{len}(x) = \sqrt{1^2 + 0^2 + (-4)^2 + 6^2} = \sqrt{53} = 7.2801\ldots$, $\text{len}(y) = \sqrt{3^2 + (-4)^2 + 1^2 + (-3)^2}$

$= \sqrt{35} = 5.9160\ldots$, and $\text{len}(x + y) = \sqrt{4^2 + (-4)^2 + (-3)^2 + 3^2} = \sqrt{50} = 7.0710\ldots$

Part (b): Using (37), we compute: $\|x\| = \max\{|1|, |0|, |-4|, |6|\} = 6$, $\|y\| = \max\{|3|, |-4|, |1|, |-3|\} = 4$, and $\|x + y\| = \max\{|4|, |-4|, |-3|, |3|\} = 4$.

These computations give experimental evidence of the validity of the triangle inequality in this special case. We now repeat these same computations using MATLAB:

```
>> x=[1 0 -4 6]; y=[3 -4 1 -3];
>> norm(x), norm(y), norm(x+y)     →ans =  7.2801      5.9161      7.0711
>> norm(x,inf), norm(y,inf),    norm(x+y,inf) →ans =  6        4       4
```

EXERCISE FOR THE READER 7.23: Show that the max norm as defined by (37) is indeed a vector norm by verifying the three vector norm axioms of (36).

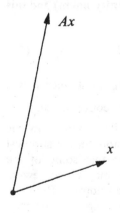

Given any vector norm, we define an **associated matrix norm** by the following:

$$\|A\| \equiv \max\left\{ \frac{\|Ax\|}{\|x\|}, \ x \neq 0 \, (\text{vector}) \right\}. \tag{38}$$

For any nonzero vector x, its norm $\|x\|$ will be a positive number (by (36A)); the transformed vector Ax will be another vector and so will have a norm $\|Ax\|$. The norm of the matrix A can be thought of as the maximum magnification factor by which the transformed vector Ax's norm will have changed from the original vector x's norm; see Figure 7.35.

FIGURE 7.35: Graphic for the matrix norm definition (38). The matrix A will transform x into another vector Ax (of same dimension if A is square). The norm of A is the maximum magnification that the transformed vector Ax norm will have in terms of the norm of x.

It is interesting that matrix norms, despite the daunting definition (38), are often easily computed from other formulas. For the max vector norm, it can be shown that the corresponding matrix norm (38), often called the **infinity matrix norm**, is given by the following formula:

$$\|A\| \equiv \max_{1 \le i \le n} \left\{ \sum_{j=1}^{n} |a_{ij}| \right\} = \max_{1 \le i \le n} \{ |a_{i1}| + |a_{i2}| + \cdots + |a_{in}| \}. \tag{39}$$

This more practical definition is simple to compute: We take the sum of each of the absolute values of the entries in each row of the matrix A, and $\|A\|$ will equal the maximum of these "row sums." MATLAB has a command that will do this computation for us:

norm(A,inf) →	Computes the infinity norm $\|A\|$ of a matrix A.

One simple but very important consequence of the definition (38) is the following inequality:

$$\|Ax\| \le \|A\|\|x\| \text{ (for any matrix } A \text{ and vector } x \text{ of compatible size)}. \tag{40}$$

To see why (40) is true is easy: First if x is the zero vector, then so is Ax and so both sides of (40) are equal to zero. If x is not the zero vector then by (38) we have $\|Ax\|/\|x\| \le \|A\|$, so we can multiply both sides of this inequality by the positive number $\|x\|$ to produce (40).

EXAMPLE 7.18: Let $A = \begin{bmatrix} 1 & 2 & -1 \\ 0 & 3 & -1 \\ 5 & -1 & 1 \end{bmatrix}$ and $x = \begin{bmatrix} 1 \\ 0 \\ -2 \end{bmatrix}$. Compute $\|x\|$, $\|Ax\|$,

and $\|A\|$ and check the validity of (40).

SOLUTION: Since $Ax = \begin{bmatrix} 3 \\ 2 \\ 3 \end{bmatrix}$, we obtain: $\|x\| = 2$, $\|Ax\| = 3$, and using (39),

$\|A\| = \max\{1+2+|-1|, \ 0+3+|-1|, \ 5+|-1|+1\} = \max\{4, 4, 7\} = 7.$ Certainly $\|Ax\| \le \|A\|\|x\|$ holds here ($3 \le 7 \cdot 2$).

EXERCISE FOR THE READER 7.24: Prove the following two facts about matrix norms: For two $n \times n$ matrices A and B:

(a) $\|AB\| \le \|A\|\|B\|$.

(b) If A is nonsingular, then $\|A^{-1}\| = \left(\min_{x \ne 0} \frac{\|Ax\|}{\|x\|} \right)^{-1}$.

With matrix norms introduced, we are now in a position to define the condition number of a nonsingular (square) matrix. For such a matrix A, the **condition number** of A, denoted by $\kappa(A)$, is the product of the norm of A and the norm of A^{-1},, i.e.,

$$\kappa(A) = \text{condition number of } A \equiv \| A \| \| A^{-1} \|. \tag{41}$$

By convention, for a singular matrix A, we define $\kappa(A) = \infty$.[14] Unlike the determinant, a large condition number is a reliable indicator that a square matrix is **nearly singular** (or **poorly conditioned**); and condition numbers will be a cornerstone in many of the error estimates for linear systems that we give later in this section. Of course, the condition number depends on the vector norm that is being used (which determines the matrix norm), but unless explicitly stated otherwise, we will always use the infinity vector norm (and the associated matrix norm and condition numbers). To compute the condition number directly is an expensive computation in general, since it involves computing the inverse A^{-1}. There are good algorithms to estimate condition numbers relatively quickly to any degree of accuracy. We will forgo presenting such algorithms, but will take the liberty of using the following MATLAB built-in function for computing condition numbers:

`cond(A, inf)` →	Computes and outputs the condition number (with respect to the infinity vector norm) of the square matrix A.

The condition number has the following general properties (actually valid for condition numbers arising from any vector norm):

$$\kappa(A) \geq 1, \text{ for any square matrix } A. \tag{42}$$

If D is a diagonal matrix with nonzero diagonal entries: $d_1, d_2, \cdots, d_n$, then $\tag{43}$

$$\kappa(D) = \frac{\max\{|d_i|\}}{\min\{|d_i|\}}.$$

If A is a square matrix and c is a nonzero scalar, then $\kappa(cA) = \kappa(A)$. $\tag{44}$

In particular, from (43) it follows that $\kappa(I) = 1$. The proofs of these identities will be left to the exercises. Before giving our error analysis results (for linear systems), we state here a theorem that shows, quite quantitatively, that nonsingular matrices with large condition numbers are truly very close to being singular. Recall that the singular square matrices are precisely those whose determinant is

---

[14] Sometimes this condition number is denoted $\kappa_\infty(A)$ to emphasize that it derives from the infinity vector and matrix norm. Since this will be the only condition number that we use, no ambiguity should arise by our adopting this abbreviated notation.

zero. For a given $n \times n$ nonsingular matrix A, we think of the distance from A to the set of all $n \times n$ singular matrices to be $\min\{\|S - A\| : \det(S) = 0\}$. (Just as with absolute values, the norm of a difference of matrices is taken to be the distance between the matrices.) We point out that $\min\limits_{\det(S)=0} \|S - A\|$ can be thought of as the distance from A to the set of singular matrices. (See Figure 7.36.)

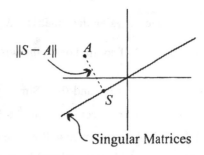

FIGURE 7.36: Heuristic diagram showing the distance from a nonsingular matrix A to the set of all singular matrices (line).

THEOREM 7.6: (*Geometric Characterization of Condition Numbers*) If A is any $n \times n$ nonsingular matrix, then we have:

$$\frac{1}{\kappa(A)} = \frac{1}{\|A\|} \cdot \min\limits_{\det(S)=0} \|S - A\| = \frac{1}{\|A\|}. \tag{45}$$

Like all of the results we state involving matrix norms and condition numbers, this one is true, in general, for whichever matrix norm (and resulting condition number) we would like to use. A proof can be found in the paper [Kah-66]. This theorem suggests some of the difficulties in trying to numerically solve systems having large condition numbers. Gaussian elimination involves many computations and each time we modify our matrix, because of roundoff errors, we are actually dealing with matrices that are close to but not the same as the actual (mathematically exact) matrices. The theorem shows that for poorly conditioned matrices (i.e., ones with large condition numbers), this process is extremely sensitive since even a small change in a poorly conditioned matrix could result in one that is singular!

We close with an example that will review some of the concepts about norms and condition numbers that have been introduced.

EXAMPLE 7.19: Consider the matrix: $A = \begin{bmatrix} 7 & -4 \\ -5 & 3 \end{bmatrix}$.

(a) Is there a (2×1) vector x such that: $\|Ax\| > 8\|x\|$? If yes, find one; otherwise explain why one does not exist.

(b) Is there a nonzero vector x such that $\|Ax\| \geq 12\|x\|$? If so, find one; otherwise explain why one does not exist.

(c) Is there a singular matrix $S = \begin{bmatrix} a & b \\ c & d \end{bmatrix}$ (i.e., $ad - bc = 0$) such that $\|S - A\| \leq 0.2$? If so, find one; otherwise explain why one does not exist.

(d) Is there a singular matrix $S = \begin{bmatrix} a & b \\ c & d \end{bmatrix}$ (i.e., $ad - bc = 0$) such that $\|S - A\| \le 0.05$? If so, find one; otherwise explain why one does not exist.

SOLUTION: Parts (a) and (b): Since $\|A\| = 7 + 4 = 11$, it follows from (38) that there exist (nonzero) vectors x with $\|Ax\| / \|x\| = 11$ or, put differently (multiply by $\|x\|$) $\|Ax\| = 11\|x\|$, but there will not be any nonzero vectors x that will make this equation work if 11 gets replaced by any larger number. (The maximum amount that matrix multiplication by A can magnify the norm of any nonzero vector x is 11 times.) Thus part (a) will have a vector solution but part (b) will not. To find an explicit vector x that will solve part (a), we will actually do more and find one that undergoes the maximum possible magnification $\|Ax\| = 11\|x\|$. The procedure is quite simple (and general). The vector x will have entries being either 1 or -1. To find such an appropriate vector x, we simply identify the row of A that gives rise to its norm being 11; this would be the first row (in general if more than one row gives the norm, we can choose either one). We simply choose the signs of the x-entries so that when they are multiplied in order by the corresponding entries in the just-identified row of A, all products are positive. In other words, if an entry in the special row of A is positive, take the corresponding component of x to be 1; if the special row entry of A is negative, take the corresponding component of x to be -1. In our case the special row of A is (the first row) $[7 \ -4]$, and so in accordance we take $x = [1 \ -1]'$. The first entry of the vector Ax is $[7 \ -4] \cdot [1 \ -1]' = 7(1) - 4(-1) = 7 + 4 = 11$, so $\|x\| = 1$ and $\|Ax\| = 11$ (actually, this shows only $\|Ax\| \ge 11$ since we have not yet computed the other component of Ax, but from what was already said, we know $\|Ax\| \le 11$, so that indeed $\|Ax\| = 11$). This procedure easily extends to any matrices of any size.

Parts (c) and (d): We rewrite equation (45) to isolate the distance from A to the singular matrices (simply multiply both sides by $\|A\|$):

$$\min_{\det(S)=0} \|S - A\| = \frac{\|A\|}{\kappa(A)}.$$

Appealing to MATLAB to compute the right side (and hence the distance from A to singulars):

```
>> A=[7 -4;-5 3];
>> norm(A,inf)/cond(A,inf)
→ ans =          0.0833
```

Since this distance is less than 0.2, the theorem tells us that there is a singular matrix satisfying the requirement of part (c) (the theorem unfortunately does not help us to find one), but there is no singular matrix satisfying the more stringent requirements of part (d). We use *ad hoc* methods to find a specific matrix S that satisfies the requirements of part (c). Note that $\det A = 7 \cdot 3 - (-5) \cdot (-4) = 1$. We will try to tweak the entries of A into those of a singular matrix $S = \begin{bmatrix} a & b \\ c & d \end{bmatrix}$ with determinant $ad - bc = 0$. The requirement of the distance being less than 0.2 means that our perturbations in each row must add up (in absolute value) to at most 0.2. Let's try tweaking 7 to $a = 6.9$ and 3 to $d = 2.9$ (motive for this move: right now A has $ad = 21$, which is one more than $bc = 20$; we need to tweak things so that ad is brought down a bit and bc is brought up to meet it). Now we have $ad = 20.01$, and we still have a perturabtion allowance of 0.1 for both entries b and c and we need only bring bc up from its current value of 20 to 20.01. This is easy— there are many ways to do it. For example, keep $c = -5$ and solve $bc = 20.01$, which gives $c = 20.01/-5 = -4.002$ (well within the remaining perturbation allowance). In summary, the matrix $S = \begin{bmatrix} 6.9 & -4.002 \\ -5 & 2.9 \end{bmatrix}$ meets the requirements that were asked for in part (c). Indeed S is singular (its determinant was arranged to be zero), and the distance from this matrix to A is

$$\|S - A\| = \left\| \begin{bmatrix} 6.9 & -4.002 \\ -5 & 2.9 \end{bmatrix} - \begin{bmatrix} 7 & -4 \\ -5 & 3 \end{bmatrix} \right\| = \left\| \begin{bmatrix} -.1 & -.002 \\ 0 & -.1 \end{bmatrix} \right\| = .102 < 0.2.$$

NOTE: The matrix S that we found was actually quite a bit closer to A than what was asked for. Of course, the closer that we wish to find a singular matrix to the ultimate distance, the harder we will have to work with such *ad hoc* methods. Also, the idea used to construct the "extremal" vector x can be modified to give a proof of identity (39); this task will be left to the interested reader as Exercise 10.

When we use Gaussian elimination to solve a nonsingular linear system (33): $Ax = b$, we will get a **computed solution** vector z that will, in general, differ from the **exact (mathematical) solution** x by the **error term** Δx :

$$z = x + \Delta x \longleftarrow \text{Error Term}$$

Computed Solution   Exact Solution

The main goal for the error analysis is to derive estimates for the size of the error (vector) term: $\|\Delta x\|$. Such estimates will give us quality control on the computed solution z to the linear system.

Caution: It may seem that a good way to measure the quality of the computed solution is to look at the size (norm) of the so-called **residual vector**:

$$r = \text{residual vector} \equiv b - Az. \tag{46}$$

Indeed, if z were the exact solution x then the residual would equal the zero vector. We note the following different ways to write the residual vector:

$$r \equiv b - Az = Ax - Az = A(x - z) = A(x - (x + \Delta x)) = A(-\Delta x) = -A(\Delta x); \qquad (47)$$

in particular, the residual is simply the (negative of) the matrix A multiplied by the error term vector. The matrix A may distort a large error term into a much smaller vector thus making the residual much smaller than the actual error term. The following example illustrates this phenomenon (see Figure 7.37).

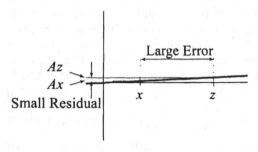

FIGURE 7.37: Heuristic illustration showing the unreliability of the residual as a gauge to measure the error. This phenomenon is a special case of the general principle that a function having a very small derivative can have very close outputs resulting from different inputs spread far apart.

EXAMPLE 7.20: Consider the following linear system $Ax = b$:

$$\begin{bmatrix} 1 & 2 \\ 1.0001 & 2 \end{bmatrix} \begin{bmatrix} x_1 \\ x_2 \end{bmatrix} = \begin{bmatrix} 3 \\ 3.0001 \end{bmatrix}.$$

This system has (unique) exact solution $x = [1, 1]'$. Let's consider the (poor) approximation $z = [3, 0]'$. The (norm of the) error of this approximation is $\|x - z\| = \|[-2, 1]'\| = 2$, but the residual vector,

$$r = b - Az = \begin{bmatrix} 3 \\ 3.0001 \end{bmatrix} - \begin{bmatrix} 1 & 2 \\ 1.0001 & 2 \end{bmatrix} \begin{bmatrix} 3 \\ 0 \end{bmatrix} = \begin{bmatrix} 3 \\ 3.0001 \end{bmatrix} - \begin{bmatrix} 3 \\ 3.0003 \end{bmatrix} = \begin{bmatrix} 0 \\ -.0002 \end{bmatrix}$$

has a much smaller norm of only 0.0002. This phenomenon is also depicted in Figure 7.37.

Despite this drawback about the residual itself, it can be manipulated to indeed give us a useful error estimate. Indeed, from (47), we may multiply left sides by A^{-1} to obtain:

$$r = A(-\Delta x) \implies -\Delta x = A^{-1} r.$$

If we now take norms of both sides and apply (40), we can conclude that:

$$\|\Delta x\| = \|-\Delta x\| = \|A^{-1} r\| \le \|A^{-1}\| \, \|r\|.$$

(We have used the fact that $\|-v\| = \|v\|$ for any vector v; this follows from norm axiom (36B) using $c = -1$.) We now summarize this simple yet important result in the following theorem.

THEOREM 7.7: (*Error Bound via Residual*) If z is an approximate solution to the exact solution x of the linear system (33) $Ax = b$, with A nonsingular, and $r = b - Az$ is the residual vector, then

$$\text{error} \equiv \|x - z\| \leq \|A^{-1}\| \, \|r\| .= \tag{48}$$

REMARK: Using the condition number $\kappa(A) = \|A\| \, \|A^{-1}\|$, Theorem 7.7 can be reformulated as

$$\|x - z\| \leq \kappa(A) \frac{\|r\|}{\|A\|}. \tag{49}$$

EXAMPLE 7.21: Consider once again the linear system $Ax = b$ of the preceding example:

$$\begin{bmatrix} 1 & 2 \\ 1.0001 & 2 \end{bmatrix} \begin{bmatrix} x_1 \\ x_2 \end{bmatrix} = \begin{bmatrix} 3 \\ 3.0001 \end{bmatrix}.$$

If we again use the vector $z = [3, 0]'$ as the approximate solution, then, as we saw in the last example, the error $= 2$ and the residual is $r = [0, -0.0002]'$. The estimate for the error provided in Theorem 9.2, is (with MATLAB's help) found to be 4:

```
>>A=[1 2; 1.0001 2]; r=[0 -.0002];
>> norm(inv(A),inf)*norm(r,inf)    →ans = 4.0000
```

Although this estimate for the error is about as far off from the actual error as the approximation z is from the actual solution, as far as an estimate for the error is concerned, it is considered a decent estimate. An estimate for the error is considered good if it has approximately the same order of magnitude (power of 10 in scientific notation) as the actual error.

Using the previous theorem on the error, we obtain the following analogous result for the relative error.

THEOREM 7.8: (*Relative Error Bound via Residual*) If z is an approximate solution to the exact solution x of the linear system (33) $Ax = b$, with A nonsingular, $b \neq 0$ (vector), and $r = b - Az$ is the residual vector, then

$$\text{relative error} \equiv \frac{\|x - z\|}{\|x\|} \leq \frac{\|A\| \, \|A^{-1}\|}{\|b\|} \|r\|. \tag{50}$$

REMARK: In terms of condition numbers, Theorem 7.8 takes on the more appealing form:

$$\frac{\|x-z\|}{\|x\|} \le \kappa(A)\frac{\|r\|}{\|b\|}. \tag{51}$$

Proof of Theorem 7.8: We first point out that $x \ne 0$ since $b \ne 0$ (and $Ax = b$ with A nonsingular). Using identity (40), we deduce that:

$$\|b\| = \|Ax\| \le \|A\|\|x\| \implies \frac{1}{\|x\|} \le \frac{\|A\|}{\|b\|}.$$

We need only multiply both sides of this latter inequality by $\|x-z\|$ and then apply (48) to arrive at the desired inequality:

$$\frac{\|x-z\|}{\|x\|} \le \frac{\|A\|}{\|b\|} \cdot \|x-z\| \le \frac{\|A\|}{\|b\|} \cdot \|A^{-1}\|\|r\|.$$

EXAMPLE 7.21: (cont.) Using MATLAB to compute the right side of (50),

```
>> cond(A,inf)*norm(r,inf)/norm([3  3.0001]',inf)
→ans = 4.0000
```

Once again, this compares favorably with the true value of the relative error whose explicit value is $\|x-z\|/\|x\| = 2/1 = 2$ (see Example 7.20).

EXAMPLE 7.22: Consider the following (large) linear system $Ax = b$, with

$$A = \begin{bmatrix} 4 & -1 & 0 & 0 & -1 & 0 & 0 & 0 & \cdots & 0 & 0 \\ -1 & 4 & -1 & 0 & 0 & -1 & 0 & 0 & \cdots & 0 & 0 \\ 0 & -1 & 4 & -1 & 0 & 0 & -1 & 0 & \cdots & & 0 \\ 0 & 0 & -1 & 4 & 0 & 0 & 0 & -1 & \ddots & & 0 \\ -1 & 0 & 0 & 0 & 4 & -1 & 0 & 0 & \ddots & & 0 \\ 0 & -1 & 0 & 0 & -1 & 4 & -1 & 0 & \ddots & & 0 \\ \vdots & & \ddots & \ddots & \ddots & \ddots & \ddots & \ddots & & & \\ & \vdots & & & & & & & & & \\ 0 & & & & \ddots & \ddots & \ddots & \ddots & \ddots & \ddots & 0 \\ 0 & 0 & \cdots & & \ddots & -1 & 0 & 0 & -1 & 4 & -1 \\ 0 & 0 & 0 & \cdots & & 0 & -1 & 0 & 0 & -1 & 4 \end{bmatrix}, \quad b = \begin{bmatrix} 1 \\ 2 \\ 3 \\ 4 \\ 5 \\ 6 \\ \vdots \\ \\ 798 \\ 799 \\ 800 \end{bmatrix}.$$

The 800×800 coefficient matrix A is diagonally banded with a string of 4's down the main diagonal, a string of –1's down each of the diagonals 4 below and 4 above the main diagonal, and each of the diagonals directly above and below the main diagonal consist of the vector that starts off with $[-1\ -1\ -1]$, and repeatedly tacks the sequence $[0\ -1\ -1\ -1]$ onto this until the diagonal fills. Such banded coefficient matrices are very common in finite difference methods for solving (ordinary and partial) differential equations. Despite the intimidating size of this

system, MATLAB's "left divide" can take advantage of its special structure and will produce solutions with very decent accuracy as we will see below:
(a) Compute the condition number of the matrix A.
(b) Use the left divide (Gaussian elimination) to solve this system, and call the computed solution z. Use Theorem 7.7 to estimate its error and Theorem 7.8 to estimate the relative error. Do not print out the vector z!
(c) Obtain a second numerical solution $z2$, this time by left multiplying the equation by the inverse A^{-1}. Use Theorem 7.7 to estimate its error and Theorem 7.8 to estimate the relative error. Do not print out the vector $z2$!

SOLUTION: We first enter the matrix A, making use of MATLAB's useful `diag` function:

```
>> A=diag(4*ones(1,800));
>> a1=[-1 -1 -1]; vrep=[0 -1 -1 -1];
>> for i=1:199, a1=[a1, vrep]; end %this is level +1/-1 diagonal
>> v4=-1*ones(1,796); %this is level +4/-4 diagonal
>> A=A+diag(a1,1)+diag(a1,-1)+diag(v4,4)+diag(v4,-4);
>> A(1:8,1:8)   %we make a quick check to see how A looks
→ans =    4   -3    0    0   -1    0    0    0
         -3    4   -3    0    0   -1    0    0
          0   -3    4   -3    0    0   -1    0
          0    0   -3    4    0    0    0   -1
         -1    0    0    0    4   -3    0    0
          0   -1    0    0   -3    4   -3    0
          0    0   -1    0    0   -3    4   -3
          0    0    0   -1    0    0   -3    4
```

The matrix A looks as it should. The vector b is, of course, easily constructed.

```
>>b=1:800; b=b';   %needed to take transpose to make b a column vector
```

Part (a).
```
>> c= cond(A,inf)   → c = 2.6257e + 003
```

With a condition number under 3000, considering its size, the matrix A is rather well conditioned.

Part (b): Here and in part (c), we use the condition number formulations (49) and (51) for the error estimates of Theorems 7.7 and 7.8.

```
>> z=A\b; r=b-A*z; errest=c*norm(r,inf)/norm(A,inf)
→ errest =    2.4875e - 010
```

```
>> relerrest=c*norm(r,inf)/norm(b,inf)
→ relerrest =    3.7313e - 012
```

Part (c):
```
>> z2=inv(A)*b; r2=b-A*z2; errest2=c*norm(r2,inf)/norm(A,inf)
→ errest2 =    6.8656e - 009
```

```
>> relerrest2=c*norm(r2,inf)/norm(b,inf)
→ relerrest2 =    1.0298e - 010
```

Both methods have produced solutions of very decent accuracy. All of the computations here were done with lightning speed. Thus even larger such systems (that are decently conditioned) can be dealt with safely with MATLAB's "left divide." The matrix in the above problem had a very high percentage of its entries being zeros. Such matrices are called sparse matrices, and MATLAB has efficient ways to store and manipulate such matrices. We will discuss this topic in the next section.

For (even moderately sized) poorly conditioned linear systems, quality control of computed solutions becomes a serious issue. The estimates provided in Theorems 7.7 and 7.8 are just that, estimates that give a guarantee of the closeness of the computed solution to the actual solution. The actual errors may be a lot smaller than the estimates that are provided. Another more insidious problem is that computation of the error bounds of these theorems is expensive, since it involves either the norm of A^{-1} directly or the condition number of A (which implicitly requires computing the norm of A^{-1}). Computer errors can lead to inaccurate computation of these error bounds that we would like to use to give us confidence in our numerical solutions. The next example will demonstrate and attempt to put into perspective some of these difficulties. The example will involve the very poorly conditioned Hilbert matrix that we introduced in Section 7.4. We will solve the system exactly (using MATLAB's symbolic toolbox),[15] and thus be able to compare estimated errors (using Theorems 7.7 and 7.8) with the actual errors. We warn the reader that some of the results of this example may be shocking, but we hasten to add that the Hilbert matrix is notorious for being extremely poorly conditioned.

EXAMPLE 7.23: Consider the linear system $Ax = b$ with

$$
A = \begin{bmatrix}
1 & \frac{1}{2} & \frac{1}{3} & \cdots & \frac{1}{48} & \frac{1}{49} & \frac{1}{50} \\
\frac{1}{2} & \frac{1}{3} & \frac{1}{4} & \cdots & \frac{1}{49} & \frac{1}{50} & \frac{1}{51} \\
\frac{1}{3} & \frac{1}{4} & \frac{1}{5} & \cdots & \frac{1}{50} & \frac{1}{51} & \frac{1}{52} \\
\vdots & & & \ddots & & \vdots & \vdots \\
\frac{1}{48} & \frac{1}{49} & \frac{1}{50} & \cdots & \frac{1}{96} & \frac{1}{97} & \frac{1}{98} \\
\frac{1}{49} & \frac{1}{50} & \frac{1}{51} & \cdots & \frac{1}{97} & \frac{1}{98} & \frac{1}{99} \\
\frac{1}{50} & \frac{1}{51} & \frac{1}{52} & \cdots & \frac{1}{98} & \frac{1}{99} & \frac{1}{100}
\end{bmatrix}, \qquad
b = \begin{bmatrix}
1 \\ 2 \\ 3 \\ \vdots \\ 48 \\ 49 \\ 50
\end{bmatrix}.
$$

Using MATLAB, perform the following computations.
(a) Compute the condition number of the 50×50 Hilbert matrix A (on MATLAB using the usual floating point arithmetic).
(b) Compute the same condition number using symbolic (exact) arithmetic on MATLAB.

---

[15] For more on the Symbolic Toolbox, see Appendix A. This toolbox may or may not be in the version of MATLAB that you are using. A reduced version of it comes with the student edition. It is not necessary to have it to understand this example.

(c) Use MATLAB's left divide to solve this system and label the computed solution as z, then use Theorem 7.7 to estimate the error. (Do not actually print the solution.)

(d) Solve the system by (numerically) left multiplying both sides by A^{-1} and label the computed solution as z2; then use Theorem 7.7 to estimate the error. (Do not actually print the solution.)

(e) Solve the system exactly using MATLAB's symbolic capabilities, label this exact solution as x, and compute the norm of this solution vector. Then use this exact solution to compute the exact errors of the two approximate solutions in parts (a) and (b).

SOLUTION: Since MATLAB has a built-in function for generating Hilbert matrices, we may very quickly enter the data A and b:

```
>> A=hilb(50);   b=1:50;b=b';
```

Part (a):   We invoke MATLAB's built-in function for computing condition numbers:

```
>> c1=cond(A,inf)
→Warning: Matrix is close to singular or badly scaled.
     Results may be inaccurate. RCOND = 3.615845e - 020.
> In C:\MATLABR11\toolbox\matlab\matfun\cond.m at line 44
→c1 = 5.9243e + 019
```

This is certainly very large, but it came with a warning that it may be an inaccurate answer due to the poor conditioning of the Hilbert matrix. Let's now see what happens when we use exact arithmetic.

Part (b):   Several of MATLAB's built-in functions are not defined for symbolic objects; and this is true for the norm and condition number functions. The way around this is to work directly with the definition (41) of the condition number: $\kappa(A) = \left\| A \right\| \left\| A^{-1} \right\|$, compute the norm of A directly (no computational difficulties here), and compute the norm of A^{-1} by first computing A^{-1} in exact arithmetic, then using the double command to put the answer from "symbolic" form into floating point form, so we can take its norm as usual (the computational difficulty is in computing the inverse, not in finding the norm).

```
>> c=norm(double(inv(sym(A))),inf)*norm(A,inf)   % "sym" declares A as
a symbolic variable, so inv is calculated exactly; double switches
the symbolic answer back into floating point form.
→c1 = 4.3303e + 074
```

The difference here is astounding! This condition number means that, although the Hilbert matrix A has its largest entry being 1 (and smallest being 1/99), the inverse matrix will have some entries having absolute values at least $4.33 \times 10^{74} / 50 = 8.66 \times 10^{72}$ (why?).   With floating point arithmetic, however,

MATLAB's computed inverse has all entries less than 10^{20} in absolute value, so that MATLAB's inverse is totally out in left field!

Part (c):
```
>> z=A\b; r=b-A*z;
→Warning: Matrix is close to singular or badly scaled.
     Results may be inaccurate. RCOND = 3.615845e - 020.
```

As expected, we get a flag about our poorly conditioned matrix.

```
>> norm(r,inf)      →ans = 6.2943e - 005
```

Thus the residual of the computed solution is somewhat small. But the extremely large condition number of the matrix will overpower this residual to render the following useless error estimate (see (49)):

```
>> errest=c1*norm(r,inf)/norm(A,inf)      →errest = 9.5437e+014
```

Since
```
>> norm(z,inf)      →ans = 5.0466e+012
```

this error estimate is over 100 times as large as the largest component of the numerical solution. Things get even worse (if you can believe this is possible) with the inverse multiplication method that we look at next.

Part (d):
```
>> z2=inv(A)*b; r2=b-A*z2;
→Warning: Matrix is close to singular or badly scaled.
     Results may be inaccurate. RCOND = 3.615845e - 020.
>>norm(r2,inf)      →ans = 1.6189e + 004
```

Here, even the norm of the residual is unacceptably large.

```
>> errest2=c1*norm(r2,inf)/norm(A,inf)      →errest = 2.2078e+023
```

Part (e):
```
>> S=sym(A);  %declares A as a symbolic matrix
>> x=S\b;  %Computes exact solution of system
>> x=double(x);  %Converts x back to a floating point vector
>>   norm(x,inf)      →ans = 7.4601e + 040
```

We see that the solution vector has some extremely large entries.

```
>> norm(x-z,inf)      →ans= 7.4601e + 040
>> norm(x-z2,inf)     →ans = 7.4601e + 040
>> norm(z-z2,inf)     →ans = 3.8429e + 004
```

Comparing all of these norms, we see that the two approximations are closer to each other than to the exact solution (by far). The errors certainly met the estimates provided for in the theorem, but not by much. The (exact arithmetic) computation of x took only a few seconds for MATLAB to do. Some comments are in order. The reader may wonder why one should not always work using exact arithmetic, since it is so much more reliable. The reasons are that it is often not necessary to do this—floating point arithmetic usually provides acceptable (and usually decent) accuracy, and exact arithmetic is much more expensive. However, when we get such a warning from MATLAB about near singularity of a matrix we must discard the answers, or at least do some further analysis. Another option (again using the Symbolic Toolbox of MATLAB) would be to use variable precision arithmetic rather than exact arithmetic. This is less expensive than exact arithmetic and allows us to declare how many significant digits with which we would like to compute. We will give some examples of this arithmetic in a few rare cases where MATLAB's floating point arithmetic is not sufficient to attain the desired accuracy (see also Appendix A).

EXERCISE FOR THE READER 7.25: Repeat all parts of the previous example to the following linear system $Ax = b$, with:

$$A = \begin{bmatrix} 1 & 1 & 1 & \cdots & 1 & 1 & 1 \\ 2^{11} & 2^{10} & 2^9 & \cdots & 4 & 2 & 1 \\ 3^{11} & 3^{10} & 3^9 & \cdots & 9 & 3 & 1 \\ \vdots & & & \ddots & & \vdots & \vdots \\ 10^{11} & 10^{10} & 10^9 & \cdots & 100 & 10 & 1 \\ 11^{11} & 11^{10} & 11^9 & \cdots & 121 & 11 & 1 \\ 12^{11} & 12^{10} & 12^9 & \cdots & 144 & 12 & 1 \end{bmatrix}, \quad b = \begin{bmatrix} 1 \\ -2 \\ 3 \\ \vdots \\ -10 \\ 11 \\ -12 \end{bmatrix}.$$

This coefficient matrix is the 12×12 Vandermonde matrix that was introduced in Section 7.4 with polynomial interpolation.

We next move on to the concepts of eigenvalues and eigenvectors of a matrix. These concepts are most easily motivated geometrically in two dimensions, so let us begin with a 2×2 matrix $A = \begin{bmatrix} a & b \\ c & d \end{bmatrix}$, and a nonzero column vector $x = \begin{bmatrix} x_1 \\ x_2 \end{bmatrix}$ $\neq \begin{bmatrix} 0 \\ 0 \end{bmatrix}$. We view A (as in Section 7.1) as a linear transformation acting on the vector x. The vector x will have a positive length given by the Pythagorean formula:

$$\text{len}(x) = \sqrt{x_1^2 + x_2^2}. \tag{52}$$

Thus A transforms the two-dimensional vector x into another vector $y = \begin{bmatrix} y_1 \\ y_2 \end{bmatrix}$ by matrix multiplication $y = Ax$. We consider the case where y is also not the zero vector; this will always happen if A is nonsingular. In general, when we graph the vectors x and y together in the same plane, they can have different lengths as well as different directions (see Figure 7.38a).

Sometimes, however, there will exist vectors x for which y will be **parallel** to x (meaning that y will point in either the same direction or the opposite direction as

x; see Figure 7.38b). In symbols, this would mean that we could write $y = \lambda x$ for some **scalar** (number) λ. Such a vector x is called an eigenvector for the matrix A, and the number λ is called an associated eigenvalue. Note that if λ is positive, then x and $y = \lambda x$ point in the same direction and $\text{len}(y) = \lambda \cdot \text{len}(x)$, so that λ acts as a magnification factor. If λ is negative, then (as in Figure 7.38b) y points in the opposite direction as x, and finally if $\lambda = 0$, then y must be the zero vector (so has no direction). By convention, the zero vector is parallel to any vector. This is permissible, as long as x is not the zero vector. This definition generalizes to square matrices of any size.

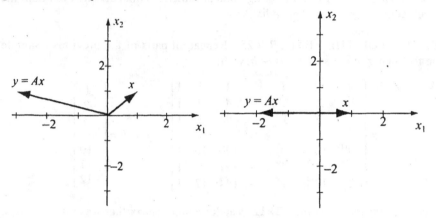

FIGURE 7.38: Actions of the matrix $A = \begin{bmatrix} -2 & -1 \\ 0 & 1 \end{bmatrix}$ on a pair of vectors: (a) (left) The (shorter) vector $x = \begin{bmatrix} 1 \\ 1 \end{bmatrix}$ of length $\sqrt{2}$ gets transformed to the vector $y = Ax = \begin{bmatrix} -3 \\ 1 \end{bmatrix}$ of length $\sqrt{10}$. Since the two vectors are not parallel, x is not an eigenvector of A. (b) (right) The (shorter) unit vector $x = \begin{bmatrix} 1 \\ 0 \end{bmatrix}$ gets transformed to the vector $y = Ax = \begin{bmatrix} -2 \\ 0 \end{bmatrix}$ (red) of length 2, which is parallel to x, therefore x is an eigenvector for A.

DEFINITION: Let A be an $n \times n$ matrix. An $n \times 1$ nonzero column vector x is called an **eigenvector** for the matrix A if for some scalar λ we have

$$Ax = \lambda x. \tag{53}$$

The scalar λ is called the **eigenvalue** associated with the eigenvector x.

Finding all eigenvalues and associated eigenvectors for a given square matrix is an important problem that has been extensively studied and there are numerous algorithms devoted to this and related problems. It turns out to be useful to know this **eigendata** for a matrix A for an assortment of applications. It is actually quite easy to look at eigendata in a different way that will give an immediate method for finding it. We can rewrite equation (53) as follows:

$$Ax = \lambda x \Leftrightarrow Ax = \lambda Ix \Leftrightarrow \lambda Ix - Ax = 0 \Leftrightarrow (\lambda I - A)x = 0.$$

Thus, using what we know about solving linear equations, we can restate the eigenvalue definition in several equivalent ways as follows:

λ is an eigenvalue of A $\Leftrightarrow$ $(\lambda I - A)x = 0$ has a nonzero solution x

(which will be an eigenvector)

$\Leftrightarrow$ $\det(\lambda I - A) = 0$

Thus the eigenvalues of A are precisely the roots λ of the equation $\det(\lambda I - A) = 0$. If we write out this determinant with a bit more detail,

$$\det(\lambda I - A) = \det \begin{bmatrix} \lambda - a_{11} & -a_{12} & -a_{13} & \cdots & -a_{1,n-1} & -a_{1n} \\ -a_{21} & \lambda - a_{22} & -a_{23} & & & \\ -a_{31} & -a_{32} & \lambda - a_{33} & & & \vdots \\ \vdots & & & \ddots & & \\ -a_{n-1,1} & -a_{n-1,2} & -a_{n-1,3} & \cdots & \lambda - a_{n-1,n-1} & -a_{n-1,n} \\ -a_{n1} & -a_{n2} & -a_{n3} & \cdots & -a_{n,n-1} & \lambda - a_{nn} \end{bmatrix},$$

it can be seen that this expression will always be a polynomial of degree n in the variable λ, for any particular matrix of numbers $A = \lfloor a_{ij} \rfloor$ (see Exercise 30). This polynomial, because of its importance for the matrix A, is called the **characteristic polynomial** of the matrix A, and will be denoted as $p_A(\lambda)$. Thus $p_A(\lambda) = \det(\lambda I - A)$, and in summary:

The eigenvalues of a matrix A are the roots of the characteristic polynomial, i.e., the solutions of the equation $p_A(\lambda) = 0$.

Finding eigenvalues is an algebraic problem that can be easily solved with MATLAB; more on this shortly. Each eigenvalue will always have associated eignvectors. Indeed, the matrix equation $(\lambda I - A)x = 0$ has a singular coefficient matrix when λ is an eigenvalue (since its determinant is zero). We know from our work on solving linear equations that a singular ($n \times n$) linear system of form $Cx = 0$ can have either no solutions or infinitely many solutions, but $x = 0$ is (obviously) always a solution of such a linear system, and consequently such a singular system must have infinitely many solutions. To find eigenvectors associated with a particular eigenvalue λ, we could compute them numerically by applying `rref` rather than Gaussian elimination to the augmented matrix $[\lambda I - A \mid 0]$. Although theoretically sound, this approach is not a very effective numerical method. Soon we will describe MATLAB's relevant built-in functions that are based on more sophisticated and effective numerical methods.

EXAMPLE 7.24: For the matrix $A = \begin{bmatrix} -2 & -1 \\ 1 & 1 \end{bmatrix}$, do the following:

(a) Find the characteristic polynomial $p_A(\lambda)$.

(b) Find all roots of the characteristic polynomial (i.e., the eigenvalues of A).

(c) For each eigenvalue, find an associated eigenvector.

SOLUTION: Part (a): $p_A(\lambda) = \det(\lambda I - A) =$

$$\det\left(\begin{bmatrix} \lambda+2 & 1 \\ -1 & \lambda-1 \end{bmatrix}\right) = (\lambda+2)(\lambda-1) + 1 = \lambda^2 + \lambda - 1.$$

Part (b): The roots of $p_A(\lambda) = 0$ are easily obtained from the quadratic formula:

$$\lambda = \frac{-1 \pm \sqrt{1 - 4 \cdot 1 \cdot (-1)}}{2} = \frac{-1 \pm \sqrt{5}}{2} \approx -1.6180, .6180.$$

Part (c): For each of these two eigenvalues, let's use rref to find (all) associated eigenvectors.

Case $\lambda = (-1 - \sqrt{5})/2$:

```
>> A=[-2 -1;1 1];  lambda=(-1-sqrt(5))/2;
>> C=lambda*eye(2)-A; C(:,3)=zeros(2,1);
>> rref(C)
→ ans =    1.0000   2.6180      0
              0        0         0
```

From the last matrix, we can read off the general solution of the system $(\lambda I - A)x = 0$ (written out to four decimals):

$$\begin{cases} x_1 + 2.6180x_2 = 0 \\ (= \text{any number}) \end{cases} \Rightarrow \begin{cases} x_1 = -2.6180t \\ x_2 = t \end{cases}, \ t = \text{any number}$$

These give, for all choices of the parameter t <u>except $t = 0$</u>, all of the associated eigenvectors. For a specific example, if we take $t = 1$, this will give us the eigenvector $x = \begin{bmatrix} -2.6180 \\ 1 \end{bmatrix}$. We can verify this geometrically by plotting the vector x along with the vector $y = Ax$ to see that they are parallel.

Case $\lambda = (-1 + \sqrt{5})/2$:

```
>> lambda=(-1+sqrt(5))/2; C=lambda*eye(2)-A; rref(C)
→ ans =    1.0000   0.3820      0
              0        0         0
```

As in the preceding case, we can get all of the associated eigenvectors. We consider the eigenvector $x = \begin{bmatrix} -.3820 \\ 1 \end{bmatrix}$ for this second eigenvalue. Since the eigenvalue is postive, x and Ax will point in the same directions, as can be checked. Of course, each of these eigenvectors has been written in format short; if we wanted we could have displayed them in format long and thus written our eigenvectors with more significant figures, up to about 15 (MATLAB's accuracy limit).

Before discussing MATLAB's relevant built-in functions for eigendata, we state a theorem detailing some useful facts about eigendata of a matrix. First we give a definition. Since an eigenvalue λ of a matrix A is a root of the characteristic polynomial $p_A(x)$, we know that $(x - \lambda)$ must be a factor of $p_A(x)$. Recall that the **algebraic multiplicity** of the root λ is the highest exponent m such that $(x - \lambda)^m$ is still a factor of $p_A(x)$, i.e., $p_A(x) = (x - \lambda)^m q(x)$ where $q(x)$ is a polynomial (of degree $n - m$) such that $q(\lambda) \neq 0$.

THEOREM 7.9: (*Facts about Eigenvalues and Eigenvectors*): Let $A = [a_{ij}]$ be an $n \times n$ matrix.
(i) The matrix A has at most n (real) eigenvalues, and their algebraic multiplicities add up to at most n.
(ii) If u, w are both eigenvectors of A corresponding to the same eigenvalue λ, then $u + w$ (if nonzero) is also an eigenvector of A corresponding to λ, and if c is a nonzero constant, then cu is also an eigenvector of A.[16] The set of all such eigenvectors associated with λ, together with the zero vector, is called the **eigenspace** of A **associated with the eigenvalue** λ.
(iii) The dimension of the eigenspace of A associated with the eigenvalue λ, called the **geometric multiplicity of the eigenvalue** λ, is always less than or equal to the algebraic multiplicity of the eigenvalue λ.
(iv) In general a matrix A need not have any (real) eigenvalues,[17] but if A is a **symmetric matrix** (meaning: A coincides with its transpose matrix), then A will always have a full set of n real eigenvalues, provided each eigenvalue is repeated according to its geometric multiplicity.

The proofs of (i) and (ii) are rather easy; they will be left as exercises. The proofs of (iii) and (iv) are more difficult; we refer the interested reader to a good linear algebra textbook, such as [HoKu-71], [Kol-99] or [Ant-00]. There is an extensive theory and several factorizations associated with eigenvalue problems. We should also point out a couple of more advanced texts. The book [GoVL-83] has become the standard reference for numerical analysis of matrix computations. The book [Wil-88] is a massive treatise entirely dedicated to the eigenvalue problem; it remains the standard reference on the subject. Due to space limitations, we will

---

[16] Thus when we throw all eigenvectors associated with a particular eigenvalue λ of a matrix A together with the zero vector, we get a set of vectors that is closed under the two linear operations: vector additon and scalar multiplication. Readers who have studied linear algebra will recognize such a set as a vector space; this one is called the **eigenspace** associated with the eigenvalue λ of the matrix A. Geometrically the eigenspace will be either a line through the origin (one-dimensional), a plane through the origin (two-dimensional), or in general, any k-dimensional hyperplane through the origin ($k \leq n$).

[17] This is reasonable since, as we have seen before, a polynomial need not have any real roots (e.g., $x^2 + 1$). If complex numbers are considered, however, a polynomial will always have a complex root (this is the so-called "Fundamental Theorem of Algebra") and so any matrix will always have at least a complex eigenvalue. Apart from this fact, the theory for complex eigenvalues and eigenvectors parallels that for real eigendata.

not be getting into comprehensive developments of eigenalgorithms; we will merely give a few more examples to showcase MATLAB's relevant built-in functions.

EXAMPLE 7.25: Find a matrix A that has no real eigenvalues (and hence no eigenvectors), as indicated in part (iv) of Theorem 7.9.

SOLUTION: We should begin to look for a 2×2 matrix A. We need to find one for which its characteristic polynomial $p_A(\lambda)$ has no real root. One approach would be to take a simple second-degree polynomial, that we know does not have any real roots, like $\lambda^2 + 1$, and try to build a matrix $A = \begin{bmatrix} a & b \\ c & d \end{bmatrix}$ which has this as its characteristic polynomial. Thus we want to choose a, b, c, and d such that:

$$\det \begin{pmatrix} \lambda - a & b \\ c & \lambda - d \end{pmatrix} = \lambda^2 + 1.$$

If we put $a = d = 0$, and compute the determinant we get $\lambda^2 - bc = \lambda^2 + 1$, so we are okay if $bc = -1$. For example, if we take $b = 1$ and $c = -1$, we get that the matrix $A = \begin{bmatrix} 0 & 1 \\ -1 & 0 \end{bmatrix}$ has no real eigenvalues.

MATLAB has the built-in function `eig` that can find eigenvalues and eigenvectors of a matrix. The two possible syntaxes of this function are as follows:

`eig(A)` →	If A is a square matrix, this command produces a vector containing the eigenvalues of A. Both real and complex eigenvalues are given.
`[V, D]=eig(A)` →	If A is an $n \times n$ matrix, this command will create two $n \times n$ matrices. D is a diagonal matrix whose diagonal entries are the eigenvalues of A, and V is matrix whose columns are corresponding eigenvectors for A. For complex eigenvalues, the corresponding eigenvectors will also be complex.

Since, by Theorem 7.9(ii), any nonzero scalar multiple of an eigenvector is again an eigenvector, MATLAB's `eig` chooses its eigenvectors to have length = 1.

For example, let's use these commands to find eigendata for the matrix of Example 7.24:

```
>> [V,D]=eig([-2 -1;1 1])
→ V =          -0.9342   0.3568
                0.3568  -0.9342

→ D =          -1.6180       0
                    0     0.6180
```

The diagonal entries in the matrix D are indeed the eigenvalues (in short format) that we found in the example. The corresponding eigenvectors (from the columns of V) $\begin{bmatrix} -0.9342 \\ 0.3568 \end{bmatrix}$ and $\begin{bmatrix} 0.3568 \\ -0.9342 \end{bmatrix}$ are different from the two we gave in that example, but can be obtained from the general form of eigenvectors that was found in that example. Also, unlike those that we gave in the example, it can be checked that these two eigenvectors have length equal to 1.

In the case of an eigenvalue with geometric multiplicity greater than 1, eig will find (whenever possible) corresponding eigenvectors that are linearly independent.[18] Watch what happens when we apply the eig function to the matrix that we constructed in Example 7.25:

```
>> [V,D]=eig([0 1;-1 0])
→ V =                    0.7071          0.7071
                         0 + 0.7071i     0 - 0.7071i

→ D =                    0 + 1.0000i     0
                         0               0 - 1.0000i
```

We get the eigenvalues (from D) to be the complex numbers $\pm i$ (where $i = \sqrt{-1}$) and the two corresponding eigenvectors also have complex numbers in them. Since we are interested only in real eigendata, we would simply conclude from such an output that the matrix has no real eigenvalues.

If you are interested in finding the characteristic polynomial $p_A(\lambda) = a_n \lambda^n + a_{n-1}\lambda^{n-1} \cdots + a_1\lambda + a_0$ of an $n \times n$ matrix A, MATLAB has a function poly that works as follows:

	For an $n \times n$ matrix A, this command will produce the vector $v = [a_n \; a_{n-1} \; \cdots a_1 \; a_0]$ of the $n + 1$ coefficients of the nth-degree characteristic polynomial $p_A(\lambda) = \det(\lambda I - A) = a_n\lambda^n + a_{n-1}\lambda^{n-1} + \cdots + a_1\lambda + a_0$ of the matrix A.
poly(A) →	

For example, for the matrix we constructed in Example 7.25, we could use this command to check its characteristic polynomial:

```
>> poly([0 1;-1 0])    →ans = 1    0    1
```

which translates to the polynomial $1 \cdot \lambda^2 + 0 \cdot \lambda + 1 = \lambda^2 + 1$, as was desired. Of course, this command is particularly useful for larger matrices where computation of determinants by hand is not feasible. The MATLAB function roots will find the roots of any polynomial:

---

[18]Linear independence is a concept from linear algebra. What is relevant for the concept at hand is that any other eigenvector associated with the same eigenvalue will be expressible as a linear combination of eigenvectors that MATLAB produces. In the parlance of linear algebra, MATLAB will produce eigenvectors that form a basis of the corresponding eigenspaces.

roots(v) →	For a vector $\begin{bmatrix} a_n & a_{n-1} & \cdots a_1 & a_0 \end{bmatrix}$ of the $n+1$ coefficients of the nth-degree polynomial $p(x) = a_n x^n + a_{n-1} x^{n-1} + \cdots + a_1 x + a_0$ this command will produce a vector of length n containing all of the roots of $p(x)$ (real and complex) repeated according to algebraic multiplicity.

Thus, another way to get the eigenvalues of the matrix of Example 7.24 would be as follows: [19]

```
>> roots(poly([-2 -1;1 1]))
→ ans          -1.6180
               0.6180
```

EXERCISE FOR THE READER 7.26: For the matrix $A = \begin{bmatrix} 2 & 1 & 0 & 0 \\ 0 & 2 & 0 & 0 \\ 0 & 0 & 1 & 0 \\ 0 & 0 & 0 & 1 \end{bmatrix}$, do the

following:
(a) By hand, compute $p_A(\lambda)$, the characteristic polynomial of A, in factored form, and find the eigenvalues and the algebraic multiplicity of each one.
(b) Either by hand or using MATLAB, find all the corresponding eigenvectors for each eigenvalue and find the geometric multiplicity of each eigenvalue.
(c) After doing part (a), can you figure out a general rule for finding the eigenvalues of an upper triangular matrix?

EXERCISES 7.6:

1.   For each of the following vectors x, find $\text{len}(x)$ and find $\|x\|$.

$x = [2, -6, 0, 3]$

$x = [\cos(n), \sin(n), 3^n]$ (n = a positive integer)

$x = [1, -1, 1, -1, \cdots, 1, -1]$ (vector has $2n$ components)

2.   For each of the following matrices A, find the infinity norm $\|A\|$.

(a) $A = \begin{bmatrix} 2 & -3 \\ 1 & 6 \end{bmatrix}$

(b) $A = \begin{bmatrix} 4 & -5 & -2 \\ 1 & 2 & 3 \\ -2 & -4 & -6 \end{bmatrix}$

(c) $A = \begin{bmatrix} \cos(\pi/4) & -\sin(\pi/4) & 0 \\ \sin(\pi/4) & \cos(\pi/4) & 0 \\ 0 & 0 & 1 \end{bmatrix}$

(d) $A = H_n$ the $n \times n$ Hilbert matrix (introduced and defined in Example 7.8)

3.   For each of the matrices A (parts (a) through (d)) of Exercise 2, find a nonzero vector x such that $\|Ax\| = \|A\|\|x\|$.

---

[19] We mention, as an examination of the M-file will show (enter type roots), that the roots of a polynomial are found using the eig command on an associated matrix.

4. For the matrix $A = \begin{bmatrix} 2 & -3 \\ -2 & 4 \end{bmatrix}$, calculate by hand the following: $\|A\|$, $\|A^{-1}\|$, $\kappa(A)$, and then verify your calculations with MATLAB. If possible, find a singular matrix S such that $\|S - A\| = 1/3$.

5. For the matrix $A = \begin{bmatrix} 1 & -1 \\ -1 & 1.001 \end{bmatrix}$, calculate by hand the following: $\|A\|$, $\|A^{-1}\|$, $\kappa(A)$, and then verify your calculations with MATLAB. Is there a singular matrix S such that $\|S - A\| = 1/1000$? Explain.

6. Consider the matrix $B = \begin{bmatrix} 2.6 & 0 & -3.2 \\ 3 & -8 & -4 \\ 1 & 2 & -1 \end{bmatrix}$.

   (a) Is there a nonzero (3×1) vector x such that $\|Bx\| \geq 13\|x\|$? If so, find one; otherwise explain why one does not exist.

   (b) Is there a singular 3×3 matrix S such that $\|S - B\| \leq 0.2$? If so, find one; otherwise explain why one does not exist.

7. Consider the matrices: $A = \begin{bmatrix} 2 & -6 \\ 11 & -5 \end{bmatrix}$, $B = \begin{bmatrix} 7 & 1 & -4 \\ 5 & -8 & -5 \\ 4 & 4 & 4 \end{bmatrix}$.

   (a) Is there a (2×1) vector X such that: $\|AX\| > 12\|X\|$?

   (b) Is there a nonzero vector X such that $\|AX\| \geq 16\|X\|$? If so, find one; otherwise explain why one does not exist.

   (c) Is there a nonzero (3×1) vector X such that $\|BX\| \geq 20\|X\|$? If so, find one; otherwise explain why one does not exist.

   (d) Is there a singular matrix $S = \begin{bmatrix} a & b \\ c & d \end{bmatrix}$ (i.e., $ad - bc = 0$) such that $\|S - A\| \leq 4.5$? If yes, find one; otherwise explain why one does not exist.

   (e) Is there a singular 3×3 matrix S such that $\|S - B\| \leq 2.25$? If so find one; otherwise explain why one does not exist.

8. Prove identities (42), (43), and (44) for condition numbers.
   Suggestion: The identity that was established in Exercise for the Reader 7.24 can be helpful.

9. (*True/False*) For each statement below, either explain why it is always true or provide a counterexample of a single situation where it is false:

   (a) If A is a square matrix with $\|A\| = 0$, then $A = 0$(matrix), i.e., all the entries of A are zero.

   (b) If A is any square matrix, then $|\det(A)| \leq \kappa(A)$.

   (c) If A is a nonsingular square matrix, then $\kappa(A^{-1}) = \kappa(A)$.

   (d) If A is a square matrix, then $\kappa(A') = \kappa(A)$.

   (e) If A and B are same-sized square matrices, then $\|AB\| \leq \|A\|\|B\|$.

   Suggestion: As is always recommended, unless you are sure about any of these identities, run a bunch of experiments on MATLAB (using randomly generated matrices).

10. Prove identity (39).
    Suggestion: Reread Example 7.19 and the note that follows it for a useful idea.

11. (*A General Class of Norms*) For any real number $p \geq 1$, the p-norm $\|\cdot\|_p$ defined for an n-dimensional vector x by the equation

$$\|x\|_p = \left(\sum_{i=1}^{n} |x_i|^p \right)^{1/p} = \left(|x_1|^p + |x_2|^p + \cdots + |x_n|^p \right)^{1/p}$$

turns out to satisfy the norm axioms ($36A - C$). In the general setting, the proof of the triangle inequality is a bit involved (we refer to one of the more advanced texts on error analysis cited in the section for details).

(a) Show that $\mathrm{len}(x) = \|x\|_2$ (this is why the length norm is sometimes called the 2-norm).

(b) Verify the norm axioms ($36A - C$) for the 1-norm $\|\cdot\|_1$.

(c) For a vector x, is it always true that $\|x\|_\infty \leq \|x\|_1$? Either prove it is always true or give a counterexample of an instance of a certain vector x for which it fails.

(d) For a vector x, is it always true that $\|x\|_\infty \leq \|x\|_2$? Either prove it is always true or give a counterexample of an instance of a certain vector x for which it fails.

(e) How are the norms $\|\cdot\|_1$ and $\|\cdot\|_2$ related?   Does one always seem to be at least as large as the other? Do (lots of) experiments with some randomly generated vectors of different sizes.
Note: Experiments will probably convince you of a relationship (and inequality) for part (e), but it might be difficult to prove, depending on your background; the interested reader can find a proof in one of the more advanced references listed in the section. The reason that the infinity norm got this name is that for any fixed vector x, we have

$$\lim_{p \to \infty} \|x\|_p = \|x\|_\infty.$$

As the careful reader might have predicted, the MATLAB built-in function for the p-norm of a vector x is norm(x,p).

12. Let $A = \begin{bmatrix} -11 & 3 \\ 4 & -1 \end{bmatrix}$, $b = \begin{bmatrix} 2 \\ 0 \end{bmatrix}$ $z = \begin{bmatrix} 1.2 \\ 7.8 \end{bmatrix}$.

(a) Let z be an approximate solution of the system $Ax = b$; find the residual vector r.
(b) Use Theorem 7.7 to give an estimate for the error of the approximation in part (a).
(c) Give an estimate for the relative error of the approximation in part (a).
(d) Find the norm of the exact error $\|\Delta x\|$ of the approximate solution in part (a).

13. Repeat all parts of Exercise 12 for the following matrices:

$$A = \begin{bmatrix} -0.1 & -9 \\ 11 & 1000 \end{bmatrix}, \ b = \begin{bmatrix} -0.1 \\ 10 \end{bmatrix} \ z = \begin{bmatrix} 11 \\ -0.2 \end{bmatrix}.$$

14. Let $A = \begin{bmatrix} 1 & 0 & 1 \\ -1 & 1 & 1 \\ -1 & -1 & 1 \end{bmatrix}, \ b = \begin{bmatrix} 1 \\ 2 \\ 3 \end{bmatrix}.$

(a) Use MATLAB's "left divide" to solve the system $Ax = b$.
(b) Use Theorems 7.7 and 7.8 to estimate the error and relative error of the solution obtained in part (a).
(c) Use MATLAB to solve the system $Ax = b$ by left multiplying the equation by the inv(A).
(d) Use Theorems 7.7 and 7.8 to estimate the error and relative error of the solution obtained in part (c).
(e) Solve the system using MATLAB's symbolic capabilities and compute the actual errors of the solutions obtained in parts (a) and (c).

15. Let A be the 60×60 matrix whose entries are 1's across the main diagonal and the last column, −1's below the main diagonal, and whose remaining entries (above the main diagonal) are zeros,

and let $b = [1 \ \ 2 \ \ 3 \ \cdots \ 58 \ \ 59 \ \ 60]'$.

(a) Use MATLAB's left divide to solve the system $Ax = b$ and label this computed solution as z. Print out only z (37).

(b) Use Theorems 7.7 and 7.8 to estimate the error and relative error of the solution obtained in part (a).

(c) Use MATLAB to solve the system $Ax = b$ by left multiplying the equation by the inv (A) and label this computed solution as z2. Print out only z2 (37).

(d) Use Theorems 7.7 and 7.8 to estimate the error and relative error of the solution obtained in part (c).

(e) Solve the system using MATLAB's symbolic capabilities, and print out x (37) of the exact solution. Then compute the norms of the actual errors of the solutions obtained in parts (a) and (c).

16. *(Iterative Refinement)* Let z_0 be the computer solution of Exercise 15(a), and let r_0 denote the corresponding residual vector. Now use the Gaussian program to solve the system $Ax = r_0$, and call the computer solution z_1, and the corresponding residual vector r_1. Next use the Gaussian program once again to solve $Ax = r_1$, and let z_2 and r_2 denote the corresponding approximate solution and residual vector. Now let

$$z = z_0, \quad z' = z + z_1, \quad \text{and} \quad z'' = z' + z_2.$$

Viewing these three vectors as solutions of the original system $Ax = b$ (of Exercise 15), use the error estimate Theorem 7.8 to estimate the relative error of each of these three vectors. Then compute the norm of the actual errors by comparing with the exact solution of the system as obtained in part (e) of Exercise 15. See the next exercise for more on this topic.

17. In theory, the iterative technique of the previous exercise can be useful to improving accuracy of approximate solutions in certain circumstances. In practice, however, roundoff errors and poorly conditioned matrices can lead to unimpressive results. This exercise explores the effect that additional digits of precision can have on this scheme.

(a) Using variable precision arithmetic (see Appendix A) with 30 digits of accuracy, redo the previous exercise, starting with the computed solution of Exercise 15(a) done in MATLAB's default floating point arithmetic.

(b) Using the same arithmetic of part (a), solve the original system using MATLAB's left divide.

(c) Compare the norms of the actual errors of the three approximate solutions of part (a) and the one of part (b) by using symbolic arithmetic to get MATLAB to compute the exact solution of the system.

Note: We will learn about different iterative methods in the next section.

18. This exercise will examine the benefits of variable precision arithmetic over MATLAB's default floating point arithmetic and over MATLAB's more costly symbolic arithmetic. As in Section 7.4, we let $H_n = \left[1/(i + j - 1)\right]$ denote the $n \times n$ Hilbert matrix. Recall that it can be generated in MATLAB using the command hilb (n).

(a) For the values $n = 5, 10, 15, \cdots, 100$ create the corresponding Hilbert matrices H_n in MATLAB as symbolic matrices and compute symbolically the inverses of each. Use tic/toc to record the computation times (these times will be machine dependent; see Chapter 4). Go as far as you can until your cumulative MATLAB computation time exceeds one hour. Next compute the corresponding condition numbers of each of these Hilbert matrices.

(b) Starting with MATLAB's default floating point arithmetic (which is roughly 15 digits of variable precision arithmetic), and then using variable precision arithmetic starting with 20 digits and then moving up in increments of 5 (25, 30, 35, ...), continue to compute the inverses of each of the Hilbert matrices of part (a) until you get a computed inverse whose norm differs from the norm of the exact inverse in part (a) by no more than 0.000001. Record (using tic/toc) the computation time for the final variable precision arithmetically computed inverse, along with the number of digits used, and compare it to the corresponding computation time for the exact inverse that was done in part (a).

19. Prove the following inequality $\|Ax\| \geq \|x\| / \|A^{-1}\|$, where A is any invertible $n \times n$ matrix and x is any column vector with n entries.

20. Suppose that A is a 2×2 matrix with norm $\|A\| = 0.5$ and x and y are 2×1 vectors with $\|x - y\| \leq 0.8$. Show that: $\|A^2 x - A^2 y\| \leq 0.2$.

21. (*Another Error Bound for Computed Solutions of Linear Systems*) For a nonsingular matrix A and a computed inverse matrix C for A^{-1}, we define the resulting **residual matrix** as $R = I - CA$. If z is an approximate solution to $Ax = b$, and as usual $r = b - Az$ is the residual vector, show that

$$\text{error} \equiv \|x - z\| \leq \frac{\|CR\|}{1 - \|R\|},$$

provided that $\|R\| < 1$.

Hint: For part (a), first use the equation $I - R = CA$ to get that $(I - R)^{-1} = A^{-1} C^{-1}$ and so $A^{-1} = (I - R)^{-1} C$. (Recall that the inverse of a product of invertible matrices equals the reverse order product of the inverses.)

22. For each of the matrices A below, find the following:
     (a) The characteristic polynomial $p_A(\lambda)$.
     (b) All eigenvalues and all of their associated eigenvectors.
     (c) The algebraic and geometric multiplicity of each eigenvalue.
     (i) $A = \begin{bmatrix} 1 & 2 \\ 2 & 1 \end{bmatrix}$ (ii) $A = \begin{bmatrix} 1 & 1 \\ 2 & 2 \end{bmatrix}$ (iii) $A = \begin{bmatrix} 1 & 2 \\ 2 & 2 \end{bmatrix}$ (iv) $A = \begin{bmatrix} -1 & 0 \\ 2 & 3 \end{bmatrix}$

23. Repeat all parts of Exercise 22 for the following matrices.
     (i) $A = \begin{bmatrix} 1 & 2 & 2 \\ 2 & 1 & 2 \\ 2 & 2 & 1 \end{bmatrix}$ (ii) $A = \begin{bmatrix} 1 & 2 & 0 \\ 2 & 1 & 2 \\ 0 & 2 & 1 \end{bmatrix}$

     (iii) $A = \begin{bmatrix} 1 & 2 & 2 \\ 2 & 1 & 2 \\ 0 & 0 & 1 \end{bmatrix}$ (iv) $A = \begin{bmatrix} 1 & 2 & 2 \\ -2 & 1 & 2 \\ -2 & -2 & 1 \end{bmatrix}$

24. Consider the matrix $A = \begin{bmatrix} 11 & 11 & 4 \\ 7 & 7 & -4 \\ -7 & -11 & 0 \end{bmatrix}$.

     (a) Find all eigenvalues of A, and for each find just one eigenvector (give your eigenvectors as many integer components as possible).
     (b) For each of the eigenvectors x that you found in part (a), evaluate $y = (2A)x$. Is it possible to write $y = \lambda x$ for some scalar λ? In other words, is x also an eigenvector of the matrix $2A$?
     (c) Find all eigenvalues of $2A$. How are these related to those of the matrix A?
     (d) For each of your eigenvectors x from part (a), evaluate $y = (-5A)x$. Is it possible to write $y = \lambda x$ for some scalar λ? In other words, is x also an eigenvector of the matrix $-5A$?
     (e) Find all eigenvalues of $-5A$; how are these related to those of the matrix A?
     (f) Based on your work in these above examples, without picking up your pencil or typing

anything on the computer, what do you think the eigenvalues of the matrix $23A$ would be? Could you guess also some associated eigenvectors for each eigenvalue? Check your conclusions on MATLAB.

25. Consider the matrix $A = \begin{bmatrix} 2 & 0 & 1 \\ 1 & -4 & 1 \\ 1 & 0 & 2 \end{bmatrix}$.

(a) Find all eigenvalues of A, and for each find just one eigenvector (give your eigenvectors as many integer components as possible).

(b) For each of the eigenvectors x that you found in part (a), evaluate $y = A^2 x$. Is it possible to write $y = \lambda x$ for some scalar λ? In other words, is x also an eigenvector of the matrix A^2?

(c) Find all eigenvalues of A^2; how are these related to those of the matrix A?

(d) For each of your eigenvectors x from part (a), evaluate $y = A^3 x$. Is it possible to write $y = \lambda x$ for some scalar λ? In other words, is x also an eigenvector of the matrix A^3?

(e) Find all eigenvalues of A^3; how are these related to those of the matrix A?

(f) Based on your work in the above examples, without picking up your pencil or typing anything on the computer, what do you think the eigenvalues of the matrix A^8 would be? Could you guess also some associated eigenvectors for each eigenvalue? Check your conclusions on MATLAB.

26. Find the characteristic polynomial (factored form is okay) as well as all eigenvalues for the $n \times n$ identity matrix I. What are (all) of the corresponding eigenvectors (for each eigenvalue)?

27. Consider the matrix $A = \begin{bmatrix} 3 & 3 & 3 \\ 3 & 2 & 4 \\ 3 & 4 & 2 \end{bmatrix}$.

(a) Find all eigenvalues of A, and for each find just one eigenvector (give your eigenvectors as many integer components as possible).

(b) For each of the eigenvectors x that you found in part (a), evaluate $y = (A^2 + 2A)x$. Is it possible to write $y = \lambda x$ for some scalar λ? In other words, is x also an eigenvector of the matrix $A^2 + 2A$?

(c) Find all eigenvalues of $A^2 + 2A$; how are these related to those of the matrix A?

(d) For each of your eigenvectors x from Part (a), evaluate $y = (A^3 - 4A^2 + I)x$. Is it possible to write $y = \lambda x$ for some scalar λ? In other words, is x also an eigenvector of the matrix $A^3 - 4A^2 + I$?

(e) Find all eigenvalues of $A^3 - 4A^2 + I$; how are these related to those of the matrix A?

(f) Based on your work in the above examples, without picking up your pencil or typing anything on the computer, what do you think the eigenvalues of the matrix $A^5 - 4A^3 + 2A - 4I$ would be? Could you guess also some associated eigenvectors for each eigenvalue? Check your conclusions on MATLAB.

NOTE: The **spectrum** of a matrix A, denoted $\sigma(A)$, is the set of all eigenvalues of the matrix A. The next exercise generalizes some of the results discovered in the previous four exercises.

28. For a square matrix A and any polynomial $p(x) = a_m x^m + a_{m-1} x^{m-1} + \cdots + a_1 x + a_0$, we define a new matrix $p(A)$ as follows:

$$p(A) = a_m A^m + a_{m-1} A^{m-1} + \cdots + a_1 A + a_0 I.$$

(We simply substituted A for x in the formula for the polynomial; we also had to replace the constant term a_0 by this constant times the identity matrix—the matrix analogue of the number

1.) Prove the following appealing formula;
$$\sigma(p(A)) = p(\sigma(A)),$$
which states that the spectrum of the matrix $p(A)$ equals the set
$$\{p(\lambda): \lambda \text{ is an eigenvalue of } A\}.$$

29.   Prove parts (i) and (ii) of Theorem 7.9.

30.   Show that the characteristic polynomial of any $n \times n$ matrix is always a polynomial of degree n in the variable λ.
       Suggestion: Use induction and cofactor expansion.

31.   (a) Use the basic Gaussian elimination algorithm (Program 7.6) to solve the linear systems of Exercise for the Reader 7.16, and compare with results obtained therein.
       (b) Use the Symbolic Toolbox to compute the condition numbers of the Hilbert matrices that came up in part (a). Are the estimates provided by Theorem 7.8 accurate or useful?
       (c) Explain why the algorithm performs so well with this problem despite the large condition numbers of A.
       Suggestion: For part (c), examine what happens after the first pivot operation.

7.7: ITERATIVE METHODS

As mentioned earlier in this chapter, Gaussian elimination is the best all-around solver for nonsingular linear systems $Ax = b$. Being a universal method, however, there are often more economical methods that can be used for particular forms of the coefficient matrix. We have already seen the tremendous savings, both in storage and in computations that can be realized in case A is tridiagonal, by using the Thomas method. All methods considered thus far have been **direct methods** in that, mathematically, they compute the exact solution and the only errors that arise are numerical. In this section we will introduce a very different type of method called an **iterative method**. Iterative methods begin with an initial guess at the (vector) solution $x^{(0)}$, and produce a sequence of vectors, $x^{(1)}, x^{(2)}, x^{(3)}, \cdots$, which, under certain circumstances, will converge to the exact solution. Of course, in any floating point arithmetic system, a solution from an iterative method (if the method converges) can be made just as accurate as that of a direct method.

In solving differential equations with so-called finite difference methods, the key numerical step will be to solve a linear system $Ax = b$ having a large and sparse coefficient matrix A (a small percentage of nonzero entries) that will have a special form. The large size of the matrix will often make Gaussian elimination too slow. On the other hand, the special structure and sparsity of A can make the system amenable to a much more efficient iterative method. We have seen that in general Gaussian elimination for solving an n variable linear system performs in $O(n^3)$-time. We take this as the ceiling performance time for any linear system solver. The Thomas method, on the other hand, for the very special triadiagonal systems, performed in only $O(n)$-time. Since just solving n independent linear equations (i.e., with A being a diagonal matrix) will also take this amount, this is the theoretical floor performance time for any linear system solver. Most of iterative methods today perform theoretically in $O(n^2)$ -time, but in practice can perform in times closer to the theoretical floor $O(n)$ -time. In recent years, iterative methods have become increasingly important and have a promising future, as increasing computer performance will make the improvements over Gaussian elimination more and more dramatic.

We will describe three common iterative methods: Jacobi, Gauss-Seidel, and SOR iteration. After giving some simple examples showing the sensitivity of these methods to the particular form of A, we give some theoretical results that will guarantee convergence. We then make some comparisons among these three methods in flop counts and computation times for larger systems and then with Gaussian elimination. The theory of iterative methods is a very exciting and interesting area of numerical analysis. The Jacobi, Gauss-Seidel, and SOR iterative methods are quite intuitive and easy to develop. Some of the more state-of-the-art methods such as conjugate gradient methods and GMRES (generalized minimum residual method) are more advanced and would take a lot more work to develop and understand, so we refer the interested reader to some references for more details on this interesting subject: [Gre-97], [TrBa-97], and [GoVL-83]. MATLAB, however, does have some built-in functions for performing such more advanced iterative methods. We introduce these MATLAB functions and do some performance comparisons involving some (very large and) typical coefficient matrices that arise in finite difference schemes.

We begin with a nonsingular linear system:

$$Ax = b. \tag{54}$$

In scalar form, it looks like this:

$$a_{i1}x_1 + a_{i2}x_2 + \cdots a_{in}x_n = b_i \quad (1 \leq i \leq n). \tag{55}$$

Now, if we assume that each of the diagonal entries of A are nonzero, then each of the equations in (55) can be solved for x_i to arrive at:

$$x_i = \frac{1}{a_{ii}}\left[b_i - a_{i1}x_1 - a_{i2}x_2 - \cdots - a_{i,i-1}x_{i-1} - a_{i,i+1}x_{i+1} - \cdots a_{in}x_n\right] \quad (1 \leq i \leq n). \tag{56}$$

The **Jacobi iteration** scheme is obtained from using formula (56) with the values of current iteration vector $x^{(k)}$ on the right to create, on the left, the values of the next iteration vector $x^{(k+1)}$. We record the simple formula:

Jacobi Iteration:

$$x_i^{(k+1)} = \frac{1}{a_{ii}}\left[b_i - \sum_{\substack{j=1, \\ j \neq i}}^{n} a_{ij}x_j^{(k)}\right] \quad (1 \leq i \leq n). \tag{57}$$

Let us give a (very) simple example illustrating this scheme on a small linear system and compare with the exact solution.

EXAMPLE 7.26: Consider the following linear system:

$$\begin{array}{rcrcrcl} 3x_1 & + & x_2 & - & x_3 & = & -3 \\ 4x_1 & - & 10x_2 & + & x_3 & = & 28 \\ 2x_1 & + & x_2 & + & 5x_3 & = & 20 \end{array}.$$

(a) Starting with the vector $x^{(0)} = [0\ 0\ 0]'$, apply the Jacobi iteration scheme with up to 30 iterations until (if ever) the 2-norm of the differences $x^{(k+1)} - x^{(k)}$ is less than 10^{-6}. Plot the norms of these differences as a function of the iteration. If convergence occurs, record the number of iterations and the actual 2-norm error of the final iterant with the exact solution.
(b) Repeat part (a) on the equivalent system obtained by switching the first two equations.

SOLUTION: Part (a): The Jacobi iteration scheme (57) becomes:

$$\begin{aligned} x_1^{(k+1)} &= (-3 - x_2^{(k)} + x_3^{(k)})/3 \\ x_2^{(k+1)} &= (28 - 4x_1^{(k)} - x_3^{(k)})/(-10) \\ x_3^{(k+1)} &= (20 - 2x_1^{(k)} - x_2^{(k)})/5 \end{aligned}.$$

The following MATLAB code will perform the required tasks:

```
xold = [0 0 0]';  xnew=xold;
for k=1:30
  xnew(1)=(-3-xold(2)+xold(3))/3;
  xnew(2)=(28-4*xold(1)-xold(3))/(-10);
  xnew(3)=(20-2*xold(1)-xold(2))/5;
  diff(k)=norm(xnew-xold,2);
  if diff(k)<1e-6
    fprintf('Jacobi iteration has converged in %d iterations', k)
    return
  end
  xold=xnew;
end
```
→ Jacobi iteration has converged in 26 iterations

The exact solution is easily seen to be $[1\ -2\ 4]'$. The exact 2-norm error is thus given by:

```
>>norm(xnew-[1 -2 4]',2)
```
→ ans = 3.9913e - 007

which compares favorably with the norm of the last difference of the iterates (i.e., the actual error is smaller):

```
>> diff(k)
```
→ans = 8.9241e - 007

We will see later in this section that finite difference methods typically exhibit linear convergence (if they indeed converge); the quality of convergence will thus depend on the asymptotic error constant (see Section 6.5 for the terminology).

Due to this speed of the decay of errors, an ordinary plot will not be so useful (as the reader should verify), so we use a log scale on the y-axis. This is accomplished by the following MATLAB command:

`semilogy(x,y)` $\rightarrow$	If x and y are two vectors of the same size, this will produce a plot where the y-axis numbers are logarithmically spaced rather than equally spaced as with `plot(x,y)`.
`semilogx(x,y)` $\rightarrow$	Works as the above command, but now the x-axis numbers are logarithmically spaced.

The required plot is now created with the following command and the result is shown in Figure 7.39(a).

```
>>semilogy(1:k,diff(1:k))
```

Part (b): Switching the first two equations of the given system leads to the following modified Jacobi iteration scheme:

$$x_1^{(k+1)} = (28 + 10x_2^{(k)} - x_3^{(k)})/4$$
$$x_2^{(k+1)} = -3 - 3x_1^{(k)} + x_3^{(k)}$$
$$x_3^{(k+1)} = (20 - 2x_1^{(k)} - x_2^{(k)})/5$$

In the above MATLAB code, we need only change the two lines for xnew(54) and xnew(55) accordingly:

```
xnew(1)=(28+10*xold(2)-xold(3))/4;
xnew(2)=-3-3*xold(1)+xold(3);
```

Running the code, we see that this time we do not get convergence. In fact, a semilog plot will show that quite the opposite is true, the iterates badly diverge. The plot, obtained just as before, is shown in Figure 7.39(b). We will soon show how such sensitivities of iterative methods depend on the form of the coefficient matrix.

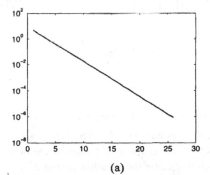
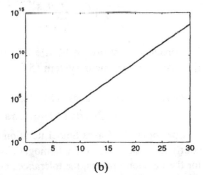

(a) (b)

FIGURE 7.39: (a) (left) Plots of the 2-norms of the differences of successive iterates in the Jacobi scheme for the linear system of Example 7.26, using the zero vector as the initial iterate. The convergence is exponential. (b) (right) The corresponding errors when the same scheme is applied to the equivalent linear system with the first two equations being permuted. The sequence now badly diverges, showing the sensitivity of iterative methods to the particular form of the coefficient matrix.

The code given in the above example can be easily generalized into a MATLAB M-file for performing the Jacobi iteration on a general system. This task will be delegated to the following exercise for the reader.

EXERCISE FOR THE READER 7.27:   (a) Write a function M-file, [x,k,diff]= jacobi(A,b,x0,tol,kmax), that performs the Jacobi iteration on the linear system $Ax = b$. The inputs are the coefficient matrix A, the inhomogeneity (column) vector b, the seed (column) vector x0 for the iteration process, the tolerance tol, which will cause the iteration to stop if the 2-norms of successive iterates become smaller than tol, and kmax, the maximum number of iterations to perform. The outputs are the final iterate x, the number of iterations performed k, and a vector diff that records the 2-norms of successive differences of iterates. If the last three input variables are not specified, default values of x0 = the zero column vector, tol = 1e-10, and kmax = 100 are used.
(b) Apply the program to recover the data obtained in part (a) of Example 7.26. If we reset the tolerance for accuracy to 1e-10 in that example, how many iterations would the Jacobi iteration need to converge?

If we compute the values of $x^{(k+1)}$ in order, it seems reasonable to update the values used on the right side of (57) sequentially, as they become available. This modification in the scheme gives the **Gauss-Seidel iteration**. Notice that the Gauss-Seidel scheme can be implemented so as to roughly cut in half the storage requirements for the iterates of the solution vector x. Although the M-file we present below does not take advantage of such a scheme, the interested reader can easily modify it to do so. Futhermore, as we shall see, the Gauss-Seidel scheme almost always outperforms the Jacobi scheme.

Gauss-Seidel Iteration:

$$x_i^{(k+1)} = \frac{1}{a_{ii}}\left[b_i - \sum_{j=1}^{i-1} a_{ij} x_j^{(k+1)} - \sum_{j=i+1}^{n} a_{ij} x_j^{(k)} \right] \quad (1 \le i \le n). \tag{58}$$

We proceed to write an M-file that will apply the Gauss-Seidel scheme to solving the nonsingular linear system (54).

PROGRAM 7.7: A function M-file,
                    [x,k,diff]=gaussseidel(A,b,x0,tol,kmax)
that performs the Gauss-Seidel iteration on the linear system $Ax = b$. The inputs are the coefficient matrix A, the inhomogeneity (column) vector b, the seed (column) vector x0 for the iteration process, the tolerance tol, which will cause the iteration to stop if the 2-norms of successive iterates become smaller than tol, and kmax, the maximum number of iterations to perform. The outputs are the final iterate x, the number of iterations performed k, and a vector diff that records the 2-norms of successive differences of iterates. If the last two input variables are not specified, default values of tol = 1e-10 and kmax = 100 are used.

```
function [x, k, diff] = gaussseidel(A,b,x0,tol,kmax)
% performs the Gauss-Seidel iteration on the linear system Ax = b.
% Inputs:   the coefficient matrix 'A', the inhomogeneity (column)
% vector 'b',the seed (column) vector 'x0' for the iteration process,
% the tolerance 'tol' which will cause the iteration to stop if the
% 2-norms of differences of successive iterates becomes smaller than
% 'tol', and 'kmax' that is the maximum number of iterations to
% perform.
% Outputs:  the final iterate 'x', the number of iterations performed
% 'k', and a vector 'diff' that records the 2-norms of successive
% differences of iterates.
% if either of the last three input variables are not specified,
% default values of x0- zero column vector, tol-1e-10 and kmax-100
% are used.

%assign default input variables, as necessary
if nargin<3, x0=zeros(size(b)); end
if nargin<4, tol=1e-10; end
if nargin<5, kmax=100; end

if min(abs(diag(A)))<eps
    error('Coefficient matrix has zero diagonal entries, iteration
                        cannot be performed.\r')
end

[n m]=size(A);
x=x0;
k=1; diff=[];

while k<=kmax
    norm=0;
    for i=1:n
        oldxi=x(i); x(i)=b(i);
        for j=[1:i-1 i+1:n]
            x(i)=x(i)-A(i,j)*x(j);
        end
        x(i)=x(i)/A(i,i);
        norm=norm+(oldxi-x(i))^2;
    end
    diff(k)=sqrt(norm);
    if diff(k)<tol
            fprintf('Gauss-Seidel iteration has converged in %d
                            iterations/r', k)
            return
    end
    k=k+1;
end
fprintf('Gauss-Seidel iteration failed to converge./r')
```

EXAMPLE 7.27: For the linear system of the last example, apply Gauss-Seidel iteration with initial iterate being the zero vector and the same tolerance as that used for the last example. Find the number of iterations that are now required for convergence and compare the absolute 2-norm error of the final iterate with that for the last example.

SOLUTION: Reentering, if necessary, the data from the last example, create corresponding data for the Gauss-Seidel iteration using the preceding M-file:

```
>> [xGS, kGS, diffGS] = gaussseidel(A,b,zeros(size(b)),1e-6);
→Gauss-Seidel iteration has converged in 17 iterations
```

Thus with the same amount of work per iteration, Gauss-Seidel has done the job in only 17 versus 26 iterations for Jacobi.

Looking at the absolute error of the Gauss-Seidel approximation,

```
>> norm(xGS-[1 -2 4]',2)          →ans = 1.4177e - 007
```

we see it certainly meets our tolerance goal of 1e-6 (and, in fact, is smaller than that for the Jacobi iteration).

The Gauss-Seidel scheme can be extended to include a new parameter, ω, that will allow the next iterate $x^{(k+1)}$ to be expressed as a linear combination of the current iterate $x^{(k)}$ and the Gauss-Seidel values given by (58). This gives a family of iteration schemes, collectively known as **SOR (successive over relaxation)** whose iteration schemes are given by the following formula:

SOR Iteration:

$$x_i^{(k+1)} = \frac{\omega}{a_{ii}} \left[b_i - \sum_{j=1}^{i-1} a_{ij} x_j^{(k+1)} - \sum_{j=i+1}^{n} a_{ij} x_j^{(k)} \right] + (1-\omega)x_i^{(k)} \quad (1 \le i \le n) \qquad (59)$$

The parameter ω, called the **relaxation parameter**, controls the proportion of the Gauss-Seidel update versus the current iterate to use in forming the next iterate. We will soon see that for SOR to converge, we will need the relaxation parameter to satisfy $0 < \omega < 2$. Notice that when $\omega = 1$, SOR reduces to Gauss-Seidel. For certain values of ω, SOR can accelerate the convergence realized in Gauss-Seidel.

With a few changes to the Program 7.7, a corresponding M-file for SOR is easily created. We leave this for the next exercise for the reader.

EXERCISE FOR THE READER 7.28:   (a) Write a function M-file, [x,k,diff]=sorit(A,b,omega,x0,tol,kmax), that performs the SOR iteration on the linear system $Ax = b$. The inputs are the coefficient matrix A, the inhomogeneity (column) vector b, the relaxation parameter omega, the seed (column) vector x0 for the iteration process, the tolerance tol, which will cause the iteration to stop if the 2-norms of successive iterates become smaller than tol, and kmax, the maximum number of iterations to perform. The outputs are the final iterate x, the number of iterations performed k, and a vector diff that records the 2-norms of successive differences of iterates. If the last three input variables are not specified, default values of x0 = the zero column vector, tol = 1e-10, and kmax = 100 are used.

(b) Apply the program to recover the solution obtained in Example 7.27.

(c) If we use $\omega = 0.9$, how many iterations would the SOR iteration need to converge?

EXAMPLE 7.28: Run a set of SOR iterations by letting the relaxation parameter run from 0.05 to 1.95 in increments of 0.5. Use a tolerance for error = 1e-6, but set kmax = 1000. Record the number of iterations needed for convergence (if there is convergence) for each value of the ω (up to 1000) and plot this number as a function of ω.

SOLUTION: We can use the M-file `sorit` of Exercise for the Reader 7.28 in conjunction with a loop to easily obtain the needed data.

```
>> omega=0.05:.05:1.95;
>> length(omega)    →ans = 39
>> for i=1:39
[xSOR, kSOR(i), diffSOR] = sorit(A,b,omega(i),zeros(size(b)),...
                1e-6,1000);
end
```

The above loop has overwritten all but the iteration counters, which were recorded as a vector. We use this vector to locate the best value (from among those in our vector `omega`) to use in SOR.

```
>> [mink ind]=min(kSOR)      →mink = 9, ind = 18
>> omega(18)                 →ans = 0.9000
```

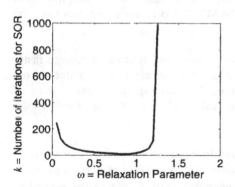

FIGURE 7.40: Graph of the number of iterations required for convergence (to a tolerance of 1e-6) using SOR iteration as a function of the relaxation parameter ω. The k-values are truncated at 1000. Notice from the graph that the convergence for Gauss-Seidel ($\omega = 1$) can be improved.

Thus we see that the best value of ω to use (from those we tested) is $\omega = 0.9$, which requires only nine iterations in the SOR scheme, nearly a 50% savings over Gauss-Seidel. The next two commands will produce the desired plot of the required number of iterations needed in SOR versus the value of the parameter ω. The resulting plot is shown in Figure 7.40.

```
>> plot(omega, kSOR),
>> axis([0 2 0 1000])
```

Figure 7.41 gives a plot that compares the convergences of three methods: Jacobi, Gauss-Seidel, and SOR (with our pseudo-optimal value of ω). The next exercise for the reader will ask to reproduce this plot.

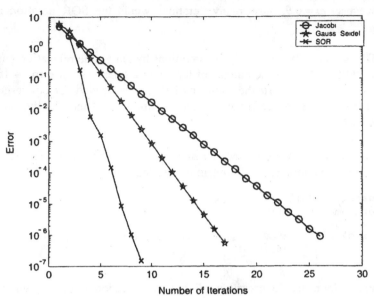

FIGURE 7.41: Comparison of the errors versus the number of iterations for each of the three iteration methods: Jacobi (o), Gauss-Seidel (*), and SOR (x).

EXERCISE FOR THE READER 7.29: Use MATLAB to reproduce the plot of Figure 7.41. The key in the upper-right corner can be obtained by using the "Data Statistics" tool from the "Tools" menu of the MATLAB graphics window once the three plots are created.

Of course, even though the last example has shown that SOR can converge faster than Gauss-Seidel, the amount of work required to locate a good value of the parameter greatly exceeded the actual savings in solving the linear system of Example 7.26. In the SOR iteration the value of $\omega = 0.9$ was used as the relaxation parameter.

There is some interesting research involved in determining the optimal value of ω to use based on the form of the coefficient matrix. What is needed to prove such results is to get a nice formula for the eigenvalues of the matrix (in general an impossible problem, but for special types of matrices one can get lucky) and then compute the value of ω for which the corresponding maximum absolute value of the eigenvalues is as small as possible . A good survey of the SOR method is given in [You-71]. A sample result will be given a bit later in this section (see Proposition 7.14).

We now present a general way to view iteration schemes in matrix form. From this point of view it will be a simple matter to specialize to the three forms we gave above. More importantly, the matrix notation will allow a much more natural way to perform error analysis and other important theoretical tasks.

To cast iteration schemes into matrix form, we begin by breaking the coefficient matrix A into three pieces:

$$A = D - L - U, \tag{60}$$

where D is the diagonal part of A, L is (strictly) lower triangular and U is the (strictly) upper triangular. In long form this (60) looks like:

$$A = \begin{bmatrix} a_{11} & & & \\ & a_{22} & & 0 \\ & & a_{33} & \\ & 0 & & \ddots \\ & & & & a_{nn} \end{bmatrix} - \begin{bmatrix} 0 & & & \\ -a_{21} & 0 & & 0 \\ -a_{31} & -a_{32} & 0 & \\ & & & \ddots \\ -a_{n1} & -a_{n2} & -a_{n3} & \cdots & 0 \end{bmatrix} - \begin{bmatrix} 0 & -a_{12} & -a_{13} & \cdots & -a_{1n} \\ 0 & -a_{23} & \cdots & -a_{2n} \\ & 0 & \ddots & \vdots \\ 0 & & \ddots & -a_{n-1,n} \\ & & & 0 \end{bmatrix}$$

This decomposition is actually quite simple. Just take D to be the diagonal matrix with the diagonal entries equal to those of A, and take L/U to be, respectively, the strictly lower/upper triangular matrix whose nonzero entries are the opposites of the corresponding entries of A.

Next, we will examine the following general (matrix form) iteration scheme for solving the system (54) $Ax = b$:

$$Bx^{(k+1)} = (B - A)x^{(k)} + b, \tag{61}$$

where B is an invertible matrix that is to be determined. Notice that if B is chosen so that this iteration scheme produces a convergent sequence of iterates: $x^{(k)} \to \tilde{x}$, then the limiting vector $\tilde{x}$ must solve (54). (Proof: Take the limit in (61) as $k \to \infty$ to get $B\tilde{x} = (B - A)\tilde{x} + b = B\tilde{x} - A\tilde{x} + b \Rightarrow A\tilde{x} = b$.) The matrix B should be chosen so that the linear system is easy to solve for $x^{(k+1)}$ (in fact, much easier than our original system $Ax = b$ lest this iterative scheme would be of little value) and so that the convergence is fast.

To get some idea of what sort of matrix B we should be looking for, we perform the following error analysis on the iterative scheme (61). We mathematically solve (61) for $x^{(k+1)}$ by left multiplying by B^{-1} to obtain:

$$x^{(k+1)} = B^{-1}(B - A)x^{(k)} + B^{-1}b = (I - B^{-1}A)x^{(k)} + B^{-1}b.$$

Let x denote the exact solution of $Ax = b$ and $e^{(k)} = x^{(k)} - x$ denote the error vector of the kth iterate. Note that $-(I - B^{-1}A)x = -x + B^{-1}b$. Using this in conjunction with the last equation, we can write:

$$e^{(k+1)} = x^{(k+1)} - x = (I - B^{-1}A)x^{(k)} + B^{-1}b - x$$
$$= (I - B^{-1}A)x^{(k)} - (I - B^{-1}A)x$$
$$= (I - B^{-1}A)(x^{(k)} - x)$$
$$= (I - B^{-1}A)e^{(k)}.$$

We summarize this important error estimate:

$$e^{(k+1)} = (I - B^{-1}A)e^{(k)}. \tag{62}$$

From (62), we see that if the matrix $(I - B^{-1}A)$ is "small" (in some matrix norm), then the errors will decay as the iterations progress.[20]   But this matrix will be small if $B^{-1}A$ is "close" to I, which in turn will happen if B^{-1} is "close" to A^{-1} and this translates to B being close to A.

Table 7.1 summarizes the form of the matrix B for each of our three iteration schemes introduced earlier. We leave it as an exercise to show that with the matrices given in Table 7.1, (61) indeed is equivalent to each of the three iteration schemes presented earlier (Exercise 20).

TABLE 7.1: Summary of matrix formulations of each of the three iteration schemes: Jacobi, Gauss-Seidel, and SOR.

Iteration Scheme:	Matrix B in the corresponding formulation (61) $Bx^{(k+1)} = (B - A)x^{(k)} + b$ in terms of (60) $A = D - L - U$
Jacobi (see formula (57))	$B = D$
Gauss-Seidel (see formula (58))	$B = D - L$
SOR with relaxation parameter ω (see formula (59))	$B = \frac{1}{\omega}D - L$

Thus far we have done only experiments with iterations. Now we turn to some of the theory.

THEOREM 7.10:   (*Convergence Theorem*) Assume that A is a nonsingular (square) matrix, and that B is any nonsingular matrix of the same size as A. The (real and complex) eigenvalues of the matrix $I - B^{-1}A$ all have absolute value less than one if and only if the iteration scheme (61) converges (to the solution of $Ax = b$) for any initial seed vector $x^{(0)}$.

---

[20] It is helpful to think of the one-dimensional case, where everything in (62) is a number. If $(I - B^{-1}A)$ is less than one in absolute value, then we have exponential decay, and furthermore, the decay is faster when absolute values of the matrix are smaller. This idea can be made to carry over to matrix situations. The corresponding needed fact is that all of the eigenvalues (real or complex) of the matrix $(I - B^{-1}A)$ are less than one in absolute value. In this case, it can be shown that we have exponential decay also for the iterative scheme, regardless of the initial iterate. For complete details, we refer to [Atk-89] or to [GoVL-83].

For a proof of this and the subsequent theorems in this section, we refer the interested reader to the references [Atk-89] or to [GoVL-83]. MATLAB's `eig` function is designed to produce the eigenvalues of a matrix. Since, as we have pointed out (and seen in examples), the Gauss-Seidel iteration usually converges faster than the Jacobi iteration, it is not surprising that there are examples where the former will converge but not the latter (see Exercise 8). It turns out that there are examples where the Jacobi iteration will converge even though the Gauss-Seidel iteration will fail to converge.

EXAMPLE 7.29: Using MATLAB's `eig` function for finding eigenvalues of a matrix, apply Theorem 7.10 to check to see if it tells us that the linear system of Example 7.26 will always lead to a convergent iteration method with each of the three schemes: Jacobi, Gauss-Seidel, and SOR with $\omega = 0.9$. Then create a plot of the maximum absolute value of the eigenvalues of $(I - B^{-1}A)$ for the SOR method as ω ranges from 0.05 to 1.95 in increments of 0.5, and interpret.

SOLUTION: We enter the relevant matrices into a MATLAB session:

```
>> A=[3 1 -1;4 -10 1;2 1 5];   D=diag(diag(A));
>> L=[0 0 0;-4 0 0;-2 -1 0];   U=D-L-A;  I = eye(3);
>> % Jacob:
>> max(abs(eig(I-inv(D)*A)))        →ans = 0.5374
>> % Gauss Seidel
>> max(abs(eig(I-inv(D-L)*A)))      →ans = 0.3513
>>    SOR omega - 0.9
>> max(abs(eig(I-inv(D/.9-L)*A)))  →ans = 0.1301
```

The three computations give the maximum absolute values of the eigenvalues of $(I - B^{-1}A)$ for each of the three iteration methods. Each maximum is less than one, so the theorem tells us that, whatever initial iteration vector we choose for any of the three schemes, the iterations will always converge to the solution of $Ax = b$. Note that the faster converging methods tend to correspond to smaller maximum absolute values of the eigenvalues of $(I - B^{-1}A)$; we will see a corroboration of this in the next part of this solution.

The next set of commands produces a plot of the maximum value of the eigenvalues of $(I - B^{-1}A)$ for the various values of ω, which is shown in Figure 7.42.

```
>> omega=0.05:.05:1.95;
>> for i=1:length(omega)
   rad(i)=max(abs(eig(I-...
   inv(D/omega(i)-L)*A)));
end
>> plot(omega, rad, 'o-')
```

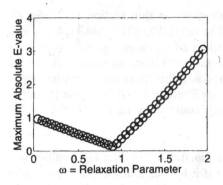

FIGURE 7.42: Illustration of the maximum absolute value of the eigenvalues of the matrix $I - B^{-1}A$ of Theorem 7.10 for the SOR method (see Table 7.1) for various values of the relaxation parameter ω. Compare with the corresponding number of iterations needed for convergence (Figure 7.40).

The above theorem is quite universal in that it applies to all situations. The drawback is that it relies on the determination of eigenvalues of a matrix. This eigenvalue problem can be quite a difficult numerical problem, especially for very large matrices (the type that we would want to apply iteration methods to). MATLAB's `eig` function may perform unacceptably slowly in such cases and/or may produce inaccurate results. Thus, the theorem has limited practical value. We next give a more practical result that gives a sufficient condition for convergence of both the Jacobi and Gauss-Seidel iterative methods.

Recall that an $n \times n$ matrix A is **strictly diagonally dominant (by rows)** if the absolute value of each diagonal entry is greater than the sum of the absolute values of all other entries in the same row, i.e.,

$$|a_{ii}| > \sum_{\substack{j=1 \\ j \neq i}}^{n} |a_{ij}|, \text{ for } i = 1, 2, ..., n.$$

THEOREM 7.11: (*Jacobi and Gauss-Seidel Convergence Theorem*) Assume that A is a nonsingular (square matrix). If A is strictly diagonally dominant, then the Jacobi and Gauss-Seidel iterations will converge (to the solution of $Ax = b$) for any initial seed vector $x^{(0)}$.

Usually, the more diagonally dominant A is, the faster the rate of convergence will be. Note that the coefficient matrix of Example 7.26 is strictly diagonally dominant, so that Theorem 7.11 tells us that no matter what initial seed vector $x^{(0)}$ we started with, both the Jacobi and Gauss-Seidel iteration schemes would produce a sequence that converges to the solution. Although we knew this already from Theorem 7.10, note that, unlike the eigenvalue condition, the strict diagonal dominance was trivial to verify (by inspection). There are matrices A that are not strictly diagonally dominant but for which both Jacobi and Gauss-Seidel schemes will always converge. For an outline of a proof of the Jacobi part of the above result, see Exercise 23.

For the SOR method (for other values of ω than 1) there does not seem to be such a simple useful criterion. There is, however, another equivalent condition in the case of a symmetric coefficient matrix A with positive diagonal entries.

Recall that an $n \times n$ matrix A is symmetric if $A = A'$; also A is **positive definite** provided that $x'Ax > 0$ for any nonzero $n \times 1$ vector x.

THEOREM 7.12: (*SOR Convergence Theorem*) Assume that A is a nonsingular (square matrix). Assume that A is symmetric and has positive diagonal entries. For any choice of relaxation parameter $0 < \omega < 2$, the SOR iteration will converge (to the solution of $Ax = b$) for any initial seed vector $x^{(0)}$, if and only if A is positive definite.

Matrices that satisfy the hypotheses of Theorems 7.11 and 7.12 are actually quite common in numerical solutions of differential equations. We give a typical example of such a matrix shortly. For reference, we collect in the following theorem two equivalent formulations for a symmetric matrix to be positive definite, along with some necessary conditions for a matrix to be positive definite. Proofs can be found in [Str-88], p. 331 (for the equivalences) and [BuFa-01], p. 401 (for the necessary conditions).

THEOREM 7.13: (*Positive Definite Matrices*) Suppose that A is a symmetric $n \times n$ matrix.
(a) The following two conditions are each equivalent to A being positive definite:
    (i) All eigenvalues of A are positive, or
    (ii) The determinants of all upper-left submatrices of A have positive determinants.
(b) If A is positive definite, then each of the following conditions must hold:
    (i) A is nonsingular.
    (ii) $a_{ii} > 0$ for $i = 1, 2, \ldots, n$.
    (iii) $a_{ii} a_{jj} > a_{ij}^2$ whenever $i \neq j$.

We will be working next with a certain class of sparse matrices, which is typical of those that arise in finite difference methods for solving partial differential equations. We study such problems and concepts in detail in [Sta-05]; here we only very briefly outline the connection.

The matrix we will analyze arises in solving the so-called Poisson boundary value problem on the two-dimensional unit square $\{(x, y) : 0 \leq x, y \leq 1\}$, which asks for the determination of a function $u = u(x, y)$ that satisfies the following partial differential equation and boundary conditions:

$$\begin{cases} -\Delta u = f(x, y), \text{ inside the square: } 0 < x, y < 1 \\ u(x, y) = 0, \text{ on the boundary: } x = 0, 1 \text{ or } y = 0, 1 \end{cases}$$

Here Δu denotes the Laplace differential operator $\Delta u = u_{xx} + u_{yy}$. The finite difference method "discretizes" the problem into a linear system. If we use the same number N of grid points both on the x- and the y-axis, the linear system $Ax = b$ that arises will have the $N^2 \times N^2$ coefficient matrix shown in (63).

In our notation, the partition lines break the $N^2 \times N^2$ matrix up into smaller $N \times N$ block matrices (N^2 of them). The only entries indicated are the nonzero entries that occur on the five diagonals shown. Because of their importance in applications, matrices such as the one in (63) have been extensively studied in the context of iterative methods. For example, the following result contains some very practical and interesting results about this matrix.

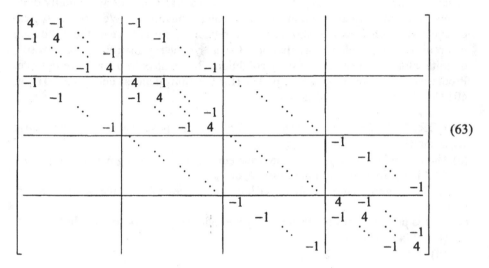

$$\text{(63)}$$

PROPOSITION 7.14: Let A be the $N^2 \times N^2$ matrix (63).

(a) A is positive definite (so SOR will converge by Theorem 7.12) and the optimal relaxation parameter ω for an SOR iteration scheme for a linear system $Ax = b$ is as follows:

$$\omega = \frac{2}{1 + \sin\left(\frac{\pi}{N+1}\right)} \, .$$

(b) With this optimal relaxation parameter, the SOR iteration scheme works on order of N times as fast as either the Jacobi or Gauss-Seidel iteration schemes. More precisely, the following quantities R_J, R_{GS}, R_{SOR} indicate the approximate number of iterations that each of these three schemes would need, respectively, to reduce the error by a factor of $1/10$:

$$R_J \approx 0.467(N+1)^2, \quad R_{GS} = \tfrac{1}{2} R_J \approx 0.234(N+1)^2, \text{ and } R_{SOR} \approx 0.367(N+1).$$

In our next example we compare the different methods by solving a very large fictitious linear system $Ax = b$ involving the matrix (63). This will allow us to make exact error comparisons with the true solution.

A proof of the above proposition, as well as other related results, can be found in Section 8.4 of [StBu-93]. Note that since A is not (quite) strictly diagonally dominant, the Jacobi/Gauss-Seidel convergence theorem (Theorem 7.11) does not apply. It turns out that the Jacobi iteration scheme indeed converges, along with SOR (and, in particular, the Gauss-Seidel method); see Section 8.4 of [StBu-93].

Consider the matrix A shown in (63) with $N = 50$. The matrix A has size 2500×2500 so it has 6.25 million entries. But of these only about $5N^2 = 12,500$ are nonzero. This is about 0.2% of the entries, so A is quite sparse.

EXERCISE FOR THE READER 7.30: Consider the problem of multiplying the matrix A in (63) (using $N = 50$) by the vector $x = [1\ 2\ 1\ 2\ 1\ 2\ \cdots\ 1\ 2]'$.

(a) Compute (by hand) the vector $b \equiv Ax$ by noticing the patterns present in the multiplication.

(b) Get MATLAB to compute $b = Ax$ by first creating and storing the matrices A and x and performing a usual matrix multiplication. Use `tic/toc` to time the parts of this computation.

(c) Store only the five nonzero diagonals of A (as column vectors): d, a1, aN, b1, bN (d stands for main diagonal, a for above-main diagonal, b for below-main diagonal). Recompute b by suitably manipulating these 5 vectors in conjunction with x Use `tic/toc` to time the computation and compare with that in part (b).

(d) Compare all three answers. What happens to the three methods if we bump N up to 100?

Shortly we will give a general development on the approach hinted at in part (c) of the above Exercise for the Reader 7.30.

As long as the coefficient matrix A is not too large to be stored in a session, MATLAB's left divide is quite an intelligent linear system solver. It has special more advanced algorithms to deal with positive definite coefficient matrices, as well as with other special types of matrices. It can numerically solve systems about as large as can be stored; but the accuracy of the numerical solutions obtained depends on the condition number of the matrix, as explained earlier in this chapter.[21] The next example shows that even with all that we know about the

---

[21] Depending on the power of the computer on which you are running MATLAB's as well as the other processes being run, computation times and storage capacities can vary. At the time of writing this section on the author's 1.6 MHz, 256 MB RAM Pentium IV PC, some typical limits, for random (dense) matrices, are as follows: The basic Gaussian elimination (Program 7.6) starts taking too long (toward an hour) when the size of the coefficient matrix gets larger than 600×600; for it to take less than about one minute the size should be less than about 250×250. Before memory runs out, on the other hand, matrices of sizes up to about 6000×6000 can be stored, and MATLAB's left divide can usually (numerically) solve them in a reasonable amount of time (provided that the condition number is moderate). To avoid redundant storage problems, MATLAB does have capabilities of storing sparse matrices. Such functionality introduced at the end of this section. Taking advantage of the structure of sparse banded matrices (which are the most important ones in numerical differential equations) will enable us to solve many such linear systems that are quite large, say up to about $50,000 \times 50,000$. Such large systems often come up naturally in numerical differential equations.

optimal relaxation parameter for the special matrix (63), MATLAB's powerful left divide will still work more efficiently for a very large linear system than our SOR program. After the example we will remedy the situation by modifying the SOR program to make it more efficient for such banded sparse matrices.

EXAMPLE 7.30: In this example we do some comparisons in some trial runs of solving a linear system $Ax = b$ where A is the matrix of (63) with $N = 50$, and the vectors x and b are as in the preceding Exercise for the Reader 7.30. Having the exact solution will allow us to look at the exact errors resulting from any of the methods.

(a) Solve the linear system by using MATLAB's left divide (Gaussian elimination). Record the computation time and error of the computed solution.

(b) Solve the system using the Gauss-Seidel program gaussseidel (Program 7.7) with the default number of iterations and initial vector. Record the computation time and error. Repeat using 200 iterations.

(c) Solve again using the SOR program sorit (from Exercise for the Reader 7.28) with the optimal relaxation parameter ω given in Proposition 7.14. Record the computation time and error. Repeat using 200 iterations.

(d) Reconcile the data of parts (b) and (c) with the results of part (c) of Proposition 7.14.

SOLUTION: We first create and store the relevant matrices and vectors:

```
>> x=ones(2500,1); x(2:2:2500,1)=2;
>> A=4*eye(2500);
>> v1=-1*ones(49,1); v1=[v1;0]; %seed vector for sub/super diagonals
>> secdiag=v1;
>> for i=1:49
     if i<49
        secdiag=[secdiag;v1];
     else
        secdiag=[secdiag;v1(1:49)];
     end
end

>> A=A+diag(secdiag,1)+diag(secdiag,-1)-diag(ones(2450,1),50)...
             -diag(ones(2450,1),-50);
>> b=A*x;
```

Part (a):
```
>> tic, xMATLAB=A\b; toc      →elapsed_time = 9.2180
>> max(xMATLAB-x)             →ans = 6.2172e-015
```

Part (b):
```
>> tic, [xGS, k, diff]=gaussseidel(A,b); toc
→Gauss-Seidel iteration failed to converge.→elapsed_time = 181.6090
>> max(abs(xGS-x))                         →ans = 1.4353

>> tic, [xGS2, k, diff] = gaussseidel(A,b,zeros(size(b)), 1e-10,200);
toc
→Gauss-Seidel iteration failed to converge.→elapsed_time = 374.5780
>> max(abs(xGS2-x)) →ans = 1.1027
```

Part (c):
```
>> tic, [xSOR, k, diff]=sorit(A,b, 2/(1+sin(pi/51))); toc
→SOR iteration failed to converge.   →elapsed_time = 186.7340
>> max(abs(xSOR-x))            →ans = 0.0031

>> tic, [xSOR2, k, diff]=sorit(A,b, 2/(1+sin(pi/51))),...
             zeros(size(b)), 1e-10,200); toc
→SOR iteration failed to converge.   →elapsed_time = 375.2650
>> max(abs(xSOR2-x))           →ans = 1.1885e - 008
```

Part (d): The above data shows that our iteration programs pale in performance when compared with MATLAB's left divide (both in time and accuracy). The attentive reader will realize that both iteration programs do not take advantage of the sparseness of the matrix (63). They basically run through all of the entries in this large matrix for each iteration. One other thing that should be pointed out is that the above comparisons are somewhat unfair because MATAB's left divide is a compiled code (built-into the system), whereas the other programs were interpreted codes (created as external programs—M-files). After this example, we will show a way to make these programs perform more efficiently by taking advantage of the special banded structure of such matrices. The resulting modified programs will then perform more efficiently than MATLAB's left divide, at least for the linear system of this example.

Using $N = 50$ in part (b) of Proposition 7.14, we see that in order to cut errors by a factor of 10 with the Gauss-Seidel method, we need approximately $0.234 \cdot 51^2 \approx 609$ additional iterations, but for the SOR with optimal relaxation parameter the corresponding number is only $.367 \cdot 51 \approx 18.7$. This corroborates well with the experimental data in parts (b) and (c) above. For Gauss-Seidel, we first used 100 iterations and then used 200. The theory tells us we would need over 600 more iterations to reduce the error by 90%. Using 100 more iterations resulted in a reduction of error by about 23%. On the other hand, with SOR, the 100 additional iterations gave us approximately $100/18.7 \approx 5.3$ reductions in the errors each by factors of 1/10, which corresponds nicely to the (exact) error shrinking from about 3e-3 to 1e-8 (literally, about 5.3 decimal places!).

In order to take advantage of sparsely banded matrices in our iteration algorithms, we next record here some elementary observations regarding multiplications of such matrices by vectors. MATLAB is enabled with features to easily manipulate and store such matrices. We will now explore some of the underlying concepts and show how exploiting the special structure of some sparse matrices can greatly expand the sizes of linear systems that can be effectively solve. MATLAB has its own capability for storing and manipulating general sparse matrices; at the end of the section we will discuss how this works.

The following (nonstandard) mathematical notations will be convenient for the present purpose: For two vectors v, w that are both either row- or column-vectors, we let $v \otimes w$ denote their juxtaposition. For the pointwise product of two vectors of the same size, we use the notation $v \odot w = [v_i w_i]$. So, for example, if $v = [v_1 \; v_2 \; v_3]$, and $w = [w_1 \; w_2 \; w_3]$ are both 3×1 row vectors, then $v \otimes w$ is the 6×1 row vector $[v_1 \; v_2 \; v_3 \; w_1 \; w_2 \; w_3]$ and $v \odot w$ is the 3×1 row vector $[v_1 w_1 \; v_2 w_2 \; v_3 w_3]$. We also use the notation $\mathbf{0}_n$ to denote the zero vector with n components. We will be using this last notation only in the context of juxtapositions so that whether $\mathbf{0}_n$ is meant to be a row vector or a column vector will be clear from the context.[22]

LEMMA 7.15: (a) Let S be an $n \times n$ matrix whose kth superdiagonal ($k \geq 1$) is made of the entries (in order) of the $(n-k) \times 1$ vector v, i.e.,

$$S = \begin{bmatrix} 0 & 0 & \cdots & 0 & v_1 & 0 & & \cdots & 0 \\ 0 & 0 & 0 & \cdots & 0 & v_2 & 0 & & \vdots \\ \vdots & \vdots & 0 & \ddots & \cdots & 0 & v_3 & \ddots & 0 \\ & & \vdots & 0 & & & \ddots & \ddots & 0 \\ & & & & \ddots & & & 0 & v_{n-k} \\ \vdots & & & & & \ddots & & & 0 \\ 0 & \vdots & \vdots & & & & 0 & 0 & 0 \\ 0 & 0 & 0 & 0 & \cdots & \cdots & 0 & 0 & 0 \end{bmatrix},$$

where $v = [v_1 \; v_2 \; \cdots \; v_{n-k}]$. (In MATLAB's notation, the matrix S could be entered as `diag(v,k)`, once the vector v has been stored.) Let $x = [x_1 \; x_2 \; x_3 \; \cdots \; x_n]'$ be any $n \times 1$ column vector. The following relation then holds:

$$Sx = (v \otimes \mathbf{0}_k) \odot ([x_{k+1} \; x_{k+2} \; \cdots \; x_n] \otimes \mathbf{0}_k). \tag{64}$$

(b) Analogously, if S is an $n \times n$ matrix whose kth subdiagonal ($k \geq 1$) is made of the entries (in order) of the $(n-k) \times 1$ vector v, i.e.,

$$S = \begin{bmatrix} 0 & 0 & \cdots & & & & \cdots & 0 & 0 \\ \vdots & 0 & 0 & \cdots & & & & & 0 \\ 0 & \vdots & 0 & 0 & \cdots & & & & \vdots \\ v_1 & 0 & \vdots & 0 & 0 & \cdots & & & \\ 0 & v_2 & 0 & & \ddots & \ddots & & & \\ \vdots & 0 & v_3 & 0 & & \ddots & \ddots & & \\ 0 & \vdots & \vdots & \ddots & \ddots & & \ddots & \ddots & \vdots \\ 0 & \vdots & \vdots & 0 & v_{n-k-1} & 0 & 0 & \cdots & 0 \\ 0 & 0 & 0 & 0 & 0 & v_{n-k} & 0 & \cdots & 0 \end{bmatrix},$$

---

[22] We point out that these notations are not standard. The symbol $\otimes$ is usually reserved for the so-called *tensor product*.

where $v = [v_1 \ v_2 \ \cdots \ v_{n-k}]$ and $x = [x_1 \ x_2 \ x_3 \ \cdots \ x_n]'$ is any $n \times 1$ column vector, then

$$Sx = (\mathbf{0}_k \otimes v) \odot (\mathbf{0}_k \otimes [x_1 \ x_2 \ \cdots \ x_k]). \tag{65}$$

The proof of the lemma is left as Exercise 21. The lemma can be easily applied to greatly streamline all of our iteration programs for sparse banded matrices. The next exercise for the reader will ask the reader to perform this task for the SOR iteration scheme.

EXERCISE FOR THE READER 7.31:   (a) Write a function M-file with the following syntax:

```
[x,k,diff] = sorsparsediag(diags, inds, b, omega, x0, tol, kmax)
```

that will perform the SOR iteration to solve a nonsingular linear system $Ax = b$ in which the coefficient matrix A has entries only on a sparse set of diagonals. The first two input variables are `diags`, an $n \times j$ matrix where each column consists of the entries of A on one of its diagonals (with extra entries at the end of the column being zeros), and `inds`, a $1 \times j$ vector of the corresponding set of indices for the diagonals (index zero corresponds to the main diagonal and should be first). The remaining input and output variables will be exactly as in the M-file `sorit` of Exercise for the Reader 7.28.   The program should function just like the `sorit` M-file, with the only exceptions being that the stopping criterion for the norms of the difference of successive iterates should now be determined by the infinity norm[23] and the default number of iterations is now 1000. The algorithm should, of course, be based on formula (59) for the SOR iteration, but the sum need only be computed over the index set (`inds`) of the nonzero diagonals. To this end, the above lemma should be used in creating this M-file so that it will avoid unnecessary computations (with zero multiplications) as well as storage problems with large matrices.
(b) Apply the program to redo part (c) of Example 7.30.

EXAMPLE 7.31:   (a) Invoke the M-file `sorsparsediag` of the preceding exercise for the reader to obtain SOR numerical solutions of the linear system of Example 7.30 with error goal 5e-15 (roughly MATLAB's machine epsilon) and compare the necessary runtime with that of MATLAB's left divide, which was recorded in that example.
(b) Next use the program to solve the linear system $Ax = b$ with A as in (63) with $N = 300$ and $b = [1 \ 2 \ 1 \ 2 \ ... \ 1 \ 2]'$. Use the default tolerance 1e-10, then, looking at the last norm difference (estimate for the actual error) use Proposition 7.14 to help to see how much to increase the maximum number of iterations to ensure convergence of the method.   Record the runtimes.   The size of A is

---

[23] The infinity norm of a vector x is simply, in MATLAB's notation, `max(abs(x))`. This is a rather superficial change in the M-file, merely to allow easier performance comparisons with MATLAB's left divide system solver.

$90,000 \times 90,000$ and so it has over 8 billion entries. Storage of such a matrix would require a supercomputer.

SOLUTION: Part (a): We first need to store the appropriate data for the matrix A. Assuming the variables created in the last example are still in our workspace, this can be accomplished as follows:

```
>> diags=zeros(2500,5);
>> diags(:,1)=4*ones(2500,1);
>> diags(1:2499,2:3)=[secdiag secdiag];
>> diags(1:2450,4:5)=-ones(2450,2);
>> inds=[0 1 -1 50 -50];

>> tic, [xSOR, k, diff]=sorsparsediag(diags, inds,b,...
              2/(1+sin(pi/51)), zeros(size(b)), 5e-15); toc
→SOR iteration has converged in 308 iterations  →elapsed_time = 1.3600
>> max(abs(xSOR-x))                             →ans = 2.3537e - 014
```

Our answer is quite close to machine precision (there were roundoff errors) and the answer obtained by MATLAB's left divide. The runtime of the modified SOR program is now, however, significantly smaller than that of the MATLAB solver. We will see later, however, that when we store the matrix A as a sparse matrix (in MATLAB's syntax), the left divide method will work at comparable speed to our modified SOR program.

Part (b): We first create the input data by suitably modifying the code in part (a):

```
>> b=ones(90000,1); b(2:2:90000,1)=2;
>> v1=-1*ones(299,1); v1=[v1;0];  %seed vector for sub/super diagonals
>> secdiag=v1;
>> for i=1:299
     if i<299
        secdiag=[secdiag;v1];
     else
        secdiag=[secdiag;v1(1:299)];
     end
   end
>> diags=zeros(90000,5);
>> diags(:,1)=4*ones(90000,1);
>> diags(1:89999,2:3)=[secdiag secdiag];
>> diags(1:89700,4:5)= [-ones(89700,1) -ones(89700,1)];
>> inds=[0 1 -1 300 -300];
>>tic, [xSORbig, k, diff]=sorsparsediag(diags, inds,b,...
              2/(1+sin(pi/301))); toc
→SOR iteration failed to converge.   →elapsed_time = 167.0320
>> diff(k-1)                         →ans = 1.3845e - 005
```

We need to reduce the current error by a factor of 1e-5. By Proposition 7.14, this means that we should bump up the number of iterations by a bit more than $5R_{SOR} \approx 5 \cdot 0.367 \cdot 301 \approx 552$. Resetting the default number of iterations to be 1750 (from 1000) should be sufficient. Here is what transpires:

```
>> tic, [xSORbig, k, diff]=sorsparsediag(diags, inds,b,...
              2/(1+sin(pi/301)), zeros(size(b)), 1e-10, 1750); toc
```

We have thus solved this extremely large linear system, and it only took about three minutes!

As promised, we now give a brief synopsis of some of MATLAB's built-in, state-of-the-art iterative solvers for linear systems $Ax = b$. The methods are based on more advanced concepts that we briefly indicated and referenced earlier in the section. Mathematical explanations of how these methods work would lie outside the focus of this book. We do, however, outline the basic concept of **preconditioning**. As seen early in this section, iterative methods are very sensitive to the particular form of the coefficient matrix (we gave an example where simply switching two rows of A resulted in the iterative method diverging when it originally converged). An invertible matrix (usually positive definite) M is used to precondition our linear system when we apply the iterative method instead to the equivalent system: $M^{-1}Ax = M^{-1}b$. Often, preconditioning a system can make it more suitable for iterative methods. For details on the practice and theory of preconditioning, we refer to Part II of [Gre-97], which includes, in particular, preconditioning techniques appropriate for matrices that arise in solving numerical PDEs. See also Part IV of [TrBa-97].

Here is a detailed description of MATLAB's function for the so-called **preconditioned conjugate gradient method**, which assumes that the coefficient matrix A is symmetric positive definite.

x=pcg(A,b,tol,kmax,M1,M2,x0) →	Performs the preconditioned gradient method to solve the linear system $Ax = b$, where the $N \times N$ coefficient matrix A must be symmetric positive definite and the preconditioner $M = M1*M2$. Only the first two input variables are required; any tail sequence of input variables can be omitted. The default values of the optional variables are as follows: tol = 1e 6, kmax = min(N,20), M1 = M2 = I (identity matrix), and x0 = the zero vector. Setting any of these optional input variables equal to [ ] gives them their default values.
[x, flag] =pcg(A,b,tol,kmax,M1,M2,x0) →	Works as above but returns additional output flag: flag = 0 means pcg converged to the desired tolerance tol within kmax iterations; flag = 1 means pcg iterated kmax times but did not converge. For a detailed explanation of other flag values, type help pcg.

We point out that with the default values $M1 = M2 = I$, there is no conditioning and the method is called the conjugate gradient method.

Another powerful and and more versatile method is the **generalized minimum residual method (GMRES)**. This method works well for general (nonsymmetric)

linear systems.   MATLAB's syntax for this function is similar to the above, but
there is one additional (optional) input variable:

x=gmres(A,b,restart,tol, kmax,M1,M2,x0) →	Performs the generalized minimum residual method to solve the linear system $Ax = b$, with preconditioner M = M1*M2. Only the first two input variables are required; any tail sequence of input variables can be omitted. The default values of the optional variables are as follows: restart = [ ] (unrestarted method) tol = 1e-6, kmax = min(N,20), M1 = M2 = I (identity matrix), and $x0$ = the zero vector. Setting any of these optional input variables equal to [ ], gives them their default values. An optional second output variable flag will function in a similar fashion as with pcg.

EXAMPLE 7.32:   (a) Use pcg to resolve the linear system of Example 7.30, with
the default settings and flag.   Repeat by resetting the tolerance at 1e-15 and the
maximum number of iterations to be 100 and then 200.   Record the runtimes and
compare these and the errors to the results for the SOR program of the previous
example.
(b) Repeat part (a) with gmres.

SOLUTION: Assume that the matrices and vectors remain in our workspace (or
recreate them now if necessary). We need only follow the above syntax
instructions for pcg:

Part (a):
```
>> tic, [xpcg, flagpcg]=pcg(A,b); toc     →elapsed_time = 3.2810
>> max(abs(xpcg-x))                        →ans = 1.5007
>> flagpcg                                 →flagpcg = 1
```

The flag being = 1 means after 20 iterations, pcg did not converge within tolerance
(1e-5), a fact that we knew from the exact error estimate.

```
>> tic, [xpcg, flagpcg]=pcg(A,b,5e-15, 100); toc →elapsed_time = 15.8900
>> max(abs(xpcg-x))                               →ans = 4.5816e - 006
>> flagpcg                                        →flagpcg = 1
```

```
>> tic, [xpcg, flagpcg]=pcg(A,b,5e-15, 200); toc →elapsed_time = 29.7970
>> flagpcg                                        →flagpcg = 0
>> max(abs(xpcg-x))                               →ans = 3.2419e - 014
```

The flag being = 0 in this last run shows we have convergence. The max norm is a
different one from the 2-norm used in the M-file; hence the slight discrepancy.
Notice the unconditioned conjugate gradient method converged in fewer iterations
than did the optimal SOR method, and in much less time than the original sorit
program.   The more efficient sorsparsediag program, however, got the
solution in by far the shortest amount of real time.   Later, we will get a more
equitable comparison when we store A as a sparse matrix.

Part (b):
```
>> tic, [xgmres, flaggmres]=gmres(A,b, [],[], 200); toc
>> tic, [xgmres, flaggmres]=gmres(A,b); toc    →elapsed_time = 2.3280
>> max(abs(xgmres-x))                            →ans = 1.5002
>> flaggmres                                     →flaggmres = 1

>> tic, [xgmres, flaggmres]=gmres(A,b, [],5e-15, 100); toc
→elapsed_time = 17.2820
>> max(abs(xgmres-x))    →ans = 6.9104e - 006

>> tic, [xgmres, flaggmres]=gmres(A,b, [],5e-15, 200); toc
→elapsed_time = 37.1250
>> max(abs(xgmres-x))    →ans = 9.2037e - 013
```

The results for GMRES compare well with those for the preconditioned conjugate gradient method. The former method converges a bit more slowly in this situation. We remind the reader that the conjugate gradient method is ideally suited for positive definite matrices, like the one we are dealing with.

Figure 7.43 gives a nice graphical comparison of the relative speeds of convergence of the five iteration methods that have been introduced in this section. An exercise will ask the reader to reconstruct this MATLAB graphic.

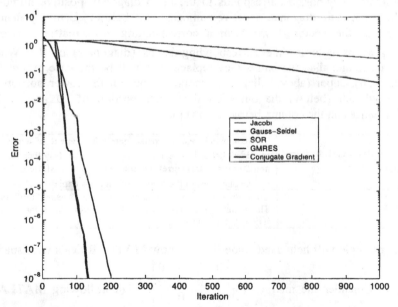

FIGURE 7.43: Comparison of the convergence speed of the various iteration methods in the solution of the linear system $Ax = b$ of Example 7.30 where the matrix A is the matrix (63) of size 2500×2500. In the SOR method the optimal relaxation parameter of Proposition 7.14 was used. Since we did not invoke any conditioning, the preconditioned conjugate gradient method is simply referred to as the conjugate gradient method. The errors were measured in the infinity norm.

In the construction of the above data, the program `sorsparsediag` was used to get the SOR data and, despite the larger number of iterations than GMRES and the conjugate gradient method, the SOR data was computed more quickly. The `sorsparsediag` program is easily modified to construct similar programs for the Jacobi and Gauss-Seidel iterations (of course Gauss-Seidel could simply be done by setting $\omega = 0$ in the SOR program), and such programs were used to get the data for these iterations. Note that the GMRES and conjugate gradient methods take several iterations before errors start to decrease, unlike the SOR method, but they soon catch up. Note also the comparable efficiencies between the GMRES and conjugate gradient methods.

We close this chapter with a brief discussion of how to store and manipulate sparse matrices directly with MATLAB. Sparse matrices in MATLAB can be stored using three vectors: one for the nonzero entries, the other two for the corresponding row and column indices. Since in many applications sparse matrices will be banded, we will explain only a few commands useful for the creation and storage of such sparse matrices. Enter `help sparse` for more detailed information. To this end, suppose that we have an $n \times n$ banded matrix A and we wish to store it as a sparse matrix. Let the indices corresponding to the nonzero bands (diagonals) of A have numbers stored in a vector d (so the size of d is the number of bands, 0 corresponds to the main diagonal, positive numbers mean above the main diagonal, negative numbers mean below). Letting p denote the length of the vector d we form a corresponding $n \times p$ matrix, Diags, containing as its columns the corresponding bands (diagonals) of A. When columns are longer than the bands they replace (this will be the case except for main diagonal), super (above) diagonals should be put on the lower portion of Diags and sub (below) diagonals on the upper portion of Diags, with remaining entries on the column being set to zero.

`S=spdiags(Diags,d,n,n)` →	This command creates a sparse matrix data type S, of size $n \times n$ provided that d is a vector of diagonal indices (say there are p), and Diags is the corresponding $n \times p$ matrix whose columns are the diagonals of the matrix (arranged as explained above).
`full(S)` →	Converts a sparse matrix data type back to its usual "full" form. This command is rarely used in dealing with sparse matrices as it defeats their purpose.

A simple example will help shed some light on how MATLAB deals with sparse data types. Consider the matrix $A = \begin{bmatrix} 0 & 1 & 0 & 0 \\ 4 & 0 & 2 & 0 \\ 0 & 5 & 0 & 3 \\ 0 & 0 & 6 & 0 \end{bmatrix}$. The following MATLAB commands will store A as a sparse matrix:

```
>> d=[-1 1];  Diags=[4 5 6 0; 0 1 2 3]';
>> S=spdiags(Diags,d,4,4)
→S =           (2,1)    4       (2,3)    2
               (1,2)    1       (4,3)    6
               (3,2)    5       (3,4)    3
```

The display shows the storage scheme. Let's compare with the usual form:

```
>> full(S)
→ ans =       0   1   0   0
              4   0   2   0
              0   5   0   3
              0   0   6   0
```

   The key advantage of sparse matrix storage in MATLAB is that if A is stored as a sparse matrix S, then to solve a linear system $Ax = b$, MATLAB's left divide operation x=S\b takes advantage of sparsity and can greatly increase the size of (sparse) problems we can solve. In fact, at most all of MATLAB's matrix functions are able to operate on sparse matrix data types. This includes MATLAB's iterative solvers pcg, gmres, etc. We invite the interested reader to perform some experiments and discover the additional speed and capacity that taking advantage of sparsity can afford. We end with an example of a rematch of MATLAB's left divide against our sorsparsediag program, this time allowing the left divide method to accept a sparse matrix. The results will be quite illuminating.

EXAMPLE 7.33: We examine the large $(10,000 \times 10,000)$ system $Ax = b$, where A is given by (63) with $N = 100$, and $x = (1\ 1\ 1\ \cdots\ 1)'$. By examining the matrix multiplication we see that

$$b = Ax = (2\ 1\ 1\ \cdots 1\ 2|1\ 0\ 0\ \cdots 0\ 1|\cdots|1\ 0\ 0\ \cdots 0\ 1|2\ 1\ 1\ \cdots 1\ 2)'.$$

We thus have a linear system with which we can easily obtain the exact error of any approximate solution.
(a) Solve this system using MATLAB's left divide and by storing A as a sparse matrix. Use tic/toc to track the computation time (on your computer); compute the error as measured by the infinity norm (i.e., as the maximum difference of any component of the computed solution with the exact solution).
(b) Solve the system using the sorsparsediag M-file of Exercise for the Reader 7.31. Compute the time and errors as in part (a) and compare.

SOLUTION: Part (a): We begin by entering the parameters (for (63)), creating the needed inputs for spdiags, and then using the latter to store A as a sparse matrix.

```
>> N=200; n=N^2; d=[-N -1 0 1 N];, dia=4*ones(1,n);
>> seed1=-1*ones(1,N-1); v1=[seed1 0];
for i=1:N-1, if i<N-1, v1 = [v1 [seed1 0]];, else, v1 = [v1 seed1];
>> end, end
>> b1=[v1 0]; a1=[0 v1]; %below/above 1 unit diagonals
>> %Next here are the below/above N unit diagonals
>> bN=[-ones(1,n-N)   zeros(1,N)];
>> aN=[zeros(1,N)   -ones(1,n-N)  ];
>> %Now we can form the n by 5 Diags matrix.
>> Diags=[bN; b1; dia; a1; aN]';
>> S=spdiags(Diags,d,n,n); %S is the sparsely stored matrix A
```

```
>> %We use a simple iteration to contruct the inhomogeneity
>> %vector b.
>> bseed1=ones(1,N);, bseed1([1 N])=[2 2]; %2 end pieces
>> bseed2=bseed1-ones(1,N); %N-2 middle pieces
>> b=bseed1; for k=1:N-2, b=[b bseed2];, end, b=[b bseed1];
>> b=b';
>> tic, xLD=S\b;, toc        →Elapsed time is 0.250000 seconds.
>> x=ones(size(xLD));
>> max(x-xLD)                → ans =1.0947e-013 (Exact Error)
```

Part (b): The syntax and creation of input variables is just as we did in Example 7.31.

```
>> d=[0 -N N -1 1];,  diags=zeros(n,5);
>> diags(:,1)=dia; diags(:,2:3)=[bN' bN']; diags(:,4:5)=[b1' b1'];
>> tic, [xSOR, k, diff]=sorsparsediag(diags, d,b,...
            2/(1+sin(pi/101))); toc
→Elapsed time is 8.734000 seconds.
>> max(x-xSOR)         → 3.9102e-012 (Exact Error)
```

Thus, now that the inputted data structures are similar, MATLAB's left divide has transcended our sorsparsediags program both in performance time and in accuracy. The reader is invited to perform further experiments with sparse matrices and MATLAB's iterative solvers.

EXERCISES 7.7

1. For each of the following data for a linear system $Ax = b$, perform the following iterations using the zero vector as the initial vector.
    (a) Use Jacobi iteration until the error (as measured by the infinity norm of the difference of successive iterates) is less than 1e-10, if this is possible. In cases where the iteration does not converge, try rearranging the rows of the matrices to attain convergence (through all $n!$ rearrangements, if necessary). Find the norm of the exact error (use MATLAB's left divide to get the "exact" solutions of these small systems).
    (b) Repeat part (a) with the Gauss-Seidel iteration.

    (i) $A = \begin{bmatrix} 6 & -1 \\ -1 & 6 \end{bmatrix}$, $b = \begin{bmatrix} 1 \\ 2 \end{bmatrix}$.   (ii) $A = \begin{bmatrix} 6 & -1 & 0 \\ -1 & 6 & -1 \\ 0 & -1 & 6 \end{bmatrix}$, $b = \begin{bmatrix} 1 \\ 2 \\ 1 \end{bmatrix}$.

    (iii) $A = \begin{bmatrix} -2 & 5 & 4 \\ 6 & 2 & -3 \\ 1 & 1 & -1 \end{bmatrix}$, $b = \begin{bmatrix} 1 \\ 2 \\ 3 \end{bmatrix}$.   (iv) $A = \begin{bmatrix} 2 & 1 & 0 & 0 \\ 2 & 4 & 1 & 0 \\ 0 & 4 & 8 & 2 \\ 0 & 0 & 8 & 16 \end{bmatrix}$, $b = \begin{bmatrix} 4 \\ -2 \\ 1 \\ 3 \end{bmatrix}$.

2. For each of the following data for a linear system $Ax = b$, perform the following iterations using the zero vector as the initial vector. Determine if the Jacobi and Gauss-Seidel iterations converge. In cases of convergence, produce a graph of the errors (as measured by the infinity norm of the difference of successive iterates) versus the number of iterations, that contains both the Jacobi iteration data as well as the Gauss-Seidel data. Let the errors go down to 10^{-10}.

    (i) $A = \begin{bmatrix} 5 & 0 \\ 2 & 4 \end{bmatrix}$, $b = \begin{bmatrix} -1 \\ 3 \end{bmatrix}$.   (ii) $A = \begin{bmatrix} 10 & 2 & -1 \\ 2 & 10 & 2 \\ -1 & 2 & 10 \end{bmatrix}$, $b = \begin{bmatrix} 1 \\ 2 \\ 3 \end{bmatrix}$.

(iii) $A = \begin{bmatrix} 7 & 5 & 4 \\ 3 & 2 & 1 \\ 2 & 8 & 21 \end{bmatrix}$, $b = \begin{bmatrix} 1 \\ 0 \\ 5 \end{bmatrix}$. (iv) $A = \begin{bmatrix} 3 & 1 & 0 & 0 \\ 1 & 9 & 1 & 0 \\ 0 & 1 & 27 & 1 \\ 0 & 0 & 1 & 81 \end{bmatrix}$, $b = \begin{bmatrix} 4 \\ 3 \\ 2 \\ 1 \end{bmatrix}$.

3. (a) For each of the linear systems specified in Exercise 1, run a set of SOR iterations with initial vector the zero vector by letting the relaxation parameter run form 0.05 to 1.95 in increments of 0.5. Use a tolerance of 1e-6, but a maximum of 1000 iterations. Plot the number of iterations versus the relaxation parameter ω.

   (b) Using MATLAB's eig function, let the relaxation parameter ω run through the same range 0 05 to 1.95 in increments of 0.5, and compute the maximum absolute value of the eigenvalues of the matrix $I - B^{-1}A$ where the matrix B is as in Table 7.1 (for the SOR iteration). Create a plot of this maximum versus ω; compare and comment on the relationship with the plot of part (a) and Theorem 7.10.

4. Repeat both parts (a) and (b) for each of the linear systems $Ax = b$ of Exercise 2.

5. For the linear system specified in Exercise 2 (iv), produce graphs of the exact errors of each component of the solution: x_1, x_2, x_3, x_4 as a function of the iteration. Use the zero vector as the initial iterate. Measure the errors as the absolute values of the differences with the corresponding components of the exact solution as determined using MATLAB's left divide. Continue with iterations until the errors are all less than 10^{-10}. Point out any observations.

6. (a) For which of the linear systems specified in Exercise 1(i)–(iv) will the Jacobi iteration converge for all initial iterates?

   (b) For which of the linear systems specified in Exercise 1(i)–(iv) will the Gauss-Seidel iteration converge for all initial iterates?

7. (a) For which of the linear systems specified in Exercise 2(i)–(iv) will the Jacobi iteration converge for all initial iterates?

   (b) For which of the linear systems specified in Exercise 2(i)–(iv) will the Gauss-Seidel iteration converge for all initial iterates?

8. (*An Example Where Gauss-Seidel Iteration Converges, but Jacobi Diverges*) Consider the following linear system:

$$\begin{bmatrix} 5 & 3 & 4 \\ 3 & 6 & 4 \\ 4 & 4 & 5 \end{bmatrix} \begin{bmatrix} x_1 \\ x_2 \\ x_3 \end{bmatrix} = \begin{bmatrix} 12 \\ 13 \\ 13 \end{bmatrix}.$$

   (a) Show that if initial iterate $x^{(0)} = [0\ 0\ 0]'$, the Jacobi iteration converges to the exact solution $x = [1\ 1\ 1]'$. Show that the same holds true if we start with $x^{(0)} = [10\ 8\ -6]'$.

   (b) Show that if initial iterate $x^{(0)} = [0\ 0\ 0]'$, the Gauss-Seidel iteration will diverge. Show that the same holds true if we start with $x^{(0)} = [10\ 8\ -6]'$.

   (c) For what sort of general initial iterates $x^{(0)}$ do the phenomena in parts (a) and (b) continue to hold?

   (d) Show that the coefficient matrix of this system is positive definite. What does the SOR convergence theorem (Theorem 7.12) allow us to conclude?
   Suggestion: For all parts (especially part (c)) you should first do some MATLAB experiments, and then aim to establish the assertions mathematically.

9. (*An Example Where Jacobi Iteration Converges, but Gauss-Seidel Diverges*) Consider the following linear system:

$$\begin{bmatrix} 1 & 2 & -2 \\ 1 & 1 & 1 \\ 2 & 2 & 1 \end{bmatrix} \begin{bmatrix} x_1 \\ x_2 \\ x_3 \end{bmatrix} = \begin{bmatrix} 1 \\ 3 \\ 5 \end{bmatrix}.$$

(a) Show that if initial iterate $x^{(0)} = [0\ 0\ 0]'$, the Jacobi iteration will converge to the exact solution $x = [1\ 1\ 1]'$ in just four iterations. Show that the same holds true if we start with $x^{(0)} = [10\ 8\ -6]'$.

(b) Show that if initial iterate $x^{(0)} = [0\ 0\ 0]'$, the Gauss-Seidel iteration will diverge. Show that the same holds true if we start with $x^{(0)} = [10\ 8\ -6]'$.

(c) For what sort of general initial iterates $x^{(0)}$ do the phenomena in parts (a) and (b) continue to hold?
Suggestion: For all parts (especially part (c)) you should first do some MATLAB experiments, and then aim to establish the assertions mathematically.
Note: This example is due to Collatz [Col-42].

10.   (a) Use the formulas of Lemma 7.15 to write a function M-file with the following syntax:

                  b = sparsediag(diags, inds, x)

The input variables are diags, an $n \times j$ matrix where each column consists of the entries of A on one of its diagonals, and inds, a $1 \times j$ vector of the corresponding set of indices for the diagonals (index zero corresponds to the main diagonal). The last input x is the $n \times 1$ vector to be multiplied by A. The output is the corresponding product $b = Ax$.
(b) Apply this program to check that $x = [1\ 1\ 1]'$ solves the linear system of Exercise 9.
(c) Apply this program to compute the matrix products of Exercise for the Reader 7.30 and check the error against the exact solution obtained in the latter.

11.   (a) Modify the program sorsparsediag of Exercise for the Reader 7.31 to construct an analogous M-file:

        [x,k,diff] = jacbobisparsediag(diags, inds, b, x0, tol, kmax)

for the Jacobi method.
(b) Modify the program sorsparsediag of Exercise for the Reader 7.31 to construct an analogous M-file:

        [x,k,diff]=gaussseidelsparsediag(diags, inds, b, x0, tol, kmax)

for the Gauss-Seidel method.
(c) Apply these programs to recover the results of Examples 7.26 and 7.27.
(d) Using the M-files of parts (a) and (b), along with sorsparsediag, recreate the MATLAB graphic that is shown in Figure 7.43.

12.   (a) Find a 2×2 matrix A whose optimal relaxation parameter ω appears to be greater than 1.5 (as demonstrated by a MATLAB plot like the one in Figure 7.42) resulting from the solution of some linear system $Ax = b$.
(b) Repeat part (a), but this time try to make the optimal value of ω to be less than 0.5.

13.   Repeat both parts of Exercise 12, this time working with 3×3 matrices.

14.   (a) Find a 2×2 matrix A whose optimal relaxation parameter ω appears to be greater than 1.5 (as demonstrated by a MATLAB plot like the one in Figure 7.42) resulting from the solution of some linear system $Ax = b$.
(b) Repeat part (a), but this time try to make the optimal value of ω to be less than 0.5.

15.   (*A Program to Estimate the Optimal SOR Parameter* ω) (a) Write a program that will aim to

find the optimal relaxation parameter ω for the SOR method in the problem of solving a certain linear system $Ax = b$ for which it is assumed that the SOR method will converge. (For example, by the SOR convergence theorem, if A is symmetric positive definite, this program is applicable.) The syntax is as follows:

$$\text{omega} = \text{optimalomega}(A, b, \text{tol}, \text{iter})$$

Of the input and output variables, only the last two input variables need explanation. The input variable tol is simply the accuracy goal that we wish to approximate omega. The variable iter denotes the number of iterations to use on each trial run. The default value for tol is 1e-3 and for iter it is 10. (For very large matrices a larger value may be needed for iter, and likewise for very small matrices a smaller value should be used.) Once this tolerance is met, the program terminates. The program should work as follows. First run through a set of SOR iterations with the values of ω running from 0.05 to 1.95 in increments of 0.05. For each value of ω we run through iter iterations. For each of these we keep track of the infinity norm of the difference of the final iterate and the immediately preceding iterate. For each tested value of $\omega = \omega_0$ for which this norm is minimal, we next run the tests on the values of ω running from $\omega_0 - .05$ to $\omega_0 + .05$ in increments of 0.005 (omit the values $\omega = 0$ or $\omega = 2$ should these occur as endpoints). In the next iteration, we single out those new values of ω for which the new error estimate is minimized. For each new corresponding value $\omega = \omega_0$ for which the norm is minimized, we will next run tests on the set of values from $\omega_0 - .005$ to $\omega_0 + .005$ in increments of 0.0005. At each iteration, the minimizing values of $\omega = \omega_0$ should be unique; if they are not, the program should deliver an error message to this effect, and recommend to try running the program again with a larger value of iter. When the increment size is less than tol, the program terminates and outputs the resulting value of $\omega = \omega_0$.

(b) Apply the above program to aim to determine the optimal value of the SOR parameter ω for the linear system of Example 7.26 with default tolerances. Does the resulting output change if we change iter to 5? To 20?

(c) Repeat part (b), but now change the default tolerance to 1e-6.

(d) Run the SOR iteration on the linear system using the values of the relaxation parameter computed in parts (a) and (b) and compare the rate of convergences with each other and with that seen in the text when $\omega = 0.9$ (Figure 7.41).

(e) Is the program in part (a) practical to run on the large matrix such as the 2500×2500 matrix of Example 7.30 (perhaps using a small value for iter)? If yes, run the program and compare with the result of Proposition 7.14.

16   (*A Program to Estimate the Optimal SOR Parameter ω for Sparse Banded Systems*) (a) Write a program that will aim to find the optimal relaxation parameter ω for the SOR method in the problem of solving a certain linear system $Ax = b$ for which it is assumed that the SOR method will converge. The functionality of the program will be similar to that of the preceding exercise, except that now the program should be specially designed to deal with sparsely banded systems, as did the program sorsparsediag of Exercise for the Reader 7.31 (in fact, the present program should call on this previous program). The syntax is as follows:

$$\text{omega} = \text{optimalomegasparsediag}(\text{diags}, \text{inds}, b, \text{tol}, \text{iter})$$

The first three input variables are as explained in Exercise for the Reader 7.31 for the program sorsparsediag. The remaining variables and functionality of the program are as explained in the preceding exercise.

(b) Apply the above program to aim to determine the optimal value of the SOR parameter ω for the linear system of Example 7.26 with default tolerances. Does the resulting output change if we change iter to 5? To 20?

(c) With default tolerances, run the program on the linear system of Example 7.30 and compare with the exact result of Proposition 7.14. You may need to experiment with different values of iter to attain a successful approximation. Run SOR on the system with this computed value for the optimal relaxation parameter, and 308 iterations. Compute the exact error and compare with the results of Example 7.31.

(d) Repeat part (c) but now with tol reset to 1e-6.

NOTE: For tridiagonal matrices that are positive definite, the following formula gives the optimal value of the relaxation parameter for the SOR iteration:

$$\omega = \frac{2}{1+\sqrt{1-\rho(D-L)^2}}, \tag{66}$$

where the matrices D and L are as in (60) $A = D - L - U$,[24] and $\rho(D - L)$ denotes the spectral radius of the matrix $D - L$.

17.   We consider tridiagonal square $n \times n$ matrices of the following form:

$$F = \begin{bmatrix} 2 & a & & & & \\ a & 2 & a & & \text{\Large 0} & \\ & a & 2 & a & & \\ & & \ddots & \ddots & \ddots & \\ & \text{\Large 0} & & \ddots & \ddots & a \\ & & & & a & 2 \end{bmatrix}.$$

(a) With $a = -1$ and $n = 10$, show that F is positive definite.
(b) What does formula (17) give for the optimal SOR parameter for the linear system?
(c) Run the SOR iteration with the value of ω obtained in part (b) for the linear system $Fx = b$ where the exact solution is $x = [1\ 2\ 1\ 2 \dots 1\ 2]'$. How many iterations are needed to get the exact error to be less than 1e-10?
(d)   Create a graph comparing the performance of the SOR of part (c) along with the corresponding Jacobi and Gauss-Seidel iterations.

18.   Repeat all parts of Exercise 17, but change a to -0.5.

19.   Repeat all parts of Exercise 17, but change n to 100. Can you prove that with $a = -1$, the matrix F in Exercise 17 is always positive definite? If you cannot, do some MATLAB experiments (with different values of n) and conjecture whether you think this is a true statement.

20.   (a) Show that the Jacobi iteration scheme is represented in matrix form (61) by the matrix B indicated in Table 7.1.
(b) Repeat part (a) for the Gauss-Seidel iteration.
(c) Repeat part (a) for the SOR iteration.

21.   Prove Lemma 7.15.

22.   (a) Given a nonsingular matrix A, find a corresponding matrix T so that the Jacobi iteration can be expressed in the form $x^{(k+1)} = x^{(k)} + Tr^{(k)}$, where $r^{(k)} = b - Ax^{(k)}$ is the residual vector for the kth iterate.
(b) Repeat part (a) for the Gauss-Seidel iteration.
(c) Can the result of part (b) be generalized for the SOR iteration?

23.   (*Proof of Jacobi Convergence Theorem*) Complete the following outline for a proof of the Jacobi Convergence Theorem (part of Theorem 7.11): As in the text, we let $e^{(k)} = x^{(k)} - x$ denote the error vector of the kth iterate $x^{(k)}$ for the Jacobi method for solving a linear system $Ax = b$, where the $n \times n$ matrix A is assumed to be strictly diagonally dominant. For each

---

[24] $D - L$ is just the iteration matrix for the Gauss-Seidel scheme; see Table 7.1.

(row) index i, we let $\mu_i = \dfrac{1}{|a_{ii}|}\displaystyle\sum_{\substack{j=1 \\ j \neq i}}^{n} |a_{ij}|$ ($1 \leq i \leq n$). For any vector v, , we let $\|v\|$ denote its

infinity norm: $\|v\|_\infty = \max(|v_i|)$.

For each iteration index k and component index i, use the triangle inequality to show that

$$|e_i^{(k)}| \leq \frac{1}{|a_{ii}|}\sum_{\substack{j=1 \\ j \neq i}}^{n} |a_{ij}|\,|e_i^{(k-1)}| \leq \mu_i \left\|e^{(k-1)}\right\|,$$

and conclude that

$$\left\|e^{(k)}\right\| \leq \|\mu\| \left\|e^{(k-1)}\right\|,$$

and, in turn, that the Jacobi iteration converges.

Chapter 8: Introduction to Differential Equations

8.1: WHAT ARE DIFFERENTIAL EQUATIONS?

Many natural phenomena are represented or modeled by functions. Such functions may depend on one or several independent variables. Choices for the independent variables are endless, but the most common ones are time and space (location) variables. Often, the explicit function is not known but rather we only know (from theory, experiments, or history) certain relations among the various rates of change (derivatives) of the function with respect to some of its independent variables. Any equation involving an unknown function along with some or all of its derivatives is called a **differential equation (DE)**. Differential equations break down into two major kinds, **ordinary differential equations (ODEs)** and **partial differential equations (PDEs)**. ODEs involve an unknown function of a single variable only, while PDEs involve an unknown function of several variables. Thus, technically an ODE falls under the umbrella of just being a special type of a PDE but the theories for these types of equations are customarily split into two different major mathematical subject areas. The derivatives of a function of several variables are called partial derivatives. We will study PDEs in Part III of this book and in this part we will focus on ODEs.

The **order** of an ODE is the order of the highest derivative of the unknown function that appears in the equation. A **solution** of an ODE is any function for which, when it (and its derivatives) are substituted for the unknown function (and the corresponding derivatives) in the ODE, the resulting equation will be an identity (i.e., always true). Our next example gives some solutions of certain ODEs. We do not assume the reader has studied differential equations, so at this point the reader should not worry about how the solutions in these examples were obtained.

EXAMPLE 8.1: For each ODE that is given, determine its order and check the given function(s) is a solution. In each of the ODEs the unknown function is written as "y" and we understand "x" to be the independent variable. Thus, "y" really means "$y(x)$".

(a) $y' = 2x$; $y = x^2 + C$ (here C is an arbitrary constant)

(b) $y' = 2xy + 1$; $y = e^{x^2} \left(\int_0^x e^{-t^2} dt \right) + e^{x^2}$

(c) $y'' + 2y' - 3y = 0$; $y_1 = e^{-3x}$, $y_2 = e^x$

(d) $y'''' + 4y''' + 3y = x$; $y_1 = x/3$; $y_2 = e^{-x} + x/3$

SOLUTION: The orders of each of these ODEs are (in order) 1, 1, 2, and 4. Checking that the functions given are actually solutions just requires some differentiation. We do only (b) (since it's a bit different) and the first function of (c), and leave the rest to the reader.   Let's begin with checking that $y_1 = e^{-3x}$ solves the ODE in (c). Since $y_1' = -3e^{-3x}$ and $y_1'' = 9e^{-3x}$, we have $y'' + 2y'$ $-3y = 9e^{-3x} + 2(-3e^{-3x}) - 3e^{-3x} = 0$, as required.

The check in part (b) will require the *fundamental theorem of calculus* (for differentiating functions defined by integrals). We recall this theorem here for convenience and future reference. It is summarized by the formula given below in which $f(t)$ is any continuous function:

$$\frac{d}{dx}\left(\int_0^x f(t)\,dt \right) = f(x).$$ (1)

(This is really just a precise statement of the fact that differentiation and integration are inverse processes.) Now using (1) together with the product rule, we obtain:

$$y' = \left(e^{x^2}\left(\int_0^x e^{-t^2}\,dt \right) + e^{x^2} \right)' = e^{x^2}(2x)\left(\int_0^x e^{-t^2}\,dt \right) + e^{x^2}\left(\int_0^x e^{-t^2}\,dt \right)' + e^{x^2}(2x)$$

$$= 2xe^{x^2}\left(\int_0^x e^{-t^2}\,dt \right) + e^{x^2}e^{-x^2} + 2xe^{x^2}$$

$$= 2x\left(e^{x^2}\left(\int_0^x e^{-t^2}\,dt \right) + e^{x^2} \right) + 1$$

$$= 2xy + 1.$$

As demonstrated by part (a) in the above example, in general, an ODE will have infinitely many solutions. The collection of all solutions of a certain ODE of order n (called the **general solution**) will involve n arbitrary constants. So to specify one such solution, we will need n **auxiliary conditions** (one for each order derivative) for the unknown function. If the independent variable is time, in many natural problems the auxiliary conditions are given at time $t = t_0 = 0$ (initially) and in this case are called **initial conditions (ICs)**. A problem that gives an ODE (of order n) along with a corresponding set of (n) ICs is called an **initial value problem (IVP)**.   This terminology still applies even when the independent variable is different from time and when $t_0 \neq 0$.

EXAMPLE 8.2: Use the information in the last example to find a solution for each of the following IVPs:

(a) $\begin{cases} (DE)\ y' = -\cos(x) \\ (IC)\ y(0) = -3 \end{cases}$ (b) $\begin{cases} (DE)\ y'' + 2y' - 3y = 0 \\ (IC)'s\ y(0) = 5,\ y'(0) = 1 \end{cases}$

SOLUTION: Part (a): The form of the DE is a familiar one from calculus: $y' = f(x)$. The general solution is the indefinite integral $y = \int f(x)\,dx$, so here

$y = \int -\cos(x)dx = -\sin(x) + C$. Substituting $x = 0$ into both sides gives (using the IC) $-3 = y(0) = -\sin(0) + C = 0 + C = C$, so $C = -3$, and $y = -\sin(x) - 3$.

Part (b): Example 8.1 gave us two solutions of this DE. Using the rules of differentiation, we observe that if we multiply either of these solutions (or any solution of the DE) by a constant it will still solve the DE (reason: Constants can be pulled out of differentiations). Also, if we add two such solutions (or any two solutions of this DE) the sum will also solve the DE (reason: Derivatives of sums are sums of the derivatives). These important facts are not true for all DEs; we will later discuss general circumstances under which they will be true. From what we have stated, it follows that for any constants C_1, C_2, the function $y = C_1 y_1 + C_2 y_2 = C_1 e^{-3x} + C_2 e^x$ will solve the DE (this actually turns out to be the general solution). If we can determine choices for C_1, C_2 that will make this function satisfy the ICs, then we can proceed. For this function, the ICs give: $5 = y(0) = C_1 + C_2$, and $1 = y'(0) = -3C_1 + C_2$. Solving these two equations gives $C_1 = 1$, $C_2 = 4$ and so a solution of the IVP is $y = e^{-3x} + 4e^x$.

From calculus we know that the solution in part (a) is **unique** (meaning: There is only one), and for both problems we saw the **existence** of a solution (meaning: There is at least one). It turns out that the solution in part (b) is also unique. Not all IVPs have such a nice existence and uniqueness phenomenon; we will give some theorems about this later.

The simplest type of ODE is like that given in part (a) of the last example where it is really just a calculus problem of finding an indefinite integral. In calculus courses, we learn that although it is possible to differentiate just about any function in sight, finding indefinite integrals is almost always impossible. Most of the integrals encountered in calculus courses are tailor-made to be evaluated using one of the techniques of integration. Yet, by the fundamental theorem of calculus, any continuous function has a definite integral (whose derivative is given by (1)). The hard fact is that in real life, chances are that the function we need to integrate will be impossible to do explicitly. By extension, more complicated differential equations are also extremely unlikely to be solvable explicitly. Much of the material in traditional courses in DEs is focused on developing methods for solving very limited classes of DEs explicitly. Thus in practice, <u>most DEs that come up cannot be solved explicitly and numerical methods are the only way to go</u>.

DEs arise in problems from practically all scientific disciplines from physics and engineering to biology and pharmacology. For each such model, certain information is known from which the DE can be formulated along with needed auxiliary conditions. In biology, for example, information might be used to set up a DE modeling an outbreak of a disease. We know the history of how the disease is spread and we will be very interested in knowing what will happen in the future. The unknown function would be the number of infected individuals, and the independent variable would be time. Biologists (and many others) would be interested in predicting the number of infected individuals in future times as well. Perhaps there are some preventative measures or vaccines that could be used. The effect of such items could further be built into the DE to help decide on the best course(s) of action to keep the disease from becoming an epidemic or preferably to wipe it out. Even people studying finance have developed a type of DE (called a *stochastic differential equation*) that can be used to model prices of stock and futures markets. This subject received a lot of attention recently, and has garnered generous funding from Wall Street tycoons who are always looking for creative new ways to turn profits. Of course, in any of the applied fields, an explicit solution is never really required. We would just like to know (within some specified tolerance for error) what the value of the unknown function will be at some values of the independent variables. For the remainder of this chapter, we will focus on (single) first-order ODEs. In the next two chapters we will extend many of our techniques to systems of several ODEs involving several unknown functions and to higher-order ODEs.

8.2: SOME BASIC DIFFERENTIAL EQUATION MODELS AND EULER'S METHOD

Let us begin with a simple example where pure mathematics alone would be quite awkward and inadequate, but MATLAB will easily come to the rescue.

EXAMPLE 8.3: Graph the solution of the IVP: $\begin{cases}(DE) \ y' = \sin(x^2) \\ (IC) \ y(0) = 1\end{cases}$, for $0 \le x \le 5$.

SOLUTION: Invoking the fundamental theorem of calculus, we obtain

$$y' = \sin(x^2) \Rightarrow y(x) = \int_0^x \sin(t^2)dt + C \ .$$

Now substituting $x = 0$ into the latter equation, the IC gives us that

$$1 = y(0) = \int_0^0 \sin(t^2)dt + C = 0 + C \Rightarrow C = 1 \ ,$$

so we now have $y(x) = \int_0^x \sin(t^2)dt + 1$. Since we cannot evaluate the indefinite integral explicitly (in fact it is impossible to do so no matter how good we are at integration by parts, substitution, etc.), pure mathematics stops here. We can now let MATLAB take over this solution. Using the numerical integrator quad (described in Chapter 3), we use a for loop to create x- and y-values of the function $y(x)$ and then plot them to obtain the desired graph. We first need to store the function to be integrated (either as an M-file or an inline function). Here is how it can all be done:

```
>> f=inline('sin(x.^2)');
>> x=0:.01:5; %This will give a very decent resolution
```

```
>> size(x)  %Need to know many components x has to create y of same
>>  % length
→   1  501

>> for i=1:501
      y(i)=1+quad(f,0,x(i));
end
>> plot(x,y)
```

NOTE: If you created f as an M-file rather than an inline function, the syntax for quad would have to be `quad('f',0,x(i))` or `quad(@f,0,x(i))`

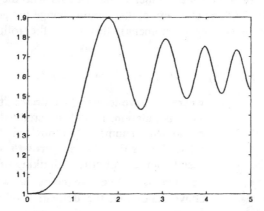

FIGURE 8.1: A plot of the solution of the IVP of Example 8.3. There exists no explicit mathematical formula for this function in terms of the standard functions of calculus.

Suppose that we wanted the numerical value of the solution when $x = 3$. We want to caution the reader that at this point we cannot just enter y(3) in our MATLAB session to get the answer. Recall that y is stored as a vector so y(3) would just be its third component, i.e., the y-coordinate when $x = 0 + 2(.01) = 0.02$. This is definitely not what we want (y when $x = 3$). Let us reiterate:

CAUTION: In the above problem, the mathematical notation $y(3)$ denotes the y-coordinate of the function when the x-coordinate equals 3, in MATLAB notation, y(3) is the third component of the vector of y-values constructed (which occurs at $x = 0.02$). Thus, depending on the context, the notation $y(3)$ could mean two different things. This problem will come up repeatedly and the reader must be made aware of it early on to avoid confusion. Remember the adage: Everything (numerical) in MATLAB is a matrix.

EXERCISE FOR THE READER 8.1: Relating to the example above, fill in the question mark: $y(3)$ (mathematical notation) = y(?) (MATLAB notation). Then find this numerical value to 4 decimals.

We now introduce our first set of differential equations on which we will begin a systematic study. They will model population growth. Here, the independent variable will be t(time)—not x. We begin with the most basic model. We will let

$P(t)$ = the number of individuals in a certain population at time t,

where the "individuals" could be humans, sharks, bacteria, etc. In the basic model, we assume that there is a constant **birth rate** = β ($\equiv$ the number of individuals born into the population per unit time per living individual) and constant **death rate** = δ ($\equiv$ the number of individuals who die per unit time per living individual). With no other effect that would change the population (e.g., no immigrations, or other such phenomena), this gives the following differential equation for $P(t)$:

$$P'(t) = (\beta - \delta)P(t) = rP(t) \quad \text{or} \quad P' = rP, \tag{2}$$

where we have let $r = \beta - \delta$ ($\equiv$ resultant **growth rate**). This population model is credited to the political economist Thomas Malthus[1] and is customarily referred to as the **Malthus growth model**. This DE is easy to solve explicitly. Thinking of a function that is its own derivative, we come up with e^t. The DE above states that the derivative of a function should equal r times the function. Since $(e^{rt})' = re^{rt}$, we see that $P(t) = e^{rt}$ will solve the DE. Also, we can multiply this solution by any constant and it will still solve the DE (why?). We have just found a collection of solutions to the DE (2) to be $P(t) = Ce^{rt}$ (where C is an arbitrary constant). It turns out that there are no other solutions (this will follow from uniqueness theorems that we give later; see also Exercise 8 of this section) and thus this is the general solution. If we substitute $t = 0$ into this general solution, we get $P(0) = Ce^0 = C$, so the constant C turns out to be the **initial population** ($\equiv$ the population at time $t = 0$). Depending on the value of r, we have three different cases for what will happen to such a population. These situations are summarized in Figure 8.3.

FIGURE 8.2: Thomas Malthus (1766–1834), English economist.

---

[1] Malthus was the first scientist to realize the power of exponential growth left unchecked. He wrote a seminal work: *Essay on the Principle of Population* (1798). In it he observed how in nature plants and animals routinely produce more offspring than can survive. and that unless family sizes were regulated, the human race would eventually become too large and poverty and famine would eventually lead to its demise. He used his models to support his claims but some of his recommendations were quite controversial and totalitarian. He proposed, for example, that poor families not be allowed to have more offspring than they can support. His social recommendations aside, Malthus's research was quite important and was even used later on by Darwin in formulating some of his famous theories on evolution.

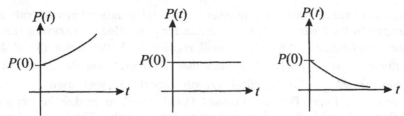

FIGURE 8.3: The three cases of the basic population model (2): (a) $r > 0$ exponential growth, (b) $r = 0$ constant population, (c) $r < 0$ exponential decay—eventual extinction.

Malthus growth left unchecked can be extreme beyond imagination. The next example puts this comment into perspective.

EXAMPLE 8.4: Under ideal conditions, a single cell of E. *coli* bacterium splits into 2 new bacteria every 20 minutes.
(a) Starting with such a single cell, estimate the population of the resulting colony after 1 day (24 hours).
(b) The average mass of an E. *coli* cell is 10^{-12} g. Compare the mass of the day-old colony to that of the Earth ($\approx 5.9763 \times 10^{24}$ kg.)

SOLUTION: Part (a): If we use hours for the unit of time, we wish to find $P(24)$. After 20 minutes, one E. *coli* cell becomes two, in 20 more minutes, these two become four, and finally after another 20 minutes (one hour), the four have become eight, resulting in the net "birth" of seven new E. *coli* cells in one hour (= 1 unit of time). So we have a growth rate $\beta = 7$. Since the death rate will not be relevant for this short-time, ideal-condition problem, this gives a growth rate of $r = 7$ and so the DE is $P'(t) = 7P(t)$. The general solution is

$$P(t) = P_0 e^{7t} = e^{7t} \text{ (since } P_0 = 1) \text{ so } p(24) = e^{7 \cdot 24} = 9.1511 \times 10^{72} \, !$$

Part (b): Using the data given, this means that the resulting population would be over 10^{33} times as massive as our planet! In his 1969 novel, *The Andromeda Strain*, Michael Crichton had made such an interesting observation. Of course, as the population starts to get big the conditions are no longer as ideal. For example, if E. *coli* has infected a certain human being, this colony will be limited to the size of the host. Also, once it is detected, human antibodies will make conditions not as hospitable, and appropriate antibiotics will make conditions so unfavorable that the whole colony can be wiped out.

More advanced models of population growth (or decay) will have variable growth rates. One useful such model is the so-called **logistical growth model**. This model takes into account factors such as limited food supply, cultural sophistication, etc., that tend to keep the population from getting too big. The DE representing this model is as follows:

$$P'(t) = rP(1 - P/k) . \tag{3}$$

Here, r and k are constants. The number r is called the **natural growth rate** and it is damped by the factor $(1 - P/k)$. The number k is called the **carrying capacity** of the environment. When P is small relative to k, the factor $(1 - P/k)$ is essentially equal to one so (3) implies that $P'(t) \approx rP$ and the growth is like Malthus growth. This logistical growth model (3) was used by Belgian mathematician Pierre François Verhulst (1804–1849) to model the population growth in Belgium.[2] We will present an example shortly. The logistical growth model has also been used in many other contexts as well. In particular, it was used in the 1970s to predict U.S. oil production. Since the logistical model can be solved explicitly, we will use it as an example to demonstrate some numerical methods for solving initial value problems.

The first method we present is due to Leonhard Euler,[3] who was the first to take action against the fact that pure mathematical methods alone were not enough to solve some important differential equations that were coming up in real-life applications. **Euler's method** applies to the following type of first-order initial value problem:

---

[2] In his 1845 paper *Recherches mathématiques sur la loi d'accroissement de la population*, Verhulst used Belgian census data to predict the parameters for his model, which he termed as *logistique*. His model predicted the population rather well all the way up into the 1990s when it began to undershoot the actual figures by only about 7%, and these discrepancies can be attributed more to immigrations which had been unanticipated in Verhulst's era. Shortly we will give a similar model for the U.S. population growth. The model comes from a 1920 paper of two demographers, Raymond Pearl and L. J. Reed, entitled: *On the rate of growth of the population of the United States since 1790 and its mathematical representation*. The latter researchers, unaware of Verhulst's work, developed a similar model using census data of the U.S. population from 1790, when it was first recorded.

[3] Leonhard Euler (pronounced "Oiler") came into this world during a very exciting time in mathematics. Calculus had recently been invented by Sir Issak Newton and Gottfried W. Leibniz and the frontier was open to apply it to solve many important problems. Euler was born and educated in Switzerland. He received his first appointment as a professor at the prestigious Saint Petersburg University in Russia at the age of 19. Euler turned out to be the most prolific mathematician of all time. His published works fill over 100 encyclopedia-sized volumes! He is considered the founder of modern pure and numerical analysis. His remarkable work touched upon every major area of mathematics, and he was able to successfully apply mathematics to numerous areas of science, such as celestial mechanics (e.g., planetary and comet motions), ship building, optics, hydrostatics, and fluid mechanics. His work in these areas led him to many differential equations that, in turn, motivated him to develop many useful methods for dealing with them. He has even done significant work in cartography and was involved in making an extensive atlas of Russia. After about six years in St. Petersburg, Euler got appointed to the Berlin Academy and eventually became its leader. Because of some quarrels with King Frederick, he was never given the official title of "President," and so after 25 years of a distinguished career in Berlin, he decided to return again to St. Petersburg. He continued to flourish there until the day of his death. During his last 17 years of life, Euler had become completely blind, but this was also one of his most productive periods! Euler had a memory that was shockingly precise. He was able to perform huge computations in his head and recite an entire novel even at age 70! At this age he could even recite the first and last sentence on each page of Vergil's *Aeneid*, which he had memorized. Once he had settled an argument in his head between two students whose answers differed in the fifteenth decimal place. We owe to Euler the notation $f(x)$ for a function (1734), e for the base of natural logs (1727), i for the square root of -1 (1777), π for pi, Σ for summation (1755), and numerous other present-day mathematical notations. Euler was also prolific in other ways such as having had 13 children. He boasted about having made his most piercing mathematical discoveries while holding one of his infants as his other children were playing at his feet.

$$(IVP) \begin{cases} y'(t) = f(t, y(t)) & (DE) \\ y(a) = y_0 & (IC) \end{cases}. \qquad (4)$$

When it is understood that t is the independent variable, the differential equation in (4) is often written more succinctly as $y' = f(t, y)$. By extension, we are allowing the initial value problem's initial condition to commence at any time $t = a$, rather than always at $t = 0$. The form of the (DE) in (4) is solved for y'. Although this is not always possible, it is a form to which most of the successful theory of differential equations can be developed. It may seem

FIGURE 8.4: Leonhard Euler (1707–1783), Swiss mathematician.

restrictive in that it seems to apply only to first-order ODEs. We will see later, however, that any higher-order ODE can be transformed into a system of first-order ODEs. Thus, the methods we will be learning about for numerically solving (4) will actually turn out to be applicable to very general ordinary differential equations (of arbitrary order) and systems of these. In order to introduce the method, we initially will only assume that the function $f(t, y)$ of the (DE) in (4) is a continuous function (in both of its variables). It turns out that this will be sufficient to guarantee the existence of a solution to (4) (at least as long as its graph stays in the region of continuity of $f(t, y)$). The condition for uniqueness is a bit more technical. We will come to this later, after we work through some more examples. It will turn out that all of the examples we consider (as well as the great majority that come up in real-life modeling) will satisfy the technical requirements to guarantee existence and uniqueness.

Euler's method is based on the tangent line approximation (special case of Taylor's theorem):

$$y(t_0 + \Delta t) \approx y(t_0) + y'(t_0)\Delta t \qquad (5)$$

The method requires specifying a **step size** $= h > 0$ (usually a small number), and will construct a sequence of y-coordinates $y_0, y_1, y_2, \cdots, y_N$ that will approximate the function at the equally spaced (by the step size h) t-coordinates $t_0 = a$, $t_1 = a + h$, $t_2 = a + 2h$, $\cdots, t_N = a + Nh$. The integer N can be as large as we want, and the t-range will cover however long an interval on which we would like to approximate the function. The goal is to get $y_n \approx y(t_n)$ for each n, $1 \le n \le N$. The initial condition (IC) in (4) lets us start off exactly: $y_0 = y(a) = y(t_0)$. Next, to get y_1, we use (4), which tells us exactly what $y'(t_0)$ is ($f(t_0, y(t_0)) = f(t_0, y_0)$) and (5):

$$y(t_1) = y(t_0 + h) \approx y(t_0) + y'(t_0)h = y_0 + hf(t_0, y_0) \equiv y_1.$$

We will next get y_2 from y_1 in the same fashion as we obtained y_1 from y_0. Things are a bit different here, though, since y_0 is exactly $y(t_0)$, whereas y_1 is only an approximation to $y(t_1)$. But since we are assuming that $f(t,y)$ is continuous, it follows that if the step size h is taken small enough, the value of the actual derivative $y'(t_1) = f(t_1, y(t_1))$ (from the (DE) in (4)) will be very close to $f(t_1, y_1)$ (since y_1 will be very close to $y(t_1)$). We use these facts and (4) to obtain y_2:

$$y(t_2) = y(t_1 + h) \approx y(t_1) + y'(t_1)h \approx y_1 + hf(t_0, y_1) \equiv y_2.$$

The subsequent y_n's are now obtained recursively in the same fashion. The actual solution is now approximated by connecting (interpolating) adjacent ordered pairs (t_n, y_n) with line segments. Recall that this is what MATLAB would do anyway if we asked it to plot the vector $y = [y_0, y_1, y_2, \cdots, y_N]$ versus $t = [t_0, t_1, t_2, \cdots, t_N]$. We can summarize Euler's method by these recursion formulas:

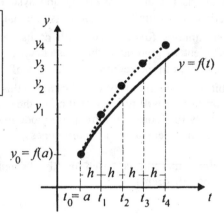

Euler's Method:
$t_0 = a,\ y_0 = y(a)$ given
h = step size
$t_{n+1} = t + h,\quad y_{n+1} = y_n + hf(t_n, y_n),$
$\qquad n = 1, 2, 3, \cdots$

FIGURE 8.5:   Illustration of Euler's method for solving the first-order IVP $\begin{cases} y'(t) = f(t, y(t)) \\ y(a) = y_0 \end{cases}$ using step size h. The approximation is the dotted graph and the actual solution is the solid graph.

As our first example, we will use Euler's method to numerically solve the historical U.S. population logistical model that was described earlier. Since the DE can be solved explicitly in this case, we will be able to compare the exact errors for Euler's method for this problem with different step sizes. Moreover, keeping in mind that the model was done near the beginning of the twentieth century, we will also be able to compare the model's predictions with some actual numbers in the U.S. populations. We will do this example by hand using MATLAB, and afterwards we will write a program that will perform Euler's method.

EXAMPLE 8.5: In the Verhulst-type population model for the U.S. population (done in 1920), the logistical population growth model was used in the initial value problem

$$\begin{cases} P'(t) = rP(1 - P/k) \\ P(0) = P_0 \end{cases},$$

using the estimates $r = 0.0318$ (growth rate), and $k = 200$ million (carrying capacity). It was known that $P(0) = 3.9$ million (where we identified $t = 0$ years with the year 1790).

(a) Use Euler's method with step size $h = 0.1$ in the Verhulst model to estimate the U.S. populations in the years 1850 ($t = 60$), 1900 ($t = 110$), and 1990 ($t = 200$).

(b) Repeat part (a) with step size $h = 0.01$.

(c) The exact solution of the logistical IVP is (from ODE methods; see Exercise 12):

$$P(t) = \frac{k}{1 + (k/P_0 - 1)e^{-rt}}.$$

In the same plane, plot the graph of the exact solution $P(t)$ along with the two Euler approximations to it for $0 \leq t \leq 200$ (1790 through 1990).

SOLUTION: For convenience, we express populations in millions. Since in part (c) we will need to plot the Euler approximations, we should create and store the whole vectors of approximations that we obtain from Euler's method in parts (a) and (b). We will need to create a function for the right side of the differential equation:

```
>> f=inline('0.0318*P*(1-P/200)');
```

Part (a): We first create the t-vector for the approximations, find out its size, and then use a for loop to create the corresponding P-coordinates of the approximations.

```
>> t=0:0.1:200; size(t)
→ 1    2001

>> P(1)=3.9; %initialize P
>> for n=1:2000
P(n+1)=P(n)+0.1*f(P(n));
end
```

In order to find out the values of this vector corresponding to the times $t = 60$ (1850), $t = 110$ (1900), and $t = 200$ (1990), we need to find the corresponding (MATLAB) indices for the vector t. This is not difficult; for example, if we use the recursion formula: $t(n + 1) = t(n) + h$, with $t(1) = 0$, we see that $t(n) = (n - 1)h = 0.1(n - 1)$. Solving for n gives $n = 10t(n) + 1$. So the indices for $t = 60, 110$, and 200 are 601, 1101, and 2001 respectively. Thus we can get the corresponding population estimates:

```
>> P(601), P(1101), P(2001)
→23.5827, 79.1281, 183.9685
```

NOTE: The first two estimates compare quite favorably with the actual U.S. populations in the corresponding years: 1850→23.2 (million), 1900→76.0, 1990→248.7. The last estimate falls rather significantly short due to an

underestimate of the carrying capacity of the United States. This is certainly
excusable, given that in the early twentieth century modern industrial technology
(e.g., skyscrapers and agricultural engineering) was not yet on the horizon of
peoples' imagination. In this respect, Verhulst's predictions for Belgium (in 1845)
were even more impressive.

Part (b): Since in part (c) we will need to compare these approximations with
those of part (a), we store both the t-vectors and P-vectors here as new vectors tb
and Pb. The constructions are analogous to those in part (a).

```
>> tb=0:.01:200; Pb(1)=3.9; size(tb)
→ 1     20001
>> for n=1:20000
Pb(n+1)=Pb(n)+0.01*f(Pb(n));
end
>> Pb(6001), Pb(11001), Pb(20001)
→23.6331, 79.3010, 183.9969
```

The indices of Pb correspond to the years 1850, 1900, and 1990 and were obtained
as explained in part (a).

Part (c): We store the exact solution as an M-file Pver.m:

```
function P=Pver(t)
P=200./(1+(200/3.9-1)*exp(-.0318*t));
```

To obtain plots of the exact solution along with the two approximate solutions of
parts (a) and (b), we must take care in plotting the two approximations with the
correct t-vectors (the vectors in a plot must be the same size).

```
>> plot(t,P), hold on, plot(tb,Pb)
>> plot(tb,Pver(tb))
>> xlabel('Years after 1790')
>> ylabel('Estimated U.S.
         population in millions')
```

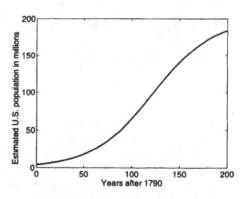

FIGURE 8.6: The graph of the logistical
U.S. population model of Example 8.5. The
function is a solution of a differential
equation and was plotted above, together
with two of the Euler approximations (with
step sizes $h = 0.1$ and $h = 0.01$). The three
graphs are indistinguishable. Compare with
Figure 8.7.

Since the three graphs are indistinguishable, we give also a plot of the errors of the
two Euler approximations. The following plot command will do it all in one line.

```
>> hold off, plot(t,abs(Pver(t)-P), tb, abs(Pver(tb)-Pb), 'o')
>> xlabel('Years after 1790')
>> ylabel('Millions')
>> title('Error Graphs')
```

FIGURE 8.7: Graphs of the absolute errors in using the Euler method to solve the Verhulst-type U.S. population initial value problem of Example 8.5. The thin curve is with step size $h = 0.1$ and the thick low curve is with step size $h = 0.01$. Note that the error appears to have decreased 10-fold as we increased the number of steps by a factor of 10.

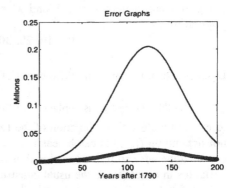

We will postpone formal error estimates and some related theory until Section 8.4. It is a simple matter to write a program for the Euler method.

PROGRAM 8.1: An M-file for Euler's method for the IVP:

$$\begin{cases} y' = f(t,y) & (DE) \\ y(a) = y_0 & (IC) \end{cases}.$$

```
function [t,y]=eulermeth(f,a,b,y0,hstep)
% M-file for applying Euler's method to solve the initial value
% problem:  (DE) y'-f(t,y),  (IC) y(a)   y0, on the t-interval [a,b]
% with step size hstep.  The output will be a vector of t's and
% corresponding y's
% input variables: f, a, b, y0, hstep
% output variables: t, y
% f is a function of two variables f(t,y)
% y(a)=y0
t(1)=a; y(1)=y0;
nmax=ceil((b-a)/hstep);
for n=1:nmax
    t(n+1)=t(n)+hstep;
    y(n+1)=y(n)+hstep*feval(f,t(n),y(n));
end
```

CAUTION: In general the function $f(t,y)$ of the (DE) in the IVP will depend on t and y. Since the above program assumes this is the case, whenever it is used to solve an IVP, the function f must be created as a function of these two variables (in this order) even if it only has one (or none) of these variables appearing in its formula. Also, in order for the final approximation produced in the program, y(nmax+1), to correspond to $y(b)$, we choose the step size h so that $(b-a)/h$ is an integer. This is what is usually done in practice. In any case, t(nmax+1) will always be within h units of b.

EXAMPLE 8.6: Using the above program in conjunction with a for loop, get MATLAB to produce a single plot that contains Euler approximations with step size $h = 0.1$ of solutions to the logistical growth model IVP:

$$\begin{cases} P'(t) = rP(1 - P/k) \\ P(0) = P_0 \end{cases}, \qquad 0 \le t \le 1.5$$

using the parameters $r = 2.2$ and $k = 100$ for each of the following initial populations:

$$P_0 = 10, 20, 30, \cdots, 190, 200.$$

Discuss the similarities and differences of this family of solutions.

SOLUTION: Here "y" is replaced by "P" and $f(t,P) = rP(1 - P/k)$. We need to explicitly store f as a function of the two variables t and P (even though it does not depend on t). This can be easily done by creating an M-file. Alternatively, it can be done with an inline function, but we must explicitly declare the two domain variables in order (since the usual construction would scan the formula and take it to be a function of one variable as in the previous example).

```
>> f=inline('2.2*P*(1-P/100)', 't', 'P')
→f = Inline function: f(t,P) = 2.2*P*(1-P/100)
```

With $f(t, P)$ thus constructed, we can obtain the desired plots very quickly with the following chain of commands:

```
>> hold on
>> for i=10:10:200
[t,yi]=eulermeth(f,0,1.5,i,0.1);
plot(t,yi)
end
```

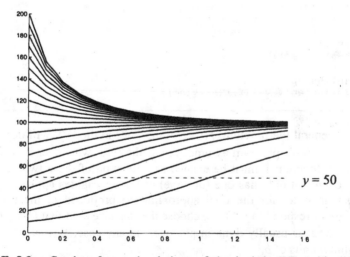

FIGURE 8.8: Graphs of several solutions of the logistic IVP (with different initial conditions); the parameters are $r = 2.2$ and $k = 100$. The green dashed line intersects solution curves in inflection points.

Notice that solutions with initial populations less than the carrying capacity will increase to it and solutions with initial populations greater than the carrying capacity will decrease to it. This behavior can be predicted from the DE since $P(t)$

$> k$ forces the right side of the DE (and hence the derivative of P) to be negative while $P(t) < k$ makes the derivative positive. Also, the net rate of change $|P'(t)|$ disintegrates to zero as $P(t)$ converges to k. When $P(t)$ is small, the DE looks more like $P'(t) \approx rP$, so we get exponential growth as in the Malthus model. Finally we note that all of the graphs that pass through $y = 50$ ($= 1/2$ of carrying capacity) have an inflection point there.

EXERCISE FOR THE READER 8.2: Use calculus to justify the statement made above regarding inflection points of solutions to the logistic DE. Are there solutions with other inflection points?

The logistic DE has the form $P' = f(P)$. To further understand such qualitative properties of the solutions of such DEs, it is useful to look at the graph of the function on the right. For the logistic DE, $f(P) = rP(1 - P/k)$ has a graph that is just a downward-opening parabola with P-intercepts at $P = 0$ and $P = k$, as shown in Figure 8.9. Since, by the chain rule, $P''(t) = (d/dt)(P') = (d/dP)(P')(dP/dt)$ $= f'(P)f(P) = r(1 - 2P/k)f(P)$, it is clear that P' will increase until P reaches $k/2$ (if it starts below this value) where its steepest slope will be (i.e., P' reaches its maximum at this point $P = k/2$ where $P'' = 0$).

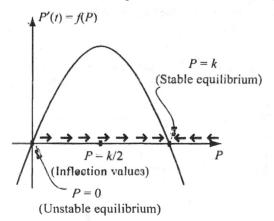

After this inflection point, P will continue to increase, but at the same time the derivative will continue to decrease. This will continue on forever, P never reaching the carrying capacity k. The two roots of $f(P)$, $P = 0$, and $P = k$, are clearly constant solutions of the logistic DE. They are called **equilibrium solutions**. The first $P = 0$ equilibrium solution is different from the second in that if an initial value were to be prescribed close to $P = 0$, the solution would continue to grow and eventually approach $P = k$ as $t \to \infty$. Thus solutions which start close to $P = 0$ will eventually diverge away from it and such equilibrium solutions are called **unstable**. If a solution started with initial condition close to $P = k$ (either greater than or less than), then as $t \to \infty$, any such solution would continue to get closer to $P = k$. Such solutions are called **stable**. Again this is clear from the graph of $P' = f(P)$ in Figure 8.9, along with the flow directions shown there: If

FIGURE 8.9: Growth function associated with the logistic equation. There are two equilibria: $P = 0$ (unstable) and $P = k$ (stable). Flow directions for P are indicated over the P-axis.

$P' = f(P) > 0$, flow is to the right (increasing P); if $P' = f(P) < 0$, flow is to the left; if $P' = f(P) = 0$, we are at an equilibrium.

Other types of growth rates result from different functions $f(P)$. The **Gomperz Law** is another such example. It is modeled by the following DE:

$$P'(t) = -sP \ln(P/k),$$

where s and k are positive constants. This DE has been proved a successful tool in clinical oncology to model tumor growth. The cells within a tumor do not have access to many nutrients and oxygen as do those on the surface, so the growth rate declines as the tumor increases in size up until the carrying capacity k (which will of course vary with the type and location of the tumor as will the constant s). Eventually the cells inside a tumor stop dividing and die, thus forming the so-called *necrotic* center. Since $\ln(x)$ is undefined at $x = 0$, this model cannot be used for very small values of P (= tumor sizes). For more details, see [EdK-87] and [Mur-03].

EXERCISE FOR THE READER 8.3: (a) Graph the right side of the Gompertz equation $g(P)$, and find all equilibrium solutions (with $P > 0$). Classify each as stable or unstable.
(b) Use MATLAB to produce graphs of solutions to the Gompertz IVPs:
$$\begin{cases} P'(t) = -sP \ln(P/k) \\ P(0) = P_0 \end{cases}, \quad 0 \le t \le 200.$$
With the parameters $s = 0.024$ and $k = 1$ create a single plot with six graphs corresponding to the initial values $P_0 = 0.1, 0.3, \cdots, 1.1$. Are there inflection points? Compare and contrast these graphs.
(c) Use calculus to show that the Gompertz DE can also be written as $P'(t) = ae^{-bt}P$, where a and b are constants.
Suggestion: For part (c), find explicitly the general solution to the Gompertz DE as follows: Introduce the new variable $y = \ln(P/k)$ to translate the Gompertz DE into a very simple (Malthus) DE for y.

EXERCISES 8.2

1. Create graphs of numerical solutions of the following IVPs on the indicated time intervals $a \le t \le b$. In each, $y = y(t)$. Also, determine your numerical approximation of $y(b)$ to five decimals.

    (a) $\begin{cases} (DE) & y' = \sqrt{1+t^5} \quad 0 \le t \le 10 \\ (IC) & y(0) = 2 \end{cases}$

    (b) $\begin{cases} (DE) & y' = \exp(\cos(2^x)) \quad 0 \le t \le 7 \\ (IC) & y(0) = 0 \end{cases}$

    (c) $\begin{cases} (DE) & y' = \cos(e^x \cos(x)) \quad 0 \le t \le 6 \\ (IC) & y(0) = 0 \end{cases}$

(d) $\begin{cases} (DE) & y' = \arctan(e^x) \\ (IC) & y(2) = -4 \end{cases} \quad 2 \le t \le 10$

2. In each part below, we assume that $y = y(t)$ has initial condition $y(0) = 1$ and satisfies the DE given. Determine $\lim_{t \to \infty} y(t)$ (i.e., as time goes to infinity, what value, if any, does the solution approach?). Also find all equilibrium solutions of the DE and classify each as stable or unstable.

    (a) $y' = y(y+1)$

    (b) $y' = y(y^2 - 4)$

    (c) $y' = y^2(y+1)(2-y)$

    (d) $y' = \sin(y)$

    (e) $y' = \cos(y)$

    (f) $y' = y\sin^2(y)$

3. For each of the DEs in Exercise 2 (parts (a) through (f)) explain when a solution will have an inflection point. What will the y-coordinates of these inflection points be?

4   For each of the DEs in Exercise 2 (parts (a) through (f)) use the Euler program in conjunction with a loop to get MATLAB to produce plots of a family of at least 15 solutions of the DE (all in the same plot) that satisfy various initial conditions $y(0) = y_0$. Choose the values of the y_0's so that your solutions will start in at least three different intervals determined by the equilibrium solution values. For each DE, specially choose (after some experimentation) appropriate time intervals $0 \le t \le b$ as well as appropriate y-ranges on the plots (via the axis command) so that the totality of your plots are effectively displayed and accurately depict the main properties of the solutions.

5. A virus culture in a host has an initial population of 10,000 and the carrying capacity of the host is known to be $k = 2$ billion. After 5 days the population grew to 24,000. Assuming logistical growth, determine the natural growth rate r.

6. (*Fishing Yields*) Suppose that for a certain species of fish in a small lake it is determined that the unencumbered annual growth rate is $r = 0.8$ and the carrying capacity of the lake is $k = 1500$ fish. The owner of the lake would like to allow harvesting of fish in this lake at the rate of $n = 200$ fish per year   Starting with the logistic equation and taking into account this annual removal, the DE for the fish population becomes:

    $$F'(t) = rF(1 - F/k) - n.$$

    (a) What initial fish populations $F(0)$ would support this fishing yield? When the yield is supported, what happens to the fish population as $t \to \infty$?

    (b) Get MATLAB to produce a good assortment (of about 15) solutions of this to the DE with several different initial conditions and put them together in a single plot. Explain some similarities and differences of your solutions.

    (c) What is the limit to the amount n of fish per year which could be harvested from this lake? Explain any problems that might arise if the fishing were to push up to this limit. What happens if the limit is exceeded?

7. (*Fishing Yields*) Redo Exercise 6 with the parameters $r = 0.66$ (slower reproducing fish), $k = 20,000$ (larger lake) and $n = 1200$ (more fishing).

8. Prove that the solutions of the Malthus growth DE $P' = rP$ (where $P = P(t)$) that satisfy an initial condition $P(0) = P_0$ are unique.

    Suggestion: Fix one such solution $P(t)$ and show that the quotient $P(t)/e^{rt}$ is a constant function.

9. Prove that the general solution of the (DE) $Q' = rQ + s$ (where $Q = Q(t)$ and r and s are nonzero numbers) is $Q(t) = Ce^{rt} - s/r$.

    Suggestion: Let $P(t) = Q(t) + s/r$. Show that $Q(t)$ solves this DE if and only if $P(t)$ solves the corresponding Malthus growth DE $P' = rP$.

10. (*Use of Predators to Keep Parasites at Bay*) On a certain island that had no cats, the mouse population doubled during the 10-year period from 1960 to 1970. In 1970 when the mouse population reached 50,000, the rulers of the island imported several cats who thereafter killed 6000 mice per year.

    (a) Letting $t = 0$ correspond to 1960, find an expression for $P(t)$ = the mouse population in the range $0 \le t \le 10$.

    (b) Find an expression for $P(t)$ for t in the range $t > 10$

    (c) What was the mouse population in 1980? In 1990?

    Suggestion: For part (b), use the result of the preceding exercise.

11. Consider the DE $y' = y^2 - t$. Use MALTAB to produce the plots of 29 solutions of this DE (approximated via the Euler method with h = 0.001) satisfying the ICs $y_0 = -14, -12, \cdots, 12, 14$. In this same plot, include the graph of the parabola $t = y^2$. Experiment with different t-intervals $0 \le t \le b$ as well as different y-ranges (via the axis command) until you obtain a plot that gives good evidence of some important behavior of these 29 solutions. Compare and contrast these different solutions. How do they behave as $t \to \infty$?

12. This exercise will show how to explicitly solve the logistical growth model equation (3) using the method of **separation of variables**.

    (a) Rewrite the equation (3): $P'(t) (= dP/dt) = rP(1 - P/k)$ so that all the "P" expressions are on the left and the "t" expressions are on the right:

    $$\frac{dP}{P(M - P)} = rdt,$$

    and now integrate both sides and use the initial condition $P(0) = P_0$ to obtain:

    $$P(t) = \frac{k}{1 + (k/P_0 - 1)e^{-rt}}$$

    Suggestion: In order to integrate the left side (if you are doing it by hand), rewrite the integrand as $1/P + 1/(k - P)$.

8.3: MORE ACCURATE METHODS FOR INITIAL VALUE PROBLEMS

Euler's method for numerically solving the IVP (4)

$$(IVP) \begin{cases} y'(t) = f(t, y(t)) & (DE) \\ y(a) = y_0 & (IC) \end{cases}$$

was based on the first-order (tangent line) approximation. If the function $f(t, y)$ on the right-hand side of the DE of (4) is sufficiently differentiable, it seems plausible that we can obtain more accurate methods by using higher-order Taylor polynomials (which can be computed from the function $f(t, y)$ using the DE (4)). This is indeed the case and we will say more about this in the next section. Computing higher derivatives can, in general, be expensive and is not always very feasible, but what is surprising is that there are methods which converge much quicker than Euler's method (and as fast as some of these higher-order Taylor methods) which only require evaluations of $f(t, y)$ (and no higher derivatives of it). In the next section, we will analyze more carefully the relative convergence speeds of the methods we introduce here compared with Euler's method (and with each other) as well as give a more detailed explanation of how these methods came about. For now we will simply introduce two such very practical methods (the improved Euler method and the Runge-Kutta method), state their relative accuracies, write codes for them, and then run them alongside each other, as well as with Euler's method, in order to see some firsthand evidence of the improvements that these methods have to offer.

Just like Euler's method, the two methods we consider require the specification of a step size h and will successively construct a sequence of y-coordinates $y_0, y_1, y_2, \cdots, y_N$ which will be approximations of the actual y-coordinates of the solution[4] of (4) at the equally spaced t-coordinates $t_0 = a, t_1 = a + h, t_2 = a + 2h, \cdots, t_N = a + Nh$. Both of these methods are so-called **one-step methods**, which means that to get from an approximation y_n to the next y_{n+1} we will use only the information y_n, h, t_n and the function $f(t, y)$. In particular, one-step methods "have no memory" of past approximations (only the current one).

We first introduce the so-called **improved Euler method** (also known as **Heun's method**). We shall briefly motivate the method as a natural extension of Euler's method. The precise error analysis will be done in the next section. Note that $t_{n+1} = t_n + h$ and by the fundamental theorem of calculus, we can write (using the DE of (4)):

$$y(t_{n+1}) = y(t_n) + \int_{t_n}^{t_{n+1}} y'(t)\, dt$$
$$= y(t_n) + \int_{t_n}^{t_{n+1}} f(t, y(t))\, dt \approx y_n + \int_{t_n}^{t_{n+1}} f(t, y(t))\, dt \tag{6}$$

In order to obtain y_{n+1}, our approximation of this value, we need to estimate the integral appearing in this formula. Euler's method can be viewed as approximating this last integral by the length of the t-interval on which we are integrating $= t_{n+1} - t_n = h$ times the approximate value of the integrand (function being integrated) evaluated at the left endpoint: $f(t_n, y(t_n)) \approx f(t_n, y_n)$. A more accurate approximation of the integral is obtained if we replace the integrand with something close to the average of the function at the two endpoints—a trapezoidal approximation; see Figure 8.10. To help approximate the value of $y'(t_{n+1})$ in the improved Euler method, we make implicit use of the Euler method as follows:

---

[4] The relevant existence and uniqueness theorem will be presented in the next section

$$y'(t_{n+1}) = f(t_{n+1}, y(t_{n+1})) \approx f(t_{n+1}, y_n + hf(t_n, y_n)).$$

Thus the improved Euler method will approximate the integral in (6) by

$$h\frac{y'(t_n) + y'(t_{n+1})}{2} = h\frac{f(t_n, y(t_n)) + f(t_{n+1}, y(t_{n+1}))}{2}$$

$$\approx h\frac{f(t_n, y_n) + f(t_{n+1}, y_n + hf(t_n, y_n))}{2}.$$

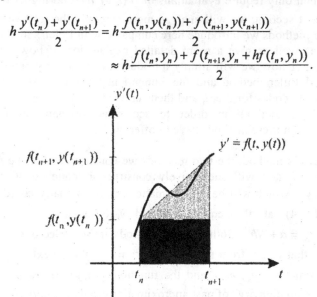

FIGURE 8.10: A graphical comparison of the philosophy of Euler's method versus the improved Euler method. Euler's method attempts to approximate the integral of y' on the indicated interval by the dark gray rectangle; the improved Euler methods attempts to use instead the area of the trapezoid determined by the values of y' at the endpoints.

In summary we have

> **The Improved Euler Method :**
> $t_0 = a,\ y_0 = y(a)$ given
> $h =$ step size
> $t_{n+1} = t + h,$
> $$y_{n+1} = y_n + h\frac{f(t_n, y_n) + f(t_{n+1}, y_n + hf(t_n, y_n))}{2}, \quad n = 1, 2, 3, \cdots$$

We will defer an example until after we also present the classical **Runge-Kutta method**. The method can also be viewed as approximating the integral in (6) with a certain weighted average of (this time four) values of the function $f(t, y)$. It is a bit more difficult to understand how this weighted average has come about and so we will not attempt to motivate it here. It can be derived using Taylor's theorem, the approach of which is given in the next section. The Runge-Kutta method, like Newton's method for rootfinding, is classical, yet highly effective and is the basis for many contemporary production-grade ODE solving programs. We present the method in the box below:

The Runge-Kutta Method:

$t_0 = a$, $y_0 = y(a)$ given,

h = step size

$t_{n+1} = t + h$,

$\quad k_1 = f(t_n, y_n)$
$\quad k_2 = f(t_n + \frac{1}{2}h, \ y_n + \frac{1}{2}hk_1)$
$\quad k_3 = f(t_n + \frac{1}{2}h, \ y_n + \frac{1}{2}hk_2)$
$\quad k_4 = f(t_n + h, \ y_n + hk_3)$

$$y_{n+1} = y_n + \frac{h}{6}(k_1 + 2k_2 + 2k_3 + k_4),$$
$$n = 1, 2, 3, \cdots$$

FIGURE 8.11a: Carle D. T. Runge (1856 –1927), German mathematician.[5]

FIGURE 8.11b: Martin W. Kutta (1867–1944), German mathematician.

In the next example we will compare these two methods with the Euler method. The example will be one where the exact solution can be computed explicitly. The solution gets very large rather quickly, making it easy to compare errors.

EXAMPLE 8.7: Use each of the three methods: Euler, improved Euler, and Runge-Kutta, to numerically solve the IVP:

$$\begin{cases} y'(t) = 2ty \\ y(1) = 1 \end{cases} \quad 1 \le t \le 3,$$

first with step size $h = 0.1$ and then with step size $h = 0.01$. Compare each of the three plots with that of the exact solution $y(t) = e^{t^2-1}$ (see Exercise for the Reader 8.5). In cases where the plots of any of the approximations are too close to compare with that of the exact solution, provide plots of the errors.

SOLUTION: We first create inline functions for both the right side of the differential equation $f(t, y)$ and the exact solution which was provided.

```
>> f=inline('2*t*y','t', 'y'); yexact=inline('exp(t.^2-1)');
```

We give the details of the MATLAB commands for part (a) ($h = 0.1$) only; the changes needed for part (b) are small and obvious. We need to create vectors corresponding to each of the approximation methods.

---

[5] The Runge-Kutta method was first developed by Runge, who was also a physicist, to help him analyze data that came up in his work in spectroscopy. Kutta, who is well known for his work in airfoil theory, extended Runge's method to systems of differential equations (we will give this in the next chapter). Runge originally started off studying literature in college. He eventually switched to mathematics and was greatly influenced by some of his teachers, who included the famous mathematical analyst Karl Weierstrass (1815–1897) and Nobel Prize–winning physicist Max Planck (1858–1947). Runge remained vigorous and prolific throughout his life and published extensively both in mathematics and in physics. During his 70th birthday party he entertained his grandchildren by doing handstands.

```
>> %Euler
>> h=0.1; [t,ye]=euler(f,1,3,1,h);
>> %we may as well use the M-file from thelast section
>> size(t) %need to know this to construct the latter approximations
→ 1   21

>> yie(1)=1; %initialize improved Euler
>> for n=1:20
       yie(n+1)=yie(n) + (h/2)*(f(t(n),yie(n)) + f(t(n+1),...
       yie(n)+h*f(t(n),yie(n))));
end

>> yrk(1)=1;   % initialize Runge-Kutta
>> for n=1:20
       k1=f(t(n),yrk(n));
       k2=f(t(n)+h/2, yrk(n)+h/2*k1);
       k3=f(t(n)+h/2, yrk(n)+h/2*k2);
       k4=f(t(n)+h, yrk(n)+h*k3);
       yrk(n+1)=yrk(n) + h/6*(k1 + 2*k2 + 2*k3 + k4);
end

>> subplot(2,1,1)   %to save space we'll use subplots
>> s=1:.01:3; plot(s,yexact(s)), hold on
>> plot(t,ye,'o',t,yie,'x', t, yrk,'+'), ylabel('y')
>> subplot(2,1,2), plot(t,abs(yexact(t)-yrk))
>> xlabel('t'), ylabel('y'), title('Runge-Kutta Error')
```

This plot is shown of Figure 8.12a. Since the Runge-Kutta plot cannot be distinguished from the exact solution, we create a separate plot (lower plot of Figure 8.12a) of just this error:[6]

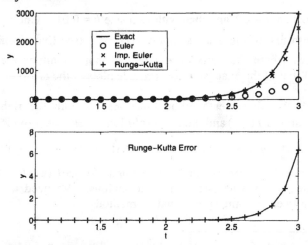

FIGURE 8.12a: Solving the initial value problem of Example 8.7. In the upper plot, the solid blue line is the exact solution. The three approximations, Euler (o o o o), improved Euler (x x x x), and Runge-Kutta (+ + + +), all used step size $h = 0.1$. The lower plot represents the error for Runge-Kutta approximation to the exact solution since the two are indistinguishable in the first plot.

---

[6] We created the legends using the "Data Statistics" menu from the "Tools" menu on the graphics window. This was first done in Chapter 7.

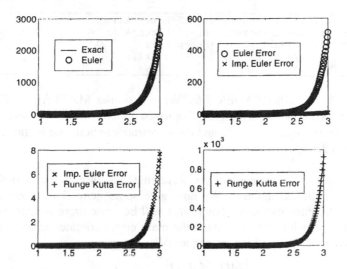

FIGURE 8.12b: In solving the initial value problem of Example 8.7 using step size $h = 0.01$, only the graph of the Euler approximation (o o o o) is distinguishable from that of the exact solution (solid graph) (Upper Left). The remaining three plots compare errors. Upper Right: Euler (o o o o) vs. improved Euler (x x x x); Lower Left: improved Euler vs. Runge-Kutta (+ + + +); and Lower Right: Runge-Kutta. Note the scale of the y-axes to see how very much smaller the errors are with the Runge-Kutta method.

EXERCISE FOR THE READER 8.4: Indicate the changes needed in the creation of the vectors ye, yie, and yrk of the above example. Also, assuming these vectors have been constructed, give MATLAB commands which would produce Figure 8.12b.

EXERCISE FOR THE READER 8.5: The DE of the previous example is separable and thus can be solved exactly by the method outlined in Exercise 12 of the last section. Use this method to derive the general solution of the DE.

It is now a simple matter to modify the codes in the last example to produce M-file programs. We do this for the Runge-Kutta method:

PROGRAM 8.2: An M-file for the Runge-Kutta method for the IVP:

$$\begin{cases} y' = f(t, y) & (DE) \\ y(a) = y_0 & (IC) \end{cases}.$$

```
function [t,y]=runkut(f,a,b,y0,hstep)
% input variables:  f, a, b, y0, hstep
% output variables:  t, y
% f is a function of two variables f(t,y).  The program will
% apply Runge-kutta to solve the IVP:  (DE):  y'=f(t,y), (IC)
% y(a)=y0 on the t-interval [a,b] with step size hstep.  The output
% will be a vector of t's and corresponding y's
t(1)=a; y(1)=y0; nmax=ceil((b-a)/hstep);
for n=1:nmax
    t(n+1)=t(n)+hstep;
    k1=feval(f,t(n),y(n));
```

```
    k2=feval(f,t(n)+.5*hstep,y(n)+.5*hstep*k1);
    k3=feval(f,t(n)+.5*hstep,y(n)+.5*hstep*k2);
    k4=feval(f,t(n)+hstep,y(n)+hstep*k3);
    y(n+1)=y(n)+1/6*hstep*(k1+2*k2+2*k3+k4);
end
```

EXERCISE FOR THE READER 8.6: Write a similar MATLAB M-file, called impeuler, which will perform the improved Euler method to solve the same IVP. The syntax, input variables, and output variables should be identical to those of Program 8.2.

The differences in accuracies of the three methods as evidenced in the above example are quite astounding. We now give some general information of the accuracy of the three methods. The results will be made more precise in the next section, but it is helpful to understand the main error estimates at this point. We say that an iterative method for solving an IVP:

$$\begin{cases} y'(t) = f(t, y) \\ y(a) = y_0 \end{cases} a \le t \le b,$$

is **of order p** (p = 1, 2, 3, ...) provided that whenever the function $f(t, y)$ is sufficiently differentiable the resulting approximations corresponding to a step size $h > 0$: $y_0 = y(x_0)$, $y_1 \approx y(x_1)$, ..., $y_N \approx y(x_N)$ (where $x_N \le b$) satisfy the error estimates:

$$| y(x_n) - y_n | \le c \cdot h^p \text{ for } n = 0, 1, 2, ..., N . \tag{7}$$

Here c is a constant which, in general, depends on the method being used as well as the function $f(t, y)$ and the interval $[a, b]$ on which the IVP is being solved.

To get a feel for differences in orders of convergence, the next example compares the effect of halving the step size in the error bounds of (7).

EXAMPLE 8.8: Suppose we had three different methods #1, #2, and #3 for solving IVPs which had orders 1, 2, and 4, respectively. Suppose also that it is known that for a certain IVP, the constant c in the right side of (7) could be taken to be 2 for all three methods. Find the resulting error bounds (using (7)) for each of the three methods using (a) step size h = 0.1 and (b) half of this, h = 0.05. Compare the results.

SOLUTION: Part (a): Using h = 0.1, and c = 2, (7) would tell us that for Method #1 the error bound is 2(0.1) = 0.2, for Method #2 it would be $2 \cdot (0.1)^2 = 0.02$, and for Method #3 it would be $2 \cdot (0.1)^4 = 0.0002$.

Part (b): Using instead h = 0.05, the resulting error bounds would now be (in the same order): 0.1, 0.005, and 0.00008. Not only were the errors smaller for higher-order methods, but the same decrease in the step size resulted in more sizeable decreases in the error bound (7) with higher-order methods. For the first-order method, halving the step size halved the error bound. For the second-order method, halving the step size resulted in an error bound equal to 1/4 of the original, and for the fourth-order method the error reduction factor was 1/16.

It turns out that Euler's method is a first-order method, the improved Euler method is a second-order method, and the Runge-Kutta method is a fourth-order method. Finding the actual constant c in (7) (or a reasonable upper bound for it) for a certain IVP using one of the methods can be extremely difficult or impossible. In fact it is very often difficult to roughly estimate c. One common practice is to solve the IVP by repeatedly decreasing the step size and comparing the differences until the discrepancy is less than or equal to the tolerance for error. To be on the safe side, one last computation is often done by halving the step size. This does not totally guarantee the desired accuracy, but for almost all well-posed IVPs that come up in practice, this method is quite reliable.

CAUTION: Of course, it is not feasible to try to get a solution with more significant digits than MATLAB can handle (about 15). Theoretically, all of these methods will reach any desired accuracy to the actual solution if the step size is sufficiently small (this follows from (7)). When h gets very small, however, the roundoff errors begin to accumulate and the solutions we get on any floating point computer system begin to loose their accuracy. So, what will happen if we continue to decrease step sizes is that the errors will get smaller and smaller, then stop getting any smaller and afterwards begin to increase (due to roundoff error accumulation)!

Our next example will deal with the problem of the free fall of an object. If we take into account air resistance, the problem becomes very difficult with conventional physics alone. In general, an object moving at a reasonable speed (e.g., a car, a baseball, a plane, a skydiver, or even a bicycle) will have a retarding air resistance force acting in the direction opposite of motion. This air resistance force will be proportional to $|v|^p$, where v denotes the velocity and the exponent p lies between 1 and 2. The exponent p, as well as the constant of proportionality, will depend on things like the size and shape of the object, the speed, as well as even the density and viscosity of the air. In general, faster speeds give larger exponents p and larger constants of proportionality.

EXAMPLE 8.9: (*Physics: Free Fall with Air Resistance*) After a skydiver jumps from an airplane and until the parachute opens, the air resistance is proportional to $|v|^{1.5}$, and the maximum speed that the skydiver can reach is 80 mph.

(a) Plot a graph of the skydiver's vertical falling velocity during the first 10 seconds of fall using the Runge-Kutta method with step size $h = 0.01$ seconds; in the same plot include the corresponding vertical fall velocity if there were no air resistance.

(b) How many seconds (to the nearest 1/100th of a second) would it take for the skydiver to break a falling speed of 60 mph?

SOLUTION: Part (a): Taking the upward vertical direction as positive, there are two forces on the diver: gravity and air resistance. By Newton's second law: $F = ma = m(dv/dt)$ and since gravity's force $= mg$ (m = mass, g = gravity constant of the earth = 32.1740 ft/ sec^2) pulls the diver down (in the negative direction) and air resistance will push the skydiver upward (against the direction of motion) in the positive direction, we arrive at the following differential equation for the velocity of the diver after jumping from the plane (valid until the parachute opens and the air resistance increases considerably)

$$v'(t) = -g + c \,|\, v(t) \,|^{1.5}.$$

The initial condition is $v(0) = 0$ (where we have let $t = 0$ correspond to the time that the skydiver jumped off the plane. Before we solve this IVP, we need to determine the constant c. We can get this by using the fact that when $v(t)$ reaches its maximum 80 mph, we must have $v'(t)=0$, so we can substitute these values into the equation and solve for c. We first arrange things so that both sides of the equation will have the same units. We change mph to ft/sec:

$$80 \frac{\text{mile}}{\text{hr}} \cdot \left(\frac{5280\,\text{ft}}{1\,\text{mile}} \right) \cdot \left(\frac{1\,\text{hr}}{60^2\,\text{sec}} \right) \approx \frac{352}{3} \frac{\text{ft}}{\text{sec}}.$$

Substituting this along with $v' = 0$ and $v = 80$ into the DE now gives $c = 32.1740/(352/3)^{1.5}$. We can now turn the problem over to MATLAB:

```
>> f=inline('-32.1740+32.1740/(352/3)^1.5*abs(v)^1.5','t','v');
>> [t,y]=runkut(f,0,10,0,0.01);
>> plot(t,y*60^2/5280) %gets the v-axis to be in mph
>> free=inline('-32.1740', 't', 'v'); %free fall DE right side
>> [t2,y2]=runkut(free,0,10,0,.01);
>> hold on
>> plot(t2,y2*60^2/5280, '-.')
>> xlabel('Time(seconds)'), ylabel('Velocity of skydiver')
```

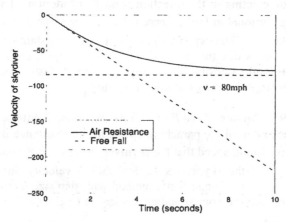

FIGURE 8.13: Comparison of the vertical free fall speed of the skydiver in Example 8.9 with air resistance (solid graph) and with no air resistance (dash-dotted graph). Speed is in mph (vertical axis) and time is in seconds (horizontal axis).

Part (b): A simple while loop will give us the index of the desired time and then we can get the time.

```
>> k=1;
>> while y(k)*60^2/5280 > -60
k=k+1;
end
>>k   →404
>>t(404)   →4.03 seconds (answer).
```

EXERCISES 8.3

1. For each IVP below, do the following: (i) Starting with step size $h = 1/2$, then $h = 1/4$, then $h = 1/8$, etc., continue to use the improved Euler method to compute $y(1)$. Stop when the computed answers for $y(1)$ differ by less than 0.001. How small a step size was needed for the process to stop? (ii) Now do the same using Euler's method.

    (a) $\begin{cases} y'(t) = \cos(ty) + 2t \\ y(0) = 0 \end{cases}$

    (c) $\begin{cases} y'(t) = \dfrac{2t + y^2}{1 + t^2 + y} \\ y(0) = 1 \end{cases}$

    (b) $\begin{cases} y'(t) = y + e^{-y} - t \\ y(0) = -1 \end{cases}$

    (d) $\begin{cases} y'(t) = ty^2 - y \\ y(0) = 0 \end{cases}$

2. Redo all parts (a) through (d) of Exercise 1 but change task (i) to use the Runge-Kutta method instead of the improved Euler method.

3   This exercise will be similar to what was done in Example 8.7. For each part, an IVP is given along with an explicit solution (so these IVPs are rare exceptions where explicit solutions exist). (i) Verify that the explicit function does indeed solve the IVP, (ii) Next, use each of the three methods: Euler, improved Euler, and Runge-Kutta, to numerically solve the IVP on the specified interval using the given step size h. Graph the approximations alongside the exact solution and plot additional error graphs as necessary.

    (a) $\begin{cases} y'(t) = t^3 - 2ty \\ y(1) = 1 \end{cases}$ $1 \le t \le 5$, Exact Solution: $y(t) = \frac{1}{2}(t^2 - 1) + e^{1-t^2}$, $h = 0.1$

    (b) $\begin{cases} r'(t) = \dfrac{r\sin(t)}{1 - \cos(t)} \\ y(0) = 4 \end{cases}$ $0 \le t \le 4$, Exact Solution: $r(t) = 4(1 - \cos(t))$, $h = 0.01$

    (c) $\begin{cases} y'(t) = t^2 \cos(t) + \dfrac{2y}{t} \\ y(2\pi) = 0 \end{cases}$ $2\pi \le t \le 10$, Exact Solution: $y(t) = t^2 \sin(t)$, $h = 0.05$

4. (*Physics: Free Fall with Air Resistance*) Redo the skydiver Example 8.9 changing the air resistance to be proportional to $|v|^{1.1}$ (a more aerodynamic skydiver), but keeping the maximum speed at 80mph. How does the graph differ from that of the example?

5. (*Physics: Free Fall with Air Resistance*) Redo the skydiver Example 8.9 changing the air resistance to be proportional to $|v|^{.9}$ (a less aerodynamic skydiver), but keeping the maximum speed at 80 mph. How does the graph differ from that of the example?

NOTE: The next three problems come from fluid dynamics. *Toricelli's Law*[7] describes how fast the level of fluid falls in a leaking tank. If the tank has cross-sectional area $A(y)$ (where y is the height fluid level measured from the bottom of the tank), and the tank has a hole of area a at its bottom, then the rate at which the fluid level drops is given by the DE

$$y'(t) = -a\sqrt{2gy} / A(y),$$

where g is the Earth's gravitational constant.

6.   (*Draining a Tank*) Suppose a cylindrical tank of radius $R = 20$ feet and height $h = 80$ feet is situated with a flat side on the ground and at the bottom there is a circular hole of radius 5 inches. Initially ($t = 0$), the tank is full with water.
        (a) What does Torricelli's DE look like for this problem?
        (b) Using the Runge-Kutta method (with step size h smaller than one minute), obtain a plot of the height of the water level (in feet) versus the elapsed time (in minutes).
        (c) Using your approximate solution, estimate how long it will take (to the nearest minute) for the tank to drain.
        (d) Redo part (c) using the Runge-Kutta approximation with half of your original step size. Did this significantly affect the answer? If did, explain what needs to be done to get a more accurate answer, if possible, and do it.
        (e) What effect would doubling the area of the drain hole have on the answer to part (a)?

7.   (*Draining a Tank*) Redo Exercise 6 with the same cylinder but this time assume that it is lying on the ground on its round (long) side (with struts to keep if from rolling away). Do you think it would take more or less time for the tank to drain if it is situated like this? Explain. (Of course, if you have done Exercise 6, you will know the answer to this last question.)

8.   (*Draining a Tank*) Redo Exercise 6 this time for a hemispherical tank of radius $R = 20$ feet which is supported with struts so the equator is at the top and the hole is circular of radius 5 inches and at the bottom.

9.   (*Ecology*) · An accidental release of 10 mongooses on a pacific island has resulted in their numbers rising and their subsequent destruction of several species of native birds. An ecologist has been tracking their numbers since their release. Since the food supply is limited and varies with the month (due to seasonal changes), the ecologist has found that the mongoose population $P(t)$ will satisfy the following DE:

$$P'(t) = rP - sP^{1.6},$$

where t is measured in months, r is the natural growth rate of the species, which she has determined to be 0.75, and the values of s (whose factor gives rise to the death rate of mongooses due to limitations in food supply) are given monthly as follows:

t	s		t	s
1 (January)	0.0084		7	0.0026
2	0.0032		8	0.0033
3	0.0014		9	0.0039
4	0.0006		10	0.0042
5	0.0005		11	0.0066
6	0.0011		12	0.0075

        (a) Using the above values for s as constants for each corresponding monthly time interval (i.e., for all of January $0 \le t \le 1$ use the value $s = 0.0084$, next for $1 < t \le 2$ use the value $s = 0.0032$,

---

[7] Evangelista Torricelli (1608–1647) was an Italian physicist who is famous for having invented the mercury barometer.

and so on), apply the Runge-Kutta method with step size $h = 0.1$ (months) to compute and plot the mongoose population for the first year.

(b) Redo part (a) with the initial mongoose population changed to be 100.

(c) In part (a) we assumed that s was constant for a whole month and then abruptly changed to a new value for the next month. Of course, it is more realistic that such changes in s would be continuous. In this part you are to redo part (a) this time using linear interpolation for s between months. Thus, over the range $0 \leq t \leq 1$, take s to be $s(t) = 0.0084 + t(0.0032 - 0.0084)$; this gives a continuous change for $s(t)$ from $s(0) = 0.0084$ to $s(1) = 0.0032$. Continue on in this fashion and then use the Runge-Kutta method as in part (a) to find the mongoose population and graph it. Can you propose an alternative, perhaps more reasonable way to interpolate the data?

10. (*Ecology*) Redo all parts of Exercise 9 with the following change to the differential equation:

$$P'(t) = rP - sP^{1.75}.$$

Use the same data for s and r as well as the same initial populations that were specified in the previous exercise.

8.4: THEORY AND ERROR ANALYSIS FOR INITIAL VALUE PROBLEMS

The hypotheses that will guarantee the initial value problem (4)

$$(IVP) \quad \begin{cases} y'(t) = f(t, y(t)) & (DE) \\ y(a) = y_0 & (IC) \end{cases},$$

to have a solution (existence) and, furthermore, for such a solution to be unique (uniqueness) are quite natural and, as indicated in the previous sections, will automatically be satisfied for most IVPs which come up as real-life models. In this section, we will state the existence and uniqueness theorems and we will also discuss and prove some error estimates for numerical methods for solving IVPs. The error estimation techniques that we introduce have a practical advantage in that they lead naturally to derivations of general one-step numerical methods for solving IVPs.

The function $f(t, y)$, when thought of as a function of the two independent variables t and y (i.e., do not think of y as a function of t), is said to satisfy a **Lipschitz condition** in the y-variable with constant L on the time interval $a \leq t \leq b$ provided that for all such t and for all y_1, y_2, we have

$$|f(t, y_1) - f(t, y_2)| \leq L |y_1 - y_2|. \tag{8}$$

The reason that this condition is a natural one is that if the partial derivative $\partial f / \partial y$ (i.e., just the ordinary derivative of $f(t, y)$ with respect to y, treating t as a constant) is bounded in absolute value by L, $|\partial f / \partial y(t, y)| \leq L$, then the Lipschitz condition (8) will hold. Conversely, if the partial derivatives $|\partial f / \partial y(t, y)|$ are not bounded for all y and for $a \leq t \leq b$ then it can be shown that $f(t, y)$ cannot satisfy a Lipschitz condition in the time interval $a \leq t \leq b$.

EXAMPLE 8.10: Which of the functions $f(t,y)$ given satisfy a Lipschitz condition in the y-variable on $0 \le t \le 1$?

(a) $f(t,y) = (1+t^2)\cos(ty)$ (b) $f(t,y) = y^{1/3}$ (c) $f(t,y) = g(t)$

SOLUTION: Part (a):   Differentiating with respect to y gives $\partial f / \partial y = (1+t^2)(-\sin(ty))t = -t(1+t^2)\sin(ty)$. Taking absolute values, we get $|\partial f / \partial y(t,y)| \le 1(1+1^2) \cdot 1 \le 2$ (since $0 \le t \le 1$) so $f(t,y)$ will satisfy the Lipschitz condition (8) with $L = 2$.

Part (b): $\partial f / \partial y = \frac{1}{3} y^{-2/3}$. This function goes to infinity as $y \downarrow 0$ so $f(t,y)$ cannot satisfy a Lipschitz condition (on any t-interval).

Part (c): $\partial f / \partial y = 0$ (no matter what the function $g(t)$ is) so the Lipschitz condition holds with $L = 0$.

We are now ready to state the existence and uniqueness theorem. We will omit the proof; it can be found in many decent ordinary differential equations textbooks (see, e.g., [Hur-90], [Arn-78], or [HiSm-97]).

THEOREM 8.1:   (*Existence and Uniqueness for Solutions of Initial Value Problems*) Consider the IVP (4) on an interval $a \le t \le b$:

$$(IVP) \quad \begin{cases} y'(t) = f(t, y(t)) & (DE) \\ y(a) = y_0 & (IC) \end{cases}.$$

(a) If the function $f(t,y)$ is a continuous function for all $a \le t \le b$ and all y, then this IVP has a solution which is valid on some interval $a \le t \le a + \delta$ ($\delta > 0$).
(b) If furthermore the function $f(t,y)$ satisfies a Lipschitz condition on $a \le t \le b$, then there is a unique solution which is valid on the whole interval $a \le t \le b$.

REMARK: The Lipschitz condition (8) is required to hold for all y with the same constant L. Often, even when $f(t,y)$ is a very simple function, the inequality (8) will only hold when the y-coordinates are bounded. In such a case it turns out that there will be a unique solution, valid not necessarily on the whole interval but on some subinterval $a \le t \le a + \delta$ ($\delta > 0$) as guaranteed by part (a). The basic pathology that can prevent the IVP from having a (unique) solution on the whole interval is that the solution can "blow up" to infinity in finite time.

EXAMPLE 8.11: (a) Apply the Runge-Kutta method with step size $h = 0.01$ to (attempt) to solve the IVP

$$\begin{cases} y'(t) = y^2 \\ y(0) = 2 \end{cases} \quad 0 \le t \le 1,$$

and plot this solution.
(b) Explain why this solution is not defined for all t in $[0,1]$.

SOLUTION: Part (a):

```
>> f=inline('y^2','t','y')
>> [t,y]=runkut(f,0,1,2,.01);
>> plot(t,y)
```

The plot of this approximation is included as Figure 8.14.

Part (b): As warned in the cautionary note above, even the simple function $f(t, y) = y^2$ of the DE has partial derivative $\partial f / \partial y = 2y$ which is not bounded for all y. The problem can be seen by looking at the exact solution of the IVP. The DE is separable (see Exercise 12 of Section 8.2) and so can be solved explicitly:

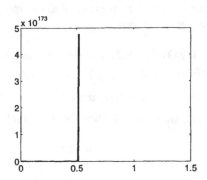

FIGURE 8.14: Plot of the Runge-Kutta approximation to the solution of the IVP of Example 8.11. Note the size of the y-coordinates.

$$y' = y^2 \Rightarrow \frac{dy}{dt} = y^2 \Rightarrow \int y^{-2} dy = \int dt \Rightarrow -y^{-1} = t + C \Rightarrow y = \frac{-1}{t + C}.$$

To get the solution of the IVP from this general solution, we substitute $t = 0$ to find (from the IC) $C = -1/2$, so the solution of the IVP is $y = 1/(0.5 - t)$, which blows up to infinity at $t = 0.5$. The Runge-Kutta method provides us with a rather accurate portrayal of this blowing-up phenomenon. This example shows the importance of checking the hypotheses in the theorem. The simple innocuous-looking expression for $f(t, y)$ in the DE actually gave rise to an explosive growth rate. The solution reached infinity in finite time. For natural phenomena this is certainly not possible. In summary then, this exact solution shows us that it is not possible to find a solution of the IVP on the whole indicated time interval [0,1] (the resulting growth rate is too explosive to allow it).

For differentiable functions $f(t, y)$ as in the example above, unless $| \partial f / \partial y |$ has a uniform upper bound L for all t in $[a, b]$, Theorem 8.1 and the subsequent remark guarantee only that the IVP has a unique **local solution** $y(t)$. This means that $y(t)$ will satisfy the IC and the DE on some time interval of positive length which starts at $t = a$: $a \le t \le a + \delta$. The last example shows that a **global solution** (defined on the whole stretch $a \le t \le b$) may not exist under such circumstances.

We now turn to error estimates for the numerical methods of approximating solutions to the IVP (4):

$$(IVP) \begin{cases} y'(t) = f(t, y(t)) & (DE) \\ y(a) = y_0 & (IC) \end{cases}.$$

For the three methods that we have so far introduced, only Euler's method has a somewhat manageable error estimate. We present this estimate in the next theorem and then proceed along more general lines to obtain error estimates for general methods.

THEOREM 8.2: (*Error Estimates for Euler's Method*) Suppose that the IVP (4) above has $f(t, y)$ satisfying $L = \max\{|\partial f / \partial y(t, y)| : a \le t \le b\} < \infty$ and $M_2 = \max\{|y''(t)| : a \le t \le b\} < \infty$. If Euler's method is used to solve the IVP (4) with step size $= h$, then for any n such that $a \le t_n \le b$, we have:

$$\text{Error} = |y(t_n) - y_n| \le \frac{hM_2}{2L}\left(e^{L(b-a)} - 1\right). \tag{9}$$

REMARK: The right side of (9) can be written as ch, where c is a constant (depending on $f(t, y)$). Thus this theorem gives a quantitative version of the result stated in the last section about the Euler method being a first-order method. Such an explicit theorem is not known for the improved Euler or Runge-Kutta methods. A proof of the Euler method result can be found, for example, in [Hur-90]. At first glance it may seem that the constant M_2 is impossible to calculate without knowing the solution, but the DE and the chain rule help with this job. The use of the theorem is illustrated in the next example.

EXAMPLE 8.12: Suppose we wish to use Euler's method to solve the IVP $\begin{cases} y' = 0.05y \\ y(0) = 10 \end{cases}$ on the interval [0,5], and seek a solution with error being less than 0.05. If we use Theorem 8.3, how small a step size h would be necessary to guarantee this desired accuracy?

SOLUTION: Here $f(t, y) = 0.05y$ so that $\partial f / \partial y = 0.05$. Using the DE twice we get $y''(t) = (0.05y)' = 0.05^2 y = 0.0025y$. So to estimate M_2 we need to get some kind of an upper bound on how large y will get. Ignoring the fact that we can get the general solution here, one could proceed as follows. From the DE and IC, $y'(0) = 0.5$. Now, as long as y is less than 20, we will have $y'(t) \le 1$, so it follows that $y(t) \le y(5) = y(0) + \int_0^5 1 \, dt \le 10 + 5 = 15$. Thus in Theorem 8.2, we can take L = 0.05 and $M_2 = 0.0025 \times 15 = 0.0375$, and the right side of (9) becomes: $\frac{h(0.0325)}{2(0.05)}\left(e^{0.05(5)} - 1\right) \approx 1.06515h$ and for this error bound to be less than the desired 0.05, we would need to take $h < 0.05/1.06515 = 0.046942$.

EXERCISE FOR THE READER 8.7: Perform the Euler approximation with step size $h = 0.046$ (the t's will not quite reach up to 5 but do not worry about this) and compare with the exact solution of this Malthus IVP to see that the resulting actual error of the approximation is < 0.004, quite a bit less than what we had needed. It is typical that the error bound of Theorem 8.3 is conservative since it is a very general result.

We now turn to another approach for estimating error bounds for general one-step methods that will lead to natural constructions of the Euler and improved Euler methods, the Runge-Kutta method, and many others of various orders. The method is focused on the so-called local truncation error, which we introduced at each step in the iterative approximation. For motivation, recall that Euler's method was based on the tangent line approximation, which we rewrite as:

$$y(t_{n+1}) = y(t_n) + hf(t_n, y(t_n)) + \varepsilon_n,$$

where ε_n is the so-called **local truncation error**. From Taylor's theorem, if y is sufficiently differentiable, we have $|\varepsilon_n| \leq Ch^2$, which turns out to give the order $p = 1$ of convergence in Euler's method. To arrive at a more general one-step method, we modify the above formula to a more general one:

$$y(t_{n+1}) = y(t_n) + G(t_n, y(t_n), h) + \varepsilon_n, \tag{10}$$

where the expression "$G(t, y(t), h)$" is allowed to depend on t, $y(t)$, and h. The resulting one-step method from (10) would simply replace the exact values $y(t_{n+1})$ and $y(t_n)$ with the corresponding approximations y_{n+1} and y_n. Efficient numerical schemes arise from intelligent choices for the expression G. The idea is to make the local truncation errors to satisfy

$$|\varepsilon_n| \leq Ch^{p+1}, \tag{11}$$

where p is as large as possible and C is some fixed constant.[8] When this can be done, we say that the one-step method arising from (10) **has local truncation error of order** p. The reason for this terminology is because the following theorem would then imply that the method will yield global errors of order p (as defined in the last section).

THEOREM 8.3: (*Error Estimates for One-Step Methods*): Suppose that the IVP (4) above has $f(t, y)$ satisfying the hypotheses of Theorem 8.1, and, furthermore, that the function $G(t, y(t), h)$ in (10) satisfies a Lipschitz condition in $y(t)$ with constant L on $a \leq t \leq b$, and that the one-step method arising from (10) has local truncation error of order p satisfying (11). If this one-step method is used to solve the IVP (4) with step size $= h$, then for any n such that $a \leq t_n \leq b$, we have:

$$\text{Error} = |y(t_n) - y_n| \leq \frac{Ch^p}{L}\left(e^{L(b-a)} - 1\right). \tag{12}$$

---

[8] Intuitively, for a method of order p, the error at each step is $O(h^{p+1})$, and there are $O(1/h)$ steps, so the total error is $O(h^{p+1}/h) = O(h^p)$. The big-O notation $f(x)$ is $O(g(x))$ will be introduced shortly in the text and means that the inequality $|f(x)| \leq C|g(x)|$ eventually is valid if x lies close enough to some specified (or understood) value. In our estimates, the understood value of h is 0.

Notice that Theorem 8.2 is a special case of this one, in case the Euler method is used. The reader is encouraged to read the proof of this theorem, which can be found in [StBu-92] or [Atk-87].

The most obvious way to improve the Euler method is to use higher-order Taylor polynomials for $G(t,y)$ in (10). We include this as our next example.

EXAMPLE 8.13: (*Higher-order Taylor Methods*) If $f(t,y)$ is p-times differentiable, then it will follow from the DE $y'(t) = f(t,y)$ that $y(t)$ is $(p + 1)$ times differentiable, and in (10) we take G to be the order-p Taylor polynomial less the first term (since it is already included in (10)):

$$G(t_n, y(t_n), h) = hy'(t_n) + \frac{h^2}{2!}y''(t_n) + \cdots + \frac{h^p}{p!}y^{(p)}(t_n).$$

From Taylor's theorem, (10) now implies that

$$|\varepsilon_n| = |R_n(t_n + h)| = \left|\frac{h^{p+1}}{(p+1)!}y^{(p+1)}(c)\right|,$$

where c is a number between t_n and $t_n + h$. Using the chain rule, under these conditions and if the partial derivatives of f (up to order p) which involve y are bounded over the indicated range, it can be shown that this $G(t, y(t), h)$ will satisfy a Lipschitz condition in the y-variable if $f(t, y)$ does (but with a larger constant). From this it now follows from Theorem 8.3 that this pth-order Taylor method is of order p. Since h is usually fixed in a given implementation of such a method, notation is sometimes abused a bit to write $G(t, y(t))$ in place of $G(t, y(t), h)$. Also, if G depends only on y, it is furthermore abbreviated as $G(y)$. Since the derivatives are, in general, expensive and awkward to compute, the method is not so widely used in practice.

EXAMPLE 8.14: Use the third-order Taylor method to solve the IVP

$$\begin{cases} y'(t) = 2ty \\ y(1) = 1 \end{cases} \quad 1 \le t \le 3$$

using step size $h = 0.01$. This was the IVP of Example 8.7 where we compared our three main methods. Use the exact solution of this IVP given in that example and plot the error of the present third-order Taylor approximation.

SOLUTION: The function $G(t_n, y_n)$ is given in the last example, but to get an explicit expression for it we need to use the specific DE for this problem to find $y''(t)$ and $y'''(t)$. This is done here (and in general) using both the DE and the chain rule repeatedly:

$$y''(t) = \left(y'(t)\right)' = \left(2ty(t)\right)' = 2y(t) + 2ty'(t) = 2y(t) + 4t^2y(t) = (2 + 4t^2)y(t),$$

$$y'''(t) = \left(y''(t)\right)' = 8ty(t) + (2 + 4t^2)y'(t) = 8ty(t) + (2 + 4t^2)2ty(t) = (12t + 8t^3)y(t).$$

Thus from the last example (with $p = 3$),

$$G(t_n, y_n) = h(2t_n y_n) + (h^2/2)(2 + 4t_n^2)y_n + (h^3/6)(12t_n + 8t_n^3)y_n.$$

We may now allow MATLAB to take over.

```
>> onestep=inline('y+2*h*t*y+h^2/2*(2+4*t^2)*y+h^3/6*...
(12*t+8*t^3)*y','t','y','h');
>> t=1:.01:3;
>> size(t)
→ 1   201
>> ytay(1)=1;
>> for n=1:200
ytay(n+1)=onestep(t(n),ytay(n),
0.01);
end
>> yexact=inline('exp(t.^2-
1)');
>> plot(t,abs(yexact(t)-ytay))
```

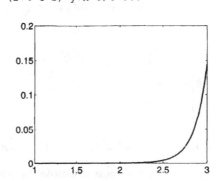

FIGURE 8.15: A plot of the error for the approximation of Example 8.14.

Upon comparing the error for this method with those in Figure 8.12 for the Euler, improved Euler, and Runge-Kutta methods, we see that the error here is significanly less than that of the improved Euler method, but is noticeably more than for the Runge-Kutta method. This makes sense since the order of this method (three) is between those of the latter two methods (two and four).

EXAMPLE 8.15: (*Derivation of a Class of Simple One-Step First- and Second-Order Methods*) Derive all first- and second-order one-step methods (10) which arise from the function $G(t, y)$ being of the following form:

$$G(t_n, y_n, h) = h[a f(t_n, y_n) + bf(t_n + ch, y_n + dhf(t_n, y_n))],$$

where a, b, c, and d are constants.

SOLUTION: The goal is to judiciously choose the parameters a, b, c, and d so that the one-step method arising from (10) will result in the pth-order (with $p = 1$ or 2) estimate (11) being valid. Each such choice will give an order-p one-step method.

We first need to express $G(t_n, y_n)$ in terms of an expression involving f (and some of its partial derivatives) evaluated at (t_n, y_n) plus some error terms involving h. This can be accomplished by repeatedly using Taylor's theorem. To make things more simple when we do this, we omit writing arguments for functions when they are (t_n, y_n). The error terms involving h will all be less than, in absolute value, some constant times a power of h: h^p. Since the individual constants that come up will be unimportant for our present purposes, we will denote each such term as $O(h^p)$. This useful notation is very often used in both pure and numerical analysis; it is commonly and affectionately called the "**big O**" **notation**. In what follows below, we first apply Taylor's theorem to $f(t, y)$ in the t-variable, and next in the y-variable.

$$G = h[af + bf(t_n + ch, y_n + dhf)] =$$

$$= h\left[af + b\left\{ f(t_n, y_n + dhf) + chf_t(t_n, y_n + dhf) + \frac{c^2 h^2}{2} f_{tt}(t_n, y_n + dhf) + O(h^3) \right\} \right]$$

$$= h\left[af + b\left\{ \begin{array}{l} f + dhff_y + \dfrac{d^2 h^2}{2} f^2 f_{yy} + O(h^3) + chf_t + cdh^2 ff_{ty} \\ \qquad\qquad\qquad + O(h^3) + \dfrac{c^2 h^2}{2} f_{tt} + O(h^3) \end{array} \right\} \right]$$

$$= h\left[(a+b)f + bh(dff_y + cf_t) + (bh^2/2)(d^2 f^2 f_{yy} + 2cdff_{ty} + c^2 f_{tt}) \right] + O(h^4).$$

In order to see how best to choose the parameters a, b, c, and d, we will compare the above expansion with that of the corresponding Taylor expansion for the unknown function $y(t)$. In obtaining the expansion below, we will be using the DE and the chain rule repeatedly. Since we wish only to approximate local errors, we assume that $y(t_n) = y_n$. The reader should keep in mind that each time we replace $y'(t)$ with $f(t, y)$ and differentiate the latter, since y is implicitly a function of t, we must use the chain rule.

$$y(t_{n+1}) = y(t_n) + hy'(t_n) + (h^2/2)y''(t_n) + (h^3/6)y'''(t_n) + O(h^4)$$
$$= y_n + hf + (h^2/2)(d/dt)f + (h^3/6)(d^2/dt^2)f + O(h^4)$$
$$= y_n + hf + (h^2/2)(f_t + ff_y) + (h^3/6)(f_{tt} + ff_{ty} + f_t f_y + f^2 f_{yy}) + O(h^4).$$

Now since y_n is already part of the one-step formula resulting from (10), we may equate coefficients of positive powers of h in the last lines of the above two expressions to minimize local truncation errors. Examination of the equations shows that it is only possible to equate the powers of h and h^2 from which we get the following conditions:

$$\text{for } h: \ a + b = 1, \qquad \text{for } h^2: \ bd = 1/2 \text{ and } bc = 1/2.$$

If we take a as an arbitrary real number and $b = 1 - a$, then we get agreement of the h coefficients and hence this leads to a first-order method for any choices of a, b, c, and d. Euler's method comes from the choice $a = 1$. If, furthermore, we require that $b \neq 0$ and $d = c = 1/2b$, then we arrive at a family of second-order methods. The improved Euler method results from the choice $b = 1/2$. When $b = 1$ we get another well-known second-order method given by the recursive formula $y_{n+1} = y_n + hf(t_n + h/2, y_n + (h/2)f(t_n, y_n))$. Of course, in order to use Theorem 8.3 to show us that these methods have the indicated orders, we must know that $G(t, y)$ satisfies a Lipschitz condition in y. This follows nicely from the fact that this is true for $f(t, y)$ and the way in which $G(t, y)$ was defined using $f(t, y)$ (see Exercise 10).

Higher-order methods can be obtained with the method of this example (using more terms, of course). It turns out that if we only use p terms involving $f(t, y)$ per step, we will be able to obtain a method of order p for $p = 1,2,3,4$, but not when $p = 5$. This partially explains the popularity of the classical Runge-Kutta method. All such methods obtainable in this way (with whatever order) are often collectively referred to as "Runge-Kutta methods."

Many popular and effective IVP solver methods used these days are based on Runge-Kutta methods, but they vary the step size. One such method is the Runge-Kutta-Fehlberg Method (abbreviated as RKF45), which is an order-5 method and requires six evaluations per step. Roughly, the way this method works is to figure out two y_{n+1}'s at each iteration: one using an order-5 Runge-Kutta-type method and the other using an order-4 Runge-Kutta-type method. If the two approximations are not close enough to each other (in comparison to the desired error goal), then these approximations are discarded, the step size is reduced, and another such pair of approximations is generated. If the approximations agree nicely, y_{n+1} is taken to be the higher-order one. If the approximations agree with much more accuracy than the desired error goal requires, then the step size is increased for the next iteration. This way, we focus the intensity of the iterations in parts of the solution where the graph is more oscillatory. When the riding is smooth we conserve energy and use large step sizes. This and other more advanced methods will be developed in more detail in the next section. MATLAB has a program which performs a more elaborate version of RKF45 which is also designed to handle systems of ODEs (which we will learn about in the next chapter). For the single IVP (4), the syntax is similar to the functions that we have built.

`[t,y] = ode45('f', [a  b], y0, options)` $\rightarrow$	Numerically solves the IVP $\begin{cases} y' = f(t,y) & (DE) \\ y(a) = y_0 & (IC) \end{cases}$ where the function $f(t,y)$ is stored as an M-file f (or an inline function entered w/out quotes), a is the initial time, b is the final time, and y0 is the value of the function at $t = a$. The last argument is optional.

You can enter `help ode45` for more details on how the program works. In particular, the default goal for the relative error is 10^{-3} and the default goal for the absolute error is 10^{-6}. In fact, MATLAB allows its users to view the actual program. To see it just enter `type ode45` and MATLAB will spit the program out for you on the command window so you can analyze it at will. MATLAB lets you view many of its programs in this way. It is a great way to expand your programming skills.

EXAMPLE 8.16: Use `ode45` to re-solve the IVP of Example 8.7 with default options and plot (only) the error graph. Next reset the default relative tolerance to 10^{-8} and compare the absolute error. Compare both with the result of Example 8.7 where our three basic methods were used to solve the same IVP.

SOLUTION: Since `ode45` does not allow inline functions, we first must store the right side of the DE as an M-file function:

```
f = inline('2*t*y','t','y')
```

Using default settings, `ode45` will now work similarly to our three basic ODE solvers (except no step size is specified).

```
>> [t,y]=ode45(f,[1 3],1);
>> yexact=inline('exp(t.^2-1)');  %from Example
>> subplot(2,1,1) %we will combine the two plots
>> plot(t,abs(yexact(t)-y))
```

Resetting the options in `ode45` takes a bit of special syntax. It is illustrated below. There are several other options that can be adjusted in a similar fashion. To see them all along with their default settings enter `odeset`.

```
>> options =odeset('RelTol',1e-8);
>> [t2,y2]=ode45(f,[1 3],1, options);
>> subplot(2,1,2)
>> plot(t2,abs(yexact(t2)-y2))
```

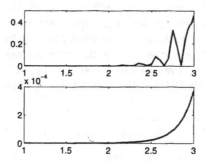

FIGURE 8.16: Plots of the errors that resulted from using MATLAB's `ode45` to solve the IVP of Example 8.7. The top plot used the default options and the latter plot set the goal for the relative error at 10^{-8}.

MATLAB has done quite well. Since (from Figure 8.11) the maximum value of the exact solution (which increases) is about 3000, this shows a relative error at $x = 3$ of about $0.5/3000 = .000167$ for the first approximation (the goal was .001) and a relative error of about 1.333×10^{-8} for the second approximation. The program is very efficient. Tests with `tic...toc` will show it nicely beats even Runge-Kutta when the same accuracy is sought. Some insight can be gained into the efficiency of `ode45` by looking at the size of the vectors constructed (= the number of iterations).

```
>> size(t), size(t2) → 45   1,   297   1
```

One disadvantage of such variable step programs is that the time vectors of each approximation are no longer uniformly spaced. It makes comparison of different plots a bit awkward. MATLAB has a vast library of ODE solver software; other examples include ode23, a lower-order version of ode45, ode113 (a variable order solver; orders from 1 to 13 can be specified), and ode15s (a good one to use if ode45 is not working well). The program ode45 represents the best all-around IVP solver.

We end this section with a few words about the **stability** of an ODE. This is an important concept which permeates many different facets of differential equations. We have seen one version of it already in this chapter. The stability of an ODE will have important effects on how errors propagate when we use a numerical method to solve an IVP. A first-order differential equation $y = f(t, y)$ which satisfies a Lipschitz condition in y on the interval $a \le t \le b$ will always give rise to a unique solution for any IC $y(a) = y_0$. Two different solution curves (for the same DE but with different initial conditions) could never cross one another (see Exercise 11). Also, any point (t, y) in the strip $\{a \le t \le b, -\infty < y < \infty\}$ must have a solution curve passing through it (see Exercise 12). Because of these two facts, the DE can be thought of geometrically as a **flow** throughout the strip $\{a \le t \le b, -\infty < y < \infty\}$. These facts remain true if we allow infinite strips $\{a \le t < \infty, -\infty < y < \infty\}$ (provided the Lipschitz condition still holds). If the strip were a flowing fluid and a small dust particle were to be dropped anywhere in the strip, then it would be carried along some solution curve and eventually reach the line $t = b$. Geometrically, the DE is called **stable** if whenever we start off with two initial conditions that are close, the two solution curves move closer and closer to one another as t advances to b (if t is allowed to go to infinity, the two curves should converge). The DE is called **neutrally stable** if the curves remain nearby but do not actually move together, and it is called **unstable** if the curves diverge away from one another with increasing time. The concept makes sense even for many DEs for which $f(t, y)$ no longer satisfies a Lipschitz condition.

More generally, we can extend this notion of stability to specific y-ranges of the strip. Examples of stable DEs include the Malthus model $y' = ry$ with $r < 0$ on any time interval (see Figure 8.3; all the solutions decay to zero), as well as the logistic DE (see Figure 8.8) on any time interval but on the y-range $y > 0$. Examples of unstable DEs include the Malthus model with $r > 0$ on any time interval and the logistic DE on the y-range $y < 0$. Any equation of the form $y' = g(t)$ is neutrally stable (the solutions $\int_a^t g(s)ds + C$ only differ by constants). The Lipschitz constant does not tell us about the stability since, for example, the two Malthus DEs $y' = ry$ and $y' = ry$ both have the same Lipschitz constant $L = |r|$, but one is stable and the other is unstable.

Note that in any one-step numerical IVP solver, at each iteration, we jump in time by a certain step size and the y-coordinate jumps, in general, to a different solution (flow) curve. The amount of vertical jump from the flow curve we were on to the new one equals what we called the local truncation error. If the equation is stable, then these local truncation errors will decay as time advances, but if it is unstable they will be amplified. Thus for stable DEs, the total error will be less than the sum of the local truncation errors and so the method performs well. For unstable DEs, the total error will exceed (sometimes greatly) the sum of the local truncation errors, so the method does not work as nicely. This phenomenon is illustrated in Figures 8.17 and 8.18.

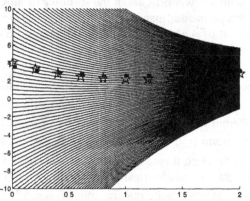

FIGURE 8.17: In the stable DE $y' = 2t - y$ the solution curves move closer to one another as time advances, making the local truncation errors in a numerical IVP solver tend to zero as time advances. An exact solution curve is followed by the pentagrams; the computed values are the heavy black segments.

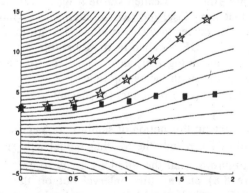

FIGURE 8.18: In the unstable DE $y' = 0.8y - 0.5y\cos(3t)$ the solutions curves move apart as time advances, making the local truncation errors in a numerical IVP solver tend to be amplified as time advances. An exact solution curve is followed by the pentagrams; the computed values are the heavy black segments.

Fortunately, there is a simple criterion for determining stability of a DE on a certain region of the form $\{a < t < b, c < y < d\}$ of the (t, y)-plane (here any of a,

b, c, d can take on infinite values). If $\partial f / \partial y < 0$ in such a region, then the DE $y' = f(t, y) =$ is stable in the region whereas if $\partial f / \partial y > 0$ then the DE is unstable in that region. In general, more negative/positive values of this partial derivative result in greater degrees of stability/instability.

EXERCISES 8.4

1. Which of the following functions satisfy a Lipschitz condition in the y-variable on the t-interval $0 \le t \le 2$? For those that do, find a corresponding Lipschitz constant L.

   (a) $f(t, y) = 6t\sin(ty) + \cos(ty)$ (c) $f(t, y) = t^2 e^{-y}$

   (b) $f(t, y) = t^3 - y^3$ (d) $f(t, y) = \cos(t^2 + y^2)$

2. For each of the following IVPs, first derive the recursion formula for Taylor's second-order method, and then use it to solve the given IVP on the indicated t-interval with the indicated step size. Plot the resulting solution along with the corresponding improved Euler solution obtained by using the same step size. In cases where the two plots are indistinguishable, provide also a plot of the absolute value of the difference of these two approximations.

   (a) $\begin{cases} y'(t) = \cos(t^2 y) \\ y(0) = 2 \end{cases}, \; 0 \le t \le 5; \; h = 0.1$

   (b) $\begin{cases} y'(t) = 2 + y/t^2 \\ y(1) = 0 \end{cases}, \; 1 \le t \le 7; \; h = 0.1$

   (c) $\begin{cases} y'(t) = e^{t/2} + \cos(y) - y \\ y(0) = 0 \end{cases}, \; 0 \le t \le 4; \; h = 0.05$

   (d) $\begin{cases} y'(t) = \dfrac{2 + y^3}{1 + ty^2}, \; 0 \le t \le 4; \; h = 0.05 \\ y(0) = -1 \end{cases}$

3. Repeat parts (b) and (c) of Exercise 2, this time using the third-order Taylor approximation, but still comparing to the improved Euler solution.

4. Repeat parts (b) and (c) of Exercise 2, this time using the fourth-order Taylor approximation, and now comparing to the Runge-Kutta method.

5. (*Simpon's Rule Is Special Case of Runge-Kutta Method*) Show that if the Runge-Kutta method is used to solve the IVP $y' = f(t), y(a) = 0$ over $[a, b]$ using $h = (b - a)/N$, it produces the formula:

$$\int_a^b f(t)dt \; (\equiv y(b)) \approx \frac{h}{6} \sum_{n=0}^{N} \left[f(t_n) + 4f(t_n + h/2) + f(t_{n+1}) \right],$$

which is known as **Simpson's rule** for approximating definite integrals.

6. For each of the DEs given below, find regions $\{a < t < b, c < y < d\}$ on which the DE is stable/unstable. Try and account for as much of the (t, y) plane as possible.

   (a) $y' = y^2 - 8y$ (b) $y' = \arctan(y)$ (c) $y' = y + 3t$ (d) $y' = 4t - t\sin(y)$

7. Provide an example of a neutrally stable DE of the form $y'(t) = f(t, y)$, where the function

$f(t,y)$ does not depend on t alone, and the neutral stability is valid on the entire region $\{0 < t < \infty, -\infty < y < \infty\}$.

8.   Here is an example of an extremely unstable differential equation: $y' = 100y - 101e^{-t} \equiv f(t,y)$.
Its general solution is given by $y(t) = e^{-t} + Ce^{100t}$. Show that the solution of the IVP
$$\begin{cases} y' = f(t,y) & (DE) \\ y(0) = 1 & (IC) \end{cases} \text{ on } 0 \le t \le 2 \oplus \text{ decays toward zero as time advances, but that solutions}$$
of the same DE with a slightly perturbed (IC): $y(0) = 1 + c$ do not. How small a step size would
the Runge-Kutta method need to solve the original IVP to within an error < 0.1? Is it possible
for MATLAB to do this or would the roundoff errors become too significant? Justify your
claims and use MATLAB to provide some numerical evidence.

9.   Is it possible for a solution curve of a DE to pass into both a region of stability and a region of
instability? Either provide an example or explain why it is not possible.

10.   Suppose that $f(t,y)$ satisfies a Lipschitz condition in the y-variable with constant L. Prove that
the function $G(t,y)$ as defined in Example 8.15 satisfies a Lipschitz condition in the y-variable
with constant $L_1 = \left[|1 - b| + |b| + \dfrac{HL}{2}\right]L$, whenever $0 \le h \le H$.

11.   Prove that if $f(t,y)$ satisfies a Lipschitz condition in the y-variable on the range $a \le t \le b$ and
$y_1(t)$ and $y_2(t)$ are solutions of the DE $y' = f(t,y)$ on $a \le t \le b$, then either these curves are
identical or they never cross.
Suggestion: Assume that the curves crossed at some value $t = c$. If $c < b$, use Theorem 8.3 to
show the curves agree also for $t > c$. If $c > a$, consider the DE $y'(t) = -f(-t,y)$ and look at
$y_1(-t)$ and $y_2(-t)$.

12.   Prove that if $f(t,y)$ satisfies a Lipschitz condition in the y-variable on the range $a \le t \le b$ and if
(t,y) is any point in the strip $\{a \le t \le b, -\infty < y < \infty\}$, then there exists a real number y_0 such
that the solution curve of the
$$\text{IVP:} \quad \begin{cases} y' = f(t,y) & (DE) \\ y(a) = y_0 & (IC) \end{cases}$$
passes through the point (t,y).

13   Verify the Lipschitz condition statement about the function $G(t, y(t), h)$ of Example 8.13.

8.5: ADAPTIVE, MULTISTEP, AND OTHER NUMERICAL METHODS FOR INITIAL VALUE PROBLEMS

In this section we briefly survey some of the more sophisticated methods for the
numerical solution of initial value problems, which serve as a basis for
contemporary production quality codes. We begin by describing adaptive methods
that will vary the step size as the iterations progress in a way that uses smaller step
sizes when needed (to reach accuracy goals) but otherwise will allow large step
sizes so as to avoid unnecessary computation. Subsequently we will describe
some implicit methods and contrast them with the explicit methods that have been

used exclusively up to this point. We will then move on to describe the idea behind a multistep method and give some typical examples. Finally, we end the section with a further discussion of stability. We will contrast the purely mathematical concept of stability, which was introduced in the last section, with numerical stability, which depends on the particular algorithm being used.

An **adaptive** initial value problem solver is any which uses some sort of check on the local truncation error at each iteration, and adjusts the step size accordingly. If the local error estimate is too large, a smaller step size is used. If it is too small, the step size is increased for the next iteration. In all other cases, the step size is maintained and the method progresses. With a constant step size, we need to set it according to the worst-case behavior of the DE; with an adaptive method, choice of a suitable step size is no longer an issue.

One rather plausible method of checking the local error for a given iteration with step size h would be to compare the result with that resulting (with the same method) from making two smaller steps of size $h/2$. A more efficient way would be to use two related schemes of different orders to approximate the next step value of the solution. An accurate estimate of the local truncation error would be the difference of these two approximations. If it is too large, the step size is cut in half and the computation is repeated. Otherwise, the higher-order approximation is accepted and we move on to the next iteration with the proviso that the step size is doubled if the measured error is very small. Because of their diversity, Runge-Kutta-type methods are often used with such schemes. Oftentimes, the Runge-Kutta methods are chosen so that the computations of each of the two different approximations share many common computations. This will be the case in the so-called **Runge-Kutta-Fehlberg method** which we now describe. This algorithm, abbreviated as **RKF45**, will be based on the following 4th- and 5th-order Runge-Kutta schemes:

The Runge-Kutta-Fehlberg Method (RKF45) for Solving the IVP: (3)

$$(IVP) \quad \begin{cases} y'(t) - f(t, y(t)) & (DE) \\ y(a) = y_0 & (IC) \end{cases}$$

$t_0 = a$, $y_0 = y(a)$ given, $h =$ initial step size, $\varepsilon =$ error tolerance

Iterative Steps: Compute

$k_1 = hf(t_n, y_n)$,

$k_2 = hf(t_n + \frac{h}{4}, y_n + \frac{1}{4}k_1)$,

$k_3 = hf(t_n + \frac{3h}{8}, y_n + \frac{3}{32}k_1 + \frac{9}{32}k_2)$,

$k_4 = hf(t_n + \frac{12h}{13}, y_n + \frac{1932}{2197}k_1 - \frac{7200}{2197}k_2 + \frac{7296}{2197}k_3)$,

$k_5 = hf(t_n + h, y_n + \frac{439}{216}k_1 - 8k_2 + \frac{3680}{513}k_3 - \frac{845}{4104}k_4)$,

$k_6 = hf(t_n + \frac{h}{2}, y_n - \frac{8}{27}k_1 + 2k_2 + \frac{3544}{2565}k_3 + \frac{1859}{4104}k_4 - \frac{11}{40}k_5)$.

From these form the order 4 Runge-Kutta approximation:

$$z_{n+1} = y_n + \frac{25}{216}k_1 + \frac{1408}{2565}k_3 + \frac{2197}{4104}k_4 - \frac{1}{5}k_5 ,$$

and also the order-5 Runge-Kutta approximation:

$$y_{n+1} = y_n + \frac{16}{135}k_1 + \frac{6656}{12,825}k_3 + \frac{28,561}{56,430}k_4 - \frac{9}{50}k_5 + \frac{2}{55}k_6 .$$

Compute the local error estimate using:

$$E = |\, y_{n+1} - z_{n+1}\, | = |\, \frac{1}{360}k_1 - \frac{128}{4275}k_3 - \frac{2197}{75240}k_4 + \frac{1}{50}k_5 + \frac{2}{55}k_6\, |$$

If $E > h\varepsilon$ (step size is too large) reduce h to $h/2$ and repeat above computations.
If $E < h\varepsilon/4$ (step size is too small) accept y_{n+1} but increase h to $2h$ for next
iteration. Otherwise (step size is good), accept y_{n+1} and continue iteration.

Some comments are in order. At each iteration, in the notation of the above
algorithm, the local truncation error of the fourth-order RK method is essentially
E/h. To see this, we let ε_n denote the local truncation error of the 4th-order
method at the nth iteration and we write:

$$|\, \varepsilon_n\, | = |\, y(t_{n+1}) - z_{n+1}\, | \approx |\, y(t_{n+1}) - y_{n+1}\, | + |\, y_{n+1} - z_{n+1}\, | = O(h^6) + E \approx E$$

The approximations hold true since $\varepsilon_n = O(h^5)$ (and the $O(h^6)$ is much smaller
for small values of h).[9]   Such RK methods as the ones implemented in the above
algorithm can be derived using Taylor's theorem (as was done in the last section
for first- and second-order methods), but the algebra gets very complicated quite
quickly as the order of the RK method increases.   See Exercises 22–24 for
examples of this type of construction.

In the criterion $E < h\varepsilon/4$ for a step size being too small, the factor $1/4$ could
have been any fraction (less than one).   The construction aims to keep the
truncation error more or less constant. It is not difficult to put the above RKF45
algorithm into a MATLAB code.   One additional feature to put in such a code is to
require that the step size does not get too large or too small (by prescribing
minimum and maximum values for the step size h).   If the program requires a step
size smaller than the minimum permitted, it should either terminate or at least give
an error message indicating what has happened.   We will leave the writing of such
an M-file as our next exercise for the reader. Producing production-level codes,
however, such as MATLAB's ode45, which is a variation of RKF45, takes a lot
more work.   For more related issues behind this and other MATLAB ODE solvers
we refer to [ReSh-97].   Up to about 1970, the standard Runge-Kutta method was
the most commonly used numerical method for solving IVPs.   Since then the

---

[9] For the fourth-order method, from the $\varepsilon_n = O(h^5)$ estimate for the local truncation error at each
iteration, we obtain a rough estimate for the global error for a time interval of unit length by: (# of
iterations) $\cdot \varepsilon_n = (1/h) \cdot O(h^5) = O(h^4)$.   This assumes stability and also that h is constant.   For the latter
assumption, think of it as an average.   The lost factor of h in going from the local truncation error to
the global error estimate is the reason that the factor of h is being multiplied by the (global) error
tolerance in the above RKF45 algorithm.   For a more detailed error analysis and general development
of Runge-Kutta-type methods, we refer to Chapter 6 of [Atk-89].

variations of the RKF45 method seem to have become the most popular general IVP numerical solvers.

EXERCISE FOR THE READER 8.8: (a) Write a function M-file for the RKF45 method for solving the IVP (4): $(IVP) \begin{cases} y'(t) = f(t, y(t)) & (DE) \\ y(a) = y_0 & (IC) \end{cases}$, which has the following syntax:

```
[t, y] = rkf45(f, a, b, y0, tol, hinit, hmin, hmax)
```

where the two output variables and first four inputs are exactly as in the runkut M-file of Program 8.2. The input variable tol is optional and specifies the error goal. The default value is tol=1e-6. The last three input variables are optional and have to do with the step sizes: hinit is the initial step size used, hmin is the minimum allowable step size, and hmax is the maximum allowable step size. The default values of these optional input variables are as follows: hinit=(b-a)/100, hmin = (b-a)/1e-5, hmax = (b-a)/2. Set it up so that if the minimum step size is reached the program will still run but will produce a message of what transpired.
(b) Run the program rkf45 to re-solve the IVP of Example 8.7 with the default settings. Compare the resultant error and number of iterations with the corresponding data for the standard Runge-Kutta method that was obtained in Example 8.7.

Adaptive methods are particularly useful for numerical solutions of DEs which undergo abrupt changes. Such DEs can often be recognized by the presence of coefficient functions which are discontinuous or vary rapidly over certain regions. DEs with discontinuous coefficients come up naturally with problems involving electric circuits as well as certain mechanical problems involving discontinuous forces, such as in tracking the velocity of a skydiver or the effects of an earthquake on a certain mechanical structure. Our next example shows how an adaptive solver such as RKF45 will automatically give more attention to such trouble areas.

EXAMPLE 8.17: Consider the following IVP: $\begin{cases} y' + b(t)y = t, & y = y(t) \\ y(0) = 1 \end{cases}$, where

the coefficient function is given by: $b(t) = \begin{cases} 1, & \text{if } 0 \leq t \leq 2 \\ 3, & \text{if } 2 \leq t \end{cases}$.

(a) Solve this IVP on the interval $0 \leq t \leq 3$ using the program rkf45 using the value of 0.0001 for tol.
(b) Count the number of time values (iterations) used in part (a), and resolve this problem using the runkut program with a step size chosen so that the number of iterations will be about equal to the number used in part (a).

(c) Compare both numerical solutions in parts (a) and (b) with the exact solution:[10]

$$y(t) = \begin{cases} t-1+2e^{-t}, & \text{for } 0 \le t \le 2, \\ \frac{1}{3}t - \frac{1}{9} + e^{-3t}(\frac{4}{9}e^6 + 2e^4), & \text{for } t > 2. \end{cases}$$

SOLUTION: Part (a): Discontinuous functions such as $b(t)$ are not well suited to be stored as inline functions, so we first create the following M-file for $f(t,y) = t - b(t)y$ (note: $f(t,y)$ is as in (4)):

```
function f = eg0817(t,y)
if (0< =t< =2)
 f=t-y;
else
 f=t-3*y;
end
```

The following commands will now solve the IVP with the specified numerical method and create a plot of the numerical solution using green circles to show the step locations.

```
>> [t,yrkf]=rkf45('eg0817',0,3,1,1e-4);
WARNING: Minimum step size has been reached; it is recommended to run the
program again with a smaller hmin and or a larger tol
>> plot(t,yrkf,'g-o'), size(t)
→ans = 1   29
```

Thus 29 steps were used. The resulting plot is shown in Figure 8.19 (the green one). Notice from the warning that our default step size (= 1e–5) has been reached. Such occurrences are quite normal for such discontinuous IVPs.

Part (b): Using 29 steps, the corresponding Runge-Kutta numerical solution is created and plotted (with red x's along with the curve of part (a)) by the following commands:

```
>> [trk,yrk]=runkut('eg0817',0,3,1,3/29);, hold on
>> plot(trk,yrk, 'r-x')
```

Part (c): So as to facilitate easy plotting we first create an M-file for the exact solution as follows:

```
function y = eg0817b(t)
for i = 1:length(t)
if (0<=t(1) & t(i)<=2)
 y(i)=t(i)-1+2*exp(-t(i));
```

---

[10] Such IVP's with discontinuous data are often amenable to solution by so-called Laplace transform methods, which are covered in any standard textbook on the analytical theory of ODEs (see, e.g., [Asm-00]). Alternatively, this one could be solved using the Symbolic Math Toolbox with the following strategy. Find the general solution of the DE using $b(t) = 1$ and use the IC to determine the unknown constant. This will be the first half of the solution, valid for $0 \le t \le 2$. Next find the general solution of the DE using $b(t) = 3$ and adjust the constant so the function matches with the first one at $t = 2$. This will be the second half of the solution, valid for $t > 2$.

```
else
  y(i)=t(i)/3-1/9+exp(-3*t(i))*(4*exp(6)/9+2*exp(4));
end
end
```

Now we may add the plot of the exact solution (in blue) onto the graph containing the two numerical plots:

```
>> plot(0:.01:3, eg0817b(0:.01:3),'b')
```

Figure 8.19 shows the end result. To compare more accurately the two numerical solutions, we next create plots of their respective errors; the results are shown in Figure 18.20.

```
>> plot(t,abs(yrkf-eg0817b(t)), 'g-o')
>> title('Error for RKF45 Solution')
>> plot(trk,abs(yrk-eg0817b(trk)), 'r-x')
>> title('Error for Runge-Kutta Solution')
```

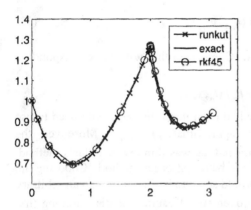

FIGURE 8.19: Comparison of the adaptive RKF45 solution and the standard Runge-Kutta solution with the exact solution of the IVP in Example 8.17. Although the number of data points of each of the numerical methods is the same (29), the adaptive RKF45 had a much more interesting and intelligent deployment of data points, concentrating them more in the area of the jump discontinuity of the data at $x = 2$. Both numerical solutions are rather good up to $x = 2$, but for $x > 2$ the standard Runge-Kutta solution is not as good—see Figure 8.20.

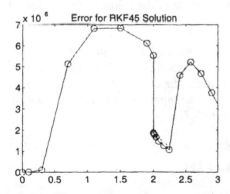
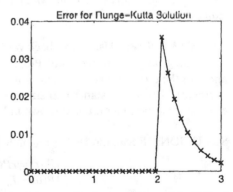

FIGURE 8.20: Comparison of the error plots for the two numerical solutions of the IVP of Example 8.17. (a) (left) The error plot for the RKF45 solution is much smaller than that for the Runge-Kutta solution in (b) (right). Indeed, by comparing y-axes scales, we see that the maximum error for RKF45 is on the order of 10^{-4} of that for the Runge-Kutta solution.

We point out that the standard Runge-Kutta method is only a fourth-order method so at first glance the above comparisons may seem a bit unfair. The reader can check, however, that the results would not be much different if we instead used the fifth-order RK method that is part of the RKF45 scheme; see Exercise 14.

We next move on to a brief discussion of implicit numerical methods for IVPs. Thus far, all of our numerical methods have been **explicit**, meaning that the value of the next approximation y_{n+1} was always expressed explicitly in terms of other known information. A numerical method for which y_{n+1} is merely expressed as the implicit solution of some equation (which is not usually analytically solvable for y_{n+1}) is called an **implicit** method. As a prototypical example of an implicit method, we describe now the so-called **backward Euler method** for solving the IVP (4) $\begin{cases} y'(t) = f(t, y(t)) & (DE) \\ y(a) = y_0 & (IC) \end{cases}$:

$$y_{n+1} = y_n + h_n f(t_{n+1}, y_{n+1}).\tag{13}$$

We have allowed for variable step sizes. Comparing this with the corresponding formula for the (original) Euler method

$$y_{n+1} = y_n + h_n f(t_n, y_n),$$

it would seem a lot less practical. Indeed, at each iteration, we would need to use some rootfinding method to compute (approximately) y_{n+1}. Moreover, the resulting benefits, if any, are unclear. Indeed, as was done with Euler's method, the backward Euler method can be shown to be a first-order method. Choosing the slope at the end of the interval (t_n, t_{n+1}) would seem to be no more than an arbitrary modification rather than a plausible improvement. Despite this discouraging first impression, implicit methods like the backward Euler method do have their merits. Before justifying this statement, we give an example of the usage of the backward Euler method.

EXAMPLE 8.18: Using the backward Euler method in conjunction with the rootfinding program newton of Program 6.2, re-solve the IVP of Example 8.5, and plot the error of this numerical solution (using the exact solution given in that example). Use a (constant) step size $h = 0.01$. Compare with the corresponding error plot for the Euler method shown in Figure 8.7.

SOLUTION: Recall the IVP of Example 8.5:

$$\begin{cases} P'(t) = rP(1 - P/k) \\ P(0) = P_0 \end{cases},$$

where $r = 0.0318$, $k = 200$, and P(0) = 3.9. Using a constant step size $h_n = h$, the Euler's backward method (13) becomes: $p_{n+1} = p_n + hrp_{n+1}(1 - p_{n+1}/k)$. Although this quadratic formula for p_{n+1} is easily solved, we will nevertheless take a more general approach using Newton's method. The M-file `newton` from Chapter 6 cannot be directly applied here since the equation needed to solve changes at each iteration. Recall the syntax of `newton` (Program 6.2):

```
[root, yval] = newton(f, df, x0, tol, nmax)
```

requires that we enter the function for which we seek a root, as well as its derivative, as inline functions. Since the function (here) changes (slightly) at each iteration, the following useful MATLAB command will allow us to incorporate such changes in the string for the function in a loop:

st=num2str(a,n) →	This command converts a number a to a string st of length at most n which represents the number a. The inputs are a number a and a positive integer n, $n \le 15$.

Using P for the variable p_{n+1} in (13), we can view the solutions of (13) as the roots of the function:

$$g(P) = P - p_n - hrP(1 - P/k).$$

We will also need the derivative of this function:

$$g'(P) = 1 - hr(1 - P/k) - hrP(-1/k) = 1 + hr(2P/k - 1).$$

The following code will now solve the IVP with the backward Euler method. Before running this code, the reader may wish to modify the M-file for `newton` so that the convergence statement is suppressed (otherwise this will be printed on the screen 2000 times!).

```
t=0:.1:200; P(1)=3.9;
%since the derivative does not change, we compute it outside the
loop
gprime=inline('1+ 0.01*0.0318*(2*P/200-1)');
for n=1:2000
    g=inline(['P-' num2str(P(n),15) '-.1*.0318*P*(1-P/200)']);
    P(n+1)=newton(g,gprime,P(n));
end
```

If we plot this function and the error as in Example 8.5, we see that the results are graphically indistinguishable from those of the ordinary Euler method (Figures 8.6 and 8.7).

In order to write a function M-file that will be able to perform the backward Euler method on an arbitrary IVP, we would need to make use of MATLAB's symbolic toolbox capabilities (so that the differentiation could be done automatically). This task will be completed in the following program.[11]

---

[11] We remind the reader that since such symbolic capabilities are not very often needed in this book (since exact arithmetic is more expensive and often unnecessary), we do not spend a lot of time explaining symbolic toolbox capabilities. The program is mainly given as an illustration, in case the reader might wish to write a similar sort of program. For more details about MATLAB's symbolic toolbox, we refer to Appendix A. We point out two useful commands that are used in this M-file.

PROGRAM 8.3: An M-file for the backward Euler method for the IVP:

$$\begin{cases} y'(t) = f(t, y(t)) & (DE) \\ y(a) = y_0 & (IC) \end{cases}.$$

```
function [t, y] = backeuler(f, a, b, y0, h)
% Performs the backward Euler method to solve the IVP y'(t) = f(t,y),
% y(a) = y0.  Cals M-file 'newton' for rootfinding and uses symbolic
% capabilities.
% Input variables:  f a function of two variables f(t,y) describing
% the ODE y' = f(t,y).  Can be an inline function or an M-file
% a, b = the left and right endpoints for the time of the IVP
%    y0 the initial value y(a) given in the initial condition
% h   the step size to be used
% Output variables:  t - the vector of equally spaced time values for
% the numerical solution, y - the corresponding vector of y
% coordinates.
syms ys
t(1)=a; y(1)=y0;
nmax=ceil((b-a)/h);
for n=1:nmax
    t(n+1)=t(n)+h;
    g=inline(char(ys-y(n)-h*f(t(n+1),ys)), 'ys');
    gp=diff(g(ys));
    gprime=inline(vectorize(char(gp)), 'ys');
    y(n+1)=newton(g,gprime,y(n));
end
```

The above M-file could be invoked to re-solve the above example; just enter:

```
>> [t, y] = backeuler(f, 0, 200, 3.9, .1);
```

Having seen firsthand all of the extra complexity needed for an implicit method, a good question to ask would be why anyone would bother using them. The answer lies in the fact that the backward Euler method (and other implicit methods) often have better numerical stability than their explicit counterparts. We have already explained the concept of (theoretical) stability for an IVP. Even when we are solving an IVP which is theoretically stable, the numerical method may not be stable but conversely, a numerical method may be numerically stable for an IVP which is not mathematically stable. Recall that an IVP is stable if small perturbations of the IC lead to solutions which converge to the desired solution as time goes on. It is unstable if small perturbations can lead to solutions which diverge away from the desired solution as time goes on. We say that an IVP is **numerically stable**[12] with respect to a certain numerical method if the following condition holds:

$$\lim_{h \to 0} \max\nolimits_{a \le t \le b} |y(t) - \tilde{y}(t)| = 0,$$

---

char(...) is used to convert expressions containing symbolic variables into strings; vectorize(...) converts a string formula into vector capability notation (i.e., the dot is inserted before any *, /, or ^). Once a symbolic expression is differentiated, any such dots that were present will disappear, so it is necessary to reinsert them

[12] We caution the reader that the word "stability" is one of the most often used words in numerical differential equations and unfortunately its definitions can vary significantly from author to author and even among different works by the same author.

where $y(t)$ is the exact solution of the IVP $\begin{cases} y'(t) = f(t, y(t)) & (DE) \\ y(a) = y_0 & (IC) \end{cases}$,

$a \leq t \leq b$ and $\bar{y}(t)$ is the numerical solution with the method using step size (at most) h. This condition basically states that we can get the numerical solution to be (uniformly) as close as we like to the actual solution by taking the step size sufficiently small. We say the method is **numerically unstable** for a particular IVP if it is not numerically stable for it. Our next example will explain this concept in the setting of a very simple IVP.

EXAMPLE 8.19: Consider the following IVP:

$$\begin{cases} y'(t) = ry \\ y(0) = y_0 \end{cases}. \tag{14}$$

We know (see Section 8.4) that this equation is (theoretically) stable if $r < 0$ and unstable if $r > 0$.
(a) With the Euler method and for $r < 0$, for which step sizes is this method numerically stable?
(b) Repeat part (a) for the backward Euler method.

SOLUTION: Part (a): For (14), the Euler method reads as follows:

$$y_{k+1} = y_k + hry_k = (1 + rh)y_k.$$

Iterating this produces the explicit formula:

$$y_k = (1 + rh)^k y_0. \tag{15}$$

Think of $(1 + rh)$ as a *magnification factor*. Recall the exact solution of (14) (see Section 8.2) is $y(t) = y_0 e^{rt}$ and this converges to 0 as $t \to \infty$. In order for the expression on the right side of (14) to also converge to zero, it is equivalent that $|1 + rh| < 1$ (unless $y_0 - 0$). Since $r < 0$ this means that $1 + rh > -1$ or equivalently, $(0 <)h < -2/r$. This range for the step size is sometimes referred to as the **region of numerical stability**.[13] Note that if the step size is outside of this range, the Euler method will diverge, even though the IVP is numerically stable. Note that for very large negative values of r, although the IVP becomes increasingly theoretically stable (the solutions converge to zero very rapidly), the region of numerical stability for the Euler method gets very small. For example if $r = -100$, Euler's method would diverge unless $h < 0.02$. This type of behavior is prototypical in what are known as **stiff** initial value problems. These problems have a solution of the form $y(t) = e^{-ct} + s(t)$, where c is a large positive constant. The term e^{-ct} is called the **transient part** of the solution and $s(t)$ is called the

---

[13] In more advanced treatments, e.g., [Atk-87], the region of stability is defined instead using the parameter $z = hr$, and r is allowed to be a complex number, so that the region of numerical stability is defined to be a subset of the complex plane.

steady-state solution. (For (14), the steady-state solution is zero.) Although the transient part will decay rapidly to zero, its derivatives $(d^n / dt^n)e^{-ct} = \pm c^n e^{-ct}$ can remain much larger, interfering with the numerical convergence.

Part (b): For the IVP (14), the backward Euler method recursion (13) reads as:

$$y_{k+1} = y_k + hry_{k+1} \implies (1-rh)y_{k+1} = y_k \implies y_{k+1} = \frac{1}{1-rh}y_k.$$

Iterating this last formula produces the following explicit formula:

$$y_k = \left(\frac{1}{1-rh}\right)^k y_0. \tag{16}$$

Since r is negative and h is positive, the magnification factor in (16), $1/(1-rh)$, is always strictly between 0 and 1 so that, regardless of the step size h, (16) will converge to zero. Thus, the region of numerical stability for the backward Euler method is $0 < h < \infty$. Such a situation is called **unconditional numerical stability**.[14] Implicit methods do not always enjoy unconditional stability, but their regions of stability are generally larger than those for the corresponding explicit methods.

The simple equation (14) is often used as a test problem for examining numerical stability of a certain numerical method. When a general single-step method is applied to the test problem (14), it will be possible to write

$$y_{n+1} = Q(hr)y_n$$

for some function Q. Numerical stability of the method will be equivalent to $|Q(hr)| < 1$, which will correspond to a certain region of numerical stability in terms of the step size h. See Exercise 10 for the analysis for the standard Runge-Kutta method. For a more detailed discussion of such stability issues we refer to Chapter 6 in [Atk-89].

EXERCISE FOR THE READER 8.9: Apply Euler's method to the IVP (14) with $r = -2$, using initial condition $y(0) = 5$ and an unstable step size to get a numerical solution on $0 \le t \le 50$ that is similar to that shown in Figure 8.21a. Solve and plot the solution on the interval $0 \le t \le 50$. Then solve it using the Runge-Kutta method with the same step size. The numerical solution should now at least converge to zero. Find a larger step size for which the Runge-Kutta method becomes unstable and the numerical solution looks like that in Figure 8.21b.

---

[14] Some numerical analysis treatments use the terms *absolutely stable* or *A-stable* for what we call unconditionally stable.

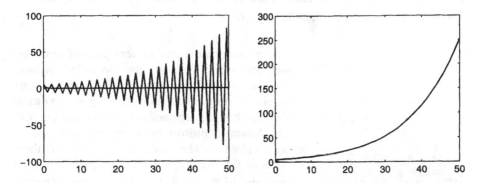

FIGURE 8.21: (a) (left) Instability of the Euler method and (b) (right) of the Runge-Kutta method in solving the simple IVP: $\begin{cases} y'(t) = -2y \\ y(0) = 5 \end{cases}$, whose exact solution $y(t) = 5e^{-2t}$ (flat) decays rapidly to zero. The numerical solutions (jagged/curved) diverge exponentially; the Euler solution does so in an oscillatory fashion, while the Runge-Kutta solution does so unilaterally. A larger step size was needed to make the Runge-Kutta method unstable.

EXERCISE FOR THE READER 8.10: (*Stability of the Trapezoid Method*) If we average the Euler and backward Euler methods, we obtain the so-called **trapezoid method** for solving IVPs:

$$y_{n+1} = y_n + h_n [f(t_n, y_n) + f(t_{n+1}, y_{n+1})]/2 .$$ (17)

Show that the trapezoid method is unconditionally stable when it is applied to the IVP (14) with $r < 0$.

Up to this point, all of the numerical iterative methods for IVPs discussed have used only the information at the present iteration (t_n, y_n) (along with, possibly some auxiliary functional evaluations) to obtain the approximation at the next iteration y_{n+1}. In moving on to the next iteration, none of the data is re-used. Such methods fall under the category of **single-step methods**. It seems reasonable that we might get better results if we were to reuse some of our previously obtained iterates to help us better determine the current iterate. Such methods are known as **multistep methods**. We will consider **linear multistep methods** with constant step size; these have the following general form:

$$y_{n+1} = \sum_{i=1}^{K} \alpha_i y_{n+1-i} + h \sum_{i=0}^{K} \beta_i f(t_{n+1-i}, y_{n+1-i}) .$$ (18)

The positive integer K is the number of steps used in the multistep method. If $\beta_0 = 0$ the method is explicit, otherwise it is an implicit method. In the creation of any multistep method, the coefficients α_i and β_i are chosen according to some polynomial interpolation (data-fitting) scheme.

We present here a pair of very popular multistep methods which lie in the so-called **Adams families** of multistep methods. These methods are usually distinguished into two types, **Adams-Bashforth** multistep methods, which are explicit, and **Adams-Moulton** multistep methods, which are implicit. The specific versions of these methods that we use will be a pair of fifth-order methods (with local truncation error $O(h^6)$), which are given below. These methods can be derived in a number of ways, we give one approach in the exercises.

FIGURE 8.22: John Couch Adams[15] (1819-1892), English mathematician.

The Adams-Bashforth 5-step method for the IVP $\begin{cases} y'(t) = f(t, y(t)) & (DE) \\ y(a) = y_0 & (IC) \end{cases}$:

$h =$ (constant) step size, $\quad t_n = a + nh$, $\quad y_0 = y(a)$ given y_1, y_2, y_3, y_4 found using a single-step method
For $n \geq 4$, (19)

$$y_{n+1} = y_n + \frac{h}{720}\big[1901 f(t_n, y_n) - 2774 f(t_{n-1}, y_{n-1}) + 2616 f(t_{n-2}, y_{n-2}) - 1274 f(t_{n-3}, y_{n-3}) + 251 f(t_{n-4}, y_{n-4})\big].$$

---

[15] Early in his youth, John Couch Adams developed a remarkable ability to perform numerical computations. He studied at St. John's College in Cambridge where he graduated as valedictorian (the term used then at Cambridge was "*Wrangler*") and it has been said that his marks were <u>double</u> those of the second-best student. His main research interest was in the motion of the heavenly bodies. As an undergraduate, he was able to predict the existence of the eighth planet (Neptune) based on his observations of irregularities in the orbit of Uranus. He passed his detailed prediction on to the director of the Cambridge Observatory but unfortunately action was not taken and subsequently credit for the discovery of Neptune was given to the French Astronomer Urbain LeVerrier, who had done a similar analysis after Adams. In 1858, Adams became a professor of mathematics at St. Andrews College but the next year he accepted a professorship at the Cambridge Observatory. Soon after moving to Cambridge, he became director of the observatory and he remained there for the rest of his career. Adams was a true scholar of many subjects. Despite his great intellect and remarkable achievements, his demeanor was always very modest. He even declined a knighthood which was offered in 1947. Adams's extensive work in planetary motion let him to seek appropriate and efficient numerical methods for solving IVP's. Francis Bashforth (1819–1912) was a classmate of Adams at St. Johns. He did extensive work in ballistics. The Adams-Bashforth methods came from a joint work published in 1883 on capillary action. Forest Ray Moulton (1872–1952) was an American mathematician who was also interested in astronomy and ballistics. He developed the so-called Adams-Moulton methods during his work for the U.S. Army in which he generalized the work of Adams and Bashforth.

The Adams-Moulton 4-step method for the IVP $\begin{cases} y'(t) = f(t, y(t)) & (DE) \\ y(a) = y_0 & (IC) \end{cases}$:

h = (constant) step size, $t_n = a + nh$, $y_0 = y(a)$ given y_1, y_2, y_3 found using a single step method

For $n \geq 3$,

$$y_{n+1} = y_n + \frac{h}{720} \left[251 f(t_{n+1}, y_{n+1}) + 646 f(t_n, y_n) \right. \tag{20}$$
$$\left. - 264 f(t_{n-1}, y_{n-1}) + 106 f(t_{n-2}, y_{n-2}) - 19 f(t_{n-3}, y_{n-3}) \right].$$

The advantage of multistep methods such as those shown above is that since they make use of previously computed and stored information, they can attain very decent accuracies with much less number crunching than comparably accurate single-step methods. It is a minor inconvenience that such methods need to use an auxiliary method to get started; usually some sort of Runge-Kutta method is used. A more serious drawback is that multistep methods are not very amenable to adaptive schemes which use nonconstant step sizes. Implicit multistep methods have, of course, the added complication of the need for some rootfinding subroutine. What is usually done in practice with multistep methods is that a pair of implicit and explicit methods of comparable order are used in conjunction with what is called a **predictor-corrector scheme**. In the context of the above Adams family pair, here is how such a scheme progresses (after having found the "seed" iterates y_1, y_2, y_3, y_4): First compute y_{n+1} using the explicit Adams-Bashforth formula (19); label this first approximation as y_{n+1}^* (the **predictor**). Next, substitute this value for y_{n+1} into the right-hand side of the Adams-Moulton formula (20) and take the resulting left side value of y_{n+1} (the **corrector**) as the approximation to $y(t_{n+1})$. It is a simple matter to convert each of the above two Adams family methods as well as the corresponding predictor-corrector scheme into MATLAB M-files. This task will be left to the next exercise for the reader.

EXERCISE FOR THE READER 8.11: (*M-files for Adams Family Methods*) This exercise asks to write M-files for two of the fifth-order Adams family methods described above to solve the IVP (4): $\begin{cases} y'(t) = f(t, y(t)) & (DE) \\ y(a) = y_0 & (IC) \end{cases}$. In each, invoke the fifth-order single-step Runge-Kutta program described earlier in this section to obtain the seed iterates.

(a) Write an M-file for the Adams-Bashforth fifth-order method (19) which has the following syntax:

```
[t, y] = adamsbash5(f, a, b, y0, h)
```

The inputs and outputs are as in the M-files for the single step (nonadaptive) methods.

(b) Write an M-file for the Adams-Bashforth-Moulton fifth-order predictor-corrector method which has the following syntax:

```
[t, y] = adamspc5(f, a, b, y0, h)
```

The inputs and outputs are as in the M-files for the single step (nonadaptive) methods.

In our next example we compare performances with the above three multistep methods.

EXAMPLE 8.20: It is easily shown that the IVP

$$\begin{cases} y'(t) = (t - 3.2)y + 8te^{(t-3.2)^2/2}\cos(4t^2) \\ y(0) = 0 \end{cases}$$

has solution $y(t) = e^{(t-3.2)^2/2}\sin(4t^2)$.[16]   Compute the numerical solutions on the interval $0 \le t \le 6$ using each of the three multistep methods: Adams-Bashforth, Adams-Moulton, and the corresponding predictor-corrector method. Plot the exact solution, and display graphically the errors for each of these three methods.

SOLUTION: After creating the needed inline function, two of the three numerical solutions are easily obtained from the M-files of the preceding exercise for the reader:

```
f=inline('(t-3.2).*y+8*t.*cos(4*t.^2).*exp((t-3.2).^2/2)', 't', 'y')
>> [t yab5]=adamsbash5(f,0,6,0,.02);
>> [t yabm]=adamspc5(f,0,6,0,.02);
```

For the Adams-Moulton solution, we need to write a code. Under the circumstances, examination of the formula (20) shows that the following code will do the job.

```
nmax=ceil((b-a)/h);
%first form the seed  iterates using single step Runge-Kutta
[t,yam]=rk5(f,a,a+4*h,y0,h);

for n=5:nmax
 t(n+1)=t(n)+h;
 g=inline(['Y-' num2str(yam(n),15) '-.02/720*251*(' num2str(t(n+1))...
   - 3.2,15) '*Y+' num2str(feval(f, t(n+1),0),15) ')-.02/720*'...
  num2str(646*feval(f, t(n),yam(n))-264*feval(f, t(n-1),yam(n-...
  1))+106*feval(f, t(n-2),yam(n-2))...
  -19*feval(f, t(n-3),yam(n-3)),15)]);
  gprime = inline(['1-.02/720*251*' num2str(t(n+1)-3.2,15)],'Y');
  yam(n+1)=newton(g,gprime,yam(n));
end
```

The exact solution can be plotted as follows (see Figure 8.23a):

```
>> s=0:.001:6; plot(s, exp((s-3.2).^2/2).*sin(4*s.^2))
```

We can add the other plots to this graph in the usual way; for example, to add the Adams-Bashforth (in green) to the existing graph we could enter:

---

[16] The solution can be obtained using MATLAB's symbolic toolbox.

```
>> hold on
>> plot(t,yab5,'g')
```

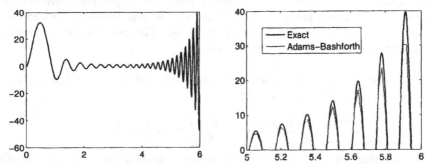

FIGURE 8.23: (a) (left) Exact solution to the IVP of Example 8.20. (b) (right) Closeup of the exact solution (dark) and the Adams-Bashforth numerical solution (light), in a problem area.

See Figure 8.23b for a closeup of where these graphs show differences. If we plot the other two numerical solutions, they will be graphically indistinguishable from the exact solution, so we create a plot comparing the errors of all three methods as follows:

```
>> plot(t,abs(yab5-exp((t-3.2).^2/2).*sin(4*t.^2)), 'g-x'), hold on
>> plot(t,abs(yabm-exp((t-3.2).^2/2).*sin(4*t.^2)), 'r')
>> plot(t,abs(yam-exp((t-3.2).^2/2).*sin(4*t.^2)), 'b')
```

The result is shown in Figure 8.24.

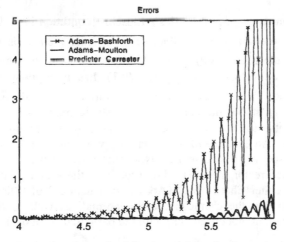

FIGURE 8.24: Error plots for each of the three Adams family methods: Adams-Bashforth (light), Adams-Moulton (dark), and predictor-corrector (medium) for the IVP of Example 8.20. The errors would not be noticeable in this graph for $t \leq 4$.

Although all methods are of the same fifth-order, notice how much better the Adams-Moulton and the predictor-corrector method are than the Adams-Bashforth method. Also despite its being much simpler (and less expensive) to use, the predictor-corrector method actually slightly beats the implicit Adams-Moulton method. These results are rather typical and this is why the predictor-corrector methods are the most popular multistep methods. The exercises will introduce some Adams family methods of different orders.

We give a brief discussion of stability for multistep methods. For a general linear K-step method $(K > 1)$ of the form (18):

$$y_{n+1} = \sum_{i=1}^{K} \alpha_i y_{n+1-i} + h \sum_{i=0}^{K} \beta_i f(t_{n+1-i}, y_{n+1-i}),$$

(we assume that either $\alpha_K \neq 0$ or $\beta_K \neq 0$ so the method is truly a K-step method) we associate the so-called **characteristic polynomial**, which is given by:

$$P(\lambda) = \lambda^K - (\alpha_1 \lambda^{K-1} + \alpha_2 \lambda^{K-2} + \cdots + \alpha_K). \tag{21}$$

The stability of a K-step method can be expressed in terms of the roots of its characteristic polynomial.

It is not difficult to show that if a K-step method is at least first-order accurate, then $\lambda = 1$ will be a root of its characteristic polynomial (see Exercise 20). If all of the other $K-1$ roots of $P(\lambda)$ (counted according to multiplicity) have absolute values less than 1,[17] then the numerical method will be numerically stable for all sufficiently small step sizes h on any initial value problem (4) $\begin{cases} y'(t) = f(t, y(t)) \\ y(a) = y_0 \end{cases}$,

provided that the function $f(t, y)$ satisfies a Lipschitz condition in y. If some root of $P(\lambda)$ has absolute value greater than one, then the method is numerically unstable, even for the basic (IVP) (14). Intermediate cases in which all roots of $P(\lambda)$ have absolute values at most one, $P(\lambda)$ has more than one root of absolute value one, but all such roots are simple are sometimes called (numerically) **weakly stable methods**. Weakly stable methods eventually will experience instability, for any step size, but it usually is less innocuous than ordinary instability. An example of weak stability will follow in the exercise for the reader below; see also Exercise 12. Note that these results do not directly apply to predictor-corrector methods. For proofs of these and related stability results we refer to either Chapter 6 of [Atk-87] or Chapter 7 of [StBu-92]. It is important to notice that the characteristic polynomial, as well as the stability results just mentioned, do not depend at all on the particular form of the IVP being solved.

---

[17] In general, the roots of a polynomial will be complex numbers; recall that the absolute value of a complex number $a + bi$ is $\sqrt{a^2 + b^2}$.

EXAMPLE 8.21: For the test problem IVP (14) $\begin{cases} y'(t) = ry \\ y(0) = y_0 \end{cases}$, with $r < 0$, classify each of the Adams-Bashforth 5-step method and the Adams-Moulton 4-step method as stable, weakly stable, or unstable.

SOLUTION: For both methods, we see that $P(\lambda) = \lambda^5 - \lambda^4 = \lambda^4(\lambda - 1)$. Hence the characteristic polynomial has a simple root $\lambda = 1$ along with a root $\lambda = 0$ of multiplicity 4, and so by the theorems mentioned above, both methods are stable. This is true not just for the test problem but for any IVP satisfying the Lipschitz assumption.

Finding out the precise regions of numerical stability for these methods is a more advanced task. For the test problem in the above example, it turns out that the Adams-Bashforth 5-step method is numerically stable if $h < -0.3/r$ and the Adams-Moulton 4-step method is numerically stable if $h < -3/r$ (see [Gea-71]).

EXERCISE FOR THE READER 8.12: Consider the following **midpoint method** for the IVP $\begin{cases} y'(t) = f(t, y(t)) \\ y(a) = y_0 \end{cases}$:

$$y_{n+1} = y_{n-1} + 2hf(t_n, y_n). \tag{22}$$

(a) Use Taylor's theorem to show the midpoint method has local truncation order $O(h^2)$ (and so this is a first-order method).

(b) Show that the midpoint method is weakly stable.

(c) Use the midpoint method to solve the IVP: $\begin{cases} y'(t) = 20 - 4y \\ y(0) = 1 \end{cases}$ for step sizes $h =$ 0.1, 0.01, 0.001, etc., until the plot looks something like that in Figure 8.25. The exact solution is $y(t) = 5 - 4e^{-4t}$.

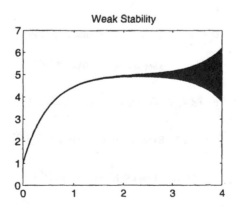

FIGURE 8.25: Illustration of weak stability of the midpoint method for IVP of Exercise for the Reader 8.12.

In general, the region of stability can change depending on the current values of t and y. This can force different constraints on the step sizes that must be taken into consideration. This is where using an implicit method may be advantageous. Although implicit methods usually require more work per iteration, often larger step sizes can be used resulting in a net reduction in the total amount of computation (over an explicit method).

EXERCISES 8.5

1. For each part, an IVP is given along with an exact solution. Solve the IVP using the following indicated numerical methods. Graph the exact solution alongside the numerical solution and in cases where the graphs are indistinguishable, graph also the error. In the multistep methods use the fifth-order Runge-Kutta method to obtain the seed iteration values, as was done in the text.
    (i) Use the RKF 45 method with tolerance = 1e-3 and then again with tolerance = 1e-6.
    (ii) Use the Adams-Bashforth 5-step method with step size $h = 0.1$, and then again with $h = 0.001$.
    (iii) Use the Adams-Bashforth-Moulton predictor-corrector method with step size $h = 0.1$, and then again with h = 0.001.

    (a) $\begin{cases} y'(t) = t^3 - 2ty \\ y(1) = 1 \end{cases}$ $1 \le t \le 5$, Exact Solution: $y(t) = \frac{1}{2}\left(t^2 - 1\right) + e^{1-t^2}$

    (b) $\begin{cases} r'(t) = \dfrac{r\sin(t)}{1 - \cos(t)} \\ y(0) = 4 \end{cases}$ $0 \le t \le 4$, Exact Solution: $r(t) = 4(1 - \cos(t))$

    (c) $\begin{cases} y'(t) = t^2 \cos(t) + \dfrac{2y}{t} \\ y(2\pi) = 0 \end{cases}$ $2\pi \le t \le 10$, Exact Solution: $y(t) = t^2 \sin(t)$

    (d) $\begin{cases} y'(t) = \dfrac{4t - ty^2}{y} \\ y(0) = 3 \end{cases}$ $0 \le t \le 4$, Exact Solution: $y(t) = \sqrt{4 + 5\exp(-x^2)}$

    (e) $\begin{cases} y'(t) = \dfrac{t^2 + ty + y^2}{t^2} \\ y(1) = 0 \end{cases}$ $1 \le t \le 3$, Exact Solution: $y(t) = t\tan(\ln t)$

2. Repeat the directions of Exercise 1 for each of the following IVPs:

    (a) $\begin{cases} y'(t) = -y\ln y \\ y(0) = 3 \end{cases}$ $0 \le t \le 4$, Exact Solution: $y(t) = e^{(\ln 3)e^{-t}}$

    (b) $\begin{cases} y'(t) = 1 - y^2 \\ y(0) = 0 \end{cases}$ $0 \le t \le 6$, Exact Solution: $y(t) = \dfrac{1 - e^{-2t}}{1 + e^{-2t}}$

    (c) $\begin{cases} y'(t) = y(2 - t) \\ y(2) = 1 \end{cases}$ $2 \le t \le 5$, Exact Solution: $y(t) = e^{-(t-2)^2/2}$

    (d) $\begin{cases} y'(t) = e^{t^2} - y/t \\ y(0) = e/2 \end{cases}$ $1 \le t \le 2$, Exact Solution: $y(t) = e^{t^2}/2t$

    (e) $\begin{cases} y'(t) = y - y/t \\ y(1) = 1/2 \end{cases}$ $1 \le t \le 5$, Exact Solution: $y(t) = e^{t-1}/2t$

3. (a) Write a function M-file, $[t,y] = rk5(f,a,b,y0,h)$, which has the same syntax, input variables, and output variables as the $runkut$ of Program 8.2, except that this one will use the 5th-order RK method of the RKF45 algorithm to solve an IVP.
(b) Apply the program to resolve the IVP of Example 8.17 and compare the resulting error plots with those in Figure 8.20.

4. Each of the following IVPs has a coefficient either with a jump discontinuity or that makes abrupt changes over a small time interval. For each one, perform the following tasks:
(i) Solve it with the fifth-order Runge-Kutta method (the one used in RKF45) by starting with a step size of $h = 1/4$ and continuing to halve the step size until the difference (in absolute value) of the current approximation and the previous one is less than 0.0001.
(ii) Repeat (i) with the Adams-Moulton predictor-corrector method.
(iii) Apply the RKF program to the problem with tolerance = 0.0001. Compare the number of grid points used with the numbers in the final approximations of (i) and (ii).

(a) $\begin{cases} y'(t) = f(t,y) \\ y(0) = 2 \end{cases}$ $0 \le t \le 4$, $f(t,y) = \begin{cases} ty, & \text{if } t < 2 \\ 1-t, & \text{if } t \ge 2 \end{cases}$

(b) $\begin{cases} y'(t) = f(t,y) \\ y(0) = 0 \end{cases}$ $0 \le t \le 4$, $f(t,y) = \begin{cases} t^2 + y^2, & \text{if } t < 2 \\ 3y, & \text{if } t \ge 2 \end{cases}$

(c) $\begin{cases} y'(t) = \sin\left(\dfrac{1}{2.01-t}\right), & 0 \le t \le 2 \\ y(0) = 2 \end{cases}$

5. Repeat the instruction of Exercise 4 with each of the following IVPs:

(a) $\begin{cases} y'(t) = f(t,y) \\ y(0) = -1 \end{cases}$ $0 \le t \le 4$, $f(t,y) = \begin{cases} 3y - 2\sin(y), & \text{if } t < 2 \\ 1 + 2y + \cos(y), & \text{if } t \ge 2 \end{cases}$

(b) $\begin{cases} y'(t) = f(t,y) \\ y(0) = 0 \end{cases}$ $0 \le t \le 4$, $f(t,y) = \begin{cases} e^{-yt}, & \text{if } t < 1 \\ -4y^{3/2}, & \text{if } t > 1 \end{cases}$

(c) $\begin{cases} y'(t) = f(t,y) \\ y(0) = 2 \end{cases}$ $0 \le t \le 4$, $f(t,y) = \begin{cases} 50\sin(50t), & 1.7 \le t \le 2.5 \\ y - 2t, & \text{otherwise} \end{cases}$

6. (*Comparison of Methods on a Very Stiff Problem*) The following IVP is very stiff: $\begin{cases} y'(t) = 101 + 100(t - y) \\ y(0) = 1 \end{cases}$ $0 \le t \le 1$. The exact solution, $y(t) = 1 + t$, comes from the general

solution $y(t) = 1 + t + ce^{-100t}$ which has an extremely fast decaying transient term.
(a) Apply the RKF45 method on this IVP with a tolerance goal of 1e-5.
(b) Compare the runtime and accuracy of this solution with that for the Adams-Bashforth-Moulton predictor-corrector method. Start with a stepsize of $h = 1/4$ and continue to halve it until the absolute value of successive numerical approximations is less than 1e-5.
(c) Compare both the above performances with that of MATLAB's built-in ode45.
(d) For stiff IVPs like this one, MATLAB has built-in solvers ode15s and ode23s whose syntax is just like that of ode45 (explained in Section 8.4). Compare the performances and runtimes of these two with all of the previous methods.

7. (*Comparison of Methods on a Mathematically Very Unstable Problem*) (a) The following IVP is very unstable: $\begin{cases} y'(t) = 100y - 101e^{-t} \\ y(0) = 1 \end{cases}$ $0 \le t \le 3$. The general solution $y(t) = e^{-t} + Ce^{100t}$ gives

rise to the specific solution $y(t) = e^{-t}$. Any numerical method will have extreme difficulty with this problem. As soon as we have a roundoff error, we are no longer on our solution curve and pick up the unwanted Ce^{100t} term, which grows explosively fast. Try to solve this problem using each of the methods in parts (a) through (d) of the previous exercise. Do not worry about the tolerances mentioned there; simply aim to get a numerical solution with error remaining less than one in absolute value on the entire interval.

(b) Repeat part (a) on the following IVP: $\begin{cases} y'(t) = 100(\sin t - x) \\ y(0) = 0 \end{cases}$ $0 \le t \le 3$. Since the exact

solution here is not given, for each numerical method, continue to solve the problem by halving the step-size (or tolerance) until successive iterates seem reasonably close, if possible. Comment on the numerical difficulties which arise and compare with the situation in the IVP of part (a). In particular, since we do not have the general solution here at our disposal, comment on the possibility of being able to predict the instability.

8.   (*Comparison of Methods on a Nonlinear Problem*) Consider the following nonlinear IVP:

$$\begin{cases} y'(t) = -300t^2 y^3 \\ y(0) = 1 \end{cases} \quad 0 \le t \le 3.$$

(a) Verify (or use separation of variables to show) that the exact solution of the DE is $y(t) = 1/(200t^3 + C)^2$.

(b) Apply the RKF45 method to solve this IVP using a tolerance of 0.0001, and compare the error with that of the exact solution.

(c)   Change the IC in the above IVP to $y(-2) = 1/1601$ and solve it with RKF45 using a tolerance of 0 0001 on the interval $-2 \le t \le 3$ and examine the error.

(d) Carefully examine how the step sizes changed in parts (b) and (c) above.

(e)   What uniform step size would be needed with the fifth-order Runge-Kutta method to achieve the same results?

(f) Repeat part (e) for the Adams-Bashforth-Moulton predictor-corrector method.

9.   (*Evaluation of an Oscillatory Integral*) The integral $I = \int_0^1 t^2 \sin(1/t)dt$ is awkward to analyze

numerically due to the oscillatory behavior of the integrand near $x = 0$. By the fundamental theorem of calculus, any integral can be viewed as the solution of an IVP (for this one $I = y(1)$, where $y(t)$ solves the IVP: $y' = t^2 \sin(1/t)$, $y(0) = 0$). This integral is proper since the integrand $t^2 \sin(1/t)$ has a limit of zero as $t \to 0$; we just need to redefine the integrand to equal zero at $t = 0$. We have already seen that the popular Simpson's rule for estimating integrals is really a special case of the Runge-Kutta method (Section 8.4, Exercise 5).

(a) Use the RKF45 method with tolerance = 1e-5 to estimate this integral. Plot the endpoints of the step intervals along with the graph of the integrand. Repeat with tolerance = 1e-10.

(b) Use the Adams-Bashforth-Moulton predictor-corrector method to estimate the integral. Start with a step size $h = 1/4$ and continue halving the step size until successive approximations differ by less than 1e-10.

(c) Apply the change of variable $u = 1/t$ to the integral I, and express I as the following convergent improper integral:

$$I = \int_1^{\infty} \frac{\sin(t)}{t^4} dt \ .$$

How large should M be so that the definite integrals $I_M = \int_1^M \frac{\sin(t)}{t^4} dt$ approximate I with total

errors less than 5e - 11?

(d) Repeat part (a) on the integral I_M of part (c) (with M appropriately large). Compare the answers and number of iterations used with those in part (a).

(e) Repeat part (b) on the integral I_M of part (c) (with M appropriately large). Compare the answers and number of iterations used with those in part (b).

(f) Repeat parts (a) through (e) on the integral $J = \int_0^1 \sin(1/t)dt = \int_1^\infty \frac{\sin(u)}{u^2} du$. Here the original

integrand is definitely not continuous at $t = 0$, but try anyway to do parts (a) and (b) if you can.

10. (*Stability of the Runge-Kutta Method*) (a) Show that when the standard Runge-Kutta method

(Section 8.3) is applied to the test IVP (14) $\begin{cases} y'(t) = ry \\ y(0) = y_0 \end{cases}$, the iteration can be expressed as:

$$y_{n+1} = (1 + rh + (rh)^2/2 + (rh)^3/6 + (rh)^4/24)y_n \, .$$

(b) For $r = -2$, what is the approximate range of step sizes for which the method is numerically stable?

(c) Repeat part (b) with $r = -10$.

11. (*Stability of the Improved Euler Method*) (a) Derive a recursion formula, analogous to that in part (a) of the preceding exercise, for the improved Euler method's (Section 8.3) solution of the

test IVP: $\begin{cases} y'(t) = ry \\ y(0) = y_0 \end{cases}$.

(b) For $r = -2$, what is the approximate range of step sizes for which the method is numerically stable?

(c) Repeat part (b) with $r = -10$.

12. (*Mathematical Analysis of the Weak Stability of the Midpoint Method*) In this exercise, we carefully examine what happens when the midpoint method (see Exercise for the Reader 8.12) is

applied to the test problem $\begin{cases} y'(t) = ry \\ y(0) = 1 \end{cases}$, whose exact solution is $y(t) = e^{rt}$.

(a) Show that for this IVP the iteration scheme becomes $y_{n+1} = y_{n-1} + 2rhy_n$. In order to leave

any blame for errors on the method itself, we use the exact value for the seed iterate

$y_1 = e^{rh}$ and we proceed (in the following outline) to explicitly solve the recursion formula in

part (a) with the form: $y_n = c_1 \rho_1^n + c_2 \rho_2^n$, where the constants c_i, ρ_i are to be determined.[18]

(b) Substitute the formula $y_n = \rho^n$ into the recursion formula of part (a) and arrive at the

equation $\rho^2 - 2hr\rho - 1 = 0$, which has roots $\rho = rh \pm \sqrt{(rh)^2 + 1}$.

(c) Using for ρ_1, ρ_2 the two values found in part (b) (with ρ_1 corresponding to the +-sign),

show that for any constants c_1, c_2 the expression $y_n = c_1 \rho_1^n + c_2 \rho_2^n$ will solve the recursion

equation of part (a). Next, determine the values of c_1, c_2 in order that this expression satisfy

also the initial conditions $y_0 = 1$, $y_1 = e^{rh}$. The resulting formula $y_n = c_1 \rho_1^n + c_2 \rho_2^n$ is the exact

solution of the numerical method.

(d) Show that the values of c_1, c_2 found in part (c) satisfy $0 < c_1 < 1$ and $c_2 < -1$. Thus the

second term of the method $c_2 \rho_2^n$ will diverge as $n \to \infty$, but rather slowly (see part (e)), which

is the nature of the weak stability.

(e) Use Taylor's theorem to show that for the values of c_1, c_2 obtained in part (c), we have

$c_1 = 1 - c_2$, and $c_2 = O([rh]^3)$.

13. (a) By mimicking the method of Exercise 12, show that the exact solution to the recursion

---

[18] The general theory of difference equations is quite vast and parallels somewhat the analytical theory of ordinary differential equations. We refer the interested reader to [Ela-99] for an introduction to this theory and to [Aga-00] for a more advanced treatment.

formula

$$y_{n+1} = y_{n-1} + \frac{h}{3}\left[f(t_{n+1}, y_{n+1}) + 4f(t_n, y_n) + f(t_{n-1}, y_{n-1})\right],$$

applied to the test IVP $\begin{cases} y'(t) = ry \\ y(0) = 1 \end{cases}$ has general solution given by $y_n = c_1 e^{rt_n} + c_2(-1)^n e^{-rt_n/3}$.

This recursion formula is known as Milne's corrector formula.

(b) Assuming exact values for the seed iterates, determine the exact solution of the recursion problem in part (a), and discuss the resulting stability for the numerical method.

(c) Use Taylor's theorem to determine the local truncation error of Milne's corrector formula.

14. (*The Hamming Predictor-Corrector Method*)   A popular predictor-corrector method in engineering fields is the following **Hamming**[19] **method** for the IVP (4) $\begin{cases} y'(t) = f(t, y(t)) \\ y(a) = y_0 \end{cases}$:

After choosing the seed iterates y_1, y_2, y_3:

Predictor explicit scheme: $y_{n+1} = y_{n-3} + \frac{4h}{3}\left[2f(t_{n-2}, y_{n-2}) - f(t_{n-1}, y_{n-1}) + 2f(t_n, y_n)\right]$

Corrector implicit scheme: $y_{n+1} = \frac{9y_n - y_{n-2}}{8} + \frac{3h}{8}\left[-f(t_{n-1}, y_{n-1}) + 2f(t_n, y_n) + f(t_{n+1}, y_{n+1})\right]$.

(a) Write a MATLAB M-file for the Hamming method having the following syntax:

$$[\texttt{t, y}] = \texttt{hamming(f, a, b, y0, h)}$$

The input and output variables are exactly as in the program adamspc5 of Exercise for the Reader 8.11.

(b) Run this program on the IVP of Example 8.20 and compare with the exact solution. Approximately how small a step size should be used to attain the same accuracy that was seen for the Adams-Bashforth-Moulton predictor-corrector method in that example?

(c) Compare the accuracy of the Hamming method for the IVP in part (b) using a step size of $h = 0.1$ with that for just the predictor explicit scheme.

(d) Compare the accuracy of the Hamming method for the IVP in part (b) using a step size of $h = 0.1$ with that for just the corrector implicit scheme.

15. (a) Write a function M-file, $[\texttt{t, y}] = \texttt{adamsmoulton(f, a, b, y0, h)}$, which has the same syntax, input variables, and output variables as the backeuler of Program 8.3, except that this one will use the 5th order Adams-Moulton multistep method (20) to solve an IVP. The fifth-order Runge-Kutta method should be used to obtain the seed iterates.

(b) Apply the program to re-solve the IVP of Example 8.20 and compare the resulting error plots with the corresponding one in Figure 8.24.

16. Consider the 2-step explicit method given by the following recursion formula for the IVP

$\begin{cases} y'(t) = f(t, y(t)) \\ y(a) = y_0 \end{cases}$: $y_{n+1} = 2y_{n-1} - y_n - 2hf(y_{n-1}, y_{n-1})$.

---

[19] Richard Wesley Hamming (1915–1998) was an American mathematician born in Illinois. He received his PhD from the University of Illinois at Urbana-Champaign in 1942. Subsequently he spent most of his career working in industry. He joined the Manhattan Project in 1945. Incidentally, the project got its name since it originated at Columbia University in New York, but later much of the research took place at Los Alamos National Laboratories in New Mexico, which is where Hamming worked. After WWII, he moved on to accept a research position at Bell Laboratories where he remained until 1976 when he moved to become chair of the computer science department at the Naval Postgraduate School in Monterey, California. Hamming wrote numerous textbooks on numerical analysis. He is best known for his research on error-correcting codes (such as the one in Exercise 19) and he has won many prestigious awards for his work. These awards include the Turing Prize from IEEE in 1968, and an award from the National Academy of Engineering in 1980.

(a) Use Taylor's theorem to find the local truncation order of this method.
(b) Discuss the stability of this method and perform some numerical experiments to justify your statements.

17. (a) Use the trapezoid method to resolve the IVP of Example 8.5 with step size $h = 0.1$. Compare the error with that for the Euler method (Figure 8.7) as well as that for the improved Euler method. (So you will also need to solve it with the improved Euler method.)
(b) Show that the trapezoid method has second-order accuracy.

18. (*Weighted Trapezoid Methods*) For a parameter σ, $0 \le \sigma \le 1$, consider the iteration scheme obtained by combining the Euler and backward Euler methods in a weighted average as follows:

$$y_{n+1} = y_n + h_n[\sigma f(t_n, y_n) + (1-\sigma)f(t_{n+1}, y_{n+1})] .$$

Note that when $\sigma = 0$ we have the backward Euler method, when $\sigma = 1$ the Euler method, and when $\sigma = 1/2$ the trapezoid rule.
(a) For each σ, $0 < \sigma < 1$ determine the region of numerical stability for this method.
(b) For each σ, $0 < \sigma < 1$ determine the order of accuracy of this method
(c) What happens to the answers in parts (a) and (b) if we allow $\sigma < 0$?
(d) What happens to the answers in parts (a) and (b) if we allow $\sigma > 1$?

19. (a) Classify the Hamming method (see Exercise 14) as either stable, unstable, or weakly stable.
(b) Use Hamming's method to re-solve the IVP in Exercise for the Reader 8.12. First use a step size of $h = 0.2$. Repeat using step sizes $h = 0.1$, $h = 0.05$, and finally with $h = 0.005$. In each case compute the solution on a large enough time interval to detect any instability. Compare with Figure 8.25 and the results of Exercise for the Reader 8.12. Compare and contrast the Hamming method and the midpoint method in terms of this example.
(c) Use Taylor's theorem to find the local truncation error for both the prediction scheme and the correction scheme in the Hamming method.

20. Show that if a K-step method ($K > 1$) $y_{n+1} = \sum_{i=1}^{K} \alpha_i y_{n+1-i} + h \sum_{i=0}^{K} \beta_i f(t_{n+1-i}, y_{n+1-i})$ is at least first-order accurate then $\sum_{i=1}^{K} \alpha_i = 1$ and hence $\lambda = 1$ will be a root of the associated characteristic polynomial.

21. (*Runge-Kutta-England Scheme*) A modification of the RKF45 scheme, introduced by England [Eng-69], uses half-steps as follows to solve the IVP $\begin{cases} y'(t) = f(t, y(t)) \\ y(a) = y_0 \end{cases}$ with a step size h:

First estimate $y(t + h/2)$ by

$$y_{n+1/2} = y_n + \frac{h}{12}(k_1 + 4k_2 + k_3) , \text{ where}$$

$k_0 = f(t_n, y_n)$
$k_1 = f(t_n + h/4, y_n + (h/4)k_0)$
$k_2 = f(t_n + h/4, y_n + (h/8)(k_0 + k_1))$
$k_3 = f(t_n + h/2, y_n - (h/2)k_1 + hk_2)$

Use this to next estimate $y(t + h)$ by:

$$z_{n+1} = y_{n+1/2} + \frac{h}{12}(k_4 + 4k_6 + k_7) , \text{ where}$$

$$k_4 = f(t_n + h/2, \ y_{n+1/2})$$
$$k_5 = f(t_n + 3h/4, \ y_{n+1/2} + (h/4)k_4)$$
$$k_6 = f(t_n + 3h/4, \ y_{n+1/2} + (h/8)(k_4 + k_5))$$
$$k_7 = f(t_n + h, \ y_{n+1/2} - (h/2)k_5 + hk_6)$$

The above method can be shown to be a fourth-order method, and furthermore with one additional functional evaluation:

$$k_8 = f(t_n + h, \ y_n + (h/12)(-k_0 - 96k_1 + 92k_2 - 121k_3 + 144k_4 + 6k_5 - 12k_6)),$$

the following estimate for $y(t + h)$ will constitute a fifth-order method:

$$y_{n+1} = y_n + \frac{h}{180}(14k_0 + 64k_2 + 32k_3 - 8k_4 + 64k_6 + 15k_7 - k_8).$$

(a) Write an M-file for an adaptive Runge-Kutta-England solver with syntax

```
[t, y] = rke45(f, a, b, y0, tol, hinit, hmin, hmax)
```

and whose input and output variables are as in the $\mathtt{rkf45}$ M-file of Exercise for the Reader 8.8.
(b) Apply your program to the IVP of Example 8.17 using a similar tolerance to what was used with $\mathtt{rkf45}$ in that example. Compare performances of the two methods.
Note: We refer the interested reader to Section 6.5 of [ShAlPr-97] for some nice ideas on how to create an effective program using the Runge-Kutta-England method.

NOTE: The general single-step Runge-Kutta (RK) method for the solution of the IVP (4):
$$\begin{cases} y'(t) = f(t, y(t)) & (DE) \\ y(a) = y_0 & (IC) \end{cases}$$ takes the following form:

$$y_{n+1} = y_n + hF(x_n, y_n, h; f),$$

where the notations for x_n, y_n and h (the step size) are as in the text. In this form, the local truncation error of the method can be expressed as $\varepsilon_n = Y(x_{n+1}) - y_{n+1}$, where $Y(x)$ is the solution of the related

local IVP: $\begin{cases} Y'(t) = f(t, Y(t)) & (DE) \\ Y(x_n) = y_n & (IC) \end{cases}$ (so, more properly, $Y(x)$ depends on n). In order that the RK

method be of order $O(h^n)$, the function F should be chosen so that this local truncation error is $O(h^{n+1})$. To facilitate construction of such function F, the following 2-dimensional version of Taylor's Theorem in two variables is very useful:

THEOREM 8.4: (*Taylor's Theorem for Functions of Two Variables*)[20] Suppose that $g(x, y)$ is a function which is continuous along with all of its partial derivatives of order up through $n + 1$ in a region containing the line segment (in the xy-plane) which joins the points (x_0, y_0) and $(x_0 + h, y_0 + k)$. There exists a number c, $0 \le c \le 1$ for which we can write:

$$g(x_0 + h, y_0 + k) = g(x_0, y_0) +$$
$$\sum_{j=1}^{n} \frac{1}{j!} \left[h \frac{\partial}{\partial x} + k \frac{\partial}{\partial y} \right]^j g(x, y) \Big|_{\substack{x = x_0 \\ y = y_0}} + \frac{1}{(n+1)!} \left[h \frac{\partial}{\partial x} + k \frac{\partial}{\partial y} \right]^{n+1} g(x, y) \Big|_{\substack{x = x_0 + ch \\ y = y_0 + ck}}.$$

---

[20] In our notation, the powers of the differential operator are to be done symbolically (just like regular binomial multiplication) and then applied to the function $g(x, y)$. For example, $\left[h \frac{\partial}{\partial x} + k \frac{\partial}{\partial y} \right]^2 g(x, y)$
is equal to $h^2 g_{xx}(x, y) + 2hk g_{xy}(x, y) + k^2 g_{yy}(x, y)$.   (Mixed partials are equal from the differentiability assumption). Actually, the ordinary Taylor theorem would also be sufficient to derive general RK methods, but the two-dimensional notation is more convenient. Indeed, the proof of the two-dimensional version of Taylor's theorem is an easy corollary of the one-dimensional Taylor theorem (Exercise 25).

The next several problems will examine such RK derivations.

22. (*Second-Order RK Formulas*) Assume that the function F in the above note has the following form:

$$F(x, y, h; f) = c_1 f(x, y) + c_2 f(x + \alpha h, y + h\beta f(x, y)),$$

where the parameters c_1, c_2, α, and β are to be determined.

(a) Use Theorem 8.4 to obtain the following expansion:

$$F(x, y, h; f) = c_1 f + c_2 \{ f + h[\alpha f_x + \beta h f f_y]$$
$$+ h^2 [\tfrac{1}{2} \alpha^2 f_{xx} + \alpha\beta f_{xy} f + \tfrac{1}{2} \beta^2 f^2 f_{yy}] \} + O(h^3)$$

(here and in what follows all evaluations of f and its partials are at (x, y) unless indicated otherwise).

(b) Using the chain rule, obtain the following expansion for the function $Y(x)$ which is defined in the preceding note:

$$Y^{(3)}(x) = f_{xx} + 2 f_{xy} f + f_{yy} f^2 + f_y f_x + f_y^2 f.$$

(c) Using the results of parts (a) and (b) obtain the following expression for the local truncation error for the RK method associated with F:

$$\varepsilon_n (\equiv Y(x_{n+1}) - y_{n+1}) = h[1 - c_1 - c_2]f + h^2[(\tfrac{1}{2} - c_2\alpha) f_x + (\tfrac{1}{2} - c_2\beta) f_y f]$$
$$+ h^3[(\tfrac{1}{6} - \tfrac{1}{2} c_2\alpha^2) f_{xx} + (\tfrac{1}{3} - c_2\alpha\beta) f_{xy} f$$
$$+ (\tfrac{1}{6} - \tfrac{1}{2} c_2\beta^2) f_{yy} f^2 + \tfrac{1}{6} f_y f_x + \tfrac{1}{6} f_y^2 f] + O(h^4),$$

where f and its derivatives are evaluated at (x_0, y_0).

(d) Observe from part (c) that the following conditions on the parameters will make $\varepsilon_n = O(h^3)$ (and hence give rise to a second-order RK method): $c_1 + c_2 = 1$, $c_2\alpha = 1/2$, $c_2\beta = 1/2$.

(e) By suitably choosing the parameters in part (d), realize both the improved Euler method and the scheme $y_{n+1} = y_n + hf(t_n + h/2, y_n + (h/2)f(t_n, y_n))$ (from Example 8.15) as special cases of this general second-order RK method.

NOTE: The RK methods developed in the last exercise were so-called **two-stage RK methods** because each iteration involved two evaluations using the function $f(x, y)$. The general (explicit) *s*-**stage RK method** will have (in the notation of the note above):

$$F(x, y, h; f) = \sum_{i=1}^{s} c_i k_i, \quad \text{where } k_i = f(x + \alpha_i h, y + \sum_{j=1}^{i-1} \beta_{ij} k_j), \quad i = 1, 2, \cdots s.$$

Third- and fourth-order RK methods can be realized as 3-stage and 4-stage RK methods respectively, but a fifth-order RK method cannot be obtained as a 5-stage RK method (6 stages are needed) and this explains the popularity of the (original) fourth-order RK method. Derivations of such formulas get extremely complicated and even computer algebra systems on a PC can only handle up to about order-5 RK methods. The minimum number of stages required for RK methods is known up to about order-8 (where 11 stages are needed). For order-9 RK methods, it is known that a minimum of somewhere between 12 and 17 stages would be required. There is a whole theory on s-stage RK methods that is quite well developed; we refer the interested reader to either the book by Butcher [But-87] or that by Lambert [Lam-91].

23. (*Third-Order RK Formulas*) Show that, in the notation of the above note, under the assumption that

$$\alpha_i = \sum_{j=1}^{i-1} \beta_{ij} \quad (i = 1, 2, 3),$$

the following conditions on a 3-stage RK method will make it into a third-order method:

$$c_1 + c_2 + c_3 = 1, \quad c_2\alpha_2 + c_3\alpha_3 = \tfrac{1}{2}, \quad c_2\alpha_2^2 + c_3\alpha_3^2 = \tfrac{1}{3}, \quad c_2\alpha_2\beta_{32} = \tfrac{1}{6}.$$

24. (*Classical RK Formulas*) Show that the classical Runge-Kutta formulas have local truncation error $\varepsilon_n = O(h^5)$, and hence result in a 4th-order method.

25. Prove Taylor's theorem in two variables (Theorem 8.4).
     Suggestion: Apply the one variable Taylor theorem (in Chapter 2) to the function $\varphi(t) = f(x_0 + th, y_0 + tk)$ on the interval $[0,1]$.

NOTE: The next three exercises explore a general method of derivation for multistep methods which is analogous to *the method of undetermined coefficients* in the analytical theory of ODE. We give a brief introduction of this method here, in the setting of deriving a 5-step explicit method for the IVP (4):
$$\begin{cases} y'(t) = f(t, y(t)) & (DE) \\ y(a) = y_0 & (IC) \end{cases}.$$ The exercises will explore more details. From the fundamental theorem of

calculus, we can write: $y(t_{n+1}) = y(t_n) + \int_{t_n}^{t_{n+1}} f(t, y(t))dt$, which leads to the approximation:

$y_{n+1} \approx y_n + \int_{t_n}^{t_{n+1}} f(t, y(t))dt$. Letting f_i denote $f(t_i, y_i)$, in a 5-step explicit method, we seek an approximation of the last integral of the form:

$$\int_{t_n}^{t_{n+1}} f(t, y(t))dt \approx h[Af_n + Bf_{n-1} + Cf_{n-2} + Df_{n-3} + Ef_{n-4}].$$

The coefficients will be determined by forcing the approximation to be exact whenever $f(t, y(t))$ is a polynomial in t of degree at most four.

26. (*Derivation of the Adams-Bashforth 5-Step Method*) Complete the following outline to derive the Adams-Bashforth 5-step explicit method (formula (19)):
     (a) For simplicity, we first assume that $h = 1$ and $t_n = 0$, so that $t_{n-1} = -1, t_{n-2} = -2, t_{n-3} = -3, t_{n-4} = -4$. It is then convenient to use the following five test polynomials (which form a basis[21] of the degree-four polynomials):
$$p_0(t) = 1,$$
$$p_1(t) = t,$$
$$p_2(t) = t(t+1),$$
$$p_3(t) = t(t+1)(t+2),$$
$$p_4(t) = t(t+1)(t+2)(t+3).$$

(Note these polynomials are chosen to have their root sets be increasing subsets of the t_i's.)

Substituting each of these polynomials $p_j(t)$ for $f(t, y(t))$ in the approximation of the above note to obtain the following:
$$\int_{t_n}^{t_{n+1}} p_j(t)dt \approx h[Ap_j(0) + Bp_j(-1) + Cp_j(-2) + Dp_j(-3) + Ep_j(-4)],$$

for $j = 0, 1, 2, 3, 4$, leads to the following linear system:

$$\begin{array}{rrrrrrr}
A & + B & + C & + D & + E & = & 1 \\
  & - B & - 2C & - 3D & - 4E & = & 1/2 \\
  &   & 2C & + 6D' & + 12E & = & 5/6 \\
  &   &   & - 6D & - 24E & = & 9/4 \\
  &   &   &   & 24E & = & 251/20.
\end{array}$$

---

[21] This means that any polynomial of degree at most 4, $p(t) = a_0 + a_1 t + a_2 t^2 + a_3 t^3 + a_4 t^4$, can be expressed as a (unique) linear combination of the five polynomials given.

Solve this linear system to obtain the coefficients of the Adams-Bashforth method.

(b) Show that the coefficients for the general case of arbitrary h and t_n are the same as those obtained in part (a), by making an appropriate change of variables in both sides of the formula

$$\int_{t_n}^{t_{n+1}} f(t, y(t))dt \approx h[Af_n + Bf_{n-1} + Cf_{n-2} + Df_{n-3} + Ef_{n-4}],$$

from the special case in part (a).

(c) Using the fact that the approximation $\int_{t_n}^{t_{n+1}} f(t, y(t))dt \approx h[Af_n + Bf_{n-1} + Cf_{n-2} + Df_{n-3} + Ef_{n-4}]$ is exact for polynomials of degree at most four, use Taylor's theorem to show that the local truncation error for the Adams-Bashforth method is $O(h^6)$, assuming that the solution of the IVP is C^6.

Suggestion: For part (c), use Taylor's theorem to express the derivative of solution using a fourth-order Taylor polynomial. Substitute this expression into the right side of the DE for $f(t, y(t))$.

27. (*Derivation of the Adams-Moulton 4-Step Method*) (a) Using the derivation for the Adams-Bashforth method in the previous exercise as a guide, derive the Adams-Moulton implicit 4-step method (formula (20)).

(b) Assuming that the solution of the IVP is C^6, show that the local truncation error for the Adams-Moulton method is $O(h^6)$.

Chapter 9: Systems of First-Order Differential Equations and Higher-Order Differential Equations

9.1: NOTATION AND RELATIONS

The previous chapter dealt quite extensively with single differential equations of order one. This chapter will extend the treatment to higher-order differential equations and their associated initial value problems, as well as to systems of differential equations. Both concepts are quite natural and have numerous applications, some of which we will introduce in this chapter. In this first section we will show a basic but important method for translating any higher-order differential equation (or any system of higher-order differential equations) into a system of first-order differential equations. In Section 9.2, we will indicate how all of the methods we introduced in the last chapter for numerically solving first-order initial value problems extend very naturally to work for systems of first-order differential equations. There is only one catch, which is that the auxiliary conditions (recall a general nth-order DE will need n auxiliary conditions to determine a unique solution) must all be specified at the same point. Section 9.3 will present some of the interesting and useful theory and geometric tools for systems of ODE. Higher-order IVPs will be dealt with in Section 9.4. There are other frequently occurring problems where we will have, say, a second-order differential equation with the auxiliary conditions specifying the unknown function at two different (boundary) points. Such problems, called **boundary value problems**, will be dealt with in the next chapter.

Suppose that $x(t)$ and $y(t)$ are unknown functions whose rates of change may depend on each other's values as well as the time t. For example, they might represent the populations of two species of animals that live in the same area and compete for resources. Thus a high population of one will in general affect the growth of the other population as well as its own (perhaps logistically, as explained in the last chapter). Thus the only way to model these two populations would be simultaneously with a pair of differential equations. In general we would like to consider first-order systems of the form:

$$\begin{cases} x'(t) = f(t, x, y) \\ y'(t) = g(t, x, y) \end{cases}.$$

$$(1)$$

The initial conditions will in general look like $x(a) = c$ and $y(a) = d$. The existence and uniqueness theory of the last chapter carries over to this setting quite analogously. A solution will exist as long as f and g are continuous functions. It will be unique if f and g satisfy (separately) Lipschitz conditions in each of the variables x and y. A precise theorem will be given in Section 9.3. There are also some nice geometric interpretations of such systems in terms of "flows" and we will say more about this later.

In general, there can be any number of unknown functions: $x_1(t), x_2(t), \cdots, x_n(t)$, which together with the associated initial conditions will solve the following initial value problem:

$$\begin{cases} x_1'(t) = f_1(t, x_1, x_2, \cdots, x_n), & x_1(a) = c_1 \\ x_2'(t) = f_2(t, x_1, x_2, \cdots, x_n), & x_2(a) = c_2 \\ \quad \cdots & \\ x_n'(t) = f_n(t, x_1, x_2, \cdots, x_n), & x_n(a) = c_n. \end{cases} \tag{2}$$

Any nth-order initial value problem of the form

$$y^{(n)}(t) = f(t, y, y', y'', \cdots, y^{(n-1)});$$
$$y(a) = c_1, y'(a) = c_2, \cdots, y^{(n-1)}(a) = c_n. \tag{3}$$

can be reformulated as a system of first-order DEs in the form (2) as follows. We introduce the functions

$$x_1(t) = y(t), x_2(t) = y'(t), x_3(t) = y''(t), \cdots, x_n(t) = y^{(n-1)}(t),$$

and then make the simple observation that they satisfy the following first-order system:

$$\begin{cases} x_1'(t) = x_2, & x_1(a) = c_1 \\ x_2'(t) = x_3, & x_2(a) = c_2 \\ \quad \cdots & \\ x_n'(t) = f(t, x_1, x_2, \cdots, x_n), & x_n(a) = c_n \end{cases}.$$

After we show how to numerically solve first-order systems such as (2), we will then be (in particular) able to numerically solve systems as above that arise from the IVP (3), and once this is done we can discard all but $x_1(t) = y(t)$ to obtain our desired solution of (3).

EXAMPLE 9.1: Express each IVP as a system of first-order IVPs:

(a) $y''(t) = \cos(ty') + 4t^2 + 6y; \quad y(0) = 2, \ y'(0) = -1$

(b) $\begin{cases} x''(t) = 6x'xy' + te^t, & x(1) = 0, \ x'(1) = 4 \\ y''(t) = t + x + y + x' + y', & y(1) = 1, \ y'(1) = 2 \end{cases}$

SOLUTION: Part (a): Upon introducing the two functions:

$$x_1(t) = y(t), x_2(t) = y'(t),$$

we obtain the equivalent system

$$\begin{cases} x_1'(t) = x_2, & x_1(0) = 2 \\ x_2'(t) = \cos(tx_2) + 4t^2 + 6x_1, & x_2(0) = -1. \end{cases}$$

Part (b): This one has a system of two second-order DEs, so it will translate to a system of four first-order DEs. Introducing the four functions $x_1(t) = x(t)$, $x_2(t) = y(t)$, $\dot{x}_3(t) = x'(t)$, $x_4(t) = y'(t)$ leads us to the following equivalent first-order system:

$$\begin{cases} x_1'(t) = x_3, & x_1(1) = 0 \\ x_2'(t) = x_4, & x_2(1) = 1 \\ x_3'(t) = 6x_3x_1x_4 + te', & x_3(1) = 4 \\ x_4'(t) = t + x_1 + x_2 + x_3 + x_4, & x_4(1) = 2. \end{cases}$$

The system in part (b) of the preceding example looks particularly daunting to attempt to solve explicitly. Indeed even a PhD who specialized in differential equations would not be able to perform this task (not to mention an Einstein). Later we will be able to plug the corresponding first-order system into a modified Runge-Kutta program to produce very satisfactory solutions (graphically or in a table) of the original problem.

EXERCISE FOR THE READER 9.1: Reformulate each of the IVPs below into a new IVP which involves a system of first-order DEs:

(a) $y''' + y'' - e^t y = \sin(3t)$; $y(0) = 1$, $y'(0) = 2$, $y''(0) = 3$

(b) $\begin{cases} R''(x) = RS + \sqrt{x^2 + 1}, & R(10) = 4, \ R'(10) = -1 \\ S'(x) = R' \cos(S), & S(10) = 1 \end{cases}$

We close this section by introducing some widely used terminology pertaining to the system (2). If the functions $f_i(t, x_1, x_2, \cdots, x_n)\ (1 \le i \le n)$ on the right sides of the DEs in (3) do not depend on the t-variable (i.e., they each look like $f_i(x_1, x_2, \cdots, x_n)$ with no "t" appearing in their formulas), then the system is called **autonomous**. The word "autonomous" means self-governing. The interpretation here is that if we have an autonomous system and specify initial conditions, the time $t = a$ at which the initial conditions are specified is irrelevant, as far as the future values of the solution(s) are concerned, since the derivatives are independent of the t-variable. When at least one of the DEs does depend on the t-variable, the system is termed **nonautonomous**.

EXERCISES 9.1:

1.   Reformulate each of the IVPs below into a new IVP involving a system of first-order DEs:

(a) $y''(t) = te^t \cos(yy')$, $y(0) = 1$, $y'(0) = 2$

(b) $y''(t) = \dfrac{2+ty}{1+(y')^2}$ $y(0) = 8, y'(0) = 4$

(c) $y'' + \cos(t)y' - \tan(t)y = 2t + 1$, $y(1) = 1, y'(1) = -5$

(d) $y''(t) = \sin^2(y) + \cos^2(y')$, $y(0) = 0, y'(0) = 1$

2.   Reformulate each of the IVPs below into a new IVP involving a system of first-order DEs:

(a) $y'''(t) = ty + y\sin(y'')$ $y(0) = 1, y'(0) = 2, y''(0) = -5$

(b) $y'''(t) = \dfrac{7y + y''}{2 + (y')^2}$, $y(0) = 0, y'(0) = 1, y''(0) = -2$

(c) $\begin{aligned} & y'''' + y''' - ty'' + \cos(t)y' - 3y = \sin(t), \\ & y(0) = 1, \ y'(0) = -5, \ y''(0) = 0, \ y'''(0) = 0 \end{aligned}$

(d) $y'''(t) = \dfrac{e^y + e^{-y'v'}}{e^{2y} - e^{-v'y''}}$, $y(0) = 0, y'(0) = 1, y''(0) = 2$

3.   Reformulate each of the IVPs below into a new IVP involving a system of first-order DEs:

(a) $\begin{cases} x'(t) = x\cos(y)y', & x(0) = 2 \\ y''(t) = t + xy, & y(0) = 0, \ y'(0) = 2 \end{cases}$

(b) $\begin{cases} x''(t) = \dfrac{1+x}{\cos^2(yy') + 2}, & x(0) = 3, \ x'(0) = -2 \\ y'' + 2y' - 3y = 2, & y(0) = 1, \ y'(0) = 0 \end{cases}$

(c) $\begin{cases} T''(t) = \sqrt{tW + T}, & T(0) = 0, \ T'(0) = 1 \\ W''(t) = W' + tT' - 4W, & W(0) = 0, \ W'(0) = 1 \end{cases}$

(d) $\begin{cases} x''(t) = xyz, & x(0) = 1, \ x'(0) = 2 \\ y''(t) = x'y'z', & y(0) = 3, \ y'(0) = 4 \\ z''(t) = xx' + yy' + zz', & z(0) = 5, \ z'(0) = 6 \end{cases}$

9.2: TWO-DIMENSIONAL FIRST-ORDER SYSTEMS

Our first example involves functions that represent the populations of two species, one of which (the **predator**) survives by consuming the other (the **prey**). This very important application was discovered by the Italian mathematician Vito Volterra[1] in the early twentieth century and independently by the Austrian mathematician and chemist Alfred Lotka. To set up the **predator-prey model**, we

---

[1] Volterra grew up in a very poor family and his father died when he was only two years old. His interest in mathematics started at a very early age. When he was 13 he worked on the (still unsolved) three-body problem concerning motion of the objects only under the influence of their interacting gravitational forces. He earned his doctorate at age 22 at the University of Pisa and became a professor there the following year. He did considerable work in the areas of functional analysis and partial differential equations. During the first world war he joined the Air Force and when he returned to civilian life was awarded a professorship in Rome. His biologist colleague Umberto D'Ancona was puzzled at why the percentage of sharks versus food fish went up so quickly during WWI, when fishing went down. This was of economic importance since sharks are not so desirable for consumption, not to mention their effect on tourism. He could not reach any reasonable conclusions with his data so he gave them to Volterra, who came up with a very powerful mathematical model for predator-prey problems. He wrote a seminal text on the subject: *Leçons sur la théorie mathématique de la lutte pour la vie* (1931). After the war the government in Italy was becoming unstable and Volterra fought hard in

FIGURE 9.1: Vito Volterra (1860–1940), Italian mathematician.

FIGURE 9.2: Alfred Lotka (1880–1949), Austrian chemist and mathematician.

will make the following assumptions:

ASSUMPTIONS: The environment has two species, the predator and the prey. The former feeds on the latter and needs it to survive. The prey feeds on a third food source which is readily available. We let:

$x(t)$ = predator population at time t, and $y(t)$ = prey population at time t.

We assume further that: (i) In the absence of predators, the prey population satisfies:

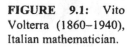

$dy/dt = cy$ $(c > 0)$ (Malthusian growth).

(ii) In the absence of prey, the predator population satisfies: $dx/dt = -ax$ $(a > 0)$ (Malthusian decay).

(iii) With both predators and prey present, the number of encounters per unit time is proportional to xy.

The Malthusian growth assumption in (i) is reasonable if there are predators since they will tend to keep the prey population at bay. If the prey population increases, then the predators will also flourish. A partial justification for assumption (iii) is that if we double the population of either species, then the number of encounters should also double. The roles of predators and prey could be played out by numerous pairs of species such as: foxes and rabbits, wasps and caterpillars, sharks and sea turtles, ladybugs and aphids, etc. The Malthusian growth rates can be determined by isolating a certain number of the species in an enclosed environment and monitoring the population changes per unit time. Similar experiments could be set up to determine exactly how the number of encounters affects the growth rate for the predators and the decay rate for the prey. In general, these assumptions thus lead us to the following system for the "unknown" functions $x(t)$ and $y(t)$, which is the general **Lotka-Volterra predator-prey model:**

---

parliament to keep the Facists at bay. In 1922, after the Facists took over Italy and abolished parliament, Volterra refused to swear an oath of allegiance to the new regime, and for this he was forced to vacate his position at the University of Rome. Volterra spent the rest of his life abroad, mostly in Paris and in Spain.

Lotka had independently discovered predator-prey models at about the same time as Volterra, and he also wrote a book on theoretical biology, which expounded on his newly discovered models. At the time, Lotka had immigrated to the United States. Soon after his arrival in America, he left academia and went to work for a New York insurance company (MetLife), which saw potential applications for Lotka's population models.

$$\begin{cases} x'(t) = -ax + bxy \\ y'(t) = cy - dxy \end{cases} \text{ where } a, b, c, d > 0, \tag{4}$$

Before moving to a specific example, we indicate how the numerical methods of the last chapter would change for systems. We present now the analogous program for the Euler method, as it applies to the general IVP resulting from the 2-dimensional system (1), and leave the corresponding programs for improved Euler and Runge-Kutta as exercises.

PROGRAM 9.1: Euler method for the two-dimensional IVP: $\begin{cases} x'(t) = f(t, x, y), & x(a) = x_0 \\ y'(t) = g(t, x, y), & y(a) = y_0 \end{cases}$.

```
function [t,x,y]=euler2d(f,g,a,b,x0,y0,hstep)
% input variables:  f, g a, b, x0, y0, hstep
% output variables:  t, x, y
% f and g are functions of three variables (t,x,y).
% The program will apply Euler's method to solve the IVP:
% (DEs):  x'=f(t,x,y), y'=g(t,x,y)  (ICs) x(a)=x0
% y(a)=y0 on the t-interval [a,b] with step size hstep.  The output
% will be 3 vectors for t-values, x-values and y-values.
x(1)=x0; y(1)=y0;
t=a:hstep:b;
[m nmax]=size(t);
for n=1:nmax-1 %This will make t have same length as x,y
    x(n+1)=x(n)+hstep*feval(f,t(n),x(n),y(n));
    y(n+1)=y(n)+hstep*feval(g,t(n),x(n),y(n));
end
```

EXERCISE FOR THE READER 9.2: Write a program `runkut2d` that extends the Runge-Kutta method to solve the two-dimensional IVP (2), so it works in a similar fashion to the above program.

EXAMPLE 9.2: Suppose that t is measured in years and that a biologist studying the interactions of sharks and sea turtles in the waters off the Northern Marianas Islands and Guam has found that the shark populations $x(t)$ (in hundreds), and the sea turtle population $y(t)$ (also in hundreds) satisfy the following IVP:

$$\begin{cases} x'(t) = -x + xy; & x(0) = 0.3 \\ y'(t) = y - xy; & y(0) = 2 \end{cases},$$

where $t = 0$ corresponds to the year 2000.

(a) Use the Euler method with step size $h = 0.1$ to solve the system for $0 \le t \le 50$ and plot the simultaneous graphs of x versus t, y versus t and do a parametric plot of y versus x.

(b) Do the same as in part (a), except use the Runge-Kutta method.

(c) Based on your results of parts (a) and (b), what do you think happens as $t \to \infty$? Do $x(t)$ and $y(t)$ approach an equilibrium or does one species die out, or what?

SOLUTION: Part (a) is made quite simple with our `euler2d` program. Although the system is autonomous, the syntax of this program requires our inputted functions be constructed as functions of the three variables (t,x,y) (in this order).

```
>> xp=inline('-x+x*y','t','x','y'); yp=inline('y-x*y','t','x','y');
>> [t,xe,ye]=euler2d(xp, yp, 0, 50, 0.3, 2, 0.01);
```

We have thus constructed the vectors for the Euler approximation to the solution. The next command below will cause MATLAB to produce a plot (in the same window) of the predator population (hundreds of sharks) *x* versus *t* in a red solid curve together with the prey population (hundreds of sea turtles) *y* versus *t* in a blue dash-dot curve. The second command will produce the corresponding parametric plot of *y* versus *x*. The two plots are reproduced in Figure 9.3.

```
>> plot(t,xe,'r',t,ye,'b-.'), xlabel('t=time in years')
>> plot(xe,ye), xlabel('x=100''s of sharks')
>> ylabel('y=100''s of sea turtles')
```

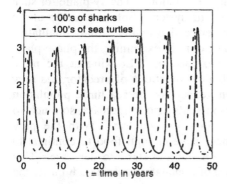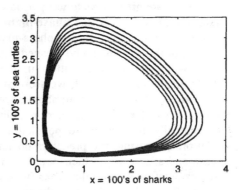

FIGURE 9.3: Using Euler's method with step size *h* = 0.01 to solve the predator-prey problem of Example 9.2 for the time range from *t* = 0 to *t* = 50 years. The first plot gives the approximations of the shark population (in hundreds) *x(t)* (= solid red graph) along with the approximation for the sea-turtle population (in hundreds) *y(t)* (= blue dash/dot graph). The second plot gives the parametric plot of *y* versus *x*.

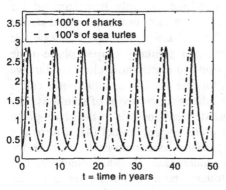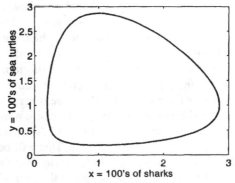

FIGURE 9.4: The plots in this figure correspond to those in the previous one (for Example 9.2), except that the Runge-Kutta method is now used (with the same step size).

Part (b) is done in the same fashion, making use of the `runkut2d` algorithm. The plots appear in Figure 9.4 and the code (without the labeling) is as follows:

```
>> [t,xrk,yrk]=runkut2d(xp, yp, 0, 50, 0.3, 2, 0.01);
>> plot(t,xrk,'r' ,t,yrk, 'b-.')
>> plot(xrk,yrk)
```

Part (c): The two methods give different plots only because Euler's method has introduced errors which have hidden the fact that the two populations are periodic functions. Even more importantly, note that in the Euler plots, successive peaks increase and successive valleys decrease for both populations. Once a population gets too low (below a fertile pair of mates or a single pregnant female), the population will soon fail to exist (extinction) so if we just looked at the Euler approximations, we might be misled to conclude that the populations will both eventually become extinct—a shockingly false conclusion! Parametric plots (the second ones) are great ways to test periodicity of functions. The xy-plane for such a system of DEs is called the **phase-plane** of the system. Solution curves of the system which are graphed in the phase-plane are called **orbits**. Many qualitative properties of the system can be gleaned from carefully examining the phase-plane and we will discuss these matters in the next section. Looking at the more accurate graphs of Figure 9.4, we see that both populations of predator and prey are periodic with cycles lasting about seven years. As the sea turtle population increases to its maximum, the shark population starts also to increase, so much so that eventually the sea turtle population begins to decrease. With the prey population decreasing, the predators no longer have enough food to continue to increase and after a while their population tops out and starts to decrease. This continues indefinitely with the peaks (valleys) of predatators lagging a bit after the corresponding peaks (valleys) for the prey.

Note further that if we set both right sides of the DEs in Example 9.2 equal to zero, we get (from the first) either $x = 0$ or $y = 1$ and (from the second) either $y = 0$ or $x = 1$. Thus there are two **equilibrium solutions** for the system, $x = 0$, $y = 0$ and $x = 1$, $y = 1$ which are constant solutions of the DE system. Only the second is interesting. Note that the phase-plane plot of the example loops around the equilibrium point, albeit in a rather peculiar way. It turns out that if we had started out with other initial conditions (which are off the phase-plane loop of the example but still with both x, $y > 0$), we would get other similarly shaped phase-plane curves which loop around the equilibrium point, no matter how close (or far) we start from the equilibrium point. Because of this the equilibrium solution is called a **vortex** (or a **center**). There are other possibilities for behavior of solutions near an equilibrium point. We will present a more detailed analysis of the phase-plane in the next section. For the Lotka-Volterra predator-prey model (4), it turns out that all solutions are periodic. This will be partially confirmed in the exercise for the reader below; see also Exercise 9 for the general case. Another interesting fact about the (periodic) solutions of the Lotka-Volterra model is that the average value of each (predator or prey) population over a cycle will always equal the corresponding equilibrium values; see Exercise 10.

EXERCISE FOR THE READER 9.3: Use MATLAB to produce a simultaneous plot of 20 different orbits in the phase-plane of the system of DEs in Example 9.2. Take your 20 different initial conditions so that some are very near the equilibrium point and some are rather far from it, but make sure the graphs are distinguishable from each other. Also, indicate for each of these orbits whether the flow is clockwise or counterclockwise.

EXERCISE FOR THE READER 9.4: If we include in the model of Example 9.2 the effect of fishing, we will see that a reduction in fishing actually will tend to reduce the food fish (prey) population and increase the shark (predator) population. (This is the phenomenon that truly puzzled D'Ancona.)[2] The amount of fishing will yield a proportional decrease in the amounts of both populations. The proportionality constant f will depend, for example, on the number of fishing boats deployed, types and numbers of nets used, etc. Incorporating this constant into the model of Example 9.2, gives the modified model.

$$\begin{cases} x'(t) = -(1+f)x + xy \\ y'(t) = (1-f)y - xy \end{cases}$$

(a) Explain why the "fishing constant" f in this model must be less than 1.
(b) Find the (only) equilibrium solution of this new system having both positive components. If f is reduced, what in turn happens to the equilibrium values of food fish and sharks?

We next turn our attention to another important model of ODE systems related to epidemiology and the spread of diseases. The models are known collectively as **SIR models**[3] and we will explain the acronym shortly.

ASSUMPTIONS: The model studies the spread of an infectious disease within a population of N subjects (humans are a good example). The subjects are separated into three classes: the **susceptibles** S, which do not have but can catch the disease; the **infectives** I, which have the disease and can pass it on to susceptibles; and the

---

[2] In Example 9.2, the prey is actually an endangered species so such fishing would not be legal (or humane). We use the data from the example for comparison only. The phenomena will remain true for any such predator-prey system, when both the predators and prey are removed at a constant rate (by hunting, fishing, etc.). For this exercise, we temporarily replace the sea turtle prey with some food fish species (say marlin).
[3] The first mathematical model for epidemiology dates back to 1760, when Swiss mathematician Daniel Bernoulli (1700–1782) investigated the effect of inoculating people with the smallpox virus to prevent the spread of the disease. The first SIR model was invented in 1927 by Kermack and McKendrick [KeMc-27], who sought to model the numbers of infected patients observed in epidemics such as the plague (London 1665–1666, Bombay 1906) and cholera (London 1865). This basic model still remains quite accurate and appropriate for analyzing numerous epidemics that spread rapidly. Subsequent modifications have been developed to accurately model different sorts of diseases and can include additional relevant aspects such as passive (inherited) immunity, vertical transmission, disease vectors, age structure, social and sexual mixing groups, vaccination, quarantine, and spacial spread, to name a few. For a recent, well-written and informative survey on the subject we cite the survey article of Heathcote [Hea-00].

removed R, who have had the disease and have either recovered with immunity, have been isolated (quarantined), or have died.

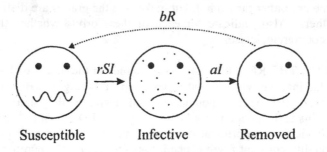

Susceptible   Infective   Removed

FIGURE 9.5: Illustration of the SIR-model for the spread of an infectious disease. The population is stratified into three subgroups and the transitions between groups are indicated with solid arrows. With the additional dotted arrow we get the SIRS model, where members of the removed class can again become susceptibles.

In particular, the removed class cannot pass the disease on to anyone in the susceptible class. Also note that since everyone (including the dead) is accounted for we have

$$S + I + R \doteq N .\tag{5}$$

We let $S(t)$, $I(t)$, $R(t)$ denote the populations of each of the three classes at time t. The rate of transfer from susceptibles to infectives is governed by the constant r, which measures the **infectivity** of a disease and has units [1/time]. The quantity $1/a$ represents the average time length of the infectious period. Thus the most dangerous diseases have a large value of r (very contagious) and a small value for a (people remain contagious for a long time). The deadly Ebola virus, which has had some isolated outbreaks in Africa starting in the 1990s, turned out not to be a major epidemic since it has only a very short infectious period. Another characteristic that can make a disease more dangerous is when infectives are contagious before showing overt symptoms, and so can pass the disease on to other unsuspecting susceptibles. HIV/AIDS is an example of such a disease. Once these parameters are understood for a disease, the effects of various control methods, such as vaccinations, quarantines, etc., can be added into the basic SIR model and appropriate public health courses of action can be prepared. The SIR model is represented by the following system:

$$\begin{cases} S'(t) = -rIS \\ I'(t) = rIS - aI \\ R'(t) = aI \end{cases}.\tag{6}$$

In light of (5), $R = N - S - I$ so we need only consider the first two equations of (6).

EXAMPLE 9.3: In 1978, a flu outbreak in a boys' boarding school in England was documented in the *British Medical Journal* and the article gave the following data for the best-fit SIR model. There were $N = 763$ boys at the school and of these 512 were confined to bed during the outbreak, which lasted a bit over two weeks. $I(0) = 1$ (only one initial infectious boy started the epidemic) and so $S(0) = N - 1 = 762$. The infectivity was $r = 2.18 \times 10^{-3}$/day and $a = 0.44036$ (so the infectivity period lasts for $1/a$ or about 2 ¼ days). ·

(a) Use the Runge-Kutta method to solve this SIR model from $t = 0$ to $t = 14$ days and in the same window, get MATLAB to plot both graphs of $S(t)$ and $I(t)$.

(b) Get MATLAB to produce a single phase-plane plot (I versus S) of the 30 solutions to the above SIR model for this flu outbreak using each of the following initial conditions: $I(0) = 1$ (the one of part (a)), $I(0) = 11$, $I(0) = 31$, ..., $I(0) = 601$. Indicate any similarities and differences of these 30 orbits.

Part (a): After creating inline functions for the right sides of the first two DEs in (6) (as functions of (t,S,T)), we perform the Runge-Kutta method and get MATLAB to produce the plots of S versus t (in black) and I versus t (in red). The plot is shown in Figure 9.6.

```
>> dS=inline('-2.18e-3*I*S','t','S','I');
>> dI=inline('2.18e-3*I*S-.44036*I','t','S','I');
>> [t,S,I]=runkut2d(dS,dI,0,14,762,1,0.01);
>> plot(t,S,'k',t,I,'r'), xlabel('t = Time in days')
```

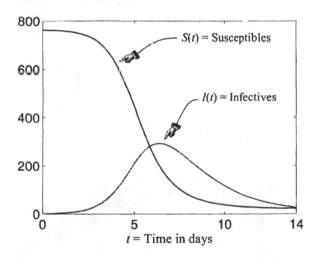

FIGURE 9.6: Plots for the sustibles and infectious puplis in the English boarding school flu outbreak of Example 9.3. Note how quickly the flu spread and in particular that only about 20 or so students (out of 763) appeared to have escaped this flu.

The model's results agree quite favorably with the actual results of this epidemic as documented in the British journal.

Part (b): The following simple for loop will allow us to obtain the desired phase portrait.

```
>> hold on
>> for i=1:20:601
        I0=i, S0=762-i;
        [t,S,I]=runkut2d(dS,dI,0,14,S0,I0,0.01);
        plot(S,I)
end
>> xlabel('S'), ylabel('I')
```

There are quite a lot of computations needed here so we intentionally left off the semicolon at the end of the I0 line within the for loop. This allowed viewing the progression of the loop, which may take a few minutes, depending on the speed of your computer. The result is shown in Figure 9.7, where the line $S + N = 763$ has been added, as well as a vertical line at which each orbit reaches its maximum I-value.

By examination of the first DE of the SIR model (6), we see that S will continue to decrease until either S or I reaches 0. The second DE of (6), when written as $dI / dt = I(rS - a)$, shows that I will increase as long as $S > a/r$ and begin decreasing after $S < a/r$. The reciprocal of important parameter $\rho = a/r$ is called the **contact rate** of the disease. If $S(0)$ starts off larger than ρ, then I will increase and there is an **epidemic**, while if $S(0)$ starts off smaller than ρ, then I will only decrease and there is no epidemic.

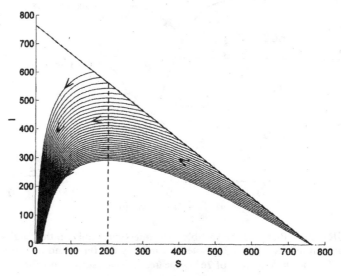

FIGURE 9.7: Phase portrait for the flu-epidemic model of Example 9.3. All initial conditions start on the diagonal solid line $S + I = N$ (flows emanate from this line). The vertical dashed line is at $S = a/r$ and this is where the values of I reach their maximum. Flow direction arrowheads were also added.

Another important parameter which takes also the size of the population into consideration is the so-called **reproduction rate** of the disease, given by:

$$R_0 = \frac{rS(0)}{a} .$$

This quantity can be viewed as the number of new infections transmitted to the suseptible population per single infected individual per unit time. If more than one new disease transmission occurs per infected individual in a unit of time ($R_0 > 1$), then it is clear that we have an epidemic. Another interesting consequence of this model is that the epidemic will evenually end because of a lack of infectives (rather than a lack of susceptibles).

EXERCISE FOR THE READER 9.5: (a) Starting with equation (6), show that in cases when $S(0) > \rho$, the maximum value reached by I is given by $N - \rho + \rho \ln(\rho / S(0))$.

(b)   From the first DE of (6), $S(t)$ is a decreasing function of t so that $S(\infty) \equiv \lim_{t \to \infty} S(t)$ exists as some number in the interval $0 \leq S \leq N$. Using the SIR DEs (6) show that $S(\infty) > 0$.

(c)   Show that $S(\infty)$ is the smallest positive root of the equation: $S(0) \exp[-(N-x)/\rho] = x$ and then use Newton's method to compute $S(\infty)$ for the flu epidemic of Example 9.3. Compare your answer to the value $S(14) = 22.0862$ of susceptibles after 14 days from the numerical solution.
Suggestion: For parts (b) and (c) use (6) to deduce that $dS/dR = -S/\rho$ and use this to write S as a function of R.

EXAMPLE 9.4: (*SIRS Model*) We modify the SIR model (6) to include the feature of temporary immunity. This will cause members of the recovered class to go back to the susceptible class at a rate bR (so $1/b$ will be the average time that the disease-imparted immunity lasts). The SIRS model is thus represented by the system:

$$\begin{cases} S'(t) = -rIS + bR \\ I'(t) = rIS - aI \\ R'(t) = aI - bR \end{cases} \tag{7}$$

Once again, from (5) we have $R = N - S - I$ so we need only solve the system resulting from the first two DEs of (7).
(a) Produce a phase portrait of about 20 well distributed orbits for I versus S of the SIRS model (7) using the following parameters: $N = 10,000$ (a small-sized city), $r = 2 \times 10^{-4}$ /yr (on average for every 5000 encounters of a susceptible with an infective there will be one new infection each year), $a = 4$ (average infection lasts for 1/4 year = 3 months), and $b = .25$ (immunity lasts for 4 years after contracting). Create the plots over a 20-year period using the Runge-Kutta method with step size = 0.01.

(b) Change the parameter a to be 1 (so now the infection lasts for a year rather than three months), but keep all other parameters the same. Create a phase portrait containing 5 well-distributed orbits over an 80-year period using the Runge-Kutta method with step size = 0.01. Compare this phase portrait with that of part (a) and interpret in terms of epidemiology.

SOLUTION: Part (a): We let the initial number of infectives range from 0 to 10,000 in increments of 500. We used the following commands to create the plot in Figure 9.8.

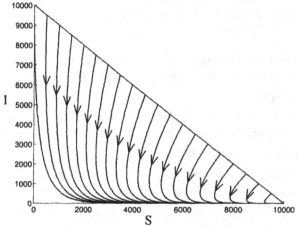

FIGURE 9.8: Orbits for the SIRS disease model of Example 9.4, part (a). Flow directions are indicated with the added arrowheads. Each initial condition emanates from the diagonal line $S + I = N$ (shown). Note that regardless of the initial conditions (even if the whole population starts off infected), the disease eventually burns itself out as the infective population converges to zero.

```
>> dS=inline('-2e-4*I*S+.25*(1e4-S-I)','t', 'S', 'I');
>> dI=inline('2e-4*I*S-4*I','t', 'S', 'I');
>> hold on
>> for k=0:500:10000
[t,S,I]=runkut2d(dS,dI,0,20,10000-k,k,0.01);
plot(S,I)
end
```

Part (b): We obtain orbits corresponding to initial infective populations from 10,000 down to 2000 in increments of 2000. The changes in the MATLAB commands are obvious and minor, so we omit them. The resulting phase portrait is given in Figure 9.9.

Notice the drastic difference in the long-term outcomes of the two very similar diseases on the same population model. The first disease eventually becomes extinct, regardless of the initial conditions, while the latter will linger on forever. Thus in order to eradicate the latter disease it would be necessary for public health officials to use supplementary measures. It is interesting to ask whether there is some borderline value for the parameter a at which the situation of the disease undergoes such a radical change. Indeed there is! In the next section we will show how to predict behaviors such as these, and in particular, for this problem we will be able to find this critical value of a. In part (b), notice that the orbits all spiral to the point $(S,I) = (5000,1000)$. Notice also that this point makes the right sides in the first two DEs of (7) equal to zero so that this point is actually an equilibrium solution.

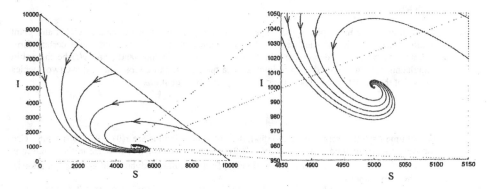

FIGURE 9.9: Orbits for the SIRS disease model of Example 9.4, part (b). Flow directions are indicated with the added arrowheads. Each initial condition emanates from the diagonal line $S + I = N$, which is shown on the left figure. Note that here the disease continues to exist. The orbits spiral to an equilibrium point (which is the same for all initial conditions). The figure plot on the right is a magnification of 100 times (note scales).

EXERCISES 9.2

1. Each of the following systems has an equilibrium solution at $(x,y) = (0,0)$ (and no others). For each of them, using the Runge-Kutta method, use MATLAB to create a representative phase portrait of the system near this equilibrium solution. Plot enough orbits so that your final plot really shows what is going on. Also, add flow directions on your orbits. Finally, from your phase portrait, can you determine whether all initial conditions which are close enough to $(0,0)$ result in solutions which always converge to $(0,0)$?

   (a) $\begin{cases} x'(t) = -x \\ y'(t) = -y \end{cases}$ (b) $\begin{cases} x'(t) = 2x \\ y'(t) = 2y \end{cases}$

   (c) $\begin{cases} x'(t) = y \\ y'(t) = -x + y \end{cases}$ (d) $\begin{cases} x'(t) = -y \\ y'(t) = x - y \end{cases}$

   (e) $\begin{cases} x'(t) = -y \\ y'(t) = x \end{cases}$ (f) $\begin{cases} x'(t) = x - 2y \\ y'(t) = x - y \end{cases}$

   Suggestion: Be careful of the fact that sometimes (depending on the particular system and initial conditions) the flow will be away from the equilibrium point, and other times the flow will be toward the equilibrium point. Thus a "good" set of initial conditions to use will vary with each problem (both in number and locations). It is best to experiment first with a few single flows before setting up a time-consuming loop to plot a bunch of them.

2. Repeat all parts of Exercise 1 for the following linear systems:

   (a) $\begin{cases} x'(t) = -2x + y \\ y'(t) = x - 2y \end{cases}$ (b) $\begin{cases} x'(t) = 2x \\ y'(t) = 4y \end{cases}$

(c) $\begin{cases} x'(t) = -y \\ y'(t) = -x \end{cases}$ (d) $\begin{cases} x'(t) = x - y \\ y'(t) = -x \end{cases}$

(e) $\begin{cases} x'(t) = y + x^2 \\ y'(t) = -x + y \end{cases}$ (f) $\begin{cases} x'(t) = x^2 - y \\ y'(t) = xy - x \end{cases}$

3. Write a program called impeul2d which performs the same task (with the same input and output variables) as the program euler2d (Program 9.1) but with the improved Euler algorithm. Use it to redo Example 9.2 (sharks/sea turtles) and compare the results with those obtained in that Example.

4. (*Agriculture*) For many plants and crops, parasites can pose serious threats to yields and plant health. The aphid is one such pest. It has a benign predator, the ladybug. In a certain farm community, we let $x(t)$ denote the population of ladybugs, in thousands, and $y(t)$ be the aphid population, also in thousands. Experimental data on growth and interaction of these two species show that they satisfy the following system of differential equations:

$$\begin{cases} x'(t) = -2x + 0.5xy \\ y'(t) = 8y - 20xy \end{cases}$$

(a) At time $t = 0$ (t is measured in months), the initial populations are $x(0) = 0.2$, $y(0) = 6.6$. Use the Runge-Kutta method with step size $h = 0.01$ to solve this system with these initial conditions for $0 \le t \le 120$ (i.e., for the subsequent 10 years, find the future ladybug and aphid populations). Plot the graph of y vs. x.
(b) Same problem as (a) but with initial conditions $x(0) = 0.5$, $y(0) = 5.0$. And then $x(0) = 0.1$, $y(0) = 12.4$. Draw separate y vs. x graphs for each and then draw a single plot, which contains all three graphs together. Do you notice anything? As time goes on, in which direction does $(x(t)$, $y(t))$ move along the curves? (Clockwise or counterclockwise?)
(c) Find the equilibrium populations x_E (for ladybugs) and y_E (for aphids) which, when used as initial conditions $x(0) = x_E$, and $y(0) = y_E$, give the positive (equilibrium) solutions = $x(t) = x_E$, and $y(t) = y_E$ (for all $t \ge 0$).
(d) Suppose an insecticide is used which kills the same proportion = 2.5 of each of the two species per month. Thus the system now becomes:

$$\begin{cases} x'(t) = -4.5x + 0.5xy \\ y'(t) = 5.5y - 20xy \end{cases}$$

Repeat the tasks of part (b) for this new system.
(e) Find the new equilibrium populations. How do they compare with those in (c) (without the insecticide)? What are your conclusions?

NOTE: The next four exercises deal with a model which is a variation of the SIR model for the spread of sexually transmitted diseases (STDs). In this model we consider 6 classes, 3 each for (promiscuous) males and females. The classes S_M, I_M and R_M denote the susceptible males, infectious males, and removed males, and S_F, I_F, and R_F denote the corresponding classes of females. The model is illustrated by the diagram:

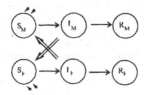

Several STDs (such as gonorrhea) do not impart immunity after an infection, so that in the above model there would be no removed classes (for males or females). In this case the model simplifies to yield the following system of differential equations:

$$\begin{cases} S_M'(t) = -rS_M I_F + aI_M, & S_F'(t) = -sS_F I_M + bI_F \\ I_M'(t) = rS_M I_F - aI_M, & I_F'(t) = sS_F I_M - bI_F \end{cases} , \tag{8}$$

where the parameters a, b, r, and s are positive numbers. Letting N_M and N_F denote the total number of (promiscuous) males and females respectively, we get that

$$S_M + I_M = N_M, \text{ and } S_F + I_F = N_F , \tag{9}$$

and from this we can reduce the fourth-order system above to a second-order one, either in S_M and S_F or in I_M and I_F. For example, in the latter two variables we get the system:

$$\begin{cases} I_M'(t) = rI_F(N_M - I_M) - aI_M \\ I_F'(t) = sI_M(N_F - I_F) - bI_F \end{cases} . \tag{10}$$

5.   Using populations $N_M = N_F = 50,000,000$ and the initial conditions $I_M(0) = 10,000$ and $I_F(0) = 2000$ of a certain STD having $a = b = 1$ (infections last one year on average) $r = 1.960 \times 10^{-8}$ and $s = 2.254 \times 10^{-8}$, do the following:

(a) Using the Runge-Kutta method with step size $h = 0.01$, obtain graphs of $I_M(t)$ and $I_F(t)$ as functions of time from $t = 0$ to $t = 500$ years. What seems to be the eventual outcome of the disease?

(b) Obtain a selection of about 20 orbits of the same system with a good selection of initial conditions. What general comments can you make about the disease from your phase portrait?

(c) Can you tell from the information given here whether the males or the females are more promiscuous in this particular model? Justify your answer.

6.   Repeat parts (a) and (b) of the preceding exercise with the same DE model (10), but the following changes in the parameters: $N_M = 5,000,000$, $N_F - 4,000,000$, $a = b - 0.5$ (so now infections last two years), $r = 5.680 \times 10^{-6}$ and $s = 4.878 \times 10^{-6}$. Also, (c) can you tell if the males in this model are more or less promiscuous than those of the other model?

7.   (a) How would the model (10) above change if we were to add in the feature of the removed class, letting c denote the rate that infective males are removed and d denote the rate that infective females are removed, and with the additional assumption that once a male or female is removed, they will not become susceptible again?

(b) Using the values $c = 0.08$ and $d = 0.05$ in this resulting model, along with the data of Exercise 5, redo part (a) of that exercise.

(c) Do part (b) of Exercise 5 using this new model.

(d) Can you think of some reasons why there might be a difference in the removal rate c for men and d for women?

8.   Redo all parts of Exercise 7 using the same values for c and d given there, but now for the data in Exercise 6. (If you have already done parts (a) and (d) you can of course skip them here.)

9.   Show that the orbits of the Lotka-Volterra predator-prey model (4) can all be expressed in the form $\dfrac{x^c y^a}{e^{dx} e^{by}} = C$, where C is some constant.

Suggestion: The separation of variables method can be used here. Use (4) to write·

$$\frac{dy}{dx} = \frac{y(c - dx)}{x(-a + by)} \Rightarrow \frac{(-a + by)}{y}\frac{dy}{dx} = \frac{c - dx}{x},$$ and then integrate. Take care not to confuse the d in the differential with the constant d in (4).

Note: This closed form solution can be used to show that each of the orbits are closed curves in the phase-plane (see [Bra-93], Lemma 1 on p. 443 for a proof) and hence all orbits of (4) are periodic.

10.   Suppose that P is the period of some pair $x(t)$, $y(t)$ of solutions to the Lotka-Volterra predator-prey model (4):

$$\begin{cases} x'(t) = -ax + bxy \\ y'(t) = cy - dxy \end{cases}.$$

(a) Show that the only equilibrium solution of (4), which has both populations being positive, is $x \equiv c/d$, $y \equiv a/b$.

(b) Show that the average values of $x(t)$ and $y(t)$ over any complete cycle are precisely a/b and c/d respectively, i.e.,

$$\frac{1}{P}\int_0^P x(t)dt = c/d, \text{ and } \frac{1}{P}\int_0^P y(t)dt = a/b.$$

Suggestion: For part (b), to show the first one, take the first equation in (4), divide it by x and integrate both sides over the interval $[0,P]$. Observe that the integrand on the left is the derivative of $\ln(x(t))$.

11.   Does the SIRS model of part (a) in the Example 9.4 have equilibrium solutions? If yes, how would you interpret them in terms of the setting of the problem?

9.3: PHASE-PLANE ANALYSIS FOR AUTONOMOUS FIRST-ORDER SYSTEMS

The last section provided us with a few glimpses of the multitude of possibilities that exist for phase-plane portraits for a given two-dimensional system. In this section we would like to try to make some general comments on how to analyze and predict properties of phase portraits for a given two-dimensional system of first-order autonomous DEs in the vicinity of an isolated equilibrium solution. Since the system is assumed to be autonomous, we will be able to ignore the independent variable t as far as predicting orbits. We begin with an example dealing with another population model, this one being for two species that compete with each other for the same food and resources but otherwise do not prey on one another. One might think of them as two different types of reef fish that feed on the same (limited) coral species or perhaps two types of squirrels that feed on the same acorns and nuts. For simplicity, in this example we will assume that the populations of the two species will grow logistically so that their populations will be limited by their own carrying capacities, but also by the total number of individuals of both species. The following system gives a simplified version of this model where both species have been treated equally:

$$\begin{cases} x'(t) = x - x^2 - rxy \\ y'(t) = y - y^2 - rxy \end{cases}. \tag{11}$$

The model has been scaled to leave only one parameter $r > 0$ as adjustable. Much can be said about the orbits without actually (numerically) solving this system. There turn out to be two different cases depending on whether $r > 1$ or $r < 1$.

We deal first with the special case $r = 2$, which will represent the first case. We should think of the system (11) as setting up a flow in the (x,y)-phase-plane. If we start with any initial condition $(x(0), y(0))$ and view it in the phase-plane, it will lie on an orbit and, as time progresses, the point will be carried along the orbit with directions and speeds determined by the DEs in (11). We would like to see what is happening to the orbit as t gets large ($t \to \infty$). That is, if we start with certain numbers for each population, what eventually happens? Do the two species establish a peaceful coexistence or does one die out? We will soon see. For a specific example, let's suppose species x had a population of 520, or $x(0) = 0.52$ (in thousands) at time zero and species y had an initial population of 500, or $y(0) = 0.5$ (also in thousands). From (11) (using $r = 2$), this means that initially $x'(0) = -.2704 (0)$ and $y'(0) = -.27$, so both populations are initially decreasing, x's a bit faster than y's. Thus, initially the orbit will move downward and to the left from the initial condition. To see what happens in the future, it is helpful to identify the so-called **nullclines**, which are the curves in the phase-plane on which either $x' = 0$ or $y' = 0$. Since $x'(t) = x(1-x-2y)$ and $y'(t) = y(1-y-2x)$, we see that the **x-nullclines** are $x = 0$ and $y = 1/2(1 - x)$ and the **y-nullclines** are $y = 0$ and $y = 1 - 2x$. At any point where an x-nullcline intersects a y-nullcline, we have an **equilibrium solution** (since if we start our initial conditions at such a point, we have both x' and $y' = 0$, so the orbit stays put). Thus, we have equilibrium solutions: $(x, y) = (0,0)$, $(x, y) = (0, 1)$, $(x, y) = (1,0)$, and $(x, y) = (1/3, 1/3)$.

The sign of either x' or y' can only change across a nullcline, so we can test x' and y' in each region determined by the nullclines and draw an appropriate arrow (or arrows) there to indicate the rough direction of flow (left or right, up or down). On the x-nullclines, the arrows will be vertical (either up or down) and on the y-nullclines the arrows will be horizontal (either left or right). The directions of these vertical and horizontal arrows can be determined by examining those in the adjacent regions. In this way, we produce the phase-plane diagram in Figure 9.10 for our system. One such computation was already done, so the figure below can be obtained by only three more such computations.

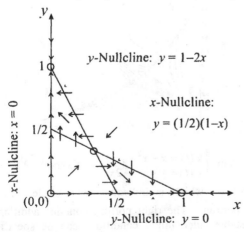

y-Nullcline: $y = 1-2x$

x-Nullcline:

$y = (1/2)(1-x)$

x-Nullcline: $x = 0$

y-Nullcline: $y = 0$

FIGURE 9.10: Phase-plane diagram for the system (11) using $r = 2$. The directions of flow are indicated by arrows. In the regions between nullclines, the arrows are meant to only indicate whether the flow is left or right and up or down. The equilibrium solutions are indicated with open circles.

From this phase diagram, we can now look to see what can

happen to the orbit with initial condition $(x(0), y(0)) = (.52, .5)$. Initially the orbit moves down and left. Either it will directly approach the equilibrium point (1/3, 1/3) or it will veer above or below. If it veers above, it will cross the nullcline $y = 1 - 2x$ into the upper triangular region. Once in this region the flow changes to upward (and still to the left). Also, it could never cross back over $y = 1 - 2x$ back into the original region it started (since the horizontal arrows on this nullcline move to the left). By the same token, it could never cross the nullcline $y = 1/2(1 - x)$ into the lower-left region. If the orbit ever did make it to the vertical x-nullcline $x = 0$, then it would have to stay on it and move vertically upward (actually, the orbit could never touch this line; see Exercise 7). In all cases, the orbit will tend to approach the equilibrium solution $(0,1)$. If the orbit initally veers off below the equilibrium point, then, as with the last possibility, we could show that the orbit would eventually approach the equilibrium point $(1,0)$.

EXAMPLE 9.5: Using the above phase-plane diagram as a guide for choosing initial conditions, get MATLAB to create a plot of about 20 orbits for the system (6) with $r = 2$ that well represent the behavior of orbits near the equilibrium solution (1/3, 1/3).

SOLUTION: The initial conditions should all be located in the upper right and lower-left regions of the phase-plane diagram in Figure 9.10. The orbits of initial conditions located in either of the two triangular regions will move away from the given equilibrium solution. Two nice spreads of such initial conditions could be taken on the parallel lines: $x + y = 1$ and $x + y = 1/3$. We create vectors for points $x1,y1$, first on the top line (nicely spread around the central point (1/2, 1/2), then vectors $x2$, $y2$ for points on the bottom line, then put them together to form single vectors $x0$, $y0$. Next, we set up an appropriate for loop to run through Runge-Kutta with time interval $0 \le t \le 20$ and step size $h = 0.01$ to plot the corresponding orbits. The following chain of commands will accomplish all of this; the result is shown in the left plot in Figure 9.11.

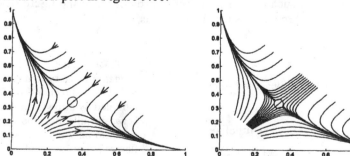

FIGURE 9.11: Phase portraits for the system $\begin{cases} x'(t) = x - x^2 - 2xy \\ y'(t) = y - y^2 - 2xy \end{cases}$ of Example 9.5 near the equilibrium solution $(x(t), y(t)) \equiv (1/3, 1/3)$, which is indicated by a circle. Flow directions have been inserted in the left portrait. The right portrait contains additional orbits. This equilibrium solution is unstable since initial conditions can be specified arbitrarily close to it and their orbits will move away from it as time goes on.

```
>> dx=inline('x-x^2-2*x*y', 't', 'x', 'y');
>> dy=inline('y-y^2-2*x*y', 't', 'x', 'y');
>> hold on
>> x1=[(1/4):(1/20):(3/4)]; y1=1-x1;
>> x2=[(1/12):(1/60):(3/12)]; y2=1/3-x2;
>> x0=[x1 x2]; y0=[y1 y2];
>> size(x0)
→1 22
>> for k=1:22
[t,x,y]=runkut2d(dx,dy,0,20,x0(k),y0(k),0.01);
plot(x,y)
end
```

The system (11) has radically different phase portraits near its central equilibrium solution, depending on whether the parameter r is chosen to be 2 or 1/2. Similar phase portraits arise in the ranges $0 < r < 1$ and $1 < r$. Recasting the results in terms of the original model, we get the following very different possible outcomes: If $r > 1$ (from Figure 9.11) we see that if the initial populations do not start off exactly equal then eventually the species with the smaller initial population will become extinct.

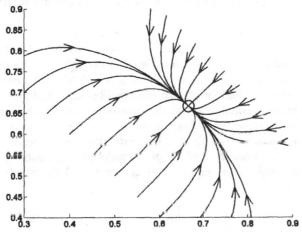

FIGURE 9.12: Phase portrait for the system $\begin{cases} x'(t) = x - x^2 - xy/2 \\ y'(t) = y - y^2 - xy/2 \end{cases}$ of Example 9.5 near the equilibrium solution $(x(t), y(t)) \equiv (2/3, 2/3)$, which is indicated by the circle. Flow directions have been inserted. This equilibrium solution is stable since any solution obtained with an initial condition close to (2/3, 2/3) will approach this equilibrium as time goes on.

EXERCISE FOR THE READER 9.6: Consider the system $\begin{cases} x'(t) = x - x^2 - xy/2 \\ y'(t) = y - y^2 - xy/2 \end{cases}$ which results from (11) by using $r = 1/2$ instead of $r = 2$ (so in this variation of the population model, the growth of one species has less effect on the growth of the other than in the first example considered).

(a) Draw by hand a phase xy-plane diagram which includes (only for $x \geq 0, y \geq 0$) all equilibrium solutions, x- and y-nullclines along with exact flow directions on the nullclines, and approximate flow directions in the regions between nullclines. (b) Use MATLAB to create a phase portrait of orbits near the equilibrium solution $(x(t), y(t)) \equiv (2/3, 2/3)$. The portrait should resemble the one in Figure 9.12.

Only in the rare case where they start off being equal will both populations survive and tend toward the central equilibrium solution. If $r < 1$ (from Figure 9.12) we see that as long as the initial populations start off close enough to the central equilibrium solution, then in the future the populations will always tend to the equilibrium solution.

Definition: An equilibrium solution $(x(t), y(t)) \equiv (x_0, y_0)$ of an autonomous system $\begin{cases} x'(t) = f(x, y) \\ y'(t) = g(x, y) \end{cases}$ is said to be **stable** if any solution resulting from initial conditions that are sufficiently near (x_0, y_0) will always approach this equilibrium solution as $t \to \infty$. If this is not the case, then the equilibrium solution is called **unstable**.

The two cases of (11) provide examples of stable and unstable equilibria. We point out that for an unstable equilibrium, we do not require that all (sufficiently near) initial conditions will result in solutions which do not converge to the equilibrium. What is required for an unstable equilibrium is that there will always exist initial conditions sufficiently near the equilibrium point whose solutions do not converge to the equilibrium. Also, the solutions need not move away from the equilibrium, just not converge to it. For example, the "vortex" equilibrium solutions of the Lotka-Volterra predator-prey model (4) are unstable (orbits with initial conditions close to the equilibrium continue to loop around it without converging to it).

Before moving on to discuss stability, we give an existence and uniqueness theorem for two-dimensional systems using some of the language we have already developed. The Lipschitz condition extends in the obvious way to a function of more variables. For example a function $f(t, x, y)$ is said to satisfy a **Lipschitz condition with constant** L in the variable x on the set R: $a \leq t \leq b$, $c \leq x \leq d$, $e \leq y \leq f$ if we have:

$$| f(t, x_1, y) - f(t, x_2, y) | \leq L | x_1 - x_2 |$$

whenever the two points being evaluated on the left lie in the set R. We can now state the existence and uniqueness theorem.

THEOREM 9.1: (*Existence and Uniqueness for Two-Dimensional Systems*) Given the initial value problem:

$$\begin{cases} x'(t) = f(t,x,y), & x(a) = x_0 \\ y'(t) = g(t,x,y), & y(a) = y_0 \end{cases},$$

where the functions f and g are assumed to be continuous near the point (a, x_0, y_0), then there will be a solution to this IVP which gives rise to an orbit. The solution will continue to exist for as long as the t-variable and the orbit stay in sets on which the functions f and g are continuous. Furthermore, if the functions f and g both satisfy a Lipschitz condition in both the x- and y-variables in a region R containing (a, x_0, y_0), then the solution of the IVP will be unique for as long as $(t,\ x(t),\ y(t))$ remains in R.

For a proof of this result we refer the reader to [Arn-78], [Hur-90], or [HiSm-97]. We remark that, in particular, if the functions f and g have first-order partial derivatives in x and y which are continuous in some region, then the Lipschitz conditions will automatically hold within any set of form $a \le t \le b,\ c \le x \le d$, $e \le y \le f$ that lies inside of this region and consequently solutions will exist and be unique for as long as their times and orbits remain in this region.

We now move on to a general discussion of the behavior of a general autonomous system:

$$\begin{cases} x'(t) = f(x,y) \\ y'(t) = g(x,y) \end{cases} \tag{12}$$

in the neighborhood of an equilibrium solution. For convenience, we assume that the equilibrium solution is at $(x,y) = (0,0)$. (This can always be achieved via translation of coordinates.) We need to assume that the equilibrium solution is **isolated**. This means that within some circle centered at $(0,0)$ in the phase-plane there are no other equilibrium solutions. We assume that the functions f and g have continuous first partial derivatives in x and y near $(0,0)$ (so in particular they will locally satisfy the required Lipschitz conditions to guarantee that unique orbits always exist when initial conditions are specified near the equilibrium solution). Because of the assumptions on the partial derivatives, we can use Taylor's theorem twice (alternatively, readers familiar with Taylor's theorem in several variables need only use this latter theorem once) to obtain the following linear approximation for $f(x,y)$ near $(0,0)$:

$$f(x,y) \approx [f(0,y)] + xf_x(\alpha,y) \approx [f(0,0) + yf_y(0,\beta)] + xf_x(\alpha,y).$$

Since $f(0,0) = 0$ (our equilibrium solution) and since the partials f_x and f_y are continuous at $(0,0)$, the above leads us to the final linear approximation:

$$f(x,y) \approx xf_x(0,0) + yf_y(0,0) \equiv ax + by,$$

where we have defined $a = f_x(0,0)$ and $b = f_y(0,0)$. In the same fashion, the linear approximation of $g(x,y)$ (near $(x,y) = (0,0)$) is $cx + dy$ where $c = g_x(0,0)$ and $d = g_y(0,0)$. From our work and experience with Taylor's theorem we know that these linear approximations work quite well near $(x,y) = (0,0)$. Thus if instead of (12), we look at the associated **linearization** (near $(x,y) = (0,0)$):

$$\begin{cases} x'(t) = ax + by \\ y'(t) = cx + dy \end{cases} \quad \text{or} \quad \begin{bmatrix} x'(t) \\ y'(t) \end{bmatrix} = \begin{bmatrix} a & b \\ c & d \end{bmatrix} \begin{bmatrix} x \\ y \end{bmatrix} \tag{13}$$

it would seem plausible that the phase portrait of (13) near the equilibrium might have some similarities to the corresponding one for the nonlinear system (12). The interesting and useful connection which we will now explain was discovered by the famous French mathematician Henri Poincaré.[4]

From the matrix of partial derivatives in (13) we will be able to get a lot of information about the phase-portrait of the original system near the equilibrium solution, so we give it a special name. It is called the **Jacobian matrix** of the system (12) (evaluated at the equilibrium solution in question). For convenience we introduce the following notations for this matrix:

$$A = J(f,g) = J(f,g)_{(0,0)} = \begin{bmatrix} f_x & f_y \\ g_x & g_y \end{bmatrix}_{(0,0)}$$

$$= \begin{bmatrix} a & b \\ c & d \end{bmatrix}.$$

FIGURE 9.13: Henri Poincaré (1854–1912), French mathematician.

If we introduce the vector notation $X(t) = \begin{bmatrix} x(t) \\ y(t) \end{bmatrix}$ and $X'(t) = \begin{bmatrix} x'(t) \\ y'(t) \end{bmatrix}$, then the

---

[4] Henri Poincaré is often called the last of the universal mathematicians. He contributed significantly to all of the major areas of mathematics. The subject has become too vast to imagine any more universal mathematicians. As a student, Poincaré excelled in all of his academic subjects and won a nationwide mathematics competition while in high school. He became a member of the French Academy of Sciences at the very young age of 32. For most of his career he was a chaired professor at *la Sorbonne* and each year he taught a different subject. His lectures were so full of insights and deep in their scope that his students who took the notes helped to formally write up his celebrated lectures, which became significant volumes in contemporary research. His eyesight was poor but he had an extremely sharp mind and was able to visualize many complicated mathematical ideas. His mind was always at work; once he even described a major mathematical breakthrough that he had the moment he was stepping onto a crowded bus in Paris. Throughout his life he won many prizes, both for his mathematical work and also for his literary works that he produced later in his life (such as *La Science et L'Hypotèse*) to help the general public understand how scientists think. Other prominent members of society belonged to the Poincaré family. His cousin Raymond Poincaré was the Prime Minister for several terms and President of the Republic from 1913 to 1920. Another cousin, Lucien Poincaré, was a high-ranking administrator at a prominent French university.

linearization (13) can be written in matrix form as:

$$X'(t) = AX \tag{14}$$

As with our original system, we assume that the equilibrium solution (0,0) for the linear system is also isolated.

EXERCISE FOR THE READER 9.7: Explain why (0,0) is an isolated equilibrium solution of (14) if and only if $\det(A) \neq 0$.

Our next theorem gives the complete story of the solution to (14) with any initial conditions. Recall that the trace of a matrix is the sum of its diagonal entries: $tr\left(\begin{bmatrix} a & b \\ c & d \end{bmatrix}\right) = a + d$. We also define the **discriminant** of the linearization (14) to be $\Delta = tr(A)^2 / 4 - \det(A)$.

THEOREM 9.2: The unique solution of (14) satisfying the initial condition $X(0) = X_0 = \begin{bmatrix} x_0 \\ y_0 \end{bmatrix}$ is given by the following formulas which involve the vector $A_0 = \begin{bmatrix} \frac{1}{2}(a-d)x_0 + by_0 \\ cx_0 + \frac{1}{2}(d-a)y_0 \end{bmatrix}$:

Case 1: $\Delta > 0$: $X(t) = \dfrac{e^{tr(A)t/2}}{2}\left[e^{\sqrt{\Delta}t}(X_0 + A_0/\sqrt{\Delta}) + e^{-\sqrt{\Delta}t}(X_0 - A_0/\sqrt{\Delta}) \right]$.

Case 2: $\Lambda = 0$: $X(t) = e^{tr(A)t/2}(X_0 + tA_0)$.

Case 3: $\Delta < 0$: $X(t) = \dfrac{e^{tr(A)t/2}}{2}\left[\cos(t\sqrt{|\Delta|})X_0 + \sin(t\sqrt{|\Delta|})A_0/\sqrt{\Delta} \right]$.

Some details of the proof of this theorem can be found in the exercises. Actually, it is not hard (just tedious) to verify that in each of the three cases, the vector function in the theorem does indeed solve the indicated IVP. There are many good books on differential equations that provide a more complete development of the solution of (14) (see, for example, [Arn-78], [HiSm-97], and [Hur-90]). Since our concern will be mostly with stability and the nature of the phase-plane near a critical point, we simply summarize the situation. The nature of the phase diagram depends only on the values of det(A) and tr(A), as summarized in the Figure 9.14 and explained in the caption below it. What is remarkable is that near an equilibrium solution the phase portrait for the nonlinear system (12) will look very much the same as that for its linearization in the most important cases (although the pictures for nonlinear systems will be distorted from the linearizations). We make this precise in the following theorem, the proof of which is referred to [Hur-90]. (Some of the phase-plane terminology of this theorem is from Figure 9.9.)

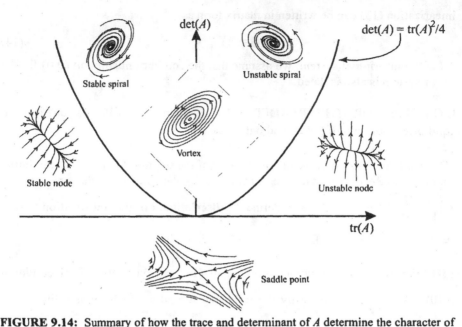

FIGURE 9.14: Summary of how the trace and determinant of A determine the character of the phase portrait of the linear system $X'(t) = AX$ near an isolated equilibrium solution. In the degenerate case where $\det(A) = tr(A)^2/4$ there will be different varieties of nodes. For a nonlinear system $\begin{cases} x'(t) = f(x,y) \\ y'(t) = g(x,y) \end{cases}$ with an isolated equilibrium solution at (α, β), and having a nonsingular Jacobian matrix $A = \begin{bmatrix} f_x & f_y \\ g_x & g_y \end{bmatrix}_{(\alpha,\beta)}$ corresponding phase portraits near (α, β) will be those indicated above, except possibly in the degenerate cases $tr(A) = 0$ or $\det(A) = tr(A)^2/4$.

THEOREM 9.3: Suppose that the functions f and g of the system $\begin{cases} x'(t) = f(x,y) \\ y'(t) = g(x,y) \end{cases}$ have continuous first-order partial derivatives near an isolated equilibrium solution $(x(t), y(t)) = (\alpha, \beta)$ (so $f(\alpha, \beta) = g(\alpha, \beta) = 0$). Let A be the corresponding Jacobian matrix $\begin{bmatrix} f_x & f_y \\ g_x & g_y \end{bmatrix}_{(\alpha,\beta)}$ and assume that $\det(A) \neq 0$.

(a) If $\det(A) < 0$, then (α, β) is a saddle point (always unstable).

(b) If $0 < \det(A) < tr(A)^2/4$, then (α, β) is a node, stable if $tr(A) < 0$, unstable if $tr(A) > 0$.

(c) If $tr(A)^2/4 < \det(A)$, then (α, β) is a spiral, stable if $tr(A) < 0$, unstable if $tr(A) > 0$.

(d) (Borderline Cases) If $\det(A) = tr(A)^2 / 4$, then (α, β) is either a node or a spiral, stable if $tr(A) < 0$, unstable if $tr(A) > 0$. If $\det(A) > 0$ and $tr(A) = 0$, then (α, β) is either a vortex or a spiral.

It turns out that without the hypothesis $\det(A) \neq 0$, there can still be isolated equilibrium solutions (α, β) and the behavior of the phase-plane near such a degenerate equilibrium solution can lead to a great variety of new phase-plane behaviors near (α, β). Some examples are examined in the exercises. As a consequence of Theorem 9.3, we obtain the following simple stability criterion.

COROLLARY 9.4: (*Stability Criterion*) Under the hypotheses of Theorem 9.3, the equilibrium solution $(x(t), y(t)) = (\alpha, \beta)$ is stable if and only if $\det(A) > 0$ and $tr(A) < 0$.

The previous theorem allows us to get a lot of useful qualitative information about the phase-plane's character (near an isolated equilibrium solution) without actually doing any numerical calculations of orbits.

EXAMPLE 9.6: For the system (11) $\begin{cases} x'(t) = x - x^2 - rxy \\ y'(t) = y - y^2 - rxy \end{cases}$ (where r can be any positive number), determine all isolated equilibrium solutions $(x(t), y(t)) = (\alpha, \beta)$. For each determine whether it is stable and also the nature of the phase-plane near (α, β).

SOLUTION: As we did earlier in this section, we find the x-nullclines of the system by setting $x' = f(x, y) = 0$, i.e., $0 = x(1 - x - ry)$. This gives the two x-nullclines: $x = 0$ and $y = (1 - x)/r$. In the same fashion we set $g(x,y) = 0$ to get the two y-nullclines: $y = 0$ and $y = 1 - rx$. The equilibrium solutions are where an x nullcline meets a y-nullcline. This gives the following isolated equilibrium solutions:

(i) (0,0), and if $r \neq 1$, (ii) (1,0), (iii) (0,1), and (solving the two sloped lines) (iv) $(1/(r + 1), 1/(r + 1))$. The Jacobian matrix for the system is

$$\begin{bmatrix} f_x & f_y \\ g_x & g_y \end{bmatrix} = \begin{bmatrix} 1 - 2x - ry & -rx \\ -ry & 1 - 2y - rx \end{bmatrix},$$

which has the following trace and determinants for each of the listed equilibrium solutions:
(i) trace = 2, det = 1, so by Theorem 9.3 (0,0) is an unstable node or spiral. It cannot be a spiral since the y-axis is an x-nullcline (a spiral orbit about (0,0) could never cross the y-axis—see Figure 9.10 if you need convincing; see also Exercise 6).
For the other three points to be isolated, we now assume $r \neq 1$.

(ii) trace $= -r$, det $= r - 1$, so by Theorem 9.3 $(1,0)$ is an unstable saddle point if r < 1. If $r > 1$, note that $tr(A)^2 / 4 - \det(A) = r^2 / 4 - r + 1 = (1 - r/2)^2 > 0$ so $\det(A)$ $< tr(A)^2 / 4$ and by the theorem we have a stable node (viz. Figure 9.11 and Figure 9.12).

(iii) $(0,1)$ has the same data and properties as $(1,0)$ (by symmetry of the system).

(iv) Here trace $= -2/(r + 1)$, det $= (1 - r)/(1 + r)$, so again by the theorem we have that $(1/(r + 1), 1/(r + 1))$ is a saddle point if $r > 1$ (cf. Figure 9.11) and a stable node if $r < 1$ (cf. Figure 9.12). The reader is encouraged to sketch phase planes for $r < 1$, and $r > 1$.

EXERCISE FOR THE READER 9.8: (a) For the SIRS model (7) of part (a) of Example 9.4, hand draw a phase-plane diagram (cf. Figure 9.10) including all nullclines (labeled), all equilibrium solutions and all flow directions (exact on nullclines and approximate between them) throughout the entire first quadrant of the *SI*-plane. (b) Use Theorem 9.3 to analyze the character of the equilibrium solution(s) obtained in part (a). (c) Next, examine what happens to the equilibrium solutions (and their character) when we allow the parameter a to decrease from 4 to zero. In the model, this corresponds to allowing the period of infection of the disease to increase from 3 months (when $a = 4$) to arbitrarily large periods.

We end this section with another famous and interesting theorem. We will present it in a very geometric fashion so we will need to first introduce a couple of concepts relating to the phase-plane of an autonomous system (12). We say that an equilibrium solution $(x(t), y(t)) = (\alpha, \beta)$ is **repelling** if any initial condition that is very close to (but not equal to) (α, β) results in a solution/orbit of (12), which will move further away as time advances. For example, any unstable spiral or node equilibrium solution is repelling. A saddle point is never repelling. A region R in the phase-plane of the system (12) is said to be a **basin of attraction,** provided that every orbit that enters R will never leave R at a later time. One handy way to confirm that a region is a basin of attraction would be to check that on the edges of R the orbit flow directions never point outside of R. For example from Figure 9.10, we can see that each of the two triangular regions between nullclines are basins of attraction. In the phase-plane a **closed orbit** is a loop which corresponds to a pair of periodic solutions $x(t)$ and $y(t)$ (as in the predator-prey problem or any vortex). We are now ready to state our theorem.

THEOREM 9.5: (*The Poincaré-Bendixson[5] Theorem*) Suppose that R is a basin of attraction for the autonomous system (12) $\begin{cases} x'(t) = f(x, y) \\ y'(t) = g(x, y) \end{cases}$ and that inside R

---

[5] Ivar Bendixson (1861–1935) was a Swedish mathematician who published an article in 1901 that expounded on some previous work of Poincaré. He was a professor at the University of Stockholm and was quite involved in public service as well. He served many years on the city council. Bendixson eventually became the president of the University of Stockholm.

there is only one equilibrium solution $(x,y) = (\alpha, \beta)$ which is repelling. Then R contains a single closed orbit which loops around (α, β). Furthermore, any initial condition which is inside R (not on the edge) and not on this loop or (α, β) will produce an orbit which will either spiral outward (if it starts inside) or inward (if it starts outside) toward the unique closed orbit loop.

NOTE: It is permissible in the theorem for the system to have equilibrium solutions on the boundary (edges) of R, just not on the inside.

To illustrate this theorem, we consider the system $\begin{cases} x'(t) = 2x(1 - x/4)/3 - xy/(1+x) \\ y'(t) = y(1 - y/x)/20 \end{cases}$, which has only one equilibrium solution $(1,1)$ which is repelling, inside the basin of attraction $R = \{0 < x < 4, 0 < y < 4\}$ (the details are left to Exercise for the Reader 9.9). The Poincaré-Bendixson theorem implies the existence of a unique periodic solution (closed orbit) to which all other orbits in R will eventually spiral. We can get a good idea of what this closed orbit looks like by picking a couple of initial conditions and running, say Runge-Kutta, for a long enough time period so that the orbits we get spiral out (or in) to the loop. The results of such a computation are illustrated in Figure 9.15.

EXERCISE FOR THE READER 9.9: Consider the system
$$\begin{cases} x'(t) = \frac{2}{3}x(1 - x/4) - xy/(1+x) \\ y'(t) = sy(1 - y/x) \end{cases},$$

where $s > 0$ is a parameter.
(a) Hand draw a phase portrait in the region $x \geq 0$, $y \geq 0$.
(b) For which values of s can we determine whether the equilibrium solution $(x,y) = (1,1)$ is repelling?
(c) Show that the region $R = \{0 < x < 4, 0 < y < 4\}$ is a basin of attraction.

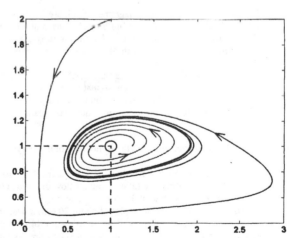

FIGURE 9.15: Illustration of the Poincaré-Bendixson theorem for the system $\begin{cases} x'(t) = 2x(1 - x/4)/3 - xy/(1+x) \\ y'(t) = y(1 - y/x)/20 \end{cases}$ with basin of attraction $R = \{0 < x < 4, 0 < y < 4\}$ containing the unique equilibrium solution (1,1) which is attracting. The equilibrium solution is located at the center of the circle shown. The outside orbit arose from the initial condition (1,2) with a time interval [0,200]; the inside orbit arose from the initial condition (1.25,1) with a time interval of [0,150].

EXERCISES 9.3

1. For each of parts (a) through (f) of Exercise 1 of Section 9.2, do the following: (i) hand draw a phase-plane diagram (cf. Figure 9.10) including all nullclines (labeled), all equilibrium solutions, and all flow directions (exact on nullclines and approximate between them) throughout the entire xy-plane. (ii) Find the Jacobian matrix evaluated at each critical point and use either the classification in Figure 9.14 (linear case) or Theorem 9.3 (nonlinear case) to describe the type of each equilibrium solution as best as possible (e.g., stable node, vortex, etc.)

2. Perform the two tasks of the preceding exercise, this time to each of parts (a) through (f) of Exercise 2 of Section 9.2.

3. Each of the autonomous systems below has an isolated equilibrium solution $(x(t), y(t)) = (0,0)$. For each one do the following: (i) Apply Theorem 9.3 to the systems at the equilibrium solution $(x(t), y(t)) = (0,0)$. What are the conclusions?
    (ii) Use MATLAB to create a plot of about 20 orbits near the equilibrium solution (0,0) chosen so that the behavior of all orbits near this equilibrium solution are well represented. After examination of these plots, can you say more (than was said in (ii)) about the nature of the behavior of the phase-plane near (0,0)? Add flow-direction arrows to your MATLAB-generated plot.

    (a) $\begin{cases} x'(t) = x + x^2 + y^2 \\ y'(t) = y - xy \end{cases}$

    (b) $\begin{cases} x'(t) = (1+x)\sin y \\ y'(t) = 1 - x - \cos y \end{cases}$

    (c) $\begin{cases} x'(t) = -x - y - 3x^2 y \\ y'(t) = y\sin x - 2x - 4y \end{cases}$

    (d) $\begin{cases} x'(t) = y \\ y'(t) = 2(x^2 - 1)y - x \end{cases}$

    (e) $\begin{cases} x'(t) = y \\ y'(t) = (1 - x^2)y - x \end{cases}$

    (f) $\begin{cases} x'(t) = \cos x - e^{y-x} \\ y'(t) = \sin(x - 2y) \end{cases}$

4. Consider each of the two nonlinear systems: (A) $\begin{cases} x'(t) = -y - x^2 \\ y'(t) = x \end{cases}$ and (B) $\begin{cases} x'(t) = -y - x^3 \\ y'(t) = x \end{cases}$.

    (a) Hand draw phase-plane diagrams (cf. Figure 9.10) for each of these two systems including all nullclines (labeled), all equilibrium solutions, and all flow directions (exact on nullclines and approximate between them) throughout the entire xy-plane.
    (b) Apply Theorem 9.3 to each of these systems at the equilibrium solution $(x(t), y(t)) = (0,0)$. What are the conclusions?
    (c) For each of the two systems, use MATLAB to create a plot of about 20 orbits near the equilibrium solution (0,0) chosen, so that the behavior of all orbits near this equilibrium solution are well represented. After examination of these plots, can you say more (than was said in part (b)) about the nature of the behavior of the phase-plane near (0,0)?

5. Here we will be working with the following nonlinear system: $\begin{cases} x'(t) = 2xy \\ y'(t) = y^2 - x^2 \end{cases}$.

    (a) Hand draw a phase-plane diagram (cf. Figure 9.10) for the system including all nullclines (labeled), all equilibrium solutions, and all flow directions (exact on nullclines and approximate between them) throughout the entire xy-plane.
    (b) Use MATLAB to create a plot of about 20 orbits near the equilibrium solution (0,0) chosen so that the behavior of all orbits near this equilibrium solution are well represented. After examination of these plots, add in flow direction arrows and describe the orbit behavior in the phase-plane near (0,0). Is the equilibrium solution stable? Why or why not?

6. Repeat both parts of Exercise 5, this time for the nonlinear system:

    $$\begin{cases} x'(t) = x^3 - 2xy^2 \\ y'(t) = 2x^2 y - y^3 \end{cases}.$$

7. (a) Explain why an orbit can never cross a vertical x-nullcline or a horizontal y-nullcline (in an area where the necessary Lipschitz conditions for Theorem 9.1 apply).
    (b) Using the existence and uniqueness Theorem 9.1, further explain why an orbit cannot merge into such a nullcline, without totally being contained in it.

8. If an autonomous system satisfies the assumptions of Theorem 9.3 at the equilibrium solution (α, β), is it possible to give conditions involving the trace and determinant of the Jacobian matrix to ensure that (α, β) is repelling?

9. For the system $\begin{cases} x'(t) = -y + x(1 - x^2 - y^2) \\ y'(t) = x + y(1 - x^2 - y^2) \end{cases}$, (a) show that $(0,0)$ is a repelling equilibrium solution and (b) find a basin of attraction for it. (c) Next, use MATLAB to plot a few orbits starting at points within this basin of attraction for long enough time intervals so that they nicely indicate the closed loop (periodic solution) guaranteed by the Poincaré-Bendixson theorem.

10. Show that the system $\begin{cases} x'(t) = 3x - y - x\exp(x^2 + y^2) \\ y'(t) = x + 3y - y\exp(x^2 + y^2) \end{cases}$ possesses a periodic solution and then get MATLAB to help find out approximately how (in the phase-plane) the corresponding orbit will look.

11. The single vector/matrix differential equation (14) $X'(t) = AX$ looks a lot like the Malthusian model from the last chapter. Recall that an eigenvalue of the matrix A is a root λ of the characteristic equation $\det \begin{bmatrix} a - \lambda & b \\ c & d - \lambda \end{bmatrix} = 0$ and each eigenvalue will have an associated (nonzero) eigenvector $v = \begin{bmatrix} v_1 \\ v_2 \end{bmatrix}$ which satisfies $Av = \lambda v$. From eigenvalues and eigenvectors we can obtain solutions of (14) as follows.
    (a) Suppose that λ is an eigenvalue of the matrix A with an associated nonzero eigenvector v. Show that the vector function $X(t) = e^{\lambda t}v$ is a solution of the DE (14).
    (b) Suppose that λ and μ are two different eigenvalues of A with associated nonzero eigenvectors v and w, respectively. Show that for any constants C and D, the vector function $X(t) = Ce^{\lambda t}v + De^{\mu t}w$ is also a solution of $X'(t) = AX$.
    (c) Show that the characteristic equation can be rewritten in the form $\lambda^2 - tr(A)\lambda + \det(A) = 0$ where $tr(A) = a + d$ is the trace of A, and $\det(A) = ad - bc$ is the determinant of A.

12. Carefully use Theorem 9.2 to geometrically interpret the solutions of a linear system in each of the cases indicated in Figure 9.14. Explain both how and why the generic pictures given are accurate, as well as why the flow directions are as indicated.

13. (a) With the aid of Theorem 9.3, discuss what changes (if any) on the parameters a and b would result in the STD model of Exercise 5 of the last section having a vortex equilibrium point (with positive populations).
    (b) Repeat part (a), this time analyzing the effect of changes on the parameters r and s on producing a vortex.

14. Reformulate Theorem 9.3 in the language of eigenvalues and positive definite matrices (see Chapter 7).

9.4: GENERAL FIRST-ORDER SYSTEMS AND HIGHER-ORDER DIFFERENTIAL EQUATIONS

All of the ODE numerical methods that we learned in the last chapter can be extended to first-order systems with any number of unknown functions. Indeed, following along the lines of the last section, we could, for example, easily build MATLAB programs called `runkut3d`, `runkut4d`, etc., that can deal with 3-dimensional systems, 4-dimensional systems and so on. Because of MATLAB's vector/matrix capabilities, however, we can build single programs that will perform any particular numerical scheme and be able to apply it to a system with any number of unknowns. We do this now for the Runge-Kutta method. This program looks quite similar to the `runkut` program of Chapter 8. The main difference is that the functions consisting of the right sides of the DEs are now stored as a single vector-valued function. Also, the solution values of the n functions $x_1, x_2, \cdots, x_n$ obtained from the method will be stored in a matrix with n columns. The first column will have the stored values of x_1, the second column will have those of x_2, and so on.

PROGRAM 9.2: Runge-Kutta method for the system (2):

$$\begin{cases} x_1'(t) = f_1(t, x_1, x_2, \cdots, x_n), & x_1(a) = c_1 \\ x_2'(t) = f_2(t, x_1, x_2, \cdots, x_n), & x_2(a) = c_2 \\ \quad \cdots \\ x_n'(t) = f_n(t, x_1, x_2, \cdots, x_n), & x_n(a) = c_n. \end{cases}$$

```
function [t,X]=rksys(vectorf,a,b,vecx,hstep)
%input variables: vectorf, a, b, vecx, hstep
%output variables: t, X
%Uses Runge-Kutta method to solve IVP of system of first-order ODE:
%x.' f1, x^' :2,... xn' fn where vectorf is the vector valued
%function def x1, x2, ..., xn which has components f1, f2, ..., fn.
%The initial conditions x1(a)=x10, x2(a)=x20,...xn(a)=xn0 are
%stored in the vector vecx, the final t-value is b, and step size
%is hstep. The output consists of the time vector t and a
%corresponding matrix X which has n columns, one for each of the
%functions x1, x2, ..., xn.
X(1,:)=vecx;
t=a:hstep:b;
[m nmax]=size(t);
for n=1:(nmax-1);
    k1=feval(vectorf,t(n),X(n,:));
    k2=feval(vectorf,t(n)+.5*hstep,X(n,:)+.5*hstep*k1);
    k3=feval(vectorf,t(n)+.5*hstep,X(n,:)+.5*hstep*k2);
    k4=feval(vectorf,t(n)+hstep,X(n,:)+hstep*k3);
    X(n+1,:)=X(n,:)+1/6*hstep*(k1+2*k2+2*k3+k4);
end
```

Thus solving such first-order IVPs numerically has no new procedural difficulties, and MATLAB's vector capabilities have made writing a simple and elegant universal program quite possible. The vector valued function `vectorf` could be stored either as an M-file or an inline function, and the usual syntax rules apply.

The existence and uniqueness theorem (Theorem 9.1) extends almost verbatim to higher-dimensional systems. Also there is an analogous theory for linear systems $X' = AX$, but because of the greater number of dimensions the results will depend not just on the two quantities (cf. Theorem 9.2 and Figure 9.14 of the last section) but, in general, on the eigenvalues of the coefficient matrix. See, for example, [Hur-90] or [HiSm-97] for a development of the linear theory. There is one stark difference, however, with nonlinear systems in two dimensions versus in higher dimensions. Unlike for two-dimensional nonlinear systems that model quite closely their linearizations near equilibrium solutions (Theorem 9.3), the behavior of nonlinear higher-dimensional systems can be truly **chaotic** in the vicinity of equilibrium points! Orbits can remain bounded near equilibrium points but wind around in different types of loops that are, for all practical purposes, unpredictable. Also, even small differences in initial conditions can lead to solutions, which differ greatly! We will see this type of phenomenon in our next example, which has the name of the **Lorenz Strange Attractor**.[6] The example results from a three-dimensional atmospheric weather model, and thus has a three-dimensional phase-plane. It will turn out to be more convenient and useful to look at two-dimensional projections and this is done in our next example.

EXAMPLE 9.7: (*The Lorenz Strange Attractor*) The Lorenz system is given by:

$$\begin{cases} x'(t) = -sx + sy \\ y'(t) = -xz + rx - y \,. \\ z'(t) = xy - bz \end{cases}$$

Using the parameters $b = 8/3$, $s = 10$, $r = 28$ and the initial conditions $x(0) = -8$, $y(0) = 8$, and $z(0) = 27$, (a) apply the Runge-Kutta method using step size $h = 0.01$ to solve the system for $0 \le t \le 50$ and then plot each of the projections z vs. x, y vs. x, and y vs. z.
(b) Perturb the initial conditions slightly to $x(0) = -8.02$, $y(0) = 7.98$ (and same $z(0)$), solve the new system with the same method, and plot the original x minus the new x versus t on the whole time interval.

SOLUTION: Part (a): To begin, we need to construct a vector-valued function for the (right sides of the) Lorenz system. Since M-files are more convenient for such functions, we will construct this one as an M-file called `lorenz`.

```
function xp=lorenz(t,xv)
x=xv(1); y=xv(2); z=xv(3);
```

---

[6] This system was named after its discoverer, Edward N. Lorenz (1917–), a meteorologist and professor at MIT. It is noteworthy that his system is very simple and it arose as a model for weather prediction. He did this work in the early 1960s and because of the lack of fully integrated and powerful computing tools, he had to work with reams of numerical output and very clumsy strip-chart plots to analyze his model. The orbits are extremely chaotic and unpredictable (see Figure 9.16) and the model is very sensitive to slight differences in initial conditions (see Figure 9.17). Such behavior is why phenomena such as weather are so difficult to predict, even for a rather short period. In his seminal paper of 1963, Lorenz commented on the sensitivity of the model to slight perturbations in initial conditions such as how a butterfly flapping its wings in Beijing could affect the weather thousands of miles away some days later.

```
xp(1)=-10*x+10*y; xp(2)=-x*z + 28*x-y; xp(3)=x*y-8/3*z;
```

Note that here the first, second, and third components of the MATLAB vectors x and xp correspond to x, y, and z respectively in the Lorenz system.

```
>> [t,X]=rksys('lorenz',0,50,[-8 8 27],0.01);
>> x=X(:,1); y=X(:,2); z=X(:,3); % back to original notation
>> plot(x,z) %the pcts resulting from this and the following
>> plot(x,y) %five plots are all shown in Figure 9.16
>> plot(y,z), plot(t,x), plot(t,y), plot(t,z)
```

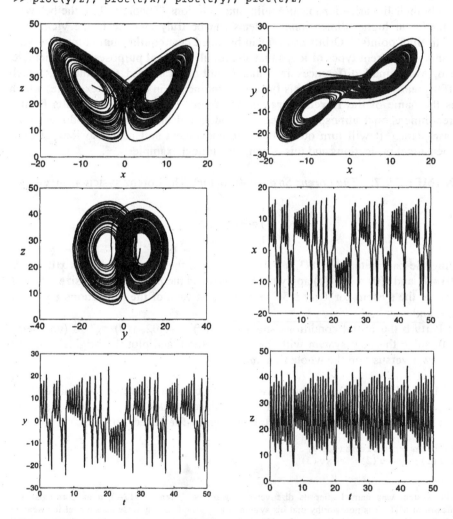

FIGURE 9.16: Views of the Lorenz Strange Attractor (Example 9.7) meteorological model. The single orbit shown (various 2-dimensional views of it) is often referred to as the "butterfly" graph. It is extremely chaotic. In fact, mathematicians conjecture that for any given sequence of positive integers: say 13, 2, 6, 22, 18, 256, 3, there will be a time when this particular orbit will make 13 loops on the left wing of the butterfly, followed by 2 loops on the right wing, then 6 on the left again, and so on.

Part (b):
```
>> [t,X2]=rksys('lorenz',0,50,[-8.02 7.98 27],0.01);
>> x2=X2(:,1);
>> plot(t,x2-x)  %plot shown in Figure 9.17
```

FIGURE 9.17: Graph of the difference of two (x-coordinates of) solutions of the Lorenz model whose initial conditions differ by only about 0.25%. Notice that there seems to be reasonable agreement for only about 3 units of time (days?) or so and after this the difference becomes as chaotic as the functions themselves. This is part of the reason that weather forecasts are usually only given for 3 or so days in advance (and they are never guaranteed).

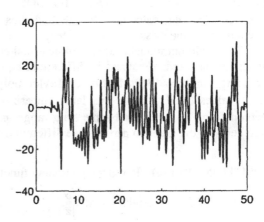

EXAMPLE 9.8: (*The Pendulum*) Figure 9.18 shows a diagram of a pendulum. The pendulum consists of a rod of length L that is connected to a hinge which is free to move back and forth in one direction. We assume the free end of the rod has a weight of mass m attached to it and that the mass of the rod is negligible in comparison. We also assume the hinge is frictionless. The position of the pendulum is recorded by the angle θ that the rod makes with the vertical. The resultant gravitational force on the weight pulls in the direction opposite but tangent to the displacement (see figure). The other component of the gravitational force is canceled off by the rod. The velocity of the mass m equals the rate of change of the arclength $s = L\theta$ that it displaces (see Figure 9.18) from its equilibrium position, so the acceleration of the mass is the second derivative of this quantity. Newton's Second Law, $F = ma$, now gives us that

$$-mg\sin\theta = mL\theta''(t) \implies L\theta''(t) + g\sin\theta = 0, \quad (15)$$

which we refer to as the **pendulum model**.

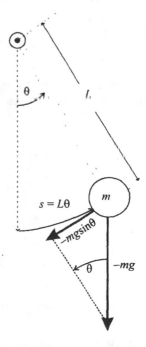

FIGURE 9.18: Illustration of a pendulum for Example 9.8.

(a)   Numerically solve the pendulum model (15) with the initial conditions $\theta(0) = \pi/6$, $\theta'(0) = 0$ (physically, this corresponds to the pendulum being held up at this angle and released at time $t = 0$) and the following parameters: $L = 1.5$ feet, $m = 200$g. Use the Runge-Kutta method with step size $h = 0.1$ and solve for the time interval $0 \le t \le 15$ [seconds] $= 1/4$ minute.   Note that the mass, although specified, does not enter into the DE and thus the motion of the pendulum is independent of the mass.

(b)   The pendulum model (15) is nonlinear and cannot be solved explicitly.   It is customary in standard courses in differential equations to linearize the model by replacing $\sin\theta$ by θ (its first-order Taylor polynomial), which is accurate for small values of θ.   Solve this latter linearization also with the Runge-Kutta method (same step size and same time range and initial conditions) and plot it together with the graph in part (a) (use different plot color/styles).

SOLUTION: Part (a): Introducing the new function $z(t) = \theta'(t)$, we can express the given IVP for the pendulum as: $\begin{cases} \theta'(t) = z, & \theta(0) = \pi/6 \\ z'(t) = -(g/L)\sin\theta, & z(0) = 0 \end{cases}$.

Using $g = 32.1740$ ft/sec^2 and putting in $L = 1.5$ ft., we can now turn the problem over to MATLAB and use our `runkut2d` program.

```
>> dth=inline('z','t','th','z');
>> dz=inline('-32.174/1.5*sin(th)','t','th','z');
>> [t,th,z]=runkut2d(dth,dz,0,15,pi/6,0,0.01);
>> plot(t,th)
```

The details for part (b) are similar and are left as an exercise.   The resulting simultaneous plot is shown in Figure 9.19.

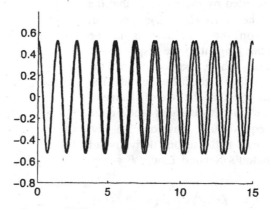

FIGURE 9.19: Graphs of the first 15 seconds of motion of the two pendulum models in Example 9.8. The left-lagging graph is the ideal pendulum model and the second one is that of its linearization. Notice that the ideal pendulum starts to lag behind the linearized model of it. Both appear (and actually are) periodic.

EXERCISE FOR THE READER 9.11: (a) Fill in the remaining details in the pendulum example necessary to obtain the second plot as shown in Figure 9.19. (b) It turns out that both pendulum models give rise to periodic motions. Can you prove this?

Newton used his calculus invention to explicitly solve the **two-body problem** where he analyzed the orbit of a single planet around the Sun. Subsequently, scientists turned to the natural next step up in difficulty: the **three-body problem** of the motion of objects subject only to the mutual forces of gravity. Much work has gone into this problem and it actually translates to an 18-dimensional first-order system! As one would expect, the problem is not explicitly solvable. In the next example, we will look at a certain restricted version of the three-body problem in which one of the objects has negligible mass (and hence negligible gravitational pull) compared with the other two. These hypotheses would be reasonable, for example, if we were tracking the motion of a space station or satellite relative to the Earth and Moon. If we set up a "rotating" coordinate system which keeps the Earth and Moon fixed and leaves the third object moving in a plane, we bring down the number of dimensions to 4. It can be shown using Newton's law of gravitation that these assumptions lead us to the following system for the position $(x(t), y(t))$ of the (relatively low mass) object at time t:

$$\begin{cases} x''(t) = 2y' + x - \dfrac{x_e(x + x_m)}{d_m^3} - \dfrac{x_m(x - x_e)}{d_e^3} \\ y''(t) = -2x' + y - \dfrac{x_e y}{d_m^3} - \dfrac{x_m y}{d_e^3}. \end{cases} \tag{16}$$

In this coordinate system, the Moon is fixed at $(-x_m, 0) = (-1/82.45, 0)$ and the Earth is fixed at $(x_e, 0) = (1 - x_m, 0)$, also, d_m and d_e denote the distances from (x, y) to the Moon and the Earth, respectively. The units in these equations have also been made large to keep the equations clean. Time is measured in years and one unit of distance equals the mean distance from the Earth to the Moon, about 380,000 km. Even in this very restricted three-body model, it is not possible to explicitly obtain the solutions. Nevertheless, it is of great practical importance, to NASA or any business wishing to send out a satellite, to be able to find initial conditions which will result in periodic orbits (otherwise, the space object could drift into outer space to be gone forever!). One set of initial conditions that will work out into a periodic orbit is the following:[7]

$$\begin{cases} x(0) = 1.2, \ x'(0) = 0 \\ y(0) = 0, \ y'(0) = -1.04935750983032. \end{cases} \tag{17}$$

---

[7] This data was computed on a supercomputer; see Chapter 6 of [ShAlPr-97].

Such initial conditions could be realized by bringing the space object to the required position and then directing appropriate thrusts to get the needed initial velocity.

EXAMPLE 9.9: (*Orbit of a Space Station*) In the restricted three-body problem presented by the IVP (16) and (17), (a) plot the orbit and (b) estimate its period.

SOLUTION: Part (a): At first, we do not know how large a time range to solve the IVP for, so it is good to do a few experiments first using Runge-Kutta with smaller step sizes until it looks like we have found a sufficiently large time range to include a complete period of the orbit. A few experiments show, however, that even with step size $h = 0.01$ we would get a very (misleading) nonperiodic orbit (try this!). As it turns out, the step size $h = 0.001$ works very well here and 10 years of time will suffice for a complete orbit. In order to apply Runge-Kutta, we first convert the system (15) into a four-dimensional system of first-order DEs. Introducing the new functions $z(t) = x'(t)$, and $w(t) = y'(t)$, we can rewrite the IVP (16), (17) as:

$$\begin{cases} x'(t) = z, \quad x(0) = 1.2 \\ z'(t) = 2w + x - \dfrac{x_e(x + x_m)}{[(x + x_m)^2 + y^2]^{3/2}} - \dfrac{x_m(x - x_e)}{[(x - x_e)^2 + y^2]^{3/2}}, \quad z(0) = 0 \\ y'(t) = w, \quad y(0) = 0 \\ w'(t) = -2z + y - \dfrac{x_e y}{[(x + x_m)^2 + y^2]^{3/2}} - \dfrac{x_m y}{[(x - x_e)^2 + y^2]^{3/2}}, \\ \qquad\qquad w(0) = -1.0493575098303 12. \end{cases}$$

We will apply our `rksys` program to solve this, but we will need first to create a vector-valued function for the right side of this system. Denoting this M-file as `threebod`, the code is as follows:

```
function xp=threebod(t,xv)
x=xv(1); z=xv(2); y=xv(3); w=xv(4);
xm=1/82.45; xe=1-xm; dm=((x+xm)^2+y^2)^(1/2); de=((x-
xe)^2+y^2)^(1/2);
xp(1)=z;
xp(2)=2*w+x-xe*(x+xm)/dm^3-xm*(x-xe)/de^3;
xp(3)=w;
xp(4)=-2*z+y-xe*y/dm^3-xm*y/de^3;

[t,X]=rksys('threebod',0,10,[1.2 0 0 -1.049357509830312],0.001);

>> x=X(:,1); y=X(:,3);
>> plot(x,y)
>> hold on
>> xm=1/82.45; xe=1-xm;
>> plot(-xm,0,'rp')  %will plot a red pentacle at moon's location
>> plot(xe,0,'go')  %will plot a green 'o' at earth's location
```

The resulting plot appears in Figure 9.20.

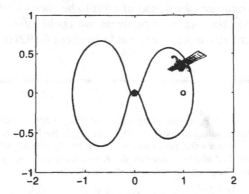

FIGURE 9.20: The orbit of the space station of Example 9.9. Each unit represents the mean distance from the Earth to the Moon. The Earth is represented by the circle on the right and the Moon by the pentacle in the center. It takes about 6.192 years for the orbit to make a complete cycle.

Part (b): One must be a bit careful here due to the errors that arise. Although the graph looks periodic, we cannot just set up a loop to see how long it takes $x(n)$ and $y(n)$ to reach their initial values exactly (since they never will). To get an idea of how we should look at coordinates, we evaluate (in format long)

```
>> x(2)=1.19999907969626    (extremely close to x(1)= 1.2)
>> y(2) -0.00104935675196    (more than 0.001 off from initial y(1) =
0).
```

We focus our attention on the y-coordinate, but we weed out all situations where x is far away from 1.2 (so the only way y can get small here is if (x,y) is near the initial point). This search can be done as follows:

```
>> n=2; %initialize
>>while x(n) < 1.19|abs(y(n))>.001
n=n+1;
end
>> n
→n = 6193
>> t(6193)
→ 6.19200000000000
```

As a further check, we look at the locations of (x,y) at nearby times:

```
>> for k=n-2:n+2
[t(k) x(k) y(k)]
end
```

→

6.19000000000000	1.19966512247978	0.00274177862485
6.19100000000000	1.19966754186654	0.00169277717629
6.19200000000000	1.19966811983163	0.00064377132264
6.19300000000000	1.19966685634965	-0.00040523438093
6.19400000000000	1.19966375141712	-0.00145423537902

Our analysis thus estimates the period of orbit to be about 6.192 years. This is as accurate as our step size (so as accurate as we could have hoped). Indeed, to greater precision, the actual orbit turns out to be about 6.192169331 years.

EXERCISES 9.4

1. Each of the following IVPs is given along with a solution $y = f(t)$. For each do the following: (i) Express the IVP as a system of first-order DEs. (ii) Apply the Runge-Kutta method to solve the IVP first with step size $h = 0.2$, then with $h = 0.1$ and finally with $h = 0.01$. (iii) Verify that $f(t)$ solves the original IVP and then compare the Runge-Kutta solutions you obtained in (ii) with $f(t)$, in cases where the graphs are indistinguishable, plot the errors.

    (a) $\begin{cases} y''(t) - 4y' + 4y = 2e^{2t} \\ y(0) = 0, \, y'(0) = 0 \end{cases}$, $0 \le t \le 2$ $\quad f(t) = t^2 e^{2t}$

    (b) $\begin{cases} y'''(t) + 3y'' + 3y' + y = e^{-t} \\ y(0) = 0, \, y'(0) = 0, y''(0) = 0 \end{cases}$, $0 \le t \le 2,$ $f(t) = t^3 e^{-t}/6$

    (c) $\begin{cases} t^3 y''' + 5t^2 y'' + 2ty' - 2y = t^4 \\ y(0) = y'(0) = y''(0) = y'''(0) = 0 \end{cases}$, $0 \le t \le 6,$ $f(t) = t^4/90$

    (d) $\begin{cases} y''' - 5y'' + 8y' - 4y = 0 \\ y(0) = 1, \, y'(0) = 4, y''(0) = 0 \end{cases}$, $0 \le t \le 5,$ $f(t) = 13e^{2t} - 10te^{2t} - 12e^t$

2. Repeat each part of Exercise 1 for the following IVPs:

    (a) $\begin{cases} y''(t) + (y')^2 = 0 \\ y(0) = 0, \, y'(0) = 1/e \end{cases}$, $1 \le t \le 10$ $\quad f(t) = \ln(x + e) - 1$

    (b) $\begin{cases} y''(t) = 2y^3 + 2y \\ y(0) = -, \, y'(0) = -2 \end{cases}$, $0 \le t \le 2,$ $f(t) = -\tan(t + \pi/2)$

    (c) $\begin{cases} t^3 y''' + t^2 y'' - 2ty' + 2y = (24 - 6t^2)/t^2 \\ y(1) = -1, \, y'(1) = 5, \, y''(1) = -12 \end{cases}$, $1 \le t \le 16,$

    $f(t) = \dfrac{-2}{t^2} + \dfrac{1}{t} - 3 + 4t - t^2$

    (d) $\begin{cases} y'' = \sqrt{y} \\ y(2) = 1/9, \, y'(0) = 2/9 \end{cases}$, $0 \le t \le 5,$ $f(t) = \dfrac{t^4}{144}$

3. MATLAB's built-in DE solver ode45, which was introduced in Section 8.4 for single ODEs, is also able to handle systems of first-order ODE. The syntax is a bit different than that of our function rksys. In this exercise, you will be repeating for each of the IVPs in Exercise 1, the following similar tasks using the ode45 program in place of the Runge-Kutta program. By experimenting with the help menu (and with what was said in Section 8.4), figure out how to use the program to solve the resulting first-order system using ode45 in the default setting. Plot the error of this approximation versus the given exact solution. Next, determine how to use the "options" to refine the accuracy to decrease the maximum error by at least 50%. Repeat once again, this time decreasing the maximum error at least another 75% (from the already decreased error).

4. Repeat Exercise 3 for each of the parts of Exercise 2.

5.   Another well-known chaotic attractor is the so-called Rossler[8] band represented by the system:

$$\begin{cases} x'(t) = -y - z \\ y'(t) = x + ay \\ z'(t) = b + z(x - c). \end{cases}$$

     (a) Using the parameters $a = 0.2$, $b = 0.2$, $c = 5.7$ (which Rossler used), initial conditions $x(0) = 0$, $y(0) = -6.78$ and $z(0) = 0.02$, and the Runge-Kutta method with step size $h = 0.01$ on the time interval [0,250], obtain graphs of y vs. x, z vs. x, z vs. y, and each of x, y, z vs. t.
     (b) Perturb slightly the initial condition $y(0)$ to equal -6.8 and solve the resulting system in the same way with the Runge-Kutta method (keeping all else the same). Plot the differences of the x's, y's, and z's (old and new) versus t. You need only do these three plots. Based on your graphs, about how long do the two different solutions seem to agree?

6.   (a) Re-solve the Rossler band model of Exercise 5, first with the Runge-Kutta method with step size $h = 0.1$ and then by successively halving the step size. Repeat this process for 6 iterations and plot the differences of successive x-coordinates versus t (so there will be a total of five plots). Based on your plots, over how large a time range does each approximation seem to be good?
     (b) MATLAB's built-in ODE solver ode45 also works for systems (see Exercise 3). Try to use it repeatedly by successively increasing the accuracy (using the "options") to obtain a solution as accurate as the final one in part (a). What is the size of the ode45 solution vector compared with the final one in part (a)?
     Suggestion: To check the accuracy of the ode45 solutions with that in part (a), it would be awkward to plot the differences (since the t-vectors in ode45 solutions are not uniformly spaced), so you should simply plot the two functions (in different colors) in the same window and visually check how long they agree.

7.   In this exercise you are to experiment with one of the parameters in the Rossler band model of Exercise 5. Keeping everything else (Runge-Kutta, step size, b, c, initial conditions, and time range) the same as in part (a), let the parameter a run from -0.4 to 0.6 in increments of 0.2 and record the y vs. x plots (only these) of the different plots. Any comments on how things are changing? (You may wish to look at some additional plots for intermediate values of the parameter a.)

8.   (a) Write a MATLAB program called eulersys which has the same syntax, input, and output variables as the program rksys in Program 9.2, except that the Euler method is used in place of the Runge-Kutta method.
     (b) Repeat part (a) of Exercise 6 but replacing the Runge-Kutta method with the Euler method and using your program eulersys.

9.   (a) Write a MATLAB program called ieulsys which has the same syntax, input, and output variables as the program rksys in Program 9.2, except that the improved Euler method is used in place of the Runge-Kutta method.
     (b) Repeat part (a) of Exercise 6 but replacing the Runge-Kutta method with the improved Euler method and using your program ieulsys.

---

[8] This model was discovered in 1976 by German Otto Rossler (1940–) in his efforts to more fully understand the Lorenz strange attractor. His model introduces a spiral type of chaos which combines a two-variable oscillator (x and y) with a switching-type subsystem (z) so that x and y switch z and, conversely, the flow of x and y depends on the switching state of z (positive or negative). To better understand how this model is working, the reader is advised to fix z at some different values and for each of these, hand draw an xy-phase-plane. Dr. Rossler appears to be quite a universal scientist; he has held professorships (at various institutes in different countries) in departments of Mathematics, Chemical Engineering, Nonlinear Studies, and Theoretical Biochemistry.

10. (*Physics Pendulum*) (a) Suppose that we wanted to build a pendulum which had a period equal to exactly one second. How long would we need to make L for this pendulum? Determine L to an accuracy so as to make the period accurate to within an error of 0.001 to exactly one second given that the initial conditions are $\theta(0) = \pi/4$, $\theta'(0) = 0$.

(b) Repeat part (a) but now for a period of exactly 10 seconds.
(c) Repeat part (a) this time using the linearized pendulum model.
(d) Repeat part (b) this time using the linearized pendulum model.

Note: Pendula are the basis for the grandfather clock. Friction and air resistance are minimized but there is a very light "kicker" mechanism which keeps them going with constant amplitude. These clocks are adjusted via a small screw near the weight end of the rod which can make slight variations in the length of the rod.

11. (*Physics Pendulum*) In Exercise 10, what happens to the answers in parts (a) and (b) if we changed the initial position $\theta(0)$ to be some other positive number in the range 0 through $\pi/2$ (exclusive)? Explain, and give some numerical evidence of your conclusions.

12. (*Physics-Damped Pendulum*) If we modify the DE (15) for the pendulum (with rod length L) to include damping (which could include such forces as air resistance on the bob and friction at the hinge mechanism), the damping force would equal $F = -c\theta'$ (proportional to speed and oppositely directed).

(a) Using Newton's Second Law $F = ma$, show that the DE for the damped pendulum becomes: $L\theta''(t) + (c/m)\theta' + g\sin\theta = 0$.

(b) For a large pendulum with $L = 20$ft, damping constant $c = 60$ (programmed to have m measured in kg.), and $m = 10$kg., use the Runge-Kutta method with $h = 0.01$ to graph θ versus t from $t = 0$ to $t = 30$ seconds. Can you determine whether successive time intervals between when $\theta = 0$ are equal?

(c) Solve and graph the corresponding linearized problem obtained by replacing $\sin\theta$ by θ in the above DE, but keeping all else the same. Graph the solutions to (a) and (b) together. Can you determine whether successive time intervals between when $\theta = 0$ are equal?

13. (a) Hand draw the $\theta - \theta'$ phase-plane for the pendulum DE (15). What are the equilibrium points? Are any of them stable?

(b) Repeat part (a) for the damped pendulum DE of the preceding exercise.

Suggestion: You will of course first need to rewrite the second-order DE as a system of first-order DEs and then proceed as in Section 9.3.

14. In the 1920s, the Dutch physicist van der Pol[9] introduced the following DE, now referred to as the **van der Pol equation**, to model electrical phenomena:

$$x''(t) + \mu(x^2 - 1)x' + x = 0,$$

where the parameter μ is assumed positive for physical reasons.
(a) Express the van der Pol DE as a system of first-order DEs.
(b) Get MATLAB to plot about 10 different orbits from an assortment of initial conditions in the square region: $-10 \le x, x' \le 10$ in the $x - x'$ phase-plane for the case $\mu = 1$.
(c) Repeat part (b) this time for the value $\mu = 1/2$ and $\mu = 2$, and then for $\mu = 1/8$ and $\mu = 8$. Any comments on the differences in phase-planes (say between the last two values of μ)?

---

[9] Balthasar van der Pol (1889–1959) first introduced his namesake equation as a model for "negative resistance in vacuum tube circuits." For different values of the parameter μ, the equation has numerous applications to electric circuits, and for a wide range of values for μ and initial conditions, the solutions very quickly spiral to periodic ones so as to appear "eventually periodic." Think of how electricity moves after you turn on a light switch.

15. (a) What does Theorem 9.3 have to say about the phase-plane of (the two-dimensional first-order system corresponding to) the van der Pol equation (Exercise 14 near the equilibrium solution $x(t) = 0$ (or $(x(t), x'(t)) = (0,0)$ in the xx' phase-plane)? How does your answer depend on the value of the positive parameter μ?

   (b) For $\mu = 5$, find a basin of attraction for the equilibrium solution $x(t) = 0$ (or $(x(t), x'(t)) =$ (0,0) in the xx' phase-plane) of the van der Pol equation, and then apply the Poincaré-Bendixson theorem to prove that this equation has a periodic solution.

   (c) For which other positive values of the parameter μ can you extend your proof of part (b) to prove the existence of a periodic solution to the van der Pol equation?

Chapter 10: Boundary Value Problems for Ordinary Differential Equations

10.1: WHAT ARE BOUNDARY VALUE PROBLEMS AND HOW CAN THEY BE NUMERICALLY SOLVED?

All of the auxiliary conditions considered so far on a differential equation (or system) were so-called initial conditions in that they all specified values of the unknown function(s) and/or derivatives at a single value of the independent variable. It is also quite common to see problems with a second-order differential equation (so it will require two auxiliary conditions) with values of the unknown function being specified at two different values of the independent variable. Examples of this include finding out how far a steel beam will bend at interior points (unknown function) given that the ends are fixed. Other examples include motion and mechanics problems where a (coordinate of the) position of an object is known at two different times and we want to find out the precise trajectory of motion during intermediate times. In this chapter we will focus our attention on second-order two-point **boundary value problems (BVPs)** of the form:

$$\text{(BVP)} \begin{cases} y''(t) = f(t, y, y') & (DE) \\ y(a) = \alpha, \ y(b) = \beta & (BCs) \end{cases} \tag{1}$$

We will present three categories of methods for numerically solving the BVP (1). The first types of these methods, called **shooting methods**, are based on our methods for solving the initial value problems (IVPs) and are quite intuitive. They work basically as follows: Introduce a corresponding IVP that has the same DE, the same first BC $y(a) = \alpha$, and a "reasonable" initial condition for $y'(a)$ (we will give some schemes later, but for now just take it to be zero). Next, solve this IVP with, say, the Runge-Kutta method, on the interval of interest $a \le t \le b$. Now check the value of $y(b)$ to see how it compares with the desired value β. If it is close enough; we accept this solution; otherwise it will be either too small or too large. Readjust the initial condition for $y'(a)$ accordingly (we will give efficient ways for this). Solve the new IVP and continue "shooting" solutions of IVPs in this way at $y(b) = \beta$ until we get close enough ("hit it").

The second type of methods we present for solving BVPs, called **finite difference methods**, are based on approximating derivatives by difference quotients. Multiple approximations are done simultaneously and so the methods will require some matrix manipulations. Finite difference methods will be seen again in later chapters to numerically solve PDEs. In the last section we introduce yet another approach called the **Rayleigh-Ritz** method. This method recasts the

399

BVP as a minimization problem for a certain functional associated with the BVP. The Rayleigh-Ritz method is the one-dimensional analogue of the very successful finite element method for solving BVPs for partial differential equations that will be discussed in Chapter 13. In contrast to the first two rather intuitive methods, the Rayleigh-Ritz method rests on many theoretical underpinnings. The shooting methods are usually more efficient than the latter two methods, but when we move on to partial differential equation BVPs, they have no analogue, but the latter two methods generalize quite well.

To compare and test efficiency of our methods, we will use the example of the deflection of a horizontal beam which is supported (at equal heights at its ends) as illustrated in Figure 10.1.

FIGURE 10.1: Deflection of a horizontal beam.

Letting $y(x)$ denote the vertical level of the beam x units from the left side (making $y = 0$ correspond to the height of the supports), we have the boundary conditions $y(0) = y(L) = 0$, where L is the horizontal distance between the supports. If y' is relatively small, and if the beam has a uniform transverse load w and tension T, it can be shown that $y(x)$ is modeled by the differential equation:

$$y''(x) = \frac{T}{EI}y + \frac{wx(x-L)}{2EI}, \quad 0 \le x \le L. \tag{2}$$

In this equation, E and I are physical constants associated with the beam and its materials (E = modulus of elasticity, I = moment of inertia). Actually the true DE has $y''(t)$ being replaced by the more complicated term $y''(t)/[1+(y')^2]^{3/2}$, which needs to be used in cases where $y'(x)$ gets large. Our reason for using (2) as our basic DE in this chapter is that the BVP can be explicitly solved. The general solution of the DE is given by

$$y(x) = C\sinh(\theta x) + D\sinh(\theta(L-x)) \tag{3}$$
$$+ wLx/2T - w/\theta^2 T - wx^2/2T,$$

where C and D are arbitrary constants.

EXERCISE FOR THE READER 10.1: (a) Verify that each function of the form (3) satisfies the DE (2), provided that $\theta^2 = T/EI$.
(b) Show further that the function (3) will satisfy the boundary conditions $y(0) = y(L) = 0$ if and only if $C = D = \dfrac{w}{\theta^2 T \sinh(\theta L)}$.

Shooting methods are more easily applied to BVPs (1) that have a **linear** differential equation. This means that the DE in (2) has the form

$$y''(t) = p(t)y' + q(t)y + r(t).$$ (4)

If, in a linear DE, the extra term $r(x)$ in (4) is zero, then we say further that the DE is **homogeneous**. There is a very important fact about homogeneous linear equations that makes them amenable to solution:

FACT: If y_1 and y_2 are two solutions of a linear homogeneous DE, and c is any constant, then $y_1 + y_2$ and cy_1 are also solutions.

Proof: Introducing the (linear) operator: $L[y] = p(t)y' + q(t)y$, the homogeneous linear DE can be rewritten as $y'' = L[y]$. Using the basic differentiation rules from calculus, we can write:

$$L[y_1 + y_2] = p(t)[y_1 + y_2]' + q(t)[y_1 + y_2]$$
$$= p(t)(y_1' + y_2') + q(t)y_1 + q(t)y_2$$
$$= p(t)y_1' + q(t)y_1 + p(t)y_2' + q(t)y_2$$
$$= L[y_1] + L[y_2],$$

which proves that $y_1 + y_2$ is also a solution. Similarly, the following computation shows that cy_1 is a solution:

$$L[cy_1] = p(t)[cy_1]' + q(t)cy_1 = p(t)cy_1' + cq(t)y_1$$
$$= c(p(t)y_1' + q(t)y_1) = cL[y_1].$$

EXERCISE FOR THE READER 10.2: Does the above fact continue to hold true for linear DEs that are not homogeneous?

EXAMPLE 10.1: For each of the second-order DEs below, indicate whether it is linear and whether it is homogeneous.

(a) $y'' + 3\cos(t)y' + \sin(t)y^2 = 0$ (b) $y'' + e'y = e'$

SOLUTION: Part (a): The equation is nonlinear due to the y^2 term (in each term, the unknown function or one of its derivatives—but not both, can only appear to the first power and must not be inserted in other functions to be linear). The homogeneity question is not applicable.

Part (b): The equation is linear (putting it in standard form it will have $p(t) = -e'$, $q(t) = 0$, and $r(t) = e'$) but not homogeneous because of the e' term.

EXERCISE FOR THE READER 10.3: For each of the second-order DEs below, indicate whether it is linear or not and (if linear) whether it is homogeneous.

(a) $y'' = tyy'$ (b) $y'' + \cos(t)y - \sin(t) = 0$

We now state a general existence and uniqueness theorem for the BVP (1). It is often a good idea to check to see if this theorem is verified before a numerical scheme is applied to any particular BVP, otherwise a solution may not be unique, or worse, may not exist and so the output could be meaningless.

THEOREM 10.1: (*Existence and Uniqueness for BVP*) If in the BVP (2)

$$\begin{cases} y''(t) = f(t, y, y') & (DE) \\ y(a) = \alpha, \ y(b) = \beta & (BCs) \end{cases},$$

the function $f(t, y, y')$ along with its two partial derivatives f_y and $f_{y'}$ are continuous in the region $R = \{a \leq t \leq b, -\infty < y, y' < \infty\}$, and furthermore if we also have satisfied on R the following two conditions: (i) $f_y(t, y, y') > 0$ and (ii) $|f_{y'}(t, y, y')| \leq M$ for some fixed positive number M, then the BVP has a unique solution.

A proof of this theorem can be found in [Kel-68]. We caution the reader that this theorem is not a definitive result on BVPs. For example, the BVP that we numerically solve in Example 10.4 of Section 10.3 violates both of the two conditions of the theorem but nevertheless does have a unique solution. In case that the DE is linear, the above theorem simplifies as follows:

COROLLARY 10.2: (*Existence and Uniqueness for Linear BVP*) If in the linear BVP

$$\begin{cases} y''(t) = p(t)y' + q(t)y + r(t) & (DE) \\ y(a) = \alpha, \ y(b) = \beta & (BC's) \end{cases},$$

the functions $p(t)$, $q(t)$ and $r(t)$ are all continuous on the interval $[a,b]$ and also $q(t) > 0$ throughout $[a,b]$, then the BVP has a unique solution.

EXAMPLE 10.2: For each of the following BVPs, indicate if it is linear or nonlinear and then determine whether the appropriate existence and uniqueness result above applies.

(a) $\begin{cases} y''(t) + 2y^2 = \cos(t) \\ y(0) = 1, \ y(1) = 2 \end{cases}$
     (b) $\begin{cases} y''(t) + 3y' - e^t y = \sin(t^2) \\ y(0) = 0, \ y(1) = -3 \end{cases}$

SOLUTION: Part (a): The DE is nonlinear because of the y^2 term. To apply Theorem 10.1, we put $f(t, y, y') = \cos(t) - 2y^2$. This function and its partial derivatives are continuous everywhere, and so certainly on the required region $R = \{0 \leq t \leq 1, -\infty < y, y' < \infty\}$. However, the partial derivative $f_y(t, y, y') = -4y$ is not always positive on the region R (just take y to be zero or any positive number). Since not all of the hypotheses are satisfied the existence and uniqueness theorem is not applicable. The BVP still may or may not have a (unique) solution but theoretically we cannot say more with what we know.

Part (b):   The DE now is clearly linear with $p(t) = -3$, $q(t) = e^t$, and $r(t) = \sin(t^2)$, and the hypotheses of Corollary 10.2 are clearly satisfied so we may conclude that the BVP has a unique solution.

EXERCISE FOR THE READER 10.4: For each of the following BVPs, indicate if it is linear or nonlinear and then determine whether the appropriate existence and uniqueness result above applies.

(a) $\begin{cases} y''(t) = y^2 + e^t y' \\ y(2) = 1,\ y(4) = 50 \end{cases}$

(b) $\begin{cases} y''(t) = ty + t\cos([y']^2) \\ y(1) = 0,\ y(3) = 2 \end{cases}$

EXERCISES 10.1

1. For each of the second-order DEs below, indicate whether it is linear or not and if it is linear whether it is homogeneous.

   (a) $y''(t)\cos t = (y + y' + 1)\sin t$

   (b) $y''(t) = 3t'y - y' + y'\cos t$

   (c) $y''(t) = \dfrac{1+y}{1+y'}$

   (d) $ty''(t) + y' = \dfrac{y}{t}$

2. For each of the following BVPs, indicate if it is linear or nonlinear and then determine whether or not the appropriate existence and uniqueness result above applies.

   (a) $\begin{cases} y''(t) = -2ty' + \cos(t) \\ y(0) = 1,\ y(2) = 1 \end{cases}$

   (b) $\begin{cases} \cos(t)y''(t) = \sin(t)y' + y + \cos t \\ y(3\pi/4) = 1,\ y(\pi) = 0 \end{cases}$

   (c) $\begin{cases} y''(t) = y'y^2 t^3 \\ y(1) = 1,\ y(2) = 2 \end{cases}$

   (d) $\begin{cases} y''(t) = y^5 + \cos(y')e^t \\ y(0) = 0,\ y(10) = 5 \end{cases}$

3. Repeat Exercise 2 for the following BVPs:

   (a) $\begin{cases} y''(t) = (t + y)^2 \\ y(0) = 1,\ y(3) = 0 \end{cases}$

   (b) $\begin{cases} y''(t) = \tan(t)y' + \arctan(t)y + \cos t \\ y(1) = 1,\ y(100) = 100 \end{cases}$

   (c) $\begin{cases} y''(t) = \dfrac{1+y'}{1-y'} + 6y \\ y(0) = 0,\ y(2) = 0 \end{cases}$

   (d) $\begin{cases} y''(t) = (1 + t + y + y')^2 \\ y(2) = 0,\ y(3) = -4 \end{cases}$

10.2: THE LINEAR SHOOTING METHOD

In case of a linear DE in (1), the shooting method will require only "two shots." More precisely, we will be able to solve the BVP (1) by solving two corresponding IVPs. An appropriate linear combination of the latter two solutions will yield a solution to the former. Passing now to the details, we begin with a linear BVP,

$$(\text{LBVP}) \quad \begin{cases} y''(t) = p(t)y' + q(t)y + r(t) & (\text{DE}) \\ y(a) = \alpha,\ y(b) = \beta & (\text{BCs}) \end{cases} \qquad (5)$$

With this we associate the following two IVPs:

$$(\text{IVP-1}) \begin{cases} y_1''(t) = p(t)y_1' + q(t)y_1 + r(t) & (DE) \\ y_1(a) = \alpha, \ y_1'(a) = 0 & (IC\text{-}1's) \end{cases}, \tag{6}$$

and

$$(\text{IVP-2}) \begin{cases} y_2''(t) = p(t)y_2' + q(t)y_2 & (DE\text{-}2) \\ y_2(a) = 0, \ y_2'(a) = 1 & (IC\text{-}2s) \end{cases}. \tag{7}$$

We point out that the DE in IVP-1 is identical to the linear DE in the original BVP while the DE-2 in IVP-2 is the **homogenization** of the original DE.

CLAIM: If $y_1(t)$ is a solution of (IVP-1) and $y_2(t)$ is a solution of IVP-2 then

$$y(t) = y_1(t) + \frac{\beta - y_1(b)}{y_2(b)} y_2(t). \tag{8}$$

is a solution of the original (BVP), as long as $y_2(b) \neq 0$.

Proof: Write $L[y] = p(t)y' + q(t)y$ so that the original DE is $y''(t) = L[y] + r(t)$ and the homogenization DE-2 is $y''(t) = L[y]$. Because of linearity, we have $L[cy_1 + dy_2] = cL[y_1] + dL[y_2]$, for any constants c and d. Using first the basic properties of differentiation and then this property we can write (keeping in mind which DEs y_1 and y_2 solve):

$$y''(t) = \left(y_1 + \frac{\beta - y_1(b)}{y_2(b)} y_2 \right)''(t)$$

$$= y_1''(t) + \frac{\beta - y_1(b)}{y_2(b)} y_2''(t)$$

$$= (L[y_1] + r(t)) + \frac{\beta - y_1(b)}{y_2(b)} L[y_2]$$

$$= L\left[y_1 + \frac{\beta - y_1(b)}{y_2(b)} y_2 \right] + r(t)$$

$$= L[y] + r(t).$$

This shows that $y(t)$ solves the DE of (LBVP) (5). The boundary conditions are easily checked using the (specially designed) initial conditions in (6) and (7):

$$y(a) = y_1(a) + \frac{\beta - y_1(b)}{y_2(b)} y_2(a) = \alpha + \frac{\beta - y_1(b)}{y_2(b)} 0 = \alpha,$$

$$y(b) = y_1(b) + \frac{\beta - y_1(b)}{y_2(b)} y_2(b) = y_1(b) + \frac{\beta - y_1(b)}{y_2(b)} y_2(b) = \beta.$$

The method just described for solving the linear BVP (5) is called the **linear shooting method**. We present now an example involving the deflection of a beam.

EXAMPLE 10.3: (a) Use the linear shooting method with $h = 0.1$ in the Runge-Kutta method to solve the following beam deflection BVP:

$$\begin{cases} y''(x) = \dfrac{T}{EI}y + \dfrac{wx(x-L)}{2EI}, & 0 \le x \le L, \\ y(0) = 0 = y(L) \end{cases}$$

having the parameters: $L = 50$ feet (length), $T = 300$ lbs (tension at ends), $w = 50$ lb/ft (vertical load), $E = 1.2 \times 10^7$ lb/ft^2 (modulus of elasticity), and $I = 4$ ft^4 (central moment of inertia). Graph the resulting numerical solution. (b) Graph the error as it compares with the exact solution as given in (3).

SOLUTION: Part (a): Comparing the generic linear BVP (5) with ours and translating the two IVPs (6) and (7) into our setting we obtain:

$$(\text{IVP-1}) \begin{cases} y_1''(x) = \dfrac{T}{EI}y_1 + \dfrac{wx(x-L)}{2EI} & (\text{DE}) \\ y_1(0) = 0,\ y_1'(0) = 0 & (\text{IC-1}) \end{cases},$$

and

$$(\text{IVP-2}) \begin{cases} y_2''(x) = \dfrac{T}{EI}y_2 & (\text{DE-2}) \\ y_2(0) = 0,\ y_2'(0) = 1 & (\text{IC-2}) \end{cases}$$

We apply the Runge-Kutta method to solve both of these problems on the interval $[0, L]$, and once this is done, we take (using (8)):

$$y(x) = y_1(x) + \frac{0 - y_1(L)}{y_2(L)} y_2(x) = y_1(x) - \frac{y_1(L)}{y_2(L)} y_2(x),$$

as our solution to the original BVP. To solve each of the second-order IVPs, we need to translate them into a system of first-order ODEs as explained in Section 9.1. Introducing the new functions $u_1 = y_1'$ and $u_2 = y_2'$, the above two IVPs translate to the following equivalent two-dimensional systems:

$$(\text{IVP-1}') \begin{cases} y_1'(x) = u_1, & y_1(0) = 0 \\ u_1'(x) = \dfrac{T}{EI}y_1 + \dfrac{wx(x-L)}{2EI}, & u_1(0) = 0 \end{cases},$$

and

$$(\text{IVP-2}') \begin{cases} y_2'(x) = u_2, & y_2(0) = 0 \\ u_2'(x) = \dfrac{T}{EI}y_2, & u_2(0) = 1 \end{cases}.$$

At this point, we turn things over to MATLAB and make use of the program `runkut2d` of the last chapter, which is ideally suited for the present purposes.

```
>> f1=inline('u', 'x','y','u');
>> g1=inline('300*y/1.2e7/4 + 50*x*(x-50)/2/1.2E7/4','x','y','u');
>> f2=f1;
>> g2=inline('300*y/1.2e7/4','x','y','u');
>> [x,y1,u1]=runkut2d(f1,g1,0,50,0,0,.1);
>> [x,y2,u2]=runkut2d(f2,g2,0,50,0,1,.1);
```

The solutions to (IVP-1) and (IVP-2) are now stored as vectors y1 and y2 respectively. We next take the indicated linear combination of these two solutions to get the solution of the BVP, which we store in the vector ybvp. To get the (MATLAB's) vector index of $x = L$, we use the `size` command, as usual.

```
>> size(x)    → 1  501
>> ybvp=y1-y1(501)/y2(501)*y2;
>> plot(x,ybvp) %See Figure 10.2.
>> xlabel('x-values'), ylabel('y-values'), title('Deflection')
```

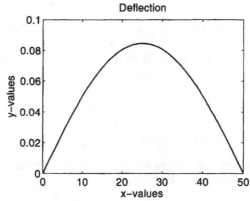

FIGURE 10.2: The graph of the vertical deflection of the horizontal beam of Example 10.3. The bending is much less severe than the graph seems to indicate (compare units on the axes); both *x*- and *y*-values are given in feet.

Part (b): Using the formula (3) for the exact solution of the DE (along with the conditions on the constants in the Exercise for the Reader 10.1 for the boundary conditions), we can easily construct the vector of the exact solution's *y*-coordinates and then plot the desired error. We construct this vector directly (since we only need it for this particular instance and an inline function construction would be a bit awkward with all of the constants).

```
>> L=50;T=300; w=50; E=1.2e7; I=4; theta=sqrt(T/E/I);
>> C=w/theta^2/T/sinh(theta*L); D=C;
>> yexact=C*sinh(theta*x)+C*sinh(theta*(L-x))+...
    w*L*x/2/T-w/theta^2/T-w*x.^2/2/T;
>> max(abs(ybvp-yexact))        →ans =7.2526e-012
```

The maximum error shows that the linear shooting method gave us quite an accurate solution.

EXERCISE FOR THE READER 10.5: Use the linear shooting method to solve the following BVP. Your IVP solver should be the Runge-Kutta method with step size $h = 0.01$. Plot your solution with a solid graph, together with the solutions of the two associated IVPs with x's and o's. Do the x- and o-plots using vectors with gaps = 0.1 to make them print nicely. Also, what is the value of the solution when $x = 1.5$?

$$\begin{cases} xy'' - y' - x^5 = 0 \ (DE) \\ y(1) = 1, \ y(2) = 4 \ (BC's) \end{cases} \quad 1 \le x \le 2, \ y = y(x)$$

EXERCISE FOR THE READER 10.6: (a) Write an M-file that will perform the linear shooting method in conjunction with the Runge-Kutta method to solve a general linear BVP (5) $\begin{cases} y''(t) = p(t)y' + q(t)y + r(t) \\ y(a) = \alpha, \ y(b) = \beta \end{cases}$. The inputs and outputs should be as follows:

```
[t, y] = linearshooting(p, q, r, a, alpha, b, beta, hstep)
```

but we intentionally leave the syntax open since the construction involves a new concept (see the suggestion below).

(b) Run this program on the BVP of Example 10.3 and compare the resulting plot with the one in Figure 10.2

Suggestion: This program is an interesting one to contemplate and the reader is encouraged to spend a decent amount of time thinking and writing such a program before consulting the solution in Appendix B. Depending on how the program is written, it will highlight the differences between different sorts of data types in MATLAB (inline objects, character strings, numbers). It is possible to write this program elegantly without resorting to the Symbolic Toolbox. Ideally, we would like to be able to write this M-file so that it will internally be able to call on the runkut2d program of Exercise for the Reader 9.2. Since the functions that need to be inputted into the latter Runge-Kutta program are constructed from (but different from) $p(t)$, $q(t)$, and $r(t)$, for this scheme to work it will be necessary to construct inline functions within the program that are built up from the inputted functions $p(t)$, $q(t)$, and $r(t)$. The syntax by which these functions are inputted should be carefully thought out to facilitate the writing of the program. Below we give a way to construct an inline function from two previously inputted strings.

Suppose that we have two strings in our workspace: s1 = x^2+1, and s2 = cos(x), and we would like to create an inline function whose formula is the following combination of these strings: s1^2+2*s2 + x, i.e., the function $f(x) = (x^2 + 1)^2 + 2\cos(x) + x$. The following construction creates this inline function directly in terms of the two previously inputted strings:

```
>> s1 = 'x^2+1'; s2 = 'cos(x)';    %enter two strings
>> f = inline(['(', s1, ')^2+2*', s2, '+x'])
→f =   Inline function: f(x) = (x^2+1)^2+2*cos(x)+x
```

EXERCISES 10.2

1. For each of the following linear BVPs, do the following (if possible): (i) Verify the DE is linear so that the linear shooting method is applicable. (ii) Write down the two associated IVPs of the linear shooting method. (iii) Introduce new variables to recast each of these IVPs into a system of first-order IVPs. (iv) Use the Runge-Kutta method with step size $h = 0.01$ to solve both IVPs. (v) Plot the solution (solid curve) together with the plots of both solutions to the IVPs done in different plotting styles.

(a) $\begin{cases} y''(t) = -2y \\ y(0) = 1, \ y(2) = 4 \end{cases}$

(b) $\begin{cases} y''(t) = -y' - 2y \\ y(0) = 1, \ y(2) = 1 \end{cases}$

(c) $\begin{cases} y''(t) = y\cos(t) + e^t \\ y(0) = 1, \ y(3) = -3 \end{cases}$

(d) $\begin{cases} y''(t) = ty - 2ty' \\ y(2) = 2, \ y(4) = 4 \end{cases}$

2. Repeat all parts of Exercise 1 for each of the following BVPs.

(a) $\begin{cases} y''(t) = t + 2y + 3y' \\ y(0) = 0, \ y(2) = -4 \end{cases}$

(b) $\begin{cases} y''(t) = -t - 2y - 3y' \\ y(0) = 0, \ y(2) = -4 \end{cases}$

(c) $\begin{cases} ty''(t) + y' - y\sin(t) = 0 \\ y(1) = 1, \ y(3) = 4 \end{cases}$

(d) $\begin{cases} 2t^2 y''(t) + ty' + y = 0 \\ y(1) = 1, \ y(3) = -1 \end{cases}$

3. Below we give a series of BVPs together with the general solution y(t) of each DE. For each part, do the following (if possible): (i) Verify the DE is linear so that the linear shooting method is applicable and also that the "general solution" given actually solves the DE. (ii) Use the linear shooting method with the Runge-Kutta method, first with step size $h = 0.1$, next with $h = 0.05$, and then again with step size $h = 0.01$ to obtain vectors (use different names) for the numerical solutions. (iii) Determine the constants in the general solution given so that it solves the given BVP. (iv) Plot the four curves in the same graph using different plot colors/styles for each. In situations where graphs are indistinguishable, plot also the errors.

(a) $\begin{cases} y''(t) + \dfrac{y'}{t} + \left(1 - \dfrac{1}{4t^2}\right)y = 0 \\ y(1) = 0, \ y(10) = 4 \end{cases}$, $y(t) = \dfrac{C\cos(t) + D\sin(t)}{\sqrt{t}}$

(b) $\begin{cases} t^2 y''(t) - ty' + y = 0 \\ y(1) = 0, \ y(3) = -2 \end{cases}$, $y(t) = Ct + Dt\ln(t) + t^2$

(c) $\begin{cases} t^2 y''(t) - ty' + 2y = 0 \\ y(1) = 0, \ y(3) = -2 \end{cases}$, $y(t) = t[C\cos(\ln t) + D\sin(\ln t)]$

4. Repeat Exercise 3 for each of the following:

(a) $\begin{cases} t^2 y''(t) + 3ty' = 2 \\ y(1) = 4, \ y(3) = 0 \end{cases}$, $y(t) = \ln t + \dfrac{C}{t^2} + D$

(b) $\begin{cases} y''(t) = 16y \\ y(0) = 1, \ y(2) = 12e^{-8} \end{cases}$, $y(t) = Ce^{4t} + De^{-4t}$

(c) $\begin{cases} y''(t) - 3y' + 2y = 3e^{-t} - 10\cos(3t) \\ y(0) = 1, \ y(2) = 4 \end{cases}$, $y(t) = Ce^t + De^{2t} - \dfrac{e^{-t}}{2} + \dfrac{7\cos(3t)}{13} + \dfrac{9\cos(3t)}{13}$

5. (Physics: *Flight of a Well-Hit Baseball*) This problem deals with the flight of a baseball in two dimensions (which we take for convenience as the xy-plane). We consider a ball that is hit so it lands 300 feet from home plate after 3 seconds. Many factors influence the flight of the ball. We assume that the air resistance acts only against the horizontal velocity, and for this particular

baseball it is proportional to the horizontal velocity. Assuming the coordinates of home plate are $(x,y) = (0,0)$, we let $x(t)$ and $y(t)$ denote the x and y coordinates of the position of the ball t seconds after it is hit. Thus, at time t, the coordinates of the baseball are $(x(t), y(t))$, $0 \le t \le 3$. The air resistance assumption and Newton's law from basic physics give the following system of second-order DEs:

$$\begin{cases} x''(t) = -cx(t) \\ y''(t) = -g \end{cases},$$

where for the ball being used the constant $c = 0.5$. The initial position of the ball is $(x(0),y(0)) = (0,3)$ <feet>. Since the ball lands after 3 seconds we also get $(x(3),y(3)) = (300,0)$.
(a) Explicitly find the function $y(t)$, just using basic calculus.
(b) Numerically find $x(t)$, by using the shooting method with step size $h = 0.01$ (implementing the Runge-Kutta method), and sketch a plot of the path of the ball. (For this part, you need not print the graph of x vs t, or give any explicit values for $x(t)$.)
(c) After how many seconds does the ball reach its maximum height? At this time what is the x-coordinate?
(d) With the same hit, how far would the ball have gone (on the x-axis), if, as in the imaginary assumptions of freshman physics courses, there was no air resistance?

NOTE: (*Mixed Boundary Conditions*) Suppose that we modify the LBVP (5):

$$\begin{cases} y''(t) = p(t)y' + q(t)y + r(t) & \text{(DE)} \\ y(a) = \alpha, \ y(b) = \beta & \text{(BCs)} \end{cases}$$

to the following:

$$\text{(LBVP)} \begin{cases} y''(t) = p(t)y' + q(t)y + r(t) & \text{(DE)} \\ y(a) = \alpha, \ y'(b) + dy(b) = \beta & \text{(BCs)} \end{cases} \tag{9}$$

The DE and first BC of (9) are identical with those of (5), the only change is that the second (BC) is now a so-called **mixed boundary condition** involving a linear combination of the values of the unknown function and its first derivative at the right endpoint $t = b$ (d is a constant).

6.   Show that if y_1 and y_2 are solutions of the associated IVPs (6) and (7) respectively, then the function

$$y(t) = y_1(t) + \frac{\beta - y_1'(b) - dy_1(b)}{y_2'(b) + dy_2(b)} y_2(t). \tag{10}$$

will solve the mixed LBVP (9), provided that $y_2'(b) + dy_2(b) \ne 0$.

7.   In each of the following parts you are given a mixed linear BVP along with the general solution for the DE. For each one, perform (if possible) the following tasks:
(i) Use the linear shooting method (as adapted in Exercise 6) with the Runge-Kutta method, first with step size $h = 0.1$, next with $h = 0.04$, and then again with step size $h = 0.01$ to obtain vectors (use different names) of numerical solutions. (ii) Determine the constants in the general solution given so that it solves the given BVP. (iii) Plot the four curves in the same graph using different plot colors/styles for each. In situations where graphs are indistinguishable, plot also the errors of the approximations with the exact solution.

(a) $\begin{cases} t^2 y''(t) + 3ty' = 2 \\ y(1) = 4, \ y'(3) + y(3) = 0 \end{cases}$, $y(t) = \ln t + \dfrac{C}{t^2} + D$

(b) $\begin{cases} y''(t) = 16y \\ y(0) = 1, \ 2y'(2) - y(2) = 12e^{-8} \end{cases}$, $y(t) = Ce^{4t} + De^{-4t}$

(c) $\begin{cases} y''(t) - 3y' + 2y = 3e^{-t} - 10\cos(3t) \\ y(0) = 1, \ y(2) - y'(2) = 4 \end{cases}$, $y(t) = Ce^t + De^{2t} - \dfrac{e^{-t}}{2} + \dfrac{7\cos(3t)}{13} + \dfrac{9\cos(3t)}{13}$

8. (*Thermodynamics*) A metal alloy cylindrical pipe at a chemical plant has a very hot fluid flowing through it. The pipe has an inner radius of one inch and an outer radius of 2 inches. The fluid inside the pipe has a temperature of $1000 °F$ and the outside temperature is $70 °F$. The temperature T within the thick pipe is a function of the radius r only and it satisfies the following heat differential equation: $rT''(r) + T' = 0$.

    (a) Solve numerically for the temperature function $T(t)$, graph it, and find $T(1.5)$.

    (b) If the pipe is insulated to minimize heat loss, the insulation will change the boundary condition at the outer radius $r = 2$ to make the derivative of the temperature function proportional to the difference of the pipe's temperature (on the outside) with the surrounding room's temperature as follows: $T'(2) = -0.068[T(2) - 70]$. Under this new insulation condition, solve numerically for the temperature function $T(t)$, graph it, and find $T(1.5)$.

    Suggestion: For part (b), see Exercise 6 and the note which precedes it. We will deal in more detail with heat equations in the chapters on partial differential equations.

9. The linear shooting method can be further modified to deal with mixed boundary conditions at both ends, i.e., with the LBVP:

    $$\text{(LBVP)} \quad \begin{cases} y''(t) = p(t)y' + q(t)y + r(t) & \text{(DE)} \\ y'(a) + cy(a) = \alpha, \ y'(b) + dy(b) = \beta & \text{(BCs)} \end{cases} \quad (11)$$

    The associated IVPs (6) and (7), however, must be modified to the following:

    $$\text{(IVP-1)} \quad \begin{cases} y_1''(t) = p(t)y_1 + q(t)y_1 + r(t) & \text{(DE)} \\ y_1(a) = 0, \ y_1'(a) = \alpha & \text{(IC-1's)} \end{cases}, \quad (12)$$

    and,

    $$\text{(IVP-2)} \quad \begin{cases} y_2''(t) = p(t)y_2' + q(t)y_2 & \text{(DE-2)} \\ y_2(a) = 1, \ y_2'(a) = -c & \text{(IC-2's)} \end{cases}. \quad (13)$$

    Prove that with these modified IVPs, the linear combination (10) will solve the LBVP (11), provided that $y_2'(b) + dy_2(b) \neq 0$.

10. In each of the following parts you are given a mixed linear BVP along with the general solution for the DE. For each one, perform (if possible) the following tasks:

    (i) Use the linear shooting method (as adapted in Exercise 9) with the Runge-Kutta method, first with step size $h = 0.1$, next with $h = 0.04$, and then again with step size $h = 0.01$ to obtain vectors (use different names) for numerical solutions. (ii) Determine the constants in the general solution given so that it solves the given BVP. (iii) Plot the four curves in the same graph using different plot colors/styles for each. In situations where graphs are indistinguishable, plot also the errors of the approximations with the exact solution.

    (a) $\begin{cases} y''(t) + \dfrac{y'}{t} + \left(1 - \dfrac{1}{4t^2}\right)y = 0 \\ y'(1) = 0, \ y'(10) + y(10) = 0 \end{cases}$, $y(t) = \dfrac{C\cos(t) + D\sin(t)}{\sqrt{t}}$

    (b) $\begin{cases} t^2 y''(t) - ty' + y = 0 \\ y'(1) = 0, \ y(3) = 0 \end{cases}$, $y(t) = Ct + Dt\ln(t) + t^2$

    (c) $\begin{cases} t^2 y''(t) - ty' + 2y = 0 \\ y'(1) + y(1) = 0, \ 2y'(3) - y(3) = -2 \end{cases}$, $y(t) = t[C\cos(\ln t) + D\sin(\ln t)]$

11. What do our existence and uniqueness theorems (of Section 10.1) say about BVPs that involve the DE: $y''(t) = y$?

12. Consider the linear DE: $y''(t) = -y$. Obviously $\sin(t)$ and $\cos(t)$ are two solutions. From these and linearity, we obtain the general solution $y(t) = C\cos(t) + D\sin(t)$ where C and D are arbitrary

constants.   To convince yourself of (prove) this, you can apply the existence and uniqueness theorem (adapted from those given in the last chapter) and show that any IVP starting at $t = 0$, say, can be solved by one of these functions. In this problem you will see that depending on the boundary conditions, different BCs for this DE can lead to nonexistence or nonuniqueness of solutions or neither. For the BVP: $\{y'' = -y,\ y(0) = 0,\ y(b) = \beta$,

(a) Exactly which values of b and β would make the BVP have no solution?

(b) Exactly which values of b and β would make the BVP have infinitely many solutions?

(c) Exactly which values of b and β would make the BVP have a unique solution?

Note: Your answers to these three parts should cover all possible values of b and β .

10.3: THE NONLINEAR SHOOTING METHOD

This method will really appear more like we are "shooting" at the solution than was the case with the linear shooting method of the last section. We once again turn to the general BVP (1):

$$\begin{cases} y''(t) = f(t,y,y') & \text{(DE)} \\ y(a) = \alpha,\ y(b) = \beta & \text{(BCs)} \end{cases}$$

Recall that when the DE was linear, we obtained the solution of the BVP as a linear combination of two solutions of (just) two specially associated IVPs. In the nonlinear case, we will solve a sequence of related IVPs:

$$(\text{IVP})_k \quad \begin{cases} y''(t) = f(t,y,y') & \text{same (DE)} \\ y(a) = \alpha,\ y'(a) = m_k & \text{(ICs)} \end{cases} \quad \Rightarrow \quad \text{solution } y_k(t) \equiv y(t,m_k) .$$

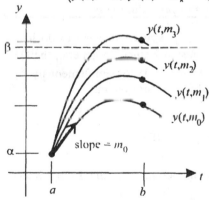

Each of the IVPs above is identical, except for the second initial condition $y'(a) = m_k$, where the parameter will be appropriately adjusted ("aimed") at each iteration. We have denoted the solution of $(\text{IVP})_k$ as $y_k(t)$ and, since it depends on m_k , we have introduced the function of two variables $y(t,m_k) \equiv y_k(t)$. The method is roughly illustrated and explained in Figure 10.3.

FIGURE 10.3:   Illustration of the nonlinear shooting method for a BVP: $\begin{cases} y''(t) = f(t,y,y') \\ y(a) = \alpha,\ y(b) = \beta \end{cases}$. The initial approximation $y_0(t) = y(t,m_0)$ is the solution of the corresponding IVP having the same DE, the same first condition, and satisfying the initial slope $y_0'(t) = m_0$, obtained numerically by methods of the last chapter. The desired second boundary condition is compared with $y_0(b)$, and, if necessary, this process is repeated with adjusted initial slopes $m_1, m_2, \cdots$ until we arrive at a solution that satisfies the second boundary condition (within a desired tolerance).

The only detail left to tend to is the important issue of how best to choose our initial slopes. It turns out to be a bit complicated; indeed, figuring out subsequent initial slopes will require solving an additional IVP. We outline the procedure now, give a specific example, and afterwards give a theoretical explanation of it.

The Nonlinear Shooting Method:

1. Start with an estimate (or guess) for the initial slope of the first IVP
   $m_0 = y_0'(a)$; a good default is the difference quotient $m_0 = \dfrac{\beta - \alpha}{b - a}$.

2. Solve the associated $(IVP)_k$ $\begin{cases} y''(t) = f(t, y, y') \\ y(a) = \alpha, \ y'(a) = m_k \end{cases}$ $(k = 0)$ on $a \le t \le b$,

   using, say, the Runge-Kutta method; denote the solution as
   $y_k(t) = y(t, m_k)$ $(k = 0)$.

3. Check for accuracy by evaluating : Diff $\equiv y(b, m_k) - \beta$. If |Diff| <
   tolerance, accept $y_k(t) = y(t, m_k)$ as solution to BVP, otherwise update

   $$m_{k+1} = m_k - \frac{\text{Diff}}{z(b, m_k)},$$

   where $z(t, m_k)$ solves the IVP:

   $$\begin{cases} z''(t) = z f_y(t, y, y') + z' f_{y'}(t, y, y') \\ z(a) = 0, \ z'(a) = 1 \end{cases}.$$

   Increase k and return to step 2 to iterate this procedure.

NOTE: To numerically solve the IVP for z in step 3, we will need to do it in conjunction with the concurrent IVP for y (y_k) in step 2 since, in general, the DE of z involves y. Thus, we will have to solve the two IVPs simultaneously by writing them into an equivalent four-dimensional first-order system.

EXAMPLE 10.4: Numerically solve the BVP:
$$\begin{cases} y''(t) = -2(yy' + ty' + y + t) & \text{(DE)} \\ y(1) = 0, \ y(2) = -2 & \text{(BCs)} \end{cases}$$

by using the nonlinear shooting method in conjunction with the Runge-Kutta method with step size $h = 0.01$.

(a) Do it first with a tolerance of 0.01. How many "shots" were required? Get MATLAB to display the totality of graphs of the functions $y_k(t) = y(t, m_k)$ ("shots") in the same plot with the final one in a different color from the rest.

(b) Next do it for a tolerance of 10^{-7}. How many "shots" were required? This time, using the subplot command, display the plots of the successive difference errors $|y_{k+1}(t) - y_k(t)|$ for $k = 0, 1, 2, 3, \ldots$

SOLUTION: We point out the DE is nonlinear (because of the yy' term) and so the linear shooting method would not be applicable. The associated initial value problems for y are

$$(\text{IVP})_k \quad \begin{cases} y''(t) = f(t, y, y') \equiv -2(yy' + ty' + y + t) \\ y(1) = 0, \ y'(1) = m_k \end{cases}.$$

By introducing the new function $yp(t) = y'(t)$ we can translate this IVP into the following equivalent system:

$$(\text{IVP})'_k \quad \begin{cases} y'(t) = yp, & y(1) = 0 \\ yp'(t) = -2(y(yp) + t(yp) + y + t), & yp(1) = m_k \end{cases}.$$

To get the IVP for the auxiliary function z, we compute

$$f_y(t, y, y') = -2y' - 2, \qquad f_{y'}(t, y, y') = -2y - 2t,$$

which brings us to the following companion IVP for z:

$$\begin{cases} z''(t) = z(-2y' - 2) + z'(-2y - 2t) \\ z(1) = 0, \ z'(1) = 1 \end{cases}.$$

By introducing the new function $zp(t) = z'(t)$ and combining this IVP with the previous one, we arrive at the following four-dimensional system:

$$\begin{cases} y'(t) = yp, & y(1) = 0 \\ yp'(t) = -2(y(yp) + t(yp) + y + t), & yp(1) = m_k \\ z'(t) = zp, & z(1) = 0 \\ zp'(t) = -2(yp + 1)z - 2(y + t)zp, & zp(1) = 1 \end{cases}.$$

For the initial slope we use the suggested default $m_0 = \dfrac{y(2) - y(1)}{2 - 1} = \dfrac{-2 - 0}{2 - 1} = -2$.

In turning the problem over to MATLAB, since we plan to make use of the rksys routine of the last chapter, we must first construct the vector-valued function corresponding to the right sides of the four-dimensional system above. We do this in the following rather generic way that can be easily mimicked for any other nonlinear shooting problem:

```
function xp=nlshoot(t,x)
xp(1)=x(2);
xp(2)=-2*(x(1)*x(2)+t*x(2)+x(1)+t);
xp(3)=x(4);
xp(4)=-2*(x(2)+1)*x(3)-2*(x(1)+t)*z(4);
```

Note that we have identified $x(1)$ with y, $x(2)$ with yp, $x(3)$ with z and $x(4)$ with zp.

Part (a): We can now perform the desired plots using the following while loop.

```
>>mk=-2; %initialize
>> while 1 %since i is true, loop will continue to execute
```

```
[t,X]=rksys('nlshoot',1,2,[0 mk 0 1],0.01);
y=X(:,1); z=X(:,3); %peel off the vectors we need
Diff=y(101)+2; %y(101) (MATLAB) corresponds to y(2) (Math)
if abs(Diff)<0.01
   plot(t,y,'r'), return
   end
plot(t,y, 'b'), hold on
n=n+1; %bump counter up one
mk=mk-Diff/z(101);   %update slope
end
```

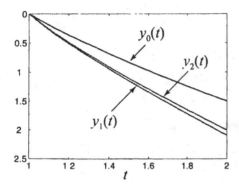

FIGURE 10.4: Illustration of the nonlinear shooting method applied to the BVP of Example 10.4(a). The successive approximations $y_k(t)$ are shown until the value of $y_k(2)$ is within a tolerance of 0.01 to -2, at which point the process grinds to a halt. The code is set up to graph the final approximation in red.

The plot shown in Figure 10.4 clearly shows that 3 iterations ("shots") were done: The first was too high, the second too low, and the third about right (within tolerance). Alternatively, by the way that the loop was set up, we could just enter n to query MATLAB to tell us how many iterations were done.

Part (b): We can easily modify the above loop to get the desired information and plots. We leave this as an exercise, but include the plot in Figure 10.5. We point out that in order to use the subplot command to get a decent plot, we first found out the number of shots needed and then ran through the loop again with an appropriately dimensioned "subplot" window. We also comment that a plot like the one in part (a) would be not quite so useful here since all approximations from the third onward are essentially indistinguishable using the graphs. This is why we look at successive differences. This is also a good way to check global errors.

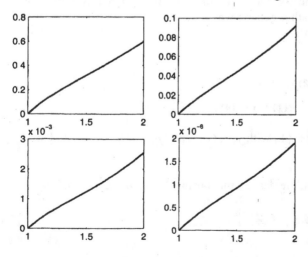

FIGURE 10.5: These graphs display the successive differences $|y_{k+1}(t) - y_k(t)|$ for nonlinear shooting method in part (b) of Example 10.4; this time the iterations continue until $y_k(2)$ gets within 10^{-7} of the desired value -2 and only 5 approximations ("shots") are needed.

EXERCISE FOR THE READER 10.7:   (a) Write a function M-file that will perform the nonlinear shooting method to numerically solve the BVP (1):
$$\begin{cases} y''(t) = f(t, y, y') & \text{(DE)} \\ y(a) = \alpha, \ y(b) = \beta & \text{(BCs)} \end{cases}$$. The inputs and outputs should be as follows:

```
[t, y, nshots] = nonlinshoot(a, alpha, b, beta, f, fy, fyp, tol,
                             hstep)
```

The input and output variables more or less correspond to the data of the problem, but we leave the syntax open (see the suggestion for Exercise for the Reader 10.7). The variable `tol` will provide a stopping criterion for the iterations: $|y(b) - \beta| < tol$.[1]   The last output variable is a positive integer giving the number of iterations (shots) that were executed.

(b) Test your program by applying it to the two sets of data of Example 10.4.

(c) Does Theorem 10.1 tell us anything about solutions for the following BVP?

$$\begin{cases} y''(t) = te^y - \sin(\cos(t))y' & \text{(DE)} \\ y(0) = 0, \ y(6) = 10 & \text{(BCs)} \end{cases}$$

Apply the nonlinear shooting method with tolerance $h = 0.01$ to solve this problem.   Then repeat with tolerance 10^{-7}.   Plot the final numerical solution and record the number of iterations.

Note: It is possible to write this program using MATLAB's Symbolic Toolbox capabilities and in this case the input variables `fy` and `fyp` could be dispensed with. The program we write in Appendix B does not use the Symbolic Toolbox; we leave such a construction to the interested reader.

As promised, we will now give a theoretical explanation of what has motivated the nonlinear shooting method.   We assume that for any initial slope m, the IVP associated with the BVP (1),

$$\begin{cases} y''(t) = f(t, y, y') & \text{same (DE)} \\ y(a) = \alpha, \ y'(a) = m & \text{(ICs)} \end{cases},$$

always has a unique solution on the time interval $[a,b]$, and we denote it by $y(t,m)$, which is a function of two variables.   Our goal is to make m be a root of the equation

$$y(b, m) - \beta = 0 . \tag{14}$$

In this equation we have held the t-variable of $y(t,m)$ to be fixed at $t = b$ so that the left side of (14) is a function of a single variable (namely m).   Assuming it is differentiable, Newton's recursion formula for rootfinding (Chapter 6) suggests that it would be a good idea to define our sequence recursively using the following scheme:

---

[1] Creation of this M-file will require features from MATLAB's Symbolic Toolbox; see Appendix A. Without the symbolic toolbox features, a similar program could be constructed but it would need more input variables, for example, $f_y(t, y, y')$ and $f_{y'}(t, y, y')$.

$$m_k = m_{k-1} - \frac{y(b, m_{k-1}) - \beta}{\partial/\partial m\{y(b, m_{k-1})\}}. \tag{15}$$

To go from one iteration to the next, after having (numerically) found $y(t, m_{k-1})$, the only difficult part of the formula (15) to obtain is the partial derivative $\partial/\partial m\{y(b, m_{k-1})\}$. This can be done (in an at first seemingly roundabout way) by finding an IVP for which the function

$$z(t) = z(t, m) \equiv \frac{\partial y}{\partial m}(t, m)$$

is a solution and then numerically solving this IVP and evaluating it at $t = b$ to get the needed partial derivative. We can get a DE for $z(t,m)$ by differentiation of the DE for $y(t,m)$ and using the chain rule as follows (wherein we reserve primes (') for differentiations in the t-variable):

$$y''(t, m) = f(t, y(t, m), y'(t, m)) \Rightarrow$$
$$\frac{\partial y''}{\partial m}(t, m) = f_t(t, y, y')\frac{\partial t}{\partial m} + f_y(t, y, y')\frac{\partial y}{\partial m}(t, m) + f_{y'}(t, y, y')\frac{\partial y'}{\partial m}(t, m).$$

Since t and m are independent variables, $\partial t/\partial m = 0$ and so the first term on the right vanishes. We now simply replace $\partial y/\partial m(t, m)$ by $z(t, m)$ and the above becomes the following DE for z:

$$z''(t, m) = f_y z + f_{y'} z'. \tag{16}$$

If we differentiate the initial conditions for $y(t,m)$: $\begin{cases} y(a, m) = \alpha \\ y'(a, m) = m \end{cases}$ we obtain

corresponding initial conditions for $z(t,m)$:

$$\begin{cases} z(a, m) = 0 \\ z'(a, m) = 1 \end{cases}. \tag{17}$$

Replacing the partial derivative in (15) by $z(b, m_{k-1})$ we see at once that (15), (16), and (17) yield the nonlinear shooting algorithm.

EXERCISES 10.3

1. For each of the linear BVPs in parts (a) through (d) of Exercise 1, Section 10.2, apply the nonlinear shooting method to solve it via the Runge-Kutta method with step size $h = 0.01$ by following the outline below (if possible):
   (i) Write down the associated IVPs both for y and for the auxiliary function z.
   (ii) Translate both IVPs for y and z into a single four-dimensional IVP system of first-order DEs.
   (iii) Use MATLAB to apply the nonlinear shooting method to solve the BVP with a tolerance of 10^{-4}. Display all of the approximations ("shots") in a single plot with the final one being displayed in a different color or plot style.

2. For each of the linear BVPs in parts (a) through (d) of Exercise 2, Section 10.2, repeat the instructions of the last exercise, but change item (iii) to: (iii'): Use MATLAB to apply the nonlinear shooting method to solve the BVP with a tolerance of 10^{-6}. How many "shots" were required? Plot the final graph and also in a separate window and using the subplot command, get MATLAB to display the plots of the successive differences of the "shots": $|y_{k+1}(t) - y_k(t)|$ for k = 0,1,2,3, ,...

3. For each of the nonlinear BVPs given, perform the following tasks (if possible):
   (i) Write down the associated IVPs both for y and for the auxiliary function z.
   (ii) Translate both IVPs for y and z into a single four-dimensional IVP system of first-order DEs.
   (iii) Use MATLAB to apply the nonlinear shooting method to solve the BVP with a tolerance of 10^{-4}. Display all of the approximations ("shots") in a single plot with the final one being displayed in a different color or plot style.
   (iv) Along with the BVP, an exact solution $f(t)$ is given; verify that this function actually solves the BVP.
   (v) Using a subplot window if you prefer, plot the errors of each of the successive shots with the exact solution given.

   (a) $\begin{cases} y'' = 12y^{5/3} \\ y(0) = 1, \ y(2) = 1/27 \end{cases}$, $f(t) = \dfrac{1}{(t+1)^3}$

   (b) $\begin{cases} y'' = -[y']^2/y \\ y(0) = 1, \ y(5) = 4 \end{cases}$, $f(t) = \sqrt{3t+1}$

   (c) $\begin{cases} y'' = y' + 2(y - \ln t)^3 \\ y(1) = 1/2, \ y(2) = 1/2 + \ln 2 \end{cases}$, $f(t) = \dfrac{1}{t} + \ln t$

4. Repeat the instructions for Exercise 3 on the following BVPs.

   (a) $\begin{cases} y'' = y'\cos t - y\sin t \\ y(0) = 1, \ y(8\pi) = 1 \end{cases}$, $f(t) = \exp(\sin t)$

   (b) $\begin{cases} y'' = y^3 - yy' \\ y(1) = 1/2, \ y(2) = 1/3 \end{cases}$, $f(t) = \dfrac{1}{t+1}$

   (c) $\begin{cases} y'' = y'(\ln t + 1) + y(1 + 1/t) \\ y(1) = 1, \ y(5) = 3125 \end{cases}$, $f(t) = t^t$

5. Use the nonlinear shooting method to solve the following BVP. Your IVP solver should be Runge-Kutta with step size $h = 0.01$. Your tolerance (for the right BC) should be 0.0001. How many iterations did this take? Plot your solution. Also, what is the value of the solution when $x = 0.4$?

$$\begin{cases} y''(t) = -t(y')^3 \\ y(0) = 0, \ y(1) = \pi/2 \end{cases}$$

6. (*Physics: Flight of a Well-Hit Baseball*) This problem deals with the flight of a baseball in two dimensions (which we take for convenience as the *xy*-plane). We consider a ball that is hit so it lands 300 feet from home plate after 3 seconds. Many factors influence the flight of the ball. We assume that the air resistance acts only against the horizontal velocity, and for this particular baseball it is proportional to the horizontal velocity with exponent 1.2. Assuming the coordinates of home plate are $(x,y) = (0,0)$, we let $x(t)$ and $y(t)$ denote the x and y coordinates of the position of the ball t seconds after it is hit. Thus, at time t, the coordinates of the baseball are $(x(t), y(t))$, $0 \le t \le 3$. The air resistance assumption and Newton's law from basic physics give the following system of second-order DEs:

$$\begin{cases} x''(t) = -cx(t)^{1.2} \\ y''(t) = -g \end{cases},$$

where for the ball being used the constant $c = 0.44$. The initial position of the ball is $(x(0), y(0))$ = $(0,3)$ <feet>. Since the ball lands after 3 seconds we also get $(x(3), y(3)) = (300,0)$.

(a) Explicitly find the function $y(t)$, just using basic calculus.

(b) Numerically find $x(t)$, by using the shooting method with step size $h = 0.01$ (and implementing the Runge-Kutta method), and sketch a plot of the path of the ball (i.e., of y vs. x). (For this part, you need not print the graph of x vs. t, or give any explicit values for $x(t)$.)

(c) After how many seconds does the ball reach its maximum height? At this time what is the x-coordinate?

(d) With the same hit, how far would the ball have gone (on the x-axis), if, as in the imaginary assumptions of physics courses, there was no air resistance?

7.   (*Civil Engineering: Deflection of a Beam*) Use the nonlinear shooting method with $h = 0.01$ in the Runge-Kutta method to solve the exact beam-deflection model BVP:

$$\begin{cases} y''(t)/[1+(y')^2]^{3/2} = \dfrac{T}{EI}y + \dfrac{wx(x-L)}{2EI}, \\ y(0) = 0 = y(L) \end{cases},$$

having the parameters: $L = 50$ feet (length), $T = 300$ lb (tension at ends), $w = 50$ lb/ft (vertical load), $E = 1.2 \times 10^7$ lb/ft^2 (modulus of elasticity), and $I = 4$ ft^4 (central moment of inertia).

(a) Graph the resulting numerical solution. (b) How does the solution compare with that obtained for the corresponding linear approximating BVP of Example 10.3?

10.4: THE FINITE DIFFERENCE METHOD FOR LINEAR BVP'S

The method we present next is philosophically quite different from the shooting methods. It immediately discretizes the BVP by approximating the derivatives with difference quotients. The problem is then translated into a linear system that is easily solved directly. This method will pave the way for the corresponding finite difference methods that we will employ in the next two chapters for solving PDEs. There are analogues of this method for nonlinear BVPs; the discretization is done in the same way but the resulting system of equations will no longer be linear.[2]

All finite difference methods are based on approximating derivatives of a function by certain difference quotients. These difference quotient formulas can always be obtained using Taylor's theorem. We will be needing them only for first and second derivatives, and we now present them in the following lemma. To describe the error bounds, we employ the "big O" notation that was introduced in Section 8.3.

LEMMA 10.3 (*Central Difference Formulas*) (a) If $f(x)$ is a function having a continuous second derivative in the interval $a - h \le x \le a + h$, then we have

$$f'(a) \approx \frac{f(a+h) - f(a-h)}{2h}, \tag{18}$$

---

[2] For the nonlinear analogues of the finite difference method, we cite the reference: [BuFa-01] (see Section 11.3 therein.)

where the error of the approximation is $O(h^2)$.

(b) If, furthermore, $f(x)$ has a continuous fourth derivative throughout $a - h \le x \le a + h$, then we also have the approximation

$$f''(a) \approx \frac{f(a+h) - 2f(a) + f(a-h)}{h^2}, \tag{19}$$

and the error of this approximation is also $O(h^2)$.

Proof of part (a): Taylor's theorem allows us to write

$$f(a+h) = f(a) + hf'(a) + h^2 f''(a)/2 + O(h^3), \text{ and}$$

$$f(a-h) = f(a) - hf'(a) + h^2 f''(a)/2 + O(h^3).$$

Subtracting the second of these equations from the first gives $f(a+h) - f(a-h)$

$= 2hf'(a) + O(h^3)$ and solving this for $f'(a)$ produces (18). We have used the facts that $O(h^3) + O(h^3) = O(h^3)$ and $O(h^3)/h = O(h^2)$. The proof of part (b) is similar and is left as the next exercise for the reader.

EXERCISE FOR THE READER 10.8: Prove part (b) of Lemma 10.3.

We now explain the finite difference method in more detail. Consider the linear BVP (5):

$$\begin{cases} y''(t) = p(t)y' + q(t)y + r(t) & \text{(DE)} \\ y(a) = \alpha, \ y(b) = \beta & \text{(BCs)} \end{cases}$$

Choose a positive integer N, and subdivide the interval $a \le t \le b$ into N equal subintervals using $N - 1$ interior grid values: $t_1 = a + h$, $t_2 = a + 2h$, $\cdots$, $t_{N-1} = a + (N-1)h$, where $h = (b-a)/N$. We also write $t_0 = a$ and $t_N = b$; see Figure 10.6. We let,

$$y_i = y(t_i) \quad (0 \le i \le N),$$

and similarly,

$$p_i = p(t_i), \ q_i = q(t_i), \ r_i = r(t_i) \quad (0 \le i \le N).$$

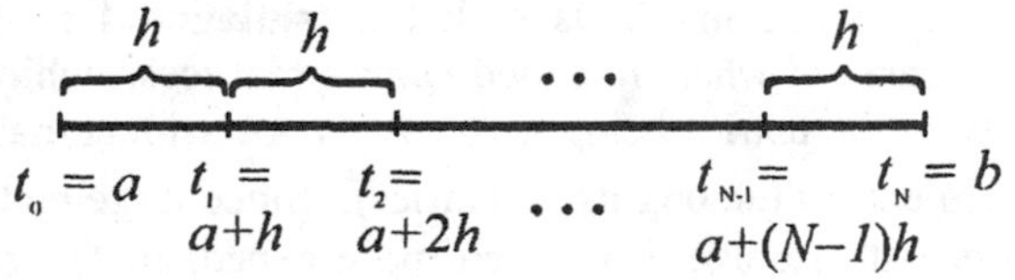

FIGURE 10.6: Grid value notation for the finite difference method for a BVP.

At each internal grid value t_i $(0 < i < N)$, we approximate the DE (5)

$$y''(t_i) = p(t_i)y'(t_i) + q(t_i)y(t_i) + r(t_i)$$

using the central difference formulas of Lemma 10.3, to obtain the approximation with local truncation error $O(h^2)$:

$$\frac{y_{i+1} - 2y_i + y_{i-1}}{h^2} = p_i \frac{y_{i+1} - y_{i-1}}{2h} + q_i y_i + r_i \quad (0 < i < N). \tag{20}$$

Multiplying by h^2, and then regrouping, we can rewrite each equation in (20) as:

$$y_{i+1} - 2y_i + y_{i-1} = hp_i(y_{i+1} - y_{i-1})/2 + h^2 q_i y_i + h^2 r_i, \text{ or}$$

$$(1 + p_i h/2)y_{i-1} - (2 + h^2 q_i)y_i + (1 - p_i h/2)y_{i+1} = h^2 r_i \quad (0 < i < N). \tag{21}$$

Since we know from the two BCs of (5) that

$$y_0 = \alpha, \text{ and } y_N = \beta,$$

the equations of (21), form a linear system in the $N - 1$ unknowns $y_1, y_2, \cdots y_{N-1}$, which, when put in matrix form $AY = C$, has

$$A = \begin{bmatrix} -(2+h^2 q_1) & 1 - p_1 h/2 & 0 & \cdots & & 0 \\ 1 + p_2 h/2 & -(2+h^2 q_2) & 1 - p_2 h/2 & \cdots & & 0 \\ 0 & \ddots & \ddots & \ddots & & \vdots \\ \vdots & & & 1 + p_{N-2}h/2 & -(2+h^2 q_{N-2}) & 1 - p_{N-2}h/2 \\ 0 & & \cdots & & 1 + p_{N-1}h/2 & -(2+h^2 q_{N-2}) \end{bmatrix}$$

and

$$C = \begin{bmatrix} h^2 r_1 - (1 + p_1 h/2)\alpha \\ h^2 r_2 \\ \vdots \\ h^2 r_{N-2} \\ h^2 r_{N-1} - (1 - p_{N-1}h/2)\beta \end{bmatrix}. \tag{22}$$

Notice the special form of the coefficient matrix A. Often in finite difference methods and in many other applications, the coefficient matrices that arise are of a similar **banded** form (i.e., nonzero entries lie entirely on a few diagonal bands). This special type of banded matrix is called a **tridiagonal matrix**. Banded matrices are special cases of what are called **sparse** matrices, which are matrices having the majority of the entries being zero. An $n \times n$ tridiagonal matrix has at most $3n - 2$ nonzero entries (among its n^2 entries). Since large matrices often eat up a lot of memory with storage, it is often more expedient to deal with sparse matrices of specialized forms using specialized methods. To solve such tri-diagonal systems, rather than Gaussian elimination, we will be using the so-called Thomas method, whose algorithm is given below:[3]

---

[3] Banded and sparse matrices were studied in Chapter 7; in particular, the Thomas method was introduced in Exercise 9 of Section 7.5.

PROGRAM 10.1: The Thomas method for solving tridiagonal systems of the form:

$$\begin{bmatrix} d_1 & a_1 & 0 & 0 & 0 & \cdots & & 0 & 0 \\ b_2 & d_2 & a_2 & 0 & 0 & \cdots & & 0 & 0 \\ 0 & b_3 & d_3 & a_3 & 0 & & & \vdots & \vdots \\ 0 & 0 & b_4 & d_4 & a_4 & 0 & & & \\ \vdots & \vdots & & \ddots & \ddots & \ddots & & \ddots & \\ 0 & 0 & & & 0 & b_{n-2} & d_{n-2} & a_{n-2} & 0 \\ 0 & 0 & \cdots & & & 0 & b_{n-1} & d_{n-1} & a_{n-1} \\ 0 & 0 & \cdots & & & & 0 & b_n & d_n \end{bmatrix} \begin{bmatrix} x_1 \\ x_2 \\ x_3 \\ \vdots \\ x_{n-2} \\ x_{n-1} \\ x_n \end{bmatrix} = \begin{bmatrix} c_1 \\ c_2 \\ c_3 \\ \vdots \\ c_{n-2} \\ c_{n-1} \\ c_n \end{bmatrix}.$$

```
function x = thomas(a,d,b,c)
%solves matrix equation Ax=c, where A is a tridiagonal matrix
%Inputs:   a=upper diagonal of matrix A a(n)=0, d=diagonal of A,
%b lower diagonal of A, b(1)-0, c-right-hand side of equation
n=length(d);
a(1)=a(1)/d(1);
c(1)=c(1)/d(1);
for i=2:n-1
   denom=d(i)-b(i)*a(i-1);
   if (denom==0), error('zero in denominator'), end
   a(i)=a(i)/denom;
   c(i)=(c(i)-b(i)*c(i-1))/denom;
end
c(n)=(c(n)-b(n)*c(n-1))/(d(n)-b(n)*a(n-1));
x(n)=c(n);
for i=n-1:-1:1
   x(i)=c(i)-a(i)*x(i+1);
end
```

EXAMPLE 10.5: Use the `thomas` program above to solve the tridiagonal system:

$$\begin{array}{rcrcrcrcl} 2x_1 & - & x_2 & & & & & = & 1 \\ -x_1 & + & 2x_2 & - & x_3 & & & = & 0 \\ & & - & x_2 & + & 2x_3 & - & x_4 & = & 0 \\ & & & & - & x_3 & + & 2x_4 & = & 1 \end{array}$$

```
>> a=[-1 -1 -1 0];d=[2 2 2 2];b=[0 -1 -1 -1]; c=[1 0 0 1];
>> format rat
>> thomas(a,d,b,c)
```

ans → 1   1   1   1   (This answer is easily checked.)

We now make some technical comments on implementing the finite difference method. In order to solve the system $AY = C$, we will need the coefficient matrix A of (22) to be nonsingular. In general this can fail, but the following theorem gives sufficient conditions to guarantee A's invertibility.

THEOREM 10.4: Suppose that the functions $p(t)$, $q(t)$, and $r(t)$ are continuous on $a \le t \le b$ and that $q(t) \ge 0$ on this time interval. Then the linear system $AY = C$ where A and C are as in (22) will have a unique solution provided that $h < 2/M$, where $M = \max\{|p(t)|: a \le t \le b\}$.

The hypotheses guarantee that the coefficient matrix A will be **strictly diagonally dominant**, which means that

$$| a_{ii} | > \sum_{\substack{j=1 \\ j \neq i}}^{n} | a_{ij} | . \tag{23}$$

In other words, the absolute value of any diagonal entry dominates the sum of the absolute values of all other entries in its row. Diagonally dominant matrices are always invertible,[4] and furthermore, it can be shown that the Thomas algorithm works very well in their presence. There are instances, however, where the Thomas method will fail for tridiagonal nonsingular matrices (e.g., this happens if $a_{11} = 0$), but there are ways to modify the method to deal with such cases.

Generally speaking, the linear shooting method is more efficient for solving a linear BVP when the former is coupled with the Runge-Kutta method. This is because the Runge-Kutta method has a local truncation error of $O(h^4)$ while that for the finite difference method is $O(h^2)$. Our main reason for introducing it here is to prime the way for its generalization to solving partial differential equations; the shooting methods do not naturally extend to the setting of PDEs.

EXERCISE FOR THE READER 10.9: Show that under the conditions of Theorem 10.4, the matrix A of (22) is diagonally dominant.

EXAMPLE 10.6: Use the finite difference method with $h = 0.1$ to solve the beam-deflection BVP of Example 10.3:

$$\begin{cases} y''(x) = \dfrac{T}{EI} y + \dfrac{wx(x-L)}{2EI}, & 0 \leq x \leq L, \\ y(0) = 0 = y(L) \end{cases}$$

having the parameters $L = 50$ feet (length), $T = 300$ lb (tension at ends), $w = 50$ lbs/ft (vertical load), $E = 1.2 \times 10^7$ lb/ft^2 (modulus of elasticity), and $I = 4$ ft^4 (central moment of inertia).
(a) Do it first for $N = 20$ subdivisions to obtain the approximate solution y1 and plot its graph.
(b) Redo it for both $N = 40$, and $N = 80$ subdivisions, to get approximate solutions y2, and y3.

---

[4] *Proof:* Suppose that A is diagonally dominant. If A were not invertible, then there would exist a nonzero vector x such that $Ax = 0$. Let k be an index so that the absolute value $| x_k |$ is as large as possible (in the norm notation from Section 7.6, this would mean $| x_k | = \| x \|_\infty$). Take the kth equation of $Ax = b$: $\sum_{j=1}^n a_{kj} x_j = 0$, divide by x_k, and solve for a_{kk} to get $a_{kk} = -\sum_{j=1, j \neq k}^n a_{kj} \cdot (x_j / x_k)$. Now take absolute values and use the triangle inequality to get $| a_{kk} | \leq \sum_{j=1, j \neq k}^n | a_{kj} \cdot (x_j / x_k) | \leq \sum_{\substack{j=1 \\ j \neq k}}^n | a_{kj} |$. What we now have contradicts diagonal dominance of the matrix A, so we have proved that A must indeed be invertible.

SOLUTION: Since the scripts are similar, we present only the one for obtaining $y1$ when $N = 20$. The script is written in such a way as to be easily modified to work for any linear BVP. The graphs of the solutions look identical, so we present only the graph of $y1$, but give plots of the differences $y1 - y2$ and $y2 - y3$.

```
%MATLAB script for finite difference method for above problem.
xa=0; xb=50; n=20;h=(xb-xa)/n;x=h:h:(xb-h);
for i=1:n-1, p(i)=0;q(i)=300/1.2e7/4;end, r=50*x.*(x-50)/2/1.2e7/4;
ya=0;yb=0; %boundary conditions
for i=1:n-1, a(i)=0;end, b=a;
a(1:n-2)=1-p(1:n-2)*h/2; %above diagonal band
d=-(2+h*h*q);  % diagonal
b(2:n-1)=1+p(2:n-1)*h/2; %below diagonal band
c(2:n-2)=h*h*r(2:n-2);
c(1)=h*h*r(1)-(1+p(1)*h/2)*ya; c(n-1)=h*h*r(n-1)-(1-p(n-1)*h/2)*yb;
y=thomas(a,d,b,c);
X=[xa x xb];
Y=[ya y yb];
plot(X,Y), grid on
```

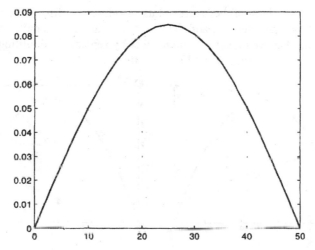

FIGURE 10.7: Graph of the solution of the beam-deflection problem of Example 10.6 using $N = 20$ subdivisions.

We can now store these x- and y-values as $x1$ and $y1$ by entering:

```
>> x1=X; y1=Y;
```

and next go on to slightly change the script to do $N = 40$ iterations and then $N = 80$ iterations and store the corresponding x and y-values as $x2$, $y2$ and $x3$, $y3$ respectively. You will notice that the graphs look quite identical. To plot the differences: $y1-y2$, and $y2-y3$, one must be a bit careful since $y1$, $y2$, and $y3$ all have different lengths. Each has $N + 1$ components ($N - 1$ grid points + the two boundary points). Here is one strategy to plot $y1-y2$ versus x.

The grid points for y2 consist of those of y1 plus one extra grid point between each adjacent pair of grid points for y1 (located at the midpoint). We must reformulate y2, only at the grid values for y1 (throwing away the extra ones at the midpoints). Let's call this "trimmed down" version of y2 by y2trim. To form y2trim in MATLAB, we could use the following line:

```
>> for i=1:21, y2trim(i)=y2(2*i-1); end  ‹alternatively:   y2trim
y2(1:2:41)
```

Now we can plot the difference of y1-y2 by simply entering:

```
>> plot(x1,y1-y2trim)
```

In a similar fashion, the following commands give the plot of the difference y2-y3:

```
>> for i=1:41, y3trim(i)=y3(2*i-1); end
>> plot(x1,y2-y3trim)
```

See Figure 10.8 for both of these error plots. Both scripts finished on the author's computer in less than a second, and the differences are quite small. We leave it to you to see what happens if one continues this by repeatedly doubling the number on subintervals N. $N = 160$, $N = 320$, $N = 640$, ...

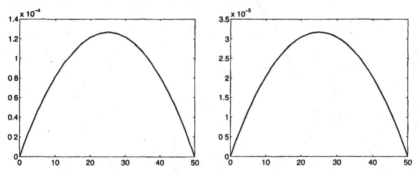

FIGURE 10.8: Plots of differences of finite difference approximated solutions to the deflected beam problem of Example 10.6. (a) The left graph is of the difference of the $N = 20$ and $N = 40$ interior grid point solutions and (b) the right one is the graph of the difference of the $N = 40$ and $N = 80$ interior grid point solutions.

EXERCISES 10.4

1 For each of the linear BVPs of parts (a) through (d) of Exercise 1 of Section 10.2, use the finite difference method with $N = 100$ to solve and then plot the solution. Whenever possible, use the Thomas method to solve the tridiagonal system. If the coefficient matrix fails to be invertible (so that errors will come up with both the Thomas and the Gaussian methods), try bumping N up to 500.

2 For each of the linear BVPs of parts (a) through (d) of Exercise 2 of Section 10.2, use the finite

difference method with $N = 100$ to solve and then plot the solution. Whenever possible, use the Thomas method to solve the tridiagonal system. If the coefficient matrix fails to be invertible (so that errors will come up with both the Thomas and the Gaussian methods), try bumping N up to 500.

3 For each of the BVPs and corresponding general solutions for the DEs given in parts (a) through (c) of Exercise 3 of Section 10.2, do the following (if possible):   (i) Use the finite difference method with $N = 100$ to solve and store the solution in vectors $t1$, $y1$. (ii) Repeat with $N = 500$ and store the solution in vectors $t2$, $y2$. (iii) Repeat once again with $N = 2500$ and store the solution in the vectors $t3$, $y3$. (iv) Determine the constants in the general solution given so that it solves the given BVP. (v) Plot the four curves in the same graph using different plot colors/styles for each. In situations where graphs are indistinguishable, plot also the errors (differences of approximations with exact solutions). Whenever possible, use the Thomas method to solve the tridiagonal system.

4 Repeat all parts of the previous exercise for each of the BVPs and general solutions given in parts (a) through (c) of Exercise 4 of Section 10.2.

5 A thin rod of length L is insulated along the lateral surface but kept at temperature $T = 0$ at both ends $x = 0$ and $x = L$.   The rod has a heat source which is proportional to the temperature at cross-section x with proportionality constant Q, The steady-state temperature function $T(x)$ $0 \le x \le L$ then satisfies the DE:

$$T_{xx} + QT = 0, \quad T = T(x), \quad 0 \le x \le L .$$

(See Section 11.2 for a derivation of more general heat equations.) Solve this DE with the given BC's $T(0) = T(L) = 0$ using the finite difference method with $N = 20$. Repeat with $N = 40$, $N = 60$, and $N = 120$. Plot these four approximations together (using different plot styles/colors). In cases where two are indistinguishable, plot the corresponding successive differences.

6 The general solution of the DE $y'' = -y$ is $y = A \sin t + B \cos t$.
(a) What restriction (on the parameters A and B) does the condition $y(0) = 0$ place?
(b) For which values of $L > 0$ does the BVP consisting of the DE and the BC's $y(0) = y(L) = 0$ have a solution (existence)? For such values of L, show that the solution is not unique.
(c) Use $L = 1$ in the BVP of part (b) and apply the finite difference method with $N = 30$. What happens? Does the Thomas algorithm work? If not, try Gaussian elimination. Is the coefficient matrix nonsingular?
(d) Repeat part (c) using $L = \pi$.

7 (a) Use Taylor's theorem to establish the following fourth-order central difference formula:

$$f'(a) \approx \frac{-f(a+2h) + 8f(a+h) - 8f(a-h) + f(a-2h)}{12h} ,$$

with error $O(h^4)$, provided that $f^{(5)}(x)$ is continuous in the interval $a - 2h \le x \le a + 2h$.

(b) In the same fashion, derive the fourth-order central difference formula

$$f''(a) \approx \frac{-f(a+2h) + 16f(a+h) - 30f(a) + 16f(a-h) - f(a-2h)}{12h} ,$$

with error $= O(h^4)$, provided that $f^{(6)}(x)$ is continuous in the interval $a - 2h \le x \le a + 2h$.

10.5: THE RAYLEIGH-RITZ METHOD

The material of this section contains much more theory than a typical section of the text. The ideas contained herein come from an important and very beautiful area of mathematics which blends linear algebra and analysis. It is fair to say that this area gave birth to the subject of functional analysis. Furthermore, the generalization of the Rayleigh-Ritz method to higher dimensions gives rise to the very important finite element method (Chapter 13) for numerical solution of PDEs. As the language in the development will indicate, many of the concepts leading to the Rayleigh-Ritz method are motivated by concepts in physics. Indeed, this was the motivational setting that led to its development. Despite the fact that the Rayleigh-Ritz method[5] dates back to the beginning of the twentieth century, it took another half century before the finite element method came to fruition. The basic idea of the Rayleigh-Ritz method is that a boundary value problem can be recast as a certain minimization problem.

FIGURE 10.9: John William Strutt (Lord Rayleigh) (1842–1919), English physicist and mathematician.

---

[5] Despite family attempts to dissuade him from vigorously pursuing a career as a full-time scientist, Lord Rayleigh (who succeeded to the title at age 30) was so intrigued by the mysteries of physics and the power of mathematics, that he made a firm commitment not to let his official diplomatic and social functions interfere too much with his dedication to scientific inquiry. For most of his life, he was financially independent, and this allowed him to set up a personal laboratory in his estate and gave him more time to focus on his research without the distraction of the other duties associated with an academic post. For the periods that he did hold academic posts at Cambridge, he took his duties with utmost conscientiousness and made some very lasting improvements in the university's scientific programs. Lord Rayleigh was a model scientist; his work touched upon and connected many areas (the Rayleigh-Ritz method is a good example inside mathematics) and was extensive (446 publications), and he won numerous prizes and recognitions for his work. Beside his scientific prowess, he was also a kind, modest, and generous man. When he won the Nobel Prize in physics in 1904, he donated his prize money to Cambridge University for the purpose of building more laboratories. In 1902, in his acceptance speech for the National Order of Merit, he stated "... the only merit of which I personally am conscious was that of having pleased myself by my studies, and any results that may be due to my researches were owing to the fact that it has been a pleasure for me to become a physicist."

Walter Ritz (1878–1909) was a Swiss/German mathematician/physicist. After entering the Polytechnic University of Zurich in an engineering program, he found that he was not satisfied with the compromises and lack of rigor in his engineering courses, so he switched to physics. He was a classmate of Albert Einstein. For health reasons, he needed to move away from the humid climate of Zürich, and went on to the University Göttingen to complete his studies. There he was influenced by the teachings of David Hilbert. Despite his short life and career, he was able to accomplish quite a lot of scientific research. Actually, Lord Rayleigh and Ritz never met. Rayleigh first developed a mathematical method for predicting the first natural frequency of simple structures by minimizing the distributed energy. Ritz subsequently extended the method to solve (numerically) associated displacement and stress functions.

Rather than strive for generality, our purpose in this section will be to understand the concepts behind the Rayleigh-Ritz method so we begin by focusing our attention on the following boundary value problem:

$$\text{(BVP)} \quad \begin{cases} -u''(x) = f(x), \ 0 < x < 1 \\ u(0) = 0, \ u(1) = 0 \end{cases}. \tag{24}$$

Here $f(x)$ is a continuous function. This problem has, by itself, numerous physical interpretations, as we have seen in previous chapters. As examples we mention the steady-state heat distribution on a thin rod (Chapter 11) with ends maintained at temperature zero, or the deflection of an elastic beam (Section 10.1) whose ends are fixed.

We introduce the **inner product** $\langle u, v \rangle$ for a pair of piecewise continuous bounded functions on $[0,1]$:

$$\langle u, v \rangle = \int_0^1 u(x)v(x)dx . \tag{25}$$

Recall that for a function $u(x)$ to be piecewise continuous on $[0,1]$, it means that the domain can be broken up into subintervals: $0 = a_0 < a_1 < \cdots < a_n = 1$ such that $u(x)$ is continuous on each open subinterval (a_{i-1}, a_i). We point out the following simple yet very important properties of this inner product. By linearity of the integral, it immediately follows that the inner product is linear in each variable, i.e.,

$$\begin{aligned} \langle \alpha u_1 + \beta u_2, v \rangle &= \alpha \langle u_1, v \rangle + \beta \langle u_2, v \rangle \\ \langle u, \alpha v_1 + \beta v_2 \rangle &= \alpha \langle u, v_1 \rangle + \beta \langle u, v_2 \rangle \end{aligned}, \tag{26}$$

where the u, u_1, v, v_1 denote arbitrary (piecewise continuous bounded) functions and α, β denote arbitrary real numbers. Even clearer is the following symmetry property:

$$\langle u, v \rangle = \langle v, u \rangle . \tag{27}$$

In light of properties (26) and (27), the inner product is said to be a **symmetric bilinear form**. Another property of the inner product is that it is **positive definite**: If $u(x)$ is a piecewise continuous function on $[0,1]$ that is not zero on some open interval (a_{i-1}, a_i), then $\langle u, u \rangle > 0$ (see Exercise 17).

We consider the following rather large class of **admissible functions** on $[0,1]$ which obey the boundary conditions of our problem (24):

$$\mathcal{A} = \{v : [0,1] \to \mathbb{R} : v(x) \text{ is continuous}, \ v'(x) \text{ is piecewise continuous} \\ \text{and bounded, and } v(0) = 0, v(1) = 0\}. \tag{28}$$

EXERCISE FOR THE READER 10.10: Show that the space $\mathcal{A}$ is closed under the operations of addition of functions and scalar multiplication. More precisely, if $v, w \in \mathcal{A}$ and α is any real number, show that the functions $v + w$, and αv also belong to $\mathcal{A}$.

For functions in this class we further define the following functional: $F : \mathcal{A} \to \mathbb{R}$ by the formula:

$$F(v) = \frac{1}{2}\langle v', v' \rangle - \langle f, v \rangle. \tag{29}$$

In the setting where (24) models the deflection of an elastic beam, certain physical interpretations can be given to some of these quantities. For a given displacement $v(x)$, the inner product $\langle f, v \rangle$ represents the so-called **load potential** and the term $\frac{1}{2}\langle v', v' \rangle$ represents the **internal elastic energy**. The functional $F(v)$ then represents the **total potential energy**. Using physics it can be proved that the solution of (24) will have minimal total potential energy over all possible admissible functions $v \in \mathcal{A}$. This fact is known as the **Principle of Minimum Potential Energy (MPE)** and we will prove it mathematically in Theorem 10.5 below. Thus, the variational problem which turns out to be equivalent to the boundary value problem (24) is the following:

(MPE)   Find $u \in \mathcal{A}$ satisfying $F(u) \le F(v)$ for all $v \in \mathcal{A}$. $\tag{30}$

Another equivalent, but very different looking problem whose equivalence to the boundary value problem is known in physics as **Principle of Virtual Work (PVW)**, is the following:

(PVW)   Find $u \in \mathcal{A}$ satisfying $\langle u', v' \rangle = \langle f, v \rangle$ for all $v \in \mathcal{A}$. $\tag{31}$

It is quite a surprising fact that the three seemingly different problems (24), (30), and (31) have equivalent solutions. The precise result is stated in the following theorem.

THEOREM 10.5 : (*Variational Equivalences of a Boundary Value Problem*) Suppose that $f(x)$ is any continuous and bounded function on $0 < x < 1$, and that $u(x)$ is an admissible function of the class $\mathcal{A}$ defined in (28). Then the following are equivalent:

(a) The function $u(x)$ is a solution of the (BVP) (24) $\begin{cases} -u''(x) = f(x), & 0 < x < 1 \\ u(0) = 0, & u(1) = 0 \end{cases}$.

(b) The function $u(x)$ is a solution of the (MPE) (30): $F(u) \le F(v)$ for all $v \in \mathcal{A}$.

(c)   The function $u(x)$ is a solution of the (PVW) (31): $\langle u',v'\rangle = \langle f,v\rangle$ for all $v \in \mathcal{A}$.
Furthermore, each of these three problems has unique solutions.

Proof: The proof is rather long, so we break it up into several pieces. The proof that (24) has a unique solution can be accomplished quite easily (see Exercise 22). We point out that Theorem 10.1 does not apply.[6]

Step 1: We first show that (b) implies (c). To this end, suppose that $u(x)$ solves the (MPE), so that $F(u) \le F(v)$ for all $v \in \mathcal{A}$. Letting ε denote any real number, we may conclude that $F(u) \le F(u + \varepsilon v)$, where $v \in \mathcal{A}$ is arbitrary. If we hold the functions u and v fixed, we can view the function on the right $\phi(\varepsilon) \equiv F(u + \varepsilon v)$ as a real-valued function of ε. Using bilinearity and then symmetry of the inner product, we may expand this function as follows:

$$\phi(\varepsilon) \equiv \frac{1}{2}\langle (u + \varepsilon v)', (u + \varepsilon v)'\rangle - \langle f, u + \varepsilon v\rangle$$

$$= \frac{1}{2}\langle u' + \varepsilon v', u' + \varepsilon v'\rangle - \langle f,u\rangle - \varepsilon\langle f,v\rangle$$

$$= \frac{1}{2}\langle u',u'\rangle + \frac{\varepsilon}{2}\langle u',v'\rangle + \frac{\varepsilon}{2}\langle v',u'\rangle + \frac{\varepsilon^2}{2}\langle v',v'\rangle - \langle f,u\rangle - \varepsilon\langle f,v\rangle$$

$$= \frac{1}{2}\langle u',u'\rangle + \varepsilon\langle u',v'\rangle + \frac{\varepsilon^2}{2}\langle v',v'\rangle - \langle f,u\rangle - \varepsilon\langle f,v\rangle.$$

Since each of the inner products in the last expression is simply a real number, the function $\phi(\varepsilon)$ is just a second-degree polynomial (in the variable ε). Since we know this function has a minimum value at $\varepsilon = 0$, we must have $\phi'(0) = 0$. Differentiating the last expression for $\phi(\varepsilon)$ in the above expansion, this gives $\langle u',v'\rangle - \langle f,v\rangle = 0$, and since $v \in \mathcal{A}$ was arbitrary, this shows (PVW).

---

[6] A general result shows that existence and uniqueness questions about general BVPs can be reduced to questions about homogeneous BVPs. The following is taken from page 197 of [Sta-79]:
Theorem: For a pair of 2×2 matrices A and B, and continuous functions $f(x)$, $g(x)$, $h(x)$ on an interval $[a,b]$, the BVP consisting of the DE $y'' = h(x)y' + g(x)y + f(x)$ ($y = y(x)$) and the general boundary conditions $A\begin{bmatrix} y(a) \\ y'(a) \end{bmatrix} + B\begin{bmatrix} y(b) \\ y'(b) \end{bmatrix} = \begin{bmatrix} \alpha \\ \beta \end{bmatrix}$ has a unique solution if and only if the corresponding homogeneous problem with $f(x) = 0$, and $\alpha, \beta = 0$ has only the trivial solution $y(x) = 0$.

For our special problem (24) we need only take $h(x) = g(x) = 0$ to get the DE and $A = \begin{bmatrix} 1 & 0 \\ 0 & 0 \end{bmatrix}$, $B = \begin{bmatrix} 0 & 0 \\ 1 & 0 \end{bmatrix}$ to get the BC's. The corresponding homogeneous problem is just: $y'' = 0$ and $y(0) = y(1) = 0$. Integrating this DE and using the BC's easily shows $y(x) = 0$ is the only solution. Thus, this theorem implies our problem (24) has a unique solution.

Step 2: We show that (c) implies (b). So assume that $u(x)$ solves the (PVW), i.e., $\langle u',v'\rangle = \langle f,v\rangle$ for all $v \in \mathcal{A}$. Fix now an admissible function $v \in \mathcal{A}$. Our task is to show that $F(v) \geq F(u)$. Setting $w = v - u$ so $v = u + w$, we may use bilinearity and symmetry as above to write:

$$F(v) = F(u+w)$$
$$= \frac{1}{2}\langle u'+w', u'+w'\rangle - \langle f, u+w\rangle$$
$$= \underbrace{\frac{1}{2}\langle u',u'\rangle - \langle f,u\rangle}_{= F(u)} + \underbrace{\langle u',w'\rangle - \langle f,w\rangle}_{= 0 \text{ by (PVW)}} + \underbrace{\frac{1}{2}\langle w',w'\rangle}_{\geq 0}$$
$$\geq F(u),$$

as desired.

Step 3: We show that (a) implies (c). We thus assume that the function $u(x)$ solves the BVP (24). From the differential equation $-u''(x) = f(x)$, $0 < x < 1$, the second derivative of $u(x)$ exists (and is continuous) so it follows that the first derivative $u'(x)$ is continuous (from calculus, differentiability implies continuity). Furthermore, since $f(x)$ is assumed to be bounded, so must be $u'(x)$, and from the boundary conditions stipulated by (24), it follows that $u(x)$ is an admissible function (i.e., $u \in \mathcal{A}$). We now fix an admissible function $v \in \mathcal{A}$, multiply both sides of the differential equation by it, and proceed to integrate by parts. Doing this and translating into inner products gives:

$$\langle f,v\rangle = \langle -u'', v\rangle = -\int_0^1 u''(x)v(x)dx = \underbrace{u'(x)v(x)\Big]_{x=1}^{x=0}}_{= 0 \text{ by (BC)}} + \int_0^1 u'(x)v'(x)dx = \langle u',v'\rangle.$$

It follows that $u(x)$ solves the (PVW), as asserted.

Up to this point we have rigorously shown the following implications for solutions of the various three problems:

$$(BVP) \Rightarrow (PVW) \Leftrightarrow (MPE).$$

We will next show that the solutions of (PVW) are unique. From this and what was already proved, it will follow that all three problems have unique solutions.

Step 4: We prove that any two solutions u_1 and u_2, both belonging to $\mathcal{A}$, of the problem (PVW) must be identical. Thus we are assuming that $\langle u_i', v'\rangle = \langle f,v\rangle$ for all $v \in \mathcal{A}$ ($i = 1, 2$). Our task is to show $u_1 = u_2$. If we use

$v = u_1 - u_2 \in \mathcal{A}$, we obtain that: $\langle u_1', [u_1 - u_2]' \rangle = \langle f, u_1 - u_2 \rangle$ and $\langle u_2', [u_1 - u_2]' \rangle$

$= \langle f, u_1 - u_2 \rangle$. Subtracting and using linearity gives us: $\langle [u_1 - u_2]', [u_1 - u_2]' \rangle = 0$,

which translates to $\int_0^1 (u_1'(x) - u_2'(x))^2 \, dx = 0$. Since the integrand is nonnegative

and piecewise continuous, it follows that it must equal zero everywhere on $[0,1]$ except, possibly, at the endpoints of the intervals making up its pieces. We have used positive definiteness of the inner product here. The same is therefore true for $u_1' - u_2' = [u_1 - u_2]'$, so it follows that the antiderivative of this latter function must be a constant. Thus we can write $u_1 - u_2 = C$ or $u_1 = u_2 + C$. But the boundary conditions $u_1(0) = 0$ then force $C = 0$ and we can conclude $u_1 = u_2$, as desired.

Step 5: *(Final Step)* We show that (PVW) implies (BVP). At this point we invoke the fact, mentioned at the outset of this proof, that the (BVP) has a solution $u(x)$ (existence). From what was already proved, this function $u(x)$ is also a solution of (PVW), but from step 4, the solution of (PVW) is unique. Consequently, any solution of (PVW) really must be the (unique) solution of (BVP), as required. QED

In order to solve the BVP (24), the above theorem allows us to focus our attention on either of the equivalent problems MPE (30) or PVW (31). The finite element method will use one of these two formulations but will replace the very large space $\mathcal{A}$ of admissible functions by a much smaller (finite-dimensional) space in each of the corresponding governing conditions.

We begin by partitioning the interval $(0,1)$ into subintervals: $\mathcal{P}: 0 = x_0 < x_1 < \cdots < x_{n+1} = 1$. We denote these intervals by $I_i = (x_i, x_{i+1})$ ($i = 0, 1, 2, \cdots, n$) and their lengths by $h_i = x_{i+1} - x_i$. Unlike with finite difference methods, we do <u>not</u> require that these lengths be equal. We define the **mesh size** $\|\mathcal{P}\|$ of this partition as the maximum of the lengths $\max_{0 \le i \le n} h_i$. Corresponding to such a partition $\mathcal{P}$ we define the following space of piecewise linear functions:

$$\mathcal{A}(\mathcal{P}) = \{ v : [0,1] \to \mathbb{R} : v(x) \text{ is continuous on } [0,1], \text{ linear on each } I_i$$
$$\text{and } v(0) = 0, v(1) = 0 \}.$$

A typical function in this space is depicted in Figure 10.10.

EXERCISE FOR THE READER 10.11: Show that the space $\mathcal{A}(\mathcal{P})$ is closed under the operations of addition of functions and scalar multiplication. More precisely, if $v, w \in \mathcal{A}(\mathcal{P})$ and α is any real number, show that the functions $v + w$, and αv also belong to $\mathcal{A}(\mathcal{P})$.

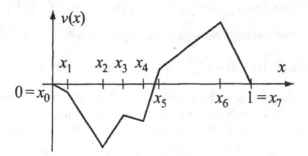

FIGURE 10.10: Illustration of a typical function in the space $\mathcal{A}(\mathcal{P})$.

Notice that a function $v \in \mathcal{A}(\mathcal{P})$ is entirely determined by its values at the interior grid points: $v(x_1), v(x_2), \cdots, v(x_n)$. This follows from linearity and continuity. We need a set of basis functions that can be used to easily describe functions in $\mathcal{A}(\mathcal{P})$. These n functions are usually chosen so that each one equals zero on most of the interval $[0,1]$, so that it will have minimum interaction with other basis functions.[7] One simple set of basis functions meeting this criterion are the so-called **hat functions** $\phi_i(x)$ $(1 \leq i \leq n)$. Each hat function $\phi_i(x)$ is that member of $\mathcal{A}(\mathcal{P})$ determined by the assignments: $\phi_i(x_i) = 1$ and $\phi_i(x_j) = 0$ for any index $j \neq i$. A typical hat function is shown in Figure 10.11.

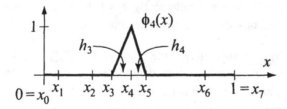

FIGURE 10.11: A typical hat function for a certain partition of $(0,1)$. Note the (possible) asymmetry.

We observe that any function $v \in \mathcal{A}(\mathcal{P})$ can be expressed in a unique way as a linear combination of the hat functions:

$$v(x) = \sum_{i=1}^{n} v(x_i)\phi_i(x). \tag{32}$$

---

[7] General Rayleigh-Ritz methods result from using any set of linearly independent functions which are continuous, piecewise differentiable and satisfy the required boundary conditions as a set of "basis functions."

(To prove this, just check that both functions agree at each partition point x_i, and then it will follow that they are always equal since both are piecewise linear.) In the language of linear algebra, we say that the n hat functions form a *basis* for the n-dimensional space $\mathcal{A}(\mathcal{P})$.

The equations of the hat functions are as follows:

$$
\phi_i(x) = \begin{cases} 0, & \text{if } 0 \le x \le x_{i-1} \text{ or } x_{i+1} \le x \le 1, \\[2mm] \dfrac{x - x_{i-1}}{h_{i-1}}, & \text{if } x_{i-1} \le x \le x_i, \\[2mm] \dfrac{x_{i+1} - x}{h_i}, & \text{if } x_i \le x \le x_{i+1}. \end{cases} \tag{33}
$$

The **Rayleigh-Ritz method** for approximating the BVP (24) is to solve the following finite-dimensional version (discretization) of it:

$$
\text{Find } u \in \mathcal{A}(\mathcal{P}) \text{ satisfying } F(u) \le F(v) \text{ for all } v \in \mathcal{A}(\mathcal{P}). \tag{34}
$$

Note that the Rayleigh-Ritz problem (34) is obtained by the corresponding (MPE) problem (30) simply by replacing $\mathcal{A}$ by $\mathcal{A}(\mathcal{P})$. We will proceed now to discuss the special Rayleigh-Ritz method for our BVP (2) using the hat functions $\phi_i(x)$ of (33). Different basis functions and, more generally, different finite dimensional spaces give rise to different versions of the Rayleigh-Ritz method. Implementations using such hat functions are often referred to as the **(piecewise) linear Rayleigh-Ritz method**. Since, as in (32), any function in $\mathcal{A}(\mathcal{P})$ can be written as $\sum c_i \phi_i$, making use of bilinearity, we may write:

$$
\begin{aligned}
F\left(\sum c_i \phi_i\right) &= \frac{1}{2}\left\langle \left[\sum c_i \phi_i\right]', \left[\sum c_j \phi_j\right]' \right\rangle - \left\langle f, \sum c_i \phi_i \right\rangle \\
&= \frac{1}{2}\left\langle \sum c_i \phi_i', \sum c_j \phi_j' \right\rangle - \sum c_i \left\langle f, \phi_i \right\rangle \\
&= \frac{1}{2}\sum_{i=1}^{n}\sum_{j=1}^{n} c_i c_j \left\langle \phi_i', \phi_j' \right\rangle - \sum c_i \left\langle f, \phi_i \right\rangle.
\end{aligned} \tag{35}
$$

The above expression can be viewed as a (quadratic) function of the variable $(c_1, c_2, \cdots, c_n) \in \mathbb{R}^n$. We can locate its minimum by setting each of the partial derivatives equal to zero. Using the product rule, we can compute as follows:

$$
\frac{\partial}{\partial c_k} F\left(\sum c_i \phi_i\right) = 0 \;\Rightarrow\; \frac{1}{2}\sum_{j=1}^{n} c_j \left\langle \phi_k', \phi_j' \right\rangle + \frac{1}{2}\sum_{i=1}^{n} c_i \left\langle \phi_i', \phi_k' \right\rangle = \left\langle f, \phi_k \right\rangle \quad (1 \le k \le n).
$$

Now using symmetry of the inner product, we can combine the two summations on the left into one:

$$\sum_{j=1}^{n} \langle \phi_k', \phi_j' \rangle c_j = \langle f, \phi_k \rangle \ (1 \le k \le n) \ . \tag{36}$$

We abbreviate this linear system as

$$Ac = b \ , \tag{37}$$

where $A = [a_{ij}] = \left[\langle \phi_i', \phi_j' \rangle \right]$ is the so-called ($n \times n$) **stiffness matrix**, and b is the

so called ($n \times 1$) **load vector**: $[b_j] = [\langle f, \phi_j \rangle]$. The terminology comes from the

model of (24) for an elastic beam.

To compute the entries of the stiffness matrix: $\langle \phi_i', \phi_j' \rangle = \int_0^1 \phi_i'(x) \phi_j'(x) dx$, we

first observe that, from the properties of the hat functions, $\phi_i'(x) \phi_j'(x) = 0$ unless

i and j are equal or are adjacent indices. Thus, the stiffness matrix is both

symmetric and tridiagonal. To compute the nonzero entries of A, there are just two

cases. We use (33) for the computations:

$$\langle \phi_i', \phi_i' \rangle = \int_{x_{i-1}}^{x_{i+1}} [\phi_i'(x)]^2 \, dx = \int_{x_{i-1}}^{x_i} [1/h_{i-1}]^2 \, dx + \int_{x_i}^{x_{i+1}} [1/h_i]^2 \, dx = \frac{1}{h_{i-1}} + \frac{1}{h_i} \ , \tag{38}$$

$$\langle \phi_i', \phi_{i+1}' \rangle = \int_{x_i}^{x_{i+1}} \phi_i'(x) \phi_{i+1}'(x) dx = \int_{x_i}^{x_{i+1}} \left[\frac{-1}{h_i} \right] \left[\frac{1}{h_i} \right] dx = \frac{-1}{h_i} \ . \tag{39}$$

EXERCISE FOR THE READER 10.12: (a) Show that the stiffness matrix A is
positive definite (i.e., show that for any $n \times 1$ vector c, we have $c'Ac \ge 0$ with
equality if and only if c is the zero vector).[8]
(b) Show that in case all grid spaces are equal, (i.e., $h_i = \|\mathscr{P}\|$ for all i), the
stiffness matrix for linear system of the linear Rayleigh-Ritz (FEM) is a constant
multiple of the coefficient matrix for the finite difference method introduced in
Section 10.3. How do the linear systems compare?

As a general rule for Rayleigh-Ritz methods (and finite element methods for
PDEs), it is usually a good idea to place more nodes where the (known) coefficient
functions in the problem undergo the most activity. Adaptive methods can be
developed in which successive refinements are used to see where to place
additional nodes. We are ready to give a numerical example of the Rayleigh-Ritz
method. In order to be able to get a check on errors, the following theorem will be
useful:

---

[8] Some general facts about positive definite matrices are that they are nonsingular and, if symmetric,
their eigenvalues are all positive. The latter is, in fact, an equivalent definition (see, e.g., Section 8.4 of
[HoKu-71] for proofs and more information on positive definite matrices). In particular, the stiffness
matrix is nonsingular, so the Rayleigh-Ritz method leads to a unique solution.

THEOREM 10.6: (*Error Estimate for Rayleigh-Ritz Approximations*)   Let $u_{\mathscr{P}}(x)$ denote the (piecewise) linear Rayleigh-Ritz approximation corresponding to a partition $\mathscr{P}$ of $[0,1]$ of the BVP (24): $\begin{cases} -u''(x) = f(x), & 0 < x < 1 \\ u(0) = 0, \ u(1) = 0 \end{cases}$, where $f(x)$ is a continuous function.   The following error estimate holds for each $x, \ 0 \leq x \leq 1$:

$$|u_{\mathscr{P}}(x) - u(x)| \leq \frac{\|\mathscr{P}\|^2}{2} \max_{0 \leq x \leq 1} |f(x)| . \tag{40}$$

The proof of this theorem involves some nice ideas from analysis; an outline is left to Exercises 18–21 (see also the note preceding Exercise 17). In fact, in this setting it is even true that $u_{\mathscr{P}}(x_i) = u(x_i)$ at each grid point and thus $u_{\mathscr{P}}$ is really the piecewise linear interpolant of u with respect to the partition $\mathscr{P}$; see Exercise 21.

EXAMPLE 10.7: Consider the (BVP) (24) $\begin{cases} -u''(x) = f(x), & 0 < x < 1 \\ u(0) = 0, \ u(1) = 0 \end{cases}$ with[9]

$$f(x) = \sin\left[\text{sign}(x - .5) \exp\left(\frac{1}{4|x - .5|^{1.05} + .3} \right) \right] .$$

$$\exp\left(\frac{1}{4|x - .5|^{1.2} + .2} - 100(x - .5)^2 \right) .$$

(a) Use the Rayleigh-Ritz method with $n = 50$ equally spaced interior grid values to solve this BVP and plot the resulting approximation.

(b) Solve the problem again with the Rayleigh-Ritz method and $n = 50$ interior grid values, but this time deploy a higher concentration of grid points where the inhomogeneity $f(x)$ is more oscillatory.

(c) Use Theorem 10.6 to find a (uniform) grid size that will guarantee that the Rayleigh-Ritz solution will be visually (without zooms) identical to the exact solution and compare both solutions of (a) and (b) with this more accurate solution.

SOLUTION: Since the BVP (24) is rather specialized, we will not bother writing here an M-file to perform the Rayleigh-Ritz method. Instead, we will go through each part directly, using MATLAB whenever convenient.

Part (a): Here we have $h_i = \|\mathscr{P}\| = 1/51$ for each i, so that from the calculations above, the stiffness matrix is given by:

---

[9] We use the notation of the "sign function" (whose MATLAB counterpart has the same name): sign(x) = 1, if $x > 0$, 0, if $x = 0$, and -1, if $x < 0$.

$$A = 51 \begin{bmatrix} 2 & -1 & & & & \\ -1 & 2 & -1 & \mathbf{0} & & \\ & -1 & 2 & -1 & & \\ & & \mathbf{0} & \ddots & \ddots & -1 \\ & & & & -1 & 2 \end{bmatrix}.$$

The entries of the load vector can be computed using MATLAB's integrator quad. The resulting system is then stored and solved using the Thomas algorithm. Note that in the case of equal grid spaces, the hat functions become symmetric and formula (33) for them can be abbreviated as (we set $h = \|\mathcal{P}\|$)

$$\phi_i(x) = \begin{cases} \dfrac{h - |x - x_i|}{h} = 1 - \dfrac{|x - x_i|}{h}, & \text{if } x_{i-1} \leq x \leq x_{i+1}, \\ 0, & \text{otherwise.} \end{cases}$$

(Verify this!) As the integrals required for the load vector entries are related, a loop will be used to compute them. The integrals will depend on a parameter. One way to compute such parameter-dependent integrals is to declare the parameter variables as global variables.

global var →	Inside the definition of a function M-file having var as a variable, this command declares this variable to be a global variable. Recall that by default, all variables appearing in an M-file are local variables. Should also be used in the command window before invoking such an M-file.

The use of this strategy is demonstrated in the remainder of this example.

The coefficients of the load vector are given by: $(1 \leq j \leq 50)$

$$b_i = \langle f, \phi_i \rangle = \int_{x_{i-1}}^{x_{i+1}} f(x)\phi_i(x)\,dx = \int_{x_{i-1}}^{x_{i+1}} \left[1 - \frac{|x - x_i|}{h} \right] f(x)\,dx$$

$$= \int_{x_{i-1}}^{x_{i+1}} \left[1 - 51 |x - x_i| \right] f(x)\,dx.$$

The integrands depend on the parameter x_i, so we will first create an M-file for them using xi, which we declare as a global variable, to represent x_i [10].

```
function y = frayritz10_7(x)
global xi;
y=(1-51*abs(x-xi)).*sin(sign(x-.5).*exp(1./(4*abs(x-
.5).^1.05+.3)))... .*exp(1./(4*abs(x-.5).^1.2+.2)-100*(x-.5).^2);
```

---

[10] A syntax note: If, after creating and storing this M-file we were to enter frayritz10_7(24), the output would be "[ ]" (the empty vector), a reasonable answer since we have not yet defined xi. If we first entered a value for xi, say xi=2 and reentered the above command, however, we would still get the empty vector as output. It is essential to first declare xi as a global variable in the command window (even though this was already done in the M-file). If this is done, and xi=2 is reentered, then entering frayritz10_7(24) would finally produce an answer (ans = −4.7321).

The load coefficients can now be created as follows: First declare our global variable and create the vector x of grid points. We remind the reader that vector indices must be positive integers so $x(1)$ represents x_0 and so on.

```
>> global xi;
>> for i=1:52
     x(i)=(i-1)/51;
end
```

With our M-file, the load coefficients are now easily created with the following loop. Notice that we have used quadl rather than quad. This former integrator works in the same syntax as quad, but uses a refined adaptive technique. It takes a bit more time to use but gives more accurate results.

```
>> for i=2:51
     xi=x(i);
     b(i)=quadl('frayritz10_7', x(i-1), x(i+1));
end
```

We have kept the indices consistent with those of the vector x, but consequently we have created a vector with one extra component $b(1) = 0$. This component must be left out when we go on to solve the linear system. In order to solve the linear system, we will use the thomas M-file, which will solve our tridiagonal system quite efficiently. We must create the appropriate vectors to meet the syntax of this M-file:

```
>> d=2*ones(50,1)*51;   %-diagonal of stiffness matrix A.
>> da=-1*ones(50,1)*51; da(51)=0; %superdiagonal (above)
>> db=-1*ones(50,1)*51; db(1)=0; %subdiagonal (below)
>> c=thomas(da,d,db,b(2:51));
```

As explained earlier, the values of the solution vector c are precisely the values of the numerical Rayleigh-Ritz solution at the interior grid points $x_1, x_2, \cdots, x_{50} = (x(24), x(25), \ldots, x(51))$. To plot the entire graph of c versus x, we need to augment the vector c to have first and last components which equal zero (from the boundary conditions). With this being done below, the resulting numerical plot is shown in Figure 10.12.

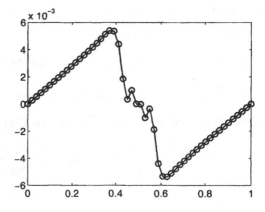

```
>> c=[0 c 0];
>> plot(x,c,'b-o')
```

FIGURE 10.12: Rayleigh-Ritz solution of the BVP in Example 10.7 using 50 equally spaced interior grid points. The grid points/values are shown with (blue) circles.

Part (b): The right-hand side of the DE $-u''(x) = f(x)$ has the graph shown in Figure 10.13.

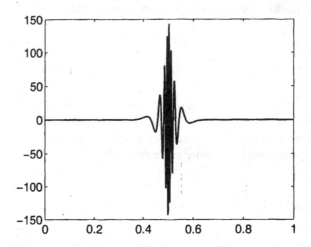

FIGURE 10.13: Graph of the right-hand side $f(x)$ of the DE $-u''(x) = f(x)$ of Example 10.7.

From Figure 10.13, we see that the inhomogeneity $f(x)$ is most oscillatory approximately on the interval [0.35, 0.65] and elsewhere is rather tame. With this perspective, it would seem that any grid that is uniformly highly dense would give rise to much wasted computation on the long intervals of inactivity. Motivated by Figure 10.13, we propose the following deployment of the 50 interior grid points.

Put 6 in each of the intervals [0, 0.35) and (0.65, 1], and put the remaining 40 in [0.35, 0.65]. We stipulate that the grid points in each of these intervals be uniformly spaced but this is by no means necessary (the Rayleigh-Ritz method is totally flexible). The linspace command will make the construction of these grid values particularly straightforward:

```
>> x2(1:7)=linspace(0,0.35,7);
>> x2(7:46)=linspace(0.35,0.65,40);
>> x2(46:52)=linspace(0.65,1,7);
```

Since the grid is no longer uniform, we need to construct a vector for the h_i :

```
>> for i=1:51, h(i)=x2(i+1)-x2(i); end
```

It is left to construct the load vector b . By (33) the coefficients are

$$b_i = \langle f, \phi_i \rangle = \int_{x_{i-1}}^{x_{i+1}} f(x)\phi_i(x)dx = \int_{x_{i-1}}^{x_i} \frac{x - x_{i-1}}{h_{i-1}} f(x)dx + \int_{x_i}^{x_{i+1}} \frac{x_{i+1} - x}{h_i} f(x)dx .$$

Employing the strategy used in part (a), we need here a pair of M-files for the two respective integrands:

```
function y = frayritz10_7a(x)
global xim;
global him;
y=(x-xim)./him.*sin(sign(x-.5).*exp(1./(4*abs(x-...
.5).^1.05+.3))).*exp(1./(4*abs(x-.5).^1.2+.2)-100*(x-.5).^2);
function y = frayritz10_7b(x)
global xip;
global hi;
y=(xip-x)./hi.*sin(sign(x-.5).*exp(1./(4*abs(x-...
.5).^1.05+.3))).*exp(1./(4*abs(x-.5).^1.2+.2)-100*(x-.5).^2);
```

The load vector is now easily constructed, and the linear tridiagonal system can be assembled and solved as before:

```
>> global xim him xip hi;
>> for i=2:51;
     xip=x2(i+1); xim=x2(i-1); hi=h(i); him=h(i-1);
     b2(i)=quadl('frayritz18_1a', x(i-1), x(i))+...
       quadl('frayritz18_1b', x(i), x(i+1));
end

>> for i=1:51, h(i)=x(i+1)-x(i); end
>> for i=2:51, d2(i)=1/h(i-1)+1/h(i); end
.. %main diagonal will be d(2:51).
>> for i=2:50, da2(i)=-1/h(i); end
>> da2(51)=0; % superdiagonal will be da(2:51).
>> for i=2:50, db2(i)=-1/h(i-1); end
>> %subdiagonal will be db(1:50)

>> c2=thomas(da2(2:51),d2(2:51),db2(1:50),b2(2:51));
```

The commands needed to plot this solution are just as in part (a), and those commands produce the plot shown in Figure 10.14(a).

Part (c): From Figure 10.12, we see that the amplitude of the solution is roughly 6e-3. Theorem 10.6 gives maximum bound for the error to be $\frac{\|\mathscr{P}\|^2}{2} \max_{0 \le x \le 1} |f(x)|$. Setting this expression to be 6e-3/100 (so the maximum error will be less than about 1/100 of the amplitude), using 150 for $\max_{0 \le x \le 1} |f(x)|$ (from Figure 10.13), and solving for $\|\mathscr{P}\|$ gives roughly 1e-4, so that if we use 10,000 interior grid points, the Rayleigh-Ritz solution should have the desired accuracy. The construction and plotting of this solution is done just as in part (a), except that instances of 50 or 51, etc. should be changed to 10,000 or 10,001, etc. The resulting graph is compared with the two obtained in parts (a) and (b) in Figure 10.14.

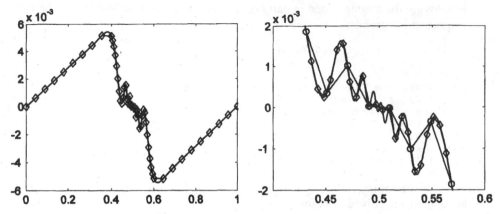

FIGURE 10.14: (a) (left) Rayleigh-Ritz solution obtained for Example 10.7(b) shown with diamonds, along with the exact solution in black. The grid used is nonuniform with more grid points (diamonds) deployed in the areas where the inhomogeneity is most active. (b) (right) Zoomed-in comparison of the Rayleigh-Ritz solutions in part (a) (circles) and part (b) (diamonds) with the exact solution (smooth curve) of Example 10.7. Note the surprising fact that the Rayleigh-Ritz solutions are exactly equal to the solution at the respective grid points, and hence the Rayleigh-Ritz solutions turn out simply to be the piecewise linear interpolants of the actual solution with respect to the associated grids (see Exercise 21 for a proof). This theorem will no longer hold in higher dimensions or even for more complicated single-variable BVPs.

We now turn to the **Galerkin method**[11] for approximating the solution of the BVP (24). In the piecewise linear setting with the finite-dimensional space $\mathcal{A}(\mathcal{P})$ in place of the space $\mathcal{A}$ of all admissible functions, this method solves the discrete analogue of the Principle of Virtual Work (31):

$$\text{Find } u \in \mathcal{A}(\mathcal{P}) \text{ satisfying } \langle u', v' \rangle = \langle f, v \rangle \text{ for all } v \in \mathcal{A}(\mathcal{P}). \tag{41}$$

In light of the bilinearity of the inner product, it is enough to check (41) for the function v running through the n (basis) functions: $\{\phi_k\}_{k=1}^n$. The discrete problem is thus to determine the coefficients $(c_1, c_2, \cdots, c_n) \in \mathbb{R}^n$ of the function

$$u = \sum_{j=1}^{n} c_j \phi_j \in \mathcal{A}(\mathcal{P}) \text{ such that } \left\langle \left(\sum c_i \phi_i \right)', \phi_k' \right\rangle = \langle f, \phi_k \rangle \ (1 \le k \le n).$$

---

[11] Like the works of Rayleigh and Ritz, the work of Russian engineer/mathematician Boris Grigorievich Galerkin (1871–1945) was motivated by physical problems. It is fair to characterize Galerkin as an applied mathematician of the purest sense. He worked many years as an engineer before his first publication at the relatively late age of 38 on longitudinal curvature. The paper was a significant extension of work of Euler and it was applied in the construction of bridges and building frames. His continued interest in structural mechanics led him to the discovery in 1915 of his most notable contribution to mathematics, what is known today as the Galerkin method. He subsequently took on some academic posts in St. Petersburg, which was the de facto mathematical capital of Russia at the time. His interests in consulting with industry and in the relevant mathematics continued until his death In 1937 he published a pivotal treatise on thin elastic plates.

Using bilinearity, these equations become $\sum_{i=1}^{n} c_j \langle \phi_j', \phi_k' \rangle = \langle f, \phi_k \rangle$, which is precisely (36). Thus, for the (BVP) (24), the Rayleigh-Ritz and Galerkin methods coincide, and this is true for any choice of basis functions (not necessarily the piecewise linear basis functions). The next exercise for the reader will use a set of basis functions that does not depend on any particular partition, but rather comes from the so-called eigenfunctions of the BVP.[12]

EXERCISE FOR THE READER 10.13: Apply the Galerkin method to re-solve the BVP of Example 10.7 using the following 50 basis functions: $\varphi_k(x) = \sin(k\pi x)$, $k = 1, 2, \cdots, 50$. Compare the accuracy with that obtained in part (a) of Example 10.7.

For general BVPs, the Rayleigh-Ritz and Galerkin methods often, but not always, coincide. For this reason the nomenclature sometimes refers to the "Rayleigh-Ritz-Galerkin method." Both methods have been developed for a great many BVPs. The formulation of the Rayleigh-Ritz method in general is a bit more involved since it entails the determination of the appropriate functional for the analogue of Theorem 10.5 to be valid. Such problems usually fall under the classical area of the *calculus of variations*. We now present a brief outline for the Rayleigh-Ritz method for solving the following more general BVP whose DE will be a prototype for the elliptic PDE problems we shall contemplate Chapter 13. All of what follows in this outline can backed up theoretically with techniques similar to those used earlier to deal with the simpler problem (24); some of the details will be left to the exercises.

$$\text{(BVP)} \quad \begin{cases} -\big(p(x)u'(x)\big)' + q(x)u(x) = f(x), \ 0 < x < 1 \\ u(0) = 0, \ u(1) = 0 \end{cases} \tag{42}$$

Different boundary conditions and more general equations can be dealt with using modified functionals. The next exercise for the reader, however, shows how BVPs

---

[12] For the BVP $-u'' = f(x)$, $u(0) = u(1) = 0$, the associated **eigenfunctions** are nontrivial solutions of the BVP $-u'' = \lambda u$, $u(0) = u(1) = 0$ for some $\lambda > 0$. It can be shown that the totality of these eigenfunctions is as follows: $u_k(x) = \sin(k\pi x)$, $k = 1, 2, \cdots$ (see Exercise 24). The eigenfunctions are pairwise orthogonal: $\langle u_k, u_\ell \rangle = \delta_{k\ell}/2$ (where $\delta_{k\ell}$ denotes the Kronecker delta equaling 1 if the indices are equal, otherwise equaling 0), as are their derivatives (Exercise 24). Moreover, the eigenfunctions have the remarkable property that any function u satisfying the same boundary conditions and satisfying reasonable regularity assumptions (say if $u \in \mathcal{A}$) can be expressed as an infinite series of these eigenfunctions: $u(x) = \sum_{k=1}^{\infty} c_k u_k(x)$. In particular, solutions of such inhomogeneous BVPs have such eigenfunction expansions. Such eigenfunction expansion theory of ODE BVPs falls under the name of *Sturm-Liouville theory*. The analog for PDE BVPs is the theory of Fourier series. Both of these analytical techniques are covered extensively in many theoretically or analytically oriented textbooks. For references we cite [Str-92] and [Sni-99]. All of these properties make finite subsets of these eigenfunctions seem like very reasonable candidates for Rayleigh-Ritz and Galerkin methods; these types of Rayleigh-Ritz methods are often referred to as *spectral methods*.

with general Dirichlet BC's can be reduced to (42) using a simple change of variables. The exercises will examine some further modifications.

EXERCISE FOR THE READER 10.14: (a) Show that the following BVP,

$$\text{(BVP)} \quad \begin{cases} -\bigl(p(x)w'(x)\bigr)' + q(x)w(x) = f(x), & 0 < x < 1 \\ w(0) = \alpha, \ w(1) = \beta \end{cases}, \tag{42a}$$

can be reduced to the form (42) by making the following change of variables/function:

$$u(x) = w(x) - (1 - x)\alpha - \beta x.$$

(b) Show that the following BVP,

$$\text{(BVP)} \quad \begin{cases} -\bigl(p(t)w'(t)\bigr)' + q(t)w(t) = f(t), & a < t < b \\ w(a) = \alpha, \ w(b) = \beta \end{cases}, \tag{42b}$$

can be reduced to (42a) by making the following change of variables/function:

$$x = (t - a)/(b - a).$$

The analogue for Theorem 10.5 (for the Rayleigh-Ritz formulation) is the following theorem whose complete proof can be found in Section 7.2 in [Sch-73].

THEOREM 10.7: (*Rayleigh-Ritz Principle for a One-Dimensional BVP*) In the BVP (42):

$$\begin{cases} -\bigl(p(x)u'(x)\bigr)' + q(x)u(x) = f(x), & 0 < x < 1 \\ u(0) = 0, \ u(1) = 0 \end{cases},$$

suppose that the (known) functions $p(x)$, $q(x)$ and $f(x)$ are continuous and additionally that $p(x)$ is differentiable on the open interval $I = [0,1]$[13]. Also assume that $p(x) > 0$ and $q(x) \geq 0$ throughout I. Under these assumptions, the BVP has a unique solution which coincides with the unique minimizer of the functional

$$F(v) = \int_0^1 \Bigl[p(x)(v'(x))^2 + q(x)(v(x))^2 - 2f(x)v(x) \Bigr] dx, \tag{43}$$

over the set of admissible functions

$$\mathcal{A} = \{ v : [0,1] \to \mathbb{R} : v(x) \text{ is continuous}, \ v'(x) \text{ is piecewise continuous}$$
$$\text{and bounded, and } v(0) = 0, \ v(1) = 0 \ \}.$$

---

[13] Actually, the theorem still works under weaker conditions stated in [Sch-73]. The most natural setting for the Rayleigh-Ritz method is in the context of Sobolev functions. This topic is rather advanced, however, so we fix our ideas on the classical formulation. The interested reader may also consult the references [StFi-73] and [AxBa-84] for more sophisticated treatments on the subject.

We remind the reader that without these hypotheses, the BVP (42), in general, may have no solution—see Exercise 12 of Section 10.2 or Exercise 24 of this section.

If we use spaces $\mathcal{A}(\mathcal{P})$ of admissible functions spanned by the hat-functions determined by a partition $\mathcal{P}$ of $[0,1]$, the Rayleigh-Ritz method seeks to minimize the functional F evaluated at a typical element $v(x) = \sum_{i=1}^{n} c_i \phi_i(x)$ of $\mathcal{A}(\mathcal{P})$ (see (32)). A similar computation to what was given above (Exercise 12) will show that if we substitute this function into (43) and set each of the partial derivatives (with respect to the parameters c_i $(1 \le i \le n)$) equal to zero, we obtain the $n \times n$ linear system

$$Ac = b,$$

where the $n \times n$ **stiffness matrix** $A = [a_{ij}]$ has coefficients given by:

$$a_{ij} = \int_0^1 \left[p(x)\phi_i'(x)\phi_j'(x) + q(x)\phi_i(x)\phi_j(x) \right] dx, \tag{44}$$

and the $n \times 1$ **load vector** b has entries given by:

$$b_j = \int_0^1 \left[f(x)\phi_j(x) \right] dx. \tag{45}$$

As before, the stiffness matrix is clearly a tridiagonal symmetric matrix that can be shown to be positive definite. Thus there will be a unique solution of the linear system and so the method will always produce Rayleigh-Ritz approximations. There is also an error estimate analogous to Theorem 10.6 which states roughly that $|u_{\mathcal{P}}(x) - u(x)| \le C \|\mathcal{P}\|^2 \max_{0 \le x \le 1} |f(x)|$. Thus, we get the same type of error estimate (proportional to $\|\mathcal{P}\|^2$) as we had in the simpler introductory BVP. The proportionality constant C will of course depend on the data $p(x)$ and $q(x)$, but not on $u(x)$ or $f(x)$ (see [StFi-73] for details). The tridiagonal coefficients of the stiffness matrix in (44) can be simplified, using (33), as was done previously, to obtain (cf. with (38), (39)):

$$a_{ii} = \frac{1}{h_{i-1}^2} \int_{x_{i-1}}^{x_i} p(x)dx + \frac{1}{h_i^2} \int_{x_i}^{x_{i+1}} p(x)dx$$

$$+ \frac{1}{h_{i-1}^2} \int_{x_{i-1}}^{x_i} (x - x_{i-1})^2 q(x)dx + \frac{1}{h_i^2} \int_{x_i}^{x_{i+1}} (x_{i+1} - x)^2 q(x)dx, \text{ for } 1 \le i \le n; \tag{46}$$

$$a_{i,i+1} = \frac{-1}{h_i^2} \int_{x_i}^{x_{i+1}} p(x)dx + \frac{1}{h_i^2} \int_{x_i}^{x_{i+1}} (x_{i+1} - x)(x - x_i)q(x)dx, \quad \text{for } 1 \le i < n. \tag{47}$$

Evaluation of the integrals in (44) through (47) can be a time-consuming process in cases of fine partitions. In such cases where the coefficient functions p, q, and f are not too wildly behaved, it is a good idea to replace each of these functions by their piecewise linear approximation (piecewise linear splines) in the integrals. By Exercise 13(b), the local errors of such approximations are $O(h^2)$ on each of the corresponding intervals, provided that the function is $\mathscr{C}^2$, and this in turn implies $O(h^3)$ estimates for each of the integrals. We do one such approximation for the fourth integral in (46); the rest are done in a similar fashion and are left as Exercise 13(a). The piecewise linear approximation $Q(x)$ to $q(x)$ relative to the partition $\mathscr{P}$ of $[0,1]$ can be expressed quite simply using the hat functions as follows:

$$Q(x) = q(x_i)\phi_i(x) + q(x_{i+1})\phi_{i+1}(x), \qquad x \in [x_i, x_{i+1}].$$

Replacing this approximation for $q(x)$ in the last integral of (46) leads us to the following estimate:

$$\frac{1}{h_i^2}\int_{x_i}^{x_{i+1}}(x_{i+1}-x)^2 q(x)dx$$

$$\approx \frac{1}{h_i^2}\int_{x_i}^{x_{i+1}}(x_{i+1}-x)^2[q(x_i)\phi_i(x) + q(x_{i+1})\phi_{i+1}(x)]dx$$

$$= \frac{1}{h_i^2}\int_{x_i}^{x_{i+1}}(x_{i+1}-x)^2\left[q(x_i)\frac{x_{i+1}-x}{h_i} + q(x_{i+1})\frac{x-x_i}{h_i}\right]dx$$

$$= \frac{q(x_i)}{h_i^3}\int_{x_i}^{x_{i+1}}(x_{i+1}-x)^3 dx + \frac{q(x_{i+1})}{h_i^3}\int_{x_i}^{x_{i+1}}(x_{i+1}-x)^2(x-x_i)dx.$$

The two latter integrals are easily evaluated explicitly, for example:

$$\int_{x_i}^{x_{i+1}}(x_{i+1}-x)^2(x-x_i)dx \underset{\substack{\text{Subst.}\\ u=x_{i+1}-x}}{=} \int_{h_i}^{0}u^2(h_i-u)(-du) = \int_{h_i}^{0}[u^3 - h_i u^2]du$$

$$= \left[\frac{u^4}{4} - h_i\frac{u^3}{3}\right]_{h_i}^{0} = \frac{h_i^4}{12}.$$

In a similar fashion we find that $\int_{x_i}^{x_{i+1}}(x_{i+1}-x)^3 dx = \frac{h_i^4}{4}$. Putting these into the above estimate gives us that:

$$\frac{1}{h_i^2}\int_{x_i}^{x_{i+1}}(x_{i+1}-x)^2 q(x)dx \approx \frac{h_i}{12}[3q(x_i) + q(x_{i+1})].$$

Similar treatments for the remaining integrals appearing in (46) through (47) (see Exercise 13(a)) result in the following estimates:

$$a_{ii} \approx \frac{1}{2h_{i-1}}\left[p(x_{i-1})+p(x_i)\right]+\frac{1}{2h_i}\left[p(x_i)+p(x_{i+1})\right]$$
$$+\frac{h_i}{12}\left[q(x_{i-1})+3q(x_i)\right]+\frac{h_i}{12}\left[3q(x_i)+q(x_{i+1})\right], \tag{48}$$

for $1 \le i \le n$, and

$$a_{i,i+1} \approx -\frac{1}{2h_i}\left[p(x_i)+p(x_{i+1})\right]+\frac{h_i}{12}\left[q(x_i)+q(x_{i+1})\right], \tag{49}$$

for $1 \le i < n$. In the same fashion, the load vector coefficients are estimated as follows:

$$b_j \approx \frac{h_{j-1}}{6}\left[f(x_{j-1})+2f(x_j)\right]+\frac{h_j}{6}\left[2f(x_j)+f(x_{j+1})\right], \tag{50}$$

for $1 \le j \le n$.

Our next example will compare performance speed and accuracy of both of the above implementations of the Rayleigh-Ritz method for a specific BVP. It is possible to solve this BVP explicitly, so we will be able to make accurate estimates for the error. The explicit solution, however, is quite a mess. It can be derived using standard methods in differential equations (undetermined coefficients). To avoid having to even write it down, we use MATLAB's Symbolic Toolbox to compute the explicit solution but suppress its output.

EXAMPLE 10.8: Consider the following problem:

$$(\text{BVP}) \begin{cases} -u''(x)+6u(x)=e^{10x}\cos(12x) & 0<x<1 \\ u(0)=0, \ u(1)=0 \end{cases}.$$

(a) Use the Rayleigh-Ritz method with $n = 500$ equally spaced interior grid values to solve this BVP and plot the resulting approximation. Keep a record of the computing time it takes to determine the load vector and stiffness matrix coefficients.
(b) Repeat part (a), this time invoking the approximations (48) thru (50) for the integrals appearing in the Rayleigh-Ritz method.
(c) Compare both solutions of (a) and (b) with the exact solution as obtained using MATLAB's Symbolic Toolbox.

SOLUTION: The BVP given indeed fits the template of (42) with $p(x)=1$, $q(x)=6$, and $f(x)=e^{10x}\cos(12x)$.

Part (a): Here we have $h_i = \|\mathscr{P}\| = 1/501$ for each i, so we must compute the tridiagonal entries of the 501×501 stiffness matrix along with the 501 load coefficients using (45)–(47). The computations are done in a similar fashion to how the load vector coefficients were done in Example 10.7. Of the $1 + 4 + 2 = 7$ integrals appearing in (45) through (47), $1 + 2 + 1 = 4$ of them will need the "global variable" strategy in conjunction with MATLAB's numerical integrator quadl. The remaining three integrals have constant integrand ($p(x) = 1$) and so will be done directly. For (45) we use the fact that since the spacing is uniform, we have $\phi_i(x) = 1 - |x - x_i| / \|\mathscr{P}\| = 1 - 501|x - x_i|$ for $x_{i-1} \le x \le x_{i+1}$. Actually, because $p(x)$ and $q(x)$ are constant functions for this problem, the approximations (48) and (49) are indeed exact. Nonetheless, we will proceed to use the quadl integrator for these integrals so as to give a good impression of the extra expense in bringing in a more sophisticated tool. For the global variables x_{i-1}, x_i, x_{i+1} we will use the MATLAB notation: xim, xi, xip (p for plus, m for minus). The four needed M-files are as follows.

```
function y = frayritz10_8load(x)
global xi;
y=(1-501*abs(xi-x)).*exp(10*x).*cos(12*x);

function y = frayritz10_8stiff1(x)
global xim;
y=(x-xim).^2*6;

function y = frayritz10_8stiff2(x)
global xip;
y=(x-xip).^2*6;

function y = frayritz10_8stiff3(x)
global xi xip;
y=(xip-x).*(x-xi)*6;
```

Note that all of the intervals of integration have length $h = \|\mathscr{P}\| = 1/501$, so that each of the integrals in (46) and (47) with integrand p equals (since $p(x) = 1$) $\|\mathscr{P}\| = 1/501$. With these M-files stored, the following loop will use (45) thru (47) to construct the needed coefficients:

```
>> x=linspace(0,1,502); h=1/501; global xi xim xip;
>> tic, for i=2:501
xi=x(i); xim=x(i-1); xip=x(i+1);
b(i)=quadl('frayritz10_8load',xim,xip);
d(i)=2/h+1/h^2*quadl('frayritz10_8stiff1',xim,xi)...
+1/h^2*quadl('frayritz10_8stiff2',xi,xip);
%d(2:501) is diagonal of stiffness matrix
da(i)=-1/h+1/h^2*quadl('frayritz10_8stiff3',xi,xip);
%da(2:501) is superdiagonal of stiffness matrix (above),
%once we set da(501)=0 (after loop).
end, toc
```

→elapsed_time = 4.0360 (seconds)

```
>> db(3:501)=da(2:500);
>> db(2)=0; da(501)=0;
>> %db is subdiagonal of stiffness matrix (below)
```

As usual, we needed to properly format the sub/superdiagonals for input into the Thomas algorithm, which we apply next.

```
>> c1=thomas(da(2:501),d(2:501),db(2:501),b(2:501));
c1=[0 c1 0];
plot(x,c1)
```

The resulting plot, which as we will see turns out to be visually indistinguishable from that of the exact solution, is shown in Figure 10.15.

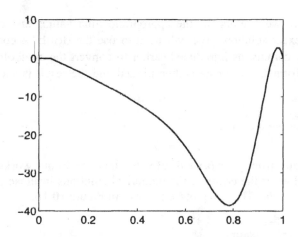

FIGURE 10.15: Plot of the solution of the BVP of Example 10.8.

Part (b): Using the estimates (48) through (50), it will be quite a simple (and quick) task to collect the needed coefficients. This can be accomplished with the following loop:

```
>> p=ones(502,1); q=6*p; x=linspace(0,1,502); f=exp(10*x).*cos(12*x);
>> h=1/501; %uniform step size

>> tic, for i=2:501
d(i)=1/(2*h)*(p(i-1)+2*p(i)+p(i+1))+h/12*(q(i-1)+6*q(i)+q(i+1));
%d(2:501) is diagonal of stiffness matrix
da(i)=-1/(2*h)*(p(i)+p(i+1))+h/12*(q(i)+q(i+1));
da(501)=0;
%da(2:501) is superdiagonal of stiffness matrix (above)
db(1)=-1/(2*h)*(p(i-1)+p(i))+h/12*(q(i-1)+q(i));
db(2)=0;
%db(2:501) is subdiagonal of stiffness matrix (below)
b(1)=h/6*(f(i-1)+4*f(i)+f(i+1));
%b(2:501) is load vector
end, toc
```

→elapsed_time = 0.0500 (seconds)

```
>> c2=thomas(da(2:501),d(2:501),db(2:501),b(2:501));
>> c2=[0 c2 0];
>>plot(x,c2)
```

The resulting plot is visually indistinguishable from the one in part (a), shown in Figure 10.15.

Part (c): The BVP is rather special in that an explicit solution can be written down. Labeling the symbolic solution as yexact and suppressing the long output, we can create it in a MATLAB session (provided the symbolic toolbox or student edition is being used) with the following command.

```
>> yexact=dsolve('-D2y+6*y=exp(10*t)*cos(12*t)', 'y(0)=0', 'y(1)=0');
```

We next create two vectors for the appropriate time values and corresponding values of the exact solution. We will need to use the double command along with the subs commands introduced earlier to convert the symbolic numbers to floating point format. The data is then plotted and the result is shown in Figure 10.15.

```
>> t=linspace(0,1,502);
>> Yexact=subs(yexact,t);
>> plot(t,Yexact)
```

With the variables from parts (a) and (b) still remaining in our workspace, we can easily obtain plots of the errors of the numerical solutions in those parts with the following commands. The two plots are shown in Figure 10.16.

```
>> plot(x,abs(c1-Yexact))
>> plot(x,abs(c2-Yexact))
```

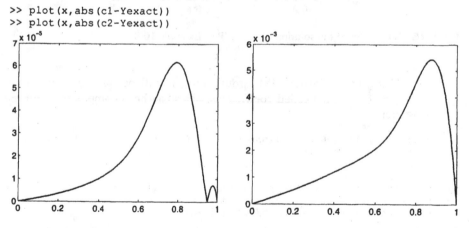

FIGURE 10.16: Plots for the errors of the two numerical solutions obtained in parts (a) (left) and (b) (right) of Example 10.8.

EXERCISE FOR THE READER 10.15: (a) Write an M-file called rayritz having the following input/output variables: [x,u]=rayritz(p, q, f, n). The program will implement the piecewise linear Rayleigh-Ritz method with (48) through (50) to solve the BVP (42):

$$\begin{cases} -\big(p(x)u'(x)\big)' + q(x)u(x) = f(x), & 0 < x < 1 \\ u(0) = 0, \ u(1) = 0 \end{cases}.$$

The first three inputs p, q, and f can represent the coefficient functions in the DE of (42), and the last input variable n denotes the number of interior grid points to use. A uniform grid is assumed. The output variables will be the domain and range vectors for the numerical solution.

(b) Starting with $n = 99$ interior grid points ($h = 1/100$), use this program to get a numerical solution $y1$ of the BVP in Example 10.8, then use 199 grid points ($h = 1/200$), getting a corresponding solution $y2$, and find the maximum absolute difference of the computed solutions on the common domain values. Now cut the gap in half again with $n = 399$, and get a corresponding solution $y3$ and look at the maximum absolute difference with the vector $y2$ at common domain points. Continue this process until the maximum absolute difference is less than 5×10^{-5}. Now (if you have access to the Symbolic Toolbox) compute the actual maximum error of this final solution compared to the exact solution in Example 10.8.

The Rayleigh-Ritz approximations we have obtained were all piecewise continuous but not differentiable. The versatility of the Rayleigh-Ritz method allows us, in fact, to use any sets of linearly independent functions as basis functions. The catch is that the resulting stiffness matrix should be reasonably well conditioned. Some different sets of basis functions will be examined in the exercises; see Exercise 6 for a problematic situation. The hat basis functions we used resulted in numerical approximations that were piecewise continuous but not differentiable. This lack of differentiability can be overcome by the use of more elaborate basis functions. One popular scheme is to use **cubic splines** for the basis functions; a typical one is shown in Figure 10.17.

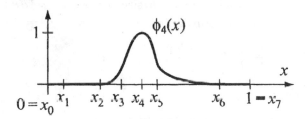

FIGURE 10.17: A cubic spline basis function. Unlike the piecewise linear hat functions of Figure 10.11, such basis functions are typically nonzero at three node points.

Each cubic spline basis function will have two continuous derivatives and thus so will the numerical approximations (since they are linear combinations of basis functions). The price we will need to pay for this extra smoothness in the numerical solutions is that the resulting stiffness matrix will typically have seven nonzero diagonals, rather than three, and the coefficients will be more complicated to compute. We proceed to outline an implementation of such cubic spline basis functions in the Rayleigh-Ritz method.

We restrict to the case of uniform grids and begin by defining the basic cubic spline function from which all other spline basis functions can be defined. This basic spline, which we denote by $BS(x)$, will be defined using the five nodes: $x_0 = -2$, $x_1 = -1$, $x_2 = 0$, $x_3 = 1$, and $x_4 = 2$ by the following requirements:

1. On each interval $[x_i, x_{i+1}]$ $(i = 0, 1, 2, 3)$, $BS(x)$ will be a polynomial of degree at most three.
2. $BS(x)$, $BS'(x)$, and $BS''(x)$ are each continuous at the node interfaces $x = x_1, x_2, x_3$.
3. $BS(\pm 2) = 0$, $BS(0) = 1$ (interpolation requirements).
4. $BS'(x)$, and $BS''(x)$ both equal zero at the endpoint nodes $x = x_0, x_4$.

EXERCISE FOR THE READER 10.16: (a) Show that the conditions (i) through (iv) above uniquely determine the function $BS(x)$ to be an even function ($BS(-x) = BS(x)$) in $\mathscr{C}^2(\mathbb{R})$ and specified by the following formula:

$$BS(x) = \begin{cases} [(2-x)^3 - 4(1-x)^3]/4, & \text{if } x \in [0,1], \\ (2-x)^3/4, & \text{if } x \in (1,2], \\ 0, & \text{if } x > 2, \\ BS(-x), & \text{if } x < 0. \end{cases} \tag{51}$$

(b) Get MATLAB to plot this function.

Using the basic spline function $BS(x)$, we can define our basis $\{\phi_i(x)\}_{i=1}^n$ functions for the BVP (42) on $[0,1]$ corresponding to a uniform grid $0 = x_0 < x_1 < \cdots < x_n < x_{n+1} = 1$ with mesh size $h = x_{i+1} - x_i = 1/(n+1)$. These functions are specified by the formulas below:

$$\phi_i(x) = \begin{cases} BS\left(\dfrac{x-h}{h}\right) - BS\left(\dfrac{x+h}{h}\right), & \text{if } i = 1, \\[2mm] BS\left(\dfrac{x-ih}{h}\right), & \text{if } i = 2,3,\cdots,n-1, \\[2mm] BS\left(\dfrac{x-nh}{h}\right) - BS\left(\dfrac{x-(n+2)h}{h}\right), & \text{if } i = n. \end{cases} \tag{52}$$

EXERCISE FOR THE READER 10.17: (a) Show that the basis functions $\{\phi_i(x)\}_{i=1}^n$, as specified in (52), form a linearly independent set of functions. Also, show that on each interval (x_i, x_{i+1}), ϕ_i is a polynomial of degree at most three and that ϕ_i, ϕ_i', ϕ_i'' are continuous at the endpoints x_i, x_{i+1}. Show that $\phi_i(x_i) = 1$, $\phi_i(x_j) = 0$ if $|i - j| \geq 2$ or $j = 0$ if $i = 1$, or $j = n + 1$ if $i = n$, and $\phi_i(x) = 0$ if there is such an x_j that lies between x_i and x.

(b) Using the value $n = 5$, get MATLAB to plot each of the five corresponding hat functions.

In order to implement these basis functions in the method, we will need to compute their derivatives. By the chain rule, these can be easily computed in terms of $BS'(x)$, which by simple computation is as specified below:

$$BS'(x) = \begin{cases} \frac{3}{4}[4(1-x)^2 - (2-x)^2], & \text{if } x \in [0,1], \\ -\frac{3}{4}(2-x)^2, & \text{if } x \in (1,2], \\ 0, & \text{if } x > 2, \\ -BS'(-x), & \text{if } x < 0. \end{cases} \tag{53}$$

Each of the " $BS(\bullet)$ " expressions in (52) will have, by the chain rule, derivative equal to $BS'(\bullet)/h$. Note also that since $\phi_i(x)$ and $\phi_i'(x)$ equal zero outside the interval $[x_{i-2}, x_{i+2}]$, it follows that the stiffness matrix entries

$$a_{ij} = \int_0^1 \left[p(x)\phi_i'(x)\phi_j'(x) + q(x)\phi_i(x)\phi_j(x) \right] dx \quad \text{(from (44))} \quad \text{will be zero if}$$

$|i-j| > 3$, and from this it follows that the stiffness matrix will be a banded matrix with (at most) seven bands. With these observations, it is a simple matter to incorporate the cubic spline Rayleigh-Ritz method into a MATLAB program. Examples will be left to the exercises. We close this section with a result on errors of the cubic spline Rayleigh-Ritz method, which shows it is often worth the extra work required over the basic piecewise linear scheme.

THEOREM 10.8: (*Errors in Cubic Spline Rayleigh-Ritz Approximations*) Suppose that the exact solution of the BVP (42)

$$\begin{cases} -\left(p(x)u'(x)\right)' + q(x)u(x) = f(x), & 0 < x < 1 \\ u(0) = 0, \ u(1) = 0 \end{cases}$$

is $\mathscr{C}^4([0,1])$ and the data $p(x)$, $q(x)$, and $f(x)$ satisfy the assumptions of Theorem 10.7. If $u_{\mathscr{P}}$ is the cubic spline Rayleigh-Ritz approximation for this problem corresponding to a partition $\mathscr{P}$ of $[0,1]$, then we have the following error estimate:

$$|u_{\mathscr{P}}(x) - u(x)| \le C \|\mathscr{P}\|^3 \max_{0 \le x \le 1} |u^{(4)}(x)| \quad \text{for each } x \text{ in } [0,1]. \tag{54}$$

The constant C is independent of $u(x)$.

For a proof of this theorem we refer to Section 7.5 of [StBu-92]. The key point is that the error estimate is proportional to $\|\mathscr{P}\|^3$, which is superior to the $\|\mathscr{P}\|^2$ estimates for the piecewise linear Rayleigh-Ritz method and for the finite difference method.

EXERCISES 10.5

1   For each of the following BVPs, perform the following tasks.
 (i) Use the piecewise linear Rayleigh-Ritz method with $n = 50$ equally spaced grid values to solve the BVP numerically and plot the results.
 (ii) Repeat part (i) with $n = 200$ equally spaced grid points.
 (iii) Repeat part (i) with $n = 500$ equally spaced grid points.
 In each part, first perform all integrals directly, and then repeat using the approximations (48)–(50) as needed. Compare performance times. When it is possible to compute the exact solution using the symbolic toolbox, or if one is given, plot the errors of each approximation obtained.

 (a) $(DE)\ u'' = e^{8x - [2(x-1)]^2} \cos(e^{8x})$ $(BC)\ u(0) = u(1) = 0$

 (b) $(DE)\ (e^{-3x}u')' - e^{-3x}u = 3\pi\cos(\pi x),\ (BC)\ u(0)$
 $= u(1) = 0;\ u_{\text{exact}}(x) = e^{3x}\cos(\pi x)$

 (c) $(DE)\ (2u')' + 12u = x^3,\ (BC)\ u(0) = u(1) = 0;\ u_{\text{exact}}(x) = (x^3 - x)/12$

2.   Repeat the instructions of Exercise 1 for each of the following BVPs:
 (a) $(DE)\ u'' = \cos(2x) + \sin(16x)/8,\ (BC)\ u(0) = u(1) = 0$

 (b) $(DE)\ ((1 + x^2)u')' = 2,\ (BC)\ u(0) = u(1) = 0;\ u_{\text{exact}}(x) = \ln(x^2 + 1)$.

 (c) $\begin{cases} (DE)\ -u'' + 400u = -400\cos^2(\pi x) - 2\pi^2\cos(2\pi x) \\ (BC)\ u(0) = u(1) = 0 \end{cases}$;
 $u_{\text{exact}}(x) = e^{20x}/(e^{20} + 1) + e^{-20x}/(e^{-20} + 1) - \cos^2(\pi x)$

3.   Repeat each part of Exercise 1 for each of the BVPs given, but this time choose the indicated number of interior nodes randomly, using the rand function.

4.   Repeat each part of Exercise 2 for each of the BVPs given, but this time choose the indicated number of nodes according to the properties of the coefficient and right-hand-side data.

5.   For each of the BVPs given below, use the piecewise linear Rayleigh-Ritz method in conjunction with the method of Exercise for the Reader 10.15 to numerically solve the BVP according to each of the following node deployments:
 (i) Use $n = 50$ equally spaced interior nodes. Repeat with each of $n = 200$ and $n = 500$.
 (ii) Use $n = 250$ randomly chosen interior nodes. Repeat with each of $n = 200$ and $n = 500$.
 (iii) Use $n = 250$ nodes deployed (without equal spaces) in a way that seems reasonable from given data. Repeat with each of $n = 200$ and $n = 500$.

 Whenever possible, graph the errors of each of these approximations.
 (a) The beam-deflection problem of Example 10.3.
 (b) The BVP of Exercise 3(a) of Section 10.2.
 (c) The BVP of Exercise 3(c) of Section 10.2.

6.   (*A Problematic Choice of Basis Functions*) Consider applying the Rayleigh-Ritz method to our
 model problem (24) $\begin{cases} -u''(x) = f(x),\ 0 < x < 1 \\ u(0) = 0,\ u(1) = 0 \end{cases}$ using the following bases: $\{\phi_i(x)\}_{i=1}^n$ where
 $\phi_i(x) = x^i(1 - x)$.
 (a) Use the Rayleigh-Ritz method with this basis and $n = 50$ to re-solve the (BVP) (24) of Example 10.7. How does the solution compare with the "exact" solution found in part (c) of that example?
 (b) Try to repeat using $n = 500$ basis functions. What happens?

(c) Show that $\langle \phi_i', \phi_j' \rangle = \dfrac{(i+1)(j+1)}{i+j+1} + \dfrac{(i+2)(j+2)}{i+j+3} - \dfrac{(i+1)(j+2)+(i+2)(j+1)}{i+j+2}$, for any i, j > 0.

(d) Make a plot of the condition numbers (use cond (A)) of the $n \times n$ stiffness matrix A as a function of n as n ranges from 2 to 100. Recall (Chapter 7) that large condition numbers make linear systems difficult to solve.

(e) How would matters change if we instead used $\phi_i(x) = x^i$ as our basis functions?

7. Repeat each part of Exercise 2 for each of the BVPs given, but this time adapt the Rayleigh-Ritz method using the basis functions $\varphi_k(x) = \sin(k\pi x)$, $k = 1, 2, \cdots, n$ of Exercise for the Reader 10.14.

8. Consider once again the (BVP) $\begin{cases} -u''(x) + 6u(x) = e^{10x}\cos(12x) \ \ 0 < x < 1 \\ u(0) = 0, \ u(1) = 0 \end{cases}$ of the Example 10.8.

    (a) Use the cubic spline Rayleigh-Ritz method with $n = 500$ equally spaced interior grid values to solve this BVP and plot the resulting approximation. Keep a record of the computing time it takes to determine the load vector and stiffness matrix coefficients.
    (b) Graph the error of the numerical solution by using the exact solution as in the last example.

9. Repeat each part of Exercise 2 for each of the BVPs given, but this time use the cubic spline Rayleigh-Ritz method. Compare the results (and errors, when possible) with the numerical solutions obtained in Exercise 2.

10. (*Natural Boundary Conditions*) This exercise will show how to develop the Galerkin method for BVPs with non-Dirichlet boundary conditions on the following model problem:
$$\begin{cases} -u''(x) = f(x), \ 0 < x < 1 \\ u(0) = 0, \ u'(1) = 0 \end{cases}.$$
For this problem we use the following for our admissible functions:
$$\mathcal{A}_1 = \{v : [0,1] \to \mathbb{R} : v(x) \text{ is continuous,}$$
$$v'(x) \text{ is piecewise continuous and bounded, and } v(0) = 0 \ \}.$$

Notice the only difference with this and the class (28) of the model problem (24) considered in the text is that this class has one less requirement: the condition $v(1) = 0$ is no longer essential.

(a) Use the DE and integrate by parts (as in step 3 of the proof of Theorem 10.5) to show that any solution of the BVP satisfies the corresponding PVW: $\langle u', v' \rangle = \langle f, v \rangle$ for all $v \in \mathcal{A}$. The converse is also true and so the PVW is equivalent to the BVP just as in Theorem 10.5. This gives rise to a Galerkin method for numerically solving the BVP, given any basis of a finite-dimensional subspace of $\mathcal{A}$. The fact that no boundary condition is required at $x = 1$ for admissible functions in this method has motivated the terminology of a **natural boundary condition** at $x = 1$ as opposed to an **essential boundary condition** like the one at $x = 0$. It is quite surprising that even though the natural boundary conditions force no conditions on the admissible functions, the solution of the PVW will automatically satisfy them.

(b) (*Piecewise Linear Galerkin Method*) Given a partition $\mathcal{P}$ of $[0,1]$, we let
$$\mathcal{A}_1(\mathcal{P}) = \{v : [0,1] \to \mathbb{R} : v(x) \text{ is continuous on } [0,1], \text{ linear on each } I_i \text{ and } v(0) = 0\}.$$

The hat functions $\phi_i(x)$ $(1 \le i \le n)$ need one more function to be added to form a basis of $\mathcal{A}_1(\mathcal{P})$. The function $\phi_{n+1}(x) \in \mathcal{A}_1(\mathcal{P})$ defined by $\phi_{n+1}(x_j) = 0, (j = 0, 1, \cdots, n)$ and $\phi_{n+1}(1) = 1$ will do the job. By substituting a linear combination of these $\sum_{i=1}^{n+1} c_i \phi_i(x)$ into the PVW, set up a linear system for the resulting Galerkin method.

(c) Apply the method using $n = 50$ equally spaced interior nodes to the BVP in case

$f(x) = e^{2x} \cos(\pi x)$. Compute the error by comparing with the exact solution (obtainable with the symbolic toolbox).

(d) Repeat part (c) with $n = 200$.

(e) What can be said in general about the stiffness matrix for this method (e.g., is it invertible, symmetric, positive definite)?

11. (*Natural Boundary Conditions*) Parts (a) through (e): Go through each part of Exercise 10 for

the BVP $\begin{cases} -u''(x) = f(x), \ 0 < x < 1 \\ u'(0) = 0, \ u(1) = 0 \end{cases}$, making changes where needed.

(f) What happens if we try to develop a similar method when both boundary conditions are natural: $u'(0) = 0, \ u'(1) = 0$?

12. Complete the justification of the approximations (48) through (50).

13. Suppose that $b - a = h$ and that $p(x)$ is a function on $[a,b]$ whose second derivative is continuous on $[a, b]$ (i.e., $p(x)$ is in the space $\mathscr{C}^2([a,b])$). Let $p_\ell(x)$ be the linear function which agrees with $p(x)$ at $x = a$ and $x = b$. Show that for any x in $[a,b]$, we have $|p_\ell(x) - p(x)| = O(h^2)$. Next use this to show that $\left| \int_a^b p_\ell(x) - p(x) \, dx \right| = O(h^3)$.

Suggestion: For Part (b), use the mean value theorem from calculus to find a number c in $[a,b]$ for which $p'(c) = p_\ell'(c)$. Next use Taylor's theorem to write $p(x) = p_\ell(x) + p''(z_x)(x-c)^2/2$ for any x in $[a,b]$, where z_x is an number between x and c. From this we get that $|p(x) - p_\ell(x)| \leq \max_{a \leq z \leq b} |p''(z)| (x-c)^2/2$ and the assertions readily follow.

14. Derive the Galerkin method for the BVP (41): $\begin{cases} -(p(x)u'(x))' + q(x)u(x) = f(x), \ 0 < x < 1 \\ u(0) = 0, \ u(1) = 0 \end{cases}$ by mimicking step 3 of the proof of Theorem 10.5. Does the method agree with the Rayleigh-Ritz method?

15. Suppose that $g(x)$ is a continuous function on $[0,1]$ that satisfies $\int_0^1 g(x)v(x)dx = 0$ for every $v \in \mathcal{A}$. Prove that $g(x) \equiv 0$ by providing more details to the following outline.

Sketch of Proof: Suppose that $g(x_0) > 0$ for some $x_0 \in (0,1)$. Then by continuity, $g(x) > 0$ for all x in some interval $(x_0 - h, x_0 + h)$. Let $v(x)$ be a hat function with $v(x_0) = 1$, $v(x_0 \pm h) = 0$. Show that $\int_0^1 g(x)v(x)dx > 0$ and this hat function is admissible. This contradiction shows that we cannot have such an $x_0 \in (0,1)$. Conclude similarly that there is no $x_0 \in (0,1)$ for which $g(x_0) < 0$.

16. The proof of Theorem 10.5 made use of one external theorem stating the existence of a solution of the (BVP) (24). In this exercise, you are to follow an outline to prove that part (c) of this theorem implies part (a) by using only the assumption that $u''(x)$ exists and is piecewise continuous and bounded whenever $u(x)$ is a solution of (PVW).

Write (PVW) in the form $\int_0^1 u'(x)v'(x)dx = \int_0^1 f(x)v(x)dx$ $(v \in \mathcal{A})$. Fix a function $v \in \mathcal{A}$ and integrate this by parts to obtain $\int_0^1 [u''(x) + f(x)]v(x)dx = 0$. Now use Exercise 16 to show the differential equation (24) must hold.

NOTE: Much error analysis for Rayleigh-Ritz methods depends on certain integral inequalities. A prototypical inequality of this sort is the so-called **Cauchy-Bunyakowski-Schwarz (CBS) inequality**, which reads as follows:

$$\left| \langle u, v \rangle \right| \le \left| \langle u, u \rangle \right|^{1/2} \left| \langle v, v \rangle \right|^{1/2} ,$$

(55)

valid for any functions u, v for which the inner product (25) is defined. In integral form (see (25)) the CBS inequality becomes:

$$\left| \int_0^1 u(x)v(x)dx \right| \le \left(\int_0^1 u(x)^2 dx \right)^{1/2} \left(\int_0^1 v(x)^2 dx \right)^{1/2} . ^{14}$$

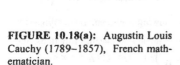

FIGURE 10.18(a): Augustin Louis Cauchy (1789–1857), French mathematician.

FIGURE 10.18(b): Viktor Yakovlevich Bunyakowsky (1804–1889), Russian mathematician.

FIGURE 10.18(c): Hermann Amandus Schwarz (1843–1921), German mathematician.

---

[14] The CBS inequality is a good example of an important mathematical result whose history is often subject to political bias. Cauchy was the first to discover a discrete version (for sums) of the inequality. Bunykowski was the first to discover, in 1859, the integral version of the CBS inequality as written above. Schwarz generalized Bunykowski's result some 25 years later to general inner products. Subsequent mathematical literature from each of the three countries usually attributes any version (from sums to general inner products) of the CBS inequality solely to their mathematical contributor. Thus, in French literature it is usually called Cauchy's inequality, etc. All three of these individuals were eminent mathematical figures in their respective countries. Cauchy began work in 1810 as a civil engineer, but his passion for mathematics kept him trying hard to land positions in mathematics. After numerous attempts he finally got one five years later. Cauchy's output was amazing—his complete works spanned over practically all areas of mathematics and filled 27 volumes. His textbooks were used for many years in most French universities. Despite his keen mathematical abilities, however, his strong religious positions and often criticism for his contemporaries made it difficult for him to retain desirable positions. Bunyakowski actually earned his doctorate under Cauchy in Paris in 1825. He then went to St. Petersburg where he spent most of his mathematical career. Schwarz originally entered what is now known as the Technical University of Berlin with the intention of earning a degree in chemistry. This school had the top German mathematics department at the time and the lessons of his mathematics teachers (including the famous analyst Karl Weierstrass (1815–1897)) led him to switch his major and eventually earn a doctorate in mathematics. Schwarz had a remarkable potential in blending analytical and geometrical methods that led him to discover many important results. After he took over Weierstrass's professorial position in 1892, however, he had already begun shifting his focus away from research being his main priority and his output decreased to a less remarkable level. At about this time, the main mathematics institute in Germany shifted from Berlin to Göttingen.

In manipulations with such integrals, it is often convenient to introduce the **norm** notation: $\|u\| = |\langle u,u \rangle|^{1/2} = \left(\int_0^1 u(x)^2 \, dx \right)^{1/2}$. Using this notation the Cauchy-Bunyakowski-Schwarz inequality takes on the more elegant form:

$$|\langle u,v \rangle| \leq \|u\| \, \|v\| .$$

For a further discussion of such concepts and, in particular, a proof of the Cauchy-Bunyakowski-Schwarz inequality, we refer the reader to any good book on analysis, for example [Ros-96] or [Rud-64]. The next few exercises will give examples of such uses of the CBS inequality.

17. Show that the function norm defined above satisfies the three vector norm axioms (see Chapter 7, equations (36A–C)). For simplicity, assume, in your proofs, that all functions are continuous on [0,1].

    (a) $\|u\| \geq 0$, $\|u\| = 0$ if and only if $u(x) = 0$ for all x in [0,1].

    (b) $\|cu\| = |c| \|u\|$, for any scalar c.

    (c) $\|u + v\| \leq \|u\| + \|v\|$ (triangle inequality).

    Suggestions: For an idea for part (a), see Exercise 15. For part (c) use the CBS inequality.
    Note: If we allow more general functions, such as piecewise continuous functions in some $A(\mathcal{P})$, then we have to change the condition in part (a) to $u(x) = 0$ for all x in [0,1] except for a possible finite set of exceptions.

18. (*Rayleigh-Ritz Approximations Have Errors of Minimum Internal Elastic Energy*) Recall that the internal elastic energy of an (admissible) function v was defined to be $(1/2)\langle v',v' \rangle =$ $(1/2)\int_0^1 (v'(x))^2 \, dx$. Follow the outline below to prove the following useful and interesting error estimate which shows that among all admissible (piecewise linear) admissible functions, the Rayleigh-Ritz approximation to the solution of the BVP (24) $\begin{cases} -u''(x) = f(x), \ 0 < x < 1 \\ u(0) = 0, \ u(1) = 0 \end{cases}$ is the best possible approximation if errors are measured by internal elastic energies. That is, if $u(x)$ is the solution of (24) and $u_{\mathcal{P}}(x)$ is the (piecewise linear) Rayleigh-Ritz approximation, both corresponding to a partition $\mathcal{P}$ of [0,1], then we have:

    $$\|(u - u_{\mathcal{P}})'\| \leq \|(u - v)'\| \quad \text{for all } v \in A(\mathcal{P}). \tag{56}$$

    (a) Show that $\langle (u - u_{\mathcal{P}})',v \rangle = 0$ for any $v \in A(\mathcal{P})$ by using the principle of virtual work..

    (b) For any $v \in A(\mathcal{P})$, note that $w \equiv u_{\mathcal{P}} - v \in A(\mathcal{P})$. Use (55) to write

    $$\langle (u - u_{\mathcal{P}})',(u - u_{\mathcal{P}})' \rangle = \langle (u - u_{\mathcal{P}})',(u - v)' \rangle .$$

    (c) Next, use the CBS inequality to obtain (56).

19. (*Comparison of Solution with Linear Interpolant*)
    (a) Let $\tilde{u}_{\mathcal{P}} \in A(\mathcal{P})$ be the (piecewise) **linear interpolant** of the solution u of (24), i.e., $v(x_i) = u(x_i)$ at each partition point (and $\tilde{u}_{\mathcal{P}}$ is linear between partition points). Let x be any number in [0,1] that lies between two partition points of $\mathcal{P}$: $x_j < x < x_{j+1}$. Use the mean value theorem from calculus to show why we can write
    $$\tilde{u}_{\mathcal{P}}(x) = u(c) + (x - c)u'(x) \quad \text{for some fixed number } c, x_j < c < x_{j+1}.$$
    (So c depends only on j, but not on the particular value of x.)
    (b) In the notation of part (a) use Taylor's theorem and only the assumption that u has a

continuous second derivative to show that for any $x, x_j < x < x_{j+1}$, we have

$$|u(x) - \tilde{u}_{\mathscr{P}}(x)| \le \frac{h_j^2}{2} \max_{x_j < x < x_{j+1}} |u''(x)|.$$

Next use the differential equation of (24) to translate this estimate to the form

$$|u(x) - \tilde{u}_{\mathscr{P}}(x)| \le \frac{\|\mathscr{P}\|^2}{2} \max_{x_j < x < x_{j+1}} |f(x)|,$$

and that this is valid for all x in $[0,1]$.

(c) By applying a similar analysis as done in parts (a) and (b) except now on the derivatives of the above two functions, obtain the following estimate for all x in $[0,1]$ (except the partition points at which $\tilde{u}_{\mathscr{P}}'(x)$ may not exist):

$$|u'(x) - \tilde{u}_{\mathscr{P}}'(x)| \le \|\mathscr{P}\| \max_{0 \le x \le 1} |f(x)|.$$

20. (*Error Estimate for Rayleigh-Ritz Approximation*) This exercise will provide an outline for using the estimates of the preceding exercises to obtain an estimate for the error of the Rayleigh-Ritz approximation. We will show that if $u_{\mathscr{P}}(x)$ is the Rayleigh-Ritz approximation corresponding to a partition $\mathscr{P}$ of $[0,1]$ of the BVP (24): $\begin{cases} -u''(x) = f(x), & 0 < x < 1 \\ u(0) = 0, & u(1) = 0 \end{cases}$, where $f(x)$ is a continuous function, then we have the following error estimate valid for all $x, 0 \le x \le 1:$ $|u_{\mathscr{P}}(x) - u(x)| \le \|\mathscr{P}\| \max_{0 \le x \le 1} |f(x)|.$

(a) Use (54) with v taken to be the linear interpolant $\tilde{u}_{\mathscr{P}}$ of Exercise 19, and then use the results of Exercise 19 to justify the following string of inequalities:

$$\|(u - u_{\mathscr{P}})'\| \le \|(u - \tilde{u}_{\mathscr{P}})'\| = \left(\int_0^1 [(u - \tilde{u}_{\mathscr{P}})'(x)]^2 dx \right)^{1/2}$$
$$\le \left(\int_0^1 [\|\mathscr{P}\| \max_{0 \le x \le 1} |f(x)|]^2 dx \right)^{1/2} \le \|\mathscr{P}\| \max_{0 \le x \le 1} |f(x)|$$

(b) Since $u_{\mathscr{P}}(0) = u(0) = 0$, we can write $u(x) - u_{\mathscr{P}}(x) = \int_0^x (u - u_{\mathscr{P}})'(x) dx$. Use this and the CDS inequality to justify the following string of inequalities, thereby completing the proof of Theorem 10.6.

$$|u(x) - u_{\mathscr{P}}(x)| \le \langle (u - \tilde{u}_{\mathscr{P}})', 1 \rangle \le \|(u - \tilde{u}_{\mathscr{P}})'\| \cdot \|1\| \le \|\mathscr{P}\| \max_{0 \le x \le 1} |f(x)|.$$

21. (*Refined Error Estimate for Rayleigh-Ritz Approximation Using Green's Functions*) In this exercise we will show that when the Rayleigh-Ritz method is to solve the BVP (24): $\begin{cases} -u''(x) = f(x), & 0 < x < 1 \\ u(0) = 0, & u(1) = 0 \end{cases}$, where $f(x)$ is a continuous function for some partition $\mathscr{P}$ of $[0,1]$, then the Rayleigh-Ritz solution $u_{\mathscr{P}}$ actually coincides with the linear interpolant $\tilde{u}_{\mathscr{P}}$ of Exercise 19. From this it will follow from the estimate of part (b) of Exercise 19 that the estimate of Theorem 10.6 is valid. The key is the introduction of so-called **Green's functions** for the BVP. For each interior node x_i of $\mathscr{P}$ the following Green's function:

$$G_i(x) = \begin{cases} (1 - x_i)x, & \text{for } 0 \le x \le x_i \\ x_i(1 - x), & \text{for } x_i \le x \le 1 \end{cases}.$$

(a) Show that $G_i(x) \in \mathcal{A}(\mathscr{P})$ and that for any $v \in \mathcal{A}$, we have: $\langle v', G_i' \rangle = v(x_i)$.

(b) Take $v = u - u_{\mathscr{P}}$ in part (a) and apply a result from one of the preceding exercises to show that $v(x_i) = 0$ and hence $u = u_{\mathscr{P}}$, as desired.

22.   Show directly that the BVP (24) $\begin{cases} -u''(x) = f(x), \ 0 < x < 1 \\ u(0) = 0, \ u(1) = 0 \end{cases}$ has a unique solution.

Suggestion: Integrate the DE once to get $u'(x) = \int_0^x -f(t)dt$ and once more to get

$u(x) = \int_0^x \int_0^t -f(s)\,ds\,dt + C$.

23.   Using the direct approach to solving the BVP (24) suggested in the preceding exercise, set up and execute a MATLAB code for solving the BVP in Example 10.7 once again. Arrange your parameters so that the total error of your numerical solution is no more than 10^{-4}.
       Suggestion: You may wish to try some different approaches using MATLAB's built-in integrator in conjunction with Simpson's rule or the trapezoidal rule (see any standard calculus textbook or [BuFa-01]) and perhaps even the Symbolic Toolbox if you have access to it.

24.   (a) Verify that the general solution of the DE $-u'' = \lambda u$ for $\lambda > 0$ is given by
       $u = C\cos(\sqrt{\lambda}x) + D\sin(\sqrt{\lambda}x)$ for arbitrary constants C and D.
       (b) Show that if we also require the boundary conditions $u(0) = u(1) = 0$, then the resulting
       BVP will only have nontrivial solutions if $\lambda = (k\pi)^2$, $k = 1, 2, \cdots$ and these (eigenfunctions)
       are $u_k(x) = \sin(k\pi x)$.
       (c) Prove the orthogonality relations for the eigenfunctions: $\langle u_k, u_\ell \rangle = \delta_{k\ell}/2$, where $\delta_{k\ell}$
       denotes the Kronecker delta.
       (d) Prove the following orthogonality relations for the derivatives of the eigenfunctions:
       $\langle u_k', u_\ell' \rangle = k\ell\pi^2 \delta_{k\ell}/2$.

Chapter 11: Introduction to Partial Differential Equations

11.1: THREE-DIMENSIONAL GRAPHICS WITH MATLAB

As mentioned in Chapter 8, **partial differential equations (PDEs)** are differential equations where the unknown function (solution) is a function of several variables (and so the derivatives will be partial derivatives). The subject of PDEs is probably the most vast branch of mathematics and we will be focusing mostly on PDEs with two independent variables. There are two main reasons for this restriction. First, many of the key aspects of the theory and numerical methods in partial differential equations are well represented in PDEs having two independent variables. Second, once we obtain (numerical) solutions of such PDE, they will be functions of two variables and we will be able to graph them using MATLAB's three-dimensional capabilities. Graphs are not feasible if there are more than two independent variables, as such graphs would require at least four dimensions. Most of our PDEs will arise from physical models, and it will be convenient to use two different sets of independent variables: either x (space) and t (time) or x and y (two space variables). The solutions of such PDEs will be functions of two variables: $z = f(x,t)$ or $z = f(x,y)$, and often the best way to understand such a function is by a graph. Such a graph will require three dimensions and it is customary to have the dependent variable's axis be vertical (just like for functions of one variable), so the two independent variables will have their axes span a two-dimensional plane that must protrude out of the paper (or screen) on which it is graphed. The graph of such a function will be a surface in the three-dimensional xyz-plane.

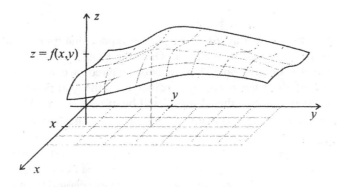

FIGURE 11.1: Mesh net graph of a function $f(x,y)$ of two variables.

In Figure 11.1, we have graphed a surface $z = f(x,y)$ with a mesh type of graph where images of equally spaced parallel lines of the form $x = c$ and $y = d$ under the function $f(x,y)$ are shown. In MATLAB, there are numerous ways of plotting and viewing the graph of a function of two variables. As with two-dimensional plots, the axis range can be specified. But to help with the third dimension, there are also other useful features such as viewing perspective (from where should we view the surface?), lighting, shading, and mesh grid lines. Color ranges can be used to vary with the height of the dependent variable. Often it is necessary to experiment with different versions of the same graph to find the best rendition of it for a particular purpose. The simplest way to understand how MATLAB does a plot of a function $z = f(x,y)$ is to consider the so-called mesh grid plots. These look a lot like the one in Figure 11.1, except only the points which are actually plotted are the grid points where one of the equally spaced horizontal lines $x = c$ meets a corresponding vertical line $y = d$—see Figure 11.2. Once the values of $f(x,y)$ are computed at these grid points, adjacent plotted points are connected by straight line segments.

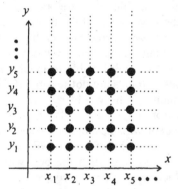

FIGURE 11.2: When MATLAB plots a function $z = f(x,y)$ of two variables, it will evaluate the function only for grid points (x,y) (the solid dots). Resolution can be improved by refining the grids (on both x- and y-axes). MATLAB has a function that will build the matrices of grid points from the corresponding vectors $x = (x_1, x_2, \cdots x_n)$ and $y = (y_1, y_2, \cdots y_m)$.

As with all things numerical in MATLAB, three-dimensional plots in MATLAB will be obtained from appropriate matrices. Once the x and y vectors (determining x- and y-coordinates of grid points) have been entered, we need to construct matrices X and Y of all of the x- and y-coordinates for the grid points. If x has n elements and y has m elements then X and Y will be of size $n \times m$. MATLAB has a convenient function meshgrid that will construct these meshed matrices for us.

`[X,Y]= meshgrid(x,y)` $\rightarrow$	For input vectors $x = (x_1, x_2, \cdots x_n)$ and $y = (y_1, y_2, \cdots y_m)$, meshgrid outputs two matrices X and Y each of size $n \times m$, which will be the corresponding x- and y-coordinates of all the points in the grid determined by the vectors x and y.

The next example will help us to understand how meshgrid is used to obtain plots of functions $z = f(x,y)$ in MATLAB. Since the three-dimensional plot functions and their options are best explained and illustrated by examples, we use the next example as a venue for illustrating several ways to plot surfaces and change the graphs. Help menus will introduce other options and features.

EXAMPLE 11.1: (a) Starting with the vectors $x = y = (-3,-2,...,2,3)$, use meshgrid to create two corresponding matrices of grid points for these vectors and then obtain a meshgrid plot of the surface

$$z = f(x,y) = \frac{3}{1+x^2+y^2} - \frac{10}{4+(x+1.5)^2+y^2}.$$

(b) Use vectors with 50 equally spaced elements over the same x- and y-ranges to obtain finer plots of this function and display plots used by several different MATLAB plotting tools.

SOLUTION: Part (a): Here are the MATLAB commands.

```
>> x=-3:3;      %x-values of grid
>> y=x;         %y-values of grid
>> [X,Y]=meshgrid(x,y) %creates matrices of grid point coordinates
```

```
X =                                Y =

  -3  -2  -1   0   1   2   3        -3  -3  -3  -3  -3  -3  -3
  -3  -2  -1   0   1   2   3        -2  -2  -2  -2  -2  -2  -2
  -3  -2  -1   0   1   2   3        -1  -1  -1  -1  -1  -1  -1
  -3  -2  -1   0   1   2   3         0   0   0   0   0   0   0
  -3  -2  -1   0   1   2   3         1   1   1   1   1   1   1
  -3  -2  -1   0   1   2   3         2   2   2   2   2   2   2
  -3  -2  -1   0   1   2   3         3   3   3   3   3   3   3
```

Corresponding entries in these grid matrices pair to give us (x,y) grid coordinates. The matrices are constructed in a way that MATLAB's plotting functions will be able to interpret, associate corresponding z-coordinates, and produce the plots.

```
>> Z=3./(1+X.^2+Y.^2)-10./(4+(X+1.5).^2+Y.^2); % construct
>>    %corresponding matrix Z of z-coordinates for grid points
>> mesh(X,Y,Z)  %produces the desired mesh grid plot.
```

The plot is shown in Figure 11.3 on the left.

Part (b): We refine the x- and y-values of the grid, reconstruct the meshgrid matrices and the corresponding Z-matrix, and then use mesh to get the higher-resolution plot on the right of Figure 11.3.

```
>> x=linspace(-3,3,50); y=x;
>> [X,Y]=meshgrid(x,y);
>> Z=3./(1+X.^2+Y.^2)-10./(4+(X+1.5).^2+Y.^2);
>> mesh(X,Y,Z)
```

CAUTION: If you just type mesh(Z) you will get the same surface graph; however, the numbers on the x- and y-axes will be the vector indices (so in this case will range from 1 to 50). It is, however, acceptable to use the original vectors x and y: mesh(x, y, Z), rather than the mesh matrices for x and y.

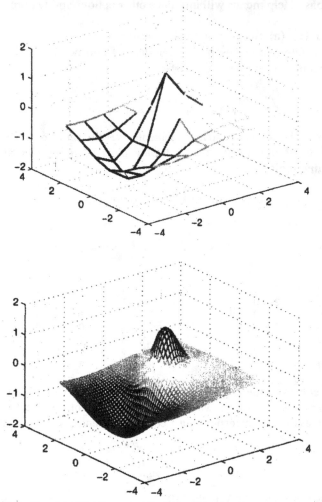

FIGURE 11.3: Two MATLAB meshgrid plots of the function
$$z = \frac{3}{1+x^2+y^2} - \frac{10}{4+(x+1.5)^2+y^2}$$
over the square $-3 \le x, y \le 3$. The first used only seven x- and y-grid coordinates, while the second used 50. The main plotting command was mesh(X, Y, Z); see Example 11.1. We point out some defaults in these plots: Colors are used with red for the highest y-values and blue for the lowest, this will come in handy for temperature plots. Also from our perspective, hidden parts of the mesh are not visible, and grid lines are shown.[1]

---

[1] To remove the grid lines simply enter grid off.

TABLE 11.1: Catalogue of several other useful plotting commands. In each we assume that x, y, X, Y, and Z have already been constructed as in Example 11.1.

MATLAB Plotting Function	Syntax and Options	Resulting Graph

meshc –>
 	
 (mesh plot with underlying contour lines)
	`>> meshc(X,Y,Z)`
	`>> axis('off')`
	%suppresses axes (and grids)	

surf →
 	
 (surface plot = mesh plot with squares between mesh lines filled in)
	`>> surf(x,y,Z)`
	`>> grid off` % leaves axes in but takes out all the extra grid lines; note that unlike with mesh plots, the mesh lines are now all black.
	
	`>> xlabel('x-axis'),`
	`ylabel('y-axis'),`
	`zlabel('z-axis')`
	%labels work like in 2 dim. plots.	

waterfall →
 	
 (surface plot using only y-grid lines; these are extended over edges of plot)
	`>> waterfall(x,y,Z)`
	`>> hidden off` %allows to see through all parts of surface
	
	`>>colormap([0 0 0])`
	%changes plot to black. 'n general the input vector [r g b] of 'colormap' has r measuring intensity of red, g intensity of green and b intensity of blue. All are numbers between 0 and 1.	

| contour →

(two-dimensional contour plot for surface) | >>c=contour(x,y,Z,12);
%we can specify the number of contour lines we want (here 12). Setting "c=" to this commands allows us to label contours with their z-values.
>>clabel(c,'manual')
%allows us to manually choose which contours to label using the mouse. | 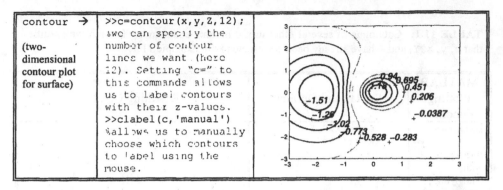 |

Other plotting commands include surfc(x,y,Z) (plots like surf and adds contour lines), surfl(x,y,Z,lvec) (surface plot with lighting; light source located at the vector lvec), and contour3(x,y,Z,n) (plots three-dimensional contour plots, the positive integer n is number of contour levels drawn). More can be learned about using these commands and their associated options by experimenting and reading on-line help menus. You can get a good general summary of MATLAB's 3D graphing tools (including some useful preprogrammed colormaps) by entering help graph3d

Once a three-dimensional plot has been created, it is often useful to view it from different perspectives. This can be accomplished using the view command:

| view(azimuth, elevation)

→ | Resets viewing angles for three-dimensional plots. The **azimuth** angle α and the **elevation** angle ε are measured as shown below in Figure 11.4. These angles are measured in degrees. The default values are $\alpha = -37.5$ and $\varepsilon = 30$.[2] |

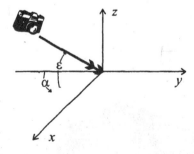

FIGURE 11.4: Measurement of the azimuth angle α and the elevation angle ε for use of the MATLAB command view(azimuth, elevation) to change perspectives on three-dimensional plots.

EXERCISE FOR THE READER 11.1: (a) Plot the function $z = \sin x \sin y \exp(-\sqrt{x^2 + y^2}/4)$ over the set $-5 \le x, y \le 5$ using 30 grid values for each of x and y and the surf command. Add labels to each of the three axes and set the grid off. Obtain views from three additional and significantly different viewing angles. Redo these plots with mesh. (b) Repeat part (a) for the "mountain pass function" $z = \sin(y + \cos x)$.

---

[2] The "3D Rotate" button ( ) on the MATLAB graphics window can be also used to change the viewing perspective.

MATLAB's three-dimensional graphics capabilities are quite vast and we will not even begin to talk about things like lighting and graphs of other types of surfaces (manifolds), as such a thorough treatment will not be needed for what we will be doing with PDEs. In closing this section, we mention two more useful plotting commands, although they are more relevant for our past work on three-dimensional orbits rather than for our subsequent work on PDEs.

plot3(x,y,z) →	Plots the curve in three-dimensional space stored by the vectors x, y, and z.

comet3(x,y,z) →	Plots the curve in three-dimensional space stored by the vectors x, y, and z in an animated fashion from start to finish. Final result is same as plot3.

This command allows us to plot orbits for three-dimensional first-order systems of ordinary differential equation which were dealt with in the last chapter or any three-dimensional parametric equations. We give an example of the latter.

EXAMPLE 11.2: Use both the plot3 and comet3 commands to plot the following decaying oscillating helix:

$$\begin{cases} x(t) = e^{-t/4} \cos(4t) \\ y(t) = e^{-t/4} \sin(4t) \quad \text{on the interval } 0 \le t \le 6\pi . \\ z(t) = \sin(t/2) + 1 \end{cases}$$

Next, use the view command to get a view from the top (i.e., the xy-plane projection of the curve).

SOLUTION:

```
>> t=0..003:6*pi;
>> x=exp(-t/4).*cos(4*t); y=exp(-t/4).*sin(4*t); z=sin(t/2)+1;
>> comet3(x,y,z)
>> plot3(x,y,z)
>> xlabel('x'), ylabel('y'), zlabel('z')
>> view(0,90)
```

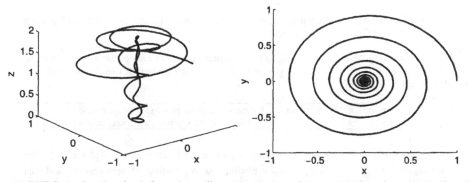

FIGURE 11.5: On the left, a three-dimensional plot of the oscillating decaying helix of Example 11.2; the right shows the top view of this decaying helix, which turns out to be just a spiral.

EXERCISES 11.1

1. (i) Using the mesh plotting tool and also using 50 grid values for both x- and y-coordinates, plot each of the following functions $z = f(x,y)$ over the indicated region. Use the colormap [0 0 0] so the display will be in black/white, and turn off the grid. Experiment with the three additional view settings: [0, 90], [90, –45], [135, 45]. (ii) Repeat these plots using 500 grid values for the x- and y-coordinates.

   (a) $z = x^2 + y^2$, $-5 \le x, y \le 5$

   (b) $z = 4x/(x^2 + y^2 + 1)$, $-4 \le x, y \le 4$

   (c) $z = \dfrac{xy(y^2 - x^2)}{x^2 + y^2}$, $-3 \le x \le 3$

   (d) $z = \exp(-x^2/2) + \exp(-2y^2)$, $-3 \le x, y \le 3$

2. (i) Using the surfc plotting tool and also using 50 grid values for both x- and y-coordinates, plot each of the following functions $z = f(x,y)$ over the indicated region. Turn off the axes. Experiment with the three additional view settings: [135 –45], [135, 0], [135, 25]. (ii) Repeat these plots using 500 grid values for the x- and y-coordinates.

   (a) $z = \sin x \sin y$, $-2\pi \le x, y \le 2\pi$

   (b) $z = \sin xy$, $-2\pi \le x, y \le 2\pi$

   (c) $z = \cos x \cos y \exp(-\sqrt{x^2 + y^2}/4)$, $-2\pi \le x, y \le 2\pi$

   (d) $z = 1/[1 + \sqrt{|y - \sin x|} + \sqrt{|x + \sin y|}]$, $-16 \le x, y \le 16$

3. (i) Using the waterfall plotting tool with the colormap [127/255 1 212/255] ("aquamarine") and with 50 grid values for both x- and y-coordinates, plot each of the following functions $z = f(x,y)$ over the indicated region. Turn off the grid and hidden defaults, and label the x-, y-, and z-axes as such. Experiment with the three additional view settings: [0 30], [0, 60], [0,90]. (ii) Repeat these plots using 500 grid values for the x- and y- coordinates.

   (a) $z = 20 - 2x^2 - 3y^2$, $-2 \le x, y \le 2$

   (b) $z = x^2 - y^2$, $-2.5 \le x, y \le 2.5$

   (c) $z = \sin\sqrt{x^2 + y^2}$, $-2\pi \le x, y \le 2\pi$

   (d) $z = \cos(x + y)\cos(3x - y) + \cos(x - y)\sin(x + 3y)\exp(-[x^2 + y^2]/8)$, $-\pi \le x, y \le \pi$

4. Redo tasks (i) and (ii) for each part (a) through (d) of Exercise 3 using the contour3 plotting tool, but this time use the default colormap.

NOTE: For analyzing three-dimensional graphics it is often useful to change the axes range of an existing plot (so to focus in on a particular portion). The axis command for two-dimensional plots extends naturally to this setting:

axis([xmin xmax ymin ymax zmin zmax]) →	Restricts a given plot to the range defined by $xmin \le x \le xmax, ymin \le y \le ymax, zmin \le z \le zmax$.

5. Obtaining first a good (or several good) surface plots of the function $z = -2y/(x^2 + y^2 + 1)$ over the region $-4 \le x, y \le 4$ (using a surface plotting tool and settings of your choice), and then making use of the axis and/or contour commands (repeatedly), find the x-, y- and z-coordinates of the maximum value of the function on this region. Your answer should be accurate to within a tolerance of 0.01 in each coordinate. Repeat for the minimum value.

6. Obtaining first a good (or several good) surface plots of the function
    $z = \cos(x + y)\cos(3x - y) + \cos(x - y)\sin(x + 3y)\exp(-[x^2 + y^2]/8)$, over the region
    $-\pi \le x, y \le \pi$ (using a surface plotting tool and settings of your choice), and then making use of
    the axis and/or contour commands (repeatedly), find the x-, y- and z-coordinates of the
    maximum value of the function on this region. Your answer should be accurate to within a
    tolerance of 0.001 in each coordinate.

7. Use plot3 and comet3 to graph each of the following space curves. Experiment with the
    view command to choose a "nice" view for your printout. Use the option axis('equal') to
    prevent axis distortion.

    (a) $x(t) = \dfrac{\cos t}{\sqrt{1 + \sin^2 2t}}$, $y(t) = \dfrac{\sin t}{\sqrt{1 + \sin^2 2t}}$, $z(t) = \dfrac{\sin 2t}{\sqrt{1 + \sin^2 2t}}$, $0 \le t \le 50$

    (b) $x(t) = \dfrac{\cos 3t}{\sqrt{1 + \sin^2 2t}}$, $y(t) = \dfrac{\sin 3t}{\sqrt{1 + \sin^2 2t}}$, $z(t) = \dfrac{\sin 2t}{\sqrt{1 + \sin^2 2t}}$, $0 \le t \le 50$

    (c) $x(t) = \dfrac{\cos 11t}{\sqrt{1 + \sin^2 5t}}$, $y(t) = \dfrac{\sin 11t}{\sqrt{1 + \sin^2 5t}}$, $z(t) = \dfrac{\sin 5t}{\sqrt{1 + \sin^2 5t}}$, $0 \le t \le 100$

    (d) $x(t) = \dfrac{\cos(\sqrt{2}t)}{\sqrt{1 + \sin^2 t}}$, $y(t) = \dfrac{\sin(\sqrt{2}t)}{\sqrt{1 + \sin^2 t}}$, $z(t) = \dfrac{\sin t}{\sqrt{1 + \sin^2 t}}$, $0 \le t \le 100$

    If you have done each of the above parts, point out some similarities and differences in the above
    plots. Which are periodic? Do you have any general comments/conjectures to make about
    parametric equations related to these?

8. Use plot3 to plot the solution of the Lorenz strange attractor of Example 9.7 in Section 9.4.
    Experiment with the view command to get one (or more) "nice" plots to include with your
    printout.

9. Redo Exercise 5 of Section 9.4 (the Rossler band), but now plot only the orbits in three
    dimensions using both comet3 and plot3. Experiment with the view command to choose a
    "nice" view for your printout.

10. The following ODE system has a very interesting orbit; an analysis of which is done in the
    article [Lan-84].
    $$\begin{cases} x'(t) = (z - 0.7)x - 3.5y, \\ y'(t) = 3.5x + (z - 0.7)y, \\ z'(t) = 0.6 + z - 0.33z^3 - (x^2 + y^2)(1 + 0.25z), \end{cases} \qquad \begin{array}{l} x(0) = 0.1 \\ y(0) = 0.03 \\ z(0) = 0.001. \end{array}$$
    Use the Runge-Kutta method with step size $h = 0.01$ and plot3 to graph the solution of the
    above system on the time range $0 \le t \le 100$. Repeat with step size $h = 0.005$. Use the
    axis(equal) command to remove any axis scale distortions.

11. Often, a three-dimensional surface plot is desired over a different shape than a rectangle, for
    example, over a circle or over an ellipse. This can be done using one of the surface plotting
    tools in MATLAB, but one needs to use polar coordinates to create the appropriate meshgrid
    matrices. For example, suppose that we wanted to plot the paraboloid $z = 1 - x^2 - y^2$ over the
    disk $x^2 + y^2 \le 4$. (a) Follow the following outline to create this plot:
    (i) Create vectors for r and theta:
    ```
    >>r=linspace(0,4,16);   theta=linspace(0,2*pi, 20);
    ```
    (ii) Create corresponding mesh matrices X and Y for these polar coordinates:
    ```
    >> X=r'*cos(theta); Y=r'*sin(theta);
    ```
    Note: Convince yourself that these will be mesh matrices.
    Optional: To see the meshgrid, type: `>>Z=zeros(size(X)); mesh(X,Y,Z)`

(iii) Create the Z-matrix and plot as usual:
```
>> Z=1-X.^2-Y.^2;   mesh(X,Y,Z)
```
Try some other options (like hidden off).

(b) Plot the graph of $z = \sqrt{\cos(x^2 + 2y^2)}$ over the disk $x^2 + y^2 \le 2$.

(c) Plot the graph of the paraboloid $z = 1 - x^2 - 3y^2$ over the ellipse $x^2 + 3y^2 \le 1$.

(d) Plot the graph of the function in part (d) of Exercise 3 over the disk $(x - \pi/2)^2 + (y - \pi/2)^2 \le 1$.

11.2: EXAMPLES AND CONCEPTS OF PARTIAL DIFFERENTIAL EQUATIONS

We begin our discussion with a natural problem about heat conduction. Consider a rod of length L that is insulated on the outer boundary but perhaps not at the ends.

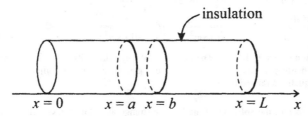

FIGURE 11.6: A one-dimensional rod of length L that is insulated along its cylindrical surface but not necessarily on the flat edges at $x = 0$ and $x = L$.

For this problem we are interested in the temperature of the rod as a function of the time t and the position x along the x-axis. We assume that the rod is very thin so that the temperature is constant for a given time throughout the cross-section. We call this function $u(x,t)$. The basic (and very intuitive) physical principle that we will use here, Fourier's Law, states that heat flows from hot places to cold places at a rate proportional to the temperature gradient. We will make this more precise shortly. The substance that the rod is made out of is relevant for knowing how quickly heat is transferred along the rod. To quantify this, we introduce

c = the **specific heat** of the substance of the rod = the heat energy needed to raise 1 unit of mass 1 unit of temperature.

The specific heat is a chemical/molecular property of the substance. Many partial differential equations in the natural sciences are based on a conservation principle, and the equation we will derive for $u(x,t)$ will be a good example of this. We let A denote the cross-sectional area of the rod and ρ denote the mass density (= mass per unit volume) of the rod. For values $x = a < x = b$ along the rod, we introduce

$$Q(t) = Q_{[a,b]}(t) = \text{the heat energy along the rod} = \int_a^b u(x,t)c\rho\,A\,dx\,. \tag{1}$$
$$\text{from } x = a \text{ to } x = b$$

By the conservation of energy, we have

$$\frac{dQ}{dt} = \text{flux term} + \text{source term.} \tag{2}$$

Letting

$F(x,t) = $ heat flux function $=$ heat energy passing through cross section at x per
unit area per unit time in the positive x-direction,

empirical physical laws imply that

$$F(x,t) = -\kappa \frac{\partial u}{\partial x}(x,t),$$

where $0 < \kappa$ is the **heat conductivity** of the material of the rod. Higher conductivities mean substances are better conductors of heat. For example, the specific heat of copper is about three times that of lead but less than half that of aluminum. We may now write,

$$\begin{aligned}
\text{flux term} &= -A[F(b,t) - F(a,t)] \\
&= A[\kappa u_x(b,t) - \kappa u_x(a,t)] \\
&= \int_a^b A[\partial/\partial x \{\kappa u_x(x,t)\}]dx,
\end{aligned}$$

where we used the fundamental theorem of calculus to write the last integral. Now, provided that u and u_t are continuous we may differentiate (1) under the integral sign (see any book on advanced calculus, e.g., [Rud-64]):

$$\frac{d}{dt}\left(\int_a^b u(x,t)c\rho\, A dx \right) = \int_a^b u_t(x,t)c\rho\, A dx.$$

Combining this with (2) and the above expression for the flux term gives:

$$\int_a^b [c\rho u_t(x,t) - \partial/\partial x \{\kappa u_x(x,t)\}]A dx = \text{source term.} \tag{3}$$

If there are no internal heat sources within the rod, then the integral in (3) must vanish for any a, b; $0 \le a < b \le L$. If the integrand is continuous, it follows (from the fundamental theorem of calculus) that the integrand must vanish identically. If κ is constant, then this leads us to the PDE:

$$u_t = k u_{xx}, \qquad u = u(x,t), \tag{4}$$

Where $k = \kappa/c\rho$ is called the **diffusivity** of the material. This equation is called the one-dimensional **heat equation** and it is also known as the one-dimensional **diffusion equation**. The reason for the latter terminology is that it more generally models the spatial (x) and temporal (t) spread of many different phenomena which obey a similar flux principle where higher concentrations spread to neighboring areas of lower concentrations at a rate proportional to the gradient. Indeed the diffusion equation has been used to successfully model spreads of populations from the molecular level (diseases) to larger-scale models of plants and animals and also the spread of other chemicals, gases, and drugs.

We will return to the one-dimensional heat equation shortly, but we first give some extensions of it and related PDEs. If the rod in Figure 11.5 has internal heat sources (e.g., chemical catalysts or electronic components), then we put

$$q_1(x,t,u) = \text{rate of production of heat energy per unit time per unit volume} \tag{5}$$
$$\text{at position } x.$$

Note that $q_1(x,t,u) > 0$ means there is a heat source at cross-section x and $q_1(x,t,u) < 0$ means there is a heat sink at cross section x. This function gives rise to the "source term" in (3) to be $\int_a^b q_1(t,x,u(x,t))A\,dx$ and so now (3) can be rewritten as

$$\int_a^b \left[c\rho u_t(x,t) - \partial/\partial x\{\kappa u_x(x,t)\} - q_1(x,t,u) \right] A\,dx = 0.$$

As before, if the integrand is continuous and κ is constant, we obtain the following PDE, known as the one-dimensional **heat equation with source term**

$$u_t = ku_{xx} + q, \qquad u = u(x,t), \tag{6}$$

where $q = q_1/c\rho$. A similar derivation would work also in two or three space dimensions to give us the corresponding two- and three-dimensional heat equations with or without source terms. For example, the **two-dimensional heat/diffusion equation** looks like:

$$u_t = k(u_{xx} + u_{yy}), \qquad u = u(x,y,t), \tag{7}$$

and similarly the three-dimensional analogue is

$$u_t = k(u_{xx} + u_{yy} + u_{zz}), \qquad u = u(x,y,z,t). \tag{8}$$

These PDEs model the spread of heat in a flat plate or a solid three-dimensional region and also population and chemical diffusions. If we are modeling a population that grows according the Malthusian or logistical model and also diffuses, we can model the temporal and spatial population distribution by the diffusion equation with the source term (ru for Malthusian growth or $ru(1 - u/K)$ for logistical growth).

The common expression appearing in the right side of each of the heat/diffusion equations ((4), (7), and (8)), obtained by adding all of the nonmixed second-order spatial partial derivatives of the function u (or just the second derivative if u has one space variable), is a very important differential operator known as the **Laplace operator** or the **Laplacian**, and is denoted Δu. It is named after the French

mathematician Pierre Simon de Laplace.[3]   Using the Laplace operator, all three of
the heat equations ((4), (7), and (8)) can be expressed in same appealing form:

$$u_t = k\Delta u. \tag{9}$$

If we consider **steady-state** heat
distributions, for which time is
no longer a relevant independent
variable, we get the so-called
Laplace equation

$$\Delta u = 0, \tag{10}$$

which could have any number of
space variables.   The Laplace
equation   arises   in   many
applications which, apart from
celestial   mechanics,   include
electromagnetism, fluid mechanics, and atomic physics. In
fact, the study of the Laplace

FIGURE 11.7(a):
Pierre   Simon   de
Laplace (1749–1827),
French mathematician
and astronomer.

FIGURE 11.7(b):
Joseph Louis Lagrange
(1736–1813), French/
Italian mathematician.

equation alone is such a fertile area of mathematics that it has its own name,
"potential theory," and this field has developed into a major area of mathematics.

To introduce some concepts, we return now to the one-dimensional heat equation
(4) (or (9)) for the heat distribution on the rod in Figure 11.5.   Just like with
ordinary differential equations, a PDE will in general have infinitely many
solutions. In fact, the variety of solutions for a PDE is usually much greater than
that for an ODE, so much so that it is often not feasible to write down the general

---

[3] Laplace and Lagrange worked on many of the problems of mechanics introduced by Euler and were
among the first mathematicians to successfully develop theories of partial differential equations
motivated by solving such celestial problems.   Laplace wrote a massive five-volume treatise,
Mécanique Céleste, which contained a plethora of results. Laplace had unfortunately neglected to
credit any of the theory in his work to any other scientists and this created some problems for him. At
one point, Napoleon had criticized him for not even having credited God as the creator of the universe
in his definitive work (something that was expected to be done in all significant scientific works during
this era). Laplace wittily replied with the now-famous retort, "Sir, I had no need for this hypothesis."
Actually in his treatise, Laplace systematically developed potential theory which has applications in
many other fields apart from mechanics. Ironically, the Laplace equation was actually first introduced
by Lagrange, but Laplace went so far with his own extensions that the equation has been named in his
honor.
  Lagrange grew up and was educated in Turin, Italy and later accepted, by a personal offer from King
Frederick the Great, a position at the Berlin Academy (which Euler vacated when he went to Saint
Petersburg). After 20 years at this post, Lagrange moved to Paris. In his famous work, *Mécanique
Analytique*, Lagrange unified the theory of mechanics in a way that made analysis much easier and
"coordinate free." Using this approach he was able to do significant work on the "three-body problem"
which won him, in 1764, the Grand Prize of the French Academy of Sciences.   Lagrange is also
credited for having invented the metric system. After his first wife died, he was remarried at age 56 to a
teenage daughter of a friend of his and he became very conscious of his health.   He never drank
alcohol and became a vegetarian. To honor his scientific accomplishments, Lagrange was buried in the
majestic Panthéon (along with literary greats Voltaire, Victor Hugo and Émil Zola) in Paris.

solution of a PDE explicitly. The next example gives a sample supply of solutions for the one-dimensional heat equation.

EXAMPLE 11.3: Show that each of the following functions $u(x,t)$ solves the heat equation (4) $u_t = ku_{xx}$, where $k > 0$ is a constant. Here a and b can be any real numbers.

(a) $u(x,t) = ax + b$

(c) $u(x,t) = e^{-kt}[ae^{\sqrt{k}x} + be^{-\sqrt{k}x}]$

(b) $u(x,t) = e^{-kt}[a\cos(\sqrt{k}x) + b\sin(\sqrt{k}x)]$

(d) $u(x,t) = \exp[-(x-a)^2/4kt]/\sqrt{t}$

SOLUTION: Each part is just a computation and comparison of partial derivatives, so we do only part (d) (the most complicated one) and leave the rest as an exercise. Writing u as $\exp[-(x-a)^2 t^{-1}/4k] \cdot t^{-1/2}$, the chain and product rules give that

$$u_t = \exp[-(x-a)^2 t^{-1}/4k](x-a)^2 t^{-2}/4k \cdot t^{-1/2}$$
$$- \exp[-(x-a)^2 t^{-1}/4k] \cdot (-1/2)t^{-3/2}$$
$$= \exp[-(x-a)^2 t^{-1}/4k] \cdot t^{-3/2}\{(x-a)^2 t^{-1}/4k + 1/2\}.$$

In the same fashion,

$$u_{xx} = \partial/\partial x\{u_x\} = \partial/\partial x\{\exp[-(x-a)^2/4kt] \cdot t^{-1/2} \cdot (-2)(x-a)/4kt\}$$
$$= \exp[-(x-a)^2/4kt] \cdot t^{-1/2} \cdot 4(x-a)^2/(4kt)^2$$
$$+ \exp[-(x-a)^2/4kt] \cdot t^{-1/2} \cdot (-2)/4kt$$
$$= \exp[-(x-a)^2/4kt] \cdot t^{-3/2}\{(x-a)^2 t^{-1}/4k^2 - 1/2k\}.$$

Comparing these two expressions we clearly have $u_t = ku_{xx}$ as desired.

Another important property of the heat/diffusion equation is that it is **linear**, as is the Laplace equation. The definition of linear a PDE is similar to that for an ODE; it is tantamount to the requirement that each term involving the unknown function (or one of its partial derivatives) must have only one such appearance of the function to the first power multiplied by some function of the independent variables (i.e., terms like $\sin u$, u^2, uu_x, $4/u_{xx}$ are not allowed). Since we will mostly be working with second-order PDEs with two independent variables, we write down the most general linear PDE of this form (with independent variables x and y):

$$a(x,y)u_{xx} + b(x,y)u_{xy} + c(x,y)u_{yy} + d(x,y)u_x$$
$$+ e(x,y)u_y + f(x,y)u = q(x,y). \tag{11}$$

Here $u = u(x,y)$ is the unknown function and a, b, c, d, e, f, q are known functions of the independent variables x and y. (Of course for the heat/diffusion equation, we would replace y with t). If the function $q(x,y)$ is zero, then the linear equation (11) is further said to be **homogeneous**. Linear equations are important because they are often more amenable to numerical methods (recall the similar situation with BVPs of Chapter 10). Also, as we had seen in Chapter 10 for linear homogeneous ODEs, linear homogenous PDEs also satisfy the following **superposition principle**.

THEOREM 11.1: (*Superposition Principle*):[4] If $u_1, u_2, \cdots$ are solutions of a linear homogenous PDE and $c_1, c_2, \cdots$ are constants, then $c_1 u_1 + c_2 u_2 + \cdots$ is also a solution of the PDE.

Sketch of Proof: We outline the proof only for the case of the second-order equation (11) with $q(x,y) = 0$ (the proof in the general case is just more writing but uses the same ideas). Furthermore, since the main ideas were already encountered in Section 10.1, we will leave similar parts of this proof as exercises. Let's rewrite the left side of (11) as the operator $L[u]$, thus (11) can be written as $L[u] = 0$. The main idea is to show that $L[u]$ is a **linear operator** in u; this means that the following two conditions hold for functions u_1 and u_2 which have second partial derivatives (so L can be computed) and for a constant c_1 :

$$\text{(i) } L[u_1 + u_2] = L[u_1] + L[u_2], \text{ and } \text{(ii) } L[c_1 u_1] = c_1 L[u_1].$$

We will leave the proofs of (i) and (ii) as exercises. From (i) and (ii) we can easily get the superposition principle; for example for a two-term sum, supposing that u_1 and u_2 are solutions of $L[u] = 0$, we have,

$$L[c_1 u_1 + c_2 u_2] = L[c_1 u_1] + L[c_2 u_2] = c_1 L[u_1] + c_2 L[u_2] = c_1 0 + c_2 0 = 0 ,$$

$$\text{by (i)} \qquad \text{by (ii)} \quad \text{since } u_1, u_2 \text{ are solutions}$$

which proves that $c_1 u_1 + c_2 u_2$ is also a solution. This proves the superposition principle with two terms. The general case now follows by induction.

EXERCISE FOR THE READER 11.2: Prove that the operator $L[u]$ defined by the left side of (11) is a linear operator, i.e., that it satisfies (i) $L[u_1 + u_2] = L[u_1] + L[u_2]$, and (ii) $L[c_1 u_1] = c_1 L[u_1]$.

Second-order PDEs are usually classified into three major types: elliptic, parabolic, and hyperbolic. Many common methods and concepts can be developed for all second-order (even nonlinear) PDEs in one of these three classes and, furthermore, PDEs of different types usually have some significant differences. We give the classification for the second-order linear PDEs of form

---

[4] The sum in this theorem is intended as a finite sum; however, with extra hypotheses the superposition remains true for certain infinite sums. This leads to Fourier series solutions of linear PDEs, which is an important topic in many theoretical PDE courses. Actually solving a PDE problem in terms of Fourier series still leaves the numerical problem of evaluating the (infinite) Fourier series to within a tolerated maximum error. Although this could be worked into MATLAB routines, there are more effective numerical schemes, so we will not go further with this approach.

(11). The classifications are entirely in terms of the coefficients a, b, c of the highest (second) order derivatives of the unknown function u: The PDE (11)

$$a(x,y)u_{xx} + b(x,y)u_{xy} + c(x,y)u_{yy} + d(x,y)u_x + e(x,y)u_y + f(x,y)u = q(x,y)$$

is said to be **elliptic** if $b^2 - 4ac < 0$,

parabolic if $b^2 - 4ac = 0$,

and **hyperbolic** if $b^2 - 4ac > 0$.

Generally speaking, elliptic PDEs describe physical processes that are in a steady-state and so do not depend on time, parabolic PDEs describe physical processes (such as diffusion of heat or a gas) which evolve toward a steady-state equilibrium, and hyperbolic PDEs describe time-dependent physical processes (such as motion of waves) which are not tending to settle into a steady-state. These terms are used throughout the subject of PDEs and have formulations for higher-order as well as nonlinear PDEs. The type of a linear PDE is entirely determined by looking at the so-called **discriminant** of the coefficients, $b^2 - 4ac$. At each point, (x,y), the PDE (11) will be of exactly one of these three types; however, it is possible for the type to change when (x,y) lies in different regions in the plane.

EXAMPLE 11.4: Show that the one-dimensional heat/diffusion equation $u_t = ku_{xx}$ is a parabolic PDE, that Laplace's equation $\Delta u = u_{xx} + u_{yy} = 0$ is elliptic, and that the one-dimensional **wave equation** $u_{tt} = c^2 u_{xx}$ is hyperbolic.

SOLUTION: If we put each of these three equations in the form (11), we see that for the heat/diffusion equation, $a = k$, $b = c = 0$, so $b^2 - 4ac = 0$ and thus it is parabolic. For Laplace's equation $a = c = 1$, $b = 0$, thus $b^2 - 4ac = -4 < 0$ so it is elliptic. Finally, for the wave equation, $a = 1$, $b = 0$, $c = -1$, so $b^2 - 4ac = 4 > 0$ and it thus is hyperbolic.

We will say more about the wave and hyperbolic equations later. We note that the Tricomi PDE $yu_{xx} + u_{yy} = 0$ has $a = y$, $b = 0$, $c = 1$ and so the discriminant $b^2 - 4ac = -4y$ shows the Tricomi PDE to be hyperbolic when $y < 0$, parabolic when $y = 0$, and elliptic when $y > 0$.

In order to specify a unique solution for a PDE, certain auxiliary conditions must also be specified. The acceptable types of auxiliary conditions that will result in existence and uniqueness theorems are varied and different for each type of PDE. If the type of auxiliary conditions given with a certain PDE will always result in existence and uniqueness of solutions for the PDE problem, we say that the PDE problem is **well-posed**. If the auxiliary conditions are either too demanding (so no solution will exist—nonexistence) or too lax (so many solutions exist—nonuniqueness), then we say that the PDE problem is **ill-posed**. Most problems that arise in applications are well posed. We proceed now to give some auxiliary conditions that will result in well-posed problems for the one-dimensional heat/diffusion equation (with or without source term) and then for the two-dimensional Laplace equation.

We begin with the heat equation and the model being the temperature of the rod in Figure 11.5. In order to know the temperature function $u(x,t)$ at all times $t \geq 0$ and at all cross sections x, $0 \leq x \leq L$, we will firstly need to know the temperature distribution on the rod at time $t = 0$ (initial temperature distribution): this is the function $u(x,0) = f(x)$ (a function of one variable). Also, we will need to know what is going on at the ends of the rod. We will call this information the **boundary conditions (BCs)**. There are a number of acceptable (and physically significant) boundary conditions which give rise (along with the heat PDE and initial temperature distribution) to well-posed problems. Table 11.2 gives some of the more important ones:

TABLE 11.2: Boundary conditions for the one-dimensional heat equation $u_t = ku_{xx}$, which, when given along with the initial temperature distribution $u(x,0) = f(x)$, result in well-posed problems.

Type of BC	Mathematical Equations	Illustration
Constant temperatures at the boundary, maintained by temperature reservoirs (heaters/coolers) at each boundary.	$u(0,t) = A$, and $u(L,t) = B$ for all $t \geq 0$	
Insulated boundaries	$u_x(0,t) = 0$ and $u_x(L,t) = 0$ for all $t \geq 0$	Same picture as above, but ends are insulators rather than temperature reservoirs.
Periodic boundary conditions. This is valid, in particular, when the rod is a loop.	$u(0,t) = u(L,t)$ and $u_x(0,t)$ $= u_x(L,t)$ for all $t \geq 0$	

To give some acceptable boundary conditions for the two-dimensional Laplace equation we will again refer to the model of $u(x,y)$ being a steady-state temperature distribution (time independent) of some thin plate in the two-dimensional xy-plane. We will denote the region as D and for simplicity here we let D be the rectangle $D = \{(x,y): \ 0 \leq x \leq a, 0 \leq y \leq b\}$. The most common boundary conditions that result in a well-posed problem are where the temperatures on the boundary are specified. This type of boundary condition is often called a **Dirichlet boundary condition** and is illustrated in Figure 11.8. What this means is that if the temperature is steady-state on a plate (does not change with time) and we know the temperature on the edges of the plate, then the temperature will be completely determined at every point inside the plate. This can be proved and is made plausible by physical or thermodynamic principles. Another common problem results from the Laplace equation on a region with the boundary being insulated. For the rectangle of Figure 11.8, this would correspond to the following boundary conditions: right edge: $u_x(a, y) = 0$; top edge: $u_y(x, b) = 0$; left edge: $u_x(0, y) = 0$; and bottom edge: $u_y(x, 0) = 0$.   Such boundary conditions are often called **Neumann boundary conditions**.   The solutions of such Neumann problems are not unique, since any constant function can be added to a solution to yield a different solution.

Look at Figure 11.8 and convince yourself of why each of these conditions corresponds to zero temperature gradients across the boundary edges. In fact, it is even permissible to stipulate that some parts of the boundary be given Dirichlet conditions and others be given Neumann conditions. Even when the interfaces of the adjacent parts of the boundary have discontinuities (breaks) in the boundary conditions, we will still have a well posed problem.   This could correspond, for example, to some parts of the boundary being insulated and others being kept at certain temperatures. Under very general circumstances, the well-posed problems that we have specified to the heat and Laplace equations also give well-posed problems for general parabolic and elliptic equations respectively.

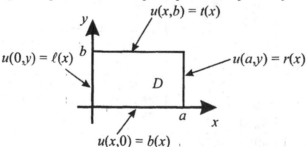

FIGURE 11.8: Dirichlet boundary conditions for the two-dimensional Laplace equation $u_{xx} + u_{yy} = 0$ are specified on each of the four sides making up the boundary of a rectangular domain in the plane. If the four functions specified on each edge are continuous, then they give a well-posed problem for the Laplace equation.

Solutions of the Laplace equation $\Delta u = 0$ (in any number of space dimensions) are known as **harmonic functions**. They include an incredibly vast collection of functions and are the subject of the very fruitful field of **potential theory**. Many aspects of potential theory have natural extensions to elliptic equations (even for nonlinear PDE). The most complete general reference on this is [GiTr-83], which is quite an advanced textbook. For example, any complex analytic function (the subject of the field of complex variables) will have harmonic functions as its real and imaginary parts. A fundamental result of potential theory is that harmonic functions satisfy the **maximum principle**, which roughly states that the maximum (or minimum) value of a harmonic function on any given (closed) bounded region in the space variables (like a rectangle) must be attained on the boundary and, furthermore, if the maximum is attained also at a point inside the boundary of the region then the harmonic function has to be a constant. In particular, applied to our steady-state temperature model for the rectangle, this says that the hottest spot on the rectangle must be on one of the edges. Again, this can be corroborated from thermodynamic principles.

EXERCISE FOR THE READER 11.3: Which of the following functions is harmonic?

(a) $u(x,y,z) = 1 + x + 2y + 3z$

(c) $u(x,y) = x^2 - y^2$

(b) $u(x,y) = x^2 + y^2$

(d) $u(x,y) = \log(r)$, $r = x^2 + y^2$

Since elliptic equations are the best understood and behaved, we will begin our numerical methods for solving PDEs with elliptic equations and this will be done in the next section. In the next chapter, we will give related methods for parabolic and hyperbolic equations and so, in particular, we will hold off until the next chapter to give some general comments about the wave equation and hyperbolic equations.

EXERCISES 11.2

1. For each of the PDEs given below, indicate its order and state whether it is a linear PDE. For those which are second-order linear and have two independent variables, state the type. If the type changes for different values of the independent variables, indicate precisely how the type varies with the independent variables.

   (a) $u_x + tu_t = e^t$, $u = u(x,t)$

   (c) $u_{xxyy} - u_{zzzz} = 0$, $u = u(x,y,z)$

   (b) $u_{xx} - 2u_{tt} + u_{tx} + t^2 u_t = u$, $u = u(x,t)$

   (d) $\Delta(\Delta u) = 0$, $u = u(x,y)$

   Note: The operator on the left side of the PDE of (d) is called the **biharmonic operator** and its solutions are called **biharmonic functions**.

2. Repeat Exercise 1 for each of the following PDEs:

   (a) $u_x + e^u = t$, $u = u(x,t)$

   (c) $u_{xxx} + u_{xx}u_{yy} = 0$, $u = u(x,y)$

   (b) $u_{xx} + u_{yy} = u_{zz}/(1+t^2)$, $u = u(x,y,z)$

   (d) $u_{xx} = (u_{yy} + u_y + u)\sin x$, $u = u(x,y)$

3. Determine which of the following functions solve the PDE $\Delta u = 2u$:

   (a) $\sin x \cos y$

   (c) $\exp(\sqrt{3}x)\sin x$

   (b) $\exp(x+y)$

   (d) $\arctan(x+y^2)y$

4. Determine which of the following functions solve the wave equation $u_{tt} = u_{xx}$:

(a) $x + 2t + 3$ (c) $\exp(x+2t) - \exp(2x-t)$

(b) $\exp(x+t) - \exp(x-t)$ (d) $f(x+t)$, f = any twice differentiable function.

5.   TRUE or FALSE?: If $u(x,y)$ and $v(x,y)$ are harmonic functions, then so is $u(x,y) + v(x,y)$. Either indicate why it is true or give a counterexample if it is false.

6.   TRUE or FALSE?: If $u(x,y)$ is a harmonic function, then so is $u(x,y)^2$. Either indicate why it is true or give a counterexample if it is false.

7.   The function $u(x,t) = \exp[-(x-a)^2/4t]/\sqrt{4\pi t}$, which is a solution of the one-dimensional heat equation $u_t = u_{xx}$ (this was seen in Example 11.3; here $k = 1$ and we are multiplying the function given there by a constant, so it will still satisfy the heat equation), is called the **fundamental solution** of the heat equation. It corresponds to the solution of the heat problem on a very long rod (mathematically think of an infinite rod) with the initial heat distribution being a very hot heat blast focused all at the section $x = a$. To get some idea of how this temperature distribution changes with time, do the following:
(a) Using $a = 1$, get MATLAB to plot snapshots of the temperature distribution at the following times: $t = 0.05$, $t = .1$, $t = .2$, $t = 1$, $t = 2$, $t = 10$, $t = 20$. Based on the graphs, what seems to be happening as time advances?
(b) For each of the above values of t, use the MATLAB function quad to integrate the snapshot at time t temperature distribution, which will be a function of x (impossible to integrate explicitly). Each of these integrals is improper (since the x-interval is unbounded), but integrate on a large enough interval so that the improper integral will be adequately approximated. What do these integrals seem to be doing? What is your interpretation?

NOTE: The problem of a very long rod for heat equation mentioned in the last exercise effectively removes the issue of the boundaries of the rod. It turns out that this problem is well posed as long as an initial temperature distribution $f(x) = u(x,0)$ is given which is continuous and decays to zero as $x \to \infty$. Furthermore, the resulting solution of this problem can be expressed as an integral involving the fundamental solution as follows:

$$u(x,t) = \int_{-\infty}^{\infty} f(s)\exp[-(x-s)^2/4t]/\sqrt{4\pi t}\,ds.$$

8.   Given the initial temperature distribution $u(x,0) = f(x)$ on a very long rod where

$$f(x) = \begin{cases} 1, & \text{if } |x| \le 1 \\ 2 - |x|, & \text{if } 1 < |x| \le 2, \\ 0, & \text{if } |x| > 2 \end{cases}$$

(a) get MATLAB to plot the initial temperature distribution $u(x,0)$ along with the following temperature distribution profiles $u(x,0.1)$, $u(x,0.2)$, $u(x,1)$, $u(x,2)$, $u(x, 10)$ all over the x-range $-10 \le x \le 10$. You can put them all in single plot (with different colors/styles) or use a subplot.
(b) Get MATLAB to plot a surface plot of the temperature function $u(x,t)$ over the range $-10 \le x \le 10$, $0 \le t \le 10$.
Suggestion: Use the integral formula of the above note and make creative use of MATLAB's quad function.

9.   Repeat both parts of Exercise 8, using instead the function

$$f(x) = \begin{cases} 5(1 - |x+5|), & \text{if } -6 \le x \le -4 \\ 10(1 - |x-5|), & \text{if } 4 \le x \le 6 \\ 0, & \text{otherwise} \end{cases}.$$

10.   In this exercise you will establish a more elaborate version of the superposition principle of

Theorem 11.1. Suppose that $L[u]$ denotes the left side of equation (11), i.e.,

$$L[u] = a(x,y)u_{xx} + b(x,y)u_{xy} + c(x,y)u_{yy} + d(x,y)u_x + e(x,y)u_y + f(x,y)u ,$$

that $u_1 = u_1(x,y)$ solves the PDE $L[u] = q_1(x,y)$, u_2 solves $L[u] = q_2(x,y)$, and so on and suppose that $c_1, c_2, \cdots$ are constants. Show that the function $c_1 u_1 + c_2 u_2 + \cdots$ solves the PDE $L[u] = c_1 q_1 + c_2 q_2 + \cdots$ (where the sums are assumed finite).

11.3: FINITE DIFFERENCE METHODS FOR ELLIPTIC EQUATIONS

We will be focusing attention mainly on variations of the **Poisson equation**

$$\Delta u = q(x,y), \quad u = u(x,y). \tag{12}$$

This linear elliptic equation specializes to the Laplace equation in case $q = 0$. Finite difference methods in general work very nicely for boundary value problems given on domains of "nice" shape, the ideal example being a rectangle. In Chapter 13 we will give a different method, called the finite element method, that works better in oddly shaped domains. The finite difference methods in this section will be based on the following **central difference formula** for approximating second derivatives:

$$f''(x) = \frac{f(x+h) - 2f(x) + f(x-h)}{h^2} + O(h^2). \tag{13}$$

We proved this formula (and others like it) in Section 10.4 using Taylor's theorem (see Lemma 10.3). Recall the "big O" notation means that the error of the approximation is less than a constant times h^2. The idea of the finite difference method for a rectangular domain can be briefly described as follows. We form a grid of points inside the rectangle, with N equally spaced x-coordinates and M equally spaced y-coordinates. We replace each of the partial derivatives in the PDE (12) by central difference quotient obtained using (13) where the terms in the quotient come from adjacent grid points. This gives a (very large) linear system of $N \cdot M$ unknowns (being the approximate values of our solution at the grid points). The boundary conditions will be enough to make the linear system well posed. We label the grid points in an efficient manner so as to make the resulting matrix for the linear system a banded matrix. We then use an efficient matrix equation solver (which may take advantage of the special form of the matrix) and solve the linear system to get an approximation of the solution to the BVP. Rather than explain the method more completely in a general fashion, we will introduce it by going through some specific examples involving Dirichlet BCs. The next section will delve into other boundary conditions for elliptic problems. Similar methods can be developed for parabolic and hyperbolic BVPs, but because stability is more often a problem for such BVPs, we put them off until the next chapter.

EXAMPLE 11.5: Use the finite difference method with $N = 4$ interior grid values on the x-axis and $M = 9$ interior grid values on the y-axis, to solve the following steady-state temperature distribution problem:

$$\begin{cases} \text{(PDE)} & \Delta u = 0, \quad u = u(x,y) \qquad\qquad 0 < x < 0.5,\ 0 < y < 1 \\ \text{(BC)} & u(x,1) = 4 \equiv t(x),\ u(x,0) = 16x^2 \equiv b(x) \\ & u(.5, y) = 4 \equiv r(y),\ u(0,y) = 4y \equiv \ell(y). \end{cases}$$

Create a surface plot of the resulting approximation to the solution.

SOLUTION: In this problem we are given the temperature reading of all of the edges of a rectangular plate and need to find the temperature at all of the interior points. Figure 11.9 summarizes graphically the given boundary conditions.

Since we will be using $N = 4$ grid points equally spaced inside the x-interval $[0, .5]$, this means that there will be a spacing of $h = (.5 - 0)/(N + 1) = .5/5 = .1$. Similarly, there will be a spacing of $k = (1 - 0)/(M + 1) = 1/10 = .1$ between the $M = 9$ grid point inserted inside the y-interval $[0,1]$. We label the x-grid points as $x_1, x_2, \cdots, x_N$ and for convenience add in the two extra endpoint grid values x_0 (left endpoint) and x_{N+1} (right endpoint), Thus the x-grid values are:

$$x_0 = 0,\ x_1 = .1(= h),\ x_2 = .2(= 2h),\ x_3 = .3(= 3h),\ x_4 = .4(= 4h),\ x_5 = .5(= 5h).$$

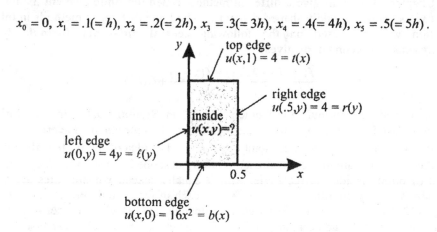

FIGURE 11.9: Illustration of the steady-state heat problem of Example 11.5. We are given the temperatures on each of the edges of the gray rectangle and we must solve for the temperatures at all of the points on the inside of the rectangle.

Doing the same for the y-grid values, we get the 11 y-grid values:

$$y_0 = 0,\ y_1 = .1(= k),\ y_2 = .2(= 2k),\ \cdots,\ x_{10} = 1(= 10k).$$

Our goal is to approximate the values $u(x_i, y_j)$ where both x_i and y_j are interior grid values (the boundary conditions give these values when either x_i or y_j is an endpoint grid value). For convenience, we employ the shorthand notation:

$$u_{i,j} = u(x_i, y_j).$$

Because of the equal spacing of the x-grid values (h = gap between adjacent x_i's), we can use the central difference formula (13) to write:

$$u_{xx}(x_i,y_j) \approx \frac{u(x_{i+1},y_j) - 2u(x_i,y_j) + u(x_{i-1},y_j)}{h^2} = \frac{u_{i+1,j} - 2u_{i,j} + u_{i-1,j}}{h^2}, \qquad (14)$$

where x_i and y_j are allowed to be any interior grid values. Here of course $h^2 = 0.01$, but we wanted to record this formula in general form for future reference. In the same fashion, we obtain the analogous approximation for the second y-partial derivatives, valid when y_j is any interior grid value:

$$u_{yy}(x_i,y_j) \approx \frac{u(x_i,y_{j+1}) - 2u(x_i,y_j) + u(x_i,y_{j-1})}{k^2} = \frac{u_{i,j+1} - 2u_{i,j} + u_{i,j-1}}{k^2}. \qquad (15)$$

If we substitute these approximations into the Laplace PDE $\Delta u = 0$ of the example, and multiply through by $h^2 = k^2$, we arrive at the following system of linear equations:

$$4u_{i,j} - u_{i+1,j} - u_{i-1,j} - u_{i,j+1} - u_{i,j-1} = 0, \quad (1 \le i \le 4, 1 \le j \le 9), \qquad (16)$$

This is a system of $M \cdot N = 4 \cdot 9 = 36$ equations in 36 unknowns. It is helpful to realize that each of the 36 equations in (16) involves values associated with the cross-shaped part of the grid shown in Figure 11.10, called the **(computational) stencil** for the finite difference method.

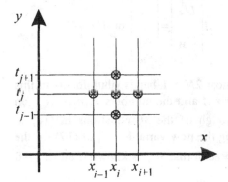

FIGURE 11.10: Illustration of the stencil for the linear equation (16).[5] These grid values form a cross centered at the $u_{i,j}$-grid value.

Although the central grid value $u_{i,j}$ will be at a point interior to the rectangle, one or two of the other four grid values may be from boundary points.

At this point the only problem that remains is to solve this large linear system. We will next put it in matrix form, but there are many ways to do this, depending on which order we choose to list the interior grid points (each interior grid point corresponds to one of the unknown 36 variables in our linear system). It is advantageous to number the interior grid points in a way that makes the resulting coefficient matrix look as simple as possible. One rather effective way to do this is to label the grid points in "reading order" starting at the upper left, proceeding right to the end of the first row of interior grid points, then moving down to the next row and continuing. This numbering scheme is illustrated in Figure 11.11. In order to write down a matrix equation for the linear system represented by the equations (16), we introduce the following notation for the variables of the system and the corresponding interior grid points gotten by following the labeling scheme of Figure 11.11:

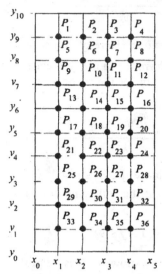

FIGURE 11.11: A generic and (as it turns out) very good way to label interior grid points in the finite difference method for solving elliptic PDEs.

$$P_k = (x_i, y_j), \quad U_k = u(P_k), \qquad (17)$$
$$(1 \le k \le 36)$$

In general, the following relationship exists between the index k and the indices i and j (Exercise 7):

$$k = i + N(M - j) \qquad (18)$$

For us, $N = 4$ and $M = 9$, so this becomes

$$k = i + 4 \, (9 - j).$$

The reader should now convince himself or herself of this using Figure 11.11. With this indexing scheme, the coefficient matrix A of the linear system,

$$A \begin{bmatrix} U_1 \\ U_2 \\ \vdots \\ U_{35} \\ U_{36} \end{bmatrix} = \begin{bmatrix} c_1 \\ c_2 \\ \vdots \\ c_{35} \\ c_{36} \end{bmatrix} \text{ or } AU = C,$$

will always be diagonally banded with at most $2N + 1$ bands (thus for us in this example, at most 9 bands). To get the matrix A and the numbers $c_1, c_2, \cdots, c_{36}$ on the right column matrix C, we rewrite enough of the 36 equations in (16) to discover the patterns. We do this now, using the new variables U_k of (17) for the unknowns (and leave these on the left side) Recall that $u_{i,j}$ is known if either $i = 0$ or 5 or $j = 0$ or 10.

TABLE 11.3: An abbreviated list of the 36 linear equations in the 36 unknowns $U_1, U_2, \cdots, U_{36}$ arising from the finite difference method for the elliptic PDE of Example 11.5. Each row represents one of the equations (16) using the notation (17) for the unknowns on the left sides of the equation and putting the known (boundary values) on the right side.

Interior vertex	Linear equation [unknowns] = [knowns]
P_1	$4U_1 - U_2 - U_5 = u_{0,9} + u_{1,10}$
P_2	$4U_2 - U_3 - U_1 - U_6 = u_{2,10}$

---

[5] This particular stencil is often referred to as the *(standard) five-point stencil for the Laplacian.*

P_3	$4U_3 - U_4 - U_2 - U_7 = u_{3,10}$
P_4	$4U_4 - U_3 - U_8 = u_{5,9} + u_{4,10}$
P_5	$4U_5 - U_6 - U_9 - U_1 = u_{0,8}$
P_6	$4U_6 - U_7 - U_5 - U_{10} - U_2 = 0$
P_7	$4U_7 - U_8 - U_6 - U_{11} - U_3 = 0$
P_8	$4U_8 - U_7 - U_{12} - U_4 = u_{5,8}$
$\vdots$	$\vdots$
P_{33}	$4U_{33} - U_{34} - U_{29} = u_{0,1} + u_{1,0}$
P_{34}	$4U_{34} - U_{35} - U_{33} - U_{30} = u_{2,0}$
P_{35}	$4U_{35} - U_{36} - U_{34} - U_{31} = u_{3,0}$
P_{36}	$4U_{36} - U_{35} - U_{32} = u_{5,1} + u_{4,0}$

From the above equations we see that A has the following appealing "diagonally banded form":

$$A = \begin{bmatrix}
4 & -1 & 0 & 0 & -1 & 0 & 0 & 0 & \cdots & & & & & \\
-1 & 4 & -1 & 0 & 0 & -1 & 0 & 0 & \cdots & & & & & \\
0 & -1 & 4 & -1 & 0 & 0 & -1 & 0 & \cdots & & & & & \\
0 & 0 & -1 & 4 & 0 & 0 & 0 & -1 & \ddots & & & & & \\
-1 & 0 & 0 & 0 & 4 & -1 & 0 & 0 & \ddots & & & & & \\
0 & -1 & 0 & 0 & -1 & 4 & -1 & 0 & \ddots & & & & & \\
\vdots & & \ddots & \ddots & \ddots & \ddots & \ddots & \ddots & \ddots & & & & & \\
& & & \ddots & \ddots & \ddots & \ddots & \ddots & \ddots & \ddots & & & & 0 \\
& & & & & \ddots & -1 & 0 & 0 & -1 & 4 & -1 \\
& & & & & & 0 & -1 & 0 & 0 & -1 & 4
\end{bmatrix},$$

i.e., A has a band of 4's down the main diagonal, two more diagonals of -1's which are 4 diagonals above/below the main diagonal, and finally two more diagonal bands, directly above and below the main diagonal, which repeat the pattern $-1, -1, -1, 0$ (with the last zero not appearing).

For simplicity in describing the entries of C, we introduce the following vectors of known boundary values (written in general form, for us, $N = 4$, $M = 9$)

$$L = [\ell(y_0), \ell(y_1), \cdots \ell(y_{M+1})] = [u_{0,0}, u_{0,1}, \cdots, u_{0,M+1}],$$
$$R = [r(y_0), r(y_1), \cdots r(y_{M+1})] = [u_{N+1,0}, u_{N+1,1}, \cdots, u_{N+1,M+1}],$$
$$B = [b(x_0), b(x_1), \cdots b(x_{N+1})] = [u_{0,0}, u_{1,0}, \cdots, u_{N+1,0}],$$
$$T = [t(x_0), t(x_1), \cdots t(x_{N+1})] = [u_{0,M+1}, u_{1,M+1}, \cdots, u_{N+1,M+1}].$$

$$(19)$$

By looking once more at the equations in Table 11.3, and using (19), we arrive at the following expression for (the transpose of) C. Please note, we are starting to use MATLAB's index notation for vectors, so, for example, $L(10)$ will mean the 10th component of the vector L that is $\ell(y_9)$ (since $L(1) = \ell(y_0)$).

$$C' = [L(10)+T(2),T(3),T(4),R(10)+T(5),L(9),0,0,R(9),L(8),0,0,R(8),$$
$$L(7),0,0,R(7),L(6),\cdots 0,0,R(6),L(5),0,0,R(5),L(4),0,0,R(4),$$
$$L(3),0,0,R(3),L(2)+B(2),B(3),B(4),R(2)+B(5)].$$

The reader is advised to convince himself or herself that this vector is correct by contemplating Figure 11.11. The column matrix C has a bit more elusive of a pattern. The beginning and end of C are a bit misleading, but the pattern of the middle terms is quite simple. The careful reader will perhaps be able to guess at what the patterns of A and C will look like in the general case (the true test would be to be able to write a MATLAB M-file that is able to do all of the above for a more general problem, not just for this specific example).

We are now ready to turn things over to MATLAB. We basically have to get MATLAB to solve the matrix equation $AU = C$ for U and then put the values of U together with the given boundary values to get the approximations for the solution on a meshgrid. Then it will be easy to plot the function. Because of the special form of A, we will be able to enter this 36×36 matrix (all 1296 entries of it) rather quickly if we take advantage of MATLAB's array operations. Let us begin by initially constructing a 36×36 matrix with 4's down the main diagonal and then add on the remaining diagonals:

```
>> A=4*eye(36);
```

We now key in the four vectors that represent the nontrivial bands above and below the diagonal (the four bands are 1 and 4 units above and below the main diagonal).

Because of its repetitive nature, $a1$ = the vector one unit above the main diagonal is easy to enter:

```
>> a1rep=[0 -1 -1 -1];
>> a1=[-1 -1 -1];
>> for i=1:8, a1=[a1 a1rep]; end
```

To check if a vector is the correct size, we can always type the size command:

```
>> size(a1)    →ans = 1   35
```

The vectors a4 and b4 (the same) are even more simple:

```
>> a4=-1*ones(1,32);
```

Now that we have the band vectors keyed in, the diag command can easily put them where we want them to be:

```
>> A=A+diag(a1,-1)+ diag(a1,1)+diag(a4,-4)+diag(a4,4);
>> x=0:.1:.5; y=0:.1:1;
>> L=4*y; B=16*x.*x; %enter left and bottom edge boundary values
```

```
>> T=4*ones(1,6);   % enter top edge boundary values
>> R=4*ones(1,11);  % enter right edge boundary values
```

Now that the vectors *L, R, T, B* are keyed in we can enter the vector *C*. Notice that since we want *C* to be a column vector, we type the transpose symbol after it (`'`) that changes it from a 1×36 row vector to a 36×1 column vector.

$$C' = [L(10) + T(2), T(3), T(4), R(10) + T(5), L(9), 0, 0, R(9), L(8), 0, 0,$$
$$R(8), L(7), 0, 0, R(7), L(6), \cdots 0, 0, R(6), L(5), 0, 0, R(5), L(4), 0, 0,$$
$$R(4), L(3), 0, 0, R(3), L(2) + B(2), B(3), B(4), R(2) + B(5)].$$

```
>> C=[L(10)+T(2) T(3) T(4) R(10)+T(5) L(9) 0 0 R(9) L(8) 0 0 R(8) ...
L(7) 0 0 R(7) L(6) 0 0 R(6) L(5) 0 0 R(5) L(4) 0 0 R(4) L(3) 0 0 ...
R(3) L(2)+B(2) B(3) B(4) R(2)+B(5)]';
```

The left divide operation will effectively solve our matrix equation:

```
>> U=A\C;
```

In order to get MATLAB to plot our solution, we need to form a matrix *Z* which is 11×6 and has all of the computed values for *u* that we just computed, as well as the boundary values that were given in the problem (and they should be placed in the corresponding spots in the matrix *Z* to where they are supposed to appear on the grid). We first contruct a 10×5 matrix *Z* that contains the just-computed boundary values (given in the vector *U*) in their correct places.

We start off by forming a matrix *Z* of zeros of the correct size.

```
>> Z=zeros(4,9);
```

Actually the correct size of the matrix will be 9×4, but we will soon take a transpose to give it the correct size. There is a useful command in MATLAB for slipping in the entries of a vector *U* (say of size 1×36) into a matrix a whose dimensions multiply to the size of the vector (say *Z*).

```
>> Z(:)=U; Z=Z';
```

The reason we needed to take the transpose of *Z* (apart from correcting the size of *Z*) is because of the order that the entries of *u* were slipped into *Z* (top-to-bottom first, rather than left-to-right first). The following example illustrates this:

```
>> E=zeros(3,2);  v=[1 2 3 4 5 6];  E(:)=v
→E =     1   4
         2   5
         3   6
```

Next, the following commands will put the correct boundary values on the top and bottom of *Z*:

```
>>Z=[T(2:5); Z; B(2:5)];
```

Note that Z will now be an 11×4 matrix; what is left to do is put the left and right boundary values on the left and right sides of Z.

Before we put the left boundary values on, it will be convenient to reverse the order of the vector L, which we can easily do by creating a new vector `Lrev` as follows:

```
>> for i=1:11, Lrev(i)=L(12-i); end
```

For R this need not be done since the components of R are all constants. We can now get the final 11×6 vector by pasting `Lrev` to the left of Z and R to the right; this can be achieved in MATLAB with the command:

```
>> Z=[Lrev; Z'; R]';
```

It is a good idea to check the matrix Z (which should represent the temperatures at the grid points):

```
>> Z        4.0000   4.0000   4.0000   4.0000   4.0000   4.0000
            3.6000   3.6782   3.7571   3.8371   3.9182   4.0000
            3.2000   3.3558   3.5131   3.6731   3.8358   4.0000
            2.8000   3.0317   3.2665   3.5065   3.7517   4.0000
            2.4000   2.7044   3.0147   3.3347   3.6644   4.0000
            2.0000   2.3711   2.7533   3.1533   3.5711   4.0000
            1.6000   2.0268   2.4741   2.9541   3.4668   4.0000
            1.2000   1.6620   2.1623   2.7223   3.3420   4.0000
            0.8000   1.2588   1.7907   2.4307   3.1788   4.0000
            0.4000   0.7825   1.3111   2.0311   2.9425   4.0000
                 0   0.1600   0.6400   1.4400   2.5600   4.0000
```

Things look pretty good (cf. Figure 11.9). We can next use the MATLAB command `surf`, to give us a graph of the surface (for the solution of our heat problem). For a technical reason in the syntax of the `surf` and other 3D plotting commands,[6] we need to use `yrev`, the reverse vector of y, for the y-axis vector.

```
>> for i=1:11, yrev(i)=y(12-i); end
>> surf(x, yrev ,Z), xlabel('x'), ylabel('y')
```

FIGURE 11.12: Plot of the heat function solution of Example 11.5. MATLAB will kindly color the "hot parts" of the surface red and the "cold parts" blue, as if it knew we were solving a heat equation. This comes from the default `colormap` setting.

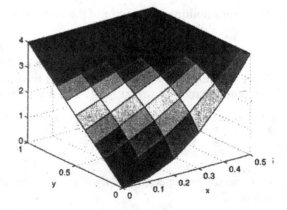

---

[6] Recall from Example 11.1 that when a meshgrid is set up, the x-coordinates are listed in the usual order but the y-coordinates go in backwards order. When we use the finite difference method, we are setting up a meshgrid for the u-values in the usual order; thus, to get correct results with `surf`, we need to feed in the y-vector in reverse order.

With the example behind us, we now turn to make some general comments on numerically solving Poisson's equation (12) $\Delta u = q(x, y)$ on a rectangular domain. The general method just needs a few technical adjustments, but the main ideas were all covered in the above example, so we indicate only the few extra changes that might be needed in general and then we comment on writing a MATLAB M-file to automate this procedure.

In general, the gap between x-grid values (h) need not be the same as the gap between y-grid values (k). (In the last example we had $h = k = 0.1$.) Also the function $q(x,y)$ need not be zero (as in our example). In this general case, letting $q_{i,j} = q(x_i, y_j)$, the central difference approximations applied to the differential equation $\Delta u = q(x, y)$ give the linear system:

$$\frac{u_{i+1,j} - 2u_{i,j} + u_{i-1,j}}{h^2} + \frac{u_{i,j+1} - 2u_{i,j} + u_{i,j-1}}{k^2} = q_{i,j}.$$

In general the gaps h and k will be small, so to keep the terms from getting too large in this system, we multiply the equation by h^2 (alternatively k^2) and regroup to obtain (for $1 \le i \le M, 1 \le j \le N$)

$$2\left(\frac{h^2}{k^2} + 1\right)u_{i,j} - u_{i+1,j} - u_{i-1,j} - \left(\frac{h^2}{k^2}\right)u_{i,j+1} - \left(\frac{h^2}{k^2}\right)u_{i,j-1} = -h^2 q_{i,j}. \tag{20}$$

We point out that when the x-mesh size h equals the y-mesh size k, (20) simplifies considerably to

$$4u_{i,j} - u_{i+1,j} - u_{i-1,j} - u_{i,j+1} - u_{i,j-1} = -h^2 q_{i,j}, \tag{21}$$

whose left side is identical with that of (16). Apart from these small differences, the procedure given in the example is very much the same as it would be to solve a general Poisson DVP on rectangle. In the next exercise for the reader, we will test the method out on a problem where the exact solution is known.

EXERCISE FOR THE READER 11.4: (a) Use the finite difference method with $N = 4$ interior grid values on the x-axis and $M = 4$ interior grid values on the y-axis, to solve the following steady-state temperature distribution problem:

$$\begin{cases} \text{(PDE)} & \Delta u = (4 - \pi^2)e^{-2y} \sin \pi x, \ u = u(x,y) \ 0 \le x \le 1, 0 \le y \le 1 \\ \text{(BC)} & u(x,1) = e^{-2} \sin \pi x \equiv t(x), \ u(x,0) = \sin \pi x \equiv b(x) \\ & u(1,y) = 0 \equiv r(y), \ u(0,y) = 0 \equiv \ell(y) \end{cases}.$$

Afterwards, graph the resulting approximation to the solution.

(b) Check that $u(x, y) = e^{-2y} \sin \pi x$ solves the above BVP and thus (since it is well posed) is the unique solution. Compare the values of this exact solution with those of the approximation you obtained in part (a). Among all interior grid points find both the maximum error and the maximum relative error of the numerical solution.

(c) Redo parts (a) and (b), this time doubling N and M to be 8.

In principle, the method described in this section for solving Poisson's equation works well for grids having up to 500 or so internal grid points. Also, if grid gaps are cut in half, we can compare the approximation with the refined grid with the approximation using the original grid (at the original internal grid points) and this can be repeated until the maximum errors (computed as in part (b) of the preceding exercise for the reader) become less than our tolerance for error so as to get a desired approximation with a confidence measurement. When the number of grid points gets larger than several hundred, the memory and storage of the matrix A starts to become a serious issue and it is better to solve the matrix equation $AU = C$ using different approaches that take advantage of the special form.

The key issue here is that for very large numbers of grid points, the percentage of entries of A which are nonzero is very small (it is less than $5n/n^2 = 5/n$); recall that such matrices are called sparse. Since our A has such a special form there are specialized algorithms (cf. the Thomas algorithm used Section 10.4) that avoid having to store the whole matrix A in its memory and solve the system much more expediently. For example, if we had 1000 internal grid points, then A would be a matrix with a million entries! But only about 5000 of these would be nonzero and it starts to become a good idea to take advantage of this structure. Most all of the matrices that arise the numerical methods that we develop for PDEs will be sparse. Readers can increase the size of problems that can be solved by utilizing sparse matrix manipulations in MATLAB as well as iterative methods. These topics were covered in Section 7.7. The same ideas work also for three (and higher)-dimensional elliptic equations but it is of course a bit more of an abstract problem to visualize the three-dimensional rectangle of internal grid points.

EXERCISE FOR THE READER 11.5: (a) Write a MATLAB function M-file called `poissonsolver` having the following syntax:

`[xgrid, ygrid, Zsol] = poissonsolver(q,a,b,c,d,h)`

which will solve the Poisson equation $\Delta u = q(x,y)$ on the rectangle $a \le x \le b$, $c \le y \le d$ with Dirichlet boundary conditions $u = 0$ on the boundary. It should use a common mesh size h and have three output variables, two vectors `xgrid`, `ygrid` and a matrix `Zsol`. More precisely:

Inputs: q (inline or M-file function for q(x,y), may be the zero function), a, b, c, d, h. The step size must divide evenly into both length and width of the rectangle.
Outputs: `xgrid`, `ygrid`, `Zsol`. From these outputs you should be able to plug them into `surf` to get a graph of the numerical solution.

Note, the `xgrid` vector will look like $[a, a + h\text{step}, a + 2h\text{step}, ..., b]$, and similarly for `ygrid`.

(b) Run your program to solve the Dirichlet problem $\Delta u = \sin(2\pi y)\{4\pi^2(x - x^3) + 6x\}$ on the unit square: $0 \le x, y \le 1$. First do it with a step size $h = 0.1$, and then repeat with $h = 0.02$. Compare these numerical solutions with the exact solution to this problem given by $u(x,y) = (x^3 - x)\sin(2\pi y)$. Plot surface graphs of the errors.

Finite difference methods are, in general, not so well suited for problems on domains that are not rectilinear. For Dirichlet boundary conditions on elliptic PDEs, however, finite difference methods can sometimes be used if we approximate the domain using a grid of squares. The general idea is illustrated in Figure 11.13.

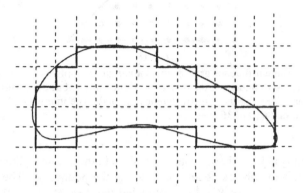

FIGURE 11.13: Approximation of a domain using a rectangular grid. Under certain regularity assumptions the solutions of Dirichlet problems for elliptic PDEs on arbitrary domains (curved) can be approximated by the corresponding solutions on the approximating rectilinear domain (segmented).

The details of such a scheme are not particularly difficult, but the creation of a general codes would be a laborious task. Such techniques work reasonably well for Dirichet problems[7] but not for boundary value problems where the BCs involve other sorts of boundary conditions (eg., Neumann or Robin). The finite element method of Chapter 13, however, is able to handle all sorts of boundary conditions and domains. Thus, we will forgo a general development, and simply outline this scheme for a particular nonrectangular domain.

EXAMPLE 11.6: Consider the problem of finding the steady-state heat distribution on the right isosceles triangular domain shown in Figure 11.14, with the boundary temperatures as indicated in the figure. Assume the short sidelengths equal one. Use the finite difference method with $h = k = 0.1$ to solve the problem numerically, and plot the solution.

---

[7] What is required is that the domain be bounded by a finite number of smooth curves, and that the coefficient of the elliptic equation are continuous functions defined in some tube about the boundary. General theorems will then show that as the mesh size gets sufficiently small, the solution of the Dirichlet problem on the approximating domain (as in Figure 11.13) will converge to the actual solution.

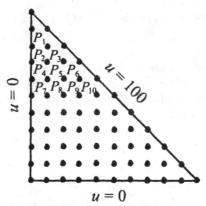

FIGURE 11.14: A finite difference grid for the domain of Example 11.6. The temperature distribution given is discontinuous at the upper and right corners, but these vertices will not enter into the linear system for the finite difference method. The interior nodes are labeled using the "reading order" (10 of 36 are shown).

SOLUTION: Being able to choose equal step sizes in the x- and y-directions helps a lot here since this distributes nodes nicely along the boundary. The number of interior nodes is $1 + 2 + \ldots + 8 = 36$. We label these as $P_1, P_2, \cdots, P_{36}$ using the "reading order" scheme of the last example, and invoke the analogues of all of the other relevant notations of that example. For the resulting 36 linear equations corresponding to (16), each interior node P_k give rises to an equation of the form:

$$4u_{i,j} - u_{i+1,j} - u_{i-1,j} - u_{i,j+1} - u_{i,j-1} = 0 ,$$

where $P_k = (x_i, y_j)$ corresponds to $u_{i,j}$, and the known values need to be moved to the right side. By looking at the figure, we see that the right sides of these equations will thus be either 200 (if k = 1, 3, 6, 10, 15, 21, 28, 36) or 0 (all other nodes). In Table 11.4, we give a few of the resulting equations in the unknowns $U_k = u(P_k)$. Although the coefficient matrix A of the linear system $AU = C$ could be easily entered by (brute) inspection, we choose to construct it instead with the following loop. The loop starts with a 36×36 diagonal matrix A and then places the -1's in appropriate places. Unlike the brute force construction, this loop easily generalizes to finer grids.

```
>> A=diag(4*ones(1,36));
border = [0 1 3 6 10 15 21 28 36];
for k=2:length(border)
  if k>2, pregap=right-left+1; end
  left=border(k-1)+1; right=border(k);
  if k<length(border)
    postgap=right-left+1;
  end
  for i=left:right
    if i<right %has right neighbor and top neighbor
      A(i,[i+1  i-pregap])=-1;
    end
```

```
   if i>left %has left neighbor
     A(i,i-1)=-1;
   end
   if k<length(border) %has bottom neighbor
     A(i,i+postgap)=-1;
   end
 end
end
```

TABLE 11.4: An abbreviated list of the 36 linear equations for the finite difference method of Example 11.6.

Interior vertex	Linear equation
P_1	$4U_1 - U_2 = 200$
P_2	$4U_2 - U_4 - U_1 - U_3 = 0$
P_3	$4U_3 - U_5 - U_2 = 200$
P_4	$4U_4 - U_2 - U_7 - U_5 = 0$
P_5	$4U_5 - U_3 - U_4 - U_6 - U_8 = 0$
P_6	$4U_6 - U_5 - U_9 = 200$
P_7	$4U_7 - U_4 - U_8 - U_{11} = 0$
P_8	$4U_8 - U_5 - U_7 - U_9 - U_{12} = 0$
$\vdots$	$\vdots$
P_{33}	$4U_{33} - U_{34} - U_{32} - U_{26} = 0$
P_{34}	$4U_{34} - U_{35} - U_{33} - U_{27} = 0$
P_{35}	$4U_{35} - U_{36} - U_{34} - U_{28} = 0$
P_{36}	$4U_{36} - U_{35} = 200$

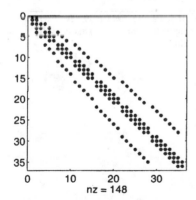

nz = 148

Since we know the numerical values of all entries of the matrix (diagonal entries = 4, off diagonal entries = 0 or –1), rather than printing the matrix A, a more practical way to view it would be using MATLAB's spy function:

spy(A) →	Produces a graphic indicating the placement of the nonzero entries of a matrix A.

Thus the command spy(A) produces the plot in Figure 11.15.

FIGURE 11.15: Spy plot of the coefficient matrix A for Example 11.6. The nz=148 indicates the number of nonzero entries of the 1296 entries of A. We know the diagonal entries all equal 4 and all other nonzero entries equal –1. Such spy plots make the general structure of such matrices quite evident.

From what was observed above, the right-side vector C can be easily constructed as follows:

```
>> C=zeros(36,1);
>> C(border(2:length(border)))=200;
```

and the system can be solved:

```
>> U=A\C;
```

We now can use the values of U to fill in the values of a matrix Z of the numerical values of the solution of the boundary value problem. We first form a 64×64 matrix for the interior grid points, temporarily putting $Z = 100$ for the values above the main diagonal.[8]

```
>> Z=100*ones(8);
>> count=1;
>> for i=1:8
     gap=border(i+1)-count+1;
     Z(i,1:gap)=U(count:(count+gap-1))';
     count=count+gap;
end
```

Next we enlarge this matrix to a 11×11 matrix which contains the given boundary values. As a compromise, we set $Z = 50$ at the interfaces of the discontinuous boundary data.

```
Z=[100*ones(1,8);100*ones(1,8);Z;zeros(1,8)];
Z=[[50 zeros(1,10)]', Z, [100*ones(1,10) 0]', [100*ones(1,10) 50]']
```

In order to get surface plot on just the triangle, we will redefine the entries of Z that are off the triangle as nan, except for those nodes which are adjacent to two nodes on the slanted side of the triangle. For these latter nodes, take the average of the values at the two neighboring nodes on the slanted edge.

nan→	When stored as an entry of a matrix or vector, nan (meaning: *not a number*) will produce a hole in any plots of this vector at the corresponding location. This is useful for three-dimensional plots over nonrectangular domains.

```
>>for i=1:11
    if i<11
       Z(i,i+1)=(Z(i,i)+Z(i+1,i+1))/2;
    end
    for j=i+2:11
        Z(i,j)=nan;
    end
end
```

Let us now check the matrix Z:
```
>>Z
→Z =
```

---

[8] This is simply a reasonable convention since we need to fill in a whole square matrix of values, even though the present problem only has actual values corresponding to the lower-left triangular part of the matrix.

50	75	NaN	NaN	NaN	NaN	NaN	NaN	NaN	NaN	NaN
0	100	100	NaN	NaN	NaN	NaN	NaN	NaN	NaN	NaN
0	60.38	100	100	NaN	NaN	NaN	NaN	NaN	NaN	NaN
0	41.52	74.90	100	100	NaN	NaN	NaN	NaN	NaN	NaN
0	30.81	58.07	80.88	100	100	NaN	NaN	NaN	NaN	NaN
0	23.66	45.68	65.44	83.26	100	100	NaN	NaN	NaN	NaN
0	18.14	35.55	51.95	67.61	83.26	100	100	NaN	NaN	NaN
0	13.36	26.44	39.19	51.95	65.44	80.88	100	100	NaN	NaN
0	8.86	17.65	26.44	35.55	45.68	58.07	74.90	100	100	NaN
0	4.43	8.86	13.36	18.14	23.66	30.81	41.52	60.38	100	75
0	0	0	0	0	0	0	0	0	0	50

Notice the symmetry of this numerical data. This should be the case because of the symmetry of the given temperature distribution.

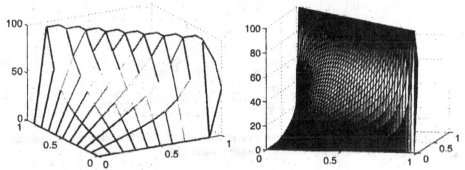

FIGURE 11.16: (a) (left) The mesh plot of the solution to the heat problem of Example 11.6. (b) (right) A surface plot of the same problem using a finer grid (Exercise for the Reader 11.6).

EXERCISE FOR THE READER 11.6: (a) Write a MATLAB function M-file that is designed precisely to solve the Dirichlet problem for the Laplace equation $\Delta u = 0$ on the special triangulular domain with vertices (0,0), (1,0), and (0,1). The syntax should be as follows:

```
[Z, x, y]=triangledirichletsolver(n,leftdata, bottomdata, slantdata)
```

where, n = the common number of interior grid values on the x- and y-axes ($h = k$), and the remaining three input variables are vectors having $n + 2$ components giving the boundary data on the three faces of the triangle: leftdata gives the boundary values at the nodes on the left face read from top to bottom, bottomdata gives the boundary values at the nodes on the bottom face read from left to right, and slantdata gives the boundary data at the nodes on the slanted face read from top to bottom. The output variables are: Z an $(n+2) \times (n+2)$ matrix of the values of the solution at the corresponding grid values of the triangle: the first column of Z should thus be the vector leftdata, the main diagonal the vector slantdata, etc. The entries of the matrix Z above the main diagonal should be NaN's. The last two output variables x and y should simply be the $(n + 2)$ vectors giving the x- and y-grid values; however, y should be given in decreasing order to facilitate plotting. The output data should be arranged so that the command surf(x, y, z) will give a plot of the numerical solution. (b) Test the program on the BVP of Example 11.6 but with $n = 49$ and obtain a graphic of the numerical solution as in Figure 11.16b. Next, use the data to obtain an isotherm (contour) plot as in Figure 11.17.

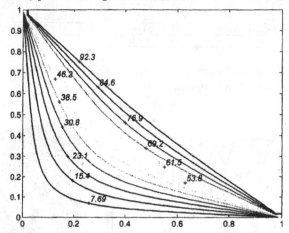

FIGURE 11.17: Isotherms (lines of constant temperature) for the heat problem in Example 11.6. This plot was obtained in Exercise for the Reader 11.6.

We close this section with some theoretical comments about the finite difference method applied to Dirichlet problems. From a purely linear algebraic perspective, it is not at all clear that a finite difference scheme will have a solution (i.e., if the coefficient matrix will be nonsingular). This turns out to be the case if we apply the method to the Poisson PDE on any rectilinear domain on which the boundary data is specified. We state this as our next theorem:

THEOREM 11.2: (*Existence and Uniqueness of the Finite Difference Method for Dirichlet Problems*) Suppose that the domain D is bounded, connected and rectilinear[9] and that the finite difference method is used to solve the Dirichlet problem for the Poisson equation:

$$\begin{cases} \Delta u = f(x, y) & \text{on } D \\ u = g(x, y) & \text{on } \partial D \end{cases} \quad u = u(x, y),$$

where that data $f(x,y)$ and $g(x,y)$ are arbitrary functions (not necessarily continuous). If any grid (with $h = x$-step and $k = y$-step not necessarily equal) is

---

[9] This means that the boundary of the domain is made up of vertical and horizontal line segments only. Such domains are allowed to have holes (e.g., an L-shaped region with some rectangular holes punched out). The connectedness assumption simply means (informally) that the domain is all in one piece. More technically, it means that any two points in the domain can be joined by an arc which lies entirely in the domain. This assumption is not at all restrictive since BVPs on nonconnected domains can be broken into separate problems on connected domains.

used for which each corner point of the boundary of D is a node, then the finite difference method will produce a unique solution.

Proof: We give the proof for the case in which $h = k$, since the ideas present themselves most elegantly in this case. The general case will be left to the exercises. First we assume that $f(x,y) = 0$ (i.e., the Laplace equation). So we can think of the BVP as a steady-state heat distribution problem. Recall that harmonic functions (solutions of $\Delta u = 0$) satisfy the maximum principle: If $u(x,y)$ attains a maximum (or minimum) in the interior of a domain (as opposed to a boundary point), then $u(x,y)$ must be a constant function. We will show that finite difference solutions to the Laplace equation also have this important property. Indeed the finite difference scheme for the Laplace equation (16) $4u_{i,j} - u_{i+1,j} - u_{i-1,j} - u_{i,j+1} - u_{i,j-1} = 0$, when rewritten as

$$u_{i,j} = \frac{1}{4}\left[u_{i+1,j} + u_{i-1,j} + u_{i,j+1} + u_{i,j-1}\right],$$

can be thought of as saying that the value of the finite difference solution at an interior point will be the average of the values of the finite difference solution at the four neighbors (right, left, top, bottom; see Figure 11.10). It follows that if this finite difference solution were to have a maximum at some interior point, then each of the four neighbors would share this maximum value (if just one was less, then so would the average and hence $u_{i,j}$). We then apply this same argument to each of the neighbor nodes that are interior nodes. By the connectedness assumption, if we continue to repeat this argument, eventually all the nodes of the domain will be accounted for and shown to have the same maximum value. This proves the maximum principle for finite difference solutions.

A slight modification of this proof will prove the analogous minimum principle: If a finite difference solution attains a minimum value at an interior point, then the finite difference solution must be a constant. From these principles, it is easy to show that finite difference solutions are unique for the Poisson equation. Indeed, if $u_{i,j}$ and $v_{i,j}$ were both finite difference solutions to the above Poisson BVP, then $w_{i,j} \equiv u_{i,j} - v_{i,j}$ would be a finite difference solution of the the Laplace equation $\Delta u = 0$ and have zero boundary data (since the boundary data of $u_{i,j}$ and $v_{i,j}$ are the same). By the maximum and minimum principles, it follows that $w_{i,j} \equiv 0$ (i.e., $u_{i,j} \equiv v_{i,j}$); this proves uniqueness. Next, any finite difference solution $u_{i,j}$ is a solution of a linear system of N equations and N unknowns ($N = $ total number of interior nodes). For such a linear system (with square coefficient matrix), existence of a solution is equivalent to uniqueness of solutions (and to the coefficient matrix being invertible); see [HoKu-71].

We caution the reader that although the finite difference method can be extended to more complicated PDEs in natural ways, care must be taken with respect to the underlying mathematical theory. Such problems may not have existence and uniqueness for mathematical solutions and in such circumstances we cannot have much hope for any numerical scheme. Mathematical existence and uniqueness theory for PDE is a vast field of contemporary research and much remains to be discovered (especially for nonlinear equations). In the theorem below we give a small sampling of some existence and uniqueness theorems for elliptic boundary value problems. In each we make the underlying assumption that the domain Ω lies in the plane and that its boundary $\partial\Omega$ is **piecewise smooth**, meaning that $\partial\Omega$ can be broken up into a finite number of pieces, each of which is the graph of a function of either x or y with continuous second derivative. We also say that a function is smooth if its second (partial) derivatives are all continuous.

THEOREM 11.3: (*Existence and Uniqueness for Some Elliptic Boundary Value Problems*) Suppose that Ω is a smooth domain in the plane.
(a) If $g(x,y)$ is a continuous function on $\partial\Omega$ then the Dirichlet problem for the Laplace equation:

$$\begin{cases} \Delta u = 0 \text{ on } \Omega \\ u = g(x,y) \quad \text{on } \partial\Omega \end{cases}$$

has a unique solution that is continuous on $\Omega \cup \partial\Omega$ and agrees with $g(x,y)$ on $\partial\Omega$.
(b) Suppose that the PDE (11):

$$a(x,y)u_{xx} + b(x,y)u_{xy} + c(x,y)u_{yy} + d(x,y)u_x + e(x,y)u_y + f(x,y)u$$
$$= q(x,y),$$

is uniformly elliptic on Ω ($b^2 - 4ac < -\delta < 0$ throughout Ω, for some positive number δ), that the coefficients have piecewise continuous partial derivatives throughout Ω, and that $a(x,y) \geq 0$ and $f(x,y) \leq 0$ throughout Ω. If $g(x,y)$ has continuous second derivatives in some tube about $\partial\Omega$, then there exists a unique solution $u(x,y)$ of the PDE (11) that satisfies the BC $u = g(x,y)$ on $\partial\Omega$.

Some of the smoothness requirements can be weakened. The result of part (a), for example, can be extended to work for very general domains, including domains with fractal boundaries. The proofs of such theorems are quite involved and are not within the scope of the text; we refer the interested reader to [GiTr-83]. In part (b), the requirement that $f(x,y) \leq 0$ is essential. Indeed, it is well known that for any such domain Ω the Laplace operator can have **eigenvalues** $\lambda < 0$ for which the PDE $\Delta u = \lambda u$ has nonzero solutions with zero boundary data. By analogy with matrix theory, these nonzero solutions are called (associated) **eigenfunctions**. As a simple example, in case Ω is the unit square $\{(x,y): 0 < x, y < 1\}$, the eigenvalues are $\lambda = -(n^2 + m^2)\pi^2$ for any integers n and m, and associated eignfunctions are $u(x,y) = \sin(n\pi x)\sin(n\pi y)$. It can be readily checked that these functions satisfy the PDE $\Delta u = \lambda u$ and have zero boundary values. Since $u(x,y) = 0$ is also (an obvious) solution of this same BVP, uniqueness is violated. Furthermore, if λ is a negative number not equal to one of these eigenvalues, it can be shown that one cannot arbitrarily assign continuous boundary data for the PDE $\Delta u = \lambda u$ and always have a solution (nonexistence).[10]

EXERCISES 11.3

1. (a) Use the finite difference method with $N = 4$ interior grid values on the x-axis and $M = 9$ interior grid values on the y-axis, to solve the following steady-state temperature distribution problem:

$$\begin{cases} (\text{PDE}) & \Delta u = 0, \quad u = u(x,y) \qquad 0 < x < 1, \, 0 < y < 2 \\ (\text{BC}) & u(x,1) = 8 \equiv t(x), \, u(x,0) = 0 \equiv b(x) \\ & u(.5,y) = y^3 \equiv r(y), \, u(0,y) = 4y \equiv \ell(y). \end{cases}$$

Afterwards, graph the resulting approximation to the solution.
(b) Repeat with $N = 9$ and $M = 19$.

2. (a) Redo part (a) of Exercise 1 using $N = 9 = M$ interior grid values on both the x- and y-axis.
(b) Repeat with $N = 19 = M$.

3. Consider a steel alloy rectangular plate that is 10 feet long and 6 feet wide. Suppose that the bottom (10-foot) edge is maintained at 400°F , the top edge is maintained at 250°F and both vertical edges are maintained at 150°F . Assume that the flat faces of the plate are insulated.
(a) Use the finite difference method to solve for the temperatures within the plate using a spacing of 1 foot $(= h = k)$. Plot the approximate location of the 300°F isothermal curve within the plate (i.e., the contour in the plate on which the temperature is constantly 300°F) and also the 200°F contour.
(b) Repeat part (a) using a 4 inch step size $(= h = k)$.

4   Consider a steel alloy plate that is 10 feet long and 6 feet wide, and is insulated on its flat surfaces. Suppose the left edge is maintained at 1000°F (very hot) and the other three edges are all maintained at 50°F .
(a) Use the finite difference method to solve for the temperatures within the plate using a spacing of 1 foot $(= h = k)$.
(b) Shade (or color, preferably in red) the part of the plate which will be over 140°F (hot part of the plate)
(c) Repeat parts (a) and (b) using a grid spacing of 4 inches $(= h = k)$.

5. (a) Use the finite difference method with $N = 4$ interior grid values on the x-axis and $M = 9$ interior grid values on the y-axis, to solve the following Poisson boundary value problem:

---

[10] There is an interesting problem about these so-called eigenvalues of the Laplacian. If the planar domain Ω is thought of as a drumhead, the (negatives of these) associated eigenvalues can be shown to be natural frequencies of vibration (see Chapter 10 of [Str-92] for details). The set of all of eigenvalues of Ω, the so-called **spectrum**, can be shown to be an infinite set satisfying: $\lambda_1 \geq \lambda_2 \geq \lambda_3 \cdots \rightarrow -\infty$.

This spectrum can thus be thought of as the totality of the range of tones which can be emmitted from a drumhead of shape Ω. In a famous 1966 paper entitled "*Can you hear the shape of a drum*" [Kac-66], it was asked that if one knows the spectrum of a given domain Ω, is the shape of Ω completely determined (up to congruence): In other words, do domains of different shapes have different spectra? The problem drew a lot of interest but remained open until 1992, when Carolyn Gordon, David Webb, and Scott Wolpert published their paper "*One cannot hear the shape of a drum*," where they found a counterexample of two noncongruent planar domains with the same spectra. Despite the fact that their domains were simple polygons, their construction actually relied on some sophisticated results from group theory.

$$\begin{cases} \text{(PDE)} & \Delta u = \sin(\pi x), \quad u = u(x,y) \qquad 0 \le x \le 1, 0 \le y \le 2 \\ \text{(BC)} & u(x,2) = 6 \equiv t(x), u(x,0) = 0 \equiv b(x) \\ & u(1,y) = 3y \equiv r(y), u(0,y) = 3\cos(\pi y) \equiv \ell(y). \end{cases}$$

Afterwards, graph the resulting approximation to the solution.
(b) Repeat with $N = 9$ and $M = 19$.

6.   (a) Use the finite difference method with $N = 9$ interior grid values on the x-axis and $M = 9$ interior grid values on the y-axis, to solve the following Poisson boundary value problem:

$$\begin{cases} \text{(PDE)} & \Delta u = x^2 + y^2, \quad u = u(x,y) \qquad 0 \le x \le 1, 0 \le y \le 2 \\ \text{(BC)} & u(x,2) = 0 \equiv t(x), u(x,0) = 0 \equiv b(x) \\ & u(1,y) = 100 \equiv r(y), u(0,y) = 100 \equiv \ell(y). \end{cases}$$

Afterwards, graph the resulting approximation to the solution.
(b) Repeat with $N = M = 19$.

7.   (a) Prove that the indexing scheme (18) $k = i + N(M - j)$ always results in the reading order of indices k for the nodes (x_i, y_j) in a rectangle.

(b) How would the formula (18) change if we wished to index the nodes so they started on the bottom row, went left to right, and then moved up one row at a time?

8.   (a) Set up the finite difference method for the steady-state heat problem for the Laplace equation $\Delta u = 0$ on the domain shown in Figure 11.18(a) with specified Dirichlet data and using a common stepsize $h = k = 1$.
(b) Get MATLAB to solve the linear system and give a surface plot of the solution.
(c) Get MATLAB to give a corresponding isotherm plot.
(d) Repeat parts (a) through (c) using a step size $h = k = 0.5$.

9.   Repeat each part of Exercise 8 on the domain of Figure 11.18(a), but change the boundary Dirichlet data as follows: $u \equiv 0$ on all four sides of the outer boundary, while on the four inner boundary sides, u is specified by: $u(x,3) = u(x,7) = 25(x - 5)^2$, $u(3,y) = u(7,y) = 100$.

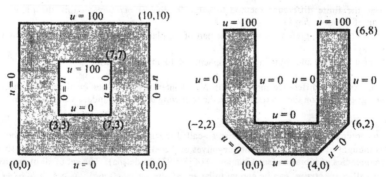

FIGURE 11.18: (a) (left) and (b): Two planar domains with boundary data for Dirichlet problems for Exercises 8–11.

10. (a) Set up the finite difference method for the steady-state heat problem for the Laplace equation $\Delta u = 0$ on the domain shown in Figure 11.18(b) with specified Dirichlet data and using a common stepsize $h = k = 1$.
(b) Get MATLAB to solve the linear system and give a surface plot of the solution.
(c) Get MATLAB to give a corresponding isotherm plot.
(d) Repeat parts (a) through (c) using a step size $h = k = 0.25$.

11. Repeat each part of Exercise 10 on the domain of Figure 11.18(b), but change the boundary Dirichlet data (only) on the four vertical sides to be linear, increasing from 0 to 100 as y increases from 2 to 8.

12. (a) Formulate a finite difference method using $N = M = 9$ interior grid values on both the x- and y-axes to solve the following elliptic boundary value problem:

$$\begin{cases} \text{(PDE)} \ (e^x u_x)_x + (e^y u_y)_y = 2e^{x+y}(e^x + e^y), \ 0 < x < 1, \ 0 < y < 1, u = u(x, y) \\ \text{(BC)} \ u(x,0) = e^x, \ u(x,1) = e^{x+1}, \ u(0, y) = e^y, \ u(1, y) = e^{y+1} \end{cases}.$$

Use MATLAB to numerically solve it and compare with the exact solution $u(x,y) = e^{x+y}$. What does the existence and uniqueness theorem (Theorem 11.3) say about this problem?
(b) Solve again with the finer grid $N = M = 19$.
(c) Repeat parts (a) and (b) for the BVP obtained by the above by changing all boundary values to zero, i.e., $u(x,y) = 0$ at all boundary points.

NOTE: (*Steady-State Fluid Flow Equation*) An important BVP that arises in applications of steady-state two-dimensional fluid flow is the following:

$$\begin{cases} \text{(PDE)} \ (au_x)_x + (bu_y)_y + cu = f(x, y) \text{ on } D \\ \text{(BC)} \ u = g(x, y) \text{ on the boundary of } D \end{cases}.$$

Here the function $u(x,y)$ denotes the pressure, a and b (which can be functions of x and y) denote fluid conductivity coefficients in the x- and y- directions. In this setting, the vector $(-au_x, -bu_y)$ turns out to be the fluid flux and $f(x,y)$ denotes the amount of fluid being added at the point (x,y). The PDE can be derived in a similar fashion to how the heat equation was done.

13. Set up a finite difference scheme for the steady-state fluid flow problem above using the parameters: $a = b = 1$, $c = -1$, and $f(x,y) = x^2$. Use $N = M = 9$ interior grid points on both the x- and y-axis on the square domain $D = \{0 < x < 1, 0 < y < 1\}$. Imposing zero boundary conditions on each side, solve the linear system and plot the resulting surface. Give also a contour plot of the isopressure lines. What does the existence and uniqueness theorem (Theorem 11 3) say about this problem?

14. Repeat all parts of exercise 13 on the problem modified as follows:
$a(x, y) = 2e^x$, $b(x, y) = x + y + 1$, $f(x, y) = 3$, if $x > 0.5$; $= 0$, if $x = 0$ 5; -3, if $x < 0.5$

NOTE: (*Nine-Point Formula for the Laplacian*) The following so-called nine-point formula

$$\Delta u(x, y) \approx \frac{1}{6h^2} \begin{bmatrix} 4u(x+h, y) + 4u(x-h, y) + 4u(x, y+h) + 4u(x, y-h) \\ u(x+h, y+h) + u(x-h, y+h) + u(x+h, y-h) \\ + u(x-h, y-h) - 20u(x, y) \end{bmatrix},$$

turns out to be extremely accurate, with truncation order $O(h^6)$ when used for the Laplace equation.[11] The next three exercises will examine its use.

---

[11] Letting $\Delta_h^{(9)} u(x, y)$ denote this approximation, it can be shown using Taylor's theorem that if u has sufficiently many continuous partial derivatives, then

$$\Delta_h^{(9)} u(x, y) - \Delta u(x, y) = \frac{1}{2} h^2 \Delta^2 u(x, y) + \frac{2}{6!} h^4 [\Delta^3 u(x, y) + 2(\Delta u)_{xxyy}(x, y)]$$

$$+ \frac{2}{8!} \frac{h^6}{3} [3\Delta^4 u(x, y) + 16(\Delta^2 u)_{xxyy}(x, y) + 20u_{xxxxyyyy}(x, y)] + O(h^7),$$

where $\Delta^2 u$ means $\Delta(\Delta u)$, etc.; see [KaKr-58]. From this it clearly follows that in case u is harmonic (Laplace equation) the approximation is $O(h^6)$, but in general it is only $O(h^2)$, and thus cannot be

15.   (a) Use the nine-point formula in a finite difference method to solve the Dirichlet problem

(PDE) $\Delta u(x, y) = 0$, $0 < x < 1$, $1 < y < 2$

(BC) $u(x, 1) = \ln(x^2 + 1)$, $u(x, 2) = \ln(x^2 + 2)$, $u(0, y) = \ln(y^2)$, $u(1, y) = \ln(y^2 + 1)$,

using $N = 4$ interior grid points on the x-axis and $M = 4$ interior grid points on the y-axis. Look at the errors by comparing with the exact solution $u(x, y) = \ln(x^2 + y^2)$. Solve also using the standard five-point finite difference formula and compare the errors.
(b) Repeat part (a), this time using $N = M = 9$.
(c) Re-solve Exercise for the Reader 11.4, this time incorporating the nine-point formula. How does the performance compare (use the exact solution provided to get the errors) with that of the standard finite difference method?

16.   Repeat each part of Exercise 8, but this time incorporating the nine-point formula.

17.   (a) Prove Theorem 11.2 in the general case of step sizes that are not necessarily equal.
(b) Prove an extension to Theorem 11.2 to the case of the following more general boundary value problem:

$$\begin{cases} au_{xx} + bu_{xx} = f(x, y) & \text{on } D \\ u = g(x, y) & \text{on } \partial D \end{cases} \qquad u = u(x, y),$$

where a and b are nonzero real numbers of the same sign. Is the result still valid if a and b are real numbers of opposite signs? How does this tie in with ellipticity?
(c) Does the extension of part (b) continue to be valid in case a and b are allowed to be (continuous) functions $a = a(x, y)$, $b = b(x, y)$ that are of the same sign?

11.4: GENERAL BOUNDARY CONDITIONS FOR ELLIPTIC PROBLEMS AND BLOCK MATRIX FORMULATIONS

Our introduction to finite difference methods in the last section centered on elliptic problems with Dirichlet boundary conditions. Allowing more general boundary conditions will lead to related methods, all of which can be very nicely expressed in the language of block matrices. The notations and concepts of this section coupled with MATLAB's ease of handling matrices will make the task of writing MATLAB codes for finite difference methods a very natural one. Furthermore, this centralized approach will carry over well into the development of finite difference methods for other sorts of PDEs, some more of which will be examined in the next chapter.

We begin with the Dirichlet problem for the Poisson equation in two space dimensions:

$$\begin{cases} (\text{PDE}) & \Delta u = f(x, y) \quad \text{on } \Omega, \quad u = u(x, y) \\ (\text{BC}) & u = g(x, y) \quad \text{on } \partial\Omega \end{cases}, \qquad (22)$$

---

much better than the usual five-point approximation. Note that the approximation is still $O(h^6)$ when used for a Poisson equation $\Delta u = f(x, y)$ whenever $\Delta f = f_{xxyy} = 0$, and so in particular whenever $f(x, y)$ is a polynomial of the form $A + Bx + Cy + Dxy + E(x^2 - y^2)$.

The domain Ω will be a rectangle in the plane, which for convenience we assume has its lower-left vertex at the origin: $\Omega = \{(x, y): 0 < x < a, \ 0 < y < b\}$. The symbol $\partial\Omega$ denotes the boundary of the domain Ω, which in this case consists of the four sides of the rectangle. Unlike for the parabolic and hyperbolic problems that we will discuss in the next chapter, the relation of horizontal to vertical step sizes is not so important for elliptic problems, so for simplicity we will use equal step sizes $(h = k)$. We assume that a step size $h > 0$ has been chosen so that $a = (N+1)h$ and $b = (M+1)h$. The x-grid points and y-grid points are then specified by:

$$x_i = ih \ (0 \leq i \leq N+1), \quad y_j = jh \ (0 \leq j \leq M+1) \tag{23}$$

and the corresponding functional values are then denoted by:

$$u_{i,j} = u(x_i, y_j) = u(ih, jh) \quad (0 \leq i \leq N+1, \ 0 \leq j \leq M+1). \tag{24}$$

We use analogous notation for the data functions of the problem: $f_{i,j}$, $g_{i,j}$ (of course, the latter is defined only for indices corresponding to boundary points).

Substituting the central difference approximations (14) and (15) into the PDE (22) results in the following discretization of the Poisson PDE:

$$\frac{u_{i+1,j} - 2u_{i,j} + u_{i-1,j}}{h^2} + \frac{u_{i,j+1} - 2u_{i,j} + u_{i,j-1}}{h^2} = f_{i,j} \ (1 \leq i \leq N, \ 1 \leq j \leq M). \tag{25}$$

Since the central difference formula employed has error $O(h^2)$, it follows that the local truncation error of the discretization (25) is $O(h^2) + O(h^2) = O(h^2)$. We rewrite (25) in the following simpler form:

$$4u_{i,j} - u_{i+1,j} - u_{i-1,j} - u_{i,j+1} - u_{i,j-1} = -h^2 f_{i,j}, \quad (1 \leq i \leq N, 1 \leq j \leq M). \tag{26}$$

This is a linear system with NM equations in the unknown interior grid values of $u(x,y)$ that we index using the scheme (18) into the components of a vector U:

$$P_k \equiv (x_i, y_j), \quad U_k \equiv u(P_k) = u_{i,j}, \quad k = i + N(M - j). \tag{27}$$

Recall that this indexing scheme results in the "reading order" labeling of the interior grid points (see Figure 11.11).

We wish to look carefully in the matrix form of this linear system:

$$AU = C. \tag{28}$$

Observe that we have relabeled only the unknown values of the $u_{i,j}$ that appear in (26); the Dirichlet boundary conditions of (22) give us the following data:

$$u_{i,0} = g_{i,0}, \quad u_{i,M+1} = g_{i,M+1} \quad (0 \le i \le N+1),$$
$$u_{0,j} = g_{0,j}, \quad u_{N+1,j} = g_{N+1,j} \quad (0 \le j \le M+1). \tag{29}$$

If we directly incorporate the indexing scheme (27) into (26), we arrive at the following:

$$4U_k - U_{k+1} - U_{k-1} - U_{k-N} - U_{k+N} = -h^2 f_k, \tag{30}$$

which, however, needs to be corrected in case any of the u-values in (26) is a known boundary value. If this happens, such values need to be substituted by the g-values in (29) and then moved to the right side. Let us carefully consider each case when such corrections are needed. It is most helpful to view such cases by thinking about the "reading order" indexing of the interior nodes $P_k = (x_i, y_j)$, $(k = i + N(M - j))$ (as in Figure 11.11) as well as the stencil for our finite difference method (Figure 11.10).

Case 1: P_k lies on the top row (so $j = M$). The value U_{k+N} is thus known and should be moved to the right side as $g_{i,M+1}$.

Case 2: P_k lies on the bottom row (so $j = 1$). The value U_{k-N} is thus known and should be moved to the right side as $g_{i,0}$.

Case 3: P_k lies on the right edge (so $i = M$). The value U_{k+1} is thus known and should be moved to the right side as $g_{N+1,j}$.

Case 4: P_k lies on the left edge (so $i = 0$). The value U_{k-1} is thus known and should be moved to the right side as $g_{0,j}$.

Note that Cases 1 and 2 cannot occur simultaneously, nor can Cases 3 and 4, but either of the last two cases can occur in conjunction with either of the first two. In light of the blocklike structure of the above cases, the $NM \times NM$ coefficient matrix A in (28) is readily expressed as an $M \times M$ **block matrix** where each block is an $N \times N$ (ordinary) matrix as indicated below:

$$A = \begin{bmatrix} T_N & -I_N & 0_N & \cdots & & \cdots & 0_N \\ -I_N & T_N & -I_N & 0_N & & & \vdots \\ 0_N & -I_N & T_N & -I_N & & & \\ \vdots & 0_N & -I_N & T_N & \ddots & & \vdots \\ & & & \ddots & \ddots & \ddots & 0_N \\ \vdots & & & & \ddots & \ddots & -I_N \\ 0_N & \cdots & & \cdots & 0_N & -I_N & T_N \end{bmatrix} \tag{31}$$

where I_N denotes the $N \times N$ identity matrix, 0_N denotes the $N \times N$ zero matrix, and T_N is the following $N \times N$ tridiagonal matrix:

$$T_N = \begin{bmatrix} 4 & -1 & & & \\ -1 & 4 & -1 & & \text{\Large 0} \\ & -1 & 4 & \ddots & \\ & & \ddots & \ddots & \ddots \\ \text{\Large 0} & & & \ddots & \ddots & -1 \\ & & & & -1 & 4 \end{bmatrix}. \tag{32}$$

The block matrix A in (31) is said to have tridiagonal block structure. Using this same decomposition into blocks, the $NM \times 1$ vectors U and C of (28) can be expressed as the following juxtapositions of M $N \times 1$ vectors:

$$U = \begin{bmatrix} U^1 \\ U^2 \\ \vdots \\ U^M \end{bmatrix}, \quad C = \begin{bmatrix} B_1 - h^2 F_1 \\ B_2 - h^2 F_2 \\ \vdots \\ B_M - h^2 F_M \end{bmatrix}, \tag{33}$$

where

$$U^J = \begin{bmatrix} U_{1+(M-j)N} \\ U_{2+(M-j)N} \\ U_{3+(M-J)N} \\ \vdots \\ U_{N-1+(M-j)N} \\ U_{(M-J+1)N} \end{bmatrix}, \quad F_J = \begin{bmatrix} f_{1+(M-j)N} \\ f_{2+(M-J)N} \\ f_{3+(M-J)N} \\ \vdots \\ f_{N-1+(M-j)N} \\ f_{(M-J+1)N} \end{bmatrix}, \tag{34}$$

and

$$B_1 = \begin{bmatrix} g_{1,M+1} + g_{0,M} \\ g_{2,M+1} \\ g_{3,M+1} \\ \vdots \\ g_{N-1,M+1} \\ g_{N,M+1} + g_{N+1,M} \end{bmatrix}, B_J = \begin{bmatrix} g_{0,M+1-J} \\ 0 \\ 0 \\ \vdots \\ 0 \\ g_{N+1,M+1-J} \end{bmatrix} (1 < J < M), B_M = \begin{bmatrix} g_{1,0} \mid g_{0,1} \\ g_{2,0} \\ g_{3,0} \\ \vdots \\ g_{N,1,0} \\ g_{N,0} + g_{N+1,1} \end{bmatrix}. \tag{35}$$

Note that the coefficient matrix A is quite sparse; indeed, it is a banded matrix with 5 bands: the main diagonal, the sub- and superdiagonals, and the diagonals that lie M units above and below the main diagonal. By Proposition 7.14, A is invertible (in fact positive definite). In Section 7.7, it was shown how the SOR method can very effectively solve linear systems having A as the coefficient matrix. Such sparse solution techniques will greatly expand the resolution that we will be able to attain in our numerical PDE solution techniques. Indeed, if we wanted to have a resolution of, say, 100 interior grid values on the x- and y-axes (with a square domain), this would mean that the linear system (28) that is needed to be solved would involve a $10,000 \times 10,000$ coefficient matrix A. Even storing such a matrix would tax most home PCs; attempting to solve the system with the general Gaussian elimination method would require on the order of $(10,000)^3 = 10^{12}$ flops.

Even at the rate of 1 million flops/second, this would take nearly two weeks to solve! MATLAB's left divide is able to take advantage of the special structure of positive definite matrices, like A, which arise in finite difference methods. This results in MATLAB's ability to solve such linear systems as long as it is possible to store the coefficient matrix. Furthermore, many of the matrices that arise in numerical differential equations are sparse, and so taking advantage of scarcity will permit the solution of even larger systems (see Section 7.7). In order to keep this chapter more accessible, we will be using this left divide solver in lieu of iterative methods; however, readers who have studied Section 7.7 are encouraged to apply some of the methods they have learned on the linear systems that come up in this and subsequent sections.

Our next example will demonstrate how the above notation will allow us to code this finite difference method into a succinct MATLAB program.

EXAMPLE 11.7: (a) Use the finite difference method to solve the following Poisson problem[12]:

$$\begin{cases} \text{(PDE)} & \Delta u = -f(x,y) \quad \text{on } \Omega = \{0 < x, y < 1\}, \quad u = u(x,y) \\ \text{(BC)} & u(x,0) = u(x,1) = 5x \\ & u(0,y) = 0, \ u(1,y) = 5 \end{cases},$$

where $f(x,y)$ is defined by:

$$f(x,y) = \begin{cases} 800, & \text{if } \frac{1}{4} < x < \frac{1}{2} \text{ and } \frac{1}{2} < y < \frac{3}{4} \\ 0, & \text{otherwise} \end{cases}.$$

Use a step size $h = 0.02$.
(b) Plot the numerical solution as a surface plot.
(c) Give a two-dimensional contour plot of the solution.

NOTE: Recall that the Poisson equation models steady-state temperature distribution with a time-independent heat source $f(x,y)$. Thus, we can view the solution to the above problem as the steady-state temperature distribution on the unit square Ω with the edges maintained at the temperatures specified by the BC and with a homogeneous heat source concentrated on the specified smaller square of sidelength 1/4. The contours of the plot in part (c) are then the isotherms of the temperature distribution.

SOLUTION: As we have been doing thus far in our development of numerically solving boundary value problems for PDEs, we will solve this problem in a way

---

[12] The reason that we used a negative coefficient in the Poisson PDE is to highlight the interpretation of the Poisson equation as a model of steady-state heat distributions with internal heat source term $f(x,y)$; cf. (6) of this chapter for the one-dimensional analogue (put $u_t = 0$). For this reason the Poisson equation is often written as $-\Delta u = f$.

that can be easily generalized to the creation of an M-file for solving more general problems.

Notice that with $h = 0.02$, from $N + 1 = 1/h$, we see that there will be $N = 49$ interior grid points on both the x- and y-axes. Thus the linear system to be solved, $AU = C$, will have a coefficient matrix of size $N^2 \times N^2$ ($N^2 = 2401$). Creating the coefficient matrix A of (32) can be done quite efficiently using MATLAB's `diag` function:[13]

```
>> N=49;
>> A=diag(4*ones(1,N^2))-diag(ones(1,N^2-N), N)-diag(ones(1,N^2-N),-
N);
>>    next create vector for sub super diagors .
>> v1=-ones(1,N-1); v=[v1 0];
>> for i=1:N-1
     if i<N-1
       v=[v v1 0];
     else
       v=[v v1];
     end
end
>> A=A+diag(v,1)+diag(v,-1);
```

In order to create the vector C of (33), we first store the given boundary values at our grid points, using some rather obvious notation:

```
>> leftdata=zeros(1,N+2); rightdata=5*ones(1,N+2);
>> xgrid=0:.02:1;
>> topdata=5*xgrid; bottomdata=topdata;
>> Bprep1=topdata(2:N+1); Bpreplast=bottomdata(2:N+1);
>> C=[];
>> Fprep=zeros(1,N);
>> h=0.02;
```

We also store the following M-file for the function on the right side of the PDE:

```
function z = squareheatsource(x,y)
if x>=.25 & x<=.5 & y>=.5 & y<=.75
     z=-800;
else
     z=0;
end

>> for j=1:N
     F=Fprep;
     for i=1:N
       F(i)=-h^2*feval('squareheatsource',h*i,1-h*j);
     end
     F(1)=F(1)+leftdata(1+j);
     F(N)=F(N)+rightdata(1+j);
     if j==1
       F=F+Bprep1;
     elseif j==N
```

---

[13] **Note:** Some of the matrix creation commands may take a second or two to execute, depending on the speed of your computer.

```
      F=F+Bpreplast;
    end
    C=[C; F'];
end
>> %now we can assemble the matrix Z of u-values
>> U=A\C;
>> Z=zeros(N,N);
>> Z(:)=U; Z=Z';
>> Z=[topdata(2:N+1); Z; bottomdata(2:N+1)];
>> Z=[leftdata' Z rightdata'];
```

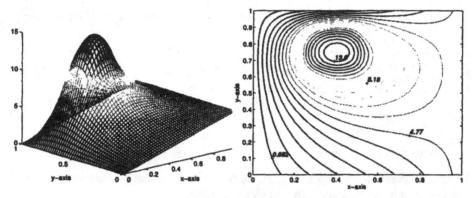

FIGURE 11.19: (a) (left) Temperature mesh plot for Example 11.7. (b) (right)
Corresponding isotherms.

```
>> size(xgrid)
→ans = 1   51
for i=1:51
ygrid(i)=xgrid(52-i);
end %as usual, we reverse the order of y-grid for plots to be
correct.
>> mesh(xgrid,ygrid,Z)
>> hidden off, xlabel('x-axis'), ylabel('y-axis')
>> c=contour(xgrid, ygrid,Z,20);
>> clabel(c, 'manual')
```

EXERCISE FOR THE READER 11.7: (a) Write a MATLAB function M-file that
is precisely designed to solve the Dirichlet problem for the Poisson equation
$\Delta u = f$ on the rectangle with vertices (0,0), (a,0), (a,b), and (0,b). The syntax
should be as follows:

```
[Z, x, y]=rectanglepoissonsolver(h, a, b, varf, leftdata, rightdata,
                         topdata, bottomdata)
```

where h = the common step size on the x- and y-axes ($h = k$) (assumed to divide
into both a and b), a and b are the dimensions of the rectangular domain, varf is
either an M-file or an inline function for the function f appearing in the PDE,
and the remaining four input variables are vectors for the boundary data. The first
two should be column vectors of length $M = b/h + 1$, with the boundary values

given from top to bottom. The last two should be row vectors of length $N = a/h +$ 1 with boundary values given from left to right. The first output variables is Z, an $(N+2) \times (M+2)$ matrix of the values of the solution at the corresponding grid values of the rectangle: The first column of Z should thus be the vector leftdata, etc. The last two output variables x and y should simply be the vectors giving the x- and y-grid values; however, y should be given in decreasing order to facilitate plotting.
(b) Test the program by re-solving Example 11.7.

We now proceed to discuss the generalized Neumann problem for the Poisson equation in two space dimensions:

$$\begin{cases} \text{(PDE)} & \Delta u = f(x,y) \quad \text{on } \Omega, \quad u = u(x,y) \\ \text{(BC)} & \partial u / \partial n = g(x,y) \quad \text{on } \partial\Omega \end{cases} \tag{36}$$

Here $\partial u / \partial n$ denotes the derivative of $u(x,y)$ in the direction of the outward-pointing normal vector n for points on the boundary. Recall that when this BVP models steady-state heat distribution (with time-independent heat source), the generalized Neumann boundary conditions specify the rate at which heat is lost ($g > 0$) or gained ($g < 0$) at the boundary point (x,y).[14]

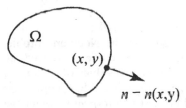

FIGURE 11.20: Illustration of the unit outward normal vector at a boundary point (x,y) of a planar domain Ω.

Unlike the Dirichlet problem for the Poisson equation, the Neumann problem requires an additional consistency hypothesis in order for a solution to exist. Also, it is clear that if $u(x,y)$ is a solution to the Neumann problem (36) and C is any constant, then $u(x,y) + C$ will also be a solution (why?). Apart from this we do have uniqueness. The following theorem makes these statements more precise.

THEOREM 11.4: (*Existence and Uniqueness for the Poisson PDE with Neumann Boundary Conditions*) Suppose that Ω is a smooth bounded planar domain and that in the BVP (36):

$$\begin{cases} \text{(PDE)} & \Delta u = f(x,y) \quad \text{on } \Omega, \quad u = u(x,y) \\ \text{(BC)} & \partial u / \partial n = g(x,y) \quad \text{on } \partial\Omega \end{cases},$$

---

[14] Traditionally, the term "Neumann boundary conditions" is reserved for the special case that $h(x,y) \equiv 0$ (insulated boundary).

$f(x,y)$ has piecewise continuous partial derivatives throughout Ω and $g(x,y)$ is piecewise continuous on $\partial\Omega$. Then the BVP has a solution if and only if the data satisfies the following **compatibility condition**:

$$\iint\limits_{\Omega} f(x,y)dxdy = \int_{\partial\Omega} g(x,y)ds. \tag{37}$$

where the latter integral is with respect to arc length along the boundary. In this case, the solution is unique up to an additive constant.

A few comments are in order. If we interpret the BVP (36) as a steady-state heat flow problem, the (BC) can be interpreted as requiring that the heat flux (net heat flow across the boundary) at (x,y) is given by $g(x,y)$. The right-side integral in (37) thus becomes the net heat flux lost through the boundary of the region. The left-side integral is the net heat produced within the region. For equilibrium (steady-state) the amount of internal heat produced must equal the net heat lost through the boundary (conservation of heat). This is why the compatibility condition is required. As for the nonuniqueness, this is plausible since the BVP is only stating the net heat production within the region and the net heat flux. There is no reference to how the temperature is being measured (Fahrenheit or Celsius?) or to how much heat energy is contained in the region, and so it is natural the additive constant must appear as a point of reference.

While for the Dirichlet problem on the square it is quite common for the boundary data to be continuous, for the Neumann problem it is typical that the boundary data is discontinuous at the corners. Indeed, at each corner point, the most typical situation would allow two values of $g(x,y)$, depending on from which side we are looking at normal derivatives. This simply corresponds to the fact that the direction of the normal vector (with respect to which we are measuring the rate of change of $u(x,y)$) takes a sharp (discontinuous) turn at each corner point and the rate of heat flow will change with the direction in which we are measuring it. To make the Neumann problem well-posed (have a unique solution) we could require an additional condition that the temperature be a certain value at a certain point in the domain. Vector calculus can be used to prove Theorem 11.4; some elements of the proof will appear in the exercises.

Turning to finite difference methods for solving the Neumann problem (36), we need to approximate the derivative BC using a finite difference formula. The following so-called forward difference and backward difference formulas appear to be rather plausible to this end:

LEMMA 11.5: (*Forward/Backward Difference Formulas*) (a), Suppose that $f(x)$ is a function having a continuous second derivative in the interval $a \le x \le a+h$; then we have the **forward difference approximation**:

$$f'(a) = \frac{f(a+h)-f(a)}{h} + O(h). \tag{38}$$

(b) Suppose that $f(x)$ is a function having a continuous second derivative in the interval $a - h \leq x \leq a$; then we have the **backward difference approximation**:

$$f'(a) = \frac{f(a) - f(a-h)}{h} + O(h). \tag{39}$$

The lemma is an easy consequence of Taylor's theorem (Exercise 14). A plausible way to set up a finite difference method for the Neumann problem (on a rectangle) would be by incorporating the boundary data with the forward difference scheme on the left and bottom edge nodes, and the backward difference scheme on the right and top edge nodes. While this could indeed be developed into a viable scheme, the drawback is that the $O(h)$ errors introduced by forward/backward difference portions of the schemes would contaminate the much better $O(h^2)$ local truncation errors that came up from the central difference approximation used in the internal node discretization of the PDE. A better approach is to use the central difference approximations both for the PDE and the boundary conditions. This can be accomplished by introducing additional nodes, so-called **ghost nodes**, as we will now explain (see Figure 11.21). The forward and backward difference schemes, however, will be of use in finite difference methods for parabolic and hyperbolic BVPs, which are studied in the next chapter.

Our next example will explain how to develop a finite difference scheme for a Neumann problem that will have local disretization error $O(h^2)$.

EXAMPLE 11.8: Use a finite difference method with common step size $h = 0.1$ to solve the following steady-state temperature distribution Neumann problem:

$$\begin{cases} \text{(PDE)} & \Delta u = 2, \quad u = u(x,y) & 0 \leq x \leq 0.5, 0 \leq y \leq 1 \\ \text{(BC)} & u_y(x,1) = -2, u_y(x,0) = 4 \\ & u_x(.5,y) = 4 - 8y, \ u_x(0,y) = 0 \end{cases},$$

with the additional requirement that $u(0,0) = 0$. Afterwards, graph the resulting approximation to the solution.

SOLUTION: The reader may check that both of the integrals in (37) have a common value of 1, so we know this Neumann problem has a solution. In contrast to the corresponding method for the Dirichlet problem (Example 11.5) the new scheme will require us to solve for the values of u at both the interior nodes as well as the boundary nodes. These nodes as well as the ghost nodes we will need are illustrated in Figure 11.21 (compare with Figure 11.11). The picture illustrates how the index scheme for the Dirichlet method should be modified. We briefly highlight the general notations that will be used:

$$x_i = (i-1)h \ (0 \leq i \leq N+1), \quad y_j = (j-1)h \ (-1 \leq j \leq M+1)$$

(note that now x_0, x_{N+1} correspond to the ghost nodes, so, as before, $x_1, \cdots, x_n$

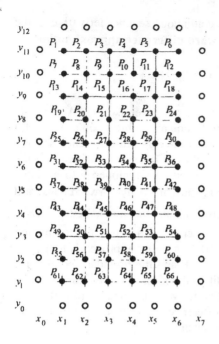

correspond to unknown function values, and similarly for y's). The indexing scheme is as before:

$$P_k \equiv (x_i, y_j), \quad U_k \equiv u(P_k) = u_{i,j},$$
$$k = i + N(M - j)$$

(ghost nodes are not indexed with k). The discretization of Poisson's PDE is just as before (26) (with equal step sizes)

$$4u_{i,j} - u_{i+1,j} - u_{i-1,j} - u_{i,j+1} - u_{i,j-1} = -h^2 f_{i,j}$$
$$(1 \le i \le N, 1 \le j \le M).$$

FIGURE 11.21: A grid for the Neumann problem of Example 11.7. The solid-labeled nodes are the ones (function values) that need to be solved for; the hollow nodes are the ghost nodes needed to set up the method.

In our example, of course, $N = 6$, $M = 11$, and $f(x, y) = -(2y)^2$, but we wish to present the general development. To allow for different Neumann data at corner points, we use the following notations for the Neumann data:

$$g_{i,j}^h = g^h(x_i, y_j) \quad \text{(horizontal side)}$$
$$g_{i,j}^v = g^v(x_i, y_j) \quad \text{(vertical side)}$$

Now we use the central difference formula to eliminate any ghost node values that appear in (26). To see how this works, let us first assume that $i = N$ (so we are dealing with a node on the right side of the rectangular domain). Then $u_{i+1,j} = u_{N+1,j}$ is a ghost value. The central difference formula gives us that

$$\frac{u_{N+1,j} - u_{N-1,j}}{2h} = g_{N,j}^v \quad \Rightarrow \quad u_{N+1,j} = u_{N-1,j} + 2hg_{N,j}^v,$$

which causes (26) to become:

$$4u_{N,j} - 2u_{N-1,j} - u_{N,j+1} - u_{N,j-1} = -h^2 f_{i,j} + 2hg_{N,j}^v \quad (1 < j < M).$$

We left out the two cases $j = 1$ and $j = M$, since these each need another ghost node to be accounted for. Analogously, these give the following:

$$4u_{N,1} - 2u_{N-1,1} - 2u_{N,2} = -h^2 f_{i,j} + 2h(g_{N,1}^v + g_{N,1}^h), \text{ and}$$
$$4u_{N,M} - 2u_{N-1,M} - 2u_{N,M-i} = -h^2 f_{i,j} + 2h(g_{N,M}^v + g_{N,M}^h).$$

Similar equations can be thus obtained for nodes on the other three sides. These considerations lead us to the linear system:

$$AU = C \tag{40}$$

where the $NM \times NM$ matrix A is given by:

$$A = \begin{bmatrix} W_N & -2I_N & 0_N & \cdots & & \cdots & 0_N \\ -I_N & W_N & -I_N & 0_N & & & \vdots \\ 0_N & -I_N & W_N & -I_N & & & \\ \vdots & 0_N & -I_N & W_N & \ddots & & \vdots \\ & & & \ddots & \ddots & \ddots & 0_N \\ \vdots & & & & \ddots & \ddots & -I_N \\ 0_N & \cdots & & & 0_N & -2I_N & W_N \end{bmatrix}. \tag{41}$$

Note that A is an $M \times M$ block matrix made up of the indicated $N \times N$ matrix blocks. Here I_N denotes the $N \times N$ identity matrix, 0_N denotes the $N \times N$ zero matrix, and W_N is the following $N \times N$ tridiagonal matrix:

$$W_N = \begin{bmatrix} 4 & -2 & & & & \\ -1 & 4 & -1 & & \mathbf{0} & \\ & -1 & 4 & \ddots & & \\ & & \ddots & \ddots & \ddots & \\ \mathbf{0} & & & -1 & \ddots & -1 \\ & & & & -2 & 4 \end{bmatrix}. \tag{42}$$

The $NM \times 1$ vectors U and C of (28) can be expressed as the following juxtapositions of M $N \times 1$ vectors:

$$U = \begin{bmatrix} U^1 \\ U^2 \\ \vdots \\ U^M \end{bmatrix}, \quad C = \begin{bmatrix} 2hB_1 - h^2 F_1 \\ 2hB_2 - h^3 F_2 \\ \vdots \\ 2hB_M - h^2 F_M \end{bmatrix}, \tag{43}$$

where

$$U^J = \begin{bmatrix} U_{1+(J-1)N} \\ U_{2+(J-1)N} \\ U_{3+(J-1)N} \\ \vdots \\ U_{N-1+(J-1)N} \\ U_{JN} \end{bmatrix}, \quad F_J = \begin{bmatrix} f_{1+(J-1)N} \\ f_{2+(J-1)N} \\ f_{3+(J-1)N} \\ \vdots \\ f_{N-1+(J-1)N} \\ f_{JN} \end{bmatrix}, \tag{44}$$

and

$$
B_1 = \begin{bmatrix} g_{1,M}^v + g_{1,M}^h \\ g_{2,M}^h \\ g_{3,M}^h \\ \vdots \\ g_{N-1,M}^h \\ g_{N,M}^v + g_{N,M}^h \end{bmatrix}, B_j = \begin{bmatrix} g_{1,M-j}^v \\ 0 \\ 0 \\ \vdots \\ 0 \\ g_{N,M-j}^v \end{bmatrix} (1 < j < M), B_M = \begin{bmatrix} g_{1,1}^v + g_{1,1}^h \\ g_{2,1}^h \\ g_{3,1}^h \\ \vdots \\ g_{N-1,1}^h \\ g_{N,1}^v + g_{N,1}^h \end{bmatrix}. \tag{45}
$$

This block matrix system shares some resemblance to the one we derived earlier in this section for the Dirichlet problem—but there is one important difference! Whereas the coefficient matrix A for the Dirichlet problem is symmetric, positive definite, and well-conditioned, the above matrix A is, in fact, singular! This can be readily verified since the sum of the entries in each row of A equals zero and therefore the vector $[1 \ 1 \ 1 \ \cdots \ 1]'$ is a solution of the homogeneous system $AU = 0$. This property of our finite difference model corresponds nicely to the property of the BVP mentioned in Theorem 11.3 that the solution of the Neumann problem (if exists) is unique only up to an additive constant. Indeed, the fact that $[1 \ 1 \ 1 \ \cdots \ 1]'$ is a solution of the homogeneous system $AU = 0$ lets us add a constant vector $c[1 \ 1 \ 1 \ \cdots \ 1]'$ to any solution of the linear system $AU = C$ and still have a solution. What is even more interesting is that the compatibility condition (37) $\int_\Omega f(x,y)dxdy = \int_{\partial\Omega} g(x,y)ds$ for the existence of a solution to the Neumann problem turns out to have the following discrete analogue for the solvability of the discrete system $AU = C$:

$$
\frac{1}{2} \sum_{P_k \text{ corner}} c_k + \sum_{P_k \text{ edge}} c_k + 2 \sum_{P_k \text{ interior}} c_k = 0. \tag{46}
$$

We leave the proofs of the facts that (46) is equivalent to $AU = C$ having a solution and that (46) can be viewed as the discrete analogue of the compatibility condition (37) to Exercises 20, and 21. For now, we proceed to verify the compatibility condition and then solve the system, $AU = C$. Since we are dealing with a singular system our usual methods cannot be applied here.

We begin coding the problem into MATLAB; as usual, we do so in a way that is amenable to the creation of more general codes.

```
>> N=6; M=11; h=0.1;
>> %next create vector for +/- N-diagonals
>> vN=-2*ones(1,N);
>> for i=2:M-1
     vN((i-1)*N+1:i*N)=-ones(1,N);
end
>> for i=1:length(vN)
     vNbott(i)=vN(length(vN)+1-i);
end
>> A=diag(4*ones(1,N*M))+diag(vN, N)+diag(vNbott,-N);

%next create vector for sub/super diagonals
>> v1=-ones(1,N-1); v1(1)=-2; v=[v1 0];
>> v2=-ones(1,N-1); v2(N-1)=-2; vbott=[v2 0];
```

```
>> for i=1:M-1
if i<M-1
   v=[v v1 0]; vbott = [vbott v2 0];
else
   v=[v v1]; vbott=[vbott v2];
end
end
>> A=A+diag(v,1)+diag(vbott,-1);
```

We next create the relevant boundary data and inhomogeneity (right-hand side) function:

```
>> xgrid=0:.1:.5; ygrid=0:.1:1;
>> leftdata=zeros(size(ygrid))'; rightdata= (4-8*ygrid)';
>> topdata=-2*ones(size(xgrid)); bottomdata=4*ones(size(xgrid));
>> f = inline('2', 'x', 'y');
```

from which we may now construct the needed vector C:

```
>> C=zeros(N*M,1);
>> for j=M:-1:1
for i=1:N
  C(i+N*(M-j))=-h^2*f(xgrid(i),ygrid(j));
   if i == 1
     C(i+N*(M-j))=C(i+N*(M-j))+2*h*leftdata(j);
   elseif i == N
     C(i+N*(M-j))=C(i+N*(M-j))+2*h*rightdata(j);
end
end
end
>> C(1:N)=C(1:N)+2*h*topdata';
>> C(M*N-N+1:M*N)= C(M*N-N+1:M*N)+2*h*bottomdata';
```

We now have constructed the known matrices A and C of the singular linear system $AU = C$. For good measure we check the validity of the discrete compatibility condition (46):

$$\frac{1}{2}\sum_{P_k \text{ corner}} c_k + \sum_{P_k \text{ edge}} c_k + 2\sum_{P_k \text{ interior}} c_k = 0.$$

We need to distinguish the indices k in these three sets. Using some MATLAB notation, these three index sets may be expressed as follows:

Corners: $k = 1, N, MN - 1, MN$,
Edges: $k = 2:(N-1), (N+1):N:(MN-1), (2N):N:((M-1)N)$
Interior nodes: all remaining indices k

The following loop will now evaluate the sum of (46):

```
>> sum=0;
for k=1:length(C)
  if ismember(k,[1 N M*N-N+1 M*N])
    sum=sum+C(k)/2;
  elseif ismember(k, [2:(N-1)   (N+1):N:(M*N)  (2*N):N:((M-1)*N) ...
    (M*N-N+1):(M*N-1)])
     sum=sum+C(k);
  else
```

```
      sum=sum+2*C(k);
   end
end
>> sum
```
→sum = 2.2204e-015

Taking into account floating point errors, this sum is zero, so the system will have a solution. To solve it numerically, we cannot use left divide. We use MATLAB's rref to put the corresponding augmented matrix $[A \mid C]$ into reduced row echelon form:[15]

```
>> Aug=[A C];  Augred=rref(Aug);
```

We now check that the row reduced augmented matrix has the expected form:

```
>> max(abs(Augred(66,:)))
```
→ans = 0 (Shows the last row is all zeros.)

If the compatibility condition (46) failed, then the last entry would not be zero, but all other entries in the last row would be zero so there would be no solution.

```
>> max(max(abs((Augred(1:65,1:65)-eye(65)))))
```
→ans =0 (Shows the upper 65×65 submatrix of Augred is the identity matrix.)

A simple solution of $AU = C$ is now obtained by setting $U_{66} = 0$ which, because of the special form observed above of Augred, simply amounts to taking U to be the last column of Augred:

```
>> U=Augred(:,67);
```

Since we would like to have $(u(0,0) =) U_{61}$ to equal zero and since constants can be added to solutions, the solution to our problem will be contained in the vector:

```
>> U=U-U(61);
```

We may now build an appropriate matrix of the u-values and plot as usual:

```
>> Z=zeros(N,M); Z(:)=U; Z=Z';

for i=1:M
y(i)=ygrid(M+1-i);
end %as usual, we reverse the order of y-grid for plots
to be correct.
>> surf(xgrid,y,Z)
>> hidden off, xlabel('x-axis'), ylabel('y-axis')
```

---

[15] We do not advocate using rref to solve singular systems, although we could get more reliable results by working with the Symbolic Toolbox. Apart from this example, we will not be needing to solve any singular linear systems in this book, so we will not delve into a serious discussion of the available numerical methods. Numerical methods for solving singular linear systems are rather sophisticated. The interested reader is encouraged to refer to Chapter 6 of [GoVL-83] or to look at the paper of A. Neumaier [Neu-98] for details.

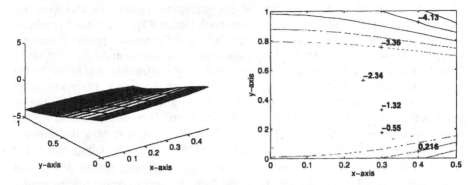

FIGURE 11.22: (a) (left) Surface plot of the solution to the Neumann problem of Example 11.7. (b) (right) Corresponding plot of isotherms.

The plot is shown in Figure 11.22 along with a corresponding isotherm plot. Note the effect of the various boundary conditions on the heat distribution near the edges. Some comments are in order. Such Neumann problems are not very stable numerically. Indeed, for the linear system to have a solution, an exact condition must hold (the discrete compatibility condition (46)); small roundoff errors can lead to a false conclusion of insolvability. Even worse, the discrete compatiblity condition (46) may not hold even when the integral version holds (37). This would happen, for example, if we changed the PDE in the above example to $\Delta u = 6y^2$ but leave the BCs intact. In this case, the reader could check that although the compatibility condition (37) is still valid, the discrete compatibility condition is no longer valid. See Exercise 22 for more details on such pathologies.

Despite the fact that Neumann problems are not very amenable to finite difference schemes (due basically to the exact requirement of the compatibility condition (37)), any other BCs on Poisson's equation (Robin, Dirichlet, or mixed, even with Neumann conditions on some—but not all of the boundary) give rise to a stable problem that can be solved by blending the methods used thus far. We state a relevant theorem and then give an example.

THEOREM 11.6: (*Existence and Uniqueness for the Poisson PDE with Mixed Boundary Conditions*) Suppose that Ω is a smooth bounded planar domain and that in the BVP

$$\begin{cases} (PDE) & \Delta u = f(x,y) \quad \text{on } \Omega, \quad u = u(x,y) \\ (BC) & p(x,y)\partial u / \partial n + r(x,y)u = g(x,y) \quad \text{on } \partial\Omega \end{cases} \quad (47)$$

where $f(x,y)$ has piecewise continuous partial derivatives throughout Ω and $g(x,y)$, $p(x,y)$, and $r(x,y)$ are piecewise continuous on $\partial\Omega$, with $p(x,y), r(x,y) \geq 0$, and $p(x,y) + r(x,y) > 0$ throughout $\partial\Omega$. If $r(x,y) > 0$ on some portion of $\partial\Omega$ of positive arclength, then the BVP (47) has a unique solution.

Like Theorem 11.5, this one can be proved using vector calculus (see for example Sections 81 and 82 of classical textbook [BrCh-93]). The last hypothesis appearing in this theorem simply ensures that the BCs are not purely Neumann (without this we we could infer nonuniqueness from Theorem 11.5). Problems with such mixed boundary conditions are usually quite amenable to solution by the finite difference methods of this section. For each boundary node involving Neumann or Robin conditions, we introduce a ghost node, while Dirichlet boundary points ($p = 0$, $r > 0$) do not require them. The next exercise for the reader requires such a mixing of methods. Even on a simple rectangular domain, for a mixed BVP (having the same type of BC on each edge) there are $3^4 = 81$ different types of BC configurations. Thus a general matrix description would not be feasible, but after working through the next exercise for the reader and some of the exercises at the end of this section, the reader should become quite adept at dealing with any sort of BCs.

EXERCISE FOR THE READER 11.8:   (a) Numerically solve the following Laplace problem with "mixed type" boundary conditions:

$$\begin{cases} \text{(PDE)} & \Delta u = 0, \quad u = u(x,y) \qquad 0 \le x \le 1, 0 \le y \le 1 \\ \text{(BC)} & u(x,0) = 0,\ u_y(x,1) = 20 \\ & u(0,y) = 100,\ u_x(1,y) = 0 \end{cases}$$

Use a common step size of $h = 0.05$. Obtain a surface graph of the solution along with an isotherm plot and interpret as a steady-state heat distribution.
(b) Repeat with $h = 0.02$. Your plots should look like those in Figure 11.23.

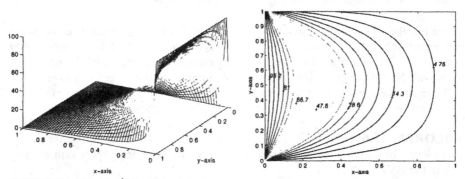

FIGURE 11.23: (a) (left) Mesh plot of the solution of the mixed BVP of Exercise for the Reader 11.8, using a common grid spacing of $h = 0.02$. Note how each of the boundary conditions are well depicted near the edges. (b) (right) Corresponding isotherm contour plot. The plots obtained for $h = 0.05$ appeared quite identical to these.

EXERCISES 11.4

1.   (*Dirichlet Problems for the Laplace Equation*) For each BVP given, do the following: (i) Set up the finite difference method for solving the problem using common x- and y-mesh size $h =$

0.1, and write down the linear system in block matrix notation. (ii) Use MATLAB to solve the resulting linear system, and produce a mesh plot of the solution surface. (iii) Obtain a contour plot of the isotherms. (iv) Repeat parts (i) through (iii) using $h = 0.05$. (v) Repeat parts (i) through (iii) using $h = 0.02$.

(a) $\begin{cases} \text{(PDE)} & \Delta u = 0 \quad \text{on } \Omega = \{0 < x < 2, 0 < y < 1\}, \quad u = u(x,y) \\ \text{(BC)} & u(x,0) = 0, \ u(x,1) = 100, \ u(0,y) = 0, \ u(2,y) = 100 \end{cases}$

(b) $\begin{cases} \text{(PDE)} & \Delta u = 0 \text{ on } \Omega = \{0 < x, y < 1\}, \quad u = u(x,y) \\ \text{(BC)} & u(x,0) = 50x, \ u(x,1) = -50x, \ u(0,y) = 0, \ u(1,y) = y^2 - 101y + 50 \end{cases}$

(c) $\begin{cases} \text{(PDE)} & \Delta u = 0 \quad \text{on } \Omega = \{3 < x < 4, 2 < y < 3\}, \quad u = u(x,y) \\ \text{(BC)} & u(x,2) = 0, \ u(x,3) = 0, \ u(3,y) = 0, \ u(4,y) = 100 \end{cases}$

(d) $\begin{cases} \text{(PDE)} & \Delta u = 0 \text{ on } \Omega = \{0 < x < 1, 0 < y < 1\}, \ u = u(x,y) \\ \text{(BC)} & u(x,0) = 20\sin(2\pi x), \ u(x,1) = 50x(1-x), \ u(0,y) = u(1,y) = 0 \end{cases}$

2.   (*Dirichlet Problems for the Poisson Equation*) Go through each of parts (i) through (v) of Exercise 1 for the following BVPs:

(a) In the BVP of Exercise 1(a), change the PDE to $\Delta u = 100(1 - (x-1)^2)$

(b) In the BVP of Exercise 1(b), change the PDE to $\Delta u = -f(x,y)$, where $f(x,y)$ is the "bump" function of Example 11.7.

(c) In the BVP of Exercise 1(c), change the PDE to $\Delta u = f(x,y)$, where

$$f(x,y) = \begin{cases} 500, & \text{if } x < 3.5 \\ 0, & \text{otherwise} \end{cases}.$$

(d) In the BVP of Exercise 1(a), change the PDE to $\Delta u = -(2x)^2 - (5y)^2$.

3.   (*Mixed Boundary Value Problems for the Laplace Equation*) Go through each of parts (i) through (v) of Exercise 1 by making the following changes in the BVP (a) of Exercise 1:

(a) Replace the corresponding BC with $u_y(x,0) = -20, \ u_y(x,1) = 40$.

(b) Replace the corresponding BC with $u_y(x,0) = u_y(x,1) = u_x(0,y) = -50$.

(c) Replace the corresponding BC with $u_y(x,0) = u_y(x,1) = u_x(0,y) = 30$.

(d) Replace the corresponding BC with $u_y(x,0) + u(x,0) = u_y(x,1) + u(x,1) = 50$.

4.   (*Mixed Boundary Value Problems for the Poisson Equation*) Go through each of parts (i) through (v) of Exercise 1 by making the following changes in the BVP (a) of Exercise 2:

(a) Replace the corresponding BC with $u_y(x,0) = 40, \ u_x(0,y) = 40y$.

(b) Replace the corresponding BC with $u_y(x,0) = -40, \ u_x(0,y) = -40y$.

(c) Replace the corresponding BC with $u_y(x,0) = u_y(x,1) = 10e^x, \ u_x(0,y) = 0$.

(d) Replace the corresponding BC with $u_y(x,0) = u_y(x,1) = 10e^x, \ u_x(0,y) + u(0,y) = 0$.

5.   (*Mixed Boundary Value Problems for the Poisson Equation*) Go through each of parts (i) through (v) of Exercise 1 by making the following changes in the BVP (c) of Exercise 2:

(a) Replace the corresponding BC with $u_y(x,3) = u_y(x,2) = u_x(3,y) = 80$.

(b) Replace the corresponding BC with $u_y(x,3) = u_y(x,2) = u_x(3,y) = -80$.

(c) Replace the corresponding BC with $u_x(3,y) + 10u(3,y) = 0, u_x(4,y) + 10u(4,y) = 40$.

(d) Replace the corresponding BC with $10u_x(3,y) + u(3,y) = 0, 10u_x(4,y) + u(4,y) = 40$.

6.   (a) Using a finite difference method with grid step sizes $h = k = \pi/5$, solve the following elliptic BVP:

$$\begin{cases} \text{(PDE)} & \Delta u + (x^2 + y^2)u = 0 \quad u = u(x,y) \qquad 0 \le x \le \pi, \, 0 \le y \le \pi \\ \text{(BC)} & u(x,\pi) = \sin(\pi x) \equiv t(x), \, u(x,0) = 0 \equiv b(x) \\ & u(\pi,y) = \sin(\pi y) \equiv r(y), \, u(0,y) = 0 \equiv \ell(y) \end{cases}$$

Plot your numerical solution and then print out the 6×6 matrix whose entries are the absolute values of the differences of the exact solution $\sin(xy)$ with the approximation at each of the grid points. What is the maximum single error occurring at a grid point?

(b) Repeat part (a) using step sizes $h = k = \pi/10$ (the grid for the second question will now be 11×11).

(c) Repeat part (a) once again using step sizes $h = k = \pi/30$.

7.   (a) Using a central difference approximation for the first-order partial derivative, use the finite difference method to solve the following elliptic boundary value problem:

$$\begin{cases} \text{(PDE)} & \Delta u - u_x = 2, \quad u = u(x,y) \qquad 0 \le x \le 1, \, 0 \le y \le 1 \\ \text{(BC)} & u(x,1) = 0 \equiv t(x), \, u(x,0) = 0 \equiv b(x) \\ & u(.5,y) = 0 \equiv r(y), \, u(0,y) = 0 \equiv \ell(y) \end{cases}$$

with equal grid step sizes $h = k = 0.2$.

(b) Repeat with $h = k = 0.1$ and compare the absolute value of the differences of this solution with that of part (a) on common grid points.

(c) Repeat again with $h = k = 0.05$ and compare the absolute value of the differences of this solution with that of part (b) on common grid points.

NOTE: The next four exercises will take advantage of sparse matrix storage and manipulations in MATLAB; this topic was discussed in Section 7.7.

8.   (a) Write an M-file whose syntax and functionality is identical to the `rectanglepoissonsolver` M-file of Exercise for the Reader 11.7 except that internally this new one will create the coefficient matrix as a sparse matrix. Call this new M-file `rectanglepoissonsolversp`.

(b) Test the new program out on the BVP of Example 11.7, and compare performance times with the original one for various step sizes. Allowing a maximum of five minutes on your computer, how many more internal nodes can your new M-file handle compared with the original?

9.   (*Dirichet Problems for the Poisson Equation*) For each of the BVPs given in Exercise 2, start off with a common step size $h = 1/4$ and solve it using the finite difference method but by storing the coefficient matrix as a sparse data type. Repeat with $h = 1/8$. Compare the two solutions at common interior grid points. Continue this halving of step sizes and comparing consecutive solutions until the maximum error falls below $1/100$ of the maximum observed amplitude of the most recent numerical solution, or the computation takes the computer more than 5 minutes.

10.   (*Mixed Boundary Value Problems for the Laplace Equation*) Repeat the instructions of Exercise 9 for each of the BVPs of Exercise 3.

11.   (*Mixed Boundary Value Problems for the Poisson Equation*) Repeat the instructions of Exercise 9 for each of the BVPs of Exercise 4.

12.   (*A Block Matrix Finite Difference Method for Unequal Step Sizes*) We apply the finite difference method to the BVP (20): $\begin{cases} \text{(PDE)} & \Delta u = f(x,y) \quad \text{on } \Omega, \\ \text{(BC)} & u = g(x,y) \quad \text{on } \partial\Omega \end{cases}$ on a rectangular domain

$\Omega = \{(x,y): 0 < x < a, \, 0 < y < b\}$; using step size h in the x-direction and k in the y-direction, the scheme is as in (20): $2\left(\dfrac{h^2}{k^2}+1\right)u_{i,j} - u_{i+1,j} - u_{i-1,j} - \left(\dfrac{h^2}{k^2}\right)u_{i,j+1} - \left(\dfrac{h^2}{k^2}\right)u_{i,j-1} = -h^2 q_{i,j}$.

(a) Using the natural ordering of grid points (Figure 11.11), what would the block matrix representation $AU = C$ look like for this scheme?

(b) Adapt the scheme of part (a) to solve the BVP of Exercise 1(a) using step sizes $h = 0.05$ and $k = 0.025$.

(c) Adapt the scheme of part (a) to solve the BVP of Exercise 1(d) using step sizes $h = 0.02$ and $k = 0.05$.

13. (*A Block Matrix Finite Difference Method for a Mixed BVP with Unequal Step Sizes*) If we apply the finite difference method to the BVP :

$$\begin{cases} \text{(PDE)} \quad \Delta u = f(x,y), \quad u = u(x,y) & 0 \le x \le a, \, 0 \le y \le b \\ \text{(BC)} \quad u(x,0) = T_b, \, u_y(x,1) = G_t \\ \qquad\quad u(0,y) = T_\ell, \, u_x(1,y) = G_r \end{cases},$$

using step size h in the x-direction and k in the y-direction, the scheme is as in (20):

$$2\left(\frac{h^2}{k^2}+1\right)u_{i,j} - u_{i+1,j} - u_{i-1,j} - \left(\frac{h^2}{k^2}\right)u_{i,j+1} - \left(\frac{h^2}{k^2}\right)u_{i,j-1} = -h^2 q_{i,j}$$

(a) Using the natural ordering of grid points and using ghost nodes as needed (cf. Figure 11.21), what would the block matrix representation $AU = C$ look like for this scheme?

(b) Adapt the scheme of part (a) to solve the BVP for the Reader 11.8 using step sizes $h = 0.05$ and $k = 0.025$.

14. Use Taylor's theorem to prove the forward and backward difference formulas (38) and (39).

NOTE: Many existence and uniqueness theorems for PDE boundary value problems, as well as the theoretical development of the finite element method, depend essentially on some integral identities known collectively as **Green's identities**. These identities are easily derived from the **divergence theorem** of vector calculus. These theorems are valid in any number of dimensions (2, 3, or more) but we develop them now in the two-dimensional setting (of this chapter). The next several exercises are thus intended for students who have studied multivariable calculus. Indeed, the divergence theorem is introduced and dealt with extensively in vector calculus courses.

DIVERGENCE THEOREM: If Ω is a bounded domain in the xy-plane that has a smooth boundary $\partial\Omega$, and $\vec{F}(x,y) = (F_1, F_2)$ is any vector-valued function with continuous first partial derivatives on $\Omega \cup \partial\Omega$, then we have $\Omega \cup \partial\Omega$

$$\iint_\Omega \text{div}\vec{F}(x,y)\,dxdy = \int_{\partial\Omega} \vec{F}(x,y)\cdot n(x,y)\,ds \tag{48}$$

where $\text{div}\,\vec{F}(x,y)$ denotes the divergence of the vector field $\vec{F}$: $\text{div}\,\vec{F}(x,y) = \partial F_1/\partial x + \partial F_2/\partial y$, $n = n(x,y)$ is the outward pointing normal vector at the point (x,y) on the boundary, and ds denotes arclength. The product on the right is the dot product.

15. (*Green's Identities*) Use the divergence theorem to prove each of the following integral identities. In each, the domain Ω is as in the divergence theorem, and the functions $u = u(x,y)$ and $v = v(x,y)$ appearing in the integrals are assumed to have continuous second partial derivatives. Also, the **gradient** operator is denoted by ∇, so, for example, $\nabla u(x,y)$ is the vector-valued function (u_x, u_y) .

(a) **Green's First Identity:** $\displaystyle\int_{\partial\Omega} v\frac{\partial u}{\partial n}\,ds = \iint_\Omega \nabla v\cdot\nabla u\,dxdy + \iint_\Omega v\Delta u\,dxdy$

(b) **Green's Second Identity:** $\int_{\partial\Omega}\left(u\dfrac{\partial v}{\partial n}-v\dfrac{\partial u}{\partial n}\right)ds=\iint_{\Omega}(u\Delta v-v\Delta u)\,dxdy$

Suggestion: For part (a), first show that $\nabla\cdot(v\nabla u)=\nabla v\cdot\nabla u+v\Delta u$ and then apply the divergence theorem. Use part (a) to prove part (b).

16. *(Proof of Uniqueness for the Dirichlet Problem for the Poisson's Equation)*   Complete the following outline to prove part of the uniqueness statements of Theorem 11.3: The Dirichlet problem $\Delta u=f(x,y)$ on Ω (smooth) and $u=g(x,y)$ on $\partial\Omega$, can have at most one solution.

   (a) Suppose that we have two solutions u_1 and u_2 of this BVP. Show that $u\equiv u_1-u_2$ is harmonic $(\Delta u=0)$ in Ω and vanishes on $\partial\Omega$.

   (b) Use Green's first identity to get that $\iint_{\Omega}|\nabla u(x,y)|^2\,dxdy=0$.

   (c) Show that $|\nabla u(x,y)|^2\equiv 0$ on Ω (see Exercise 14 of Section 10.5).

   (d) Show that u, having vanishing gradient on Ω, must be constant on each component (piece) of Ω. By the zero boundary conditions for u, we must have in fact $u\equiv 0$ and hence $u_1\equiv u_2$ on Ω.

17. *(Proof of Uniqueness for the Neumann Problem for the Poisson's Equation)* Using the outline of the previous exercise as a guide, prove that if u_1 and u_2 both solve the Neumann problem $\Delta u=f(x,y)$ on Ω (smooth) and $\partial u/\partial n=g(x,y)$ on $\partial\Omega$, then $u_2\equiv u_1+C$ throughout Ω, for some constant C.

18. *(Proof of Uniqueness for the Robin Problem for the Poissons Equation)* Using the outline of the Exercise 16 as a guide, prove that if u_1 and u_2 both solve the Robin problem $\Delta u=f(x,y)$ on Ω (smooth) and $\partial u/\partial n+ru=0$ on $\partial\Omega$, where $r>0$ throughout $\partial\Omega$, then $u_2\equiv u_1$ throughout Ω.

19. *(Dirichlet's Principle)*   Consider the Dirichlet problem for the Laplace equation on a smooth domain Ω: $\Delta u=0$ on Ω and $u=g(x,y)$ on $\partial\Omega$. For any "admissible" function v that has continuous partial derivatives on Ω and satisfies the boundary condition $v=g(x,y)$ on $\partial\Omega$, we define the **energy** of v by:

$$E(v)=\iint_{\Omega}|\nabla v(x,y)|^2\,dxdy\,.$$

Dirichlet's Principle states that the solution of the Dirichlet problem has the lowest possible energy (physicists refer to this as the "ground state") among all admissible functions. In other words, if u is the solution of the Dirichlet problem and v is any other admissible function, then $E(v)\geq E(u)$. Follow the outline below to prove Dirichlet's principle:

   (a) Consider the difference function $w=v-u$, which has zero boundary data. Show that we can expand:

$$E(v)=E(u)+\iint_{\Omega}\nabla u\cdot\nabla w\,dxdy+E(w).$$

   (b) Use Green's first identity to show that the middle term (the double integral) in the above expansion is zero. Since each of the three energies is nonnegative, deduce Dirichlet's principle.

20. *(The Discrete Compatibility Condition for the Neumann Problem)*
   (a) Consider the smallest possible discretization of a Neumann problem that actually has interior nodes: a 3×3 grid of nodes. The problem then has nine nodes, as shown in Figure 11.24. Show that the discretization $AU=C$ of the Neumann problem (36)

$$\begin{cases} \text{(PDE)} & \Delta u = f(x,y) \quad \text{on } \Omega, \quad u = u(x,y) \\ \text{(BC)} & \partial u / \partial n = h(x,y) \quad \text{on } \partial\Omega \end{cases}$$

has a solution if and only if the condition (46) (specialized to the present setting):

$$\tfrac{1}{2}c_1 + c_2 + \tfrac{1}{2}c_3 + c_4 + 2c_5 + c_6 + \tfrac{1}{2}c_7 + c_8 + \tfrac{1}{2}c_9 = 0$$

holds.

(b) Generalize your proof in part (a) to show that a general discretization $AU = C$ of the Neumann problem (36) will have a solution if and only if (46)

$$\frac{1}{2} \sum_{P_k \text{ corner}} c_k + \sum_{P_k \text{ edge}} c_k + 2 \sum_{P_k \text{ interior}} c_k = 0$$

holds.

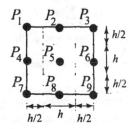

FIGURE 11.24: A small grid for easy visualization of some properties of finite difference schemes.

Suggestion: For part (a), use (46) to formulate the following sequence of elementary row operations to perform to the last row of the augmented matrix $[A \mid C]$

$$R_9 \to \tfrac{1}{2}R_9 + R_8 + \tfrac{1}{2}R_7 + R_6 + 2R_5 + R_4 + \tfrac{1}{2}R_3 + R_2 + \tfrac{1}{2}R_1 .$$

Verify that this will clear out the last row of A, and leave the expression in (46) in the last entry on the other side of the partition line.

21. (*Interpretation of the Discrete Compatibility Condition for the Neumann Problem*)
(a) When specialized to the case of a 3×3 grid on a square as in Figure 11.24, with $a = b = 1$, and $h = 1/2$, interpret the condition (46) (specialized to the present setting) $\tfrac{1}{2}c_1 + c_2 + \tfrac{1}{2}c_3 + c_4$

$+2c_5 + c_6 + \tfrac{1}{2}c_7 + c_8 + \tfrac{1}{2}c_9 = 0$ as a discrete version of the compatibility condition (37)

$\int_\Omega f(x,y) dx dy = \int_{\partial\Omega} g(x,y) ds$.

(b) Generalize your proof in part (a) to interpret condition (46) $\frac{1}{2} \sum_{P_k \text{ corner}} c_k + \sum_{P_k \text{ edge}} c_k$

$+2 \sum_{P_k \text{ interior}} c_k = 0$ as a discrete version of the compatibility condition.

Suggestion: For part (a), write out the vector C as defined by (43), (44), (45):

$$C = [2h(g_{1,3}^v + g_{1,3}^h) - h^2 f_{11}, \ 2hg_{1,3}^h - h^2 f_{12}, \ 2h(g_{2,3}^v + g_{2,3}^h) - h^2 f_{13}, \ 2hg_{2,1}^v - h^2 f_{21}, \cdots]'.$$

Interpret each individual term of (46) as an approximation to a subintegral of one of the integrals in (37); for example, for the first term $\tfrac{1}{2}c_1$ of (46), we can write:

$$\tfrac{1}{2}c_1 = h(g_{1,3}^v + g_{1,3}^h) - \tfrac{1}{2}h^2 f_{11}$$
$$\approx 2 \int_{3/4}^{1} g^v(0,y) dy + 2 \int_{0}^{1/4} g^h(x,1) dy - 2 \int_{0}^{1/4} \int_{3/4}^{1} f(x,y) dy dx$$

(simply approximate the functions on the sets of integration by the corresponding constants on the left). Once this is done, it will be apparent that the expression (46) is an approximation to

$$2\left(\int_{\partial\Omega} g(x,y)ds - \int_{\Omega} f(x,y)dxdy\right)$$ and thus setting this latter expression equal to zero produces (37).

22. In Example 11.8, suppose that the PDE is changed to $\Delta u = 6y^2$ but the boundary conditions are left the same.

(a) Show that the compatibility condition (37) $\int_{\Omega} f(x,y)dxdy = \int_{\partial\Omega} g(x,y)ds$ holds and thus this Neumann problem has a solution.

(b) Using the same grids as were employed in Example 11.8, show that the discrete compatibility condition (46) $\frac{1}{2}\sum_{P_k \text{ corner}} c_k + \sum_{P_k \text{ edge}} c_k + 2\sum_{P_k \text{ interior}} c_k = 0$ fails. Thus, for this BVP, although it has a solution, the associated linear system (from the finite difference method) does not have a solution.

(c) In the language of Exercise 21, explain how these two compatibility conditions are not consistent.

(d) In the language of Exercise 21, explain why the discrete approximations of (46) correspond exactly to the integrals of (37) for the original BVP of Example 11.8.

Suggestion: For part (c), the discrete approximations corresponding to the boundary integral $\int_{\partial\Omega} g(x,y)ds$ are exact, but those for the domain integral $\int_{\Omega} f(x,y)dxdy$ are not.

23. Construct a Neumann problem $\begin{cases} (PDE) & \Delta u = f(x,y) \quad \text{on } \Omega, \quad u = u(x,y) \\ (BC) & \partial u/\partial n = g(x,y) \quad \text{on } \partial\Omega \end{cases}$ on a rectangular

domain along with a grid of nodes for the finite difference method having the property that the compatibility condition (37) $\int_{\Omega} f(x,y)dxdy = \int_{\partial\Omega} g(x,y)ds$ fails and thus this Neumann problem does not have a solution, but such that the corresponding discrete compatibility condition (46) $\frac{1}{2}\sum_{P_k \text{ corner}} c_k + \sum_{P_k \text{ edge}} c_k + 2\sum_{P_k \text{ interior}} c_k = 0$ does hold. Explain your example in the context of Theorem 11.6 and Exercises 20 and 21.

Chapter 12: Hyperbolic and Parabolic Partial Differential Equations

12.1: EXAMPLES AND CONCEPTS OF HYPERBOLIC PDE'S

In the last chapter, we discussed in some detail the heat and Laplace's equations, which are prototypes for parabolic and elliptic PDEs, respectively. We would like now to introduce some concepts and theory for the wave equation, which is the prototype for hyperbolic equations. The wave equation models many natural phenomena, including gas dynamics (in particular, acoustics), vibrating solids and electromagnetism. It was first studied in the eighteenth century to model vibrations of strings and columns of air in organ pipes. Several mathematicians contributed to these initial studies, including Taylor, Euler, and Jean D'Alembert, about whom we will say more shortly. Subsequently in the nineteenth century, the wave equation was used to model elasticity as well as sound and light waves, and in the twentieth century, it has been used in quantum mechanics and relativity and most recently in such fields as superconductivity and string theory. In general, the **wave equation** has a time variable t and any number of space variables x, y, z,... and takes the form

$$u_{tt} = c^2 \Delta u = c^2 (u_{xx} + u_{yy} + \cdots), \tag{1}$$

where c is a positive constant and the Laplace operator on the right is with respect to all of the space variables. Modifications of this equation have been successfully used to model numerous physical waves and wavelike phenomena. In two space variables, for example, allowing for a variable wave speed due to depth differences in an ocean, the PDE: $u_{tt} = \nabla \cdot [H(x, y, t) \nabla u] + H_{tt}$ has been used to model large destructive ocean waves.[1] In such an application, the function H is the depth of the ocean at space coordinates (longitude and latitude) (x, y) and at time t. The latter term corresponds to the changes in depth due to underwater landslides. For more on this and other applications of this variable media wave equation, we mention the text [Lan-99].

---

[1] The symbol ∇, read as "nabla" or "del," is used to represent the gradient operator, which is the vector of all partial derivatives of a function. Thus for a function of two variables $f(x, y)$, $\nabla f = \nabla f(x, y) \equiv (f_x(x, y), f_y(x, y))$. The large dot represents the vector dot product, so in long form: $\nabla \cdot [H(x, y, t) \nabla u] = (\partial_x, \partial_y) \cdot (Hu_x, Hu_y) = \partial_x (Hu_x) + \partial_y (Hu_y)$. In particular, when $H \equiv 1$ we have $\nabla \cdot [\nabla u] = \partial_x (u_x) + \partial_y (u_y) = u_{xx} + u_{yy} = \Delta u$, another way to write the Laplacian of u. Such notations are very common in the literature for partial differential equations involving several space variables.

Much of the general theory of hyperbolic PDEs is well represented by that for the **one-dimensional wave equation** ($u = u(x,t)$ depends on time t and one space variable x), so we proceed now to introduce it through its historical model of a vibrating string and present some of the theory.   At the end of the section we indicate some differences and similarities of higher-dimensional waves to one-dimensional waves.

We consider a small segment of taut string having length Δs and uniform tension T that is acted on by a vertical force q, as shown in Figure 12.1.

We assume that the string is displaced only in the vertical (transverse) direction, and let $u(x,t)$ denote the y-coordinate of the string at horizontal coordinate x at the time t. If we let ρ denote the mass density (mass per unit length) of the string (assumed constant), then Newton's second law ($F = ma$) gives us that

$$-T\sin\theta + T\sin(\theta + \Delta\theta) + q\Delta s = \rho\Delta s u_{tt}(x,t),$$

where the first two terms represent the vertical component of the internal elastic forces acting on the segment of string.

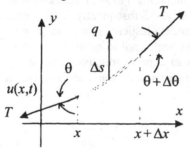

FIGURE 12.1: A segment of a uniformly taut string having tension T and external load q. The string is displaced vertically only, and $u(x,t)$ is the vertical level of the string at time t and horizontal position x.

For   small   deflections   in   the   string,   we   have $\Delta s \approx \Delta x$ and   also $\sin(\theta) \approx \theta \approx u_x(x,t)$. In the limit as $\Delta s \rightarrow 0$, this brings us to

$$Tu_{xx} + q = \rho u_{tt}, \quad u = u(x,t), \tag{2}$$

which is the **one-dimensional wave equation with external load term** q. In case $q = 0$, this reduces to the one-dimensional wave equation (1) with $c = (T/\rho)^{1/2}$. It turns out that this parameter c is the speed at which the wave (i.e., any solution of the equation) propagates. This will be made clear shortly. Intuitively, it makes sense that the speed of any disturbance on a string should increase along with the tension and decrease for heavier strings. For a derivation of wave equations for strings under more general hypotheses we refer to the article by S. Antman [Ant-80] or Chapter 3 of the textbook by Kevorkian [Kev-00].

FIGURE 12.2: Jean Le Rond D'Alembert (1717–1783), French mathematician.

The general solution of the one-dimensional wave equation was first derived by the French mathematician Jean D'Alembert.[2] D'Alembert's derivation is simple and elegant and the form of the solution will give many insights into qualitative aspects of wave equations. It begins by introducing the new variables:

$$\xi = x - ct, \quad \eta = x + ct. \tag{3}$$

We may now think of u as either a function of (x,t) or of (ξ, η). When we use the chain rule to translate the wave equation (1) into a PDE with respect to the new variables (ξ, η), something very nice will happen. The resulting PDE will be extremely easy to solve for the general solution. Applied using (3), the chain rule gives the following:

$$u_x = u_\xi \xi_x + u_\eta \eta_x = u_\xi + u_\eta$$
$$u_t = u_\xi \xi_t + u_\eta \eta_t = -cu_\xi + cu_\eta \tag{4}$$

In the same fashion, if we differentiate once again, we arrive at

$$u_{xx} = u_{\xi\xi} + 2u_{\xi\eta} + u_{\eta\eta}, \quad u_{tt} = c^2(u_{\xi\xi} - 2u_{\xi\eta} + u_{\eta\eta}). \tag{5}$$

When we substitute equations (5) into the one-dimensional wave equation (1), we obtain the following version of the wave equation in the new variables (ξ, η):

$$u_{\xi\eta} = 0. \tag{6}$$

This PDE is very easy to solve, by "integrating" twice. Since it says that $\partial / \partial \eta(u_\xi) = 0$, we can integrate with respect to η to get $u_\xi = F(\xi)$, where

---

[2] Jean D'Alembert was born in Paris as an illegitimate child of a former nun while the father was out of the country. Unable to support her son, his mother left him on the steps of a church. The infant was quickly found and taken to an orphanage. He was baptized as Jean Le Rond, after the name of the church where he was found. When the infant's father returned to Paris, he arranged for Jean to be adopted by a married couple, who were friends of his. His adoptive parents brought him up well. He studied law and earned a law degree. He soon decided that mathematics was his true passion and studied it on his own. Although mostly self-taught, D'Alembert became an eminent mathematician and scholar in the same league with the likes of Euler, Laplace, and Lagrange. He made significant contributions to partial differential equations and his elegant methods, including his solution to the wave equation, very much impressed Euler. Frederick II (King of Prussia) offered D'Alembert the presidency of the prestigious Berlin Academy, a position which he declined. He was quite an eloquent and well-rounded scholar and he made significant contributions to Diderot's famous encyclopedia. Apparently, D'Alembert was prone to argumentation and his disputes with other contemporary mathematicians caused him some professional difficulties on several occasions

$F(\xi)$ is an arbitrary function of ξ. Next we integrate again, this time with respect to ξ, to conclude that

$$u(\xi,\eta) = f(\xi) + g(\eta), \tag{7}$$

where $f(\xi)$ and $g(\eta)$ are arbitrary functions of the indicated variables. (Note $f(\xi)$ is an antiderivative of $F(\xi)$.) Translating back to the original variables using (3) gives us the following general solution of the wave equation:

$$u(x,t) = f(x-ct) + g(x+ct), \tag{8}$$

where f and g are arbitrary functions (with continuous second derivatives). We point out that each term in (8) represents a wave propagating along the x-axis with speed c. For example, $f(x-ct)$ is constant on lines of the form $x = ct$. As time t advances, values of x must also increase to maintain the same value of f (disturbance). Thus the first term represents a wave that propagates in the positive x-direction with speed c (right traveling wave). Similarly, the term $g(x+ct)$ represents a left-traveling wave. Both waves travel without distortion (i.e., the profile of either one of them t units of time later will be the exact same profile, but shifted to the left or right ct units along the x-axis.)

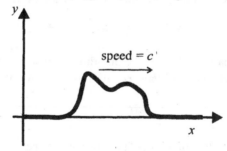

FIGURE 12.3: A right-propagating pulse $f(x-ct)$. The general solution (8) of the one-dimensional wave equation $u_{tt} = c^2 u_{xx}$ also includes a left-propagating pulse. Both wavefronts propagate without distortion.

D'Alembert went on further with his general solution (8), formulating and solving a well-posed problem for the one-dimensional wave equation. We consider a very long string and so consider the one-dimensional wave equation on the space range $-\infty < x < \infty$, and the time range $0 \le t < \infty$. Unlike with the heat equation, it is quite clear from (8) that merely specifying the wave profile $u(x,0)$ at time $t = 0$ is not sufficient to determine a unique solution. Indeed, the initial wave could come from a single left-moving wave, a single right-moving wave, or more generally could be made up as a superposition of two waves each moving in different directions. If we specify both the initial wave profile $u(x,0)$ and its initial velocity $u_t(x,0)$, then this together with the wave equation will give a well-posed problem. These initial boundary conditions are often referred to as **Cauchy boundary conditions** (or **Cauchy boundary data**). Thus the **Cauchy problem for the wave equation** is summarized as follows:

$$\begin{cases} (PDE) & u_{tt} = c^2 u_{xx}, \quad -\infty < x < \infty, \ 0 < t < \infty, \quad u = u(x,t) \\ (BC's) & u(x,0) = \varphi(x), \ u_t(x,0) = v(x) \quad -\infty < x < \infty, \ 0 \le t < \infty \end{cases} \quad (9)$$

This highlights an important general difference between elliptic PDEs versus hyperbolic PDEs. Recall from the last chapter that for elliptic PDEs, simply specifying the value of the solution on the boundary of the domain (Dirichlet boundary conditions) resulted in a well-posed problem. For hyperbolic PDEs, more information is needed for the problem to be well posed. We now state d'Alembert's solution of this Cauchy problem:

THEOREM 12.1: (*D'Alembert's Solution of the Cauchy Problem*)[3] Suppose that the function $\varphi(x)$ has a continuous second derivative and $v(x)$ has a continuous first derivative on the whole real line. Then the Cauchy problem (9) for the one-dimensional wave equation has the unique solution given by

$$u(x,t) = \frac{1}{2}[\varphi(x+ct) + \varphi(x-ct)] + \frac{1}{2c} \int_{x-ct}^{x+ct} v(s)ds. \quad (10)$$

Proof: Substitution of the general solution (8) into the BCs of (9) produces (put $t = 0$):

$$\varphi(x) = f(x) + g(x), \quad \text{and} \quad v(x) = -cf'(x) + cg'(x).$$

Integrating the second equation and dividing by c gives: $(1/c)\int_0^x v(s)ds$ $= g(x) - f(x)$. (Since $f(x)$ and $g(x)$ are arbitrary functions we can assume that the constant of integration is zero.) This last equation together with the first of the original pair are easily solved to give:

$$f(x) = \frac{1}{2}\left[\varphi(x) - (1/c)\int_0^x v(s)ds\right], \quad g(x) = \frac{1}{2}\left[\varphi(x) + (1/c)\int_0^x v(s)ds\right].$$

---

[3] In applications, it is convenient to allow functions $\varphi(x)$ and $v(x)$ for initial data which may violate the technical assumptions of having the required derivatives at all values of x. Often there are a finite set of values (**singularities**) of x at which either $\varphi(x)$ or $v(x)$ may not even be defined or their derivatives may not exist. Such singularities do not pose any serious problems for d'Alembert's solution, but they will give rise to corresponding singularities in the solution at all future time values. See, for example, the initial profile of Figure 12.4 ($\varphi(x)$) for a wave problem. This function has singularities at the three points where there are sharp corners in the graph $x = -1, 0, 1$. Future profiles in the solution shown in Figure 12.5 show also the presence of such singularities. Recall that for solutions of heat equations that were seen in the last chapter, singularities arising from discontinuities in an initial temperature distribution or its derivative immediately got smoothed out as time advanced. This is one of the major distinguishing features between hyperbolic versus parabolic PDE's. In the former, singularities are preserved and propagate, while in parabolic PDE's, initial singularities disappear as soon as time becomes positive.

Substituting these formulas into (8) now lets us write the solution as:

$$f(x-ct)+g(x+ct)=\frac{1}{2}[\varphi(x-ct)+\varphi(x+ct)]$$

$$+\frac{1}{2c}\left[-\int_b^{x-ct}v(s)ds+\int_b^{x+ct}v(s)ds\right],$$

which equals the expression in (10).

We emphasize that the foregoing analysis was only for one-dimensional waves on an infinite string. Of course, infinite strings do not exist, but for long strings, or for modeling disturbances on finite strings for limited time intervals, the above analysis can lead to useful insights. It is rare to have such an explicit analytical general solution. Soon we will consider boundary conditions that will require nonanalytical numerical methods, and finite-difference methods will be employed as in the last chapter. For now, let us get some hands-on experience with traveling waves. In the following example, we will get MATLAB to create a series of snapshots of a solution of a natural wave problem.

EXAMPLE 12.1: (*A Plucked Infinite String*) Consider what happens to a long string that is plucked with three fingers as shown in Figure 12.4 and then released (at time $t = 0$). Assume that the units are chosen so that wave speed $c = (T/\rho)^{1/2}$ equals 1. Using d'Alembert's solution, get MATLAB to create a series of snapshots of the wave profiles for each of the seven times starting with time $t = 0$ and advancing to $t = 3$ in increments of 0.5.

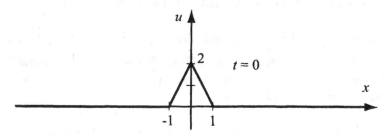

FIGURE 12.4: Initial profile for the plucked string of Example 12.1.

SOLUTION: In the Cauchy problem (9), we put $c = 1$, and $v(x) = 0$ (since at time $t = 0$, the three-finger plucked string is released with no initial velocity). From Figure 12.4, we can write the initial profile of the string as $\varphi(x) = \begin{cases} 2-2|x|, & \text{for } |x| \le 1 \\ 0, & \text{for } |x| \ge 1 \end{cases}$. It is not too difficult to analyze the resulting wave propagation analytically using Theorem 12.1, but a MATLAB code can be easily written to produce snapshots and/or movies of this and more complicated waves. Since an inline function construction is not appropriate for functions whose formulas change, we first construct an M-file for the function $\varphi(x)$:

```
function y = EX12_1(x)
if abs(x)<1, y=2-2*abs(x);
else y=0;
end
```

Using this M-file in the following code, we create relevant vectors to produce the snapshots, and we use the `subplot` command to conveniently collect all of the profiles in a single figure. The resulting MATLAB plot window is reproduced in Figure 12.5.

```
>> x=-5:.01:5;
>> counter =1;
>> for t=0:.5:3;
     x1=x+t; x2=x-t;
     for i=1:1001
       u(i)=.5*(EX12_1(x1(i))+EX12_1(x2(i)));
     end
     subplot(7,1,counter)
     plot(x,u)
     hold on
     axis([-5 5 -1 3]) %We fix a good axis range.
     counter=counter+1;
end
```

EXERCISE FOR THE READER 12.1: Following the procedure for making a movie in Section 7.2, get MATLAB to create a movie of the solution of the wave problem of Example 12.1 for the time range $0 \le t \le 4$. Play it back at varying speeds (and perhaps with varying repetitions).

EXERCISE FOR THE READER 12.2: (a) Write a function M-file:

```
function [] = dalembert(c,step, finaltime, phi, nu, range),
```

for creating a series of snapshots for the solution of the one-dimensional wave problem (9). The inputs should be: a positive number c for the wave speed, a positive number step for the time steps of the snapshots, and another positive number finaltime for the time limit of the snapshots. Also, the initial data of the problem will be inputted as two inline or M-file functions phi and nu. The last input variable is a 4×1 vector range for the xy-axis range to use in the snapshots. There will be no output variables, but the program will produce a graphic of snapshots of the Cauchy problem (9) starting at time $t = 0$ and continuing in increments of step until finaltime is exceeded.
(b) Run your program using the data of Example 12.1.
(c) Run your program on the "hammer blow" problem that consists of the Cauchy problem (9) (for the wave equation) with $c = 1$, $\varphi(x) = 0$, and

$$v(x) = \begin{cases} 1, & \text{for } |x| \le 1 \\ 0, & \text{for } |x| > 1 \end{cases}.$$ Create a series of snapshots of the solution from $t = 0$ to t

$= 5$ in increments of $t = 0.5$.
(d) Use your program to help you estimate the length of time it takes for the disturbances of the waves of both parts (b) and (c) above to reach an observer at

position $x = 10$. How do your answers fit in with the previously mentioned fact that the waves in making up d'Alembert's general solution of the wave equation travel at speed c (here $c = 1$)?

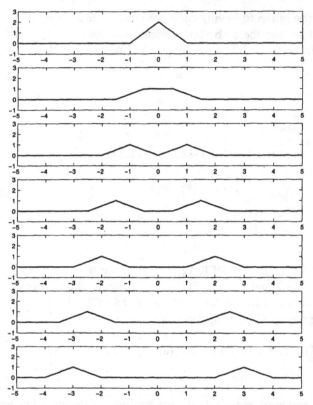

FIGURE 12.5: Progressive snapshots of the solution of the Cauchy problem for the plucked string of Example 12.1, at times $t = 0$, $t = 0.5$, $t = 1$, ..., $t = 3$. Note that the initial disturbance separates into two disturbances that eventually take on the same shape but each having half the size of the original. The function $u(x,t)$ could also be graphed in three dimensions as a function of two variables. The snapshots, which are merely "slices" of the three-dimensional graphs, are often more useful than the latter.

We now introduce a concept that will help us to highlight another important difference between parabolic and hyperbolic PDEs. Note that from d'Alembert's solution of the wave initial value problem (9), the solution is made up of two waves propagating at speed c and traveling in opposite directions. The actual disturbances can travel at speeds less than but not exceeding c (see part (d) of Exercise for the Reader 12.2). It also follows from d'Alembert's solution that the value of the solution u of (9) at a certain point (x,t), i.e., the vertical disturbance of the string at location x and at time t, can only be affected by the initial data φ, v over the interval $[x - ct, x + ct]$. This interval is called the interval of dependence of the "space-time" point (x,t); and the corresponding triangle (see Figure 12.6) in the space-time plane is called the **domain of dependence** of (x, t).

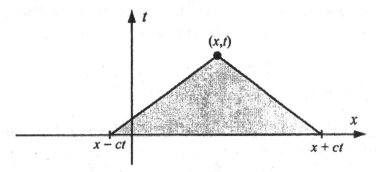

FIGURE 12.6: Illustration of the interval of dependence $[x - ct, x + ct]$ (on the x-axis) for the wave equation on a line. The shaded triangle in the space-time plane (xt-plane) is called the domain of dependence. The values of the initial condition functions $\varphi(x)$ and $v(x)$ outside of the interval of dependence for (x,t) are irrelevant to the determination of $u(x,t)$.

Although d'Alembert's solution of the wave equation on the infinite string makes it possible to analyze analytically most of the properties of the solution, the next variation of a Cauchy problem for the one-dimensional wave equation that we consider will give rise to analytical formulas that are extremely complicated and intractable. We now study the wave equation on a string of finite length, which is fixed at both ends. The precise Cauchy problem that we work with is as follows:

$$\begin{cases} (PDE) \quad u_{tt} = c^2 u_{xx}, \qquad 0 < x < L, \, 0 < t < \infty, \; u = u(x,t) \\ (BC's) \quad \begin{cases} u(x,0) = \varphi(x), \; u_t(x,0) = v(x) \\ u(0,t) = u(L,t) = 0 \end{cases}, 0 < x < L, \, 0 \le t < \infty \end{cases} \qquad (11)$$

A model to help visualize this Cauchy problem would be the motion of a guitar string of length L that is fixed at both ends. What makes a nice analytical formula impossible here is the fact that once the disturbances reach the ends of the string, they will bounce back, and things will continue to get more complicated as time goes on.

Theoretically, we can solve (11) by using d'Alembert's solution for the infinite string in a clever way. The useful artifice that will be used is called the **method of reflections**. We first extend the functions $\varphi(x)$ and $v(x)$ to be functions on the whole real line, based on their values in the interval $0 < x < L$. Labeling these extensions as $\hat{\varphi}(x)$ and $\hat{v}(x)$, respectively, they will be created so that they are odd functions across both of the boundary values $x = 0$ and $x = L$. Analytically, this means that

$$\hat{\varphi}(-x) = -\hat{\varphi}(x) \quad \text{and} \quad \hat{\varphi}(2L-x) = -\hat{\varphi}(x), \qquad -\infty < x < \infty, \tag{12}$$

and the corresponding identities for $\hat{v}(x)$. It can be easily verified (Exercise 14) that the following formula gives such an extension $\hat{\varphi}(x)$ of $\varphi(x)$[4].

$$\hat{\varphi}(x) = \begin{cases} \varphi(x), & \text{if } 0 < x < L \\ -\varphi(-x), & \text{if } -L < x < 0 \\ \text{Extend to be periodic of period } 2L \end{cases}, \qquad -\infty < x < \infty. \tag{13}$$

See Figure 12.7 for a graphical depiction of this construction. An analogous formula is used to construct $\hat{v}(x)$.

EXERCISE FOR THE READER 12.3: (*Constructing an M-file for a Periodic Function*) (a) For the function $\varphi(x) = 1 - |1-x|$ on the interval $[0, 2]$ ($L = 2$). Write an M-file, called   y=phihat(x) that extends the given function to $-\infty < x < \infty$ by the rule of (13).   Try to write your M-file so that it does not use any loops.
(b)   Get MATLAB to plot the graph of your phihat(x) on the interval $-6 \leq x \leq 6$.

If we solve the corresponding Cauchy problem (9) on the whole real line using as data the extended functions $\hat{\varphi}(x)$ and $\hat{v}(x)$ for boundary data, the function $\hat{u}(x,t)$ that arises will, in fact, extend the solution of (11).

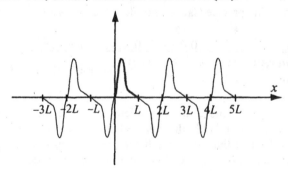

FIGURE 12.7: Illustration of the extension (13) of a function $\varphi(x)$ defined from $x = 0$ to $x = L$ (heavy graph portion) to a function $\hat{\varphi}(x)$ that is odd about each of the endpoints $x = 0$ and $x = L$.

---

[4] Technically, this definition does not define $\hat{\varphi}(x)$ for $x = 0, \pm L, \pm 2L, \cdots$.   The original function $\varphi(x)$ was also not defined at the endpoints $x = 0$, and $x = L$. This was only for notational convenience in the boundary conditions of (11). The boundary conditions corresponding to the ends of the string being fixed would force $\varphi(0) = \varphi(L) = 0$ so we extend the definition of $\hat{\varphi}(x)$ to all real numbers by specifying $\hat{\varphi}(0) = \hat{\varphi}(\pm L) = \hat{\varphi}(\pm 2L) = \cdots = 0$. The resulting function will be continuous (otherwise the string would be broken).

Some parts of this assertion are clear. Defining $u(x,t) = \hat{u}(x,t)$ for $0 \le x \le L$, and $t \ge 0$ (i.e., take u to be the function $\hat{u}$ restricted to the domain of the problem (11)), it is clear that $u(x,t)$ satisfies the wave equation and the first two (initial) boundary conditions since $\hat{u}$ does. Because of the odd extension properties of $\hat{\varphi}(x)$ and $\hat{v}(x)$, it also follows that $u(0,t) = u(L,t) = 0$ for all $t \ge 0$ (the reader should verify this using (10)). Thus this function u does indeed furnish a solution to the Cauchy problem (11).

EXAMPLE 12.2: (*A Plucked Guitar String*) Consider what happens to a guitar string of length 4 units that is plucked with one finger as shown in Figure 12.8 and then released (at time $t = 0$). Assume that the units are chosen so that wave speed $c = (T/\rho)^{1/2}$ equals 1. By using the method of reflections, get MATLAB to create a series of snapshots of the wave profiles for each of the 12 times starting with time $t = 0$ and advancing to $t = 6$ in increments of 0.5.

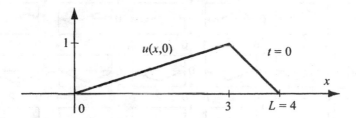

FIGURE 12.8: The initial profile of the plucked guitar string of Example 12.2.

SOLUTION: Looking at Figure 12.8, we can write:

$$\varphi(x) \ (\equiv u(x,0)) = \begin{cases} x/3, & \text{for } 0 \le x \le 3 \\ 4 - x, & \text{for } 0 \le x \le 3 \end{cases}.$$

Also $v(x)(\equiv u_t(x,0)) = 0$ since the string is released without velocity. We first create an M-file for $\hat{\varphi}(x)$ using a similar construction as was done in the solution of Exercise for the Reader 12.3.

```
function y = EX12_2phihat(x)
if (0 <= x)&(x <= 3), y=x/3;
elseif (x >=3)&(x<=4), y=4-x;
elseif (x<0)&(x>=-4), y = -EX12_2(-x);
else q=floor((x+4)/8); y=EX12_2phihat(x-8*q);
end
```

We can now use MATLAB to create the desired snapshots. To make for a convenient single graphic of all 41 plots, we use the `subplot` command to partition the plot window into smaller pieces.

```
>> counter=1;
>> x=0:.01:4;
```

```
>> z=zeros(size(x));    %will be used to add axes to plots
>> clf    %freshen up the plot window
>> for t=0:.2:8
     x1=x+t; x2=x-t;
     for i=1:401
       u(i)=.5*(EX12_2phihat(x1(i))+EX12_2phihat(x2(i)));
     end
     subplot(7,6,counter), plot(x,u), hold on
     plot(x,z,'k') %adds a central axis to each plot
     axis([0 4 -1 1])
     counter=counter+1;
     hold off
end
```

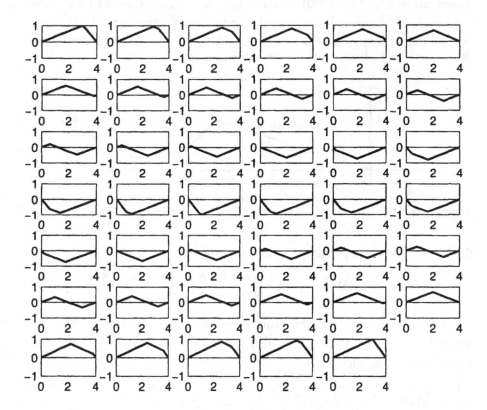

FIGURE 12.9: Snapshots of the plucked guitar string of Example 12.2. (To be read from left to right, and then top to bottom.) The speed of the wave is taken to be one unit length per unit time. Each successive square represents an increment of 0.2 units of time. Notice that the last frame corresponds to eight units of time and is exactly the initial profile.

Analytically, the waves that result on such finite strings are quite messy to describe. Physically, what is happening is that two waves are still moving in opposite directions at speeds equal to c. Each is constantly bouncing off the ends, reflecting and superimposing with the other. To get a better idea of the properties of the solution, it is a good idea to create a MATLAB movie for it (Exercise 4). Further details in this area can be found in Section 3.2 of [Str-92].

EXERCISE FOR THE READER 12.4: Prove that the solution of the wave problem on the finite string (11) is always periodic in the time variable with period L/c.

Suggestion: Use the solution arising from the method of reflections.

EXERCISE FOR THE READER 12.5: (*Single Pulse Wave on a Finite String*) Consider the wave problem (11) with $c = 2$, and initial profile $\varphi(x)$ given as in Figure 12.10. Obtain a series of snapshots from time $t = 0$ through $t = 10$ in increments of 0.5 of the solution of the problem (11) under the hypotheses that:
(a) The impulse is moving to the right initially with speed 2 units per unit of time.
(b) The impulse is moving to the right initially with speed 1 unit per unit of time.
(c) The impulse is moving to the right initially with speed 4 units per unit of time.

You need not worry about finding an extremely accurate analytical formula to model the initial profile $\varphi(x)$; you can simply use polynomial interpolation (as in Section 7.4) with or without derivative conditions.

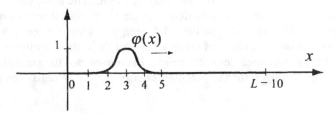

FIGURE 12.10: Initial profile for the impulse wave of Exercise for the Reader 12.5. The impulse is moving to the right.

Waves (i.e., solutions of the wave equation) satisfy a conservation of energy principle that is very important in physics. We demonstrate this principle for the one-dimensional wave equation written in physical form: $\rho u_{tt} = T u_{xx}$, where, we recall, ρ is the mass density of the string and T is the tension. From physics, the **kinetic energy** of a mass m, which is moving at a velocity v, is defined to be $\frac{1}{2}mv^2$. Breaking the wave into infinitesimal segments, this gives rise to the definition:

$$KE(t) = \frac{1}{2}\rho \int_{-\infty}^{\infty} u_t(x,t)^2 dx \qquad t \geq 0, \tag{14}$$

for the kinetic energy of the string at time t. This improper integral will converge under most reasonable physical assumptions. For example, if both of the initial condition functions $\varphi(x)$, $v(x)$ vanish outside a finite interval, so will the integrand (but with a larger interval determined by the intervals of dependence). If we differentiate this kinetic energy function with respect to t, we may differentiate under the integral sign to obtain:[5]

$$\frac{d}{dt}KE(t) = \rho \int_{-\infty}^{\infty} u_t u_{tt}\,dx \qquad t \ge 0.\tag{15}$$

Using the PDE to substitute Tu_{xx} for ρu_{tt} in the above integral, and then integrating by parts, we obtain:

$$\frac{d}{dt}KE(t) = T \int_{-\infty}^{\infty} u_t u_{xx}\,dx = Tu_t u_x \Big]_{-\infty}^{\infty} - T \int_{-\infty}^{\infty} u_{tx}u_x\,dx = -T \int_{-\infty}^{\infty} u_{tx}u_x\,dx,$$

the last equation being valid since the integrated term vanishes off a finite interval. Since $u_{tx}u_x = \partial/\partial t(\tfrac{1}{2}u_x^2)$, we may write (again using the differentiation under the integral sign rule):

$$\frac{d}{dt}KE(t) = -\frac{d}{dt}\int_{-\infty}^{\infty}\frac{1}{2}Tu_x^2\,dx, \qquad t \ge 0.\tag{16}$$

In basic physics, the **potential energy** of an object of mass m located at height h is defined to be mgh, where g is the gravitational constant. The basic conservation of energy principle in elementary mechanical physics states that if no external forces other than gravity are present, then the total energy = kinetic energy + potential energy remains constant. (Think of when an object falls, its velocity increases so its kinetic energy increases and its height decreases so its potential energy decreases.) The analogue for the potential energy for the string is the following integral:

$$PE(t) = \frac{1}{2}T \int_{-\infty}^{\infty} u_x(x,t)^2\,dx \qquad t \ge 0,\tag{17}$$

and, correspondingly, the **total energy** is defined to be

$$E(t) = KE(t) + PE(t) = \frac{1}{2}\int_{-\infty}^{\infty}\left[\rho u_t^2 + Tu_x^2\right]dx \qquad t \ge 0.\tag{18}$$

---

[5] Such differentiations are permissible under general circumstances. Here is a relevant theorem: Suppose that $f(x,t)$ is a continuous function of two variables in some rectangular region in the xt-plane: $a \le x \le b$, $c \le t \le d$. Suppose also that the partial derivative $f_t(x,t)$ is continuous in this same region. Then the following identity is valid for any t, $c \le t \le d$: $\dfrac{d}{dt}\int_a^b f(x,t)dx = \int_a^b f_t(x,t)dx$. Note that although the integral in (14) is over the whole real line, if $\varphi(x)$, $v(x)$ vanish outside a finite interval, the integral can be evaluated over a finite interval and the theorem can be applied. The theorem can even be extended to certain improper integral settings and in cases where the continuity assumptions break down at isolated singularities. See any good book on advanced calculus for details on this theorem and related results, for example, [Rud-64], [Ros-96], or [Apo-74].

The identity (16) states that $\dfrac{d}{dt}KE(t) = -\dfrac{d}{dt}PE(t)$, and it follows from (18) that $E'(t) = 0$ (i.e., the total energy in the wave remains constant). This is the **conservation of energy**. It is extremely important and noteworthy! Regardless of how long we let the string propagate, the total energy E of the configuration will remain unchanged.

EXAMPLE 12.3: (a) Compute the total energy of the plucked infinite string of Example 12.1, and (b) of the plucked guitar string of Example 12.2.

SOLUTION: In light of the conservation of energy, we may simply use the initial conditions to evaluate $E(0)$ in each case. In both cases, $u_t(x,0) = v(x) = 0$ and $u_x(x,0) = \varphi'(x)$.

Part (a): $E = E(0) = \dfrac{T}{2}\displaystyle\int_{-\infty}^{\infty} u_x^2 dx = \dfrac{T}{2}\displaystyle\int_{-\infty}^{\infty}[\varphi'(x)]^2 dx = \dfrac{T}{2}2^2 \cdot 2 = 4T$. The tension T is

not specified in the example, so this is as far as we can take this answer.

Part (b): Here, since the string is finite, we similarly obtain:

$$E = E(0) = \frac{T}{2}\int_0^4 u_x^2 dx = \frac{T}{2}\int_0^4 \varphi'(x)dx = \frac{T}{2}\left[(1/3)^2 \cdot 3 + 1^2 \cdot 1\right] = 2T/3.$$

There are some interesting similarities and differences of waves in one, two, three, and higher dimensions. We first point out that future profiles of one-dimensional waves will inherit symmetries in the initial conditions. Such results can be obtained from d'Alembert's formula (see Exercise 10). The analogue in higher dimensions of such symmetry would be **radially symmetric waves**. In n space dimensions such a wave would be a solution of the wave equation (1):

$$u_{tt} = c^2\Delta u = c^2(u_{x_1 x_1} + u_{x_2 x_2} + \cdots + u_{x_n x_n}), \quad u = u(x_1, x_2, \cdots, x_n, t).^{6}$$

which is expressible in the form $u(r,t)$, where $r = \sqrt{x_1^2 + x_2^2 + \cdots + x_n^2}$ is the distance to the origin. Thus, a radially symmetric n-dimensional wave is not really a function of $n + 1$ variables (as a general such wave might be) but actually just a function of two variables. There are analytical techniques for finding formulas for radially symmetric waves, but they involve special mathematical functions (such as Bessel functions) and the analysis can get a bit complicated. See, for example, [Str-92] for a nice treatment on radially symmetric waves. In two dimensions, water ripples provide a nice and telling example of radially symmetric waves. In three dimensions, sound waves and electromagnetic (e.g., radio) waves provide prototypical examples. If a pebble is dropped in water, the water ripples continue to propagate and reproduce themselves. In general, disturbances resulting from

---

[6] Most interesting applications of the wave equation occur in one, two or three space dimensions in which cases the customary choices x, y, and z are used in place of x_1, x_2 and x_3.

two-dimensional waves continue to propagate at a given point of space, once they have reached this point. In one and three dimensions, once the disturbance of a wave passes by a certain point, the wave is finished there and moves on. In three dimensions, however, there is an important difference from one-dimensional waves. The intensity of the wave decreases as we move away from the source. This can be proved from the conservation of energy. (Once a disturbance from a three-dimensional radially symmetric wave reaches a distance R from the source, it must cover an entire sphere with the same amount of energy that the wave packed on much smaller spheres, and the intensity will be decreased at each point on these larger spheres. This argument can be made into a rigorous proof.) In higher than three dimensions, radially symmetric waves turn out to have the same distorted properties of two-dimensional waves. These facts make it clear that we are very fortunate to live in a three-dimensional world. Indeed, if the dimension of our world were two or higher than three, than anytime someone spoke, we would never stop hearing them. In a one-dimensional world, anytime anyone spoke or a noise was made, everyone would hear it and with the same intensity regardless of how far away from the source they were! For a rigorous proof that radially symmetric distortion-free waves are only possible in one and three dimensions, and that only in one dimension are radially symmetric waves possible without loss of intensity, we refer the reader to the article (with a rather presumptuous title) by Morley [Mor-85] and [Mor-86].

EXERCISES 12.1

1. (*Making Snapshots of Vibrating Strings*) For each of the following initial data sets, create a series of snapshots of the solution of the wave problem (9):

$$\begin{cases} \text{(PDE)} & u_{tt} = u_{xx}, \qquad -\infty < x < \infty, \, 0 < t < \infty, \; u = u(x,t) \\ \text{(BCs)} & u(x,0) = \varphi(x), \, u_t(x,0) = v(x) \quad -\infty < x < \infty, \, 0 \le t < \infty \end{cases}$$

with c (wave speed) $= 1$.

(a) $\varphi(x) = \begin{cases} \sin(x), & \text{for } 0 \le x \le 2\pi \\ 0, & \text{otherwise} \end{cases}$, $v(x) = 0$.

(b) $\varphi(x) = \begin{cases} \sin(2x), & \text{for } 0 \le x \le 2\pi \\ 0, & \text{otherwise} \end{cases}$, $v(x) = 0$.

(c) $\varphi(x) = 0$, $v(x) = \begin{cases} 1, & \text{for } 6\pi \le x \le 8\pi \\ 0, & \text{otherwise} \end{cases}$.

(d) $\varphi(x) = \begin{cases} \sin(x), & \text{for } 0 \le x \le 2\pi \\ 0, & \text{otherwise} \end{cases}$, $v(x) = \begin{cases} 1, & \text{for } 6\pi \le x \le 8\pi \\ 0, & \text{otherwise} \end{cases}$

Obtain snapshots for the time range $0 \le t \le 14$ in increments of $\Delta t = 2$, and choose the axes range so that the plots show all disturbances of the wave in an informative fashion.

2. (*More snapshots of vibrating strings*) For each of the following initial data sets, create a series of snapshots of the solution of the wave problem (9):

$$\begin{cases} \text{(PDE)} & u_{tt} = u_{xx}, \qquad -\infty < x < \infty, \, 0 < t < \infty, \; u = u(x,t) \\ \text{(BCs)} & u(x,0) = \varphi(x), \, u_t(x,0) = v(x) \quad -\infty < x < \infty, \, 0 \le t < \infty \end{cases}$$

with c (wave speed) $= 1$.

(a) $\varphi(x) = \begin{cases} \sin(x), & \text{for } 0 \le x \le \pi \\ 0, & \text{otherwise} \end{cases}$, $\quad v(x) = \begin{cases} -\cos(x), & \text{for } 0 \le x \le \pi \\ 0, & \text{otherwise} \end{cases}$.

(b) $\varphi(x) = \begin{cases} \sin(x), & \text{for } 0 \le x \le \pi \\ 0, & \text{otherwise} \end{cases}$, $\quad v(x) = \begin{cases} -2\cos(x), & \text{for } 0 \le x \le \pi \\ 0, & \text{otherwise} \end{cases}$.

(c) $\varphi(x) = \begin{cases} \sin(x), & \text{for } 0 \le x \le \pi \\ 0, & \text{otherwise} \end{cases}$, $\quad v(x) = \begin{cases} -0.5\cos(x), & \text{for } 0 \le x \le \pi \\ 0, & \text{otherwise} \end{cases}$,

(d) $\varphi(x) = \begin{cases} \sin(x), & \text{for } 0 \le x \le \pi \\ 0, & \text{otherwise} \end{cases}$, $\quad v(x) = \begin{cases} -4\cos(x), & \text{for } 0 \le x \le \pi \\ 0, & \text{otherwise} \end{cases}$,

Obtain snapshots for the time range $0 \le t \le 20$ in unit increments. Physically, explain how the four sets of initial conditions are related.

3. (*Making Movies of Vibrating Strings*) For each of the vibrating string problems ((a) through (d)) of Exercise 1, create a MATLAB movie of the vibrating string on the time range $0 \le t \le 14$. View each at various speeds and repetitions.

4. (*More Movies of Vibrating Strings*) For each of the vibrating string problems ((a) through (d)) of Exercise 2, create a MATLAB movie of the vibrating string on the time range $0 \le t \le 20$. View each at various speeds and repetitions.

5. (a) Create a MATLAB movie for the guitar string wave of Example 12.2 from time $t - 0$ till time $t = 24$. View it at various speeds and repetitions.
    (b) Create a MATLAB movie for the single impulse wave of Exercise for the Reader 12.5 from time $t = 0$ till time $t = 40$. View it at various speeds and repetitions.

6. (*Snapshots of Vibrating Finite Strings*) For each of the following initial data sets, create a series of snapshots of the solution of the wave problem (11):

$$\begin{cases} (\text{PDE}) & u_{tt} = u_{xx}, \qquad\qquad 0 < x < L, \ 0 < t < \infty, \ u - u(x,t) \\ (\text{BCs}) & \begin{cases} u(x,0) = \varphi(x), \ u_t(x,0) = v(x) \\ u(0,t) = u(l,t) - 0 \end{cases}, \quad 0 < x < L, \ 0 \le t < \infty \end{cases}$$

with c (wave speed) $= 1$.

(a) $\varphi(x) = \begin{cases} \sin(x), & \text{for } 0 \le x \le 2\pi \\ 0, & \text{otherwise} \end{cases}$, $\quad v(x) = 0, \ L = 4\pi$.

(b) $\varphi(x) = \begin{cases} \sin(x), & \text{for } 0 \le x \le 2\pi \\ 0, & \text{otherwise} \end{cases}$, $\quad v(x) = 0, \ L = 3\pi$.

(c) $\varphi(x) = \begin{cases} \sin(x), & \text{for } 0 \le x \le \pi \\ 0, & \text{otherwise} \end{cases}$, $\quad v(x) = \begin{cases} -2\cos(x), & \text{for } 0 \le x \le \pi \\ 0, & \text{otherwise} \end{cases}$, $L = 3\pi$.

(c) $\varphi(x) = \begin{cases} \sin(x), & \text{for } 0 \le x \le 2\pi \\ 0, & \text{otherwise} \end{cases}$, $\quad v(x) = \begin{cases} 1, & \text{for } 6\pi \le x \le 8\pi \\ 0, & \text{otherwise} \end{cases}$, $L = 10\pi$

Obtain snapshots for the time range $0 \le t \le 40$ in increments of $\Delta t = 2$.

7. (*Making movies of Vibrating Finite Strings*) For each of the vibrating string problems ((a) through (d)) of Exercise 6, create a MATLAB movie of the vibrating string on the time range $0 \le t \le 60$. View each at various speeds and repetitions.

8. Compute the total energies of each of the vibrating infinite strings in Exercise 1.

9. Compute the total energies of each of the vibrating finite strings in Exercise 6.

10. (*Symmetry of Waves on an Infinite String*) Consider the solution of the wave problem (9):

$$\begin{cases} \text{(PDE)} \ \ u_{tt} = c^2 u_{xx}, & -\infty < x < \infty, \ 0 < t < \infty, \ u = u(x,t) \\ \text{(BCs)} \ \ u(x,0) = \varphi(x), \ u_t(x,0) = v(x) & -\infty < x < \infty, \ 0 \le t < \infty \end{cases}.$$

Use d'Alembert's formula to prove the following symmetry inheritance results.
(a) If both of the initial data are even functions of x (i.e., $\varphi(-x) = \varphi(x)$ and $v(-x) = v(x)$ for all x), then so will be the wave profile at any future time: $u(-x,t) = u(x,t)$ for all x and $t \ge 0$.
(b) If both of the initial data are odd functions of x (i.e., $\varphi(-x) = -\varphi(x)$ and $v(-x) = -v(x)$ for all x), then so will be the wave profile at any future time: $u(-x,t) = -u(x,t)$ for all x and $t \ge 0$.

11.   (*Waves on a Semi-infinite String*) Consider the solution of the following wave problem similar to the finite-string problem (11) except that only one end of the string is held fixed.

$$\begin{cases} \text{(PDE)} \ \ u_{tt} = u_{xx}, & 0 < x < \infty, \ 0 < t < \infty, \ u = u(x,t) \\ \text{(BCs)} \ \begin{cases} u(x,0) = \varphi(x), \ u_t(x,0) = v(x) \\ u(0,t) = 0 \end{cases}, & 0 < x < \infty, \ 0 \le t < \infty \end{cases}.$$

(a) Making use of d'Alembert's formula and an appropriate "method of reflections" technique similar to that used in the text for the finite string, develop a program for solving this problem. We point out that such a method will not be a numerical method, per se, since it will simply use the computer to perform analytical computations (and the only errors are due to roundoff).
(b) Obtain snapshots of profiles of the solution to the above problem using the following initial

conditions: $\varphi(x) = \begin{cases} \sin(x), & \text{for } 0 \le x \le 2\pi \\ 0, & \text{otherwise} \end{cases}$, $v(x) = 0, \ c = 1$.

(c) Obtain snapshots of profiles of the solution to the above problem using the following initial conditions:

$$\varphi(x) = \begin{cases} \sin(x), & \text{for } 0 \le x \le 2\pi \\ 0, & \text{otherwise} \end{cases}, \ v(x) = \begin{cases} 1, & \text{for } 6\pi \le x \le 8\pi \\ 0, & \text{otherwise} \end{cases}, L = 10\pi$$

(d) Create a MATLAB movie of the propagation of the wave in part (b).
(e) Create a MATLAB movie of the propagation of the wave in part (c).

12.   (*A Maximum Principle for the Wave Equation*) (a) Suppose that the hypotheses of d'Alembert's theorem are satisfied for the Cauchy problem (9):

$$\begin{cases} \text{(PDE)} \ \ u_{tt} = c^2 u_{xx} & -\infty < x < \infty, \ 0 < t < \infty, \ u = u(x,t) \\ \text{(BCs)} \ \ u(x,0) = \varphi(x), \ u_t(x,0) = v(x) & -\infty < x < \infty, \ 0 \le t < \infty \end{cases},$$

and that $|\varphi(x)| \le M$ for all x and that $\left| \int_a^b v(s)ds \right| \le L$ for all numbers a and b. Show that the

solution $u(x,t)$ of the Cauchy problem satisfies the inequality: $|u(x,t)| \le M + L/2c$ for all x and t in the domain.
(b) Under what general circumstances can you conclude that the maximum amplitude of the wave is attained at time zero (i.e., $|u(x,t)| \le \max\{|u(x,0)| : -\infty < x < \infty\}$)?

12.2: FINITE DIFFERENCE METHODS FOR HYPERBOLIC PDE'S

We begin this section by developing finite difference schemes for the numerical solution of the one-dimensional wave problem (11):

$$\begin{cases} \text{(PDE)} \ \ u_{tt} = c^2 u_{xx}, & 0 < x < L, \ 0 < t < \infty, \ u = u(x,t) \\ \text{(BCs)} \ \begin{cases} u(x,0) = \varphi(x), \ u_t(x,0) = v(x) \\ u(0,t) = u(L,t) = 0 \end{cases}, & 0 \le x \le L, \ 0 \le t < \infty \end{cases}$$

on a finite string. Our development works in general if we allow c to be a function of t and/or x: $c = c(x,t)$. Physically, this corresponds to modeling a vibrating string where its characteristics can change depending on time and space. We have already shown that in case c is a constant, d'Alembert's Theorem 12.1 coupled with the method of reflections can lead to a practical numerical method for solving this problem. Since the method simply evaluates the theoretical solution, it is relatively error free and so completely adequate for solving (11) with any sets of data. This will allow us to compute the errors of the numerical solutions we obtain from the finite difference methods. D'Alembert's solution, however, is specific to the wave equation, while the finite difference methods that we introduce can be easily adapted to work for more general hyperbolic PDE problems.

At first glance, the similarity of the wave and Laplace's equation would make it seem quite plausible that the same general finite difference discretization would work nicely, as we witnessed in the case for elliptic boundary value problems. The boundary conditions in (11), however, are different in two major ways: (i) The region $0 \le x \le L$, $0 \le t < \infty$ is no longer a bounded rectangle, but rather a half strip extending to infinity in the positive t-direction. (ii) There are two boundary conditions on the lower side of the strip rather than one. We will indeed discretize the PDE in the analogous fashion to what was done to Laplace's equation (replace each second derivative with its central difference approximation), but because of (i), we will not be able to set up the problem as a finite linear system (there are infinitely many nodes). Instead, we will do what is called a *marching scheme*, where the nodal approximations are computed one time level at a time, moving up from $t = 0$. At first glance, this may seem like a better situation, since the number of variables and the size of the linear systems will be much smaller than if we were to do it all at once, as with the elliptic method. For the most part it is true that the computations will generally move faster, but one new issue that we will need to confront with such marching schemes is the issue of stability. At each step, the local truncation errors will still be very small, but they can compound quite quickly to make the numerical solutions meaningless. Fortunately, there are some stability criteria that give easy ways to arrange the relative step sizes so that the schemes will be stable.

All finite difference methods require that the variables be restricted to finite intervals,[7] so we will need to restrict time to some specified range, $0 \le t \le T$. As in Chapter 11 (see Figure 11.11), we begin by introducing a grid of equally spaced x- and t-coordinates for the rectangular region $0 \le x \le L$, $0 \le t \le T$:

$$0 = x_0 < x_1 < x_2 < \cdots < x_{N+1} = L, \quad \Delta x_i \equiv x_i - x_{i-1} = h,$$
$$0 = t_0 < t_1 < t_2 < \cdots < t_{M+1} = T, \quad \Delta t_i \equiv t_i - t_{i-1} = k. \tag{19}$$

---

[7] Thus, in terms of the original variables of the PDE, the regions on which finite difference methods can be used to solve problems must be rectangular (if there are two variables, or n-dimensional box shapes if there are n variables). Coordinate transforms (such as polar coordinates) can allow for other sorts of shapes. One of the key advantages of the finite element methods that we will introduce in the next chapter is that they allow the solution of PDE problems on more complicated geometrical configurations.

By using the central difference formulas (see Lemma 10.3) in the wave equation, we get the following discretization of it:

$$
\frac{u(x_i,t_{j+1})-2u(x_i,t_j)+u(x_i,t_{j-1})}{k^2}
$$
$$
= c^2 \frac{u(x_{i+1},t_j)-2u(x_i,t_j)+u(x_{i-1},t_j)}{h^2} . \tag{20}
$$

We recall that the truncation errors here are $O(k^2)$ and $O(h^2)$, respectively. Using the notation:

$$
u_{i,j} = u(x_i,t_j),
$$

and introducing the parameter

$$
\mu = ck/h, \tag{21}
$$

we may express (20) in the following simplified form:

$$
u_{i,j+1} - 2u_{i,j} + u_{i,j-1} - \mu^2 \left[u_{i+1,j} - 2u_{i,j} + u_{i-1,j} \right] = 0. \tag{22}
$$

We next solve this equation for the unique term corresponding to the highest time value to obtain:

$$
u_{i,j+1} = 2(1-\mu^2)u_{i,j} + \mu^2 \left[u_{i+1,j} + u_{i-1,j} \right] - u_{i,j-1}, \tag{23}
$$

for $i = 1, 2, \ldots, N$, and $j = 1, 2, \ldots, M$. The endpoint boundary conditions tell us that:

$$
u_{0,j} = 0 = u_{N,j}, \quad \text{for all } j. \tag{24}
$$

It follows from (22) and (23) that we may represent the time level $j + 1$ functional values in terms of the previous two time level functional values by means of the following tridiagonal linear system:

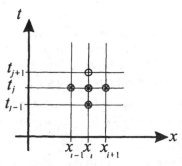

FIGURE 12.11: Illustration of the computational stencil for the discretization (20), (21) of the wave equation. The single point with largest time coordinate is emphasized, since the finite difference method will solve for it using the values of the solution at the previously found lower time grid points.

$$
\begin{bmatrix} u_{1,j+1} \\ u_{2,j+1} \\ \vdots \\ u_{N-1,j+1} \\ u_{N,j+1} \end{bmatrix} = \begin{bmatrix} 2(1-\mu^2) & \mu^2 & 0 & \cdots & 0 \\ \mu^2 & 2(1-\mu^2) & \mu^2 & \ddots & \vdots \\ 0 & \ddots & \ddots & \ddots & 0 \\ \vdots & \ddots & & & \mu^2 \\ 0 & \cdots & 0 & \mu^2 & 2(1-\mu^2) \end{bmatrix} \begin{bmatrix} u_{1,j} \\ u_{2,j} \\ \vdots \\ u_{N-1,j} \\ u_{N,j} \end{bmatrix} - \begin{bmatrix} u_{1,j-1} \\ u_{2,j-1} \\ \vdots \\ u_{N-1,j-1} \\ u_{N,j-1} \end{bmatrix} \tag{25}
$$

Such a scheme is referred to as an **explicit three-level scheme**, explicit since the highest ($t = (j + 1) k$) level values are explicitly solved in terms of the lower level values; three-level simply means that the nodal values involved in the scheme span over three time levels ($t = (j -1)k, jk, (j + 1)k$). This scheme will progress by iterating as we march upward in time. In order to start this recursion, we will need the functional values at the first two time levels $t = 0$ and $t = k\,(= t_1)$. These are the two column vectors on the right when $j = 1$. At time $t = 0$, these values are specified by the initial condition $u(x,0) = \varphi(x)$ (initial wave profile) of (11) and this gives us:

$$
u_{i,0} = \varphi(x_i) \text{ for } i = 1, 2, \ldots, N. \tag{26}
$$

In order to get the required next time level functional values we will need to make use of the initial wave velocity condition of (11): $u_t(x,0) = v(x)$. The fact that this extra information is actually required (unlike in the elliptic case) is consistent with the fact that the wave problem (11) is well posed. To use this initial velocity to approximate the time level $t = k$ functional values, we will need another difference formula for approximation of derivatives; either the forward or backward difference formulas (Lemma 11.5) will give us what we need. For reasons that will soon be apparent, we choose to use the forward difference formula here.

For a fixed value of x, and treating $u(x,t)$ as a function of t, the forward difference formula implies that:

$$
v(x) = u_t(x,0) \approx (u(x,k) - u(x,0))/k \Rightarrow u(x,k) \approx u(x,0) + kv(x)
$$

(this is nothing more than the usual tangent line approximation). In terms of our grid functional values this translates to:

$$
u_{i,1} \approx u_{i,0} + kv(x_i) \text{ for } i = 1, 2, \ldots, N. \tag{27}
$$

Note that (viz. Lemma 11.5) the error of this approximation is $O(k)$, which is of lower order and hence potentially much greater than the $O(h^2 + k^2)$ local truncation error for (23). Thus, this lower quality estimate for the $t = k$ time level values (needed to start (23)) could contaminate the overall quality of (23). This problem can be avoided since the approximation can be improved to have error $O(k^2)$ (thus matching those in the foregoing development) if we furthermore assume that the wave equation is valid on the initial line and is sufficiently differentiable.   Indeed, based on the differentiability assumption, (the one-variable) Taylor's theorem from Chapter 2 allows us to write:

$$u(x_i,k) = u(x_i,0) + ku_t(x_i,0) + \frac{k^2}{2}u_{tt}(x_i,0) + \frac{k^3}{6}u_{ttt}(x_i,\hat{k}),$$

where $\hat{k}$ is a number between 0 and k. The assumption that the wave equation is valid on the initial line tells us that $u_{tt}(x,0) = c^2 u_{xx}(x,0) = c^2\varphi''(x)$, and we are led to the following approximation

$$u_{i,1} \approx u_{i,0} + kv(x_i) + \frac{c^2 k^2}{2}\varphi''(x_i), \text{ for i} = 1, 2, ..., N, \tag{28}$$

with error bound $O(k^2)$.[8]   To avoid computation of derivatives, we may approximate $\varphi''(x_i)$ using the central difference formula: $\varphi''(x_i) \approx [\varphi(x_{i+1})$ $-2\varphi(x_i) + \varphi(x_{i-1})]/h^2$ (with error $O(h^2)$). Installing this approximation into (28) produces the following practical approximating formula for the time level $t = 1$ functional values:

$$u_{i,1} = (1-\mu^2)\varphi(x_i) + \frac{\mu^2}{2}[\varphi(x_{i+1}) + \varphi(x_{i-1})] + kv(x_i), \text{ for } i = 1, 2, ..., N \tag{29}$$

which has local truncation error $O(h^2 + k^2)$.   The next exercise for the reader gives us another way to arrive at the above $O(k^2)$ approximations for $u_{i,1}$ and show it to be valid under slightly different assumptions.

EXERCISE FOR THE READER 12.6: (a) Use Taylor's theorem to establish the following **centered difference approximation**: Suppose that $f(x)$ is a function having a continuous third derivative in the interval $a-h \leq x \leq a+h$, then the following approximation for $f'(x)$ is valid:

$$f'(x) = \frac{f(x+h) - f(x-h)}{2h} + O(h^2). \tag{30}$$

Note that this is a second-order approximation to $f'(x)$, whereas the forward and backward difference approximation are only first-order approximations.
(b) Using the artifice of ghost nodes (introduced in Section 11.4), obtain estimate (29) (with local truncation error $O(h^2 + k^2)$) under the assumption that the solution $u(x,t)$ of the Cauchy problem (11) extends to have a continuous third-order time derivative for $t \geq -k$.

---

[8] This is the reason we choose to adopt the forward over the backward difference method. We would not have been able to make such a local error truncation if we had used the backward difference method.

Suggestion: For part (b), introduce a line of nodes at level $t = -k$ and denote the ghost values of u on these nodes by $u_{i,-1}$. The centered difference approximation gives the estimate $u_{i,1} - u_{i,-1} \approx 2kv(x_i)$ that has error $O(k^2)$.

Even with all of the above attention to detail in developing a finite difference method with $O(h^2 + k^2)$ local truncation error, stability issues can seriously corrupt the method. The next example will give good evidence of how badly things can go.

EXAMPLE 12.4: *(Illustration of Instability)* Consider the following Cauchy problem of a long plucked string:

$$\begin{cases} \text{(PDE)} & u_{tt} = 4u_{xx}, & -\infty < x < \infty,\ 0 < t < \infty,\ u = u(x,t) \\ \text{(BCs)} & u(x,0) = \varphi(x),\ u_t(x,0) = 0, & -\infty < x < \infty,\ 0 \le t < \infty \end{cases}$$

where $\varphi(x)$ is as in Example 12.1 (Figure 12.4). Note this problem is identical to the problem of Example 12.1 except that the wave speed has changed from $c = 1$ to $c = 2$. By D'Alembert's solution (Theorem 12.1), we know the exact solution of this problem is given by (from (10)):

$$u(x,t) = \frac{1}{2}\left[\varphi(x + 2t) + \varphi(x - 2t)\right],$$

and the solution will look just like the one shown in Figure 12.5, except now the speed is doubled. Thus, for any specified range of time values, we can view this problem as taking place on a finite string centered at $x = 0$ of sufficiently large length.
(a) Apply the above finite difference scheme (22) with $h = k = 1$ up to time level $t = 5k (= 5)$, and $-12 \le x \le 12$. To isolate just the effectiveness of (22), use the exact values for the time level $t = k$ values, as determined by D'Alembert's solution, in place of (28). Examine the u-values and compare with those of the exact solution (cf. Figure 12.5).
(b) Repeat Part (a) with $h = k = 0.1$ up to time level $t = 50k (= 5)$.
(c) Repeat Part (a) with $h = 1$, $k = .5$ up to time level $t = 10k (= 5)$.

SOLUTION: Part (a): With $c = 2$, we have, by (21), $\mu = ck/h = 2$, so that (23) becomes:

$$u_{i,j+1} = -6u_{i,j} + 4\left[u_{i+1,j} + u_{i-1,j}\right] - u_{i,j-1}.$$

Since $h = 1$, we get $x_0 = -12$, $x_{12} = 0$, $x_{25} = 12$ so by (10) (exact solution), we may write $u_{i,0} = \begin{cases} 2, & i = 12 \\ 0, & \text{otherwise} \end{cases}$ and $u_{i,1} = \begin{cases} 1, & i = 12 \pm 2 \\ 0, & \text{otherwise} \end{cases}$. Note that by (22), the set of indices with nonzero u-values can advance only one index to the left/right with each new time level. The following MATLAB loop will produce the needed nonzero u-values up to time level $t = 5k$. The instability is so severe that it is convenient to view the matrix of values. In creating the 6×25 matrix of nodal values, we let the bottom row correspond to the time level zero values and so the top row corresponds to the $t = 5$ values. Note this requires us to modify the indexing of (23) accordingly in our MATLAB code below:

```
>> U=zeros(6,25); U(6,12)=2; U(5,[10 14])=1;
>> for j=5:-1:2
     for i=2:24
       U(j-1,i)=-6*U(j,i)+4*[U(j,i+1)+U(j,i-1)]-U(j+1,i);
end
end
```

The nonzero matrix values are shown below. Note that the actual solution has two pulses of height 1 moving from left to right at speed two. The numerical solution below is totally off and unstable, it oscillates out of control. Also, the disturbances only propagate at speed one.

256	-1536	4432	-8048	10373	-10688	10424	-10688	10373	-8048	4432	-1536	256
0	64	-288	616	-812	776	-710	776	-812	616	-288	64	0
0	0	16	-48	67	-56	44	-56	67	-48	16	0	0
0	0	0	4	-6	4	-2	4	-6	4	0	0	0
0	0	0	0	1	0	0	0	1	0	0	0	0
0	0	0	0	0	0	2	0	0	0	0	0	0

Part (b): Since c and μ are still 2, (22) takes the same form as in part (a), but since $x_0 = -12$, $x_{120} = 0$, $x_{241} = 12$, and we have

$$u_{i,0} = \begin{cases} 2-|120-i|/5, & 110 \le i \le 130 \\ 0, & \text{otherwise} \end{cases}.$$

To get $u_{i,1}$, we note that (10) and the initial values give us (since $h = k = 0.1$) that $u_{i,1} = u_{i+2,0} + u_{i-2,0}$. It is most simple to use a MATLAB loop to compute these values before entering into the main loop based on (23). Using the matrix conventions of part (a), the construction of the matrix of values can be accomplished in MATLAB with the following commands:

```
>> U=zeros(51,251);
for i=110:130
  U(51,i)=2-abs(i-120)/5;
end
  for i=108:132
  U(50,i)=U(51,i+2)+U(51,i-2);
end
for j=50:-1:2
  for i=2:250
    U(j-1,i)=-6*U(j,i)+4*[U(j,i+1)+U(j,i-1)]-U(j+1,i);
  end
end
```

To see that the numerical solution is still badly unstable, we need only look at the middle portion of the last six rows of the matrix (corresponding to the time range: $0 \le t \le .5$):

-45.2	416.8	-842.8	1091.2	-1008.8	920	-1008.8	1091.2	-842.8	416.8	-45.2
5	-19.6	71	-84.8	77.8	-50.8	77.8	-84.8	71	-19.6	5
4	4.8	-0.8	12.8	-0.4	7.2	-0.4	12.8	-0.8	4.8	4
3	3.6	4.2	3.2	4.6	4.4	4.6	3.2	4.2	3.6	3
2	2.4	2.8	3.2	3.2	3.2	3.2	3.2	2.8	2.4	2
1	1.2	1.4	1.6	1.8	2	1.8	1.6	1.4	1.2	1

Indeed, this shows that even at time level $t = 0.5$, the profile oscillates rapidly between ± 1000.

Part (c): Since k is now half of h, we have $\mu = 1$, so that (23) takes the following form:

$$u_{i,j+1} = u_{i+1,j} + u_{i-1,j} - u_{i,j-1}.$$

Since $h = 1$, we get $x_0 = -12$, $x_{12} = 0$, $x_{25} = 12$ as in part (a), and from (10) (exact solution), we may write $u_{i,0} = \begin{cases} 2, & i = 12 \\ 0, & \text{otherwise} \end{cases}$ and $u_{i,1} = \begin{cases} 1, & i = 12 \pm 1 \\ 0, & \text{otherwise} \end{cases}$. Note that by (23), the set of indices with nonzero u-values can advance only one index to the left/right with each new time level. The construction of the 11×25 matrix of u-values is done as before and the relevant entries are displayed below.

```
U=zeros(11,25);U(11,12)=2; U(10,[11 13])=1;
for j=10:-1:2
  for i=2:24
    U(j-1,i)=U(j,i+1)+U(j,i-1)-U(j+1,i);
  end
end
```

1	0	0	0	0	0	0	0	0	0	0	0	0	0	0	0	0	0	0	0	0	0	0	0	1
0	1	0	0	0	0	0	0	0	0	0	0	0	0	0	0	0	0	0	0	0	0	0	1	0
0	0	1	0	0	0	0	0	0	0	0	0	0	0	0	0	0	0	0	0	0	1	0	0	0
0	0	0	1	0	0	0	0	0	0	0	0	0	0	0	0	0	0	0	0	1	0	0	0	0
0	0	0	0	1	0	0	0	0	0	0	0	0	0	0	0	0	0	0	1	0	0	0	0	0
0	0	0	0	0	1	0	0	0	0	0	0	0	0	0	0	0	0	1	0	0	0	0	0	0
0	0	0	0	0	0	1	0	0	0	0	0	0	0	0	0	0	1	0	0	0	0	0	0	0
0	0	0	0	0	0	0	1	0	0	0	0	0	0	0	0	1	0	0	0	0	0	0	0	0
0	0	0	0	0	0	0	0	1	0	0	0	0	0	1	0	0	0	0	0	0	0	0	0	0
0	0	0	0	0	0	0	0	0	1	0	1	0	0	0	0	0	0	0	0	0	0	0	0	0
0	0	0	0	0	0	0	0	0	0	2	0	0	0	0	0	0	0	0	0	0	0	0	0	0

Note the rather surprising results! The result of part (c) quite well represents the actual solution (up to the resolution on the x-grid). The results of parts (a) and (b) were totally unstable, and despite the fact that the grid of part (b) was much finer (in both variables) than that for part (c), the grid for part (c) turned out to give a much more stable method. It turns out that the relative ratio of h and k, not their actual sizes, is what will make or break stability. Such remarkable phenomena did not occur when we applied finite difference methods to elliptic problems in the last chapter.

The finite difference methods shown above can be proved to converge to the exact solution of the wave problem (11) (as the partitions become more and more refined) provided that, in addition to the required differentiability assumptions, the following **Courant-Friedrichs-Levy (CFL) condition** holds:

$$\mu \equiv ck/h \le 1. \tag{31}$$

If this condition is violated (i.e., if $\mu > 1$), examples can be constructed where (as in Example 12.4), although all other differentiability assumptions are satisfied, the finite difference approximations will not converge to the exact solution, even as the mesh sizes of both variables tend to zero! In fact when $\mu > 1$, the method is **unstable** in the sense that errors made at each time stage of the process can significantly affect the subsequent time numerical values. For complete details and proofs on these matters we refer to Section 9.3.1 of [IsKe-66]. The exercises will include an outline of the proof and in the next section we will give some details of the analogous theory for the heat equation.

We give here a nontechnical explanation of why such instability can arise. From (23), the numerical values at a new time level at x_i depend on those of previous two levels, which lie at most one horizontal step to the left and right of x_i. From this we can determine the "numerical interval of dependence" of a grid point (x_i, t_{j+1}), analogously to how we defined the interval of dependence of the exact solution (see Figure 12.12). From Figure 12.12, we can see how violation of the CFL condition can lead to instability of the numerical method. What this amounts to is that the size of the t-steps $(= k)$ is too large relative to the size of the x-steps $(= h)$. This means that the numerical interval of dependence (shown by the black double arrowed segment) for (x_i, t_{j+1}) is smaller than the theoretical interval of dependence (green arrowed segment). We know from the theory in the last section that the numerical interval of dependence thus does not take enough information into account to properly formulate the approximations. This can lead to catastrophic results, as we have seen. The diagonally upward black arrows indicate flows of information in the finite difference scheme.

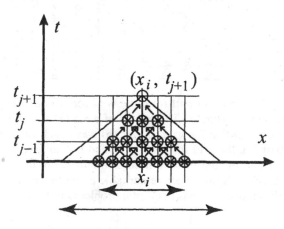

FIGURE 12.12: Illustration of the problem when the Courant-Friedrichs-Levy (CFL) condition is violated.

In both parts (a) and (b) of Example 12.4, we had $h = k$, so that (since $c = 2$), $\mu = 2$ and the Courant-Friedrichs-Levy condition is violated. We can see that the numerical wave profiles propagate at only h units (to left and right) for each k unit time level increase. Thus, the numerical profiles cannot keep up with the actual wave propagation (two h units left and right of space for each k unit of time) and the scheme goes haywire. In part (c), however, $k = h/2$ and now the numerical scheme can keep up, and it does so quite well in that example. It turns out that when the problem is a smooth one, taking step sizes so that $\mu = 1$ can greatly enhance the accuracy of the scheme. It seems quite surprising that for a given (stable) choice of h and k, fixing h and decreasing k can sometimes have a detrimental effect on the numerical solution. Shortly, we will introduce an implicit scheme that has better stability properties.

Recall that the finite difference scheme for elliptic PDEs that we used in the last chapter was implicit and very stable; also, in Part II, we saw that implicit schemes for ODE problems, although more difficult to work with, had better stability properties than explicit schemes. This is a general rule: Implicit schemes are more stable than explicit schemes in numerical differential equations. One advantage of explicit schemes, however, is that many are easily adapted to effectively solve nonlinear problems (provided stability requirements are met). Although we will not enter into any detailed discussion of stability issues for nonlinear PDEs, we will occasionally try to adapt some of our linear schemes to solving nonlinear problems. Often this is what is done in practice. Indeed, a nonlinear problem, when looked at locally (in a small portion of the domain), can be approximated by a linear problem and the latter one dealt with according to linear schemes. For more on nonlinear PDEs, we cite the reference [Log-94]. A more advanced treatment is given in [Smo-83].

Numerical methods for nonlinear PDEs is an extremely active area of mathematical research. We caution the reader that many reasonable-looking finite difference schemes may do poorly for a given nonlinear problem. In general those that are based on conservation laws (physical principles) are the most successful. This seems to imply that a purely mathematical approach to the numerical solution of nonlinear PDEs is not sufficient; an additional requirement is a certain knowledge of the physical principles governing the phenomena that are modeled by the PDEs. For a detailed investigation of such issues, we refer to the two-volume set [Tho-95a], [Tho-95b]. The book [Dur-99] gives a detailed treatment of various numerical methods for wave (hyperbolic) problems. A particular nonlinear one-dimensional wave equation with a conservation law based finite difference method is nicely examined in [StVa-78].

We now proceed to write a function M-file that will apply the above finite difference scheme to solve a more general version of the wave problem (11) which allows a certain nonlinearity in the PDEs. Specifically, we allow the ends of the wave to have time-dependent variable heights and we allow the coefficient c (wave speed) to depend on t, x, and/or u. The former conditions mean that we allow forced control on each of the string ends; the more general assumption on c corresponds physically to having a string whose characteristics are not uniform in x (e.g., it could be thicker in some places than in others), are time dependent (e.g., it could weaken or strengthen with time), and even depend on the current position and slope of the string (e.g., the properties of the string may weaken, depending on its composition, in areas where there is a steep slope stretch.)

In Program 12.1, the main change will be that when we use (29), we need take note of the fact that $\mu = ck/h$ is now no longer (necessarily) constant: $\mu = \mu_{i,j} = c(t_j, x_i, u_{i,j}, (u_x)_{i,j})k/h$. In the fourth argument of c we use the centered difference approximation: $(u_x)_{i,j} \approx [u_{i+1,j} - u_{i-1,j}]/2h$. The resulting scheme (23) will still be an explicit one.

PROGRAM 12.1: Function M-file for solution of the following wave problem by the finite difference method,[9]

$$\begin{cases} \text{(PDE)} & u_{tt} = c(t,x,u,u_x)^2 u_{xx}, \ 0 < x < L, 0 < t < \infty, \ u = u(x,t) \\ \text{(BCs)} & \begin{cases} u(x,0) = \varphi(x), \ u_t(x,0) = v(x) \\ u(0,t) = A(t), \ u(L,t) = B(t) \end{cases}, 0 < x < L, 0 \le t < \infty \end{cases} \tag{32}$$

This program uses the improved approximation (29) for the level-one time values that work better under greater differentiability hypotheses on the initial data. A more basic program, which uses (27) in place of (29), is left to the following exercise for the reader; it is recommended over this one in case the initial conditions possess singularities. The program assumes that the Courant-Friedrichs-Levy condition has been checked to be satisfied in the region under consideration.

```
function [x, t, U] = onedimwave(phi, nu, L, A, B, T, N, M, c)
% solves the one-dimensional wave problem u_tt - c(t,x,u,u_x)^2*u_xx
% Input variables:  phi-phi(x) - initial wave profile function
% nu=nu(x) = initial wave velocity function, L = length of string, A
% = A(t) height function of left end of string u(0,t)=A(t), B=B(t) =
% height function for right end of string u(L,t)-B, T- final time for
% which solution will be computed, N - number of internal x-grid
% values,  M - number of internal t-grid values, c -c(t,x,u,u_x)-
% speed of wave.  Functions of the indicated variables must be
% stored as (either inline or M-file) functions with the same
% variables, in the same order.
% Output variables: t - time grid row vector (starts at t-0, ends at
% t-T, has M+2 equally spaced values),  x - space grid row vector, U-
% (M+2) by (N+2) matrix of solution approximations at corresponding
% grid points; y grid will correspond to first (row) indices of U, x
% grid values to second (column) indices of U.
% CAUTION:  For stability of the method, the Courant-Friedrichs-Levy
% condition should hold:  c(x,t,u,u_x)(T/L)(N+1)/(M+1) <1.

h = L/(N+1);  k = T/(M+1);
```

---

[9] Although we have not yet made explicit mention of the incorporation of boundary conditions into finite difference schemes for the wave equation, this is a rather obvious extension of ideas presented in the previous chapter. Program 12.1 provides an example of such a feature.

```
U=zeros(M+2,N+2); x=0:h:L; t=0:k:T;
% Recall matrix indices must start at 1. Thus the indices of the
% matrix will always be one more than the corresponding indices that
% were used in theoretical development.

%Assign left and right Dirichlet boundary values.
U(:,1)=feval(A,t)'; U(:,N+2)=feval(B,t)';

%Assign initial time t=0 values and next step t=k values.
for i=2:(N+1)
    U(1,i)=feval(phi,x(i));
    mu(i)=k*feval(c,0,x(i),U(1,i),(feval(phi,x(i+1))-feval(phi,x(i-
1)))/2/h)/h;
    U(2,i)=(1-mu(i)^2)*feval(phi,x(i))+mu(i)^2/2*(feval(phi,x(i-1))+
...
            feval(phi,x(i+1))) + k*feval(nu,x(i));
end

%Assign values at interior grid points
for j=3:(M+2)
for i=2:(N+1)
    mu(i)=k*feval(c,t(j),x(i),U(j-1,i),(U(j-1,i+1)-U(j-1,i-1))/2/h)/h;
    %First form needed tridiagonal matrix
    Tri = diag(2*(1-mu(2:N+1).^2)) + diag(mu(3:N+1).^2, -1) +
diag(mu(2:N).^2, 1);
    %Now perform the matrix multiplications to iteratively obtain
    % solution values for increasing time levels.
    U(j,2:(N+1))=(Tri*(U(j-1,2:(N+1))'))'-U(j-2,2:(N+1));
    U(j,2)=U(j,2)+mu(2)^2*feval(A,t(j-1));
    U(j,N+1)=U(j,N+1)+mu(N+1)^2*feval(B,t(j-1));
end
end
```

As was implicit in Program 12.1, we point out that the Courant-Friedrichs-Levy (CFL) condition can be expressed using the input parameters in the above M-file in the following way:

$$\mu \equiv c(t,x,u,u_x)\frac{T}{L}\left(\frac{N+1}{M+1}\right) < 1 \tag{33}$$

The actual M-file is quite short. In the next example we will test both the accuracy and runtime of this program with a wave problem with nicely smooth input data and whose exact solution is available to compute errors. It will also demonstrate some interesting pathologies when played against the Courant-Friedrichs-Levy condition.

EXAMPLE 12.5: Use Program 12.1 to solve the following wave problem on the time interval $0 \le t \le 2$:

$$\begin{cases} \text{(PDE)} & u_{tt} = u_{xx}, & 0 < x < \pi, 0 < t < \infty, \ u = u(x,t) \\ \text{(BCs)} & \begin{cases} u(x,0) = \sin x, \ u_t(x,0) = 0 \\ u(0,t) = 0, \ u(\pi,t) = 0 \end{cases}, & 0 < x < \pi, 0 \le t < \infty \end{cases}$$

using the following grid sizes. In each case, compare the results with the actual solution $u(x,t) = \cos t \sin x$ on the indicated time levels. If the graphs are too close to discern differences, compute the maximum error numerically.
(a) $N = 10$, $M = 15$. Note that this set of parameters slightly violates the Courant-Friedrichs-Levy condition. Compare the numerical solution with the exact solution at time levels $t = 0.5$, $t = 1$, $t = 1.5$, $t = 2$.
(b) $N = 10$, $M = 29$. Note that this set of parameters satisfies the Courant-Friedrichs-Levy condition. Compare numerical solution with exact solution at time levels $t = 4$ and $t = 8$.
(c) $N = 100$, $M = 15$. Note that this set of parameters strongly violates the Courant-Friedrichs-Levy condition (31) ($\mu = 16.0746$).

SOLUTION: We create three sets of data for each of the three sets of parameters and label them differently for future use.

We first create inline functions for the initial data of this wave problem:

```
>> phi = inline('sin(x)');
>> nu= inline('0'); A=nu; B=A; c=inline('1','t','x','u','ux');
→nu = Inline function:
     nu(x) = 0
```

We now create the numerical solutions for each of the three parts:

```
>> [x1, t1, U1] = onedimwave(phi, nu, pi, A, B, 8, 10, 15, c);
>> [x2, t2, U2] = onedimwave(phi, nu, pi, A, B, 8, 10, 29, c);
>> [x3, t3, U3] = onedimwave(phi, nu, pi, A, B, 8, 100, 15, c);
```

Part (a): To produce the desired snapshots, we take note of the general relationships between t and j: $k = \dfrac{2}{M+1}$, $t_j = jk = \dfrac{2j}{M+1}$, so $j = \dfrac{(M+1)t_j}{2}$.
Thus, when $M = 15$, we have $j = 8t_j$, so the values $t = 0.5$, 1.0, 1.5, and 2.0 correspond respectively to the indices $j = 4$, $j = 8$, $j = 12$ and $j = 16$. Since the MATLAB indices are one greater than these actual indices, we may create and plot the desired numerical snapshots as follows:

```
>> subplot(1,4,1)
>> plot(x1, U1(5, :)), axis([0 pi -1  1]),hold on
>> plot(x1, cos(2)*sin(x1),'r')
>> subplot(1,4,2)
>> plot(x1, U1(9, :)), axis([0 pi -1  1]),hold on
>> plot(x1, cos(4)*sin(x1),'r')
>> subplot(1,4,3)
>> plot(x1, U1(13, :)), axis([0 pi -1  1]),hold on
>> plot(x1, cos(6)*sin(x1),'r')
>> subplot(1,4,4)
>> plot(x1, U1(17, :)), axis([0 pi -1  1]),hold on
>> plot(x1, cos(8)*sin(x1),'r')
```

We have set the axes to an appropriate setting for comparisons and used the horizontal stacking of the subplot so as to make the vertical errors more detectable. The resulting graphic is shown in Figure 12.13.

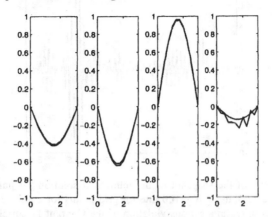

FIGURE 12.13: Comparison of the computed finite difference solution's snapshots (jagged) with the actual solution's snapshots (smooth) for the wave problem of Example 12.5. The four plots correspond to snapshots at time levels $t = 0.5$, $t = 1$, $t = 1.5$, $t = 2$, respectively. The numerical solution was obtained using $N = 10$ interior grid points for x and $M = 15$ interior grid points for t, which resulted in a violation of the Courant-Friedrichs-Levy condition (31) with $\mu = 1.75... > 1$. All except the last profile show the numerical snapshots to be reasonably decent with only small errors that are barely visible to the naked eye. At $t = 2$, however, the numerical graph starts to break its pattern and relative errors reach orders of magnitude of 100%. Time levels (from left to right) are $t = 0.5$, $t = 1$, $t = 1.5$, and $t = 2$.

Part (b): In this case, if we plot (as in part (a)) and compare the numerical solution with the actual solution, the results are indistinguishable at both time levels $t = 1$ and $t = 2$. To compute the maximum absolute values of the differences, using again the index relation $j = \dfrac{(M+1)t}{2}$ (and adding one to j to get MATLAB's indices) we enter the following commands:

```
>> max(max(abs(U2(:,16)'-cos(4)*sin(x2))))
→ans = 0.00131359012313

>> max(max(abs(U2(:,31)'-cos(8)*sin(x2))))
→ans = 0.00343796574347
```

The results are rather accurate considering the somewhat large step sizes (especially in t).

Part (c): Plotting the numerical solution's snapshot at time level $t = 2$ is accomplished with the command:

```
>>plot(x3, U3(17, :))
```

The rather surprising result is shown in Figure 12.14.

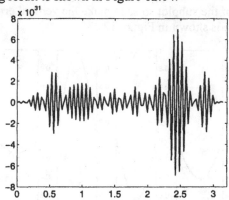

FIGURE 12.14: Plot of the snapshot of the numerical solution for part (c) of the wave problem of Example 12.5, using $N = 100$ interior grid points for x and $M = 15$ interior grid points for t, which resulted in a serious violation of the Courant-Friedrichs-Levy condition (31) with $\mu = 16.07... > 1$.   Note the amplitude of the graph is 31 orders of magnitude greater than the actual solution, so the result is quite meaningless. Note also that the grid used was actually finer than that used in part (a) (which gave much better results). Thus, blindly refining grids can lead to disastrous results that use more computing time, unless the Courant-Friedrichs-Levy condition is respected.

The next exercise for the reader will show how, with a bit finer of a grid on the time axis (and keeping the same grid on the x- axis) we can arrive at numerical solutions with the above program that are numerically indistinguishable from the actual solution. We also point out that the above program is able to handle grids for both x and t with over 1000 points in a reasonable amount of time (a few minutes). This is quite different from the situation for the finite difference methods for elliptic PDEs discussed in the previous chapter. Recall that in the algorithm for elliptic PDEs, it was required to solve a linear system of order roughly $N \cdot M$ to simultaneously solve for the numerical solution at all interior grid values. Numerically solving parabolic equations will also take far fewer computations than do elliptic equations with similar grids, and this is another reason we have grouped hyperbolic and parabolic PDEs together in this chapter.

EXERCISE FOR THE READER 12.7: (a) Modify Program 12.1 into one that uses (27) in place of (29) for the approximation of the function on the level $t = k$ time line. Call this modified function onedimwavebasic.
(b) Starting with $N = 10$ interior x-grid points, begin with $M = 30$ interior t-grid points and re-solve the wave problem of Example 12.5, with both the onedimwave program and your newly constructed onedimwavebasic program. Compare with the exact solution at $t = 8$. Continue to double M (keeping N fixed) until you have completed nine doublings of M. Collect the graphs of the resulting errors (at $t = 8$) in a separate 5×2 partitioned (by

subplot) window.   Repeat with $N = 40$. Compare and contrast the graphical results, and comment on any observed instability.

Physically, both hyperbolic and parabolic problems model time-dependent phenomena.   The main difference between them is that while parabolic phenomena are dissipative, solutions to hyperbolic PDEs are conservative.   In particular, initial singularities are smoothed out and lost with time under a parabolic PDE and are preserved and propagated under a hyperbolic PDE. Since finite difference schemes tend to average things out (we saw a good example of this in the previous chapter when we showed maximum principles hold for elliptic finite difference schemes), this suggests that finite difference schemes may run into problems for hyperbolic problems with discontinuous data. This is indeed the case, and for this reason, there are other methods that are more suitable for hyperbolic problems with discontinuous data. Examples of alternative methods suitable for hyperbolic problems with singularities include the method of characteristics (the D'Alembert method of the previous section is a special case), and the method of lines. More can be found on such methods in [Abb-66], [Dur-99], and [Ame-77].   Discontinuous data is very natural in hyperbolic problems modeling events such as shocks, explosions, or earthquakes. Our next example will show some typical pathologies that can occur when the finite difference method is applied to a discontinuous problem.

EXAMPLE 12.6: Consider the following wave problem:

$$
\begin{cases}
(\text{PDE}) \quad u_{tt} = c(x)^2 u_{xx}, \qquad 0 < x < 5, \, 0 < t < \infty, \quad u = u(x,t) \\
(\text{BCs}) \quad \begin{cases} u(x,0) = 0, \ u_t(x,0) = 0 \\ u(0,t) = A(t), \ u(L,t) = 0 \end{cases} 0 < x < 5, \, 0 \le t < \infty
\end{cases}
$$

where $c(x) = \begin{cases} 2, & \text{if } 0 \le x \le 3 \\ 1, & \text{if } x > 3 \end{cases}$ and $A(t) = \begin{cases} \sin(5t), & \text{if } 0 \le t \le \pi/5 \\ 0, & \text{if } t > \pi/5 \end{cases}$. Physically this

wave problem can be thought of as that of a string of length 5 that is made up of two materials which are glued together at $x = 3$. The right portion is more dense than the left (recall c is inversely proportional to $1/\sqrt{\rho}$, ρ = mass density of string). The string is initially at rest, and for a short period of time the left end is "moved" upward and then back down to the original position and held there. The right end of the string remains pinned down.   Use the above program onedimwave, respecting the Courant-Friedrichs-Levy condition, to numerically solve this problem on the time range $0 \le t \le 5$ using finer and finer grids until the solutions become visually consistent   Plot a series of snapshots of the wave profiles.

SOLUTION: In cases where the wave speed c is nonconstant, we must replace c with its maximum value (worst-case scenario) in the CFL condition (33). This tells us that we should have

$$M + 1 \ge 2(N + 1).$$

We create M-files for $c(x)$ and $A(t)$, and inline functions for the remaining data:

``
function y =
c_EG12_5(t,x,u,ux)
for i = 1:length(x)
  if (0<=x(i))&(x(i)<=3)
  y(i)=2;
  elseif (3<x(i))&(x(i)<=5)
  y(i)=1;
  else
  y(i)=0;
  end
end
``
	``
	function y = Apulse_EG12_5(t)
	for i = 1:length(t)
	  if (0<=t(i))&(t(i)<=pi/5)
	  y(i)=sin(5*t(i));
	  else
	  y(i)=0;
	  end
	end
	``

```
>> phi=inline('0'); nu=phi; B=nu;
```

After some experimentation with increasing resolution, the patterns of the solutions become clear. The code below produced the series of snapshots shown in Figure 12.15.

```
>> [x, t, U] = onedimwave(phi, nu, 5, @Apulse_EG12_5, B, 5, 350,...
               800, @c_EG12_5);
>> count=1;
>> for k=1:80:800
      subplot(10,1,count)
      plot(x,U(k,:)), hold on, axis([0 5 -1.1 1.1]);
      text(5.1,0, ['t = ',num2str((k-1)/800*5),])
      count=count+1;
end
```

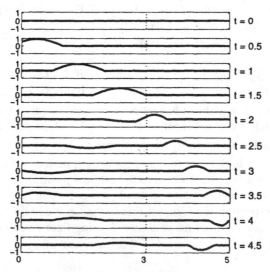

FIGURE 12.15: Series of snapshots for the solution of the "glued string" problem of Example 12.6. The interface at $x = 3$ (where the heavier string on the right meets the lighter one on the left) is emphasized with a grid. Note that when the wave reaches the heavier portion of the string, a secondary reflection takes place, and the original wavefront gets smaller and slows down.

We point out the very large numbers of $N = 350$ and $M = 500$ that were used. The discontinuities in the data necessitated this. Indeed, Figure 12.5 shows a single plot of a snapshot of the above numerical solution at a late time `plot(x,U(780,:))`. This is in sharp contrast to the results of part (b) of the Example 12.5 (a wave problem with smooth data), where we got excellent accuracy with a very rough grid on the finite difference method. The general rule is that for hyperbolic problems with discontinuous data, finite difference methods will require a lot of work to get decent results. Even (many) nonlinear hyperbolic problems take less work if the data is smooth. The exercises will further explore these issues.

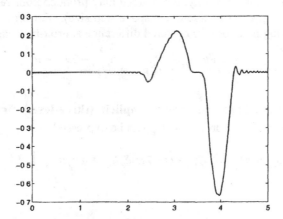

FIGURE 12.16: Wave profile near the end of the time interval $0 \le t \le 5$ of the numerical solution of Example 12.6. Note the small (secondary) oscillations. This type of numerical noise is not part of the exact solution; it arises since the finite difference method tries to smooth out certain discontinuities in the data of the problem.

EXERCISE FOR THE READER 12.8: The reader may observe that because of the discontinuities in the data, it would have been more appropriate to use the less restrictive version of the finite difference method `onedimwavebasic` of Exercise for the Reader 12.7. Check that with this M-file, the results will be quite identical to those of the above example.

We turn now to briefly describe implicit finite difference methods. For simplicity, we describe them only for the basic wave equation $u_{tt} = c^2 u_{xx}$ of (11). Their main advantage of allowing a more flexible choice of time step sizes is not so crucial here since the Courant-Friedrichs-Levy condition does not pose too stringent (small) a step size requirement on t. When we move on to parabolic equations in the next section, however, we will see that the stability of explicit methods requires a much smaller time step and so implicit methods will be a more attractive alternative. A family of such methods can be obtained by approximating u_{tt} with the centered difference: $u_{tt}(x_i,t_j) \approx [u_{i,j+1} - 2u_{i,j} + u_{i,j-1}]/k^2$, but for u_{xx} we

use a weighted average of the centered difference approximations at the three
levels: $t = t_{j-1}, t_j, t_{j+1}$:

$$u_{xx}(x_i, t_j) \approx \omega[u_{i+1,j-1} - 2u_{i,j-1} + u_{i-1,j-1}]/h^2$$
$$+(1-2\omega)[u_{i+1,j} - 2u_{i,j} + u_{i-1,j}]/h^2 + \omega[u_{i+1,j+1} - 2u_{i,j+1} + u_{i-1,j+1}]/h^2, \tag{34}$$

where the parameter ω satisfies $0 < \omega < 1$. The choices $\omega = 1/2$ and $\omega = 1/4$ are
the most popular since they lead to symmetric stencils. It can be shown that this
scheme is unconditionally stable as long as $\omega \geq 1/4$. We caution the reader that
despite the unconditional stability, the scheme may produce poor results if we use
too large a time step size (relative to the space step size). For brevity, it is helpful
to use the following notations for centered difference approximations:

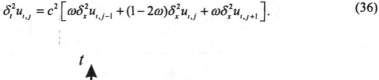

$$\delta_x^2 u_{i,j} \equiv \frac{u_{i+1,j} - 2u_{i,j} + u_{i-1,j}}{h^2}, \quad \delta_t^2 u_{i,j} \equiv \frac{u_{i,j+1} - 2u_{i,j} + u_{i,j-1}}{k^2}. \tag{35}$$

Thus with these notations, the general **implicit (three-level) finite difference
scheme** for the wave equation $u_{tt} = c^2 u_{xx}$ can be expressed as:

$$\delta_t^2 u_{i,j} = c^2 \left[\omega \delta_x^2 u_{i,j-1} + (1-2\omega)\delta_x^2 u_{i,j} + \omega \delta_x^2 u_{i,j+1} \right]. \tag{36}$$

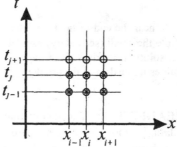

FIGURE 12.17: Stencil the implicit finite difference method (36) for the wave equation.
At each iteration in the upward marching scheme, the hollow nodal values need to be
determined. In the case where the parameter ω is 1/2, the central row of nodes (time level
t_j) is not present in the stencil.

When translated into a linear system, at each iteration, the scheme (36) leads to a
tridiagonal system in the variables $u_{i,j+1}$ $(1 \leq i \leq M)$ so the Thomas method can be
used.

EXERCISE FOR THE READER 12.9: (a) Write a function M-file for applying
the implicit finite difference method (36) with parameter $\omega = 1/4$. The syntax
should be as follows:

```
[x, t, U] = onedimwaveimpl_4(phi, nu, L, A, B, T, N, M, c)
```

where the variables and functionality should be just as with Program 12.1.

(b) Do some experiments on the problem of Example 12.6 comparing the performance of the implicit finite difference method of part (a) with the explicit method of Program 12.1 using values of N and M that would make the latter unstable.

(c) How fine a resolution is needed in the implicit finite difference method of part (a) to obtain graphical results of the quality of the numerical solution created in Example 12.6? Do the results seem better when the CFL condition is satisfied? Do some experiments.

A wave problem, or for that matter, any PDE problem with two space variables (and the time variable) would require four dimensions to represent the graph of the solution. Snapshots, however, obtained by fixing the time at a certain level, can be graphed in three dimensions, and from these movies of wave propagation can be put together and viewed. Since such graphical investigations are particularly useful for understanding wave propagation, we proceed now to develop a finite difference method for solving wave equations in two space variables.

We consider the following two-dimensional wave problem on a rectangular domain:

$$
\begin{cases}
\text{(PDE)} \ u_{tt} = c^2(u_{xx} + u_{yy}), \ 0 < x < a, 0 < y < b, 0 < t < \infty, \ u = u(x,y,t) \\
\text{(BCs)} \begin{cases}
u(x,y,0) = \varphi(x,y), \quad 0 \le x \le a, \ 0 \le y \le b, 0 \le t < \infty \\
u_t(x,y,0) = v(x,y), \quad 0 \le x \le a, \ 0 \le y \le b, 0 \le t < \infty \\
u(x,y,t) = 0, \quad \text{for all } (x,y) \text{ on the boundary of the rectangle:} \\
\qquad\qquad R = \{0 \le x \le a, 0 \le y \le b\}
\end{cases}
\end{cases} \tag{37}
$$

For simplicity we have kept the edges of the wave fixed at height zero. We may visualize (37) as a problem for the vibrations of a flexible membrane of elastic material that has been stretched over the edges of the rectangular frame R, much like a drumhead. The initial conditions will produce a vibration of the drumhead, which will be governed by the wave PDE in (37). With this interpretation, it can be shown using physical principles that $c^2 = T/\rho$, where T is the tension of the membrane[10] and ρ is the mass per unit area of the membrane.

We restrict time to be bounded on some fixed interval: $0 \le t \le T$ (apologies for using the letter T for two different purposes; the distinction should be clear from the context). We introduce a grid for each variable under the following notation:

$$
0 = x_0 < x_1 < x_2 < \cdots < x_{N_x+1} = a, \quad \Delta x_i \equiv x_i - x_{i-1} = h,
$$

$$
0 = y_0 < y_1 < y_2 < \cdots < y_{N_y+1} = b, \quad \Delta y_i \equiv y_i - y_{i-1} = h \tag{38}
$$

---

[10] The stretching of the membrane results from the boundary forces. It is assumed that the membrane is stretched uniformly in all directions.

$$0 = t_0 < t_1 < t_2 < \cdots < t_{M+1} = T, \quad \Delta t_i \equiv t_i - t_{i-1} = k.$$

We have specialized to using the same step size h for x- and y-coordinates. This, of course, is not always possible, depending on the dimensions of the rectangle R, but it will keep the notation more manageable.   To keep the notation somewhat streamlined, we use superscripts to denote indices for time levels, and subscripts to indicate indices of space variables:

$$u_{i,j}^{\ell} = u(x_i, y_j, t_\ell).$$

Using this notation and applying the central difference formulas for approximating the PDE in (37), as we did in the one-dimensional case, we arrive at the following discretization of the two-dimensional wave equation:

$$u_{i,j}^{\ell+1} - 2u_{i,j}^{\ell} + u_{i,j}^{\ell-1} - \mu^2 \left[u_{i+1,j}^{\ell} + u_{i-1,j}^{\ell} + u_{i,j+1}^{\ell} + u_{i,j-1}^{\ell} - 4u_{i,j}^{\ell} \right] = 0, \qquad (39)$$

where, once again we have let $\mu = ck/h$. Solving for the highest level time term produces:

$$u_{i,j}^{\ell+1} = 2(1 - 2\mu^2)u_{i,j}^{\ell} + \mu^2 \left[u_{i+1,j}^{\ell} + u_{i-1,j}^{\ell} + u_{i,j+1}^{\ell} + u_{i,j-1}^{\ell} \right] - u_{i,j}^{\ell-1}, \qquad (40)$$

We are now faced with a new difficulty. Namely, in order to store all of the functional values of our numerical solution, standard matrices are no longer feasible.   Fortunately, MATLAB can handle higher-dimensional arrays or matrices. We digress momentarily to introduce them. A three-dimensional matrix will have three coordinates that specify each of its entries. For example, to create a three-dimensional matrix of size $2 \times 2 \times 2$ all of whose entries are zeros, we could enter:

```
>> A=zeros(2,2,2)
```

This object can be thought of as two 2×2 matrices stacked on top of one another. When displaying such a matrix, as above, MATLAB will display each "layer" matrix in order starting from the lowest (final) index and moving on. All of the matrix manipulation tricks that we have learned, as well as all of the MATLAB functions that make sense for such higher-dimensional matrices, can be used for these general arrays.   There is no limit to the dimension of the matrices that MATLAB can handle. For example, we could create a $3 \times 2 \times 2 \times 2$ all of whose entries are 5's as follows:

```
>> A=5*ones(3,2,2,2)
```

```
→ A(:,:,1,1) =   5  5     A(:,:,2,1) =   5  5
                 5  5                     5  5
                 5  5                     5  5

  A(:,:,1,2) =   5  5     A(:,:,2,2) =   5  5
                 5  5                     5  5
                 5  5                     5  5
```

Geometrically, we can visualize a three-dimensional matrix as simply a three-dimensional array of numbers, as, for example, a discretization of the heat distribution of a three-dimensional object. Algebraically, it is helpful to think of higher-dimensional matrices as simply being indexed sets of ordinary (two-dimensional) matrices. This is how MATLAB treats them; the first two indices will always denote the indices of the displayed two-dimensional constituent matrices of a higher-dimensional matrix. Higher-dimensional matrices are ideal for storing numerical values for functions of more than two variables.

As in the one-dimensional case, the Courant-Friedrichs-Levy condition now takes on the form $\mu^2 \equiv (ck/h)^2 \le 1/2$.[11] Under the sufficient differentiability assumptions, Taylor's theorem (Chapter 2) can be used, once again, to produce the following $O(h^2 + k^2)$ approximation for the time level $t = k$ values:

$$u_{i,j}^1 \approx (1 - 2\mu^2)\varphi(x_i, y_j) + \frac{\mu^2}{2}[\varphi(x_{i+1}, y_j) + \varphi(x_{i-1}, y_j) + \varphi(x_i, y_{j+1})$$

$$+\varphi(x_i, y_{j-1})] + kv(x_i, y_j), \quad \text{for } i = 1, 2, \cdots, N_x \text{ and } j = 1, 2, \cdots, N_y. \tag{41}$$

EXERCISE FOR THE READER 12.10: Justify the $O(h^2 + k^2)$ error quality of the approximation (41).

In order to turn the above formulas into an M-file, it is not feasible to incorporate (or even try to make sense of) higher-dimensional matrix multiplication. MATLAB's versatile features for manipulating arrays, however, will allow us to again write a rather short (and efficient) M-file for the above finite difference scheme. For smooth problems, it will no longer be true that taking μ to be its maximal value for stability will yield zero truncation errors (see Exercises 19 and 22 and the note preceding Exercise 19), but nonetheless, for smooth problems, this choice is usually a good one and further it simplifies the formulas. In our M-file below, we thus take $\mu = 1/2$.

PROGRAM 12.2: Function M-file for solution of the two-dimensional wave problem (37) by the finite difference method:

$$\begin{cases} \text{(PDE)} & u_{tt} = c^2(u_{xx} + u_{yy}), \quad 0 < x < a, 0 < y < b, 0 < t < \infty, \quad u = u(x, y, t) \\ \text{(BCs)} & \begin{cases} u(x, y, 0) = \varphi(x, y), & 0 \le x \le a, 0 \le y \le b, 0 \le t < \infty \\ u_t(x, y, 0) = v(x, y), & 0 \le x \le a, 0 \le y \le b, 0 \le t < \infty \\ u(x, y, t) = 0, & \text{for all } (x, y) \text{ on the boundary of the rectangle:} \\ & R = \{0 \le x \le a, 0 \le y \le b\} \end{cases} \end{cases}$$

```
function [x, y, t, U] = twodimwavedirbc(phi, nu, a, b, T, h, c)
% solves the two-dimensional wave problem u_tt = c^2(u_xx+u_yy)
% on the rectangle {0<= x <= a,  0<= y <= b}, with u(x,y)=0
```

---

[11] In general, if we wanted to allow different step sizes h_x, h_y for the x- and y-grids, the Courant-Friedrichs-Levy stability condition takes the form: $\mu^2 \equiv c^2 k^2 (h_x^{-2} + h_y^{-2}) \le 1$, see [IsKe-66].

```
% on the boundary of the rectangle.
% Input variables:  phi -initial wave profile function
% nu initial wave velocity function, both should be functions of
% (x,y).  a= right endpoint of x, b = upper endpoint of y,
% T = final time solution will be computed.  h=common gap on
% x, y-grids,  c - speed of wave.
% Output variables: x = row vector for first space variable, y -
% row vector for second space variable,  t = time grid row vector
%(starts at t=0, ends at t=T, has Nt equally spaced values),
% U = (Nx)by(Ny)by(Nt) matrix of solution approximations at
% corresponding grid points (where Ny - number of y-grid points)
% x grid will correspond to first entries of U, y grid
% values to second entries of U, and t grid to third entries of U.
% CAUTION:  This method will only work if n is chosen so that the x
% and y grids can have a common gap size, i.e., if h = a/h and
% b/h must be integers.
% The time grid gap is chosen so that mu^2 - 1/2; this guarantees the
% Courant-Friedrichs-Levy condition holds and simplifies the
% main finite difference formula.

k=h/sqrt(2)/c; %k is determined from mu^2 = 1/2
MaxRatio=max([b/h a/h 1]);
if (abs(b/h-round(b/h))>MaxRatio*eps)|(abs(a/h-
round(a/h))>MaxRatio*eps)
fprintf('Space grid gap h must divide evenly into both a and b \r')
fprintf(' Either try another input or modify the algorithm')
error('M-file will exit')
end
Nx = round(a/h)+1; %number of points on x-grid
Ny = round(b/h)+1; %number of points on y-grid
Nt = floor(T/k)+1; %number of points on t-grid
U=zeros(Nx, Ny, Nt); x=0:h:a; y=0:h:b; t=0:k:T;
% Recall matrix indices must start at 1.  Thus the indices of the
% matrix will always be one more than the corresponding indices that
% were used in theoretical development.
% Note that by default, zero boundary values have been assigned to
% all grid points on the edges of the rectangle (and, for the time
% being, at a l grid points).

%Assign initial time t=0 values and next step t=k values.
for i=2:(Nx-1)
    for j=2:(Ny-1)
    U(i,j,1)=feval(phi,x(i),y(j));
    U(i,j,2)=.25*(feval(phi,x(i-1),y(j))+...
    feval(phi,x(i+1),y(j))+feval(phi,x(i),y(j-1))+...
    feval(phi,x(i),y(j+1)))+ k*feval(nu,x(i),y(j));
end
end
% Assign values at interior grid points
for ell=3:Nt %letter ell looks too much like number one

    U(2:(Nx-1),2:(Ny-1), ell) = ...
    +.5*(U(3:Nx,2:(Ny-1), ell-1)+ U(1:(Nx-2),2:(Ny-1), ell-1)...
    + U(2:(Nx-1),3:Ny, ell-1) + U(2:(Nx-1),1:(Ny-2), ell-1))...
    - U(2:(Nx-1),2:(Ny-1), ell-2);
end
```

EXAMPLE 12.7: (a) Make a (color) movie of the wave that solves the following problem:

$$\begin{cases} \text{(PDE)} \quad u_{tt} = u_{xx} + u_{yy}, \quad 0 < x < \pi, 0 < y < \pi, 0 < t < \infty, \quad u = u(x,y,t) \\[2mm] \text{(BCs)} \begin{cases} u(x,y,0) = \sin 2x \sin 2y, \quad 0 \le x \le \pi, \ 0 \le y \le \pi, 0 \le t < \infty \\ u_t(x,y,0) = 0, \quad 0 \le x \le \pi, \ 0 \le y \le \pi, 0 \le t < \infty \\ u(x,y,t) = 0, \quad \text{for all } (x,y) \text{ on the boundary of the rectangle:} \\ \qquad\qquad\qquad\qquad R = \{0 \le x \le \pi, \ 0 \le y \le \pi\} \end{cases} \end{cases},$$

for the time interval $0 \le T \le 4$. Divide the x- and y-ranges into 25 equally spaced intervals for the grid.

(b) The exact solution of this problem is $u(x,y,t) = \cos(2\sqrt{2}t)\sin(2x)\sin(2y)$, as is easily verified. Measure the maximum value of the errors of the computed solution above versus the exact solution at four time values that are close to the values $t = 1, 2, 3, 4$.

SOLUTION: We first construct inline functions for the boundary conditions. It is important that they be made functions of x and y (in this order).

```
>> phi=inline('sin(2*x)*sin(2*y)', 'x', 'y')
→phi =Inline function:
        phi(x,y) = sin(2*x)*sin(2*y)

>> nu=inline('0','x', 'y')
→nu = Inline function:
        nu(x,y) = 0
```

The remaining input parameters for the above M-file are as follows: $a = b = \pi$, $T = 4$, $h = \pi/25$, and $c = 1$.

```
>>[x, y, t, U] = twodimwavedirbc(phi, nu, pi, pi, 4, pi/25, 1)
>> size(U)
→ano = 26   26   40

>> for ell=1:26
surf(x,y,U(:,:,ell));
axis([0 pi 0 pi -1.5 1.5]);
M(:,:,ell)=getframe;  ¹²%see footnote below
end
>> movie(M, 2, 4)
```

Several representative snapshots of the movie are displayed in Figure 12.18; the reader is urged to run the code on his or her machine and view the actual movie.

Part (b): We first create an inline function for the exact solution:

```
>> exact = inline('cos(2*sqrt(2)*t).*sin(2*x).*sin(2*y)','x', 'y',
't')
→exact = Inline function:
        exact(x,y,t) = cos(2*sqrt(2)*t).*sin(2*x).*sin(2*y)
```

---

¹² Note that the movie matrix M for three-dimensional graphics is set up as a three-dimensional array. For versions earlier than Version 7 in MATLAB, the usual (two-dimensional) syntax `M(:,ell)=getframe;` should be used.

By viewing the vector t, we see that a good set of representative times would be:

```
>> t([13 24 35 46])
→ans =1.0663   2.0437   3.0212   3.9986

>> for i=1:26
for j=1:26
Uexact1(i,j)=exact((i-1)*pi/25, (j-1)*pi/25,t(13));
end
end

>> max(max(U(:,:,13)-Uexact1))
→ans =1.3323e-015
```

Repeating this for the other three listed time levels gives the following errors (in order): 1.7764e-015, 1.6653e-015, 1.8319e-015! The results are astoundingly pleasing; not only is the numerical solution graphically indistinguishable from the exact solution, but they are still identical up to MATLAB's working precision! In general, whenever the solution of the wave problem (11) or (37) has a "smooth" exact solution, then the finite difference solution (with appropriate value of μ) coincides with the exact solution! See Exercise 19 for a precise statement and outline of the proof in the case of (11).

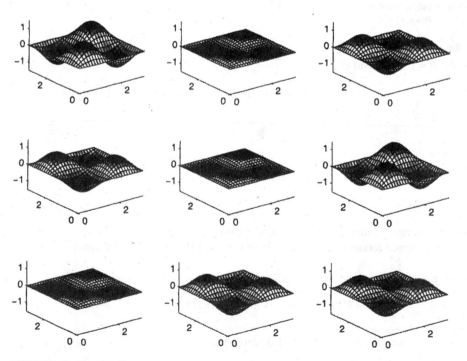

FIGURE 12.18: Slides from the movie of Example 12.7 of a wave on a square membrane. MATLAB's default colormap is set up so that highest values of the graph are colored red (hot) and lowest values are colored blue (cold). The movie must be seen!

EXERCISES 12.2

1. (a) Use finite difference methods to create a series of snapshots of the solution of the following vibrating string problem:

$$\begin{cases} \text{(PDE)} \ u_{tt} = 2u_{xx}, & 0 < x < \pi, \ 0 < t < \infty, \ u = u(x,t) \\ \text{(BCs)} \ \begin{cases} u(x,0) = \sin x, \ u_t(x,0) = 0 \\ u(0,t) = u(\pi,t) = 0 \end{cases}, & 0 < x < \pi, \ 0 \le t < \infty \end{cases}$$

Have your snapshots range from $t = 0$ through $t = 6$.

(b) Compute the maximum errors of the snapshots at six time levels close to $t = 1, 2, 3, 4, 5, 6$ by comparing with the exact solution $u(x,t) = \cos(\sqrt{2}t)\sin x$.

(c) Are there moderate values of N (= number of interior x-grid points) and M(= number of interior t-grid points) for which the finite difference solution would be accurate essentially to machine precision? Take moderate to mean less than 100, and machine precision to be approximately 10^{-15}.

(d) Do the answers to these questions change significantly depending on whether we use Program onedimwave as the solver or the program onedimwavebasic of Exercise for the Reader 12.7?

2. Make a movie, as distortion-free as possible, of the wave in Exercise 1 in the range $t = 0$ through $t = 6$.

3. Repeat all parts of Exercise 1 for the following vibrating string problems:

(a) $$\begin{cases} \text{(PDE)} \ u_{tt} = u_{xx}, & 0 < x < \pi, \ 0 < t < \infty, \ u = u(x,t) \\ \text{(BCs)} \ \begin{cases} u(x,0) = \sin(x), \ u_t(x,0) = \sin(x) \\ u(0,t) = u(\pi,t) = 0 \end{cases}, & 0 < x < \pi, \ 0 \le t < \infty \end{cases}$$

The exact solution is $u(x,t) = \sin(x)(\cos(xt) + \sin(t))$.

(b) $$\begin{cases} \text{(PDE)} \ u_{tt} = 9u_{xx}, & 0 < x < \pi, \ 0 < t < \infty, \ u = u(x,t) \\ \text{(BCs)} \ \begin{cases} u(x,0) = 0, \ u_t(x,0) - \sin^3(x) \\ u(0,t) = u(\pi,t) = 0 \end{cases}, & 0 < x < \pi, \ 0 \le t < \infty \end{cases}$$

The exact solution is $u(x,t) = [9\sin x \sin 3t - \sin 3x \sin 3at]/36$.

(c) $$\begin{cases} \text{(PDE)} \ u_{tt} = 4u_{xx}, & 0 < x < 1, \ 0 < t < \infty, \ u = u(x,t) \\ \text{(BCs)} \ \begin{cases} u(x,0) = x(x-1), \ u_t(x,0) = 0 \\ u(0,t) = u(\pi,t) = 0 \end{cases}, & 0 < x < \pi, \ 0 \le t < \infty \end{cases}$$

The exact solution is $u(x,t) = \dfrac{8}{\pi^3} \sum_{n=1}^{\infty} \dfrac{1}{(2n-1)^3} \sin[(2n-1)\pi x]\cos[(4n-2)\pi t]$.

Suggestion: For part (c), when computing the exact solution, use a finite sum that approximates the infinite sum to within machine precision. See Section 5.3 for related approximations.

4. Make movies, as distortion-free as possible, for each of the waves in parts (a) through (c) of Exercise 3. Use the time range $0 \le t \le 6$.

5. Consider the moving wave problem modeled by the basic wave equation $u_{tt} = u_{xx}$ on the string $0 \le x \le 10$ with initial profile $u(x,0) = \varphi(x) = \begin{cases} x/2, & \text{if } 0 \le x \le 2 \\ (x-3)^2, & \text{if } 2 \le x \le 3 \text{ and moving to the right at} \\ 0, & \text{otherwise} \end{cases}$

speed one; see Figure 12.17. (a) Use the finite difference method (with sufficiently fine grids) to solve this problem from $t = 0$ through $t = 20$, and plot some snapshots of the numerical solution.

Do the results change significantly depending on whether we use Program 12.1 onedimwave as the solver or the program onedimwavebasic of Exercise for the Reader 12.7? (b) Create a MATLAB movie of this wave motion. (c) Use the implicit finite difference method (of Exercise for the Reader 12.9) so re-solve part (a) with grids comparable to those that were used in part (a). How do the results compare? (d) Make a movie of the motion of the wave, from $t = 0$ through t = 12.

Suggestion: To get the initial velocity $u_t(x,0)$, use D'Alembert's theorem to get the analytical solution (for t less than 7) and differentiate with respect to t). Note that this initial velocity will be discontinuous.

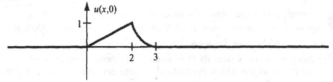

FIGURE 12.19: Initial wave profile for Exercise 5.

6. The wave problem of Exercise 5 involved a nonsmooth wave. Repeat all parts of Exercise 5 for the corresponding wave problem with initial pulse $u(x,0) = BS(x-2)$, where $BS(x)$ is the cubic spline given by equation (51) of Chapter 10. The initial pulse is still moving to the right at speed one. Because of the smoothness of the data, the finite difference methods should perform much better than for Exercise 5.

7. Use finite difference methods to solve the following vibrating string problem where one end is in sustained motion:

$$\begin{cases} \text{(PDE)} & u_{tt} = u_{xx}, \qquad\qquad 0 < x < \pi, \, 0 < t < \infty, \ u = u(x,t) \\ \text{(BCs)} & \begin{cases} u(x,0) = 0, \ u_t(x,0) = 0 \\ u(0,t) = \sin t, \ u(\pi,t) = 0 \end{cases} 0 < x < \pi, \, 0 \le t < \infty \end{cases}$$

(a) Display the results graphically as a series of snapshots.
(b) Create a MATLAB movie of this vibrating string.
(c) According to your data, does the solution eventually become a periodic function? If it seems so, can this be precisely confirmed?
(d) Compare the performance of explicit finite difference methods versus the implicit method of Exercise for the Reader 12.9 (using grids that obey the CFL stability criterion).

8. For each wave problem below, do the following: (i) Use an explicit finite difference method to create a series of snapshots of the wave propagation from time $t = 0$ up through (at least) $t = 10$. Try successively refining the resolutions until the numerical results stabilize (graphically at least). (ii) Create a MATLAB movie of the solution of (i). (iii) Repeat (i) using the implicit finite difference method of Exercise for the Reader 12.9.

(a) $$\begin{cases} \text{(PDE)} & u_{tt} = (1 + x/2)u_{xx}, \qquad 0 < x < \pi, \, 0 < t < \infty, \ u = u(x,t) \\ \text{(BCs)} & \begin{cases} u(x,0) = \sin(x), \ u_t(x,0) = \sin(x) \\ u(0,t) = u(\pi,t) = 0 \end{cases}, 0 < x < \pi, \, 0 \le t < \infty \end{cases}$$

(b) $$\begin{cases} \text{(PDE)} & u_{tt} = (1 + x^2)^{1/2}u_{xx}, \quad 0 < x < \pi, \, 0 < t < \infty, \ u = u(x,t) \\ \text{(BCs)} & \begin{cases} u(x,0) = 0, \ u_t(x,0) = 0 \\ u(0,t) = \sin(4x), \ u(\pi,t) = 0 \end{cases}, 0 < x < \pi, \, 0 \le t < \infty \end{cases}$$

(c) $$\begin{cases} \text{(PDE)} & u_{tt} = (1 + x^2)^{1/2}u_{xx}, \qquad 0 < x < \pi, \, 0 < t < \infty, \ u = u(x,t) \\ \text{(BCs)} & \begin{cases} u(x,0) = 0, \ u_t(x,0) = 0 \\ u(0,t) = \sin(4x), \ u(\pi,t) = (1/4)\sin(10x) \end{cases}, 0 < x < \pi, \, 0 \le t < \infty \end{cases}$$

9   (a) Write a function M-file for applying the implicit finite difference method (36) with parameter $\omega = 1/2$. The syntax should be as follows:

```
[x, t, U] = onedimwaveimpl_2(phi, nu, L, A, B, T, N, M, c)
```

where the variables and functionality should be just as in the program `onedimwaveimpl_4` of Exercise for the Reader 12.9.

(b) Do some experiments on the problem of Example 12.6 comparing the performance of the implicit finite difference method of part (a) with the explicit method of Program 12.1, using values of N and M that would make the latter unstable. Compare with the performance of `onedimwaveimpl_4` as was seen in Exercise for the Reader 12.9.

(c) How fine a resolution is needed to get the implicit finite difference method of part (a) to obtain graphical results of the quality of the numerical solution created in Example 12.6? Do the results seem better when the CFL condition is satisfied? Do some experiments. Compare with the performance of `onedimwaveimpl_4` as was seen in Exercise for the Reader 12.9.

10.   (*Another M-file for Two-Dimensional Waves*) (a) Write another M-file with the following syntax:

```
[x, y, t, U] = twodimwavedirbcV2(phi, nu, a, b, T, h, k, c)
```

that is designed to solve the two-dimensional wave problem (34). The variables and functionality are similar to the `twodimwavedirbc` M-file of Program 12.2 but there is one new input variable, k, which is the time step size. Thus this new program gives more flexibility in choosing time step in that it is no longer determined by forcing $\mu^2 = 1/2$ (the maximum value allowed in the stability condition).

(a) Run the program on the problem of Example 12.7 to reproduce the results of that example. In other words, choose k to be the value that was internally computed $= t(1)$ by the previous M-file, and check if the U matrix is identical, modulo roundoff, to the one in obtained in Example 12.7. (They should be if your program is correctly written.)

(b) Keeping h the same as in part (c), successively halve the k-step size from its value there, rerun the program, and compute the errors (at the same four t-values) of the numerical solution of your program with that in part (a). Do things get better, worse, or stay about the same? Comment on your findings.

(c) Can you find a two dimensional wave DVP where the `twodimwavedirbcV2` seems (with appropriate choices of k) to do a better job than `twodimwavedirbc`? This may take a good deal of numerical experimentation.

11   For each of the following two-dimensional wave problems do the following: (i) Create a series of snapshots of the wave motion. (ii) Create a movie of the wave motion. (iii) If an exact solution is given, compute the (maximum) errors of the finite difference solutions at several different time values. The problems all fall under the umbrella of the following Dirichlet BVP:

$$\begin{cases} \text{(PDE)} \quad u_{tt} = c^2(u_{xx} + u_{yy}), \quad 0 < x < a, 0 < y < b, 0 < t < \infty, \quad u = u(x,y,t) \\ \\ \text{(BCs)} \quad \begin{cases} u(x,y,0) = \varphi(x,y), \quad 0 \le x \le a, \ 0 \le y \le b, \ 0 \le t < \infty \\ u_t(x,y,0) = v(x,y), \quad 0 \le x \le a, \ 0 \le y \le b, \ 0 \le t < \infty \\ u(x,y,t) = 0, \quad \text{for all } (x,y) \text{ on the boundary of the rectangle:} \\ \qquad\qquad\qquad\qquad R = \{0 \le x \le a, \ 0 \le y \le b\} \end{cases} \end{cases}$$

(a) (*Two-Dimensional Hammer Blow*) Take $a = b = 3, c = 1, \quad \varphi(x,y) = 0$, and $v(x,y) = \begin{cases} 5, \ 1 \le x, y \le 2 \\ 0, \text{ otherwise} \end{cases}$. Run your computations on the time interval $t = 0$ through $t = 5$.

(b) Take $a = b = 1, \quad c = 2, \quad \varphi(x,y) = x(1-x)y(1-y)$, and $v(x,y) = 0$. Exact solution:

$$u(x,y,t) = \frac{64}{\pi^6} \sum_{n=0}^{\infty} \sum_{m=0}^{\infty} \frac{1}{(2n+1)^3(2m+1)^3} \cdot$$
$$\sin((2n+1)\pi x)\sin((2m+1)\pi y)\cos(\pi t \sqrt{(2n+1)^2 + (2m+1)^2}$$

(c) Data as in part (b) but $v(x,y) = 2\sin(2\pi x)\sin(\pi y)$. Exact solution: Solution of part (b) plus $(2/\sqrt{5})\sin(2\pi x)\sin(\pi y)\sin(\sqrt{5}\pi t)$.

Suggestion: The exact solutions of parts (b) and (c), so-called *double Fourier series*, can be estimated by finite sums of the form $u(x,y,t) = \frac{64}{\pi^6} \sum_{n=0}^{K} \sum_{m=0}^{K} \ldots$, where the integer K is chosen sufficiently large so that the series is accurate to MATLAB precision (i.e., maximum error $< 10^{-15}$). Try to estimate mathematically how large K should be for this accuracy. See Section 5.3 for similar estimates.

12. (a) (*Pinch-Gripped Membrane*) Suppose a membrane that occupies the square $0 \le x, y \le 4$ has initial position given by a pyramid with height 1 at $(x,y) = (2,2)$, and initial velocity equal to zero; see Figure 12.20(a). Create a movie of the resulting motion of the membrane if the wave speed is $c = 1$.

(b) (*Smooth Bumped Membrane*) Suppose a membrane that occupies the square $-2 \le x, y \le 2$ has initial position given by the graph $z = BS(r)$; where $r = \sqrt{x^2 + y^2}$ is the distance to the origin and BS is the cubic spline function (51) of Chapter 10, see Figure 12.20(b). The initial velocity is equal to zero. Create a movie of the resulting motion of the membrane if the wave speed is $c = 1$.

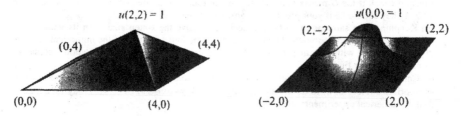

FIGURE 12.20: Initial membrane profiles for the wave problems of Exercise 12 (a) (left) and (b) (right).

13. Write a function M-file for solution of the following two-dimensional wave problem by the finite difference method.

$$\begin{cases} \text{(PDE)} \quad u_{tt} = c^2(u_{xx} + u_{yy}), \quad 0 < x < a, 0 < y < b, 0 < t < \infty, \;\; u = u(x,y,t) \\ \text{(BCs)} \begin{cases} u(x,y,0) = \varphi(x,y), \quad 0 \le x \le a, \; 0 \le y \le b, 0 \le t < \infty \\ u_t(x,y,0) = v(x,y), \quad 0 \le x \le a, \; 0 \le y \le b, 0 \le t < \infty \\ u(0,y,t) = L(y,t), \;\; u(a,y,t) = R(y,t), \\ u(0,x,t) = B(x,t), \;\; u(b,x,t) = T(x,t) \end{cases} \end{cases}$$

The syntax should be as follows:

```
[x,y,t,U] = twodimwavedirbc2(phi, nu, a, b, T, h, c, L, R, B, T)
```

where the variables (all but the last four input variables) and functionality are as in Program 12.2. The last four input variables are the boundary value functions.

(a) Use the program to solve the problem above with the following data:

$$c = 1, a = b = \pi, \varphi = v = L = T = R = 0, \;\; B(x,t) = \sin t \sin(x).$$

Create both snapshots and a movie of the wave propagation.

(b) Repeat part (a) but change $T(x,t) = (1/4)\sin(5t)\sin(3x)$.

NOTE: The next exercise will outline a proof of the necessity of the Courant-Friedrichs-Levy condition for the stability of the finite difference method in solving the wave equation. The proof will rely on finding exact solutions of the difference equation. As was the case with difference schemes for ODE in Part II, the theory of finding a solution of a finite difference scheme for a PDE closely parallels the theory of analytical solutions. Since we do not discuss this analytical theory, we will only start with a suitable form for a solution of the difference equation without discussing the motivation for this choice. The proofs will run more smoothly if we use complex numbers and, in particular, Euler's identity: $e^{i\theta} = \cos(\theta) + i\sin(\theta)$, where $i = \sqrt{-1}$ is the complex unit.

14. Use Euler's identity to show that $\cos(\theta) = (e^{i\theta} + e^{-i\theta})/2$ and $\sin(\theta) = (e^{i\theta} - e^{-i\theta})/2i$.

15. (*Stability Analysis*) This exercise will outline a proof that the Courant-Friedrichs-Levy condition (31) $\mu \equiv ck/h \le 1$ is necessary for the stability of the finite difference scheme (23)

$$u_{n,j+1} = 2(1-\mu^2)u_{n,j} + \mu^2\left[u_{n+1,j} + u_{n-1,j}\right] - u_{n,j-1}$$

for the wave equation $u_{tt} = c^2 u_{xx}$. Since the proof will invoke complex number notation, we have changed the index i to n in (23) to avoid any confusion.

(a) We begin by looking at functions (of n and j) of the form: $U_{n,j} = \alpha^j e^{i\beta nh}$ (here α and β are parameters and h is the grid spacing on the x-axis). Substitute this function into (23) and show that it solves it if and only if $\alpha + 1/\alpha - 2 = \mu^2(\eta - 2 + 1/\eta)$.[13]

(b) Show that the equation obtained in part (a) can be rewritten in the form $\alpha^2 + 2(2\mu^2 \sin^2(\beta h/2) - 1)\alpha + 1 = 0$. The reason that we wrote the equation as a quadratic in the variable α (rather than emphasizing β) is that in the solution form $U_{n,j} = \alpha^j e^{i\beta nh}$, the factor depending on β always has bounded absolute value, in fact $|e^{i\beta nh}| = 1$ regardless of the values of β, n, and h. The α-dependent factor α^j can blow up (cause instability) if $|\alpha| > 1$, but will remain stable if $|\alpha| \le 1$.

(c) Introduce the parameter $Q = 2\mu^2 \sin^2(\beta h/2) - 1$ so the equation in part (b) can be expressed as $\alpha^2 + 2Q\alpha + 1 = 0$. Use the quadratic formula to show the roots are $\alpha = -Q \pm \sqrt{Q^2 - 1}$. Show that both roots will have absolute values less than one if $Q \le 1$.

(d) Show that if $\mu \le 1$, then both roots of the equation in part (b) have absolute values less than or equal to one, and hence conclude the stability of the finite difference method, i.e., conclude that the solutions of part (a) will remain bounded independent of j and n.
Suggestions: For part (b) use a half angle formula from trig.

NOTE: If we impose Neumann and Robin boundary conditions at the ends of a wave problem (for a finite string):

$$u_{tt} = u_{xx}, \quad a < x < b, \, 0 < t < \infty, \quad u = u(x,t),$$

the result will be a well-posed problem. The standard Neumann boundary condition (at the right end $x = b$) $u_x(b,t) = 0$ corresponds to the end of the string being a **free end**, perhaps being glued to a

---

[13] For readers who are familiar with the method of separation of variables for finding solutions of PDEs, this form of the discrete problem can be derived by a similar discrete approach, i.e., we assume that the solution of the discrete equation can be separated as a product $A(j)B(n)$, substitute it into the difference equation, and then determine the form of A and B. For more details on this interesting analogy (as well as a full account of the relevant analytical theory) we refer the reader to [DuCZa-89].

massless ring that is free to move up and down on a frictionless vertical rod. The Robin boundary condition ($\alpha u(b,t) + \beta u_x(b,t) = \gamma$) can be viewed physically as (Hooke's Law) having a spring attached to the end of the string (with one end of the spring fixed). In general, the wave problem on a finite string will be well posed if there is specified any combination of Dirichlet, Neumann, or Robin boundary conditions at the two ends. This is physically quite reasonable; for a mathematical proof, see [Wei-65].

16. (*M-file for a Dirichlet-Neumann mixed Wave Problem*) (a) Write a MATLAB function M-file that will employ the finite difference method to solve the following wave problem:

$$\begin{cases} \text{(PDE)} & u_{tt} = c(t,x,u,u_x)^2 u_{xx}, & 0 < x < L, 0 < t < \infty, \ u = u(x,t) \\ \text{(BCs)} & \begin{cases} u(x,0) = \varphi(x), \ u_t(x,0) = v(x) \\ u(0,t) = A(t), \ u_x(L,t) = B(t) \end{cases}, \ 0 < x < L, 0 \leq t < \infty \end{cases}$$

The syntax of the M-file should be similar to that of Program 12.1:

```
[x, t, U] = onedimwavedirneu(phi, nu, L, A, B, T, N, M, c)
```

(b) Use the program of part (a) to solve the BVP above with the following data: $c = 1$, $L = 5$, $\varphi = v = B = 0$, and $A(t)$ as in Example 12.6. Display the solution as a series of snapshots.

(c) Repeat part (b), but change c to be as in Example 12.6..

17. (*Conservation Laws*) A one-dimensional PDE that models numerous physical conservation phenomena is the following first-order PDE: $u_t + c(t,x,u)u_x = 0$, $u = u(x,t)$. For applications, such as to highway traffic flow and fluid flow, we refer to [DuCZa-89] or [Log-94].
(a) Write a function M-file that will use finite difference methods to solve the above PDE on a finite segment $0 \leq x \leq L$ with initial profile $u(x,0) = \varphi(x)$ and Dirichlet boundary conditions $u(0,t) = A(t)$, $u(L,t) = B(t)$ (because the equation is only first-order we do not need to specify the initial wave velocity to make the problem well-posed). The syntax and functionality should be as in Program 12.1. Use centered difference (second-order) schemes on both the time and space derivative discretizations.
(b) Use your M-file of part (a) to solve the above BVP with data: $c(t,x,u) = x$, $L = 6$, $\varphi(x) = \exp(-2(x-2)^2)$ and $A = \varphi(0)$, $B = \varphi(L)$. Plot a series of snapshots of the solution over the time range $0 \leq t \leq 10$. Experiment with different resolution parameters until the numerical solutions tend to stabilize.
(c) In case $c(t,x,u) = c$ is constant, the resulting PDE $u_t + cu_x = 0$, $u = u(x,t)$ is often known as the **one-way wave equation**. Show that the solution of the resulting Dirichlet problem is given by $u(x,t) = \varphi(x - ct)$, and interpret in terms of moving waves.
(d) Apply the program of part (a) to solve the one-way wave problem with data $c = 1$, $\varphi(x)$ as in Figure 12.19, $L = 10$. How fine a resolution is needed so as to get a decent approximation to the exact solution as specified in part (c)?

18. (*Some Nonlinear Shock Wave Problems*) For each nonlinear wave problem do the following: (i) Use explicit finite difference methods to create a series of snapshots of the wave propagation from time $t = 0$ up through (if possible) $t = 10$. Try successively refining the resolutions until the numerical results stabilize (graphically at least). (ii) Create a MATLAB movie of the solution of (i). (iii) Repeat (i) using the implicit finite difference method of Exercise for the Reader 12.9.

(a) $\begin{cases} \text{(PDE)} & u_{tt} = (1+u^2)^{1/2} u_{xx}, & 0 < x < \pi, 0 < t < \infty, \ u = u(x,t) \\ \text{(BCs)} & \begin{cases} u(x,0) = \sin(x), \ u_t(x,0) = 0 \\ u(0,t) = u(\pi,t) = 0 \end{cases}, \ 0 < x < \pi, 0 \leq t < \infty \end{cases}$

(b) $\begin{cases} \text{(PDE)} & u_{tt} = (1+u^2)^{1/2} u_{xx}, & -10 < x < 10, 0 < t < \infty, \ u = u(x,t) \\ \text{(BCs)} & \begin{cases} u(x,0) = f(x), \ u_t(x,0) = 0 \\ u(0,t) = 0, \ u(\pi,t) = 0 \end{cases}, \ -10 < x < 10, 0 \leq t < \infty \end{cases}$

where $f(x) = \begin{cases} 1 - |x|, & \text{if } |x| < 1 \\ 0, & \text{otherwise} \end{cases}$.

(c) $\begin{cases} \text{(PDE)} & u_{tt} = (1 + u^2)^{1/2} u_{xx}, \\ \text{(BCs)} & \begin{cases} u(x,0) = 0, \ u_t(x,0) = 0 \\ u(0,t) = \sin(4t), \ u(\pi,t) = 0 \end{cases} \end{cases}$ $\quad \begin{array}{l} 0 < x < 6\pi, \ 0 < t < \infty, \ u = u(x,t) \\ 0 < x < 6\pi, \ 0 \le t < \infty \end{array}$

Note: The PDE here represents a wave equation where the speed of the wave depends on the amplitude, with higher (in absolute value) portions moving faster than lower portions. It would thus seem that high parts of the wave would catch up and overpass the lower parts that are ahead. This would seem to indicate that eventually the profiles would no longer be functions of x. (Think of the surface of an ocean wave as it begins to break.) See Figure 12.21 for an illustration. Such nonlinear BVPs thus do not have ordinary single-valued functions as solutions but rather what are called *multivalued wave forms*. For more on this interesting phenomenon see Chapter 3 of [Log-94]. Of course, the finite difference methods we developed are not set up to deal with such multivalued wave forms. In this exercise you should simply carry out the finite difference methods for a time interval stretching until the results no longer seem meaningful. Do your numerical results allow you to detect such shocking phenomena?

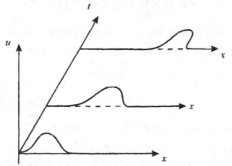

FIGURE 12.21: The shock-wave phenomena of the nonlinear wave equation of Exercise 18. Higher parts of the wave propagate faster than lower parts, eventually causing the wave profile to "break" from being a function of x.

NOTE: A general mathematical principle of partial differential equations roughly states that if the "data" of a boundary value problem has a certain amount of "smoothness," then the solution of the problem will enjoy the same amount of "smoothness."[14] Often, numerical methods require certain smoothness assumptions on the solution that is not usually known, so such smoothness results can be of great practical value in deciding in advance on a numerical method and predicting its success. The next exercise gives a wonderful yet rare situation where the numerical method gives the exact result.

19.   In this exercise we will show that if the solution to the wave problem (11)

$\begin{cases} \text{(PDE)} & u_{tt} = c^2 u_{xx}, \\ \text{(BCs)} & \begin{cases} u(x,0) = \varphi(x), \ u_t(x,0) = v(x) \\ u(0,t) = u(L,t) = 0 \end{cases} \end{cases}$ $\quad \begin{array}{l} 0 < x < L, \ 0 < t < \infty, \ u = u(x,t) \\ 0 \le x \le L, \ 0 \le t < \infty \end{array}$

is infinitely differentiable, then the finite difference scheme (23):

$$u_{i,j+1} = 2(1 - \mu^2)u_{i,j} + \mu^2 \left[u_{i+1,j} + u_{i-1,j} \right] - u_{i,j-1},$$

---

[14] The language here is admittedly quite vague, since it is not feasible to rigorously state a general result. To clarify things a bit: "Data" simply refers to the known functions in the problem, i.e., functions appearing as coefficients in the PDE or in the boundary conditions. "Smoothness" has to do with the order of differentiability of a given function. Such theorems are referred to as **regularity theorems**, and often require quite advanced mathematical methods to prove (and even to precisely state).

will be exact provided that we take $\mu = 1$ (the maximum value allowed by the Courant-Friedrichs-Levy stability condition). Thus in applying this explicit finite difference scheme to the wave problem (11), the only errors that will arise will be either roundoff errors or errors in the (needed) level $t = k$ values $u_{i,1}$ (perhaps obtained from (29)). Proceed through the following outline to establish this result. The method being used here is the so-called **bootstrapping technique**.

(a) Use Taylor's theorem to obtain the following expansions of finite difference quotients:

$$\Delta_t \equiv \frac{u(x_i,t_{j+1}) - 2u(x_i,t_j) + u(x_i,t_{j-1})}{k^2}$$

$$= u_{tt}(x_i,t_j) + 2\left[\frac{k^2}{4!}\partial_t^4 u(x_i,t_j) + \frac{k^4}{6!}\partial_t^6 u(x_i,t_j) + \cdots \right],$$

and

$$\Delta_x \equiv \frac{u(x_{i+1},t_j) - 2u(x_i,t_j) + u(x_{i-1},t_j)}{h^2}$$

$$= u_{xx}(x_i,t_j) + 2\left[\frac{h^2}{4!}\partial_x^4 u(x_i,t_j) + \frac{h^4}{6!}\partial_x^6 u(x_i,t_j) + \cdots \right].$$

(The symbols Δ_x, Δ_y have been introduced as shorthand for what remains.)

(b) Use part (a) and the fact that u satisfies the PDE of (11) to show that:

$$\Delta_t - c^2 \Delta_x = \frac{2}{4!}\left[k^2 \partial_t^4 u(x_i,y_j) - c^2 h^2 \partial_x^4 u(x_i,y_j) \right]$$

$$+ \frac{2}{6!}\left[k^4 \partial_t^6 u(x_i,y_j) - c^2 h^4 \partial_x^6 u(x_i,y_j) \right] + \cdots.$$

(c) In the expansion of part (b), use the PDE and to show that

$$k^2 \partial_t^4 u(x_i,y_j) = k^2 \partial_t^2 [\partial_t^2 u(x_i,y_j)] = k^2 \partial_t^2 [c^2 \partial_x^2 u(x_i,y_j)]$$

$$= c^2 k^2 \partial_x^2 [\partial_t^2 u(x_i,y_j)] = c^2 k^2 \partial_x^2 [c^2 \partial_x^2 u(x_i,y_j)] = k^2 c^4 \partial_x^4 u(x_i,y_j).$$

Use this to show that under the assumption $\mu = 1$, the first term on the right side of the expansion in part (b) is zero.

(d) Repeating the argument in part (c), show that the second term on the right side of the expansion in part (b) is zero.

(e) Go on to show that all of the remaining (infinitely many) terms on the right side of the expansion in part (b) are zero, and hence the local truncation error of the finite difference scheme (23) is zero.

20.   To what extent (if any) does the result of Exercise 19 still continue to hold if we allow time dependent Dirichlet boundary conditions:

$$u(0,t) = A(t), \quad u(L,t) = B(t), \quad t \geq 0,$$

assuming the functions $A(t)$ and $B(t)$ are infinitely differentiable?

21.   To what extent (if any) does result of Exercise 19 still continue to hold if we allow the wave-speed in the PDE to depend on any of the variables t, x, u? Assume (if you need to) that the function c is infinitely differentiable.

22.   (a) Explain why the argument of Exercise 19 does not hold for the corresponding wave problem in two variables (34) with $\mu = 1/2$ being the maximal allowable value for stability.

(b) Choose from this section (or one of the previous exercises) a particular instance of the problem (34) where an exact solution is known. Use the exact solution to determine the seed values $u_{i,j}^1$, and do some numerical experiments with the finite difference scheme given in the text (with $\mu = 1/2$) to demonstrate that the scheme is not exact.

12.3: FINITE DIFFERENCE METHODS FOR PARABOLIC PDE'S

As a prototypical problem for this section, we will use the one-dimensional heat equation on a finite interval with (possible) internal heat source and (possibly time-dependent) Dirichlet boundary conditions at both ends:[15]

$$\begin{cases} \text{(PDE)} & u_t = \alpha u_{xx} + q(x,t), \quad 0 < x < L, 0 < t < \infty, \quad u = u(x,t) \\ \text{(BCs)} & \begin{cases} u(x,0) = \varphi(x), \\ u(0,t) = A(t), \ u(L,t) = B(t) \end{cases} \quad 0 < x < L, 0 \le t < \infty \end{cases} \quad (42)$$

As was explained in Section 11.2, this boundary value problem models the heat distribution $u(x,t)$ on a thin rod of length L whose ends are maintained at the specified temperatures $A(t)$ and $B(t)$, whose initial temperature distribution is specified by $\varphi(x)$, and with an internal heat source $q(x,t)$. Note that for simplicity we initially assume the diffusivity α is constant, but this restriction can be easily lifted later. We will numerically solve this problem for all time values up to some predetermined value T. With our experience of finite difference methods for elliptic and hyperbolic problems behind us, we have developed essentially all of the methods required for parabolic problems. As with hyperbolic problems, stability will be a serious issue here. For explicit methods, stability puts a rather severe restriction on the size of the time steps, so implicit methods tend to be more practical for parabolic problems. Because of their dissipative nature, parabolic problems do tend to be more amenable to finite difference schemes than were hyperbolic problems, especially when discontinuous data is concerned.

Following the notation of our previous finite difference schemes, we introduce grids of equally spaced x- and t-coordinates for the rectangular region $0 \le x \le L$, $0 \le t \le T$:

$$\begin{aligned} 0 = x_0 < x_1 < x_2 < \cdots < x_{N+1} = L, \quad \Delta x_i \equiv x_i - x_{i-1} \equiv h, \\ 0 = t_0 < t_1 < t_2 < \cdots < t_{M+1} = T, \quad \Delta t_i \equiv t_i - t_{i-1} \equiv k. \end{aligned} \quad (43)$$

If we discretize the PDE of (42) by using the forward difference formula (Lemma 11.5) for u_t and the central difference formulas (Lemma 10.3) for u_{xx}, we get the **forward-time central-space scheme**:

$$\frac{u(x_i, t_{j+1}) - u(x_i, t_j)}{k} = \alpha \frac{u(x_{i+1}, t_j) - 2u(x_i, t_j) + u(x_{i-1}, t_j)}{h^2} + q(x_i, t_j), \quad (44)$$

---

[15] Notice that we have changed the notation a bit from Chapter 11. The diffusivity constant used to be labeled as "k", but we have changed this parameter to "α" so that we may continue to use "k" as the time step for finite difference methods.

the stencil for which is shown in Figure 12.23(a). We recall that the truncation errors here are $O(k)$ and $O(h^2)$, respectively, and so the local truncation error of this method is $O(k+h^2)$. Using the notation:

$$u_{i,j} = u(x_i,t_j), \quad q_{i,j} = q(x_i,t_j)$$

and introducing the parameter

$$\mu = \alpha k / h^2$$

allows us to express (44) in the following simplified form:

$$u_{i,j+1} = (1-2\mu)u_{i,j} + \mu\left[u_{i+1,j} + u_{i-1,j}\right] + kq_{i,j}. \tag{45}$$

For $i = 1, 2, ..., N$, and $j = 1, 2, ...M$. The stencil for (45) is shown in Figure 12.23(a). It can be shown that the forward-time central-space method is stable, provided that the following **stability condition** is met:

$$\mu \equiv \alpha\frac{k}{h^2} \le \frac{1}{2}. \tag{46}$$

Recall that stability means that any errors introduced in any particular stage of (44) (e.g., truncation errors) remain under control in all future iterations and that if the initial data is sufficiently differentiable, then the global error of the method will have the same quality bound as the local truncation error: $O(k+h^2)$. A complete proof of this stability result can be found in Section 9.4 of [Kels-66]; some elements will be given in the exercises. To see that (46) often makes the method impractical, for example, with $\alpha = 1$, if we wanted to use a space step size of $h = 0.05$, the stability condition (46) would force us to take a much smaller time step size $k \le 0.00125$.

By using instead the backward-time discretization (Lemma 11.5), we get the following **backward-time central-space scheme** (see Figure 12.23(b)), which is implicit but unconditionally stable:

$$\frac{u(x_i,t_j)-u(x_i,t_{j-1})}{k} = \alpha\frac{u(x_{i+1},t_j)-2u(x_i,t_j)+u(x_{i-1},t_j)}{h^2} + q(x_i,t_j). \tag{47}$$

Using the parameter $\mu = \alpha k / h^2$, this scheme can be expressed as:

$$-\mu u_{i-1,j+1} + (1+2\mu)u_{i,j+1} - \mu u_{i+1,j+1} = u_{i,j} + kq_{i,j}. \tag{48}$$

By the underlying finite difference approxim-ations, both of the above schemes have local truncation errors of order $O(k+h^2)$, so unless k is $O(h^2)$ (i.e., small enough so that the stability condition (46) holds), then the local truncation error will be first-order ($\approx O(k)$). Experience might lead us to believe using a centered difference (second-order) discretization of u_t would lead to a more effective scheme. The resulting finite difference method, known as **Richardson's method** (see Exercises 12 and 13), indeed has local truncation order $O(k^2 + h^2)$ but unfortunately has serious stability problems.

An interesting second-order implicit scheme can be obtained by averaging the forward-time central-space scheme and the backward-time central-space scheme (at corresponding time levels). This scheme is known as the **Crank-Nicolson method**[16] and runs as follows:

FIGURE 12.22(a): John Crank (1916–), English mathematician.

FIGURE 12.22(b): Phyllis Nicolson (1917–1968), English mathematician.

$$\frac{u_{i,j+1} - u_{i,j}}{k} = \frac{\alpha}{2}\left[\frac{u_{i+1,j} - 2u_{i,j} + u_{i-1,j}}{h^2} + \frac{u_{i+1,j+1} - 2u_{i,j+1} + u_{i-1,j+1}}{h^2}\right]$$
$$+ \frac{1}{2}\left[q(x_i, t_j) + q(x_i, t_{j+1})\right]. \tag{49}$$

EXERCISE FOR THE READER 12.11: Show that the local truncation error for the Crank-Nicolson method is $O(k^2 + h^2)$ provided that the solution is sufficiently differentiable.[17]

The stencil for the method is shown in Figure 12.23(c). As before, we let $\mu = \alpha k / h^2$, , which allows us to rewrite (51) in the form:

$$-\mu u_{i-1,j+1} + 2(1+\mu)u_{i,j+1} - \mu u_{i+1,j+1} = \mu u_{i-1,j} + 2(1-\mu)u_{i,j} + \mu u_{i+1,j}$$
$$+ k[q_{i,j} + q_{i,j+1}] \tag{50}$$

---

[16] The English school was quite involved with finite difference methods for evolution equations. Back in 1910, Lewis Richardson (1881–1953) had introduced his method and used it and other finite difference methods in numerous applications ranging from meteorology to eddy-diffusion in the atmosphere, and even (at the start of the Cold War) foreign policy and arms control. The stability problems with Richardson's method were not really recognized until the 1940s, when Crank, Nicolson, and others, with the aid of simple mechanical desk computing machines, performed lengthy computations. Crank and Nicolson's work culminated in a 1947 paper [CrNi-47] in which they introduced their unconditionally stable second-order method. This history brings up an important point regarding finite difference methods. It is easy to invent finite difference methods for any PDE. Getting efficient methods which actually converge with good stability and small truncation errors often requires a much more detailed investigation.

[17] The proof is rather technical, so this exercise for the reader may well be skipped by less theoretically oriented readers.

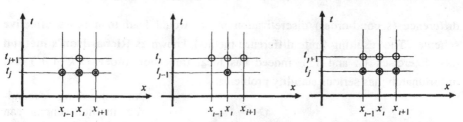

FIGURE 12.23: Stencils for three finite difference methods for the parabolic BVP (42): (a) (left) The forward-time central-space scheme, an explicit method. (b) (middle) The backward-time central-space scheme, an implicit unconditionally stable method. (c) (right) The Crank-Nicolson scheme.

We will revisit stability issues in more detail later in this section as well as in the exercises; for now we pass to the task of developing M-files for the above methods and then illustrate and compare them with some examples.   The M-file constructions are similar to those in the last section; we outline now the construction only for the Crank-Nicolson method and leave the other two as exercises.

We can convert (50) into matrix form quite naturally. Before we do this, let us first observe what happens in (50) when one of the terms involves a boundary value (i.e., either when $i = 1$ or $i = N$), In case $i = 1$, we then have $u_{i-1,j(+1)} = u(0, t_{j(+1)}) = A(t_{j(+1)})$,   and   similarly   in   case $i = N$, $u_{i+1,j(+1)} = u(L, t_{j(+1)}) = B(t_{j(+1)})$. Thus, for example, in case $i = 1$, (50) should be rewritten in the form:

$$2(1+\mu)u_{i,j+1} - \mu u_{i+1,j+1}$$
$$= 2(1-\mu)u_{i,j} + \mu u_{i+1,j} + \mu(A(t_j) + A(t_{j+1})) + k[q_{i,j} + q_{i,j+1}].$$

Borrowing from MATLAB's notation by letting $U(:, j)$ denote the jth column vector of U: $[u_{1,j} \ u_{2,j} \ \cdots \ u_{N,j}]'$, the equation (50) can be written in the form:

$$TU(:, j+1) = SU(:, j) + \mu V_j + kQ_j ,\qquad (51)$$

where the T and S are the following tridiagonal matrices:

$$T = \begin{bmatrix} 2(1+\mu) & -\mu & 0 & \cdots & 0 \\ -\mu & 2(1+\mu) & -\mu & \ddots & \vdots \\ 0 & \ddots & \ddots & \ddots & 0 \\ \vdots & & \ddots & & -\mu \\ 0 & \cdots & 0 & -\mu & 2(1+\mu) \end{bmatrix},$$

$$S = \begin{bmatrix} 2(1-\mu) & \mu & 0 & \cdots & 0 \\ \mu & 2(1-\mu) & \mu & \ddots & \vdots \\ 0 & \ddots & \ddots & \ddots & 0 \\ \vdots & & \ddots & & \mu \\ 0 & \cdots & 0 & \mu & 2(1-\mu) \end{bmatrix},$$

and the vectors V_j and Q_j are: $V_j = [A(t_j) + A(t_{j+1}), 0, \cdots, 0, B(t_j) + B(t_{j+1})]'$, and $Q_j = [q_{1,j} + q_{1,j+1}, \cdots, q_{N,j} + q_{N,j+1}]'$. Making use of MATLAB's matrix handling capabilities will now make coding the Crank-Nicolson method a simple task. The following program does this on a slightly more general version of the BVP (42).

PROGRAM 12.3: Function M-file for solution of the following parabolic BVP by the Crank-Nicolson method.[18]

$$\begin{cases} \text{(PDE)} & u_t = \alpha(x,t,u)u_{xx} + q(x,t), \ 0 < x < L, \ 0 < t < \infty, \quad u = u(x,t) \\ \text{(BCs)} & \begin{cases} u(x,0) = \varphi(x), \\ u(0,t) = A(t), \ u(L,t) = B(t) \end{cases} \quad 0 < x < L, \ 0 \le t < \infty \end{cases} \tag{52}$$

The Thomas algorithm is used to solve the tridiagonal systems which arise.

```
function [x, t, U] = cranknicolson(phi, L, A, B, T, N, M, alpha,q)
% solves the one-dimensional heat problem
% u_t = alpha(t,x,u)*u_xx+q(x,t)
% using the Crank-Nicolson method.
% Input variables: phi=phi(x) - initial wave profile function
% L - length of rod, A -A(t) - temperature of left end of rod
% u(0,t)-A(t), B=B(t) - temperature of right end of rod u(L,t)=B(t),
% T- final time for which solution will be
% computed, N = number of internal x-grid values, M = number
% of internal t-grid values, alpha =alpha(t,x,u)= diffusivity of rod.
% q - q(x,t) - internal heat source function
% Output variables: t   time grid row vector (starts at t=0, ends at
% t=T, has M+2 equally spaced values), x - space grid row vector,
% U = (M+2) by (N+2) matrix of solution approximations at
% corresponding grid points. x grid will correspond to second
% (column) entries of U, y grid values to first (row) entries of U.
% Row 1 of U corresponds to t = 0.

h = L/(N+1);  k = T/(M+1);
U=zeros(M+2,N+2); x=0:h:L; t=0:k:T;

% Recall matrix indices must start at 1.  Thus the indices of the
% matrix will always be one more than the corresponding indices that
% were used in theoretical development.

%Assign left and right Dirichlet boundary values.
U(:,1)=feval(A,t)'; U(:,N+2)=feval(B,t)';

%Assign initial time t=0 values.
for i=2:(N+1)
    U(1,i)=feval(phi,x(i));
```

---

[18] Since we are now allowing α to be variable, (49) takes on the following slightly more general form:

$$\frac{u_{i,j+1} - u_{i,j}}{k} = \left[\frac{\alpha_{i,j}}{2} \frac{u_{i+1,j} - 2u_{i,j} + u_{i-1,j}}{h^2} + \frac{\alpha_{i,j+1}}{2} \frac{u_{i+1,j+1} - 2u_{i,j+1} + u_{i-1,j+1}}{h^2} \right] + \frac{1}{2} \left[q(x_i,t_j) + q(x_i,t_{j+1}) \right],$$

where $\alpha_{i,j} = \alpha(x_i, t_j, u(x_i, t_j))$. The obvious problem, of course, is that we will not be able to evaluate the coefficients $\alpha_{i,j+1}$ in cases where α depends on u (or u_x). We circumvent this problem in such cases by setting $\alpha_{i,j+1} \approx \alpha(x_i, t_{j+1}, u(x_i, t_j))$. The other formulas are modified accordingly, and this is how the M-file is constructed; see Exercise 16.

```
end
%Assign values at interior grid points
for j=2:(M+2)
   for i=2:(N+1)
      mu(i)=k*feval(alpha,t(j-1),x(i),U(j-1,i))/h^2;
      mu2(i)=k*feval(alpha,t(j),x(i),U(j-1,i))/h^2;
      q1(i)=feval(q,x(i),t(j-1));  q2(i)=feval(q,x(i),t(j));
   end

% First form needed vectors and matrices, because we will be using
% the thomas M-file, we do not need to construct the coefficient
% matrix T.

   S = diag(2*(1-mu2(2:N+1))) + diag(mu2(3:N+1), -1) + diag(mu(2:N),
1);
   V = zeros(N,1);  V(1)=mu(2)*U(j-1,1)+mu2(2)*U(j,1);
   V(N)=mu(N+1)*U(j-1,N+2)+mu2(N+1)*U(j,N+2);
   Q = k*(q1(2:N+1)+q2(2:N+1))';

%Now perform the matrix multiplications to iteratively obtain
solution values for increasing time levels.

   c=S*((U(j-1,2:(N+1)))')+V+Q;
   a=-mu2(2:N+1); b=a; a(N)=0; b(1)=0;
   U(j,2:N+1)=thomas(a,2*(1+mu2(2:N+1)),b,c);
end
```

EXERCISE FOR THE READER 12.12: (a) Write a corresponding M-file for the problem (52) that applies the forward-time central-space scheme. The syntax should be as follows:

```
[x, t, U] = fwdtimecentspace(phi, L, A, B, T, N, M, alpha,q)
```

where the input variables, output variables, and functionality are just as in Program 12.3. If possible, avoid any square matrix multiplication in your program.
(b) Write a corresponding M-file for the problem (52) that applies the backward-time central-space scheme. The syntax should be as follows:

```
[x, t, U] = backwdtimecentspace(phi, L, A, B, T, N, M, alpha,q)
```

where the input variables, output variables, and functionality are just as in Program 12.3.

The reader is encouraged to compare the performances of the above two programs with the Crank-Nicolson method results of the next example. Another such comparison will be called for in Exercise for the Reader 12.13.

EXAMPLE 12.8: Consider the heat problem:

$$\begin{cases} \text{(PDE)} \ u_t = u_{xx}, & 0 < x < \pi, 0 < t < \infty, \ u = u(x,t) \\ \text{(BCs)} \ \begin{cases} u(x,0) = \varphi(x), \\ u(0,t) = u(L,t) = 0 \end{cases} & 0 < x < \pi, 0 \le t < \infty \end{cases},$$

where $\varphi(x) = \begin{cases} x, & \text{if } x < \pi/2 \\ \pi - x, & \text{if } x > \pi/2 \end{cases}$.

(a) Use the Crank-Nicolson method to create a mesh plot of the solution for $0 \le t \le 3$ with 25 interior space grid points and 92 interior time grid points. Then obtain a simultaneous two-dimensional plot of several temperature profiles of this numerical solution for various times in this range.

(b) The exact solution is given by $u(x,t) = \sum_{n=1}^{\infty} \frac{(-1)^{n+1}}{\pi(2n+1)^2} \sin[(2n+1)x]e^{-(2n+1)^2 t}$

(see, e.g., [Asm-00]). How large a value of N can be used so that the corresponding partial sum

$$u_N(x,t) \equiv \sum_{n=1}^{N} \frac{(-1)^{n+1}}{\pi(2n+1)^2} \sin[(2n+1)x]e^{-(2n+1)^2 t},$$

approximates the exact solution with error less than 10^{-15}, i.e., $|u(x,t) - u_N(x,t)| < 10^{-15}$ for all values of x and t?

(c) Use the "exact solution" of part (b) to plot a surface graph of the error of the Crank-Nicolson solution of part (a).

(d) Using the same number of space grid points, use two different numbers of interior time grid points to illustrate the stability condition (46) for this example with the forward-time central-space method.

SOLUTION: Part (a): We first create an M-file for the initial temperature distribution and then use Program 12.3.

```
function y = phi_EG12_8(x)
for i = 1:length(x);
   if (0<=x(1))&(x(1)<=pi/2)
       y(i)=x(i);
   else
      y(i)=pi-x(i);
   end
end
>> alpha=inline('1','x','t','u');
>> A=inline('0'); B=A; q=inline('0', 'x','t');
>> [x, t, UCN] = cranknicolson(@phi_EG12_8, pi, A, B, 3, 25, 92,
alpha,q);
```

To create the surface plot we simply enter:

```
>> surf(x,t,UCN)
>> xlabel('space'), ylabel('time'), zlabel('temperature')
```

The result is shown in Figure 12.24(a). The temperature profile plot shown in Figure 12.24(b) can be obtained by continuing to enter commands similar to the following:

```
>> plot(x,UCR(1,:)), hold on
>> xlabel('space'), ylabel('temperature')
>> gtext('t=0') %use mouse to place text
>> plot(x,UCR(5,:)), gtext('t=.323') %continue
```

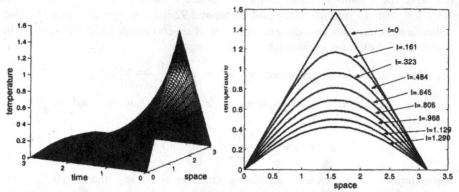

FIGURE 12.24: (a) (left) Surface graph of the (Crank-Nicolson) solution to the heat problem of Example 12.8. (b) (right) Individual temperature profiles of this solution for some particular values of t. The individual curves in (b) are simply "slices" of the surface in (a) with planes perpendicular to the time axis.

(b) We estimate the error by using a standard idea from calculus (see Chapter 5):

$$|u(x,t) - u_N(x,t)| < \left| \sum_{n=N+1}^{\infty} \frac{(-1)^{n+1}}{\pi(2n+1)^2} \sin[(2n+1)x]e^{-(2n+1)^2 t} \right|$$

$$< \sum_{n=N+1}^{\infty} \frac{1}{\pi(2n+1)^2} \leq \frac{1}{\pi}\int_{2N+2}^{\infty}\frac{du}{u^2} = \frac{1}{8\pi(N+1)^3}.$$

Thus, the error will always be less than 1e-15 (regardless of x and t) provided that $1/(8\pi(N+1)^3) < 1e\text{-}15$. Solving for N gives $N > 34{,}139$. Thus, we may take the finite sum $u_N(x,t)$ as the exact solution provided $N \geq 34{,}140$.

(c) We may now create a matrix Uexact, of exact solution values on the space-time grid of part (a). Rather than perform the entire sum for each point, the following code takes advantage of the separated nature of the summands:

```
>> Uexact=zeros(92+2,25+2);
>> for n=0:34140
   V=sin((2*n+1)*x); W=exp(-(2*n+1)^2*t);

   for i=1:length(t)
     Uexact(i,:)=Uexact(i,:)+(-1)^(n)*4/pi/(2*n+1)^2*V*W(i);
   end
end
>> mesh(x,t,Uexact-UCN)
>> xlabel('space'), ylabel('time'), zlabel('temperature')
```

The result is shown in Figure 12.25.

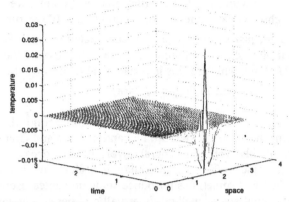

FIGURE 12.25: Surface plot of the error for the Crank-Nicolson approximation in part (a) of Example 12.8. Notice how the initially large error (due to the singularity of the boundary data) tends to dissipate very rapidly as time increases.

Part (d): For future reference, we first rewrite the stability condition (46) in terms of the input variables for our M-file solvers:

$$\mu \equiv \frac{T(N+1)^2}{L^2(M+1)} \leq \frac{1}{2}. \tag{53}$$

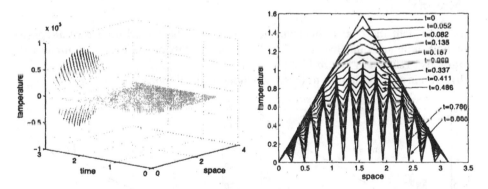

FIGURE 12.26: Illustration of instability when the forward-time central-space method is used with $\mu \approx .513...$ in the heat problem of Example 12.8. (a) (left) Mesh plot of the forward-time central-space showing the large-scale instability. (b) (right) Simultaneous plots of some heat profiles showing small-scale evolution of instability.

The parameter in part (a) would give a value $\mu \approx 2.23...$ which would result in the forward-time central-space method being quite unstable. We will look at what happens with the forward-time central-space scheme with some values of μ that are much closer to the stability barrier. Using the same number of internal space grid points ($N = 25$), and with $M = 400$ internal time grid points, we would have $\mu \approx .513...$ which slightly violates the stability condition. If we run the M-file fwdtimecentspace, just as in part (a), we can analogously obtain the mesh and two-dimensional plots shown in Figure 12.26, which illustrate the instability. If instead we used $M = 500$, which results in a stable value of $\mu \approx .410...$, we get a surface graph identical to the one in Figure 12.24(a).

The initial temperature in the last example was continuous but not differentiable, and for such problems, both the backward-time central-space and Crank-Nicolson methods perform admirably well. In general, errors tend to decay with time as they did in the last example. When the initial data is discontinuous, however, the Crank-Nicolson method will sometimes introduce some unwanted oscillations and the backward-time central-space method is usually a better choice for such problems. Of course, the oscillations can be mitigated by using smaller time step sizes, but this might entail significantly more computation than with the backward-time central-space method. For some theoretical explanations of such pathologies, see Section 9.1 of [Epp-02]. The next exercise for the reader gives an illustration.

EXERCISE FOR THE READER 12.13: Consider the following heat problem:

$$\begin{cases} \text{(PDE)} \ \ u_t = 3u_{xx}, & 0 < x < 4, \ 0 < t < \infty, \ \ u = u(x,t) \\ \text{(BCs)} \ \ \begin{cases} u(x,0) = \varphi(x), \\ u(0,t) = 0, \ u(4,t) = 100(1-e^{-t}), \end{cases} & 0 < x < 4, \ 0 \le t < \infty \end{cases},$$

where $\varphi(x) = \begin{cases} 100, & \text{if } 1 \le x \le 2 \\ 0, & \text{otherwise} \end{cases}$.

(a)   Run both the Crank-Nicolson method and the backward-time central-space method with an equivalent set of grids to obtain plots of the resulting numerical solutions as shown in Figure 12.27.

 (b) Obtain also mesh plots of these two numerical solutions.

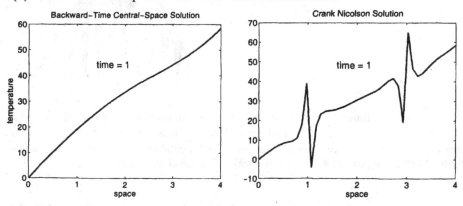

FIGURE 12.27: With any given grid sizes, the (a) (left) backward-time central-space method usually outperforms the (b) (right) Crank-Nicolson method when the parabolic problem has discontinuous data, as is demonstrated from these two snapshots of the two methods on the problem of Exercise for the Reader 12.13. The oscillations in (b) are not part of the actual solution; using finer time grids will cause them to fade away.

We next move on to boundary conditions involving derivatives. Previously we have separated these sorts of boundary conditions into two kinds: Neumann and Robin. For a physical interpretation in terms of our heated-rod problem, it is helpful to consider both of these conditions as special cases of the following general condition:

$$\partial u / \partial n = -\gamma(u - \eta). \tag{54}$$

In general the parameters γ and η are allowed to be functions of other variables, e.g., $\eta = \eta(x,t,u)$, but we have written them as constants for simplicity. In case $\gamma = 0$, we get the basic Neumann boundary condition corresponding to an insulated end of the rod. The general Neumann BC is gotten by taking $\eta = \eta(x) = u(x) + 1$ (so (54) becomes $\partial u / \partial n = \gamma$), which corresponds to heat being lost or absorbed by the end of the rod at a specified rate. The general form of (54) with $\gamma \neq 0$ gives Robin BCs that physically correspond to heat being radiated into or out of the rod at a rate proportional to the difference of u and some specified temperature η.

EXAMPLE 12.9: Consider the heat problem:

$$\begin{cases} \text{(PDE)} & u_t = u_{xx}, \quad 0 < x < 1, 0 < t < \infty, \quad u = u(x,t) \\ \text{(BCs)} & \begin{cases} u(x,0) = 100, \\ u_x(0,t) = u(0,t), \ u_x(1,t) = -u(1,t) \end{cases} \quad 0 < x < 1, 0 \leq t < \infty \end{cases}$$

This corresponds to a laterally insulated rod that is initially uniformly heated to a temperature of 100 and for which heat is being radiated from both ends at a rate equal to the current temperature of the end. Adapt the forward-time central-space method to solve this problem on the time range $0 \leq t \leq 1$ using a stable grid, and give plots of some temperature profiles.

SOLUTION: Part (a): In order to maintain the $O(h^2)$ quality portion of the local truncation error, we should approximate both derivative boundary conditions using (second-order) central differences. This will require the use of "ghost nodes" at both left and right ends. We modify our indexing scheme for the space variables accordingly as follows:

$$x_i = (i-1)h \quad (0 \leq i \leq N+1).$$

Thus the unknown grid values are still indexed as $x_1, \cdots, x_N$, and $x_0 = -h$ and $x_{N+1} = L + h$ correspond to the ghost nodes, but now $L = (N-1)h$. In this notation, we discretize the left boundary condition (at each time level) with the central difference approximation:

$$[u_{2,j} - u_{0,j}] / 2h = u_{1,j},$$

where $u_{0,j}$ is the ghost node value. We can eliminate the ghost node in the system (45) by assuming that this discretization is valid at the left end, i.e.,

$$u_{1,j+1} = (1-2\mu)u_{1,j} + \mu\left[u_{2,j} + u_{0,j}\right].$$

Eliminating the ghost node value using the two preceding equations leads us to

$$u_{1,j+1} = (1-2\mu)u_{1,j} + 2\mu\left[u_{2,j} - hu_{1,j}\right].$$

In the same fashion, we obtain the following formula for $u_{N,j+1}$ corresponding to the right boundary:

$$u_{N,j+1} = (1-2\mu)u_{N,j} + 2\mu\left[u_{N-1,j} - hu_{N,j}\right].$$

To move to the next time level we use these in conjunction with (45) to compute all of the remaining $u_{i,j+1}$ $(1 < i < N)$:

$$u_{i,j+1} = (1-2\mu)u_{i,j} + \mu\left[u_{i+1,j} + u_{i-1,j}\right].$$

We can specialize the code of fwdtimecentspace with the above modifications to create a matrix of numerical values for the numerical solution as follows. Below we are using $h = 1/20$ and $k = 1/1850$. This gives a stable value for $\mu = \alpha k / h^2 = 1 \cdot (1/1850)/(1/20)^2 = 0.2162... < 1/2$.

```
h=1/20; k=1/1850; mu=k/h^2;
N=21; M=1851;
U=zeros(M+1,N); x=0:h:1; t=0:k:1;
%Assign initial time t=0 values and next step t=k values.
for i=1:N, U(1,i)=100; end
%Assign values at interior grid points
for j=2:M+1
    U(j,2:N-1)=(1-2*mu)*U(j-1,2:N-1)+mu*(U(j-1,3:N)+U(j-1,1:N-2));
    U(j,1)= (1-2*mu)*U(j-1,1)+2*mu*(U(j-1,2)-h* U(j-1,1));
    U(j,N)= (1-2*mu)*U(j-1,N)+2*mu*(-h*U(j-1,N)+U(j-1,N-1));
end
```

The simultaneous plots of some temperature profiles, shown in Figure 12.28(a), are now obtained just as in the previous example.

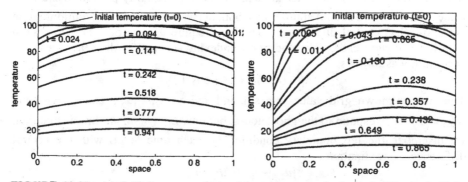

FIGURE 12.28: (a) (left) Some snapshots of the temperature in the rod of Example 12.9, where heat is lost through equal radiation at both ends. (b) (right) Snapshots for the analogous problem of Exercise for the Reader 12.14. The latter problem has a greater radiation heat loss on the left end rather than on the right, perhaps due to differences in insulation.

EXERCISE FOR THE READER 12.14: (a) Redo the Example 12.9 when the left BC is changed to $u_x(0,t) = u(0,t)^{1.5}$; your plots should look similar to those of Figure 12.28(b).

(b) Will the temperature on the right end eventually be lower than that on the left end? Try to answer this question on physical grounds first, then attempt to use MATLAB to back up your answer. Does the BC of this example have any physical models?

(c) If instead the left BC of the BVP of Example 12.9 was changed to $u_x(0,t) = -u(0,t)$, what do you think would eventually happen to the temperature in the rod? Try to answer this question on physical grounds first, then attempt to use MATLAB to back up your answer.

Similar methods can be developed for any other kinds of boundary conditions, and everything can be done likewise for the Crank-Nicolson or the purely implicit method. There will be some exercises to give more practice with this.

EXERCISE FOR THE READER 12.15: (a) Write an M-file with the following syntax:

```
[x, t, U] = cranknicolsonRobinLR(phi, L, A, B, T, N, M, alpha,q)
```

that will numerically solve the following BVP:

$$\begin{cases} (PDE) \ u_t = \alpha(x,t,u)u_{xx} + q(x,t), & 0 < x < L, 0 < t < \infty, \ u = u(x,t) \\ (BC's) \ \begin{cases} u(x,0) = \varphi(x), \\ u(0,t) = au_x(0,t) + b, \ u_x(L,t) = cu(0,t) + d \end{cases} & 0 < x < L, 0 \le t < \infty \end{cases}$$

where a, b, c, and d are constants. The input variables and functionality are similar to those of Program 12.3, except here the input variables A and B represent the vectors $[a \ b]$ and $[b \ c]$, respectively.

(b) Apply the program to re-solve the problem of Example 12.9 (using the same values for M and N) and compare plots of the corresponding numerical temperature profiles with those of Figure 12.28(a) (obtained using the explicit method).

We end this section with some theoretical developments for a heat problem and associated finite difference methods. For simplicity, we work with a basic Dirichlet problem for the one-dimensional heat equation of the following theorem. Many of the results, ideas, and concepts generalize to other parabolic BVPs.

THEOREM 12.2: (*Existence and Uniqueness for a Heat Problem*) Suppose that L and T are positive numbers and $A(t)$, $B(t)$, and $\varphi(x)$ are continuous functions. The following BVP (42) has a unique solution

$$\begin{cases} (PDE) \ u_t = \alpha u_{xx} + q(x,t), & 0 < x < L, 0 < t < \infty, \ u = u(x,t) \\ (BCs) \ \begin{cases} u(x,0) = \varphi(x), \\ u(0,t) = A(t), \ u(L,t) = B(t) \end{cases} & 0 < x < L, 0 \le t < \infty \end{cases}$$

Elements of the proof of this result can be found in the exercises; see also [Asm-00] (in particular, see Sections 3.5 and 3.10 therein).   We now give a formal definition of stability of a finite difference scheme for this (BVP) (42).

Definition: A finite difference scheme for the BVP (42) that satisfies the hypotheses of Theorem 12.2 is **unconditionally stable** if there exists a constant $C > 0$ (depending only on the data α, q, A, B of the BVP) such that for any initial profile function $\varphi(x)$, any grid of the space interval [0, L] and any grid of the time interval [0, T], we have

$$\max_{i,j} |u_i^j| \leq C \max_{0 \leq x \leq L} |\varphi(x)|. \tag{55}$$

In words, this roughly states that the numerical values do not get too much larger than the initial data.   The method is **conditionally stable** if (54) holds provided that the uniform time step size (k) and space step size (h) satisfy some specified relationship.

Note that if we used a stable finite difference method to solve the problem (42) with perturbed initial heat profile $\tilde{\varphi}(x)$ with $|\tilde{\varphi}(x) - \varphi(x)| < \varepsilon$ where ε is a very small number, and if $\tilde{u}_i^j$ denotes the resulting numerical values, then $|\tilde{u}_i^j - u_i^j| < C\varepsilon$.   (Proof: Apply the stability inequality (55) to the BVP (42) with initial heat profile changed to $\tilde{\varphi}(x) - \varphi(x)$.)   The advantage of stable schemes is that any roundoff errors propagate only in an additive fashion.   In unstable schemes, as we have seen, the errors usually propagate exponentially.

Using the so-called *von Neumann approach*, we will establish the stability results that have been stated thus far in the section as well as others in the context of the (42).[19]   To help make the ideas more transparent, we assume the equation and the boundary conditions are homogeneous: $q(x,t) = 0$, $A(t)$, $B(t) = 0$.   Let us begin with the (homogeneous) explicit method (45):

$$u_{i,j+1} = (1 - 2\mu)u_{i,j} + \mu\left[u_{i+1,j} + u_{i-1,j}\right].$$

For the homogeneous problem, in addition to the guaranteed existence and uniqueness from the above theorem, we also have the **maximum principle**: The solution satisfies $|u(x,t)| \leq \max|\varphi(x)|$.   Physically, this simply says that the maximum temperature of the rod is attained initially.[20]   We assume that the

---

[19] This approach was used in the exercises of the last section.   There we used complex notation which afforded a more efficient analysis; the approach in the text above will use only real numbers, but the exercises will revisit the complex number approach.

[20] The maximum principle holds more generally in the presence of time-dependent Dirichlet boundary conditions $u(0,t) = A(t)$, $u(L,t) = B(t)$, in which case the inequality changes to $|u(x,t)| \leq M$ where M is the largest of $\max|\varphi(x)|$, $\max|A(t)|$, and $\max|B(t)|$ (if these maxima exist).   For a proof, see

(discrete) solution of the finite difference method has the following separated form:

$$u_{i,j} = X_i T_j.$$ (56)

Substitution of (56) into the above finite difference scheme and dividing by $X_i T_j$ produces:

$$\frac{T_{j+1}}{T_j} = 1 - 2\mu + \mu \frac{X_{i+1} - X_{i-1}}{X_i}.$$ (57)

Since the left sequence depends only on j while that on the right depends only on i, it follows that both sides must equate to the same constant. Calling this constant ξ, the temporal (time) portion of (57) yields $T_{j+1} = \xi T_j$ and hence

$$T_j = \xi^j T_0.$$ (58)

The spatial portion of (57): $1 - 2\mu + \mu \dfrac{X_{i+1} - X_{i-1}}{X_i} = \xi$, will take a bit more effort

to solve. We assume that X_i takes the form: $X_i = A\cos i\theta + B\sin i\theta,$ [21] where the parameters A and B and θ are to be determined. The left (homogeneous) boundary condition forces $A = 0$. The right (homogeneous) boundary condition now implies that $(N+1)\theta = n\pi$ for some positive integer n. Combining this with the relation $(N + 1)h = L$ (h = spatial step size), produces $\theta = n\pi h/L$. The (nonzero) value of the parameter B may now be arbitrarily specified (since it cancels out of the spatial equation and the boundary conditions are already guaranteed, note that we have not yet dealt with the boundary condition). Using B = 1 gives the following candidate solution to the spatial equation:

$$X_i = \sin(in\pi h/L).$$ (59)

Substituting this into the spatial equation gives:

$$1 - 2\mu + \mu \frac{\sin((i+1)n\pi h/L) + \sin((i-1)n\pi h/L)}{\sin(in\pi h/L)} = \xi.$$

Section 3.10 of [Asm-00]. The maximum principle for the heat equation ceases to be valid in case of heat sources; see Exercise 21.

[21] Here i is an index (not a complex number). For readers who are familiar with the analytic theory of ODE, the choice of the form of X_i was motivated by viewing the spatial equation

$\mu \dfrac{X_{i+1} - X_{i-1}}{X_i} + 1 - 2\mu = \xi$ as a discretization of the second-order ODE: $\mu X''(s) + 1 - 2\mu = \xi.$

If we apply the trig addition formulas (e.g., $\sin((i+1)n\pi h/L) = \sin(in\pi h/L)\cos(n\pi h/L) + \sin(n\pi h/L)\cos(in\pi h/L)$) we can convert the above equation to the following form:

$$\xi = \xi(n) = 1 - 2\mu(1 - \cos(n\pi h/L)).. \tag{60}$$

Since the resulting separated solutions $\xi^j T_0 \sin(in\pi h/L)$ of the finite difference scheme exist for any integer n, and since we know (from the maximum principle) the actual solution to the BVP is bounded, it follows that we must have $|\xi(n)| \le 1$ because otherwise this finite difference solution will grow exponentially. Since n can be arbitrary (because for stability h can be arbitrary) the $\cos(n\pi h/L)$ can get arbitrarily close to -1, so (59) shows us that in order for $|\xi(n)| \le 1$, we must have $2\mu \le 1$ or $\mu \le 1/2$, which is the stability condition (46).

This von Neumann method can be applied to obtain stability results for many finite difference schemes for a wide range of BVPs. For example, the following finite difference scheme (for the same homogeneous problem considered above):

$$\frac{u_{i,j+1} - u_{i,j}}{k} = \frac{\alpha}{2}\left[(1-\sigma)\frac{u_{i+1,j} - 2u_{i,j} + u_{i-1,j}}{h^2} \right. $$
$$\left. + \sigma\frac{u_{i+1,j+1} - 2u_{i,j+1} + u_{i-1,j+1}}{h^2} \right], \tag{61}$$

where $0 \le \sigma \le 1$, includes as special cases the explicit method ($\sigma = 0$), the backward-time central-space method (49) ($\sigma = 1$), and the Crank-Nicolson method ($\sigma = 1/2$). Note that this method is always implicit when $\sigma > 0$. The von Neumann method can be used to show this method is unconditionally stable whenever $1/2 \le \sigma \le 1$ and when $0 \le \sigma < 1/2$ the method will be stable if the following condition holds:

$$\mu \equiv \alpha k/h^2 < 1/(2-4\sigma). \tag{62}$$

There are other approaches to stability theory of finite difference methods. One of these, known as the spectral method, looks at eigenvalues of matrices associated with the finite difference methods. The exercises will delve deeper into stability theory. Some references include [Smi-85], [RiMo-67], and [IsKe-66] and the more elementary [DuCZa-89].

EXERCISES 12.3

NOTE: For convenience in these exercises, we will refer to the forward-time central-space method (44) simply as the "explicit method" and the backward-time central-space method as the "implicit method."

1   Use the explicit method to solve the following BVP on $0 \le t \le 1$ using the indicated grids:

$$\begin{cases} \text{(PDE)} & u_t = u_{xx}, \qquad\qquad\qquad 0 < x < 1, 0 < t < \infty, \;\; u = u(x,t) \\ \text{(BCs)} & \begin{cases} u(x,0) = \sin(\pi x)(1 + 2\cos(\pi x)), \\ u(0,t) = 0, \; u(1,t) = 0 \end{cases}, \quad 0 < x < 1, 0 \le t < \infty \end{cases}$$

In each case, compare the results with the actual solution $u(x,t) = \sin(\pi x)e^{-\pi^2 t} + \sin(2\pi x)e^{-4\pi^2 t}$
at the indicated time levels.
(a) $N = 19$, $M = 39$. Note that this set of parameters violates the stability condition, by (P8),
$\mu = 1$ Compare the numerical solution with the exact solution at time levels $t = .25$, $t = .5$, $t = .75$, $t = 1$.
(b) $N = 39$, $M = 99$. Note that this set of parameters satisfies the stability condition with $\mu = 0.4$.
Compare numerical solution with exact solution at time levels $t = .5$, $t = 1$, $t = 1.5$, $t = 2$.

2.   (a) Re-solve the BVP of Exercise 1 using the Crank-Nicolson method with $N = 39$ and $M = 99$
and give a mesh plot of the error of this numerical solution.
(b) Repeat part (a) this time using the implicit method.

3.   For each of the following BVPs, do the following: (i) Use the explicit method to solve the
problem on the indicated time interval. Begin with 10 interior space node values and a
corresponding number of interior time node values that results in a stable scheme. (ii) Continue
halving the space step size and using a smaller (stable) time step that evenly divides into the
previous time step. Compare common temperature profiles of adjacent numerical solutions.
Keep track of the maximum discrepancy. Continue until the maximum discrepancy is smaller
than 10^{-3} or the numerical computations take more than two minutes on MATLAB. (iii) For
your final numerical solution, plot a (three-dimensional) mesh plot of the solution. (iv) For your
final numerical solution, plot (and label) several two-dimensional temperature profiles in the
same graph.

(a) $\begin{cases} \text{(PDE)} & u_t = u_{xx}, \qquad\qquad 0 < x < 1, 0 < t \le 1 \;\; u = u(x,t) \\ \text{(BCs)} & \begin{cases} u(x,0) = 50, \\ u(0,t) = 0, \; u(1,t) = 100 \end{cases}, \quad 0 < x < 1, 0 \le t \le 1 \end{cases}$

(b) $\begin{cases} \text{(PDE)} & u_t = u_{xx}, \qquad\qquad 0 < x < 1, 0 < t \le 1, \;\; u = u(x,t) \\ \text{(BCs)} & \begin{cases} u(x,0) = \begin{cases} 2x, & \text{if } 0 \le x \le 1/2 \\ 1, & \text{if } 1/2 \le x \le 1 \end{cases} \\ u(0,t) = 0, \; u(1,t) = 0, \; 0 \le t \le 1 \end{cases} \end{cases}$

(c) $\begin{cases} \text{(PDE)} & u_t = u_{xx}, \qquad\qquad 0 < x < \pi, 0 < t \le 1, \;\; u = u(x,t) \\ \text{(BCs)} & \begin{cases} u(x,0) = 100x(\pi - x), \\ u(0,t) = 50\sin(\pi x), \; u(\pi,t) = 0 \end{cases}, \quad 0 < x < \pi, 0 \le t \le 1 \end{cases}$

(d) $\begin{cases} \text{(PDE)} & u_t = 2u_{xx}, \qquad\qquad 0 < x < \pi, 0 < t \le 1, \;\; u = u(x,t) \\ \text{(BCs)} & \begin{cases} u(x,0) = \begin{cases} 100\pi x \\ 100\pi^2 - 100\pi x \end{cases}, \quad 0 < x < \pi, 0 \le t \le 1 \\ u(0,t) = 50\sin(\pi x), \; u(\pi,t) = 0 \end{cases} \end{cases}$

4.   Parts (a) through (d): For each of the BVPs Exercise 3, repeat the directions of that problem,
this time using the Crank-Nicolson method.   Try using your temporal step size k to
satisfy $h/2 \le k \le h$, where h is the spatial step. Try redoing some of these numerical solutions
using a much smaller temporal step size (as stipulated by the stability condition (46) for the
explicit method).   Does this seem like a better strategy than using roughly the same step sizes?
In answering this latter question, you should, of course, weigh in the extra work needed for a
given spatial step size.

5.   Parts (a) through (d): For each of the BVPs of the Exercise 3, repeat the directions of that
      problem, this time using the implicit method. Try using your temporal step size k to satisfy
      $h/2 \le k \le h$, where h is the spatial step. Try redoing some of these numerical solutions using
      a much smaller temporal step size (as stipulated by the stability condition (46) for the explicit
      method). Does this seem like a better strategy than using roughly the same step sizes? In
      answering this latter question, you should, of course, weigh in the extra work needed for a given
      spatial step size.

6.   Consider the following heat problem:

$$\begin{cases} \text{(PDE)} \ u_t = u_{xx}, & 0 < x < 1, \ 0 < t < \infty, \ u = u(x,t) \\ \text{(BCs)} \ \begin{cases} u(x,0) = \sin(\pi x), \\ u(0,t) = t, \ u(1,t) = 0, \end{cases} & 0 < x < 1, \ 0 \le t < \infty \end{cases}$$

      This can be interpreted as the modeling of a rod of length one with initial heat distribution as
      specified, right end being maintained at temperature zero, and left end being constantly heated
      up so has to have a temperature of t at time t.

      (a) Modify the explicit method to estimate the time t^* it takes for the temperature at the
      midpoint $t = 1/2$ to first reach a value of $u = 2$. Run the new program to estimate the time t^* by
      trying out several different grid choices.
      (b) Repeat part (a), this time using the implicit method.
      (c) Repeat part (a), this time using the Crank-Nicolson method.

7.   For each of the following BVPs, do the following: (i) Use the explicit method to solve the
      problem on the indicated time interval. Begin with 10 interior space node values and a
      corresponding number of interior time node values that results in a stable scheme. (ii) Continue
      halving the space step size and using a smaller (stable) time step that evenly divides into the
      previous time step. Compare common temperature profiles of adjacent numerical solutions.
      Keep track of the maximum discrepancy. Continue until the maximum discrepancy is smaller
      than 10^{-3} or the numerical computations take more than two minutes on MATLAB. (iii) For
      your final numerical solution, plot a (three-dimensional) mesh plot of the solution. (iv) For your
      final numerical solution, plot (and label) several two-dimensional temperature profiles in the
      same graph.

(a) $$\begin{cases} \text{(PDE)} \ u_t = \alpha(x)u_{xx}, & 0 < x < 1, \ 0 < t \le 1 \ \ u = u(x,t) \\ \text{(BCs)} \ \begin{cases} u(x,0) = 50, \\ u(0,t) = 0, \ u(1,t) = 100, \end{cases} & 0 < x < 1, \ 0 \le t \le 1 \end{cases}$$

      where $\alpha(x) = \begin{cases} 1, \text{ if } x \le 1/2 \\ 4, \text{ if } x > 1/2 \end{cases}$.

(b) $$\begin{cases} \text{(PDE)} \ u_t = \alpha(x)u_{xx}, & 0 < x < 1, \ 0 < t \le 1, \ u = u(x,t) \\ \text{(BCs)} \ \begin{cases} u(x,0) = \begin{cases} 100, & \text{if } 1/4 \le x \le 3/4 \\ 0, & \text{otherwise} \end{cases} \\ u(0,t) = 0, \ u(1,t) = 0, \ 0 \le t \le 1 \end{cases} \end{cases}$$

      where $\alpha(x)$ is as in part (a).

      (c) Same BVP as (a) but change $\alpha(x)$ to $\alpha(u) = (1/25)\sqrt{u^2 + 1}$.
      (d) Same BVP as (b) but change the PDE to $u_t = \alpha(x)u_{xx} + q(x,t)$, where

$$q(x,t) = \begin{cases} 200\sin(3\pi t/2), & \text{if } 0 \le t \le 2/3 \text{ and } x \in [0,1/8] \cup [7/8,1] \\ 0, & \text{otherwise} \end{cases}$$

8.   Parts (a) through (d): For each of the BVPs of Exercise 8, repeat the directions of that problem,
      this time using the Crank-Nicolson method. Try using your temporal step size k to
      satisfy $h/2 \le k \le h$, where h is the spatial step. Try redoing some of these numerical solutions

using a much smaller temporal step size (as stipulated by the stability condition (46) for the explicit method). Does this seem like a better strategy than using roughly the same step sizes? In answering this latter question, you should, of course, weigh in the extra work needed for a given spatial step size.

9   Parts (a) through (d): For each of the BVPs of Exercise 7, repeat the directions of that problem, this time using the implicit method. Try using your temporal step size k to satisfy $h/2 \leq k \leq h$, where h is the spatial step. Try redoing some of these numerical solutions using a much smaller temporal step size (as stipulated by the stability condition (46) for the explicit method). Does this seem like a better strategy than using roughly the same step sizes? In answering this latter question, you should, of course, weigh in the extra work needed for a given spatial step size.

10.   Rewrite Program 12.3 (for the Crank-Nicolson method) so that it avoids the creation of any square matrices, but is otherwise the same program; in particular, the input and output variables and functionality of the two programs should be identical. Find a problem and corresponding input parameters where your modified program noticeably outperforms Program 12.3.

11.   (*Some Related Neumann Problems*) For each of the following BVPs, set up an appropriate finite difference scheme and numerically solve the problem. Continue to re-solve the problem with a decreasing set of space steps and corresponding decreasing time steps (in a stable way), compare consecutive numerical solutions (at common grid values), and continue until the maximum error becomes less than 0.001 or the computations take more than two minutes. Comment on the stability (or lack thereof) of your method on the problem. In case of stability, plot your final solution, both as a three-dimensional surface plot and as a two-dimensional plot of several superimposed time level profiles.

(a) $\begin{cases} \text{(PDE)} & u_t = u_{xx}, \qquad\qquad\qquad 0 < x < 1, 0 < t \leq 1 \quad u = u(x,t) \\ \text{(BCs)} & \begin{cases} u(x,0) = \sin(x), \\ u_x(0,t) = 0, \, u_x(1,t) = 0 \end{cases}, \quad 0 < x < 1, 0 \leq t \leq 1 \end{cases}$

(b) $\begin{cases} \text{(PDE)} & u_t = u_{xx} + u, \qquad\qquad\quad 0 < x < 1, 0 < t \leq 1 \quad u = u(x,t) \\ \text{(BCs)} & \begin{cases} u(x,0) = \sin(x), \\ u_x(0,t) = 0, \, u_x(1,t) = 0 \end{cases}, \quad 0 < x < 1, 0 \leq t \leq 1 \end{cases}$

(c) $\begin{cases} \text{(PDE)} & u_t = u_{xx} + 2y, \qquad\qquad\quad 0 < x < 1, 0 < t \leq 1 \quad u = u(x,t) \\ \text{(BCs)} & \begin{cases} u(x,0) = \sin(x), \\ u_x(0,t) = 0, \, u_x(1,t) = 0 \end{cases}, \quad 0 < x < 1, 0 \leq t \leq 1 \end{cases}$

(d) $\begin{cases} \text{(PDE)} & u_t = u_{xx} + u_x, \qquad\qquad\quad 0 < x < 1, 0 < t \leq 1 \quad u = u(x,t) \\ \text{(BCs)} & \begin{cases} u(x,0) = \sin(x), \\ u_x(0,t) = 0, \, u_x(1,t) = 0 \end{cases}, \quad 0 < x < 1, 0 \leq t \leq 1 \end{cases}$

12.   For each of the following BVPs, set up an appropriate finite difference scheme and numerically solve the problem. Continue to re-solve the problem with a decreasing set of space steps and corresponding decreasing time steps (in a stable way), compare consecutive numerical solutions (at common grid values), and continue until the maximum error becomes less than 0.001 or the computations take more than two minutes. Comment on the stability (or lack thereof) of your method on the problem. In case of stability, plot your final solution, both as a three-dimensional surface plot and as a two-dimensional plot of several superimposed time level profiles.

(a) $\begin{cases} \text{(PDE)} & u_t = u_{xx}, \qquad\qquad\qquad\quad 0 < x < 1, 0 < t \leq 1 \quad u = u(x,t) \\ \text{(BCs)} & \begin{cases} u(x,0) = 100, \\ u_x(0,t) = -20, \, u_x(1,t) = 0 \end{cases}, \quad 0 < x < 1, 0 \leq t \leq 1 \end{cases}$

(b) $\begin{cases} \text{(PDE)} & u_t = u_{xx}, \qquad\qquad\qquad\qquad\quad 0 < x < 1, 0 < t \leq 1 \quad u = u(x,t) \\ \text{(BCs)} & \begin{cases} u(x,0) = 100, \\ u_x(0,t) = -20, \, u_x(1,t) = u - 90 \end{cases}, \quad 0 < x < 1, 0 \leq t \leq 1 \end{cases}$

(c) $\begin{cases} \text{(PDE)} & u_t = 2xu_{xx}, \qquad\qquad\qquad 0 < x < 1,\ 0 < t \le 1 \ \ u = u(x,t) \\ \text{(BCs)} & \begin{cases} u(x,0) = 100, \\ u_x(0,t) = -20,\ u_x(1,t) = u - 90 \end{cases}, \quad 0 < x < 1,\ 0 \le t \le 1 \end{cases}$

(d) $\begin{cases} \text{(PDE)} & u_t = u_{xx}, \qquad\qquad\qquad 0 < x < 1,\ 0 < t \le 2 \ \ u = u(x,t) \\ \text{(BCs)} & \begin{cases} u(x,0) = 100, \\ u_x(0,t) = 90 - u,\ u_x(1,t) = u - 90 \end{cases}, \quad 0 < x < 1,\ 0 \le t \le 1 \end{cases}$

13. *(Richardson's Method–Experimental Instability)* Recall that Richardson's method for the heat equation $u_t = \alpha u_{xx}$ uses centered difference approximations for both the time and space derivative terms. Thus it takes the following form:

$$\frac{u(x_i, t_{j+1}) - 2u(x_i, t_j) - u(x_i, t_{j-1})}{k^2} = \alpha \frac{u(x_{i+1}, t_j) - 2u(x_i, t_j) + u(x_{i-1}, t_j)}{h^2}.$$

Richardson's method turns out to be unstable for any choice of space and time steps (therefore it is an unconditionally unstable method).

(a) Perform Richardson's method on a BVP (of your choice) to demonstrate this instability for several values of $\mu = \alpha k / h^2$.

(b) Write an M-file that will perform Richardson's method on the BVP of Program 12.3; the syntax, and input and output variables should be just as in Program 12.3, i.e.,

    [x, t, U] = richardson(phi, L, A, B, T, N, M, alpha, q).

Run your program to reproduce the results of part (a). Also, run your program on the BVP of Example 12.7 (with similar step sizes) and compare the performance with that witnessed for the Crank-Nicolson method in that example.

14. *(Richardson's Method–Theoretical Instability)* (a) Perform a von Neumann stability analysis using real numbers (as in this section) or complex arithmetic to show that Richardson's method (Exercise 13) is always unstable for the BVP:

$$\begin{cases} \text{(PDE)} & u_t = \alpha u_{xx}, \qquad\qquad 0 < x < L,\ 0 < t < \infty,\ \ u = u(x,t) \\ \text{(BCs)} & \begin{cases} u(x,0) = \varphi(x), \\ u(0,t) = 0,\ u(L,t) = 0 \end{cases} \quad 0 < x < L,\ 0 \le t < \infty \end{cases}$$

Suggestion: Although the real number notation used in the text will work, complex notation, as used in the exercises of the last section, will yield a more succinct proof.

15. In this exercise we will consider the BVP

$$\begin{cases} \text{(PDE)} & u_t = \alpha u_{xx}, \qquad\qquad 0 < x < L,\ 0 < t < \infty,\ \ u = u(x,t) \\ \text{(BCs)} & \begin{cases} u(x,0) = \varphi(x), \\ u(0,t) = 0,\ u(L,t) = 0 \end{cases} \quad 0 < x < L,\ 0 \le t < \infty \end{cases}$$

along with the following finite difference scheme:

$$\frac{u(x_i, t_{j+1}) - 2u(x_i, t_j) - u(x_i, t_{j-1})}{k^2} = \alpha \frac{u(x_{i+1}, t_j) + u(x_{i-1}, t_j) - u(x_i, t_{j+1}) - u(x_i, t_{j-1})}{h^2}.$$

Like Richardson's method (Problem 13), this one used a centered difference approximation for the time derivative, but a different sort of space derivative approximation.

(a) Use this method to solve the BVP of Example 12.8 using the same step sizes that were used in that example (with the Crank-Nicolson method), and compare errors using the "exact" solution given in that example.

(b) Show that the local truncation error of this method is $O(h^2 + k^2)$.

(c) Perform a von Neumann stability analysis to show that this method is unconditionally stable.

16. (a) Show that with the general formulation of the Crank-Nicolson method given in the footnote for Program 12.3, equation (50) takes on the following form:

$$-\mu_{i,j+1}u_{i-1,j+1} + 2(1+\mu_{i,j+1})u_{i,j+1} - \mu_{i,j+1}u_{i+1,j+1}$$
$$= \mu_{i,j}u_{i-1,j} + 2(1-\mu_{i,j})u_{i,j} + \mu_{i,j}u_{i+1,j} + (k/2)[q_{i,j} + q_{i,j+1}]\,,$$

where $\mu_{i,j} = \alpha_{i,j}k/h^2$.

(b)   In cases where α depends on u (and/or u_x), the approximation on which our cranknicolson program is based is rather Spartan. Experiment with some particular BVPs where $\alpha = \alpha(u)$ and modify this program to incorporate the first-order (rather than zeroth-order) approximation, until you find one in which the latter method gives improved results.

$$\alpha_{i,j+1} \approx \alpha(x_i,t_{j+1},u(x_i,t_j),[u(x_{i+1},t_j)-u(x_{i-1},t_j)]/2h)$$
$$+\partial\alpha/\partial u(x_i,t_{j+1},u(x_i,t_j),[u(x_{i+1},t_j)-u(x_{i-1},t_j)]/2h)\Big[u(x_i,t_j)-u(x_i,t_{j-1})\Big].$$

Suggestion: Unless you can find such a problem with a known analytic solution, you should judge the success of the two methods by checking to see how much the numerical results change when both temporal and spatial steps are cut in half.

17.   (*Finite Difference Schemes for Heat Flow in Two Space Dimensions*)   As was done with hyperbolic equations in Section 12.2, the finite difference methods of this section can be extended to deal with the heat equation (and other parabolic equations) in two-space variables. This exercise outlines the procedure for the following Dirichlet problem on a rectangle:

$$\begin{cases} \text{(PDE)} \ u_t = \alpha(u_{xx}+u_{yy}), \ \ 0<x<a, 0<y<b, 0<t<\infty, \ \ u=u(x,y,t) \\ \text{(BCs)} \ \begin{cases} u(x,y,0)=\varphi(x,y), \ \ \ 0\le x\le a, \ 0\le y\le b, 0\le t<\infty \\ u(x,y,t)= A(x,y), \ \ \text{for all } (x,y) \text{ on the boundary of the rectangle:} \\ \qquad\qquad\qquad\qquad R=\{0\le x\le a, \ 0\le y\le b\} \end{cases} \end{cases}$$

(a) For a fixed time interval $0\le t\le T$ introducing a grid as in (40) and letting $u_{i,j}^\ell = u(x_i,y_j,t_\ell)$, derive the following forward-time central-space finite difference approximation of the PDE:

$$u_{i,j}^{\ell+1} = \mu\Big[u_{i+1,j}^\ell + u_{i-1,j}^\ell + u_{i,j+1}^\ell + u_{i,j-1}^\ell\Big] + (1-4\mu)u_{i,j}^\ell,$$

where $\mu = \alpha k/h^2$.   (Recall in (38) we have assumed equal time steps for the two space variables.) The stability condition for this scheme is $\mu \le 1/4$.

(b)   Derive a similar finite difference scheme in which different step sizes h_x and h_y are permitted for the x- and y-directions. The stability condition in this general setting becomes $\alpha k/(h_x^2 + h_y^2)\le 1/8$.

(c)   By analogy with (61) derive a family of finite difference schemes, indexed by the parameter σ ($0\le\sigma\le1$) that take the form:

$$u_{i,j}^{\ell+1} = \mu(1-\sigma)\Big[u_{i+1,j}^\ell + u_{i-1,j}^\ell + u_{i,j+1}^\ell + u_{i,j-1}^\ell\Big]$$
$$+ \mu\sigma\Big[u_{i+1,j}^{\ell+1} + u_{i-1,j}^{\ell+1} + u_{i,j+1}^{\ell+1} + u_{i,j-1}^{\ell+1}\Big] + (1-4\mu)u_{i,j}^\ell.$$

This scheme is uniformly stable if $1/2\le\sigma\le1$ and otherwise the stability condition becomes $\mu \equiv \alpha k/h^2 < 1/(4-8\sigma)$. When $\sigma = 1/2$, it is referred to as the Crank-Nicolson method since it naturally generalizes the method in one space variable.
Note: The stability assertions can be established by the von Neumann method; see [RiMo-67] or [IsKe-66].

18.   (*Cooling of a Uniformly Heated Slab*) Consider the following heat problem:

$$\begin{cases} \text{(PDE)} \ u_t = \alpha(u_{xx}+u_{yy}), \ \ 0<x<a, 0<y<b, 0<t<\infty, \ \ u=u(x,y,t) \\ \text{(BCs)} \ \begin{cases} u(x,y,0)=T_0, \ \ \ \ 0\le x\le a, \ 0\le y\le b, 0\le t<\infty \\ u(x,y,t)=0, \ \ \text{for all } (x,y) \text{ on the boundary of } \{0\le x\le a, \ 0\le y\le b\} \end{cases} \end{cases}$$

Physically, this problem can be thought of as the cooling of a thin rectangular slab with insulated lateral surfaces and whose edges are maintained at temperature of 0. Initially, the temperature is T_0. We have left the diffusivity as variable.

(a) Use the forward-time, central-space explicit method of Exercise 17(a) to numerically solve this problem on the time interval $0 \le t \le 1$ using 10 interior space grid nodes in both the x- and y-directions. Use the following data: $a = b = 1$, $T_0 = 100$, $\alpha = 1$ and perform your solution on the time interval $0 \le t \le 1$. Give three-dimensional snapshots of temperature profiles at the times (in the time grid as close as possible to) $t = 0, .2, .4., .6, .8, 1$.

(b) Repeat part (a), this time using the Crank-Nicolson method of Exercise 17(c). Do it first with the grid that would be suitable for part (b), then repeat with a grid with approximately the same total number of internal nodes (spatial-temporal) but where the step sizes are the same for each of the three variables (x, y, and t).

(c) The exact solution of this BVP can be expressed as:

$$u(x,y,t) = \frac{16T_0}{\pi^2} \left\{ \sum_{n=0}^{\infty} \frac{1}{2n+1} \sin\big((2n+1)\pi x/a\big) \exp(-\pi^2(2n+1)^2 \alpha t/a^2) \right\} \times$$
$$\left\{ \sum_{m=0}^{\infty} \frac{1}{2m+1} \sin\big((2m+1)\pi y/b\big) \exp(-\pi^2(2m+1)^2 \alpha t/b^2) \right\},$$

see Section 3.7 of [Asm-00]. Find a positive integer N so that the partial sum product:

$$u_N(x,y,t) = \frac{16T_0}{\pi^2} \left\{ \sum_{n=0}^{N} \frac{1}{2n+1} \cdots \right\} \times \left\{ \sum_{m=0}^{N} \frac{1}{2m+1} \cdots \right\},$$

approximates the exact solution with an error less than 10^{-6} uniformly for all x, y, and t between 0 and 1.
How much better will the accuracy of this approximation be for each of the temperature profiles corresponding to the time values (in the time grid as close as possible to) $t = .2, .4., .6, .8, 1$?

(d) Use the "exact" solution of part (c) to obtain three-dimensional mesh plots of the errors of each of the snapshots obtained in part (a), and then for each of those obtained in part (b).

(e) Using the numerical solution of part (a), estimate the time it takes for the <u>maximum temperature</u> on the slab to decrease to 50. Repeat with the numerical solutions of part (b) and finally with the "exact" solution of part (c).

(f) For a particular grid on the x- and y-axes, the <u>average temperature</u> on the plate at time level t_0 can be discretely defined by $\dfrac{1}{N_x N_y} \sum_{i=1}^{N_x} \sum_{j=1}^{N_y} u(x_i, y_j, t)$. Using the numerical solution of part

(a), estimate the time it takes for the average temperature on the plate to decrease to 50. Repeat with the numerical solutions of part (b) and finally with the "exact" solution of part (c).
Suggestions: For parts (a) and (b) use some of the MATLAB techniques introduced in Example 12.7, Program 12.2 and the development that precedes it. In part (b), to find (approximately) the correct step size h_b, solve the equation $h_b^3 = h_a^2 k_a$, where h_a and k_a denote the spatial and temporal step sizes, respectively, that were used in part (a). For part (c), use the ideas from Example 12.8(b). For part (d), much computation can be saved by making use of the separated nature of the exact solution.

19.　Repeat all parts of Exercise 18, keeping everything the same, except now use the value $\alpha = 2$.

20.　Repeat all parts of Exercise 18, keeping everything the same, except now use the value $\alpha = 4$.

21.　(*Cooling of a Half-Heated Slab*) Consider the following heat problem:

$$\begin{cases} \text{(PDE)} & u_t = u_{xx} + u_{yy}, & 0 < x < 1, 0 < y < 1, 0 < t < \infty, \ u = u(x,y,t) \\ \text{(BCs)} & \begin{aligned} & u(x,y,0) = \varphi(x,y), & 0 \le x \le a, \ 0 \le y \le b, 0 \le t < \infty \\ & u(x,y,t) = 0, \ \text{for all } (x,y) \text{ on the boundary of } \{0 \le x \le a, 0 \le y \le b\} \end{aligned} \end{cases},$$

where $\varphi(x,y) = \begin{cases} 100, & \text{if } y \leq x \\ 0, & \text{otherwise} \end{cases}$.

(a) Use the backward-time, central-space purely implicit method of Exercise 17(c) with $\sigma = 1$ to numerically solve this problem on the time interval $0 \leq t \leq 1$ using 25 interior grid values in each of the x- and y- and t-directions. Give three-dimensional snapshots of temperature profiles at the times (in the time grid as close as possible to) $t = 0, .2, .4., .6, .8, 1$. Repeat using 50 interior grid values for each variable.

(b) Repeat part (a), this time using the Crank-Nicolson method of Exercise 17(c).

(c) The exact solution of this BVP can be expressed as:

$$u(x,y,t) = 100 \sum_{n=1}^{\infty} \sum_{m=1}^{\infty} C_{n,m} \sin(n\pi x)\sin(m\pi x)\exp(-\pi^2(n^2 + m^2)t),$$

where $C_{n,m} = \begin{cases} 4((-1)^n((-1)^m - 1)n^2 + ((-1)^n - 1)m^2)/[nm(n^2 - m^2)\pi^2], & \text{if } n \neq m \\ 2((-1)^n - 1)^2 /[n^2\pi^2], & \text{if } n = m \end{cases}$

(see Section 3.7 of [Asm-00]). Find a positive integer N so that the partial sum:

$$u_N(x,y,t) = 100 \sum_{n=1}^{N} \sum_{m=1}^{N} C_{n,m} \cdots$$

approximates the exact solution with an error less than 10^{-6} uniformly for all x, y, and t between 0 and 1.

How much better will the accuracy of this approximation be for each of the temperature profiles corresponding to the time values $t = .2, .4., .6, .8, 1$?

(d) Use the "exact" solution of part (c) to obtain three-dimensional mesh plots of the errors of each of the snapshots obtained in part (a), and then for each of those obtained in part (b).

(e) Repeat part (e) of Exercise 18 for the BVP of this problem.

(f) Repeat part (f) of Exercise 18 for the BVP of this problem.

22. Repeat all parts of Exercise 21, keeping everything the same, except now change the PDE to
$u_t = 2(u_{xx} + u_{yy})$.

23. Repeat all parts of Exercise 21, keeping everything the same, except now change the PDE to
$u_t = 4(u_{xx} + u_{yy})$.

24. (*Failure of the Maximum Principle for Heat Problems with Sources*) Consider the following heat problem:

$$\begin{cases} \text{(PDE)} & u_t = u_{xx} + 2(t+1) + x(1-x), \quad 0 < x < 1, 0 < t < \infty, \quad u = u(x,t) \\ \text{(BCs)} & \begin{cases} u(x,0) = x(1-x), \\ u(0,t) = 0, \, u(1,t) = 0, \end{cases} \quad 0 < x < 1, 0 \leq t < \infty \end{cases}$$

(a) Show that $u(x,t) = (t+1)x(1-x)$ solves this BVP and violates the maximum principle stated in the section in that the internal temperature of the rod can exceed the boundary and initial temperature values.

(b) Apply each of the three methods, implicit, explicit, and Crank-Nicolson, to this problem with comparable grids and compare the numerical results with the solution of part (a).

Chapter 13: The Finite Element Method

13.1: A NONTECHNICAL OVERVIEW OF THE FINITE ELEMENT METHOD

The **Finite Element Method (FEM)** is actually a large collection of numerical methods for solving PDEs. It was first devised as a numerical tool by mathematician Richard Courant[1] in a 1943 paper on torsion problems [Cou-43], and is based on analogous techniques and principles to those developed in the early twentieth century for one-dimensional boundary value problems, as were presented in Section 10.5. The method was extensively elaborated on during the 1950s and 1960s by engineers as a practical approach for solving various PDEs in structural engineering. In the 1960s and 1970s, mathematicians worked to give the method a solid theoretical basis and extended it as a tool for solving many different PDE problems. Active research on this method continues today and it has become the most commonly used numerical method for partial differential equations. As a very pertinent example, in MATLAB's PDE Toolbox, all of the programs for solving PDEs use FEMs. Writing even somewhat general programs for FEMs is a very complicated task. Our goal in this chapter will be to explain the method and to write some programs to implement it in several specific instances. This should be sufficient for readers needing to delve deeper into FEMs to be able to extend the programs into more general ones. MATLAB's PDE Symbolic Toolbox programs are open to its users to read and modify. So, in principle, after reading this chapter, readers could modify some of the FEM programs in MATLAB's toolbox to suit their exact needs (if not already met).

---

[1] Richard Courant was an exceptionally influential mathematician. He grew up in Germany and had a rather difficult childhood, having to work to help support his family while going to school. He eventually entered the Univerisity of Breslau (now Wroclaw, Poland) as an undergraduate and was lured to major in mathematics by the exciting lectures in his classes. He went on to Göttingen for his graduate studies where he worked with Hilbert and later became a professor there. His education was interrupted by military service for Germany in WWI, where he developed an effective electronic communications system that was implemented for the troops. Despite the important contributions he was making to the University of Göttingen, not to mention his important military service to his country, when the Nazis came to power in the early 1930s, he was forced to resign his professorship. The Nazis had decreed that any "non-Aryan" civil servant was to be terminated and having one Jewish grandparent was sufficient to make someone "non-Aryan." There was supposed to be an exemption for individuals who gave Germany military service in WWI, but despite ardent efforts on the part of the university to keep him, Courant was still "retired." He subsequently accepted an offer at New York University. The transition was very difficult for him. Coming from a world-renowned institute and having been surrounded by top-notch mathematicians, when he got to NYU, he found his colleagues to be very weak and the students likewise poor. He made use, however, of his extensive contacts and was able to hire a large group of strong new faculty. Today, NYU's mathematics department, also known as the *Courant Institute*, is considered by many as the premiere applied mathematics institute in the world.

Figure 13.1: Richard
Courant (1888–1972),
German mathematician.

The FEM methods basically split up the domain of the problem into small pieces, called **elements**, that have simple structure. There are many different ways to perform such decompositions and the geometry certainly changes with the dimension of the space. A common approach for two-dimensional domains is to **triangulate** the domain into small triangles. The **triangulation** must be done in such a way that whenever two triangles touch, they will have either an entire common edge (and thus two common vertices) or just a common vertex. The reason for this is that the FEM approximate solution for the PDE will be made up of separate "pieces" on the various elements and they need to connect up (interpolate) together in a neat fashion. When a domain has a curved boundary, the sizes of the triangles can be made small enough so that the triangulation approximates the domain reasonably well. An example of such a triangulation is shown in Figure 13.2. In three dimensions, tetrahedra (pyramids) are commonly used; but the process is still known as triangulation.

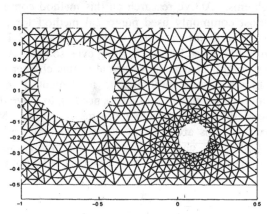

FIGURE 13.2: A triangulation of a planar domain consisting of a rectangle with two circles deleted. The circular boundary portions are thus approximated by polygons (as shown). This triangulation was created using MATLAB's PDE Toolbox.

Even in two dimensions, of course, there are numerous ways to perform such a triangulation. We point out some important features of the triangulation in Figure 13.2. First, notice that the triangles in the mesh all seem to be roughly the same size. This uniformity is not necessary, but for a general problem on a given domain it is usually the best generic triangulation scheme. Another important property is that none of the sidelengths of any triangle in the mesh is much shorter than the other two sides of the same triangle. Another way to describe this property is that the area of the inscribed circle of any of these triangles is not much smaller than the area of the triangle. This feature is essential in order that the finite element method be stable.

Just writing a good program to create such generic triangulations is already an arduous task. It must be thought out how the geometry of the original domain should be inputted (as a matrix) and then the triangles must be created and stored, usually by their vertices. Additionally, the vertices of the triangulation will need to be numbered and it will be helpful for the numbering to be done in such a way that the numbers of the three vertices of any triangle are reasonably close.

Like finite difference methods, finite element methods will discretize the PDE into a linear system. The nature of the discretization, however, is very different for FEMs. Mathematically, the PDE is first converted to a so-called **variational problem**. This is usually done in one of two ways: the **Rayleigh-Ritz method**, where the solution of the PDE is recast as the solution of a certain minimization problem among a large class of functions, or the **Galerkin method**, where the solution is recast as a certain unique representing function. Although different in philosophy, the two approaches often turn out to be equivalent. With either method, the approximate solution is found by restricting attention to a certain finite-dimensional space of so-called **admissible functions** determined by **basis functions** corresponding to each of the elements. Even with the type of elements being specified, there are numerous choices for the basis functions. The simplest choice would be constant functions, but these do not blend together well. The next simplest choice would be to have linear functions on each element. For two-dimensional domains with triangulations, this type of basis function turns out to be quite effective. Over each element, the graph of such a basis function will be the triangular portion of a plane (sitting over the two-dimensional triangle). Since three points determine a plane, these basis functions will be flexible enough to accommodate specifying values at the three vertices of their triangle, and mesh triangles that have common vertices or edges will have their graphs coinciding at common points. The resulting approximating functions will thus be continuous over the original domain, but in general will not be differentiable at common edges or vertices of different triangles. More complicated spline-type basis functions can be used to overcome this differentiability problem, but, of course, the limited benefits of using such more complicated basis functions would have to be weighed against the resulting increase in technical difficulties. For most applications on two-dimensional domains with triangular elements, such linear basis functions are sufficient and most commonly used. MATLAB's PDE Toolbox, for example, uses such basis functions.[2]

---

[2] The programs in MATLAB's PDE toolbox are designed only to handle PDEs with two space variables, and so, for example, they cannot solve three-dimensional elliptic problems such as steady-state heat distributions and structure of materials. The programs can, however, accommodate a time variable in hyperbolic or parabolic equations. The reason for this is that FEMs for parabolic and hyperbolic problems invoke finite difference schemes for the time derivatives and thus can accommodate two space variables in addition to the time variable.

Focusing only on (linear combinations of) the basis functions, the FEM will solve a linear system to determine the best candidate to solve the variational problem (Rayleigh-Ritz or Galerkin) and this will be the approximation to the actual solution. This approximation turns out to be simply the projection of the actual solution onto the finite-dimensional space of admissible functions or, informally, the best admissible function for solving the variational problem. Figure 13.3 shows the FEM solution for the following PDE problem on the domain Ω of Figure 13.2:

$$\begin{cases} \Delta u = 0 \text{ on } \Omega, \quad u = u(x,y) \\ \partial u / \partial n = 0, \text{ on outer rectangle} \\ u = 40, \text{ on small circle, } u = 500, \text{ on large circle} \end{cases} \tag{1}$$

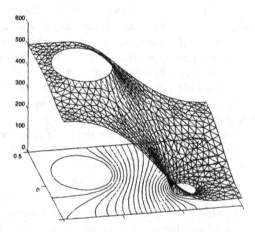

FIGURE 13.3: A FEM solution of the steady-state heat problem (1) on the domain and triangulation as pictured in Figure 13.2. Contour lines have been added. This solution was created using MATLAB's PDE Toolbox.

The problem can be thought of as the steady-state heat distribution on the rectangular region Ω (from the Laplace equation $\Delta u = 0$). The first boundary condition on the edge of the outer rectangle: $\partial u / \partial n = 0$, is a Neumann boundary condition (n denotes the unit outward normal vector for the domain), stating that heat does not flow out of or into the rectangle (i.e., the boundary is insulated). The two constant temperatures on the interior circular boundaries are Dirichlet boundary conditions specifying certain temperatures that are fixed on each. The problem may be thought of as a basic version of the cooling of a nuclear reactor within some enclosed region (the rectangle); the large very hot circle denoting the reactor and the small circle denoting the cooling source (usually a stream of fresh water).

In the triangulation process, it is not always efficient to make the triangles be all of essentially the same size. Indeed, at places where the solution varies drastically, smaller triangles should be used and in areas of small variation larger ones can be and should be used. More triangles entail more work so we should use very small ones only where they are needed. Of course, not knowing the solution ahead of time can make it difficult to predict where the solution will be varying wildly. Sharp corners or curves in the domain, as well as areas where the coefficients of the PDE (if variable) rapidly change, are usually problem areas. There are more sophisticated so-called **adaptive methods** of the FEM that iteratively take into account all available information so as to refine the elements accordingly in a way that aims to reach the best possible accuracy with specified constraints (such as operating time, number of triangles, etc.). Figure 13.4 shows such an adaptive triangulation for the boundary value problem (1). Triangulation of domains is an art!

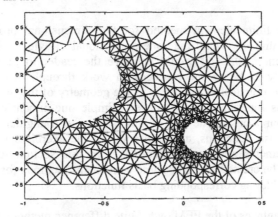

FIGURE 13.4: A triangulation of the planar domain of Figure 13.2 that was obtained using an adaptive FEM to solve the steady-state heat problem (1). Compare with the triangulation of Figure 13.2 and, in particular, notice how the triangles closer to the boundaries of the circles (facing inward) are much smaller than the farther away triangles. This triangulation was created using MATLAB's PDE Toolbox.

An outline for the rest of this chapter is as follows: We will be focusing on linear elliptic boundary value problems on planar domains. Our FEM will use piecewise linear basis functions on triangulations of the domains. Section 13.2 will introduce some practical techniques for producing effective triangulations of planar domains and explain how to construct and manipulate basis functions. Section 13.3 will explain the complete program of using a FEM to solve quite general boundary value problems on arbitrary planar domains and boundary conditions. The most time-consuming step of the FEM is the construction of the linear system whose solution will give the values of the approximate solution. This process, known as "the assembly process," is broken up into an element-by-element computation involving the calculation of certain double integrals and (depending on the boundary data) line integrals. We will demonstrate with examples that it is not efficient to use MATLAB's integration tools in the assembly process. Indeed, if the elements are sufficiently small (depending on the coefficients of the problem), it turns out to be perfectly adequate to use some simple numerical quadrature formulas (for triangles and line segments). This will allow us to attain essentially the same accuracy as with the more elaborate integrators but at a very small fraction of the time. In numerical differential equations, much can be learned from experimentation, and this chapter provides numerous opportunities in this area. The committed reader can gain a great deal by exploring some of the more advanced topics that are introduced in the exercises.

13.2: TWO-DIMENSIONAL MESH GENERATION AND BASIS FUNCTIONS

Theoretically, the finite element method for two-dimensional problems shares many common threads with the one-dimensional Rayleigh-Ritz methods introduced in Section 10.5. It would behoove the reader to glance over that section periodically as he or she proceeds to work through this and the next section. The major practical difference is in the geometry of the two-dimensional elements and basis functions versus the very simple one-dimensional elements. We will restrict our focus in the text of this chapter to triangular elements and piecewise linear basis functions, although some of the exercises will delve into other sorts of elements and basis functions. In this section we show how triangulations can be created and give some convenient methods for constructing, storing, and manipulating corresponding basis functions.

The two main advantages of the FEM over finite difference methods are its ease in dealing with more complicated domains than simply rectangular ones, and its flexibility in dealing with many sorts of boundary conditions. To illustrate construction of the basis functions, we use the simple triangulation of the hexagonal domain shown in Figure 13.5.

FIGURE 13.5: A simple triangulation of a hexagonal domain. The eight nodes are labeled in red and the eight triangles are also labeled. The ordering is somewhat arbitrary. It is just a coincidence that the number of nodes coincides with the number of triangles. At this point we left out x- and y-coordinates so as to emphasize the element numbering.

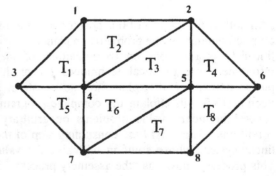

The **nodes** in a triangulation are simply the vertices of the triangles. As in the one-dimensional method, each node gives rise to a basis function that takes on the value 1 at its corresponding node and zero at all other nodes. Piecewise linear functions work well here since a linear function is completely determined on a triangle once its values are specified on the three vertices. Furthermore, two linear functions so determined on triangles that share an edge will agree on the common edge. The resulting basis function will be the unique piecewise linear function on the hexagon having the property that it is linear on each element and takes on the value 1 at its associated node and 0 at all other nodes. In this context the basis functions are sometimes known as **pyramid functions**. The pyramid function for the node #4 of Figure 13.5 is illustrated in Figure 13.6.

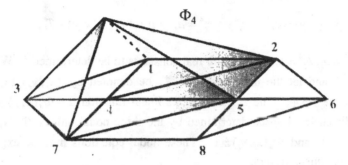

FIGURE 13.6: A graph of the piecewise linear basis or pyramid function $\Phi_4 = \Phi_4(x, y)$ for node #4 in the triangulation of the hexagonal domain of Figure 13.5. The function takes on the value 1 at node #4, zero at all other nodes, and is linear on each triangle. Thus, on the unshaded triangles, the pyramid function is identically zero.

To get formulas for the pyramid functions, we need to introduce coordinates. For the purpose of an example, we assign coordinates as follows: node #3 will be put at the origin $(0,0)$, node #4 will have coordinates $(1,0)$, and the coordinates of nodes #1 and #7 are $(1,1)$, and $(1,-1)$, respectively. The corresponding coordinates of nodes #2, #5, and #8 are obtained by adding 3/2 to the first coordinates of the last three, and node #6 has coordinates $(7/2,0)$. Knowing these coordinates, all of the information of the triangulation can be represented by the following two matrices, N(odes) and T(riangles):

$$
N = \begin{bmatrix} 1 & 1 \\ 2.5 & 1 \\ 0 & 0 \\ 1 & 0 \\ 2.5 & 0 \\ 3.5 & 0 \\ 1 & -1 \\ 2.5 & -1 \end{bmatrix}, \quad
T = \begin{bmatrix} 1 & 3 & 4 \\ 1 & 2 & 4 \\ 2 & 4 & 5 \\ 2 & 5 & 6 \\ 3 & 4 & 7 \\ 4 & 5 & 7 \\ 5 & 7 & 8 \\ 5 & 6 & 8 \end{bmatrix}.
$$

The eight rows of N give the coordinates of the corresponding numbered nodes, the first column entry gives the x-coordinates, and the second column entry the y-coordinates. The eight rows of T give the node numbers of the corresponding numbered triangles, in order (see Figure 13.5). Such matrices will be needed in writing programs for the FEM.

EXAMPLE 13.1: Write down a formula for the basis function $\Phi_4 = \Phi_4(x, y)$ shown in Figure 13.6.

SOLUTION: From its piecewise linearity, on each of the eight triangles $T_\ell (1 \le \ell \le 8)$, $\Phi_4(x, y)$ will be a linear function and so can be written as

$$\Phi_4(x, y) = a_\ell^4 x + b_\ell^4 y + c_\ell^4 = a_\ell x + b_\ell y + c_\ell, \quad (x, y) \in T_\ell, \qquad (2)$$

where $a_\ell^4 = a_\ell$, $b_\ell^4 = b_\ell$, $c_\ell^4 = c_\ell$ are real constants to be determined.[3] We now fix an index ℓ and let the three nodes of T_ℓ be denoted by (x_r, y_r), (x_s, y_s), and (x_t, y_t) where $t = 4$. The graph of such a linear function $z = \Phi_4(x, y)$ is a plane in three-dimensional space determined by the three nodal values $\Phi_4(x_r, y_r) = 0$, $\Phi_4(x_s, y_s) = 0$, and $\Phi_4(x_t, y_t) = 1$. These nodal equations may be expressed as the following linear system:

$$\begin{cases} a_\ell x_r + b_\ell y_r + c_\ell = 0 \\ a_\ell x_s + b_\ell y_s + c_\ell = 0 \\ a_\ell x_t + b_\ell y_t + c_\ell = 1 \end{cases} \qquad (3)$$

Putting (3) in matrix notation gives:

$$MA = Z, \text{ where } M = \begin{bmatrix} x_r & y_r & 1 \\ x_s & y_s & 1 \\ x_t & y_t & 1 \end{bmatrix}, A = \begin{bmatrix} a_\ell \\ b_\ell \\ c_\ell \end{bmatrix}, \text{ and } Z = \begin{bmatrix} 0 \\ 0 \\ 1 \end{bmatrix}. \qquad (4)$$

Geometrically, since three noncollinear points determine a unique plane, it follows that the linear system (3)/(4) will have a unique solution as long as the three nodes are not collinear. This is certainly the case for any triangulation. We mention one further important point that the system will be well conditioned provided that the area of the triangle T_ℓ is not much greater than that of the inscribed circle. This is a quantitive way of saying that the three nodes of T_ℓ should not be close to being collinear (convince yourself of this!). This can be analytically verified using the following explicit formulas for the determinant of M and M^{-1}:

$$|\det(M)| = 2 \cdot \text{area}(T_\ell), \qquad (5)$$

---

[3] The superscipts, although technically necessary, can be omitted in this example since there is only one basis function under consideration.

$$M^{-1} = \frac{1}{\det(M)} \begin{bmatrix} (y_s - y_t) & (y_t - y_r) & (y_r - y_s) \\ (x_t - x_s) & (x_r - x_t) & (x_s - x_r) \\ (x_s y_t - x_t y_s) & (x_t y_r - x_r y_t) & (x_r y_s - x_s y_r) \end{bmatrix}. \tag{6}$$

For a proof of the interesting equation (5), the reader is referred to Exercise 22. Using (5), equation (6) can be proved by a direct (albeit tedious) verification. Equations (5) and (6) plainly show that the matrix is well conditioned as long as the triangle is not too long and thin. Apart from this, the explicit formula (6) is useful to build into large-scale FEM programs where such matrices need to be inverted large numbers of times in constructing the basis functions. We continue with this example in a way that will help us later when we need to write general programs. We begin by entering the node matrix N and the triangle matrix T into our MATLAB session:

```
>> N=[1 1;5/2 1;0 0;1 0;5/2 0;7/2 0;1 -1;2.5 -1];
>> T=[1 3 4;1 2 4;2 4 5;2 5 6;3 4 7;4 5 7;5 7 8;5 6 8];
```

Since Φ_4 vanishes over triangles #4, #7, and #8, the coefficients a, b, c are all zero for these triangles. The following loop will give us what we need and is easily modified to function in general FEM routines. It will store the needed coefficients of Φ_4 on the remaining triangles in a four-column matrix A: The first column gives the triangle number; the remaining three give the corresponding coefficients of Φ_4 as in (2). Since the coefficients are all fractions, we display the output in rational format.

```
>> format rat, counter=1;
>> for L=1:8
if ismember(4,T(L,:))==1
%checks to see if 4 is a node of triangle #L
%if yes, next two commands reorder the vector T(L,:) to
%construct a vector "nv" of length 3
%of nodes of triangle #L with 4 appearing last
   index=find(T(L,:)==4);
   nv=[T(L,1:2) 4]; nv(index)=T(L,3);
   xr=N(nv(1),1); yr=N(nv(1),2); xs=N(nv(2),1); ys=N(nv(2),2);
   xt=N(nv(3),1);yt=N(nv(3),2);
   M=[xr yr 1;xs ys 1;xt yt 1]; %matrix M of (4)
Minv=[ys-yt yt-yr yr-ys; xt-xs xr-xt xs-xr; xs*yt-xt*ys xt*yr-xr*yt
xr*ys-xs*yr]/det(M);
   % inverse matrix M from (6)
   abccoeff=Minv*[0;0;1]; %coefficents of basis function on triangle#L
   A(counter,:)=[L abccoeff'];
   counter=counter+1;
end
end
>> A
```

→A =

1	1	-1	0
2	0	-1	1
3	-2/3	0	5/3
5	1	1	0
6	-2/3	1	5/3

From this matrix, we can write down the explicit formula for Φ_4:

$$\Phi_4(x,y) = \begin{cases} x - y, & \text{if } (x,y) \in T_1, \\ -y+1, & \text{if } (x,y) \in T_2, \\ -\frac{2}{3}x \quad +\frac{5}{3}, & \text{if } (x,y) \in T_3, \\ x+ y, & \text{if } (x,y) \in T_5, \\ -\frac{2}{3}x+ y+\frac{5}{3}, & \text{if } (x,y) \in T_6, \\ 0, & \text{otherwise.} \end{cases}$$

The reader should verify that these formulas indeed possess the required values at the nodes and hence (by linearity) on each element.

EXERCISE FOR THE READER 13.1: Find formulas for Φ_3 and Φ_5 analogous to that found for Φ_4 in the above example.

The careful reader may have realized that we can further cut our computation time down in the solution of the system (4) if we always agree to set it up so that (x_t, y_t) is the vertex on which the value of the local basis function equals 1. Since the inverse of the coefficient matrix M of (4) is explicitly known (6), the matrix product MZ will simply be the third column of M^{-1} so that in the notation of (4), we have (using (5)):

$$\begin{bmatrix} a_\ell \\ b_\ell \\ c_\ell \end{bmatrix} = \frac{1}{2 \cdot \text{area}(T_\ell)} \begin{bmatrix} y_r - y_s \\ x_s - x_r \\ x_r y_s - x_s y_r \end{bmatrix}. \tag{6a}$$

Up to now, most of our plots for functions of two variables have been over rectangular domains. Thus it will be important for us to learn how to get MATLAB to plot functions, such as the above basis functions, that are piecewise linear and continuous on a set of triangular finite elements. Such functions are determined entirely by their nodal values. MATLAB can accommodate us quite nicely for this task with the following command:

`trimesh(T,x,y,z,C)` →	Given a 3-column matrix T whose rows are node numbers for a triangulation, vectors x and y of the coordinates of the numbered nodes, and corresponding z coordinates of a piecewise linear function on the nodes, this command will produce a plot of the resulting piecewise linear function. The last argument C is an optional *rgb* vector that can be used to specify color (see Section 7.2). The default edge coloring is proportional to the edge height as in Chapter 11. The vector z can be omitted to produce a two-dimensional plot of the triangulation.

We are nicely set up to have MATLAB construct a plot of Φ_4.

```
>> x=N(:,1); y=N(:,2);
>> z=zeros(8,1); z(4)=1;
>> trimesh(T,x,y,z)
>> hidden off %allows hidden edges to appear
```

The resulting MATLAB plot is shown in Figure 13.7. Since there are only two heights for the edges, the coloring is not very elaborate. With finer triangulations and more complicated functions, the resulting trimesh plots can be quite useful and attractive, as the one shown in Figure 13.2.

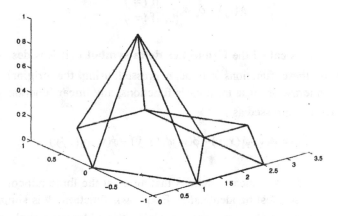

FIGURE 13.7: MATLAB's graphical rendition of the basis function $\Phi_4(x, y)$ of Example 13.1.

Given a triangulation of a domain and any function or data f defined on the nodes $N_1, N_2, \cdots, N_m$, the **finite element interpolant** of this function/data is given by:

$$\sum_{j=1}^{n} f(N_j)\Phi_{N_j}(x, y).$$

We point out that the graph of this interpolant is most easily obtained by simply using the trimesh command directly on the triangle matrix and corresponding values of f; the calculations for the basis functions are not necessary. This will not be the case for more general elements (see Exercise 26).

We stress that in the determination of the hat functions Φ_j, we really split up the problem into determining Φ_j on each element (triangle). On each element T_ℓ, $\Phi_j(x, y) = a_\ell^j x + b_\ell^j y + c_\ell^j$, is a linear combination of the three functions x, y, and 1. These three functions are a basis for the set of all linear functions on T_ℓ. We refer to them as a **local basis,** to distinguish them from the (global) basis functions Φ_j. Although these local basis functions are quite natural and have simple formulas, there is another local basis that often has theoretical advantages. To simplify notation, we fix an element $T = T_\ell$, and denote its three vertices by

v_1, v_2, and v_3 (the exact ordering is unimportant, but let's assume they are numbered in counterclockwise order). The corresponding three **standard local basis** functions ϕ_1, ϕ_2, and ϕ_3 are the linear functions determined (exactly) by the following conditions:

$$\phi_i(v_j) = \delta_{ij} \equiv \begin{cases} 1, & \text{if } i = j, \\ 0, & \text{if } i \neq j. \end{cases} \tag{7}$$

(The symbol δ_{ij} is called the **Kronecker delta symbol**.) It was described earlier how each of these functions can be expressed using the original local basis functions. In terms of these local basis functions, any linear function $\phi(x, y)$ can be conveniently expressed as:

$$\phi(x, y) = \phi(v_1)\phi_1(x, y) + \phi(v_2)\phi_2(x, y) + \phi(v_3)\phi_3(x, y). \tag{8}$$

(*Proof:* Both sides are linear functions that agree at the three noncollinear points v_1, v_2 and v_3 and so must be identical.) Each basis function Φ is simply made up of pieces of corresponding elements containing the node associated with Φ, and each of these pieces is a linear combination (8) of the above local basis functions for its element. The previous example thus gave efficient strategies for computing all of the local basis functions as well as the corresponding basis functions.

EXERCISE FOR THE READER 13.2: (a) Explain why piecewise linear basis functions could not be used if rectangular elements (with sides parallel to the axes) were used in place of triangular ones.
(b) Give an example of a simple type of basis function that could be used for such rectangular elements. Make sure that your construction will ensure that any given basis function will be continuous across element edges.

As mentioned Section 13.1, triangulation is an art, and as such there has been a notable amount of research in the development of efficient and effective triangulation and mesh generation schemes. One particularly successful and often used method in this area is that of the **Delaunay triangulation**, relative to a given finite set of points in the plane. This triangulation will result in a set of triangles whose vertices coincide with the given finite set (of nodes) and with the further property that the circumcircle of each triangle in the collection contains only nodes that are vertices of that triangle. This condition favors well-rounded triangles over thin ones, which are better for the FEM.

Definitions: Suppose that we have a set of distinct points $P = \{p_1, p_2, \cdots, p_n\}$ in the plane $\mathbb{R}^2$. The **Delaunay triangulation** relative to P consists of all triangles connecting three noncollinear points $p_i, p_j, p_k \in P$ with the property that there exists a point $a \in \mathbb{R}^2$ which lies equidistant to each of the points p_i, p_j, p_k and closer to these three than to any other point $p_\ell \in P$, $\ell \neq i, j, k$.

Figure 13.8 illustrates the Delaunay triangulation for a very small set of four points. It can be shown that the Delaunay triangulation of a finite set of points will always be a triangulation for the **convex hull**[4] of this set. The Delaunay triangulation has the important property that the minimum angle of any of its triangles is as large as possible for any triangulation of the same set of points (see Section 1.2 of [Ede-01]). This makes the Delaunay triangulation very suitable for the FEM. There is a dual notion of the Delaunay triangulation which leads to an equivalent formulation. We give the relevant definitions:

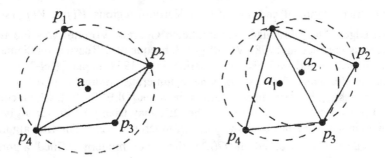

FIGURE 13.8: Two triangulations are shown for the same 4-point set $\{p_1, p_2, p_3, p_4\}$. (a) (left) The first one violates the Delaunay condition since p_3 lies in the circumcircle of the larger triangle. (b) (right) The second gives the Delaunay triangulation. Circles and centers are drawn in to show the validity of the condition.

Definitions: Suppose that we have a set of distinct points $P = \{p_1, p_2, \cdots, p_n\}$ in the plane $\mathbb{R}^2$ (write $p_i = (x_i, y_i)$). Relative to this set, for each $p_i(1 \leq i \leq n)$ we define the **Voronoi region** $V(p_i)$ as:

$$V(p_i) = \{p \in \mathbb{R}^2 : |p - p_i| \leq |p - p_\ell| \text{ for each } p_\ell \in P, \ell \neq i\}. \tag{9}$$

---

[4] The convex hull of a set of points is the smallest convex set which contains each of the points. There is a degenerate case in which some four of the points lie on a common circle (with no other points inside this circle). Here no three of the points will lie any closer to the center of the circle than the fourth. In algorithms for the Delaunay triangulation what is usually done in degenerate cases is that one of the points is slightly perturbed (moved). Since the area of a circle is zero, the probability that a fourth point will lie on the circle determined by three points is zero. In degenerate cases "the" Delaunay triangulation is not unique. The whole subject of triangulation and more general mesh generation has become quite an important discipline in itself. Good references are Chapter 13 of [Ros-00] and [Ede-01].

Here absolute values denote the Euclidean distance (this coincides with the 2-norm introduced in Chapter 7). The **Voronoi diagram** for P is simply an illustration of the totality of all of the Voronoi regions.

It is not difficult to show that each Voronoi region is a convex set which, if bounded, is a polygon (Exercise 27). In words, the Voronoi region $V(p_i)$ is simply the set of all points in the plane whose closest element of the set P is p_i. If a school district wished to minimize bussing times and costs, and if the points p_i represented locations of schools, the Voronoi region of a given school would roughly include all households whose children would be sent to that school.

The duality result states that two points $p_i, p_j \in P$ are joined by an edge in the Delaunay triangulation if and only if their Voronoi regions $V(p_i)$, $V(p_j)$ share a common edge. The Ukrainian mathematician Georges Voronoi was the first to introduce his concept in 1908 [Vor-08]. Subsequently, Russian mathematician Boris Delaunay introduced his triangulation in a 1934 paper [Del-34] that he dedicated to Voronoi. These concepts have numerous applications; details of the rich and interesting history can be found in the book [OkBoSu-92], which contains over 600 references. Construction of the Delaunay triangulation for a given set of n points in the plane has been the focus of much research. The first algorithms that were discovered worked in $O(n^4)$ time, but modern refined algorithms perform in $O(n \log n)$ time. Some survey articles of this area are: [SuDr-95] and [BeEp-92], see also Chapter 13 of [Ros-00]. We will make use of MATLAB's built-in functions that will perform both of the Delaunay triangulation and the Voronoi diagrams, so the task of triangulations will thus be reduced to the more simple problem of node deployment.

We proceed now to introduce the relevant MATLAB functions:

`tri = delaunay(x,y)` →	If x and y are vectors of the same length n giving the coordinates of n (noncollinear) points in the plane, this command will output an $n \times 3$ matrix `tri` whose rows contain the indices (rel. to the x and y vectors) of the triangles in the Delaunay triangulation.
`voronoi(x,y)` →	If x and y are as above, this command will result in a graphic of the Voronoi diagram[5] for the set of points corresponding to x and y.

Once created, the Delaunay triangulation can be used, just like any other triangulation for the FEM. To view the Delaunay triangulation, we could use the above `trimesh` command. We illustrate by having MATLAB compute both

---

[5] Actually, the `voronoi` command will show only the bounded Voronoi regions (i.e., those that have finite areas). There is an easy way to get MATLAB to show all of the regions; see Exercise for the Reader 13.2.

objects for the set of node values that we used for the diagram in Figure 13.5. The following commands result in the plot shown in Figure 13.9(a).

```
>> N=[1 1;5/2 1;0 0;1 0;5/2 0;7/2 0;1 -1;2.5 -1];
>> x=N(:,1); y=N(:,2);
>> tri=delaunay(x,y), trimesh(tri,x,y)
>> axis([-1 4.5 -1.5 1.5]), hold on
>> plot(x,y,'ro')
```

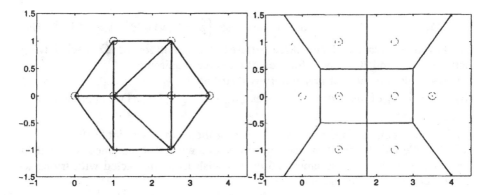

FIGURE 13.9: (a) (left) MATLAB plot of the Delaunay triangulation of the set of data points indicated by circles. (b) (right) MATLAB plot of the dual Voronoi diagram for the same set of data points.

The Voronoi diagram in Figure 13.9(b) was obtained using the minor modification of the MATLAB's voronoi program that appears in the following exercise for the reader.

EXERCISE FOR THE READER 13.3: (a) Write an M-file called voronoiall(x,y) that will function just like MATLAB's voronoi, except that it will show the unbounded Voronoi regions (not all of them, of course) with a reasonable axis view. (b) Use your program to recreate the plot of Figure 13.9(b).

Some comments are in order. First notice that the Delaunay triangulation that MATLAB gave us coincides with the one we used previously. Also notice that this example demonstrates that the Delaunay triangulation is not unique (so it really should not have been called "the" Delaunay triangulation). Indeed, the two diagonal edges in the center could have been reversed (i.e., reflect the triangulation horizontally) to result in another triangulation that also meets the Delaunay criteria, or the duality theorem's criterion. (The reader should convince himself or herself of these assertions.) The Voronoi regions are of course uniquely defined and so the Voronoi diagram is unique.

Having the delaunay function to work with makes it a lot easier to do a triangulation; we need only specify the node points. This should be done in a way that will give rise to a Delaunay triangulation whose triangles do not get too thin.

A good general rule is to deploy node points in more or less squarelike configurations. The sizes of adjacent squares should not change too abruptly. Of course, when approaching the boundary, special care must be exercised. For boundary value problems, nodes need to be put on the boundary as well. It is also possible to increase the density of nodes in certain parts of the domain in regions where coefficients of the PDE are more active (oscillatory). The next example will create three different triangulations for the same domain, a disk.

EXAMPLE 13.2: Let Ω denote the unit disk $\{p = (x, y) \in \mathbb{R}^2 : \|p\|_2 \leq 1\}$.[6] Use MATLAB to create and plot three different triangulations of Ω each having between 1000 and 2000 nodes for each of the three requirements:
(a) The nodes are more or less uniformly distributed.
(b) The density of the nodes increases as $\|p\|_2$ increases, i.e., as we approach the boundary.
(c) The distribution of the nodes increases near the boundary point $(1,0)$.
NOTE: We left the exact number of nodes somewhat flexible since we want to stress node deployment schemes and do not wish to be distracted with trying to use a precise number of nodes.

SOLUTION: Part (a): We will give two different strategies for this part.

Method 1: We use a squarelike configuration for the nodes. For the most part, this will be quite a simple scheme, but near the boundary circle $\|p\|_2 = 1$, things get a bit awkward. The square $S = \{p = (x, y) \in \mathbb{R}^2 : -1 \leq x, y \leq 1\}$ includes the disk Ω as its inscribed circle. Since the ratio of the areas of Ω to S is $\pi \cdot 1^2 / (2 \cdot 2) = \pi / 4 = 0.785...$, it follows that if we uniformly distribute a large number of nodes in the interior of S, roughly 78.5% of them will be in the interior of Ω. Since it is a simple matter to uniformly distribute any (square) number of nodes in S, we will begin by uniformly distributing about 2000 nodes in S, and let δ denote the square side length that is used. Of these nodes we will keep all of them that lie inside of Ω but at a distance of at least $\delta / 2$ from the boundary circle $\|p\|_2 = 1$. Then we add on a set of nodes on the boundary circle, which are uniformly spaced with gaps about equal to δ. The total amount of nodes thus constructed for Ω will be (well) over 1000 and certainly less than 2000.

To begin, since $\sqrt{2000} = 44.721...$, we will first construct a square grid of $N_0 = 45^2 = 2025$ interior nodes in S. The horizontal and vertical gaps should be

---

[6] We are using here the norm notation from Chapter 7: $\|p\|_2 = \|(x, y)\|_2 = \sqrt{x^2 + y^2}$ is the 2-norm which is simply the (planar) Euclidean distance from $p = (x, y)$ to the origin $(0,0)$.

$\delta = 2/(45+1)$.   The following MATLAB commands will create these nodes, and store them in two vectors x0, y0.

```
>> delta =2/46; counter=1;
for i=1:45
for j=1:45
  x0(counter)=-1+i*delta; y0(counter)=-1+j*delta;
  counter=counter+1;
end
end
```

Next, from these two vectors we extract those components corresponding to points which lie within the slightly smaller circle $\|p\|_2 = 1 - \delta/2$; the newly created vectors will be labeled as x and y.

```
>> counter=1;
>> for i=1:2025
    if norm([x0(i) y0(i)],2) < 1-delta/2
      x(counter)=x0(i); y(counter)=y0(i);
      counter=counter+1;
    end
end
```

Finally, since $2\pi/\delta = 144.5133...$, we tack onto the existing vectors x and y an additional 145 entries corresponding to 145 equally spaced points on the unit circle $\|p\|_2 = 1$.   Figure 13.10(a) shows a plot of this node set.

```
>> for i=1:145
    x(i+1597)=cos(2*pi/145*i); y(i+1597)=sin(2*pi/145*i);
end
>> plot(x,y,'bo'), axis('equal')
```

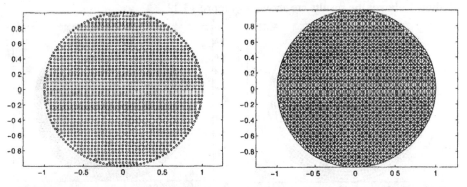

FIGURE 13.10: (a) (left) Grid of nodes from Method 1 of Example 13.2(a), essentially a square pattern except near the boundary.   There are 1742 nodes.   (b) (right) A corresponding Delaunay triangulation that has 3337 triangles.

The corresponding Delaunay triangulation will result from the following MATLAB commands and is shown in Figure 13.10b.

```
>> tri=delaunay(x,y); trimesh(tri,x,y), axis('equal')
```

Method 2: Here we will deploy nodes on circles of increasing radii. The gaps between nodes on a given circle and the gaps between radii of adjacent circles of deployment should be all about equal (uniformity). The final circle will be the boundary of Ω : $\|p\|_2 = 1$. The only mathematical preliminaries are to decide how many circles to deploy. Letting δ denote the common gap size, since the radii of the circles increase steadily from 0 to 1, the average radius will be about $1/2$, which means the average circumference will be about $2\pi(1/2) = \pi$. Thus the average number of nodes on a circle will be π / δ. Likewise, the number of circles of deployment is about $1/\delta$, so that the total number of nodes will be, roughly, $(\pi / \delta) \cdot (1/\delta) = \pi / \delta^2$. Setting this equal to 1800, say (we want it close to 2000, but to insure the actual number of nodes remains under 2000 we play it a bit safe), and solving for delta gives $\delta = \sqrt{\pi / 1800} \doteq 0.04177...$. We may now turn the node deployment over to MATLAB with this scheme:

```
>> delta=sqrt(pi/1800); x(1)=0;   y(1)=0;
>> nodecount=1; ncirc=floor(1/delta); minrad=1/ncirc;
>> for i=1:ncirc
     rad=i*minrad;
     nnodes=floor(2*pi*rad/delta);
     anglegap=2*pi/nnodes;
     for k=1:nnodes
       x(nodecount+1)=rad*cos(k*anglegap);
       y(nodecount+1)=rad*sin(k*anglegap);
       nodecount = nodecount+1;
     end
   end
```

The plotting of the nodes and then the Delaunay triangulation is done just as in Method 1 above; the results are shown in Figure 13.11.

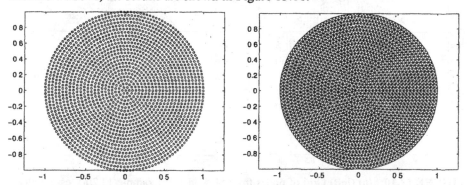

FIGURE 13.11: (a) (left) Grid of nodes from Method 2 of Example 13.2(a). There are 1887 nodes. (b) (right) A corresponding Delaunay triangulation that has 3438 triangles. Both the node distribution as well as the Delaunay triangulation take on an aesthetically more appealing pattern than those of Method 1, since this method better respected the symmetry of the domain.

Part (b): The requirement is rather vague. We will use a deployment scheme similar to that of Method 2 in part (a). The new difficulty is that there will need to be more circles of nodes of larger radii so it will take more work to estimate the total number of nodes. Such an estimate will depend first on how we plan to distribute the radii for the circles of nodes. Here is (but) one scheme. We start off with a single node at the origin $(0,0)$. Then we move to the circle $\|p\|_2 = 1/2 = \text{rad}(1) = 1 - 1/2$ and deploy 8 (equally spaced) nodes on this circle. Our next circle is $\|p\|_2 = 3/4 = \text{rad}(2) = 1 - 1/4$ on which we deploy $2 \cdot 8 = 16$ nodes. After this we deploy $2 \cdot 16 = 32$ nodes on the circle. We continue this pattern, so that at the nth circle of deployment will be $\|p\|_2 = \text{rad}(n) = 1 - 1/2^n$ on which we will deploy 2^{n+2} nodes. This will continue until the number of remaining nodes is still greater than the number of most recently installed nodes (on the last circle of deployment). The final step will be to put all of the remaining nodes on the unit circle $\|p\|_2 = 1$. This plan will create exactly 2000 nodes. Here now is the MATLAB code needed to create such a set of nodes.

```
>> xb(1)=0; yb(1)=0; rnodes=1999; %remaining nodes
>> newnodes=8; %nodes to be added on next circle
>> radcount=1; %counter for circles
>> oldnodes=1; %number of nodes already deployed
>> while newnodes < rnodes/2
   rad = 1 - 2^(-radcount);
   for i=1:newnodes
     xb(oldnodes+i)=rad*cos(2*pi*i/newnodes);
     yb(oldnodes+i)=rad*sin(2*pi*i/newnodes);
   end
   oldnodes=oldnodes + newnodes; %update oldnodes
   rnodes = rnodes - newnodes; %update rnodes
   radcount=radcount+1; %update radcount
   newnodes = 2*newnodes; %update newnodes
end
% now deploy remaining nodes on boundary
>> for i=1:rnodes
     xb(oldnodes+i)=cos(2*pi*i/rnodes);
     yb(oldnodes+i)=sin(2*pi*i/rnodes);
end
```

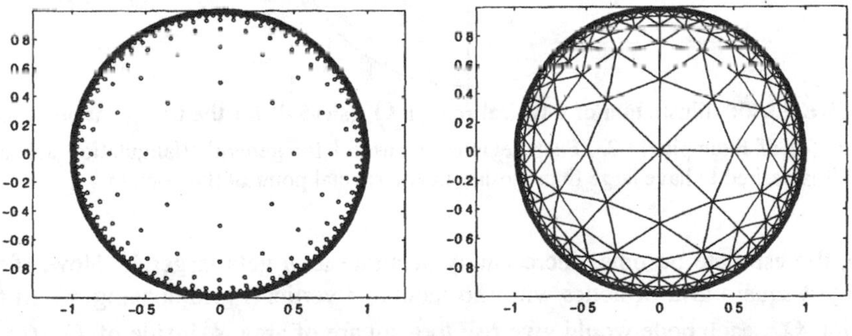

FIGURE 13.12: (a) (left) Grid of nodes for Example 13.2(b). There are 2000 nodes. (b) (right) A corresponding Delaunay triangulation that has 3015 triangles. Such a triangulation is useful for BVPs which are particularly sensitive to boundary data.

The plotting of the nodes and creation of the Delaunay triangulation is obtained in the same fashion as in part (a). The results are shown in Figure 13.12.

Part (c): The way we will deploy nodes is to first partition Ω into subsets determined by the regions between pairs of circles centered at $(1,0)$. For each positive integer n, we define the following subset $\Omega_n \subseteq \Omega$:

$$\Omega_n = \{(x,y) \in \Omega : 1/2^n < \text{dist}((x,y),(1,0)) < 2 \cdot (1/2^n)\}.$$

In words, Ω_n is just the portion of points inside Ω that lie in between the two concentric circles with center $(1,0)$, and radii $1/2^n$ and $2 \cdot (1/2^n)$. A typical such region is illustrated in Figure 13.13. The idea will be to deploy roughly a fixed number of nodes (we will use about 100) in each of these regions, up to a certain value of the index. We will also need to put some nodes in the remaining part of Ω (which actually can be written as Ω_0); here we again put roughly the same number, about 100 nodes. To decide how to put the nodes on the boundary circle of Ω as well as in the interior of the domains Ω_n, we use the following estimates. From Figure 13.13, it is clear that Ω_n is always contained in the half annulus (enclosed between the two dotted circles in the figure), and consequently,

$$\text{Area}(\Omega_n) \leq \tfrac{1}{2}[\pi \cdot (2 \cdot 2^{-n})^2 - \pi \cdot (2^{-n})^2] = \tfrac{3\pi}{2} 2^{-2n}.$$

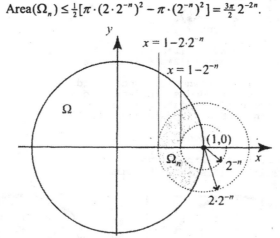

FIGURE 13.13: Illustration of a typical region Ω_n (shaded) for the triangulation scheme of part (c) of Example 13.2. Such regions are useful for general triangulation schemes when it is desired to have large finer meshes near a special point of the domain.

Also, the estimate becomes increasingly accurate as n gets larger. Now, if we deploy a square grid of nodes with (horizontal = vertical) grid spacing $= s$ in the interior Ω_n, each node would give rise to a square of area s^2 inside of Ω_n (to be specific, let's say that the node gets associated with the square of side length s having the node as its lower left vertex). Thus, if we were to put a grid of 100 such nodes in the interior of Ω_n, the area bound above would yield the following estimate for s:

$$100 \cdot s^2 \le \text{Area}(\Omega_n) \le \tfrac{3\pi}{2} 2^{-2n} \Rightarrow s \le \sqrt{3\pi} 2^{-n} /(10\sqrt{2}).$$

We use this as a scheme for the horizontal/vertical grid gap to put between nodes that lie inside each Ω_n. The actual number of nodes on each deployment will be less than 100 because the above estimates are inequalities. Since we will essentially be placing 100 nodes at each iteration on Ω_n (and the corresponding portion of the boundary circle adjacent to Ω_n) starting with $n = 0$, it follows that we should let n run up to about 15 in this scheme. For deploying nodes on the boundary circle that lie adjacent to Ω_n, we also use s as the gap (this time the circular arclength gap) between boundary nodes. Since the boundary circle has radius one, angles are equal to the corresponding boundary arclengths. We will create the nodes using nested loops. On each iteration for n (master loop), the loops will first create and store the corresponding value of s, determined by using an equality in the above inequality for s.

Next, a double loop will be set up that will run through a horizontal and vertical grid that will cover the domain Ω_n and have (horizontal = vertical) grid gap = s. For this part, note (again from Figure 13.13) that the domain Ω_n is always contained within the rectangle: $R_n = \{(x, y) : 1 - 2 \cdot 2^{-n} \le x \le 1, -2 \cdot 2^{-n} \le y \le 2 \cdot 2^{-n}\}$ Grid points lying in the interior of Ω_n are added as nodes. Once this double nested loop has been executed and interior nodes have been added, the same master loop will then move on to install nodes on the two portions of the boundary of Ω_n that lie on the unit circle (= boundary of the main domain Ω). We will need to compute the angles (made from (0,0) to the positive x-axis) of the two endpoints of the top boundary arc of Ω_n on the unit circle. (The node deployment on the bottom symmetric boundary arc can be gotten by simply negating the y-coordinates of the nodes in the upper arc.) It is easily shown using the law of cosines that these two angles θ_1 and θ_2 (which technically should be denoted by $\theta_{1,n}$ and $\theta_{2,n}$ to indicate their dependence on Ω_n) satisfy: $\cos(\theta_1) = 1 - 2/2^{2n}$ and $\cos(\theta_2) = 1 - 2^{2n}/2$. The following MATLAB code is an implementation of the scheme described above.

```
>> n=0; nodecount=1;
>> while n<16
  s=sqrt(3*pi/2)/10/2^n; hgrid=2/s/2^n; vgrid=4/s/2^n;
%these will be sufficient horizontal and vertical grid counts to
%create a rectangular grid (with gap size =s) that will cover the
%domain Omega_n
  for i=1:hgrid
  for j=1:vgrid
  xnew=1-2/2^n+i*s; ynew=-2/2^n+j*s;
  pij = [xnew ynew]; p=[1 0];
 if norm(pij,2)<1-s/2 & norm(pij-p,2)<2/2^n & norm(pij-p,2)>1/2^n+s/2
%The three conditions here check to see if the node should be added.
%The first says that the node should be in the unit circle (with a
%safe distance to the boundary to prevent interior nodes from getting
%too close to boundary nodes which will be added later). The second
```

```
%and third state that the distance from the node to the special
%boundary point (1,0) should be between the two required radii.  The
%last condition has a safety term added to the lower bound to prevent
%nodes from successive iterations from getting too close.
   x(nodecount)=xnew; y(nodecount)=ynew; nodecount=nodecount+1;
  end
 end
 end
%The next part of the loop puts nodes on the boundary.
 theta1=acos(1-2/2^(2*n)); theta2=acos(1-2^(-2*n)/2);
 if n==0, theta1=theta1-s; end
 for theta = theta1:-s:(theta2+s/2)
  x(nodecount)=cos(theta); y(nodecount)=sin(theta);
  x(nodecount+1)=cos(theta); y(nodecount+1)=-sin(theta);
  nodecount=nodecount+2;
 end
 n=n+1;
end
%We need to put a node at the special unsymmetric point (-1,0).
x(nodecount)=-1; y(nodecount)=0; nodecount=nodecount+1;
%Finally we put nodes in the portion of the domain between (1,0)
%and the last Omega_n, and then on the boundary.
%We need first to bump n back down one unit.
n=n-1;
for i=1:hgrid
for j=1:vgrid
  xnew=1-2/2^n+i*s; ynew=-2/2^n+j*s;
  pij = [xnew ynew]; p=[1 0];
  if norm(pij,2)<1-s/2 & norm(pij-p,2)< 1/2^n
   x(nodecount)=xnew; y(nodecount)=ynew; nodecount=nodecount+1;
  end
end
end
for theta = -theta2:s:theta2
  x(nodecount)=cos(theta); y(nodecount)=sin(theta);
  nodecount=nodecount+1;
end
```

Plotting of the nodes as well as the corresponding triangulation is accomplished exactly as it was done in the above two parts. The results are shown in Figures 13.14 and 13.15.

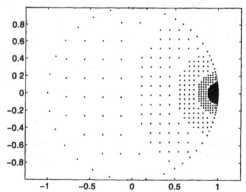

FIGURE 13.14: Node distribution from the solution of part (c) of Example 13.2. The 1457 nodes are constructed in clusters with each cluster getting its grid gap size cut in half as we move in towards the special boundary point (1,0). The exercises will examine some related schemes for this domain where there is a smoother transition in gap sizes of nodes as we progress toward (0,1).

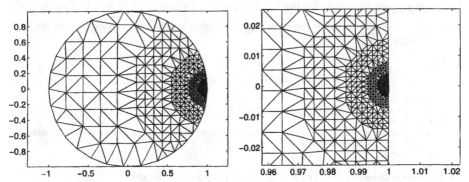

FIGURE 13.15: (a) (left) The Delaunay triangulation corresponding to the network of nodes of Figure 13.14 that has 2733 triangles. (b) (right) A $20\times$ magnification of the triangulation of (a) near the point of focus (1,0).

On the node sets that were constructed in the last example, the Delaunay triangulation worked quite nicely because the domain Ω was convex. In general, the Delaunay triangulation of a set of nodes will triangulate the convex hull of this set. Delaunay triangulation can still be used to triangulate a nonconvex domain Ω. This is usually done either by breaking up the domain into convex pieces, triangulating each piece, and merging these triangulations or simply by triangulating the convex hull and deleting triangles that are not part of the domain.[7] With either strategy, some sort of (global) reindexing will be necessary when constructing the final triangulation. Our next example will illustrate the latter strategy and the exercise for the reader which follows will require also the former strategy.

EXAMPLE 13.3: Let Ω denote the annular domain $\left\{ p = (x, y) \in \mathbb{R}^2 : 1 \le \|p\|_2 \right.$ $\left. \le 2 \right\}$. Use MATLAB to create and plot a triangulation of Ω having between 200 and 400 nodes that are more or less uniformly distributed.

SOLUTION: We use a node deployment strategy that is based on that of Method 2 of part (a) of the last example, distributing nodes on concentric circles starting at $\|p\|_2 = 1$ and ending at $\|p\|_2 = 2$. Letting, as before, δ denote the (approximate) common gap size between nodes (and the circles of node deployment), the average radius will be (roughly) 3/2, so that the average circumference will be $2\pi(3/2) = 3\pi$. The average number of nodes on a circle of deployment will thus be $3\pi/\delta$ and the number of such circles will be (roughly) $1/\delta$. This gives the

---

[7] Of course, the convex hull of a set of nodes for a domain will not always coincide with the domain even when the domain is convex (e.g., a disk) just as the triangulation will not coincide with the domain. But if the mesh is finer these approximations will become indistinguishable from the true objects. ·

following approximation for the total number of nodes: $(3\pi / \delta) \cdot (1 / \delta) = 3\pi / \delta^2$. Setting this equal to 350 and solving for delta gives us a good value to use: $\delta = 0.164097....$

```
>> delta=sqrt(3*pi/350);
nodecount=1; ncirc=floor(1/delta); radgap=1/ncirc;

for i=0:ncirc
   rad=1+i*radgap; nnodes=floor(2*pi*rad/delta); anglegap=2*pi/nnodes;
   for k=1:nnodes
   x(nodecount)=rad*cos(k*anglegap); y(nodecount)=rad*sin(k*anglegap);
     nodecount = nodecount+1;
   end
end
>> tri=delaunay(x,y); trimesh(tri,x,y), axis('equal')
>> size(x)
→ans = 1   399
```

We are just under the desired upper bound on the number of nodes (if we had gone over, we could just increase δ a bit and try again. The resulting plot of the triangulation is shown in Figure 13.16(a).

Locating and deleting the unwanted triangles in Figure 13.16(a) would be an arduous task. The problem could be greatly simplified if we were to throw in an extra "ghost node" at (0,0). This is done by simply entering

```
>> x(400)=0; y(400)=0;
>> tri=delaunay(x,y); trimesh(tri,x,y), axis('equal')
```

The resulting triangulation shown in Figure 13.16(b) is much easier to work with. The triangles that need to get deleted are simply those that have (0,0) (node #400) as one of their vertices. So we simply delete from the triangulation matrix all rows that contain the entry 400. It is very simple to tell MATLAB how to do this, after checking there are 722 triangles (elements). We will make use of the following "set difference" command:

c = setdiff(a,b) →	If a and b are vectors, the output of this "set difference" command will be another vector c whose elements consist of the different values of a that do not occur in b.

Here is a simple example:
```
>> setdiff([3 1 2 3], [2]) →ans = 1   3
```

Now back to our problem; the following series of commands will produce the final triangulation of Figure 13.16(c).

```
>> badelcount=1;
for ell=1:722
   if ismember(400,tri(ell,:))
   badel(badelcount)=ell;
   badelcount=badelcount+1;
   end
end
```

```
>> tri=tri(setdiff(1:722,badel),:);
>> x=x(1:399); y=y(1:399);
>> trimesh(tri,x,y), axis('equal')
```

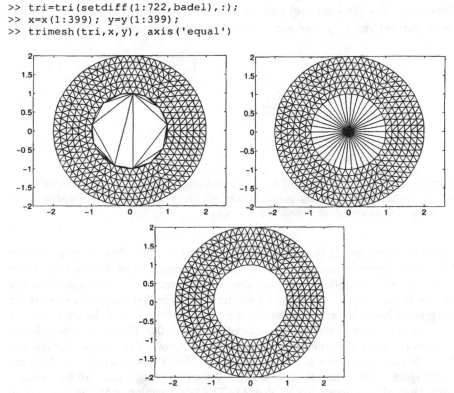

FIGURE 13.16: (a) (upper left) The Delaunay triangulation obtained from a set of nodes in the annular domain $\left\{ p = (x, y) \in \mathbb{R}^2 : 1 \le \|p\|_2 \le 2 \right\}$ of Example 13.3. (b) (upper right) The Delaunay triangulation for the same set with one additional "ghost node" added at (0,0). (c) Resulting triangulation for the annulus.

The schemes introduced in the previous examples can be combined in various fashions to give a decent collection of strategies for triangulation of planar domains that will be sufficient for our purposes. The topic of mesh generation has been receiving a great deal of attention beginning in the 1990s. We will see later in this chapter that boundary value problems on domains with obtuse ($> \pi$) interior angles (see Figure 13.17 for two examples of such domains) usually require special attention with the numerical methods at corresponding boundary points. The next exercise for the reader asks the reader to construct suitable triangulations for such domains.

EXERCISE FOR THE READER 13.4: Using a scheme similar to that of the solution of part (c) of Example 13.2, get MATLAB to create and plot triangulations each having between 500 and 1000 nodes for the two domains illustrated in Figure 13.17(a), (b), respectively. In each, arrange it so that the distribution of nodes increases near the special boundary point p indicated in the illustration. For the domain of part (a), take the exterior angle to be 60°; for the one of part (b), make up your own coordinates and dimensions.

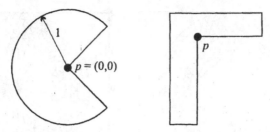

FIGURE 13.17: (a), (b) A pair of domains possessing a boundary point p with an obtuse interior angle. Such boundary points usually require extra care when boundary value problems on them are solved numerically.

In each of the triangulations done above, we tried to create them to meet most of the desired properties that were mentioned at the beginning of the chapter. There is one notable exception, however, that we did not even contemplate in our constructions. Namely, we made no efforts to arrange that the node numbers for any given triangle in the triangulations were reasonably close. (The constructions were already quite complicated without this and the Delaunay triangulation program left parts of the construction out of our control). The reason why this is a desirable property to have is that the resulting stiffness matrix will be banded (and hence sparse and easier to deal with). To get a rough idea of the relative numbering of the nodes, recall that MATLAB's function spy, introduced in Chapter 11, can gives us a "graph" of the nonzero entries of a matrix. For convenience, we give here a quick example reviewing its syntax.

```
>> d=ones(1,6); b=2*d(1:5);
>> A=diag(d)+diag(b,-1)
```

```
A =        1  0  0  0  0  0
           2  1  0  0  0  0
           0  2  1  0  0  0
           0  0  2  1  0  0
           0  0  0  2  1  0
           0  0  0  0  2  1
```

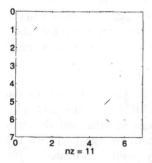

nz = 11

```
>> spy(A, 'rx')  %mark nonzero entries with red
>> %                 x's; or use spy(A)
```

FIGURE 13.18: A simple spy plot of a banded 6×6 matrix. The locations of nonzero entries are indicated by x's. The total number of nonzero entries ($nz = 11$) is indicated below the graph. The spy command is a useful tool for obtaining a quick understanding of the structure of a matrix, and, in particular, allows for quick detection of sparse and banded matrices.

The triangulation that we created in the solution of Example 13.2(c) had 1457 nodes. The corresponding stiffness matrix A would thus be 1457×1457. As in the one-dimensional FEM, we will see in the next section that the a_{ij} entry (corresponding to node numbers #i and #j) of the stiffness matrix will be given by a certain integral involving products of (gradients of) the corresponding basis functions Φ_i and Φ_j. Throughout the text of this chapter, we will be restricting our attention to piecewise linear basis functions, and for such a basis function, say Φ_i, it will be zero except on those elements that have node #i as one of their vertices. It follows that a_{ij} will be zero unless nodes #i and #j are both vertices of the same triangle. In the following example, we will use this fact to find out all possible nonzero entries of the stiffness matrix, draw a spy diagram, and list the total number of possible nonzero entries. The way we will form this matrix is to simply put a positive integer at all entries that are possibly nonzero.

EXAMPLE 13.4: Let A denote the 1457×1457 stiffness matrix for the triangulation obtained in Example 13.2(c) and with piecewise linear basis functions. Using the information above, construct a matrix M that will have positive integer entries where the corresponding entries of the stiffness matrix are zero, and zero entries where the stiffness matrix has zero entries.

SOLUTION: The way we will construct M will be similar to the so-called "assembly" method that we will use in the next section to build stiffness matrices. The construction will proceed element by element. More precisely, we begin with M being a 1457×1457 matrix of zeros. We then run down the list of triangles/elements (all 2733 of them), and for each one we change the corresponding entry of the of the matrix M to equal 1 (these are the only entries of that stiffness matrix A that could be nonzero). If an element is represented by the three vertices: [i j k], the entries we will bump up by one for this element would be the following nine entries $a_{\alpha\beta}$ where α and β run through i, j, and k. This is a much more efficient scheme rather than constructing the nonzero elements directly (in which case for each one a search would need to be done over all elements to see if the corresponding pair of nodes share a common element).

Assuming the matrix `tri` obtained in the last example (for the Delaunay triangulation of the nodes in part (c)), the following commands will "assemble" a suitable matrix M, and then create a `spy` diagram of it (and hence also of the stiffness matrix). The spy diagram is shown in Figure 13.19.

```
>> M=zeros(1457); for c=1:2733
   E=tri(c,:);
   for i=1:3
   for j=1:3
   M(E(i),E(j))=M(E(i),E(j))+1;
   end
   end
end
>> spy(M,'b+') %or use spy(M) to use default '.' markers
```

>> 9835/1457^2 →ans = 0.0046

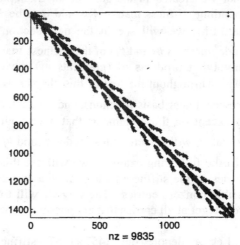

FIGURE 13.19: A spy diagram of the stiffness matrix for Example 13.4. The possible nonzero entries account for only 0.46% of all of the entries, so this stiffness matrix is sparse as stiffness matrices usually are. The fuzzy patterns (top and bottom) correspond to the boundary nodes being added after the interior nodes. The number of such patterns is the number of master iterations in the node construction.

The last ratio is simply that of the nonzero entries to the total entries of M. Thus at most 0.46% of the entries in the stiffness matrix can be nonzero, and the matrix is indeed sparse.

Can the reader explain the isolated four markers in the upper-right and lower-left of Figure 13.19?

EXERCISES 13.2

1. (a) For the hexagonal domain of Figure 13.5 with node coordinates as given by the matrix N following Figure 13.6, deploy between 1000 and 2000 nodes more or less uniformly throughout the domain and boundary and plot the nodal configuration. You should of course include the boundary nodes of Figure 13.5 but not necessarily the interior nodes (#4, #5).
    (b) Create a corresponding Delaunay triangulation and plot.

2. Repeat parts (a) and (b) of Exercise 1, but this time let the nodes increase in density as one moves toward the boundary.

3. Repeat parts (a) and (b) of Exercise 1, but this time let the nodes increase in density as one approaches the exterior node #6.

4. Letting Ω denote the unit disk $\left\{ p = (x, y) \in \mathbb{R}^2 : \|p\|_2 \le 1 \right\}$ of Example 13.2, use MATLAB to create and plot a triangulation of Ω having between 1000 and 2000 nodes for which the nodes increase in density near the segment $-\pi/4 \le \theta \le \pi/4$ on the boundary, that is, near the (smaller) circular arc connecting the points $(\sqrt{2}/2, \pm\sqrt{2}/2)$ on the boundary. Let the node distribution

elsewhere in the disk be more or less uniform.
Suggestion: Use the solution to part (c) of Example 13.2 for some relevant ideas. First deploy some nodes in a circle centered at (0,0), say $\|p\|_2 \leq 0.5$, then use a loop to deploy nodes in the annuli $A_n = \{p : 1 - 2^n \leq \|p\|_2 \leq 1 - 2^{n+1}\}$, $n = 1, 2, \ldots$. As an indicator of closeness to circular arc (for points (x,y)) in A_n use the inequality $\tan(y/x) \leq 1 + 2 \cdot 2^n$.

5. Repeat Exercise 4 with the modification that node distribution should increase near the boundary. For the portion of the boundary complementary to $-\pi/4 \leq \theta \leq \pi/4$, make the rate of increase in node density to be roughly 10% compared to the rate of increase near the special portion.

6. (a) Write an M-file, call it [x y tri]=circtri(angle1, angle2, maxnodes) that will do the following. The input variables angle1 and angle2 denote two angles on the unit circle such that $0 \leq$ angle2 $-$ angle1 $< 2\pi$. The M-file will create a set of nodes, stored in the output variables x and y for the triangulation of the unit disk $\{p = (x, y) \in \mathbb{R}^2 : \|p\|_2 \leq 1\}$ in a way analogous to the one explained in Exercise 4 (for the special case angle1 $= -\pi/4$ and angle2 $= \pi/4$) but with the total number of nodes deployed being between the input variable maxnodes and half of this variable. Thus maxnodes should be a positive integer, at least equal to, say, 20. The final output variable tri will be a three-column matrix corresponding to the Delaunay triangulation of the node set. Note that the syntax includes the possibility that angle1 $=$ angle2, in which case a triangulation similar to that done in Example 13.2(c) is required.
(b) Use your program to redo Exercise 4.
(c) Run your program, and plot the nodes and resulting triangulations for each of the following sets of input variables:
(i) angle1 $= \pi/2$, angle2 $= 5\pi/6$, maxnodes $= 500$
(ii) angle1 $= -\pi$, angle2 $= \pi/2$, maxnodes $= 1500$
(iii) angle1 $= 7\pi/6$, angle2 $= 11\pi/6$, maxnodes $= 1200$

7. (a) Write an M-file, call it [x y tri]=circtri2(angle1, angle2, maxnodes, r), that has the same syntax as that explained in Exercise 6, except that there is an additional input variable r which is to be a positive number less than 1 and the triangulation will be performed as explained in Exercise 5 (for the special case angle1 $= -\pi/4$ and angle2 $= \pi/4$, and $r = 0.1$). The parameter r will denote the relative density that nodes are increasing as we near the complementary arc compared to when we near the arc angle1 $< \theta <$ angle2.
(b) Use your program to redo Exercise 5.
(c) Run your program, and plot the nodes and resulting triangulations for each of the following sets of input variables:
(i) angle1 $= \pi/2$, angle2 $= 5\pi/6$, maxnodes $= 500$, $r = 0.25$
(ii) angle1 $= -\pi$, angle2 $= \pi/2$, maxnodes $= 1500$, $r = 0.05$
(iii) angle1 $= 7\pi/6$, angle2 $= 11\pi/6$, maxnodes $= 1200$, $r = 0.025$.

8. (*Triangulating General Convex Polygons*) (a) Write an M-file, call it [x y tri]=unipolytri(xv, yv, maxnodes), that will do the following. The input variables xv and yv denote the vectors corresponding to the x- and y-coordinates of vertices of a convex polygon which are assumed to be ordered in counterclockwise fashion around the boundary. The first vertex should also be the last vertex (to close the polygon). The last variable maxnodes denotes a positive integer, say, at least 20. The program will create a set of nodes for a triangulation of the polygon and its boundary that will be stored in the output variables x and y. The nodes are to be configured in a square pattern (cf., Method 1 of the solution to Example 13.2(a)) throughout the polygon and its boundary. The number of nodes deployed should be somewhere between maxnodes and half of this number. The third output

variable `tri` denotes a 3-column matrix corresponding to the Delaunay triangulation for the node set which is constructed.

(b) Use your program to redo Exercise 1.

(c) Run your program, and plot the nodes and resulting triangulations to obtain triangulations for each of the following convex polygons using between 200 and 400 nodes for each.

(i) The rectangle with vertices $(\pm 1, \pm 10)$.

(ii) The triangle with vertices $(0,0)$, $(1,0)$, $(0,8)$.

(iii) A regular octagon unit sidelength.

(iv) The septagon with vertices $(0,0)$, $(2,0)$, $(16,1)$, $(16,4)$, $(13,5)$, $(11,4)$, $(1,3)$.

Suggestion: One way to view a convex polygon is that its set of points can be described as the intersection of all points in the plane which simultaneously lie on the correct side of each of its edges. Each such edge requirement can be written mathematically in the form $ax + by \leq c$. Set up a grid of about maxnodes nodes in the rectangle $R = \{(x,y): \min(xv) \leq x \leq \min(xv)$, $\min(yv) \leq y \leq \min(yv)\}$, then use each of the edge requirements (put in form $ax + by < c$ to save boundary points for later) to decide with a loop which of these points should be interior points. Finally put nodes on the boundary with appropriate density. Make sure that there are no interior nodes that are too close to the boundary. That the number of nodes put in the polygon will satisfy the required bounds follows from the fact that the area of the polygon is at least half the area of the rectangle R (why?).

NOTE: (*Triangulating General Polygons*) Since any polygon can be decomposed into convex pieces, the program `unipolytri` of Exercise 8 can be used to essentially uniformly triangulate general polygons. For example, the polygon of Figure 13.17(b) is not convex but can be written as a union of two (convex) rectangles R_1 and R_2, that have corresponding areas A_1 and A_2. (There are a couple of ways to do this.) Suppose we wish to triangulate the region using somewhere between 500 and 1000 nodes. We could run the program `unipolytri` on R_1 using maxnodes to be about $1000A_1/(A_1 + A_2)$ and then on R_2 using maxnodes to be about $1000A_2/(A_1 + A_2)$. The ratios attempt at allocating an appropriate number of nodes to each piece. We can pretty much juxtapose these two node sets to arrive at a node set for the original polygon (after reindexing and deleting some nodes at the common interior interface boundary). This idea, and its extensions to greater numbers of convex pieces, is explored in the following three exercises. In particular, these exercises require the reader to have completed Exercise 8(a).

9.   Use the idea of the above note to redo Exercise for the Reader 13.4(b).

10.   Use the idea of the above note to triangulate, using between 400 and 800 nodes, the decagon that has the following vertices: $(\pm 2, 0)$, $(\pm 1, \pm 1)$, $(\pm 2, \pm 2)$. Plot both the node diagram as well as the triangulation.

11.   Consider the symmetric (nonconvex) polygon consisting of the rectangle with vertices: $(\pm 1, -1)$, $(\pm 1, 0)$ with two (left and right) triangles with vertices: $(\pm 1, 1)$, $(\pm 1, 0)$, $(0, 0)$ joined on top.

(a) Apply the method in the above note to triangulate this region by splitting it into the left and right halves (which are each convex). Display the node configuration and corresponding triangulation.

(b) Apply the method in the above note to triangulate this region by splitting it into the following three convex pieces: the bottom rectangle, and the two triangles. Display the node configuration and corresponding triangulation. Does the density appear uniform? If not, explain, and adjust the ratios of node densities to correct the problem.

The next three exercises will involve triangulations of the domains having domains with curved boundaries illustrated in Figure 13.20.

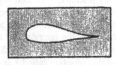

FIGURE 13.20: Four domains with curved boundary portions: (from left) (a) An ellipse, (b) a square with two circular holes, (c) a disk with a square hole, (d) an airfoil removed from a rectangle.

12. Let the elliptical region Ω of Figure 13.20(a) have equation (for its boundary): $x^2 + 4y^2 = 4$.

Use MATLAB to create and plot triangulations of Ω having between 400 and 800 nodes and with the following additional properties:
(a) The nodes are more or less uniformly distributed with essentially a square grid (as in Method 1 of part (a) in the solution of Example 13.2).
(b) The nodes are deployed on concentric ellipses of the same eccentricity as the boundary ellipse (cf. Method 2 of part (a) in the solution of Example 13.2).
(c) The nodes are deployed in concentric ellipses (as in part (b)) but the density increases as we near the boundary (cf. part (b) in the solution of Example 13.2).
(d) The density of the nodes increases as we approach the interior point $(x, y) = (1, 0)$ and such that between 20 and 30 nodes are deployed on the boundary (cf. Method 2 of part (a) in the solution of Example 13.2).

13. Let the region Ω of Figure 13.20(b) be specified as follows: The square (outside) boundary has equations: $x = 0, 2$, $y = 0, 2$ and the removed circles have the following centers and radii: upper left circle: center = $(0.5, 1.5)$, radius = 0.25; lower right circle: center = $(1.5, 0.5)$, radius = 0.1. Use MATLAB to create and plot triangulations of Ω having between 400 and 800 nodes and with the following additional properties:
(a) The nodes are, more or less, uniformly distributed with essentially a square grid (cf. Method 1 of part (a) in the solution of Example 13.2).
(b) The density of the nodes increases as we near each of the two interior circle boundary portions and such that the square boundary has between 20 and 30 nodes.

14. Let the region Ω of Figure 13.20(c) be specified as follows: The outside circle has: center = $(0, 0)$ and radius = 2; the inside square has equations: $x = \pm 1, y = \pm 1$. Use MATLAB to create and plot triangulations of Ω having between 400 and 800 nodes and with the following additional properties:
(a) The nodes are more or less uniformly distributed with essentially a square grid (cf. Method 1 of part (a) in the solution of Example 13.2).
(b) The nodes are deployed on concentric circles (to the outer boundary circle) and more or less uniformly distributed.
(c) The density of the nodes increases as we near any of the four corner points on the inside square boundary and the outside circle will have between 20 and 30 nodes.

15. Let Ω denote the region of Figure 13.20(d). (a) Use MATLAB to plot an airfoil (the inside boundary of Ω) by setting up a set of points on the boundary of the foil. Then enter (as different vectors) the four vertices of an appropriate rectangle for the outer boundary of Ω. Your foil does not have to be identical with the one in the figure, but should more or less resemble it. We are more concerned here with triangulations rather than aerodynamics. Use MATLAB to create and plot triangulations of Ω having between 400 and 800 nodes and with the following additional properties:
(b) The nodes are more or less uniformly distributed with essentially a square grid (cf. Method 1 of part (a) in the solution of Example 13.2).
(c) The density of the nodes increases as we near the airfoil (inside) part of the boundary and the outside rectangle will have between 20 and 30 nodes. See Figure 13.21 for examples of

some related triangulations.

Suggestions: To find appropriate x and y vectors for the foil, it is probably easiest to copy the figure down on graph paper and record a set of ordered (so it will plot correctly) vertices on the foil that is dense enough so as to render a decent plot. A more elaborate scheme would be to build up the boundary in terms of piecewise cubic splines whose derivatives match up on the interfaces (see the end of Section 10.5). MATLAB has a useful command for some of the tasks of this problem called inpolygon that will test if a given point lies within a given polygon:

test = inpolygon(x,y, xpoly, ypoly) $\rightarrow$	If xpoly and ypoly are the x- and y-coordinates of a set of vertices defining a polygon and x and y are the coordinates of any point in the plane, the output test will be 1 if the point (x,y) lies inside or on the boundary of the polygon and 0 otherwise. If x and y are vectors for a set of points, the output test will be a corresponding vector of 1's and/or 0's.

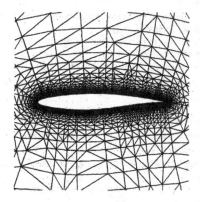 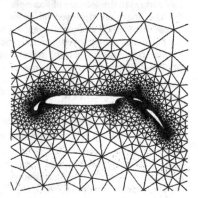

FIGURE 13.21: Two triangulations of airfoils. (a) (left) A single component airfoil similar to that in Figure 13.20(d). The triangulation is structured using lines normal to the surface (with increasing density as we near the boundary). (b) A more complex airfoil with flaps. The triangulation is done in a way that the node density increases as we near crucial portions of the configuration.[8]

NOTE: (*Rectangular Elements*) For domains whose boundaries are made up of only vertical and horizontal segments, rectangular elements are often a popular choice for the FEM. A typical rectangular element is illustrated in Figure 13.22(a). If we use just the four vertices as the nodes of each rectangular element, then each local basis function has four degrees of freedom, so linear functions (whose graphs are planes) are no longer permissible to use as local basis functions. Popular choices for basis functions in this case are piecewise bilinear functions:

$$axy + bx + cy + d.$$

These functions reduce to linear functions on any of the four edges of the rectangle so that continuity is assured across boundaries when elements are put together. Note that this would not be the case if the element were an arbitrary quadrilateral (if not all four sides are parallel to one of the axes). The next four exercises look more closely into rectangular elements.

---

[8] These two triangulations were created by Tim Barth (at the NASA Ames Research Center) and we thank him for his kind permission to include them in this text. Such triangulations of airfoils coupled with the FEM are used to model aerodynamics and design space and air vessels.

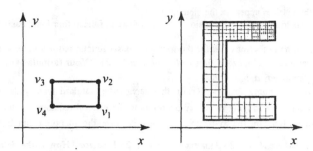

FIGURE 13.22: (a) (left) Illustration of a typical rectangular element with its four nodes consisting of its vertices. (b) (right) Tessellation of a domain into rectangular elements.

16. For the domain in Figure 13.22(b), let the outer vertices be $(1,1)$, $(7,1)$, $(7,3)$, $(3,3)$, $(3,6)$, $(7,6)$, $(7,8)$, and $(1,8)$. Tessellate the domain with square elements having unit sidelength (so there should be 30 elements).
    (a) Write down a formula for the basis function $\Phi_{(2,2)}(x,y)$ corresponding to the interior node $(2,2)$.
    (b) Use MATLAB to draw a three-dimensional graph of this basis function.
    (c) Repeat parts (a) and (b) for the basis function $\Phi_{(1,1)}(x,y)$ corresponding to the interior node $(1,1)$.
    (d) Are these basis functions differentiable (smooth) across all edges of adjacent elements? (It was already pointed out that they are continuous across edges, and this should be evidenced from the graphs.)

17. Let the domain in Figure 13.22(b) have the vertices and tessellation of the last exercise. On this domain, consider the following function:

$$f(x,y) = \begin{cases} [(x-1)/2]^2, & \text{if } y \le 3, \\ [(x-1)/2]^2(2/3)|y-9/2|, & \text{if } 3 \le y \le 9/2, \\ -[(x-1)/2]^2(2/3)|y-9/2|, & \text{if } 9/2 \le y \le 6, \\ -[(x-1)/2]^2, & \text{if } y > 6, \end{cases}$$

(a) Use MATLAB to draw a three-dimensional graph of this function.
    (b) Use MATLAB to draw a three-dimensional graph of the finite element interpolant to this function using the basis functions (Exercise 16) for the square elements of the tessellation. Note that this approximation is simply the function:

$$f(1,1)\Phi_{(1,1)}(x,y) + f(1,2)\Phi_{(1,2)}(x,y) + \cdots,$$

where each term of the sum corresponds to a node of the tessellation.
    (c) Create and plot a corresponding approximation to $f(x,y)$ that arises from the triangulation of the domain using 60 triangles, each square element giving rise to two triangular elements via the diagonal from lower left to upper right.
    (d) Repeat part (b) except this time use squares of sidelength 1/4 in the tessellation. (So there will be 16 times as many elements.)
    Suggestion: In parts (b) and (d), use the `meshgrid` command for each element and use the `hold on` command.

18. The **standard rectangular element** has vertices $(\pm 1, \pm 1)$. (a) Show that the corresponding four local basis functions (viz. (7)) are given by the following formulas (the ordering of the nodes is as in Figure 13.22a):

$$\rho_3(x,y) = (1/4)(1-x)(1+y), \quad \rho_2(x,y) = (1/4)(1+x)(1+y),$$
$$\rho_4(x,y) = (1/4)(1-x)(1-y), \quad \rho_1(x,y) = (1/4)(1+x)(1-y).$$

(The local basis function ρ_i corresponds to the vertex v_i and they are written with the same orientation as the vertices appear in the element.)

(b) Use MATLAB to draw three-dimensional graphs of each of these four local basis functions.

19. (a) Find formulas, as in Exercise 18, for the four local basis functions for a general rectangular element with vertices: $(a,b), (a+h,b), (a+h,b+k), (a,b+k)$. Your formulas will depend, of course, on the parameters a, b, h, and k.

(b) Find an affine mapping $(x,y) = F(\tilde{x}, \tilde{y})$ that carries the standard rectangular element of Exercise 18 (thought of as lying in the $\tilde{x}\tilde{y}$-plane) onto the general rectangular element of part (a) (thought of as lying in the xy-plane). In matrix form the mapping can be written as $\begin{bmatrix} x \\ y \end{bmatrix} = A \begin{bmatrix} \tilde{x} \\ \tilde{y} \end{bmatrix} + v$, where A is a 2×2 matrix and v is a 2×1 vector. How is the determinant of the matrix A related to the areas of the two rectangular elements?

Suggestion: For part (a), try first by trial and error for some simple specific parameters; let them get more general and look for patterns. For example, you might start with $a = -1$, $h = 2$, $b = -1$, $k = 1$. These parameters are very close to those of the standard element with one difference (h is 2 instead of 1). Next try changing h to 3, keeping all else fixed. Then use $a = -1$, $h = 1$, $b = -1$, $k = 2$; finally change a and b to other values, etc. Alternatively, part (a) can be done quite elegantly using part (b). See Exercise 23 for the relevant idea.

20. We define a planar domain Ω to be *horizontally blocklike* if it has the following form:
$$\Omega = \{(x,y): a \le x \le b, \ 0 \le y \le \lambda(x)\},$$
where $\lambda(x)$ is a positive-valued step function on $a \le x \le b$, i.e., there is a partition of $a \le x \le b$: $a = a_0 < a_1 < a_2 < \cdots < a_{n+1} = b$ into $n+1$ subintervals $I_i = [a_{i-1}, a_i]$ $(1 \le i \le n+1)$ and corresponding positive numbers c_i $(1 \le i \le n+1)$, such that $\lambda(x)$ can be written as: $\lambda(x) = c_i \Leftrightarrow x \in I_i$ $(1 \le i \le n+1)$. A typical horizontally blocklike domain is illustrated in Figure 13.23.

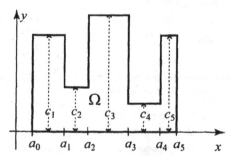

FIGURE 13.23: Illustration of a typical horizontally blocklike domain Ω for Exercise 20.

(a) Write an M-file, $[x\ y\ \text{nodes}\ \text{elems}] = \text{recttess\_hbd\_basic}(a,c,h)$, that will perform a basic rectangular tessellation of a horizontally blocklike domain in the following fashion. The input parameters are firstly two vectors a and c that contain the defining parameters of the horizontally blocklike domain to be tessellated. It is assumed of course that c has one less component than a, the components of a are increasing, and the components of c are positive (otherwise they would not define a horizontally blocklike domain).

The final input variable, h, will be the (approximate) sidelength of each of the rectangles used in the tessellation. More specifically, each of the elements (rectangles) in the tessellation should have its length (l) and width (w) lying within the interval: $\frac{1}{2}h \le w, l \le 2 \cdot h$. The tessellation will be a basic one in the sense that it will be completely determined by a single set of horizontal and vertical grid lines. (Note: This is not the case for the tessellation of Figure 13.22(b), since different sets of vertical lines are used for the grids in the upper and lower passages.) The first two output variables x and y are vectors of the values of the corresponding vertical lines and horizontal lines defining the tessellation. The third output variable, nodes, is a 2-column matrix giving all of the nodes of the tessellation. The fourth and final output variable, elems, is a 4-column matrix giving the node numbers of the each of the elements, where the ordering of the elements starts at the lower left, moves all the way up, then back down to the bottom to the next element on the right, and so on.

(b) Run your program using the following sets of input variables:

(i) $a = [1\ 3\ 4\ 7\ 9\ 10], c = [8\ 4\ 12\ 2\ 10], h = 3,$

(ii) $a = [1\ 3\ 4\ 7\ 9\ 10], c = [8\ 4\ 12\ 2\ 10], h = 1,$

(iii) $a = [1\ 3\ 4\ 7\ 9\ 10], c = [8\ 4\ 12\ 2\ 10], h = 0.13,$

and for each of these for which a tessellation is created, plot the tessellation. For (iii), your plot should look like the one shown in Figure 13.24.

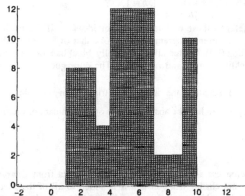

FIGURE 13.24: A tessellation of a horizontally blocklike domain obtained using a program of Exercise 16 using the input data of part (b)(iii). There are 4694 nodes and 4432 rectangular elements.

Suggestions: Part (a): First use the sort command to create a vector cvals of the values of c (in increasing order), with zero appended as the smallest value. Now find the minimum gaps occurring in the vectors cvals and a. The inputted sidelength should not be too small relative to the smaller of these two gaps. If the sidelength exceeds, say, twice this minimum gap value, have the function exit with an error flag (and no tessellation). Now move on to defining a vector x for the vertical gridlines of the tessellation. Use a loop, running through each of the gaps determined from the values of a. If the size of a certain gap is less than twice the sidelength, let x simply contain the values of a at the ends of this gap (no interior grid values); otherwise, use $k - 1$ interior and equally spaced gridlines within the gap, where k=ceil(gap/sidelength). (You need to verify that this will result in elements having horizontal sidelengths within the desired bounds.) In a similar fashion, define a vector y for the horizontal gridlines of the tessellation. Next, use the vectors a, c, x, and y to define the matrix nodes. This can be done with a double loop, but note that you will need to set it up so the larger c value is used in cases where x(i) lies on an interface of two blocks. Finally use the vectors x, y and the matrix nodes to create the matrix elems of elements. You should set things up so that for a given element (row of the elems matrix) the nodes progress, say, counterclockwise, around the element. With this being done, plotting of the tessellations (part (b)) can easily be accomplished using the following simple loop:

```
hold on, [el  e2]=size(elems);
for i=1:el
R=nodes(elems(i,:),:);
xr=R(:,1); xr(5)=xr(1);yr=R(:,2); yr(5)=yr(1);
plot(xr,yr)
end
```

MATLAB's find command can be useful for many parts of this program.

21. (a) Referring to Exercise 20, formulate the definition of the corresponding concept of a *vertically blocklike domain*.
(b) Write an M-file:

$$\texttt{[x y nodes elems]=recttess\_vbd\_basic(a,c,sidelength)}$$

that will perform a basic rectangular tessellation of a vertically blocklike domain in a similar syntax and fashion to the program of part (a) of Exercise 16. Here, the input variable a is an increasing vector corresponding to the y-values of endpoints of the blocks, and the vector c (length one less than y) gives the corresponding horizontal lengths of the blocks.
(c) Run your program using the following sets of input variables:
(i) $a = [2\ 3\ 5\ 6\ 8\ 10]$, $c = [6\ 8\ 7\ 4\ 2]$, $h = 3$,
(ii) $a = [2\ 3\ 5\ 6\ 8\ 10]$, $c = [6\ 8\ 7\ 4\ 2]$, $h = 1$,
(iii) $a = [2\ 3\ 5\ 6\ 8\ 10]$, $c = [6\ 8\ 7\ 4\ 2]$, $h = 0.13$.
Suggesions: Refer to those of Exercise 20 for ideas. If the reader has already completed Exercise 20(a), the current program could invoke that of Exercise 20 along with a rotation of axes (viz. Section 7.2). After all, a vertically blocklike domain is simply a rotation of a horizontally blocklike domain and vice versa. The same goes for corresponding tessellations.

22. Prove identity (5) equating the area of a triangle in the plane with vertices: (x_r, y_r), (x_s, y_s), and (x_t, y_t) to half the absolute value of determinant of the matrix

$$M = \begin{bmatrix} x_r & y_r & 1 \\ x_s & y_s & 1 \\ x_t & y_t & 1 \end{bmatrix}.$$

Suggestion: First use some properties of determinants from Chapter 7 to observe that the determinant will not change if a constant X is added to all of the x-coordinates (first column) and/or a constant Y is added to all the y-coordinates (second column). The corresponding effect on the triangle is simply a shift, leaving the area unchanged. Thus, we may assume that the triangle lies in the first quadrant. Furthermore, reduce to a configuration such as that shown in Figure 13.25. Express the area of the triangle shown as the difference of the sum of the areas of the two trapezoids between the top two edges of the triangle and the x-axis, less the area of the trapezoid between the bottom edge of the triangle and the x-axis. Compare this expression with the det(M). Recall that the area of a trapezoid with base b and heights h_1, h_2 equals $(b/2)(h_1 + h_2)$.

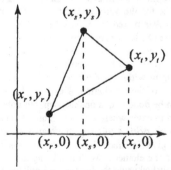

FIGURE 13.25: Geometric diagram for the proof in Exercise 22.

23. To gain a deeper understanding of elements, it is often convenient to work with a so-called **standard element**, which is essentially equivalent to all elements. For our triangular elements, with three nodes at the vertices, we will use the standard element $\tilde{T}$ that has vertices $v_1 = (1,0)$,

$v_2 = (0,1)$, and $v_3 = (0,0)$. This standard element is illustrated in Figure 13.26.

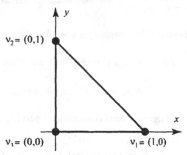

FIGURE 13.26: Illustration of the standard element for all triangular elements with three nodes at the vertices.

(a) Show that the standard local basis functions (viz. (7)) for the standard element of Figure 13.26 are given by:

$$\phi_1(x, y) = x, \ \phi_2(x, y) = y, \ \text{and} \ \phi_3(x, y) = 1 - x - y.$$

(b) For any triangular element T with specified vertices $v_1 = (x_r, y_r)$, $v_2 = (x_s, y_s)$, and $v_3 = (x_t, y_t)$ (labeled in counterclockwise order), show that the following affine mapping $(x, y) = F(\tilde{x}, \tilde{y})$ (see Section 7.2) will transform the standard basis element $\tilde{T}$ onto T and map corresponding nodes onto one another:

$$x = (x_r - x_t)\tilde{x} + (x_s - x_t)\tilde{y} + x_t, \text{ or } \begin{bmatrix} x \\ y \end{bmatrix} = \begin{bmatrix} x_r - x_t & x_s - x_t \\ y_r - y_t & y_s - y_t \end{bmatrix} \begin{bmatrix} \tilde{x} \\ \tilde{y} \end{bmatrix} + \begin{bmatrix} x_t \\ y_t \end{bmatrix}.$$
$$y = (y_r - y_t)\tilde{x} + (y_s - y_t)\tilde{y} + y_t,$$

For clarity, we have used two different sets of coordinates, $(\tilde{x}, \tilde{y})$ for the coordinates of the plane for $\tilde{T}$ and (x, y) for the coordinates of the plane of T. The action of this affine transformation is illustrated in Figure 13.27.

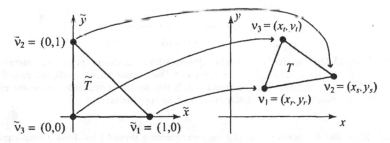

FIGURE 13.27: Illustration of the action of the affine mapping of Exercise 23(b) that takes the standard element $\tilde{T}$ onto an arbitrary element T.

(c) Discover and then prove a relationship for the determinant of the 2×2 matrix of the affine transformation of part (b) and the areas of the two elements T, $\tilde{T}$. If you are not sure of the relationship, do some experiments using MATLAB. For the proof, cf., Exercise 22.

(d)   Writing $A = \begin{bmatrix} x_r - x_t & x_s - x_t \\ y_r - y_t & y_s - y_t \end{bmatrix}$ for the matrix of the affine transformation of part (b)

observe first that the inverse affine mapping is given by: $(\tilde{x}, \tilde{y}) = F^{-1}(x, y) = A^{-1} \cdot \left(\begin{bmatrix} x \\ y \end{bmatrix} - \begin{bmatrix} x_t \\ y_t \end{bmatrix} \right)$,

and show that the standard local basis elements for T are related to those (of part (a)) for $\tilde{T}$ as follows:

$$\phi_i(x, y) = \tilde{\phi}_i(F^{-1}(x, y)), \quad i = 1, 2, 3.$$

Note: Here we have used the notational convention of part (b), so that the $\tilde{\phi}_i$'s are the standard local basis elements corresponding to $\tilde{T}$, while ϕ_i are those corresponding to T.   To prove this relation one simply needs to observe that both sides are linear functions of (x,y) and compare them on the vertices v_1, v_2, v_3.

24.   (*Quadratic Basis Functions on Triangular Elements*) For some BVPs it is desirable to use basis functions Ψ_j which are piecewise quadratic rather than the piecewise linear basis functions that were used in the text. Thus, on each element T, such a basis function will have its general formula written as:

$$\Psi_j(x, y) = ax^2 + bxy + cy^2 + dx + ey + f,$$

where, in order to simplify notation, we have omitted subscripts and superscripts on the six coefficients a, b, c, d, e, f. Since we now have six local basis functions (for each term), we will need to correspondingly have six nodes on each element in order that the coefficients be uniquely determined. A very natural (and as it turns out effective) way to do this is to put three extra nodes on the midpoints of each of the edges of the elements. The corresponding standard element (see Exercise 23), is shown in Figure 13.28.

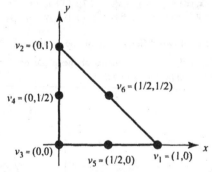

FIGURE 13.28: The standard triangular element with six nodes for piecewise quadratic FEM. The three additional nodes from the piecewise linear standard elements are placed at the midpoints of the segments, the numbering is as before for the vertex nodes, while the midpoint nodes are numbered counterclockwise in order of the opposite vertices.

(a)   Show that (by analogy with (7)) the corresponding standard local basis functions for the standard element of Figure 13.26 are given by:

$$\psi_1(x, y) = \phi_1(x, y) \cdot (2\phi_1(x, y) - 1),$$
$$\psi_2(x, y) = \phi_2(x, y) \cdot (2\phi_2(x, y) - 1),$$
$$\psi_3(x, y) = \phi_3(x, y) \cdot (2\phi_3(x, y) - 1),$$
$$\psi_4(x, y) = 4\phi_2(x, y) \cdot \phi_3(x, y),$$
$$\psi_5(x, y) = 4\phi_1(x, y) \cdot \phi_3(x, y),$$
$$\psi_6(x, y) = 4\phi_1(x, y) \cdot \phi_2(x, y),$$

where the ϕ_j's are the piecewise linear standard local basis functions of Exercise 23(a).

(b) Do the six identities of part (a) continue to remain valid when the local basis functions correspond to an arbitrary element?

(c) Show that the affine mapping $(x,y) = F(\tilde{x},\tilde{y})$ of Exercise 23(b) maps the standard triangular element of Figure 13.28 (viewed in the $\tilde{x}\tilde{y}$-plane) onto the corresponding triangular element with midpoint nodes (viewed in the xy-plane) such that the node correspondence is maintained.

(d) Now letting $\tilde{\psi}_j(x,y)$ $j = 1,...,6$, denote the standard local basis functions of part (a) and $\psi_j(x,y)$ denote the corresponding local basis functions for an arbitrary element, prove that

$$\psi_i(x,y) = \tilde{\psi}_i(F^{-1}(x,y)), \quad i = 1,...,6, \quad \text{where } F \text{ is the affine mapping of part (c).}$$

25.   (*Quadratic Basis Functions on Triangular Elements, Cont*)   Let $\psi(x,y) = ax^2 + bxy + cy^2$ $+ dx + ey + f$ be a quadratic function on a triangular element T with six nodes (vertices and midpoints). Assume that $\psi(x,y) = 0$ on an entire line segment of T that is opposite to vertex v_j. Prove that $\psi(x,y)$ can be factored as $\psi(x,y) = \phi_j(x,y) \cdot \phi(x,y)$ where $\phi_j(x,y)$ is the standard linear local basis function for the vertex v_j and $\phi(x,y)$ is some other linear function $\phi(x,y) = ax + by + c$.

Suggestion: First prove this for the case of the standard element and $j = 2$, then use affine mapping (see Exercises 23 and 24) to translate your proof to work for a general triangular element.

26.   (*Quadratic Basis Functions on Triangular Elements, Cont.*) (a) Use MATLAB to create three-dimensional plots of each of the standard local basis functions ψ_j ($j = 1,...,6$) for the standard element of Figure 13.28. The graphs of two of these functions are roughly depicted in Figure 13.29.

Suggestion: One way to get high-resolution plots over such a triangular region is to triangulate it into much smaller elements and then use the `trimesh`.

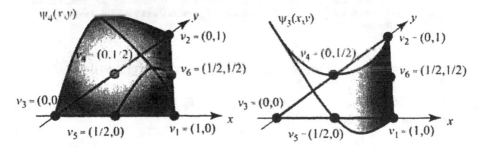

FIGURE 13.29: Graphical illustration of two of the quadratic local basis functions for the standard triangular element of Figure 13.28.

(b) Write down a formula for the (nonlocal) basis function $\Psi_4 = \Psi_4(x,y)$ corresponding to the interior node #4 of the triangulation of Figure 13.4 using the nodal parameters given below Figure 13.5.

Suggestion: The midpoint nodes will need to be numbered. This can be done systematically as in Example 13.1, but the linear systems will of course be larger.

(c) Use MATLAB to plot the (nonlocal) basis function $\Psi_4 = \Psi_4(x,y)$.

(d) Repeat parts (b) and (c) for the midpoint node between the numbered nodes 4 and 5.

27.   Given any finite set $P = \{p_1, p_2, \cdots, p_n\}$ of distinct points in the plane, show that each of the
corresponding Voronoi boxes $V(p_i)$ is a convex set.
Suggestion: Observe that a Voronoi box is an intersection of half-planes.

28.   As mentioned in the text, when a triangulation is created for a given domain to use in the FEM,
it is usually desirable to have the angles of each of the elements not get too small. In this
exercise you will be creating an M-file that will be able to perform a check for this on a given
triangulation and locate any "problem elements."
(a)   Write an M-file, called theta = minangle(v1, v2, v3) whose three input
variables v1, v2, v3 are 2×1 matrices giving the coordinates of the three vertices of a triangle
in the xy-plane, and whose output theta is the smallest (interior) angle of this triangle,
measured in degrees.
(b)   Write an M-file called theta = minanglemesh(x,y,tri) whose inputs are two
vectors x, y of the same size giving the coordinates of the nodes of a triangulation, and tri, a
3-column matrix having as its rows the node numbers of the elements in the triangulation. The
output theta will be the minimum angle (measured in degrees) of any angle of any element of
the triangulation. Run this program on the triangulations of Figure 13.5 (with parameters given
in the matrices N and T preceding Example 13.1), as well as each of the triangulations created in
Example 13.2 of the unit disk.
(c)   Write an M-file called [badelems, thetas] = minanglemesh(x,y,tri,tol)
that, along with the input variables of part (b), has the additional input variable tol that will be
a positive number denoting the smallest desired angle (measured in degrees) to be tolerated in a
triangulation.   There are two output variables: badelems, which will give the element
numbers (corresponding to row numbers of tri) whose minimum angles are less than tol, and
thetas, which is a vector of the same size as badelems gives the corresponding offending
minimum angles of the bad elements. Additonally, a graphic will be produced that will graph
only the elements corresponding to badelmens. This will allow for appropriate measures to
be taken to modify the triangulation, if necessary. Run this program on the triangulations of
Figure 13.5 (with parameters given in the matrices N and T preceding Example 13.1), as well as
each of the triangulations created in Example 13.2 of the unit disk, using three different values
for tol for each: The first one chosen so that there are no offending elements, the second one
chosen so that there are a few offending elements (if possible), and the third chosen so there are
a lot of offending elements.

13.3: THE FINITE ELEMENT METHOD FOR ELLIPTIC PDE'S

In this section we will present versions of the FEM for solving the following
general type of BVP on a domain $\Omega \subset \mathbb{R}^2$

$$\begin{cases} \text{(PDE)} \quad -\nabla \cdot (p\nabla u) + qu = f & \text{on } \Omega \\ \text{(BCs)} \qquad\qquad\qquad u = g & \text{on } \Gamma_1 \,. \\ \qquad\qquad\quad \vec{n} \cdot \nabla u + ru = h & \text{on } \Gamma_2 \end{cases} \qquad (10)$$

The data functions: p, q, f, g, r, h are allowed to be functions of (x, y), defined on
their respective indicated sets. The boundary $\partial\Omega$ is decomposed into the portions
Γ_1 and Γ_2, $\partial\Omega = \Gamma_1 \cup \Gamma_2$. On the first portion Γ_1 there are Dirichlet boundary
conditions $u = g$, and on the complementary portion Γ_2 we are assuming

generalized Neumann boundary or **Robin** boundary conditions: $\vec{n} \cdot (p\nabla u) + ru = h$. Here $\vec{n} = \vec{n}(x,y)$ denotes the outward unit normal vector defined on the $\partial\Omega$, and $\nabla u = (\partial u / \partial x, \partial u / \partial y)$ is the gradient of $u(x,y)$. Thus, from multivariable calculus, the dot product $\vec{n} \cdot \nabla u(x,y)$ is just the partial derivative of u in the direction of the outward pointing normal vector $\vec{n} = \vec{n}(x,y)$ at any point (x,y) on the boundary. If $r(x,y) \equiv 0$, the BCs on Γ_2 generalize the usual Neumann boundary conditions.[9] We allow for the possibility that either $\Gamma_1 = \partial\Omega$ (so $\Gamma_2 = \varnothing$) and the boundary conditions are purely of Dirichlet form, or that $\Gamma_2 = \partial\Omega$ with boundary conditions being entirely of Robin form. The PDE in (10) is written in the so-called **divergence form**. This is the most general form for linear elliptic PDEs on which the standard FEM is applicable, and indeed this is the most general elliptic PDE to which MATLAB's symbolic toolbox is applicable. A great many elliptic boundary value problems can be expressed in the form (10). Sometimes, it will be convenient for us to write the PDE in (10) in expanded form:

$$-\partial / \partial x[pu_x] - \partial / \partial y[pu_y] + qu = f \qquad \text{on } \Omega.$$

The reason for the negative signs will become clear once the FEM is introduced. This PDE is the natural generalization to two space variables of the ODEs that were considered in Section 10.5. We begin by outlining the FEM for the BVP (10) in the case of purely Dirichlet boundary conditions (i.e., $\Gamma_2 = \varnothing$). The FEM will look quite similar to the one-dimensional version presented in Section 10.5. The proofs of the underlying results will not be included in this text. They share many common elements with the one-dimensional theory presented in Section 10.5, but for technical reasons, the higher dimensional analogues require some more advanced mathematical machinery (including, for example, some elements of Sobolev spaces). The interested reader can consult one of the following references: [Cia-02], [AxBa-84], [StFi 73], or [Joh 87] for more details on the theory.

In cases of purely Dirichlet boundary conditions and when the data the BVP (10) satisfy: p, q, f are piecewise continuous on Ω, along with the first partial

---

[9] Let us briefly review the physical significance of the three types of BCs in the context of a steady-state heat distribution BVP (a prototypical BVP). The Dirichlet boundary condition $u = g$ means that (on the portion of the boundary where the condition holds) the boundary is being maintained (by some coolant or heat reservoir) at a specified temperature. The Neumann boundary condition $\vec{n} \cdot \nabla u = 0$ means that the boundary is insulated (no heat loss or transfer). The Robin boundary condition (after dividing through by p, which will always be assumed positive): $\vec{n} \cdot \nabla u + ru = h$ when written in the form $\vec{n} \cdot \nabla u = -r(u - \bar{h})$ looks like the usual Newton's law of cooling where the net heat transfer (out of the region) is proportional to the difference of the inside temperature (u) and the outside temperature ($\bar{h}$).

derivatives of p and q, g is piecewise continuous on $\partial\Omega$ and $p(x,y) > 0$, $q(x,y) \geq 0$, the BVP can be shown to be equivalent to the minimization problem:

Minimize the functional:

$$F[u] = \iint_\Omega [\tfrac{1}{2} pu_x^2 + \tfrac{1}{2} pu_y^2 + \tfrac{1}{2} qu^2 - fu]\,dxdy, \tag{11}$$

over the following set of admissible functions:

$$\mathcal{A} = \{v : \Omega \to \mathbb{R} : v(x) \text{ is continuous}, v'(x) \text{ is piecewise continuous}$$
$$\text{and bounded, and } v(x,y) = g(x,y) \text{ on } \partial\Omega\}. \tag{12}$$

The concept of piecewise continuity on Ω (or $\partial\Omega$) simply means that the domain (or boundary) can be broken up into finitely many elements (arcs) on each of which the given function reduces to a continuous function.

Analogous to the one-dimensional method presented in Section 10.5, the FEM will solve a corresponding finite-dimensional minimization problem where the functional $F[u]$ of (11) is kept the same, but the set of admissible functions is reduced to an approximating smaller set that is determined by the basis functions of the triangulation. Thus we will be looking for minimizers of the functional F among functions of the form $v = \sum_{i=1}^{m} c_i \Phi_i$, where the $\Phi_i = \Phi_i(x,y)$ are the basis functions. The basis functions corresponding to nodes on the boundary will have their coefficients determined by the Dirichlet boundary conditions; it is the remaining coefficients (corresponding to interior nodes) that need to be determined. We now briefly outline the FEM for BVPs with purely Dirichlet BCs. We follow this outline with some additional details and then give examples.

FEM FOR THE BVP (10) IN CASE OF PURELY DIRICHLET BC'S ($\Gamma_2 = \varnothing$):

Step #1: **Decompose the domain into elements, and represent the set of nodes and elements using matrices. Separate the nodes N_i into the internal nodes: $N_1, N_2, \cdots, N_n$ (that lie in Ω), and the boundary nodes $N_{n+1}, N_{n+2}, \cdots, N_m$ (that lie on $\partial\Omega$). Denote the basis function Φ_{N_i} corresponding to node N_i simply by Φ_i.**

Step #2: **Use the Dirichlet BCs $u(x,y) = g(x,y)$ on $\partial\Omega$ to determine the coefficients of the boundary node basis functions of an admissible function: $v = \sum_{i=1}^{m} c_i \Phi_i$, i.e., $c_i = g(N_i)$ for each $i = n+1, n+2, \cdots, m$.**

Step #3: Assemble the $n \times n$ stiffness matrix A and load vector b needed to determine the remaining coefficients $c_1, c_2, \cdots, c_n$ that work to solve the discrete minimization problem corresponding to the BVP.

Step #4: Solve the stiffness equation $Ac = b$, and obtain the FEM solution

$$v = \sum_{i=1}^{m} c_i \Phi_i.$$

The first step was examined in detail in the last section for triangular elements with piecewise linear basis functions. Such elements and basis functions are the ones that will be used exclusively in the text of this section. The exercises will consider some other sorts of elements and/or basis functions. Step #2 is rather clear. Step #3 will be accomplished by a so-called **assembly** technique where the entries of the stiffness matrix and load vectors are built by looking at the contributions of each element.

If we substitute the expression $v = \sum_{i=1}^{m} c_i \Phi_i$ for u into the functional $F[u]$, and then differentiate with respect to c_k (under the integral sign), we arrive at the following equation (Exercise 17):

$$\frac{\partial}{\partial c_k} F\left[\sum_{i=1}^{m} c_i \Phi_i \right] = \iint_{\Omega} [p \sum_{i=1}^{m} c_i \partial_x(\Phi_i)\partial_x(\Phi_k) + p \sum_{i=1}^{m} c_i \partial_y(\Phi_i)\partial_y(\Phi_k)$$
$$+ q \sum_{i=1}^{m} c_i \Phi_i \Phi_k - f\Phi_k]dxdy. \tag{13}$$

Keeping in mind that the values of c_k for $k > n$ will have been computed in Step #2, since we seek a critical point of F, we set the above equations equal to zero for $1 \le k \le n$ to obtain the following $n \times n$ linear system for the unknown coefficients:

$$Ac = b, \tag{14}$$

where the c represents the (column) vector of the unknown (internal node) coefficients: $c = [c_1 \quad c_2 \quad \cdots \quad c_n]'$. The entries of the stiffness matrix $A = [a_{ij}]$ are given by (Exercise 17):

$$a_{ij} = \iint_{\Omega} [p\nabla\Phi_i \cdot \nabla\Phi_j + q\Phi_i\Phi_j]dxdy \quad (1 \le i,j \le n), \tag{15}$$

and the entries of the load vector $b = [b_j]$ are given by:

$$b_j = \iint_{\Omega} f\Phi_j \, dxdy - \sum_{s=n+1}^{m} \iint_{\Omega} [p\nabla\Phi_s \cdot \nabla\Phi_j + q\Phi_s\Phi_j]dxdy \quad (1 \le j \le n). \tag{16}$$

We point out that the coefficients c_s ($s > n$) are known from Step #2. Note that from (15) (since the dot product is commutative: $\vec{v} \cdot \vec{w} = \vec{w} \cdot \vec{v}$) it follows that $a_{ij} = a_{ji}$, i.e., the stiffness matrix is a symmetric matrix.

Keeping in mind that each of the basis functions is made up of its linear "pieces" on each of the elements, it is more efficient to compute the stiffness entries a_{ij} and load entries b_j by running through each of the elements and adding up contributions. Assuming that the nodes and elements have been stored in a 2-column matrix N and a 3-column matrix E, respectively (as in the last section, but then we labeled the element matrix as T), we now outline the assembly process:

ASSEMBLY PROCESS FOR THE FEM FOR (10) IN CASE OF PURELY DIRICHELT BC'S ($\Gamma_2 = \varnothing$):

Step #1: Initialize $n \times n$ stiffness matrix A and $n \times 1$ load vector b with all zero entries.
Step #2: Let ℓ run from 1 to $L =$ the number of elements (= number of rows of the matrix E whose ℓth row gives the node numbers of the ℓth element T_ℓ). For each index ℓ, we create the 3×3 element stiffness matrix $A^\ell = [a_{\alpha\beta}^\ell]$ ($1 \leq \alpha, \beta \leq 3$) for the element T_ℓ and the corresponding 3×1 element load vector $b^\ell = [b_\alpha^\ell]$ by restricting the integrals in formulas (15) and (16) from Ω to T_ℓ:

$$a_{\alpha\beta}^\ell = \iint_{T_\ell} [p \nabla \Phi_{i_\alpha} \cdot \nabla \Phi_{i_\beta} + q \Phi_{i_\alpha} \Phi_{i_\beta}] \, dxdy \quad (1 \leq \alpha, \beta \leq 3), \tag{15'}$$

and

$$b_\alpha^\ell = \iint_{T_\ell} f \Phi_{i_\alpha} \, dxdy - \sum_{s=n+1}^{m} c_s \iint_{T_\ell} [p \nabla \Phi_s \cdot \nabla \Phi_{i_\alpha} + q \Phi_s \Phi_{i_\alpha}] \, dxdy \quad (1 \leq \alpha \leq 3). \tag{16'}$$

(Here, the index i_α denotes the global node number corresponding to the αth vertex of T_ℓ, i.e., $i_\alpha = E(\ell, \alpha)$, whereas the local node number α for a vertex of T_ℓ is just the corresponding column number of the index α in the ℓth row of the element matrix E.) We then transplant these contributions into the appropriate places of the (global) stiffness matrix and load vector:

$$A(E(\ell, \alpha), E(\ell, \beta)) = A(E(\ell, \alpha), E(\ell, \beta)) + a_{\alpha\beta}^\ell \quad (1 \leq \alpha, \beta \leq 3), \tag{17}$$

and

$$b(E(\ell, \alpha)) = b(E(\ell, \alpha)) + b_\alpha^\ell \quad (1 \leq \alpha \leq 3). \tag{18}$$

We point out that formulas $(15')$ and $(16')$ need only be carried out when the indices α and/or β correspond to interior nodes.[10] Also, the integrands in summation of $(16')$ will vanish on the element T_ℓ unless the corresponding exterior node (number s) is a vertex of T_ℓ.

We turn now to a simple example involving the Poisson PDE and constant (Dirichlet) boundary conditions on the hexagonal domain of the last section with only eight triangular elements (Figure 13.5). MATLAB will be able to help us with general multiple integrals, and we will explain how this can be done after this introductory example. The integrals that will need to get done in the course of this example will be simple enough to do by hand, and all will be evaluated using the results of the following exercise for the reader:

EXERCISE FOR THE READER 13.5: Let T denote any (convex) triangle in the plane having vertices $v_1 = (x_r, y_r)$, $v_2 = (x_s, y_s)$, and $v_3 = (x_t, y_t)$ (Figure 13.30), and let $\phi = \phi_3$ denote the local basis function for T corresponding to the vertex v_3, i.e., $\phi(x, y)$ is the linear function determined by the equations $\phi(v_i) = \delta_{i3}$ ($i = 1, 2, 3$). Establish the following formulas:

(a) The gradient vector $\vec{\nabla}\phi$ points in the direction of the altitude $\vec{a}$ and has magnitude $\| \vec{\nabla}\phi \| = 1 / \| \vec{a} \|$.

(b) $\iint_T \phi(x, y)\,dxdy = \frac{1}{3}\,\text{Area}(T)$.

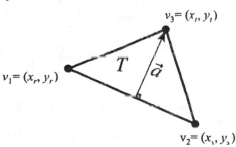

FIGURE 13.30: A typical (convex) triangular element whose local basis function $\phi = \phi_3$ is analyzed in Exercise for the Reader 13.5. Since the element is convex, the (blue) altitude vector $\vec{a}$ shown will lie inside the triangle.

EXAMPLE 13.5: Let Ω be the hexagonal domain of Figure 13.5 with eight nodes (as labeled) given by: #1: (1,1), #2: (2.5,1), #3: (0,0), #4: (1,0), #5:

---

[10] Otherwise the entries are meaningless. Thus, technically, the element stiffness matrices A^ℓ (load vectors b^ℓ) will not be complete $3 \times 3 (3 \times 1)$ matrices in cases where the element T_ℓ has some of its vertices on the boundary (in Example 13.5, this will be the case for all of the elements).

(2.5,0), #6: (3.5,0), #7: (1, −1), and #8: (2.5, −1).   Consider the following
Poisson BVP for this domain:

$$\begin{cases}\text{(PDE)} & -\Delta u = f(x,y) & \text{on } \Omega \\ \text{(BC)} & u = 1 & \text{on } \partial\Omega\end{cases},$$

where the "load" $f(x, y)$, is given by:

$$f(x,y)=\begin{cases}0, & \text{if } x \leq 2.5, \\ -1, & \text{if } x > 2.5.\end{cases}$$

Using the triangulation of Figure 13.5 and the corresponding piecewise linear
basis functions of the last section, apply the FEM to solve this BVP.

SOLUTION: In this problem the BC is purely Dirichlet, so we may follow the
above procedure.

   The numbering of the nodes in Figure 13.5 has one drawback in that it does not
conform to our current notation where the interior nodes are numbered first. We
could redo the numbering to conform but instead will work around the numbering
that was already set up. The corresponding matrices N(odes) and E(lements) are
reproduced here:

$$N = \begin{bmatrix} 1 & 1 \\ 2.5 & 1 \\ 0 & 0 \\ 1 & 0 \\ 2.5 & 0 \\ 3.5 & 0 \\ 1 & -1 \\ 2.5 & -1 \end{bmatrix}, \quad E = \begin{bmatrix} 1 & 3 & 4 \\ 1 & 2 & 4 \\ 2 & 4 & 5 \\ 2 & 5 & 6 \\ 3 & 4 & 7 \\ 4 & 5 & 7 \\ 5 & 7 & 8 \\ 5 & 6 & 8 \end{bmatrix}.$$

Keep in mind that there are $m = 8$ nodes here of which $n = 2$ are interior (nodes #4
and #5).   Thus an admissible function (for the FEM) $v = \sum_{i=1}^{8} c_i \Phi_i$ will have all but
two of the coefficients (c_4, c_5) determined by the Dirichlet boundary conditions.
Since (in the notation of (10)) $g(x, y) = 1$, we have that $c_i = g(N_i) = 1$ for $i \neq 4, 5$,
and so the FEM solution will have form: $v = c_4 \Phi_4 + c_5 \Phi_5 + \sum_{s \neq 4,5} \Phi_s$ and the rest of
the problem is to compute these remaining two coefficients.

We are now at the assembly stage of the FEM. Note that since (in the notation of
(10)), $p = 1$ and $q = 0$, and $c_s = g(N_s) = 1$ ($s \neq 4, 5$), equations $(15')$ and $(16')$
simplify to:

$$a_{\alpha\beta}^{\ell} = \iint_{T_{\ell}} \nabla\Phi_{i_{\alpha}} \cdot \nabla\Phi_{i_{\beta}}\, dxdy \quad (1\le \alpha,\beta \le 3),$$

and

$$b_{\alpha}^{\ell} = \iint_{T_{\ell}} f\Phi_{i_{\alpha}}\, dxdy - \sum_{s\ne 4,5} \iint_{T_{\ell}} \nabla\Phi_s \cdot \nabla\Phi_{i_{\alpha}}\, dxdy \quad (1\le \alpha \le 3),$$

respectively (we have incorporated the change needed to accommodate the node numbering scheme).

We initialize a 2×2 stiffness matrix A of zeros and the corresponding 2×1 initial load vector b and pass now to a detailed calculation of the first iteration of the assembly loop: $\ell = 1$ corresponding to the first element T_1 of Figure 13.5. Figure 13.31 shows this element and its corresponding element stiffness matrix A^1.

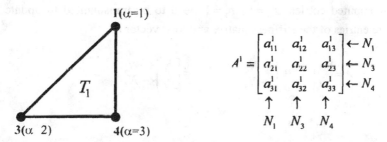

FIGURE 13.31: (a) (left) Illustration of the first element T_1 of Figure 13.5 with the global node numbers (from Figure 13.5) as well as the local node numbers from the matrix T. (b) (right) The corresponding element stiffness matrix A^1 along with a labeling of the corresponding nodes.

Of the nodes for T_1, only node #4 ($\alpha = 3$) is an interior node so we need only compute the single entry:

$$a_{33}^1 = \iint_{T_1} \nabla\Phi_4 \cdot \nabla\Phi_4\, dxdy.$$

From the formula obtained in Example 13.1 for Φ_4, we know that on T_1, $\Phi_4(x,y) = x - y$, so that $\nabla\Phi_4 = (1,-1)$ (this also follows from the preceding exercise for the reader), and $\nabla\Phi_4 \cdot \nabla\Phi_4 = 2$. Consequently,

$$a_{33}^1 = \iint_{T_1} \nabla\Phi_4 \cdot \nabla\Phi_4\, dxdy = \iint_{T_1} 2\, dxdy = 2\cdot \text{Area}(T_1) = 2\cdot(1/2) = 1.$$

Similarly, we have only to compute the single load entry:

$$b_3^1 = \iint_{T_1} f\Phi_4 \, dxdy - \sum_{s=1,3} \iint_{T_1} \nabla\Phi_s \cdot \nabla\Phi_4 dxdy.$$

Since the load $f(x, y)$ vanishes throughout T^1, only the latter two integrals need to be computed. Both integrands are constants and so the integrals can be simply evaluated as the preceding one. We need the gradients of Φ_1 and of Φ_3 on T^1. Using part (a) of the preceding exercise for the reader, we compute $\nabla\Phi_1 = (0,1)$ and $\nabla\Phi_3 = (-1,0)$ and so the corresponding dot products with $\nabla\Phi_4 = (1,-1)$ are both -1. Hence,

$$b_3^1 = -\iint_{T_1} \nabla\Phi_1 \cdot \nabla\Phi_4 dxdy - \iint_{T_1} \nabla\Phi_3 \cdot \nabla\Phi_4 dxdy = -(-\text{Area}(T_1) - \text{Area}(T_1)) = 1.$$

The just-computed entries $a_{33}^1 = 1$, $b_3^1 = 1$ need to be transplanted to update the appropriate entries of the stiffness matrix and load vector b:

$$A = \begin{bmatrix} a_{11} & a_{12} \\ a_{21} & a_{22} \end{bmatrix} \begin{matrix} \leftarrow N_4 \\ \leftarrow N_5 \end{matrix} \qquad b = \begin{bmatrix} b_1 \\ b_2 \end{bmatrix} \begin{matrix} \leftarrow N_4 \\ \leftarrow N_5 \end{matrix}$$
$$\begin{matrix} \uparrow & \uparrow \\ N_4 & N_5 \end{matrix}$$

Since the local index $\alpha = 3$ corresponds to the internal node N_4, the corresponding index for the (global) stiffness matrix and load vector is 1, and we update: $a_{11} = a_{11} + a_{33}^1 = 0 + 1 = 1$, and $b_1 = b_1 + b_3^1 = 0 + 1 = 1$. In summary, after the first iteration of the assembly process ($\ell = 1$), our updated stiffness matrix and load vector are as follows:

$$A = \begin{bmatrix} 1 & 0 \\ 0 & 0 \end{bmatrix}, \quad b = \begin{bmatrix} 1 \\ 0 \end{bmatrix}.$$

The treatment for the next iteration $\ell = 2$ is quite similar since the element T_2 also has one interior node (#4) and two boundary nodes (#1, #2). To prepare for the computations, we note that $\text{Area}(T_2) = 3/4$ and on T_2:

$$\nabla\Phi_1 = (-2/3, 1), \quad \nabla\Phi_2 = (2/3, 0), \quad \nabla\Phi_4 = (0, -1).$$

We have used Exercise for the Reader 13.5. Actually, with less work, the needed gradient vectors here and in all other computations of this example can be gleaned from the explicit formula for Φ_4 obtained in Example 13.1 by comparing relevant triangles.

From the second row of the element matrix E, we see that the three vertices of T_2, nodes #1, #2, and #4, have local node numbers $\alpha = 1, 2,$ and 3, respectively, so that the node correspondence for the element stiffness matrix A^2 is as follows:

$$A^2 = \begin{bmatrix} a^2_{11} & a^2_{12} & a^2_{13} \\ a^2_{21} & a^2_{22} & a^2_{23} \\ a^2_{31} & a^2_{32} & a^2_{33} \end{bmatrix} \begin{matrix} \leftarrow N_1 \\ \leftarrow N_2 \\ \leftarrow N_4 \end{matrix}$$
$$\begin{matrix} \uparrow & \uparrow & \uparrow \\ N_1 & N_2 & N_4 \end{matrix}$$

Since only node N_4 is internal, we need only compute the entry a^2_{33} and the corresponding element load vector entry b^2_3, and since $f(x,y)$ again vanishes on T_2, these computations can be carried out just as before, using the above gradients and area:

$$a^2_{33} = \iint_{T_2} \nabla\Phi_4 \cdot \nabla\Phi_4 \, dxdy = \iint_{T_2} 1 \, dxdy = \text{Area}(T_2) = 3/4,$$

$$b^2_3 = -\iint_{T_2} \nabla\Phi_1 \cdot \nabla\Phi_4 \, dxdy - \iint_{T_2} \nabla\Phi_2 \cdot \nabla\Phi_4 \, dxdy = -(-\text{Area}(T_2) + 0) = 3/4.$$

Transplanting these results into the appropriate places in the stiffness matrix and load vector results in the following updates:

$$A = \begin{bmatrix} 1+3/4 & 0 \\ 0 & 0 \end{bmatrix} - \begin{bmatrix} 7/4 & 0 \\ 0 & 0 \end{bmatrix}, \quad b - \begin{bmatrix} 1+3/4 \\ 0 \end{bmatrix} = \begin{bmatrix} 7/4 \\ 0 \end{bmatrix}.$$

Proceeding now to $\ell = 3$, the situation is a bit different in that the element T_3 has two internal nodes. This will mean that we will need to compute a total of six entries (four for the element stiffness matrix A^3 and two for the corresponding element load vector b^3). We obtain, as before, the area $\text{Area}(T_3) = 3/4$, and the gradient vectors on T_3,

$$\nabla\Phi_2 = (0, 1), \quad \nabla\Phi_4 = (-2/3, 0), \quad \nabla\Phi_5 = (2/3, -1).$$

From the third row of the element matrix E, we see that the three vertices of T_3: nodes #2, #4, and #5 have local node numbers $\alpha = 1, 2,$ and 3, respectively, so that the node correspondence for the element stiffness matrix A^3 is as follows:

$$A^3 = \begin{bmatrix} a_{11}^3 & a_{12}^3 & a_{13}^3 \\ a_{21}^3 & a_{22}^3 & a_{23}^3 \\ a_{31}^3 & a_{32}^3 & a_{33}^3 \end{bmatrix} \begin{matrix} \leftarrow N_2 \\ \leftarrow N_4 \\ \leftarrow N_5 \end{matrix}$$
$$\begin{matrix} \uparrow & \uparrow & \uparrow \\ N_2 & N_4 & N_5 \end{matrix}$$

The computations of the needed entries of A^3 and b^3 are now done just as before. We briefly summarize them:

$$a_{22}^3 = \iint\limits_{T_3}\nabla\Phi_4\cdot\nabla\Phi_4\,dxdy = \frac{4}{9}\cdot\frac{3}{4}=1/3, \quad a_{23}^3 = a_{32}^3 = \iint\limits_{T_3}\nabla\Phi_4\cdot\nabla\Phi_5\,dxdy = -\frac{4}{9}\cdot\frac{3}{4}=-1/3,$$

$$a_{33}^3 = \iint\limits_{T_3}\nabla\Phi_5\cdot\nabla\Phi_5\,dxdy = \frac{13}{9}\cdot\frac{3}{4}=13/12, \quad b_2^3 = -\iint\limits_{T_3}\nabla\Phi_2\cdot\nabla\Phi_4\,dxdy = 0,$$

and $b_3^3 = -\iint\limits_{T_3}\nabla\Phi_2\cdot\nabla\Phi_5\,dxdy = -(-\text{Area}(T_3)) = 3/4.$

Transplanting these results into the appropriate places in the stiffness matrix and load vector results in the following updates:

$$A = \begin{bmatrix} 7/4+1/3 & 0-1/3 \\ 0-1/3 & 0+13/12 \end{bmatrix} = \begin{bmatrix} 25/12 & -1/3 \\ -1/3 & 13/12 \end{bmatrix}, \quad b = \begin{bmatrix} 7/4+0 \\ 0+3/4 \end{bmatrix} = \begin{bmatrix} 7/4 \\ 3/4 \end{bmatrix}.$$

In the next iteration, $\ell = 4$ and $f(x,y)$ no longer vanishes on the element. Since $f(x,y)$ is constant throughout T_4, however, we will still be able to use Exercise for the Reader 13.5 to evaluate the new integral that arises. The nodes of T_4: N_2, N_5, N_6 have local node numbers (from the fourth row of E) $\alpha = 1, 2, 3$, respectively. The needed element area is $\text{Area}(T_4) = 1/2$, and the gradient vectors on T_4:

$$\nabla\Phi_2 = (0, 1), \quad \nabla\Phi_5 = (-1, -1), \quad \nabla\Phi_6 = (1, 0).$$

As only one of the nodes is internal, we have only two entries to compute:

$$a_{22}^4 = \iint\limits_{T_4}\nabla\Phi_5\cdot\nabla\Phi_5\,dxdy = \iint\limits_{T_2}2\,dxdy = 1, \quad \text{and}$$

$$b_2^4 = \iint\limits_{T_4}f\Phi_5\,dxdy - \iint\limits_{T_4}\nabla\Phi_2\cdot\nabla\Phi_5\,dxdy - \iint\limits_{T_4}\nabla\Phi_6\cdot\nabla\Phi_5\,dxdy$$
$$= -\tfrac{1}{3}\text{Area}(T_4) - (-\text{Area}(T_4) - \text{Area}(T_4)) = 5/6.$$

(In the last calculation we use Exercise for the Reader 13.5(b).) The updated stiffness matrix and load vectors now become:

$$A = \begin{bmatrix} 25/12 & -1/3 \\ -1/3 & 13/12+1 \end{bmatrix} = \begin{bmatrix} 25/12 & -1/3 \\ -1/3 & 25/12 \end{bmatrix}, \quad b = \begin{bmatrix} 7/4 \\ 3/4+5/6 \end{bmatrix} = \begin{bmatrix} 7/4 \\ 19/12 \end{bmatrix}.$$

Each of the remaining four iterations is done almost identically to one of the four that has just been done. We summarize each remaining iteration only by the stiffness matrix and load vector updates:

$$\ell = 5: \quad A = \begin{bmatrix} 25/12+1 & -1/3 \\ -1/3 & 25/12 \end{bmatrix} = \begin{bmatrix} 37/12 & -1/3 \\ -1/3 & 25/12 \end{bmatrix}, \quad b = \begin{bmatrix} 7/4+1 \\ 19/12 \end{bmatrix} = \begin{bmatrix} 11/4 \\ 19/12 \end{bmatrix}$$

$$\ell = 6:$$
$$A = \begin{bmatrix} 37/12+13/12 & -1/3-1/3 \\ -1/3-1/3 & 25/12+1/3 \end{bmatrix} = \begin{bmatrix} 25/6 & -2/3 \\ -2/3 & 29/12 \end{bmatrix}, b = \begin{bmatrix} 11/4+3/4 \\ 19/12 \end{bmatrix} = \begin{bmatrix} 15/4 \\ 19/12 \end{bmatrix}$$

$$\ell = 7: \quad A = \begin{bmatrix} 25/6 & -2/3 \\ -2/3 & 29/12+3/4 \end{bmatrix} = \begin{bmatrix} 25/6 & -2/3 \\ -2/3 & 19/6 \end{bmatrix}, \quad b = \begin{bmatrix} 15/4 \\ 19/12+3/4 \end{bmatrix} = \begin{bmatrix} 15/4 \\ 7/3 \end{bmatrix}$$

and finally,

$$\ell = 8: \quad A = \begin{bmatrix} 25/6 & -2/3 \\ -2/3 & 19/6+1 \end{bmatrix} = \begin{bmatrix} 25/6 & -2/3 \\ -2/3 & 25/6 \end{bmatrix}, \quad b = \begin{bmatrix} 15/4 \\ 7/3+5/6 \end{bmatrix} = \begin{bmatrix} 15/4 \\ 19/6 \end{bmatrix}.$$

With the stiffness matrix and load vector now "assembled," the remaining coefficients c_4, c_5 are simply the solutions of the linear system:

$$Ac = b \Leftrightarrow \begin{bmatrix} 25/6 & -2/3 \\ -2/3 & 25/6 \end{bmatrix}\begin{bmatrix} c_4 \\ c_5 \end{bmatrix} = \begin{bmatrix} 15/4 \\ 19/6 \end{bmatrix} \Rightarrow \begin{bmatrix} c_4 \\ c_5 \end{bmatrix} = \begin{bmatrix} 1277/1218 \\ 565/609 \end{bmatrix}.$$

With this small system (solved on MATLAB) exact arithmetic was feasible. The FEM solution $v = c_4\Phi_4 + c_5\Phi_5$ can now be plotted quite easily using the trimesh command as in the last section. We need to make sure we have the node matrix N and the element matrix E stored, and then assign the values for c_4, c_5 to nodes #4, #5 and values of one for the remaining nodes (from the Dirichlet BCs):

```
>> N=[1 1;5/2 1;0 0;1 0;5/2 0;7/2 0;1 -1;2.5 -1];
>> E=[1 3 4;1 2 4;2 4 5;2 5 6;3 4 7;4 5 7;5 7 8;5 6 8];
>> x=N(:,1); y=N(:,2);
>> z=ones(8,1); z(4)= 1277/1218; z(5)= 565/609;
>> trimesh(E,x,y,z)
>> hidden off, xlabel('x-values'), ylabel('y-values')
```

The resulting plot is shown in Figure 13.32.

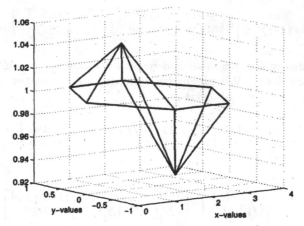

FIGURE 13.32: Plot of our first FEM solution to the BVP of Example 13.5. Only 8 elements and 2 internal nodes were used, so the plot is rather coarse.

EXERCISE FOR THE READER 13.6: If in the BVP of Example 13.5 we change the BC to $u \equiv 2$ on $\partial\Omega$, but leave all else the same, how would the exact solution of this modified problem compare with that of the original? Perform the FEM on this modified problem (with the same triangulation) and compare the numerical solution with that of the original problem.

   The resolution used in the last example was made deliberately coarse so that we could focus on the various facets of the FEM. We now move on to apply the FEM to a problem with a much more elaborate triangulation of the domain. The added complexity will force us to write some MATLAB loops to make the FEM feasible. The BVP we choose, the Laplace equation with Dirichlet boundary conditions on the unit disk, is rather special in that an explicit solution is available. We will thus be able to compare our FEM solution with the exact solution. Such examples are important as an aid for creating and testing production-level FEM codes. We state as a theorem this beautifully explicit result due to Poisson.[11]

---

[11] After his secondary education, Siméon-Denis Poisson went to work as a surgeon's apprentice with an uncle in Fontainbleau, a small city not far from Paris. His lack of coordination forced him to abandon his pursuit of this profession and he subsequently went to the local École Central for undergraduate studies in search of a new career. His mathematical ability was noticed by his instructors who encouraged him to take the entrance exams at the premiere École Polytechnique in Paris. Despite his relatively minor training, he placed at the very top and was admitted in 1798. His talents were quickly noticed and further cultivated by his teachers Laplace and Lagrange. Although his lack of manual dexterity precluded him from doing well in certain subjects (such as descriptive geometry), he excelled in subjects where drawing diagrams was not needed and at age 18 wrote a seminal memoir on finite differences which was well received. After graduation from École Polytechnique he was offered a position there, a rare honor which he accepted. He spent the remainder of his career there and led a very productive life of contributions both to mathematics and physics. He cared deeply for mathematics and for maintaining the quality and sanctity of the École Polytechnique. He was able to stop a group of politically active students at the École from publishing a lampooning attack on Napoleon's leadership, fearing that this could do harm to the École. He was elected to the physics section of the prestigious national Institute (a corresponding position in the mathematics section was

Figure 13.33: Siméon-Denis Poisson (1781–1840), French mathematician.

THEOREM 13.1: (*Poisson's Integral Formula*) Suppose that $f(\theta)$ is a continuous function (given in polar coordinates) on the circle $x^2 + y^2 = R^2$ (θ is the polar coordinate angle). If Ω is the disk inside this circle, $\Omega = \left\{ p = (x, y) \in \mathbb{R}^2 : \|p\|_2 < R \right\}$, then the solution of the Dirichlet problem:

$$\begin{cases} \text{(PDE)} & \Delta u = 0 & \text{on } \Omega \\ \text{(BC)} & u(R, \theta) = g(\theta) & \text{on } \partial\Omega \end{cases} \quad (19)$$

is unique and is given by:

$$u(r, \theta) = \frac{R^2 - r^2}{2\pi} \int_0^{2\pi} \frac{g(\phi)\, d\phi}{R^2 - 2Rr\cos(\theta - \phi) + r^2} \quad (20)$$

Here, (r, θ) denotes the polar coordinates of any point inside Ω, ($r < R$).

We omit the proof of this result (an enlightening one can be found in Section 4.6 in the textbook [Ahl-79]). The result and proof actually extends to higher dimensions; see Section 7.5 in [Zau-89] for the three-dimensional analogue. It turns out as well that the result remains valid for more general boundary data $f(\theta)$. For example, if $f(\theta)$ is only piecewise continuous, then (20) will still solve the Dirichlet problem (19), and the solution will be continuous at all points on $\Omega \cup \partial\Omega = \left\{ p = (x, y) \in \mathbb{R}^2 : \|p\|_2 \leq R \right\}$ except at those points on the boundary at which $f(\theta)$ is discontinuous (see again Section 4.6 in the textbook [Ahl-79]). This beautiful formula is one very rare instance where a general BVP has an explicit and practical solution. Recall that solutions of the Laplace PDE in (19) are called harmonic functions (Chapter 11). The BVP (19) can be viewed, for example, as finding the steady-state heat distribution of a circular plate whose temperature on the boundary is maintained with a certain known distribution ($f(\theta)$).

---

not available; due to the limit set on membership a death of a member had to occur for a new slot to open). His name permeates many areas of mathematics and physics, which apart from differential equations (Poisson bracket and integral formulas), include probability (Poisson distribution), harmonic analysis (Poisson·summation formula), and elasticity (Poisson's ratio). During his career he wrote over 300 research papers, but he was known never to work on more than one project at a time. He was extremely methodical and well-organized; if an idea for a new project would cross his mind while working on one paper he would write a brief note about it and place it in his wallet. After finishing one paper, he would then pull out all of the notes from his wallet to decide on the best topic for his next project.

We will be able to use (20) to get MATLAB to run through a sufficiently fine set of nodes in the disk to obtain a plot of the exact solution. The nodes could be chosen to be those used in a FEM approximation so that the errors of the FEM solution could be examined. All of this will be done in Example 13.7.[12]

Example 13.5 was intentionally set up so as to avoid the problem of having to numerically integrate functions of two variables. In more general examples, we will need to show how to deal with such integrals. MATLAB has an integrator to perform double integrals in floating point arithmetic. Such integrals can be time consuming depending on the oscillatory behavior of the integrand. Triangulations can be made finer in parts of the domain where the data functions have larger variations, and thus the integrals become less difficult to evaluate numerically. In practice, however, rather than using general integration programs or symbolic integrators, well-known quadrature approximations are employed. Such approximation schemes take advantage of the special structure of elements to approximate an integral over an element by a certain weighted average among certain special points of the element. We will present both approaches below. The first method will be to use MATLAB's numerical integrator. To facilitate general codes, we will appeal to some of the Symbolic Toolbox capabilities. The second method will utilize special quadrature formulas. The performance accuracy and times of both approaches will be compared and contrasted with an example where the exact solution can be obtained (and in which the FEM integrals will be quite simple). After presenting both methods, we will discuss some of the underlying theory. Particular readers may wish to cover only one method. Readers who do not have access to or wish to avoid using the Symbolic Toolbox may wish either to skip Method 1, or to be prepared to recode those parts of it which appeal to symbolic functionality. In our numerical example (as we will see below), Method 2 ran about 200 times faster than Method 1 and gave the same quality of results. Such results are typical and this is why we recommend Method

---

[12] As an aside, we point out here some related facts. A celebrated result in the theory of complex variables (which can be found in [Ahl-79], the classic treatise on the subject) known as the Riemann mapping theorem, states that any simply connected planar domain $D \subset \mathbb{R}^2$ can be mapped conformally onto the unit disk $U = \left\{ p = (x, y) \in \mathbb{R}^2 : \|p\|_2 < 1 \right\}$. Simply connected means roughly that the domain has no holes inside, i.e., if γ is any closed path in the domain, then the interior of γ contains only points in the domain; see [Ahl-79] for more details. A conformal mapping is a one-to-one function (of two variables) F such that $F(D) = U$. Conformal mappings have the property that they preserve angles and have many beautiful properties (see [Ahl-79]). One particularly useful property of conformal mappings is that they preserve harmonic mappings, i.e., if $u(x, y)$ is a harmonic function on the domain U and $F : D \to U$ is a conformal mapping, then $v = u(F(x, y))$ is a harmonic function on D. This result means that for any simply connected domain in the plane, there is a corresponding Poisson integral formula for solutions of the Dirichlet problem gotten by changing variables to the disk. This is quite a satisfying and complete result, theoretically, at least. The practical problem for a given simply connected domain thus reduces to computing explicity a function which conformally maps it to the disk U. This problem has been extensively studied and there are many situations where the mappings have been found. This approach has led to numerous applications to physical BVPs involving the Laplace equation, including also steady-state fluid flow, and electrostatics. See [BrChSi-03] for more on conformal mapping with an emphasis on applications.

2. We include Method 1 only for comparison purposes; for readers interested in practical codes, it may be skipped altogether.

NUMERICAL APPROXIMATION TO DOUBLE INTEGRALS— METHOD 1: USING MATLAB's NUMERICAL INTEGRATOR dblquad:

MATLAB's numerical integrator for double integrals has a syntax that requires the integration to be performed over a rectangle. We explain its functionality below and then show how it can be adapted to perform integrations over more general regions.

dblquad(fun,xmin,xmax, ymin,ymax) →	Assume that fun is an inline function of x and y.[13] This command will numerically compute the integral $$\int_{x\,min}^{x\,max} \int_{y\,min}^{y\,max} fun(x,y)dydx,$$ using a double iteration with the single variable function integrator quad and with a default tolerance for error being 1e-6. As with quad, the syntax of dblquad requires that we make the integrand fun(x,y) able to input a vector argument for (the first variable) x and return a vector of the same size.
dblquad(fun,xmin,xmax, ymin,ymax, tol,@quadl,p1,p2,...) →	Optional extra inputs: tol allows specification of an error tolerance, @quadl specifies that the more refined quadl integrator be used in the iterations, the last inputs p1, p2, ... represent numerical values to assign in case fun depends on additional parameters: fun = fun(x,y,p1,p2,…).

The following simple example will illustrate the syntax requirement on fun:

To evaluate the integral of x^2y^2 over the rectangle $R=[0,2]\times[1,2]$:

$$\int_R x^2 y^2\,dxdy = \int_0^2\int_1^2 x^2 y^2\,dydx,$$

we could simply enter:

```
>> dblquad(inline('x.^2.*y.^2','x', 'y'), 0, 2, 1, 2)   →ans = 6.2222
```

The vector syntax requirement on the first variable x is automatically satisfied since this variable appears in the single term for the integrand. If, however, we wanted (for testing purposes) to compute the area of the rectangle R, the corresponding command:

```
>> dblquad(inline('1','x', 'y'), 0, 2, 1, 2)
```

---

[13] As usual, if instead "fun" has been stored as an M-file, it should be written with single quotes: dblquad('fun',...) or preceded with the "@" symbol: dblquad(@fun,...).

gives a series of error messages:
```
??? Index exceeds matrix dimensions.

Error in ==> C:\MATLAB6p5\toolbox\matlab\funfun\quad.m
On line 67   -=> if ~isfinite(y(7))
(more...)
```

The syntax can be adjusted accordingly as follows:
```
>> dblquad(inline('1*ones(size(x))','x', 'y'), 0, 2, 1, 2)
→ans = 2
```

which (as we know) gives the correct answer. A similar syntax note was pointed out in Chapter 3 for quad.

In order to use dblquad to integrate over regions other than rectangles the following identity will be useful:

$$\int_T \text{fun}(x,y)dxdy \equiv \int_{xmin}^{xmax} \int_{ylow(x)}^{ytop(x)} \text{fun}(x,y)dydx$$

$$= \int_{xmin}^{xmax} \int_0^1 \text{fun}(x, ylow(x) + u(ytop(x) - ylow(x)))[ytop(x) - ylow(x)]dudx,$$

(21)

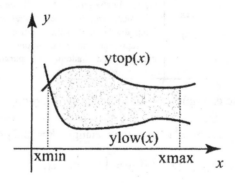

FIGURE 13.34: Illustration of a typical planar region on which integrals can be computed using (21).

Here, the region T need not be a triangle, but rather any region in the plane bounded below by the curve $ylow(x)$ and above by the curve $ytop(x)$ and over the range [xmin, xmax]; see Figure 13.34.

The identity (21) is easily established by a simple variable substitution; see Exercise 20. Using this identity, we may use dblquad to compute any double integral. Since all of our integrals in the text proper of this section will be over triangles, the next example will present some more or less typical evaluations of double integrals over triangles.

EXAMPLE 13.6: Let the triangle T of Figure 13.30 have the following vertices: $v_1 = (1,3)$, $v_2 = (5,1)$, and $v_3 = (4,6)$. Use MATLAB's dblquad to numerically compute the following integrals:

(a) $\int_T 2xy^2 dxdy$

(b) $\int_T \sin(xy\sqrt{y}) dxdy$

In each, decrease the tolerance or change to quadl, as needed, until the answers agree to four decimals.

SOLUTION: We need first to express the integrals as double integrals. Letting x be the outer integration variable, the x-range of T is $1 \le x \le 5$. Over this range, the lower function ylow of x will be the line segment from v_1 to v_2 (see Figure 13.30). Writing this line segment as a function of x yields: $\text{ylow}(x) = -\frac{1}{2}x + \frac{7}{2}$. The corresponding upper function ytop of x splits up into two formulas determined by the two segments $v_1 v_3$ and $v_3 v_2$. Writing each of these segments as a function of x yields the following formula for ytop:

$$\text{ytop}(x) = \begin{cases} x+2, & \text{if } x \le 4 \\ -5x+26, & \text{if } x > 4 \end{cases}.$$

Part (a): Using the above functions, we can rewrite the integral in the following iterated form:

$$\int_T 2xy^2 dxdy = \int_1^5 \int_{\text{ylow}(x)}^{\text{ytop}(x)} 2xy^2 dydx = \int_1^4 \int_{\text{ylow}(x)}^{x+2} 2xy^2 dydx + \int_4^5 \int_{\text{ylow}(x)}^{26-5x} 2xy^2 dydx.$$

The latter form is a more convenient one to implement on MATLAB. The code given below is written in a way that will make it easy to adapt to handle the general computation of such integrals and to this end it is more convenient to use some Symbolic Toolbox capabilities.

```
>> syms x y u
>> ylow = -.5*x+3.5; ytop1=x+2; ytop2=-5*x+26;
>> fun=2*x*y^2;
>> ynew1=ylow+u*(ytop1-ylow);
>> funprep1=subs(fun,y,ynew1)*(ytop1-ylow);
>> ynew2=ylow+u*(ytop2-ylow);
>> funprep2=subs(fun,y,ynew2)*(ytop2-ylow);
>> funnew1=vectorize(inline([char(funprep1),'*ones(size(u))'],...
                                    'u','x'));
>> %we needed to convert the symbolic expression back into a
>> %character string for construction of an inline function.
>> funnew2=vectorize(inline([char(funprep2),'*ones(size(u))'],...
                                    'u','x'));
>> dblquad(funnew1,0,1,1,4)+ dblquad(funnew2,0,1,4,5)
```

→ans = 724.8000

Using a smaller tolerance (than the default 10^{-6}) gives the same result:

```
>> dblquad(funnew1,0,1,1,4,1e-7)+ dblquad(funnew2,0,1,4,5,1e-7)
→ans =724.8000
```

Part (b) Implementing the same strategy, we obtain:

```
>> fun=sin(sin(x)*y);
>> funprep1=subs(fun,y,ynew1)*(ytop1-ylow);
>> funprep2=subs(fun,y,ynew2)*(ytop2-ylow);
>>
funnew1=vectorize(inline([char(funprep1),'*ones(size(u))'],'u','x'))
>>
funnew2=vectorize(inline([char(funprep2),'*ones(size(u))'],'u','x'))
>> dblquad(funnew1,0,1,1,4)+ dblquad(funnew2,0,1,4,5)
→ ans = 0.1397
```

There is agreement when we reduce the tolerance as above.

The numerical integration(s) of part (b), unlike that in part (a), took a noticeable amount of time. This is due to the fact that the integrand in part (b) is very oscillatory over the domain. In general, double integrals can take a lot of work to evaluate effectively since, if the integrals cannot be done explicitly, any method basically has to iterate evaluations of a one-variable integral on numerous slices (the number goes up when more accuracy is desired). When performing the FEM to solve a given BVP, the triangulation can and should be done so as to use smaller elements in areas of high oscillation of the given data. This will assure that the integrals that arise in the assembly process will be numerically quite tame and easy to compute. MATLAB's symbolic integrator int can also be used to evaluate double integrals, and although the syntax is a bit simpler than for dblquad, the M-files we introduce below will help to make dblquad more convenient to use. Also, the extra computing time needed for int to attempt to find exact antiderivatives, which is usually not possible in general, is not worth the occasional extra precision in the answers.

To save on having to go through the above complicated syntax each time a numerical integral is encountered, we give here an M-file that is essentially a user-friendly version of dblquad. It is a simple modification of the code employed in the last example.

PROGRAM 13.1: User-friendly M-file for numerically computing double integrals over planar regions bounded between two functions of x, as in Figure 13.34. Integrand fun is entered as a function of the symbolic variables x and y.

```
function nint= quad2d(fun,xmin,xmax,ylow,ytop)
% numerically computes a double integral of a function 'fun' on a
% region over the interval minx<x<maxx and between the functions of
% x: ylow<y<ytop.
% INPUTS:   fun = a function of the symbolic variables x and y
% minx = minimum x-value for region
% maxx - maximum x-value for region
% ylow = function of symbolic variable for lower boundary of region
% ytop - function of symbolic variable for upper boundary of region
```

```
% OUTPUT:  nint - numerical approximation of integral using the
% integrator 'dblquad' in conjunction with the default settings.
% x and y should be declared symbolic variables before this M-file is
% used.
syms u x y
ynew=ylow+u*(ytop-ylow);
funprep=subs(fun,y,ynew)*(ytop-ylow);
funnew=vectorize(inline([char(funprep),'*ones(size(u))'],'u','x'));
nint = dblquad(funnew,0,1,xmin,xmax);
```

EXERCISE FOR THE READER 13.7: Use the above program to numerically compute the following double integrals:

(a) $\int_S xy^2 \, dxdy$, where S is the circular sector $\{(r,\theta): 0 \le r \le 1, 0 \le \theta \le \pi/4\}$.

(b) $\int_U \exp(1-x^2-2y^2) \, dxdy$, where U is the region enclosed between the curves

$y = e^x$, $y = x^2 - 1$ and $y = 0$.

EXERCISE FOR THE READER 13.8: (a) Write an M-file, integ=triangquad2d (fun, v1, v1, v3) whose inputs are a function of x,y (written as a symbolic expression), and three 2×1 matrices v1, v2, v3 which are vertices (in any order) of triangle in the xy-plane. If we denote this triangle by T, the output integ will be the numerical integral $\int_T \text{fun}(x,y) \, dxdy$, computed with quad2d as in Example 13.6.

(b) Use your function in Part (a) to reevaluate the integrals of Example 13.6, and also to compute the following integrals in which T_1 is the triangle with vertices $(0,0), (6,0), (12,2)$, and T_2 is the triangle with vertices $(1,3), (3,2),$ and $(2,5)$.

(i) $\int_{T_1} 1 \, dxdy = 6$,   (ii) $\int_{T_2} 1 \, dxdy = 5/2$,   (iii) $\int_{T_1} 2x^1 \, dxdy = 504$,   and

(iv) $\int_{T_2} \sin(x^2) \, dxdy \approx -0.2998$

Suggestions: Branch your program off into two cases: Either the triangle has a vertical side or the three x-coordinates of the vertices are distinct. Draw lots of pictures of triangles as you are proceeding.

EXAMPLE 13.7: Consider the Dirichlet problem (19) on the unit disk $\Omega = \{(x,y): x^2 + y^2 < 1\}$,

$$\begin{cases} \text{(PDE)} & \Delta u = 0 & \text{on } \Omega \\ \text{(BC)} & u(1,\theta) = g(\theta) & \text{on } \partial\Omega\text{'} \end{cases}$$

(we put $R = 1$ in (19)), where $g(\theta) = \begin{cases} 2\theta^2, & \text{if } 0 \le \theta \le 2 \\ 8, & \text{if } 2 < \theta \le 3 \\ 0, & \text{if } 3 < \theta \le 2\pi \end{cases}$.

(a) Use the FEM with a triangulation of the disk involving between 50 and 100 nodes deployed on circles of increasing radii but more or less uniformly (as in Method 2 of the solution to Example 13.2(a) of the last section) to solve this BVP and plot the FEM solution.

(b) Use the Poisson integral formula (20) to numerically compute the exact solution at each of the nodes in part (a), and plot it. Compare with the plot obtained in part (a), and compute the maximum error (at the interior nodes).

(c) Repeat both parts (a) and (b), this time using between 500 and 1000 nodes.

SOLUTION: Part (a): The triangulation can be done in exactly the same fashion as was done in Method 2 of part (a) of the solution of Example 13.2 (simply change the value of delta=sqrt(pi/90); everything else is the same). The code is thus omitted here; the nodes were stored in vectors x and y and the triangulation in the matrix tri. The triangulation is shown in Figure 13.35.

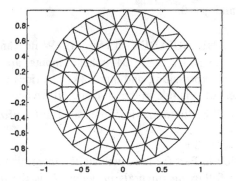

FIGURE 13.35: Triangulation for the FEM solution of Example 13.7(a). There are 99 nodes and 163 triangular elements.

By the way in which the nodes were created, the numbering scheme conforms to that of the procedure outline (the boundary nodes are indexed last). In the notation of the procedure, $m = 99$ (= total number of nodes), as seen by entering size(x). We can use a simple MATLAB loop to compute n (= number of interior nodes):

```
>> n=1;
>> while x(n)^2+y(n)^2<1-eps
     n=n+1;
end
>> n=n-1   →n = 66
```

(Note: We used eps (= machine epsilon) to safeguard the inequality from roundoff errors.) Thus there are $n = 66$ interior nodes.

We now use the boundary data to assign the corresponding coefficients c_i $(i > n)$ of the basis functions for thefor the FEM solution $v = \sum_{i=1}^{m} c_i \Phi_i$. To facilitate this, we will create an M-file for the boundary data function $g(\theta)$. Since the function will eventually need to be integrated (in part (b) when we use the Poisson integral formula), and the function is defined by cases, we will implement the special vector construction for this M-file that was explained in Chapter 4:

```
function y = EX13_7_bdydata(x)
for i = 1:length(x)
if (0<=x(i))&(x(i)<=2)
    y(i)=2*x(i)^2;
elseif (2<x(i))&(x(i)<=3)
    y(i)=8;
else
    y(i)=0;
end
end
```

Now, since the boundary data function is a function of the angle θ, and the nodes are stored as ordered pairs of xy-coordinates, in order to use this function to assign the node coefficients, we must compute and input the corresponding angles for each node. MATLAB has the following built-in functions for such coordinate changes:

`[th,r]=cart2pol(x,y)` →	If (x,y) denote the cartesian coordinates of a point in the plane, the output `[th,r]` will be the corresponding polar coordinates, where the angle `th` is chosen in the interval $(-\pi,\pi]$, and the radius `r` is nonnegative.
`[x,y]=pol2cart(r,th)` →	Inputs a set of polar coordinates (r,th) and outputs the corresponding cartesian coordinates.

The following loop will now store the boundary node coefficients:

```
for i=67:99
  th=cart2pol(x(i),y(i));
  if th<0, th=th+2*pi; end
  %need to ensure th is in domain of boundary data function
  c(i)=EX13_7_bdydata(th);
end
```

We are now ready to move on to the assembly process. We first observe that since (in the notation of (10)), $q \equiv 0$, $f \equiv 0$ and $p \equiv 1$, equations $(15')$ and $(16')$ simplify to:

$$a_{\alpha\beta}^\ell = \iint_{T_\ell} \nabla\Phi_{\iota_\alpha} \cdot \nabla\Phi_{\iota_\beta} \, dxdy \quad (1 \leq \alpha, \beta \leq 3),$$

and

$$b_\alpha^\ell = -\sum_{s=n+1}^{m} c_s \iint_{T_\ell} \nabla\Phi_s \cdot \nabla\Phi_{\iota_\alpha} \, dxdy \quad (1 \leq \alpha \leq 3),$$

respectively. Also, of the 33 possible indices s in the b_α^ℓ formulas, only those (at most two) corresponding to boundary nodes of the element T_ℓ need to be considered.   Since each gradient appearing in the above integrals is of a linear function on an element, the integrands are all constants, and so the corresponding integrals will be simply the constant times the area of the underlying element. We will use the M-file of Exercise for the Reader 13.8 to evaluate each of these integrals (within a loop).

We will make use of MATLAB's `setdiff` built-in function, which was introduced in the last section, but with an optional second output variable.

`[d,ind] = setdiff(a,b)` →	The first output variable was explained in the last section. The optional second output variable will be the indices of a which produce the vector d.

Here is a brief usage example:

```
>> a = [1 2 3];.b = [2 4];
>> d = setdiff(a,b)      →d = 1   3
>> [d,ind] = setdiff(a,b);   ind  →ind = 1   3
>> a = [3 2 1];
>> [d,ind] = setdiff(a,b)  →d = 1   3,  ind = 3   1
```

As usual, we first initialize the $n \times n$ ($n = 66$) stiffness matrix A of zeros and the corresponding $n \times 1$ initial load vector b and create a program that will completely perform the assembly. Here is the complete code for the assembly process for Example 13.7.

```
>> N=[x' y'];
>> E=tri;
>> n=66; m=99; syms x y
>> A=zeros(n); b=zeros(n,1);
>> [L cL]=size(E);
>> for ell=1:L
  nodes=E(ell,:);
  intnodes=nodes(find(nodes<=n));
  bdynodes=nodes(find(nodes>n));
  %find gradients [a b] of local basis functions
  % ax ' by 'c; distinguish between int node
  %local basis functions and bdy node local basis
  %functions

  for i=1:length(intnodes)
   xyt=N(intnodes(i),:); %main node for local basis function
   onodes=setdiff(nodes,intnodes(i));
   %two other nodes (w/ zero values) for local basis function
   xyr=N(onodes(1),:);
   xys=N(onodes(2),:);
   M=[xyr 1;xys 1;xyt 1]; %matrix M of (4)
   abccoeff=[xyr(2)-xys(2); xys(1)-xyr(1); xyr(1)*xys(2)-...
   xys(1)*xyr(2)]/det(M);%coefficents of basis function on triangle#L
   %See formula (6a)
   intgrad(i,:)=abccoeff(1:2)';
  end

  for j=1:length(bdynodes)
   xyt=N(bdynodes(j),:); %main node for local basis function
```

```
  onodes=setdiff(nodes,bdynodes(j)); %two other nodes
  % (w/ zero values) for local basis    function
  xyr=N(onodes(1),:);
  xys=N(onodes(2),:);
  M=[xyr 1;xys 1;xyt 1]; %matrix M of (4)
  abccoeff=[xyr(2)-xys(2); xys(1)-xyr(1); xyr(1)*xys(2)-...
  xys(1)*xyr(2)]/det(M); %coefficents of basis function on
triangle#L
  bdygrad(j,:)=abccoeff(1:2)';
  end

  %update stiffness matrix
  for i1=1:length(intnodes)
  for i2=1:length(intnodes)
  fun = sym(intgrad(i1,:)*intgrad(i2,:)'); %integrand for (15ell)
  integ=triangquad2d(fun,xyt,xyr,xys);
  A(intnodes(i1),intnodes(i2))=A(intnodes(i1),intnodes(i2))+integ;
  end
  end

  %update load vector
  for i=1:length(intnodes)
  for j=1:length(bdynodes)
  fun = sym(intgrad(i,:)*bdygrad(j,:)'); %integrand for part of
(16ell)
  integ=triangquad2d(fun,xyt,xyr,xys);
  b(intnodes(i))=b(intnodes(i))-c(bdynodes(j))*integ;
  end
  end
end
sol=A\b;
c(1:n)=sol';
```

The result is now easily plotted using the `trimesh` function of the last section:

```
>> x=N(:,1); y=N(:,2);
>> trimesh(E,x,y,c), xlabel('x-values'), ylabel('y-values')
>> hidden off
```

The resulting plot is shown in Figure 13.36.

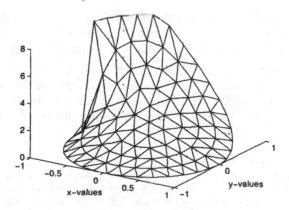

FIGURE 13.36: Plot of the FEM solution of the Dirichlet problem of Example 13.7.

Part (b): The following simple loop will implement the Poisson integral formula (20) to determine the value of the exact solution at each of the interior nodes c_i ($i \leq n$). As has been the convention thus far, we continue to use MATLAB's numerical integrators for one-dimensional integrals; in this case we use `quadl`. We will leave the already assigned values at the boundary nodes c_i ($i > n$). We first create and store an M-file for the integrand in the Poisson integral formula (20) using the boundary data of the current example. Since this function will be integrated (with `quadl`) we will need to construct it as shown in Chapter 4 so that it will appropriately handle vector inputs.

```
function y = EX13_7_poisson(phi,r,th)
for i = 1:length(phi)
if (0<=phi(i))&(phi(i)<=2)
    y(i)=2*phi(i)^2*(1-r^2)/2/pi/(1-2*r*cos(th-phi(i))+r^2);
elseif (2< phi(i))&( phi(i)<=3)
    y(i)=8*(1-r^2)/2/pi/(1-2*r*cos(th-phi(i))+r^2);
else
    y(i)=0;
end
end
```

```
>> cp=c;  %initialize node values for Poisson integral method.
>> for i=1:n
[th, r]=cart2pol(N(i,1),N(i,2)); %polar coors for node #i
cp(i)= quadl(@EX13_7_poisson,0,3,[],[],r,th);
%since integrand vanishes on (3, 2*pi) we can reduce the interval of
%integration.
end
```

The plot of the exact solution just obtained[14] will be quite similar to that of our FEM approximation in Figure 13.36. The resulting error plot is now easily obtained by the following commands, and the plot is shown in Figure 13.37.

```
>> trimesh(E,x,y,abs(c-cp))
>> hidden off
>> xlabel('x-values'), ylabel('y-values')
```

Part (c): The code in parts (a) and (b) is written in a way so that just one small change in one line of the code is required to do part (c). In the creation of the nodes (as in Method 2 of in the solution of Example 13.2(a)) we only need to change the paratmer δ to $\sqrt{\pi}/900$. The resulting node set contains $m = 897$ nodes of which the first $n = 791$ are interior nodes, and the Delaunay triangulation contains 1686 elements. The main loop took close to an hour on the author's computer. See Figure 13.38 for plots of the FEM solution and error.

---

[14] Of course, the Poisson integral formula, as mentioned, is exact. The only errors will be the errors that arise from the numerical integration. By default, the accuracy goal will have error < 1e-6, and the integrand is well-behaved so such errors will not be relevant for our present comparison purposes. In case they do become relevant (with a much finer mesh, say), we could always set a new accuracy goal for `quadl`.

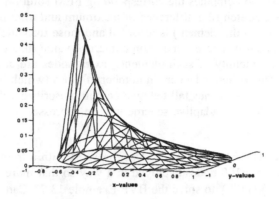

FIGURE 13.37: Plot of the error of the FEM solution of Example 13.7, obtained by comparing it to the exact solution over the same grid from the Poisson integral formula.[15]

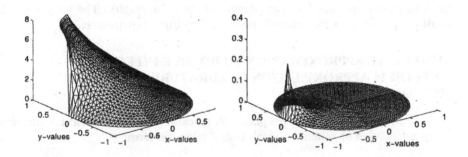

FIGURE 13.38: (a) (left) Plot of the FEM solution of the Dirichlet problem of Example 13.7(b). There are 897 nodes and 1686 triangular elements. (b) (right) Plot of the corresponding error. Notice that in the solution there is one distinguished element (near $(x, y) = (\cos(3), \sin(3))$) whose z range stretches all the way from 0 to the maximum value of 8. This is inevitable since the boundary data has a jump discontinuity from $z = 8$ to $z = 0$ at $(x, y) = (\cos(3), \sin(3))$. The error, although quite small over most of the domain, has a distinguished spike near $(x, y) = (\cos(3), \sin(3))$.

Looking at the errors of the last two solutions, we would be led to conjecture that better FEM solutions could be obtained (for the same numbers of nodes) if we were to concentrate more nodes near the boundary point $(x, y) = (\cos(3), \sin(3))$ at which there is a jump discontinuity. The next exercise for the reader will explore this. This example also motivates the concept of adaptive methods for FEM. One scheme for such a method begins with a more or less uniform node distribution

---

[15] More preciselely, this plot is the difference between the FEM solution and the piecewise linear interpolant of the exact solution.

(and triangulation) and computes the corresponding FEM solution of a BVP. For each element, the z-stretch (the difference of maximum and minimum z-values of FEM solution over just the element) is recorded and those for which this stretch is in, say the largest 10% or exceeds a certain numerical value (this can be adjusted) are flagged. In the vicinity of such elements, extra nodes are added and a new mesh is created. This is iterated a certain number of times (which can be adjusted) or until the maximum z-streches fall below a certain prescribed value (which also can be adjusted). Such an adaptive scheme will be addressed in the exercises of this section.

EXERCISE FOR THE READER 13.9: Use the FEM with a triangulation of the disk involving roughly 100 nodes deployed in a way so that more nodes are used near $(x, y) = (\cos(3), \sin(3))$ to solve the BVP Example 13.7. Can you triangulate in such a way that the maximum error is smaller than that obtained in the solution of part (b) of Example 13.7 (when 897 nodes were used)? Plot the error (as computed above using the Poisson integral formula).
Suggestion: Try several different schemes with the main goal being to minimize the maximum total error (i.e., the z-height of the error graph). The node sets are small enough so that CPU time will not hinder multiple experiments.

NUMERICAL APPROXIMATION TO DOUBLE INTEGRALS— METHOD 2: APPROXIMATION QUADRATURE FORMULAS (RECOMMENDED):

Suppose that T is a region in the plane. A so-called **Gauss quadrature** formula for approximation of general integrals over T takes the form:

$$\int_T f(x, y)dxdy \approx w_1 f(\xi_1) + w_2 f(\xi_2) + \cdots + w_n f(\xi_n), \qquad (22)$$

where the **weights** $w_1, w_2, \cdots, w_n$ are specified real numbers and the **sampling points** $\xi_1 = (x_1, y_1)$, $\xi_2 = (x_2, y_2)$, $\xi_1 = (x_1, y_1)$, $\xi_2 = (x_2, y_2)$, $\cdots$, $\xi_n = (x_n, y_n)$ are specified points in T. In general, these formulas are developed with the goal that they be exact for polynomials (in two variables) up to a specified degree. If such a formula was exact for polynomials of degree up to p, Taylor's theorem in two variables could then be used to show that if the integrand has continuous partial derivatives up to order $p + 1$ then the error of the approximation (22) is $O(h^{p+1})$, where h is the diameter of T (Exercise 28). For each sampling point there are three degrees of freedom (the weight, and the coordinates of the sampling point). For example, when T is a triangle with vertices V_1, V_2, and V_3 it can be shown (Exercise 29) that the following formula is exact for any polynomial of degree at most one:

$$\int_T f(x, y)dxdy \approx \frac{\text{Area}(T)}{3}\{f(V_1) + f(V_2) + f(V_3)\}. \qquad (23)$$

This may be interpreted as a two-dimensional generalization of the trapezoidal rule. With the same number of sample points we can do better: If we choose them to be the midpoints of the edges of the triangle, rather than the vertices, we arrive at the following formula that turns out to be exact for polynomials of degree at most 2:

$$\int_T f(x,y)\,dxdy \approx \frac{\text{Area}(T)}{3}\{f([V_1+V_2]/2)+f([V_1+V_3]/2)+f([V_2+V_3]/2)\}. \quad (24)$$

For a brief but enlightening introduction on how such formulas are derived, see Section 5.2 of [ZiMo-83]. More details can be found in the article [Cow-73]; see also [Kry-62].

EXERCISE FOR THE READER 13.10: (a) Write an M-file for the Gaussian quadrature formula (24) having the following syntax:

```
int = gaussianintapprox(f, V1, V2, V3)
```

The input variables are: f, an inline function or an M-file, and V1, V2, and V3, the vertices of a triangle in the plane (listed as row vectors of length two). The output int is a number corresponding to the integral approximation of (24).
(b) Run through the MATLAB codes of part (c) of Example 13.7 on your own computer, and take note of the time it takes for the main finite element part of the code (after the triangulation). Then rewrite this part of the code to use the M-file of part (a) of this exercise in place of dblquad, and compare the resulting error and runtime.

Example 13.7 gives a nice demonstration of how refinements of the mesh will reduce the errors of the FEM approximations of the actual solution. In general, if the data for the BVP (10) satisfy: $p, q,$ and f are piecewise continuous on Ω, the first partial derivatives of p q, and g are piecewise continuous on Γ_1, r and h are piecewise continuous on Γ_2, and $p(x,y) \ge p_0 > 0, q(x,y) \ge 0$, then it can be shown that with the above FEM scheme (as well as the one below for more general boundary conditions), the error of the FEM approximation is of order δ, where δ is the maximum diameter of any of the (triangular) elements. This result can be roughly expressed by the following inequality:

$$\|u - \hat{u}\| \le C\delta. \quad (25)$$

Here u represents the exact solution of the BVP, $\hat{u}$ is the FEM solution (corresponding to a triangular mesh with h defined as above), and C is a constant that depends on the problem but not on δ. The norm on the left can be any of several norms to measure the errors. The order of the errors can be upgraded from δ to higher powers of δ by using basis functions that are locally polynomial of higher degree (some examples of such elements were given in the exercises of the previous section). To give more specific results would require a deeper theoretical discussion involving some functional analysis. We refer the interested reader to Chapter 4 of [Joh-87]; see also [Cia-02] and [StFi-73]. We caution the reader that the situation of the Dirichlet problem on a disk for the Laplace PDE is very atypical in that an explicit solution is available (Theorem 13.1). The next two exercises for the reader will involve slight variations of this PDE; the first one deals with the same type of BVP but on another domain, while the second deals with a slightly different PDE (Poisson's) on the disk. It may come as a surprise that for such mild variations, no explicit solution techniques are known.

From our experience with both methods of numerical quadrature applied to the same problem of Example 13.7, we see that in any FEM program, the amount of time devoted to numerical quadrature is a crucial consideration. The relevant theorem on how numerical quadrature schemes affect the order of convergence of an FEM depends on the maximum degree of general polynomials for which the quadrature scheme integrates exactly. Stated roughly, if the error of an FEM approximation is of order δ^k, i.e., $\|u - \hat{u}\| \le C\delta^k$ (cf. (25)), and if the Gaussian quadrature formula (of form (22)) used is exact for all polynomials (in x and y) of degree at most $2k - 2$, then the same order of accuracy (perhaps with a different value of C) will hold for the solution resulting from the FEM with the numerical quadrature formula being used. For details, see Chapter IV of [CiLi-89]; see also Section 4.3 of [StFi-73]. In our case, $k = 1$, so that this theorem really only requires that constant functions be integrated exactly by the numerical quadrature formula being used. Nonetheless, the extra precision of our quadrature formula (24) comes at very little cost and will imrove the accuracy of our method. In the following two exercises for the reader, this quadrature formula is to be invoked in the FEM.

EXERCISE FOR THE READER 13.11: Using the grid of Example 13.3, apply the FEM to solve the following Dirichlet problem on the annulus $\Omega = \{(x, y) : 1 \le x^2 + y^2 \le 4\}$:

$$\begin{cases} \text{(PDE)} & \Delta u = 0 & \text{on } \Omega \\ \text{(BC)} & u \equiv 2 & \text{on } x^2 + y^2 = 1 \, . \\ & u(2, \theta) = \cos(2\theta) & \text{on } x^2 + y^2 = 4 \end{cases}$$

Plot your FEM solution and indicate the number of nodes, internal nodes, and elements. Your solution plot should look like that in Figure 13.39(a).

EXERCISE FOR THE READER 13.12: (a) Use the FEM to solve the following Poisson-Dirichlet problem on the unit disk $\Omega = \{(x, y) : x^2 + y^2 \le 1\}$ with the triangulation of Method 2 of the solution to Example 13.2(a)

$$\begin{cases} (PDE) & \Delta u = f & \text{on } \Omega \\ (BC) & u = 0 & \text{on } \partial\Omega \, , \end{cases}$$

where the load $f(x, y)$ equals 100 on the (small) disk of radius $r = 0.125$ and center $(x, y) = (0, 0.5)$, and zero elsewhere. Plot your FEM solution and indicate the number of nodes, internal nodes, and elements. Your solution plot should look like that in Figure 13.39(b).

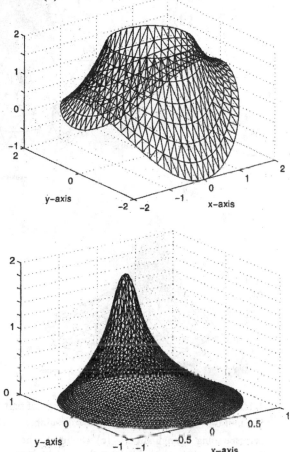

FIGURE 13.39: (a) (top) Plot of the FEM solution to the BVP of Exercise for the Reader 13.11. (b) (bottom) Plot of the FEM solution to the BVP of Exercise for the Reader 13.12(a).

(b) The triangulation of part (a) was rather uniform and had 1795 nodes. In this part we try to work with a smaller number of nodes but deploy them in a strategy that concentrates more of them near where the inhomogeneity $f(x, y)$ has most of its action.   Construct a triangulation using between 500 to 1000 nodes and distributed in the four subregions of Figure 13.40(a) as follows: Roughly 50% of the nodes are to be deployed in Ω_1, 25% in Ω_2, 15% in Ω_3, and only 10% in Ω_4. Each of these regions is simply the intersection of the whole domain with the insides of the circles with center $(0, \frac{1}{2})$ having radii: 1/4, 1/2, 1, and 3/2, respectively. The distribution should be more or less uniform in each subregion. Obtain and plot the FEM solution. Your solution should look something like the one shown in Figure 13.40(c).

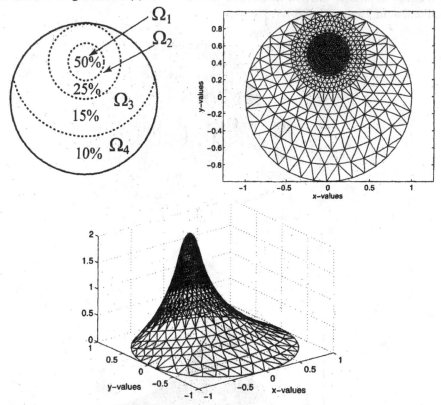

FIGURE 13.40: (a) (left) Diagram for node deployment strategy of part (b) in Exercise for the Reader 13.12. The unit disk Ω is split up into four subregions: $\Omega_1, \Omega_2, \Omega_3,$ and Ω_4. (b) (top right) A corresponding triangulation. (c) (bottom) The corresponding FEM solution; it appears graphically indistinguishable from the one obtained in part (a).

We now move on to describe the FEM for the general case of the BVP (10):

$$\begin{cases} \text{(PDE)} & -\nabla\cdot(p\nabla u) + qu = f & \text{on } \Omega \\ \text{(BCs)} & u = g & \text{on } \Gamma_1 \\ & \vec{n}\cdot\nabla u + ru = h & \text{on } \Gamma_2 \end{cases}.$$

Under the assumptions indicated in the theoretical discussion earlier in this section, this BVP can be shown to be equivalent to the following minimization problem:

Minimize the functional:

$$F[u] = \iint_{\Omega} [\tfrac{1}{2} pu_x^2 + \tfrac{1}{2} pu_y^2 + \tfrac{1}{2} qu^2 - fu]\,dxdy + \int_{\Gamma_2} [\tfrac{1}{2} ru^2 - hu]\,ds, \qquad (26)$$

over the following set of admissible functions:

$$A = \{v : \Omega \to \mathbb{R} : v(x) \text{ is continuous}, \ v'(x) \text{ is piecewise continuous}$$
$$\text{and bounded, and } v(x,y) = g(x,y) \text{ on } \Gamma_1\}. \qquad (27)$$

Note that the class of admissible functions requires only the Dirichlet boundary conditions (on Γ_1). The Robin boundary conditions (on Γ_2), are accounted for in the functional (26) and will be automatically satisfied by the solution.

Analogous to the one-dimensional method presented in Section 10.5, the FEM will solve a corresponding finite-dimensional minimization problem where the functional $F[u]$ of (26) is kept the same, but the set of admissible functions is reduced to an approximating smaller set determined by the basis functions of the triangulation. Thus we will be looking for minimizers of the functional F among functions of the form $v = \sum_{i=1}^{m} c_i \Phi_i$, where the $\Phi_i = \Phi_i(x,y)$ are the basis functions.

The basis functions corresponding to nodes on the boundary portion Γ_1 will have their coefficients determined by the Dirichlet boundary conditions; it is the remaining coefficients (corresponding to interior nodes and nodes on the boundary portion Γ_2) that need to be determined. We now briefly outline the FEM for this general BVP. We follow this outline with some additional details and then give examples.

FEM FOR THE BVP (10) —GENERAL CASE:

Step #1: Decompose the domain into elements, and represent the set of nodes and elements using matrices. Separate the nodes N_i **into the internal nodes and non-Dirichlet boundary nodes:** $N_1, N_2, \cdots, N_n$ **(that lie in** $\Omega \cup \Gamma_2$ **), and the Dirichlet boundary nodes** $N_{n+1}, N_{n+2}, \cdots, N_m$ **(that lie on** Γ_1 **). Denote the basis function** Φ_{N_i} **corresponding to node** N_i **simply by** Φ_i. **It is important that nodes be placed at all endpoints (interfaces) of** Γ_1 / Γ_2 **and that these endpoints be counted as Dirichlet boundary nodes (i.e., grouped with those in** Γ_1 **).**

Step #2: Use the Dirichlet BCs $u(x,y) = g(x,y)$ **on** Γ_1 **to determine the coefficients of the Dirichlet boundary node basis functions of an admissible function:** $v = \sum_{i=1}^{m} c_i \Phi_i$, **i.e.,** $c_i = g(N_i)$ **for each** $i = n+1, n+2, \cdots, m$.

Step #3: Assemble the $n \times n$ stiffness matrix A and load vector b needed to determine the remaining coefficients $c_1, c_2, \cdots, c_n$ which work to solve the discrete minimization problem corresponding to the BVP.
Step #4: Solve the stiffness equation $Ac = b$, and obtain the FEM solution

$$v = \sum_{i=1}^{m} c_i \Phi_i.$$

As before, the coefficients $c_1, c_2, \cdots, c_n$ will eventually be determined as the solution vector $c = [c_1 \ c_2 \ \cdots \ c_n]'$ of a linear system (14) $Ac = b$. The stiffness matrix A and load vector b will, in general, have entries given as follows:

$$a_{ij} = \iint_{\Omega} [p \nabla \Phi_i \cdot \nabla \Phi_j + q \Phi_i \Phi_j] \, dxdy + \int_{\Gamma_2} r \Phi_i \Phi_j \, ds \ (1 \leq i, j \leq n), \tag{28}$$

and

$$b_j = \iint_{\Omega} f \Phi_j \, dxdy + \int_{\Gamma_2} h \Phi_j \, ds$$

$$- \sum_{s=n+1}^{m} c_s \left\{ \iint_{\Omega} [p \nabla \Phi_s \cdot \nabla \Phi_j + q \Phi_s \Phi_j] \, dxdy + \int_{\Gamma_2} r \Phi_s \Phi_j \, ds \right\} \ (1 \leq j \leq n). \tag{29}$$

In these formulas the integrals over Γ_2 are with respect to arclength (i.e., positively oriented line integrals). These can be derived in a similar fashion to what was done in the purely Dirichlet BC case (see the development of (13)). As before, we observe that the stiffness matrix is a symmetric matrix. The time-consuming part of the FEM is still the assembly process. The mechanics are as in the purely Dirichlet case (just replace (15′) and (16′) with their analogs for (28) and (29)).

The assembly process can be coded much like the way we did it for Example 13.7. The only new feature here is the presence of the line integrals. Before entering into the MATLAB code, we give a brief outline of how such line integrals can be evaluated. We show how to numerically evaluate integrals of the form $\int_{\Gamma_2 \cap T_\ell} F \, ds$, where F is any function on Γ_2 in the setting of an assembly code. Any such integral can be broken up into a sum of corresponding integrals over line segments. Let L denote a typical such line segment, connecting nodes N_1 and N_2 of T_ℓ. Letting $\vec{v} = N_2 - N_1$, we can write:

$$\int_L F \, ds = \int_0^1 F(N_1 + s\vec{v}) \|\vec{v}\| \, ds, \tag{30}$$

from the definition of line integrals. Such integrals could be done with MATLAB's quad (or quadl)—but in a general FEM code, it would be awkward to combine the M-files for the integrand "F" with the needed change in variables unless we resort to symbolic variables. We avoid this dilemma and maintain consistency with the recommended method for approximating double integrals by invoking the following numerical quadrature approximation for ordinary integrals:

$$\int_0^1 f(x)dx \approx (1/6)\{f(0)+4f(1/2)+f(1)\}. \tag{31}$$

This formula, known as the **Newton-Coates formula** with three equally spaced points, is exact for polynomials up to degree three (Exercise 24). For more on such one-dimensional quadrature formulas, see Chapter 5 of [ZiMo-83], or see any good book on numerical analysis. What is most pertinent is that the accuracy of this approximation makes it feasible to use in the FEM; the underlying theory can be found in the references mentioned above. Combining (30) and (31) yields the following quadrature approximation, which is easily incorporated in FEM codes:

$$\int_L F\, ds \approx (\| N_1 - N_2 \|/6)\{F(N_1)+4F([N_1 + N_2]/2)+F(N_2)\}, \tag{32}$$

EXERCISE FOR THE READER 13.13: (a) Write an M-file lineint = bdyintapprox(fun, tri, redges) that works as follows: The inputs will be fun, an inline (or M-file) function of the variables x and y, a 3×2 matrix tri of nodes of a triangle in the plane, and a 2-column matrix redges, possibly empty ([ ]). The rows of redges consist of the corresponding increasing node indices (from 1 to 3 corresponding to their row in tri) of nodes that are endpoints of segments of the triangle that are part of the "Robin" boundary (for an underlying FEM problem). Thus the rows of redges can include only the following three vectors: [1 2], [1 3], and [2 3]. The output, lineint, will be the approximation of the corresponding line integral of fun over the Robin segments of the triangle, by using formula (32).
(b) Test the accuracy of your M-file in computing the following line integrals over the indicated edge sets of the triangle with vertices $N_1 = (0,0)$, $N_2 = (2,0)$, $N_3 = (0,3)$, and then on the triangle with vertices $N_1 = (0,0)$, $N_2 = (.2,0)$, $N_3 = (0,.3)$. In the notation used below, ε_{ij} denotes the edge of this triangle joining N_i to N_j $(i \neq j)$. The line integrals are given below for the larger triangle:

$$\int_{\varepsilon_{12}\cup\varepsilon_{23}} 4\, ds = 8+4\sqrt{13}, \quad \int_{\varepsilon_{12}\cup\varepsilon_{13}} \cos(\pi x/4 + \pi y/2)ds = 2/\pi.$$

In our next example, we will solve a BVP over an odd-shaped region. The problem is carefully constructed so that the exact solution will be available for comparison purposes. In Exercise for the Reader 13.14, the reader will be asked to solve another such problem on the same region for which an exact solution is not available.

EXAMPLE 13.8: Use the finite element method to solve the following mixed BVP over the parabolically shaped domain $\Omega = \{(x, y) : 0 \le x \le 10, \ 0 \le y \le x(10 - x)\}$:

$$\begin{cases} \text{(PDE)} & -\Delta u = -1/25 & \text{on } \Omega \\[2mm] \text{(BCs)} & u = 0 & \text{on } x\text{-axis} \\[2mm] & \bar{n} \cdot \nabla u = \dfrac{y}{25(101 - 40x + 4x^2)^{1/2}} & \text{on } y = x(10 - x) \end{cases}$$

(a) Use first a triangulation with between 300 and 500 nodes that are more or less uniformly distributed. Compare with the exact solution $u(x, y) = y^2 / 50$.

(b) Repeat part (a) this time using a similar triangulation with between 1000 and 2000 nodes.

Before we begin to solve this example, we leave the reader to perform the following:

EXERCISE FOR THE READER 13.14: Verify that the exact solution provided really solves the BVP in the above example.

SOLUTION TO EXAMPLE 13.8: Part (a): To decide on the linear gap distance between nodes, we first find the area of Ω:

$$\text{area}(\Omega) = \int_0^{10} x(10 - x)dx = [5x^2 - x^3/3]_0^{10} = 500/3$$

If we place the nodes in small square configurations (cf. Method 1 of Example 13.2(a)), then, roughly, each node would account for an area δ^2. Thus, if m denotes the number of nodes we use, then we would have (approximately) $m\delta^2 \approx \text{area}(\Omega)$, the left side being a bit larger due to boundary nodes. This gives the estimate

$$\delta \approx \sqrt{\text{area}(\Omega)/m}, \tag{33}$$

for the gap size we should use if we want to deploy m nodes. This formula can be used in the creation of squarelike nodegrids on any two-dimensional region with smooth boundary curves. Using (33) with the above area and $m = 350$, we arrive at the value $\delta = \sqrt{500/3/350} \approx 0.6901....$

We begin by using MATLAB to deploy nodes in the interior of Ω, maintaining a safe distance close to δ away from the boundary and placing them in a square grid configuration with sidelength δ:

```
>> bdyf= inline('x.*(10-x)');
```

```
>> delta=sqrt(500/3/350);
>> nodecount=1;
>> for i=1:10/delta
for j=1:bdyf(i*delta)/delta
  xtemp=i*delta; ytemp=j*delta;
  if (bdyf(xtemp-delta/2)>ytemp)&...
  (bdyf(xtemp)>ytemp+delta/2)&(bdyf(xtemp+delta/2)>ytemp)
%These conditions assure that the parabolic portion of the boundary
%does not get too close to the candidate (xtemp, ytemp) for an
%internal node.
    x(nodecount)=xtemp; y(nodecount)=ytemp;
    nodecount = nodecount+1;
  end
end
end
```

We would like to assign the boundary nodes in such a way that the distance gap between nodes is approximately δ. This is quite simple to do on the straight portion of the boundary. For the curved portion, we now introduce a general method to accomplish such node deployment. Recall, the arclength formula for the graph of a function $f(x)$ over an interval $[a,b]$: $L = \int_a^b \sqrt{1+[f'(x)]^2}\,dx$. Now the parabolic boundary graph function has $f'(x) = 10 - 2x$ so that the largest that the integrand $\sqrt{1+[f'(x)]^2}$ will be over $[0, 10]$ is $\sqrt{101} \approx 10$. (This maximum change in arclength occurs near the endpoints where the parabola is most steep.) Since we will place a node on the parabola at $x = 0$, $y = 0$ (call this the "most recent node"), and then continue advancing x by $\delta/3 \cdot 10$ (so the corresponding arclength of the parabola will advance by no more than about $\delta/3$), as soon as the arclength from the most recent node exceeds δ, we create a new node. Since we will place a node also at $(10,0)$, we will place a safeguard to prevent the nodes on the parabola from getting too close to this one. The code below is set up so that the Dirichlet nodes are indexed last.

```
>> arcint = inline('sqrt(101-40*x+4*x.^2)');
>> xref1=0; xref2=delta/30; cumlen=quad(arcint,xref1,xref2);
>> while xref1<10
  while cumlen<delta
   xref2=xref2+delta/30;;
   cumlen=quad(arcint,xref1,xref2);
  end
  if xref2<10-delta/40
   x(nodecount)=xref2; y(nodecount)=bdyf(xref2);
   nodecount = nodecount+1;
  end
  xref1=xref2; xref2=xref2+delta/30;
  cumlen=quad(arcint,xref1,xref2);
end
if quad(arcint,xref1,10)>delta/2
  nodecount = nodecount+1;
end
>> x(nodecount)=10; y(nodecount)=0;
>> nodecount = nodecount+1;
```

```
>> %finally put nodes on interior of horizontal segment
>> num = floor(10/delta); delta2=10/num; xref=10-delta2;
>> while xref>delta2/4
   x(nodecount)=xref; y(nodecount)=0;
   nodecount = nodecount+1; xref=xref-delta2;
end
>> x(nodecount)=0; y(nodecount)=0; %last node
>> nodecount = nodecount+1; tri=delaunay(x,y);
>> trimesh(tri,x,y), axis('equal') %Plots the triangulation
```

The triangulation is shown in Figure 13.41(a). From the way the nodes were constructed, the boundary nodes come after the interior nodes and the first boundary node is on the parabolic portion of the boundary. We can thus find the key indices by:

```
>> nint=min(find(abs(y-bdyf(x))<10*eps))-1 %number of interior nodes
→nint = 307
>> n=find(x==10&y==0)-1 %number of interior/Robin nodes
→n = 373
>> m=length(x) %number of nodes
→m = 388
>> size(tri)
→ans = 693    3 (So there are 693 elements.)
```

We give special names to the node numbers of the endpoints of the segment (interface with Robin/Dirichlet nodes):

```
>> dir1 = m; %node (0,0)
>> dir2 = nint+1; %node (10,0)
```

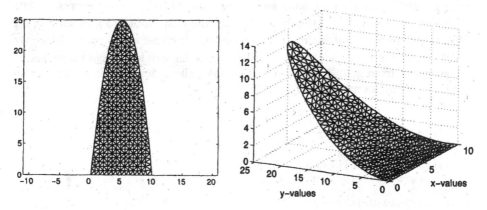

FIGURE 13.41: (a) (left) The triangulation of the parabolic region for the BVP of Example 13.8(a). There are 388 nodes and 693 elements. With this resolution the curved boundary is rather well represented by the element boundaries, except near the top where the curvature of the parabola is most extreme. (b) (right) The FEM solution of Example 13.8(a). The exact solution is graphically indistinguishable, the maximum relative error being less than 1%.

From the key node indices found above, we conclude:

Interior nodes: 1:307
Robin nodes (on interior of parabola): 308: 373
Dirichlet nodes (on line segment): 374:388

Notice we have created nodes at the interfaces of the two boundary portions (Dirichlet meets Robin) and these interface nodes will be assigned the Dirichlet conditions, as required. Since the Dirichlet boundary conditions are zero, simply creating a 388-length vector c of zeros will take care of assigning the Dirichlet nodes their appropriate values:

```
>> c= zeros(m,1);
```

The crucial index here is $n = 373$, the number of interior nodes added to the number of Robin boundary nodes; this is how many coefficients need to be determined. Since, in (10), we have $p \equiv 1$, $q \equiv 0$, $f = -1/25$, $g \equiv 0$, $r \equiv 0$, and since $c_s = 0$ $(s > n)$, the element matrix analogues of (28) and (29) (cf. (15$'$) and (16$'$)) are as follows:

$$a'_{\alpha\beta} = \iint_{T_\ell} [\nabla \Phi_{i_\alpha} \cdot \nabla \Phi_{i_\beta}] \, dx \, dy \quad (1 \le \alpha, \beta \le 3),$$

and

$$b'_\alpha = (-1/25) \iint_{T_\ell} (\Phi_{i_\alpha}) \, dx \, dy + \int_{\Gamma_2 \cap T_\ell} h(x,y) \Phi_{i_\alpha} \, ds \quad (1 \le \alpha \le 3),$$

where $h(x,y) = \dfrac{y}{25(101 - 40x + 4x^2)^{1/2}}$.

For each element index ℓ, these coefficients need to be computed only when the nodes i_α and/or i_β are interior or Robin nodes (i.e., $i_\alpha, i_\beta \le n \equiv 373$) corresponding to vertices of the corresponding element. The assembly code will invoke the M-file gaussianintapprox of Exercise for the Reader 13.10 for approximating the double integrals, and the M-file robinbdyint of Exercise for the Reader 13.13 for numerically evaluating line integrals. Here is the assembly code:

```
N=[x' y']; E=tri; A=zeros(n); b=zeros(n,1); [L cL]=size(E);
for ell=1:L
   nodes=E(ell,:); %global node indices of element
   intnodes=nodes(find(nodes<=n)); %global interior/Robin node indices
%find coefficients [a b c] of local basis functions
% ax + by +c; for int/robin nodes
   for i=1:length(intnodes)
     xyt=N(intnodes(i),:); %main node for local basis function
     onodes=setdiff(nodes,intnodes(i));
%global indices for two other nodes (w/ zero values) for local basis
%function
     xyr=N(onodes(1),:);
```

```
    xys=N(onodes(2),:);
    M=[xyr 1;xys 1;xyt 1]; %matrix M of (4)
%local basis function coefficients using (6B)
    abccoeff=[xyr(2)-xys(2);xys(1)-xyr(1);xyr(1)*xys(2)-
xys(1)*xyr(2)]/...
    det(M);
    intgrad(i,:)=abccoeff(1:2)';
    abc(i,:)=abccoeff';
end

% determine if there are any Robin edges
marker=0; %will change to 1 if there are Robin edges.
roblocind=find(nodes==dir1|nodes==dir2|(nodes<=n& nodes >=(nint+1)));
%local indices of nodes for possible robin edges
if length(roblocind)>1
    elemnodes = N(nodes,:);
%now find robin edges and make a 2 column matrix out of their local
%indices.
    rnodes=nodes(roblocind); %global indices of robin nodes
    count=1;
    for k=(nint+1):n
      if ismember(k,rnodes) & ismember(k+1,rnodes)
        robedges(count,:)=[find(nodes==k)   find(nodes==k+1)];
        count=count+1; marker =1;
      end
    end
end

%update stiffness matrix
for i1=1:length(intnodes)
for i2=1:length(intnodes)
    if intnodes(i1)>=intnodes(i2) %to save some computation, we use
%symmetry the stiffness matrix.
    fun1 = num2str(intgrad(i1,:)*intgrad(i2,:)',10);
%integrand a-integral
    fun=inline(fun1,'x', 'y');
    integ=gaussianintapprox(fun,xyt,xyr,xys);
A(intnodes(i1),intnodes(i2))=A(intnodes(i1),intnodes(i2))+integ;
end
end
end

%update load vector
for i1=1:length(intnodes)
    ai1 = num2str(abc(i1,1),10);
    bi1 = num2str(abc(i1,2),10);
    ci1 = num2str(abc(i1,3),10);
    fun=inline([ai1,'*x+',bi1, '*y+', ci1],'x','y');
    integ=-1/25*gaussianintapprox(fun,xyt,xyr,xys);
    b(intnodes(i1))=b(intnodes(i1))+integ;
%now add Robin portion, if applicable
%robin edges were computed above
    if marker==1
prod=inline(['y./(25.*sqrt(101-
40.*x+4.*x.^2))','*(',ai1,'*x+',bi1,...
 '*y+', ci1,')'],'x','y');
b(intnodes(i1))=b(intnodes(i1))+bdyintapprox(prod, elemnodes,...
robedges);
end
```

```
end
clear roblocind rnodes robedges
end
A=A+A'-A.*eye(n); %Use symmetry to fill in remaining entries of A.


sol=A\b; c(1:n)=sol'; c(n+1:m)=0;

%The result is now easily plotted using the 'trimesh' function of the
%last section:

x=N(:,1); y=N(:,2);
trimesh(E,x,y,c), hidden off
xlabel('x-axis'), ylabel('y-axis')
```

The following commands will plot the error using the exact solution provided. The result is shown in Figure 13.42(a).

```
cexact = zeros(m,1);
for i=1:length(x),cexact(i)=y(i)^2/50; end
trimesh(E,x,y,abs(c-cexact))
```

Part (b) is done in exactly the same fashion. In fact, the above code is designed in such a way that only the second line (defining delta) in the node deployment needs to be adjusted (change 350 to 1800). With this change, the above code will produce a numerical solution with error plot shown in Figure 13.42(b).[16]

The various examples done so far contain all of the necessary techniques needed to apply the FEM to general BVPs of form (10). The next two exercises for the reader contain two more examples.

EXERCISE FOR THE READER 13.15: Consider the following steady-state heat distribution problem on the parabolic region $\Omega = \{(x,y).0 \leq x \leq 10, 0 \leq y \leq x(x-10)\}$ of the preceding example:

$$\begin{cases} \text{(PDE)} & -\Delta u = f & \text{on } \Omega \\ \text{(BCs)} & u = 0 & \text{on } x\text{-axis} \\ & \bar{n} \cdot \nabla u + 2u = 40 & \text{on } y = x(10-x) \end{cases},$$

where the source function is given by: $f(x,y) = \begin{cases} 200, & \text{if } 4 \leq x \leq 6 \text{ and } 10 \leq y \leq 15 \\ 0, & \text{otherwise.} \end{cases}$

The region Ω along with the boundary conditions and the support of the source function f is illustrated in Figure 13.43(a).

(a) Compute and plot the numerical solution of this BVP using the triangulation of the solution to part (a) of Example 13.8.

---

[16] For the convenience of the reader, the entire MATLAB codes for this example (and other longer examples and exercises for the reader of this chapter) are included as downloadable text files on the ftp site for this book (see the preface for the URL of this ftp site). These codes can easily be pasted directly into the MATLAB window, and they can be modified to solve other FEM problems.

(b) Compute and plot the numerical solution of this BVP using the triangulation of the solution to Part (b) of Example 13.8. Your plot should look like the one in Figure 13.43(b).

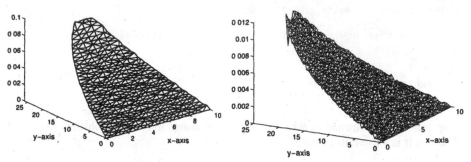

FIGURE 13.42: Error plots for the FEM solution of Example 13.8. (a) (left) Using the triangulation of part (a), which had 693 elements, the actual error was less than 1%. (b) (right) Using the triangulation of part (b), which had 3587 elements, the actual error was less than 0.1%.

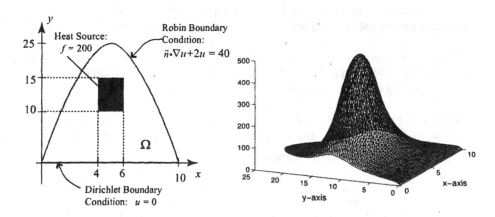

FIGURE 13.43: (a) (left) Illustration of the domain and boundary conditions for the steady-state heat distribution problem of Exercise for the Reader 13.15. The parabolically shaped plate has an internal rectangular heat source (temperature = 200) shown by a dark rectangle. The bottom (flat) edge is maintained at temperature 0 and the curved part of the boundary has a Robin boundary condition. (b) (right) An FEM solution of this problem.

EXERCISE FOR THE READER 13.16: (a) Contruct a squarelike grid and then a corresponding triangulation with between 2000 and 3000 nodes for the domain of Figure 13.44(a).

(b) Use the FEM with your triangulation to solve the Laplace problem $\Delta u = 0$ on this domain with boundary conditions as shown in Figure 13.44(a) and then plot your solution. Your plot should look like the one shown in Figure 13.43(b).

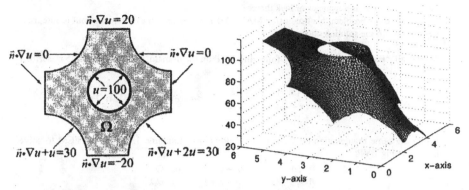

FIGURE 13.44: (a) (left) Illustration of the domain and boundary conditions for the BVP problem of Exercise for the Reader 13.16. The circular (inner) boundary portion has Dirichlet boundary conditions; the remaining (outer) boundary portions have the indicated Neumann or Robin conditions. (b) (right) Plot of the FEM solution for the Laplace problem having the indicated boundary data of (a). The triangulation used had 2655 nodes and 5024 elements.

EXERCISES 13.3

1. (a) Using the exact method of Example 13.5 (with Exercise for the Reader 13.4), solve the following BVP on the same hexagonal domain and triangulation of that example:

$$\begin{cases} \text{(PDE)} & -\Delta u = 0 \quad \text{on } \Omega \\ \text{(BC)} & u = x + y \quad \text{on } \partial\Omega' \end{cases}$$

and plot the resulting numerical solution.
(b) Check that $u(x,y) = x + y$ is the exact solution of the BVP; compare the numerical solution with this exact solution.

2. (a) Using the exact method of Example 13.5 (with Exercise for the Reader 13.4), solve the following BVP on the same hexagonal domain and triangulation of that example:

$$\begin{cases} \text{(PDE)} & -\Delta u = 0 \quad \text{on } \Omega \\ \text{(BC)} & u = x^2 + y^2 \quad \text{on } \partial\Omega' \end{cases}$$

and plot the resulting numerical solution.
(b) Check that $u(x, y) = x^2 + y^2$ is the exact solution of the BVP; compare the numerical solution with this exact solution.

3. (a) Using the exact method of Example 13.4 (with Exercise for the Reader 13.4), apply the FEM on the triangular domain Ω of Figure 13.45 with the triangulation shown there to solve the following BVP:

$$\begin{cases} \text{(PDE)} & -\Delta u = 2 \quad \text{on } \Omega \\ \text{(BC)} & u = x \quad \text{on } \partial\Omega' \end{cases}$$

The vertical/horizontal distance between adjacent nodes is one, and node #7 has coordinates (0,0) (so, for example, node #1 is at (0,3) and node #10 is at (3,0)). Plot the resulting numerical solution.
(b) Solve this problem again with the FEM but this time use the Gauss quadrature formula (24) (or the gaussianintapprox M-file) to evaluate integrals. Compare with the solution obtained in part (a);

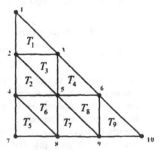

FIGURE 13.45: Triangular domain with basic triangulation for Exercise 3.

comment on any discrepancies or lack thereof.

(c) Re-solve the problem this time using MATLAB's dblquad to evaluate all double integrals. Compare with the solution obtained in part (a).

4.   Repeat all parts of Exercise 3 for the following BVP on the domain Ω of Figure 13.45 described there.

$$\begin{cases} (PDE) & -\Delta u = f(x,y) & \text{on } \Omega \\ (BC) & u = (x+y)^2 & \text{on } \partial\Omega \end{cases}, \quad \text{where } f(x,y) = \begin{cases} 0, & \text{if } y \le 1, \\ 1, & \text{if } 1 < y \le 2 \\ 2, & \text{if } 2 < y \end{cases}.$$

5.   Using the same triangulation on the hexagonal domain of Example 13.5, use the FEM to solve the following BVP:

$$\begin{cases} (PDE) & -\Delta u + u = 3 & \text{on } \Omega \\ (BC) & u = x + y & \text{on } \partial\Omega \end{cases}.$$

Note: Since the quadrature formula (24) is exact for polynomials up to second degree, it can be used to exactly evaluate all of the integrals that arise.

6.   Using the same triangulation on the triangular domain of Exercise 3, use the FEM to solve the following BVP:

$$\begin{cases} (PDE) & -\Delta u + u = f(x,y) & \text{on } \Omega \\ (BC) & u = x^2 & \text{on } \partial\Omega \end{cases}, \quad \text{where } f(x,y) = \begin{cases} 0, & \text{if } x \le 1, \\ -2, & \text{if } x > 1 \end{cases}.$$

Note: Since the quadrature formula (24) is exact for polynomials up to second degree, it can be used to exactly evaluate all of the integrals that arise.

7.   Consider the Dirichlet problem (19) on the unit disk $\Omega = \{(x,y): x^2 + y^2 < 1\}$:

$$\begin{cases} (PDE) & \Delta u = 0 & \text{on } \Omega \\ (BC) & u(1,\theta) = \cos^2(\theta) & \text{on } \partial\Omega \end{cases}.$$

(a) Use the FEM with the triangulation of part (c) of Example 13.7 to compute the numerical solution of the problem, performing the double integrals as in Example 13.7. Keep track of the time needed to perform the main assembly (using tic...toc). Use Poisson's integral formula (Theorem 13.1) to compute the "exact solution" to this problem at the nodes of the triangulation and plot solution and the error of the FEM solution obtained.

(b)   Repeat part (a), but this time use the Gauss quadrature formula (24) (or the gaussianintapprox M-file) to compute the double integrals in the FEM. Compare and contrast the FEM numerical solutions of parts (a) and (b).

(c)   Use the FEM as in part (b) to find the numerical solution of this problem using a triangulation of the circle having between 3000 and 4000 nodes. Plot the error against the corresponding "exact solution" from Poisson's integral formula.

(d) Repeat each of the above parts for the Dirichlet problem identical to the above but with the boundary condition being changed to $u(1,\theta) = \sin^2(\theta/2)$.

8.   Consider the following Dirichlet problem (19) on the disk $\Omega = \{(x,y): (x-1)^2 + (y-3)^2 < 5\}$:

$$\begin{cases} (PDE) & \Delta u = 0 & \text{on } \Omega \\ (BC) & u(x,y) = \ln x + 2y & \text{on } \partial\Omega \end{cases}.$$

(a)   Use the FEM with a triangulation having between 500 and 1000 nodes to compute the numerical solution of the problem, performing the double integrals as in Example 13.7. Keep track of the time needed to perform the main assembly (using tic...toc). Use Poisson's integral formula (Theorem 13.1) to compute the "exact solution" to this problem at the nodes of the triangulation and plot solution and the error of the FEM solution obtained.

(b)   Repeat part (a), but this time use the Gauss quadrature formula (24) (or the gaussianintaprrox M-file) to compute the double integrals in the FEM. Compare and contrast the FEM numerical solutions of parts (a) and (b).

(c)   Use the FEM as in part (b) to find the numerical solution of this problem using a triangulation of the circle having between 3000 and 4000 nodes. Plot the error against the corresponding "exact solution" from Poisson's integral formula.

(d) Repeat each of the above parts for the Dirichlet problem identical to the above but with (i) the boundary condition being changed to $u(x,y) = 2x + e^y$ and then (ii) with the PDE changed to

$-\nabla \cdot (e^{x+y}u) + u = y$ but all else as in the original problem.

9. Consider the following Robin problem for the Laplacian on the unit disk $\Omega = \{(x,y): x^2 + y^2 < 1\}$:

$$\begin{cases} \text{(PDE)} & \Delta u = 0 & \text{on } \Omega \\ \text{(BC)} & \vec{n} \cdot \nabla u + u = 3 & \text{on } \partial\Omega \end{cases}$$

(a) Use the FEM to solve this problem using a triangulation having between 500 to 1000 nodes and plot your numerical solution.

(b) Create a triangulation having between 1500 and 2000 nodes containing the node set of your triangulation of part (a). Re-solve the BVP with the FEM on this triangulation. Plot the new solution, compare it with that of part (a), and finally plot the difference of the two solutions on the common node set.

(c) Create a triangulation having between 3000 and 3500 nodes containing the node set of your triangulation of part (b). Re-solve the BVP with the FEM on this triangulation. Plot the new solution, compare it with that of part (b), and finally plot the difference of the two solutions on the common node set.

(d) Repeat each of parts (a) through (c) for the BVP with the same Robin boundary conditions of the above problem, but with the PDE changed to:

(i) $\nabla \cdot (e^{x+y}u) = 0$, (ii) $\nabla \cdot (e^{x+y}u) = 3$, (iii) $\nabla \cdot (e^{x+y}u) = -3$, (iv) $-\nabla \cdot (e^{x+y}u) + u = 3$.

10. Consider the following BVP on the annulus $\Omega = \{(x,y): 1 \le x^2 + y^2 \le 4\}$ of Exercise for the Reader 13.11:

$$\begin{cases} \text{(PDE)} & \Delta u = e^{x^2/2} & \text{on } \Omega \\ \text{(BCs)} & \vec{n} \cdot \nabla u \equiv 10 & \text{on } x^2 + y^2 = 1 . \\ & u(2,\theta) = 50 & \text{on } x^2 + y^2 = 4 \end{cases}$$

(a) Use the FEM to solve this problem using a triangulation having between 500 to 1000 nodes and plot your numerical solution.

(b) Create a triangulation having between 1500 and 2000 nodes containing the node set of your triangulation of part (a). Re-solve the BVP with the FEM on this triangulation. Plot the new solution, compare it with that of part (a), and finally plot the difference of the two solutions on the common node set.

(c) Create a triangulation having between 3000 and 3500 nodes containing the node set of your triangulation of part (b). Re-solve the BVP with the FEM on this triangulation. Plot the new solution, compare it with that of Part (b), and finally plot the difference of the two solutions on the common node set.

(d) Repeat each of parts (a) through (c) for the BVP with the same Robin boundary conditions of the above problem, but with the PDE changed to:

(i) $\nabla \cdot (e^{x+y}u) = 0$, (ii) $\nabla \cdot (e^{x+y}u) = 3$, (iii) $\nabla \cdot (e^{x+y}u) = -3$, (iv) $\nabla \cdot (e^{x+y}u) + u = 3$.

11. This exercise will use the FEM to solve the heat problem (1)

$$\begin{cases} \Delta u = 0 \text{ on } \Omega, & u = u(x,y) \\ \partial u / \partial n = 0, & \text{on outer rectangle} \\ u = 40, & \text{on small circle}, u = 500, \text{ on large circle} \end{cases}$$

from the introductory section. Take the domain (see Figure 13.2) to be the rectangle: $-1 < x < 0.5, \ -0.5 < y < 0.5$ with the following two disks deleted: larger circle: center = $(-0.65, 0.15)$, radius = 0.25 and smaller circle: center = $(0.1, -0.2)$, radius = 0.1. In each case you are to plot your results.

(a) First use a triangulation with between 300 and 500 nodes, more or less uniformly spaced.

(b) Repeat part (a) using a triangulation with between 1500 and 2000 nodes.

(c) (i) Repeat parts (a) and (b) on the BVP gotten from (1) by changing the BC on the outer rectangle to be $\partial u / \partial n + u = 40$, but keeping all else the same. (ii) Do this again using instead the BC on the larger circle to be $\partial u / \partial n + 4u = 40$. (iii) Repeat using instead the BC on the larger circle to be $\partial u / \partial n + u = 80$.

(d) (i) Repeat parts (a) and (b) on the BVP gotten from (1) by changing the PDE to be $\Delta u = f(x, y)$, where $f(x,y) = -100$ on the circle with center center $= (0.3, 0.25)$, radius $= 0.1$, and $f(x,y) = 0$ elsewhere (but keeping all else the same). (ii) Do this again but change the PDE to $\Delta u + 2u = f(x, y)$.

12. *(Comparison of the FEM and the Finite Difference Method for a Certain Mixed BVP)* (a) Use the FEM to solve the BVP of Exercise for the Reader 11.8. For the triangulation, let the node set correspond to that in part (a) of that exercise for the reader, i.e., nodes are uniformly spaced in a squarelike grid with horizontal and vertical gap size equaling h = 0.05. Let MATLAB's delaunay produce the actual triangulation once you create the node set. Plot your solution and compare it with Figure 11.23(a). Produce also a contour plot for your FEM solution and compare it with Figure 11.23(b).

(b) Repeat part (a), but using the finer grid with horizontal/vertical gap size $h = 0.02$.

13. Repeat both parts of Exercise 11 on the BVP with the same boundary conditions but with the PDE changed from the Laplace equation to $-\nabla \cdot ([x^2 + y^2 + 1]u) + u = \cos(xy)$.

14. *(Determination of Maximum Tolerable Heat)* Consider the domain of Figure 13.46:

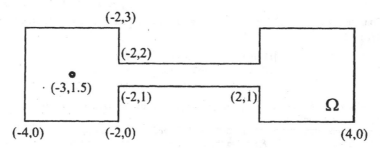

FIGURE 13.46: A domain consisting of two squares joined by a rectangular neck.

(a) In this domain, an observer at location (-3, 1.5) (left side) cannot tolerate a temperature greater than 50. All edges except for the right edge are kept insultated $\vec{n} \cdot \nabla u = 0$ while the right edge will be maintained at a certain temperature $u = T_{hot}$. What is the maximum value of T_{hot} so the observer's requirement is met? Try to get your answer accurate to at least two decimals. For the PDE in the domain use the basic Laplace equation $\Delta u = 0$.

(b) How would the answer in part (a) change if the rectangular length were to be doubled in length?

(c) How would the answer in part (a) change if the rectangular length were to have only half of its height?

(d) How would the answer in part (a) change if the square on the right were to have its sidelength doubled (but the left square is still kept the same)?

(e) How would the answer in part (a) change if the hot edge of the square on the right were the top edge rather than the right edge?

15. Let Ω be the domain shown in Figure 13.47, with the deleted disk having center (2.5, 2.5) and radius, 0.75.

(a) Create triangulation of Ω having between 300 and 400 (essentially equally spaced) nodes.

(b) Create a triangulation of Ω having between 1500 and 2000 nodes.

(c) Use the FEM with the triangulation of part (a) to solve the following BVP on Ω :

$$\begin{cases} \Delta u = 0 \text{ on } \Omega \\ \vec{n} \cdot \nabla u = 10, \text{ on triangle .} \\ u = 100, \text{ on circle} \end{cases}$$

Plot your result and then repeat with the triangulation of part (b).

(d) Repeat part (c) on the modified BVP gotten by changing the PDE to be (i), $-\Delta u = x^2/2$, but keeping all else the same; and then to (ii)

$-\Delta u + e^{x/2}u = x^2/2$.

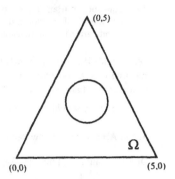

FIGURE 13.47: Triangular domain with basic triangulation for Exercise 3.

16. (a) Triangulate the domain of Figure 13.48 using between 400 and 800 nodes, more or less equally spaced.

(b) Repeat part (a) but this time use between 2000 and 2500 nodes.

(c) Use the FEM and the triangulation of part (a) to solve the heat problem on the domain of Figure 13.48 governed by the Laplace equation $\Delta u = 0$ and the boundary conditions shown in the figure. Plot your numerical solution.

(d) Repeat part (c), this time using the triangulation of part (b).

(e) Repeat both parts (c) and (d) on the modified BVP gotten by changing the boundary conditions on Γ_1^3 and Γ_1^3 to be $\vec{n} \cdot u = 0$

(insulated), but keeping all else the same.

(f) Repeat both parts (c) and (d) on the modified BVP gotten by changing the boundary conditions on $\Gamma_1^?$ and $\Gamma_1^?$ to be the Robin

conditions: $\vec{n} \cdot u = 20$ and $\vec{n} \cdot u = -20$, respectively, but keeping all else the same.

(g) Repeat both parts (c) and (d) on the

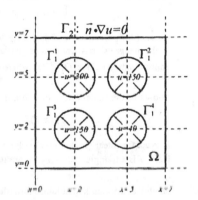

FIGURE 13.48: Boundary conditions for the heat problem of Exercise 11. The outer square boundary Γ, is insulated, while the four circular inner boundary portions Γ_1^i, $1 \le i \le 4$, are each maintained at the indicated temperatures.

modified BVP gotten by changing the boundary conditions on Γ_1^2 and Γ_1^3 to be the Robin

conditions: $\vec{n} \cdot u + u = 20$ and $\vec{n} \cdot u + u = 20$, respectively, but keeping all else the same.

(h) Repeat both parts (c) and (d) on the modified BVP gotten by changing the PDE to be $\nabla \cdot (([2^{x/2} + y]u) + u = 10$, but keeping all else the same.

17. Let Ω be the domain between the x-axis and the graph of $y = e^x$ from $x = 0$ to $x = 4$.

(a) For the function $u(x, y) = \sin(x/(y+1)) + x^2 y/25$, determine functions $g(x), f(x,y)$ and $h(x,y)$ so that $u(x,y)$ solves the following BVP:

$$\begin{cases} \text{(PDE)} & -\Delta u + u = f(x, y) & \text{on } \Omega \\ \text{(BCs)} & u = g(x) \text{ on x-axis}; \quad \vec{n} \cdot \nabla u + u = h(x, y) & \text{on curved portion of } \partial\Omega \end{cases}$$

(b) Construct a triangulation of Ω having between 300 and 500 nodes. Compute the corresponding FEM solution and use the exact solution to plot the error.
(c) Repeat part (b) this time using between 1500 and 2000 nodes.

18. Write an M-file, integ=quadint(fun,v1,v2,v3, v4), whose inputs are fun an inline function of x,y, and four 2×1 matrices v1, v2, v3, v4 that are vertices (in any order) of quadrilateral (four sided polygon) in the xy-plane. If we denote this quadrilateral by Q, the output integ should be the numerical integral $\int_Q \text{fun}(x,y)\,dxdy$, computed using dblquad, MATLAB's numerical integrator.

19. Derive formulas (13) through (18) for the FEM for BVPs with purely Dirichlet BCs.

20. (a) Establish the integral formula (21) for general planar regions of Figure 13.34.
     (b) Derive a similar integration formula for regions between functions of y.
     Suggestion: For part (a), in the last integral, make the following substitution $y = y\text{low}(x) + u(y\text{top}(x) - y\text{low}(x))$.

21. Suppose that a Gauss quadrature formula (22):
$$\int_T f(x,y)\,dxdy \approx w_1 f(\xi_1) + w_2 f(\xi_2) + \cdots + w_n f(\xi_n)$$
is exact for polynomials of degree up to p. Use Taylor's theorem in two variables to show that if the integrand has continuous partial derivatives up to order $p + 1$, then the error of the approximation (22) is $O(h^{p+1})$, where h is the diameter of the triangle T.

22. Show that the Gauss quadrature formula (23):
$$\int_T f(x,y)\,dxdy \approx \frac{\text{Area}(T)}{3}\{f(V_1) + f(V_2) + f(V_3)\}$$
is exact for linear (first-degree) polynomials, but not for quadratic (second degree) polynomials. Here, T is a triangle and V_1,V_2,V_3 are its vertices.

23. Show that the Gauss quadrature formula (24):
$$\int_T f(x,y)\,dxdy \approx \frac{\text{Area}(T)}{3}\{f([V_1 + V_2]/2) + f([V_1 + V_3]/2) + f([V_2 + V_3]/2)\},$$
is exact for quadratic (second-degree) polynomials. Here, T is a triangle and V_1,V_2,V_3 are its vertices.
     Suggestion: First work with the standard triangle with vertices (0,0), (1,0), and (0,1). You need only verify it for the basis polynomials and use of linearity. Once this is done use affine maps to get the result for general triangles (see Exercise 19 of the previous section).

24. Show that the Newton-Coates quadrature formula (30):
$$\int_0^1 f(x)\,dx \approx (1/6)\{f(0) + 4f(1/2) + f(1)\}$$
is exact when $f(x)$ is polynomial of degree at most three.

NOTE: The next four exercises will introduce the reader to some refinement and adaptive implementations of the FEM. These are based on the simple refinement scheme of splitting an element into four similar elements by introducing a new node at the midpoint of each edge; see Figure 13.49.

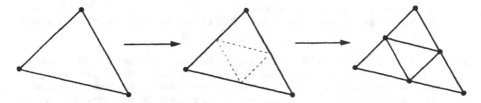

FIGURE 13.49: A refinement scheme for triangular meshes that is easily programmed into FEM routines.

25. (*M-file for Automatic Mesh Refinement*) The scheme of Figure 13.49 gives rise to a very natural refinement scheme that can be repeated for any number of iterations. Simply start off with any triangulation $\mathcal{T}_0$ of a given domain. For the first refinement $\mathcal{T}_1$ of $\mathcal{T}_0$, the node set will be the node set of $\mathcal{T}_0$, along with the midpoints of all element edges. Each element of $\mathcal{T}_0$ gives rise to four elements of $\mathcal{T}_1$ as in Figure 13.49. This procedure can be iterated to get a sequence of successively finer triangulations $\mathcal{T}_0$, $\mathcal{T}_1$, $\mathcal{T}_2$, ... We point out two very nice properties: (i) the node set of $\mathcal{T}_n$ is contained in the node set of $\mathcal{T}_{n+1}$ (making it simple to compare FEM solutions on successive triangulations), and (ii) the minimum angle of any element of $\mathcal{T}_{n+1}$ equals the minimum angle of any element of $\mathcal{T}_n$ (this keeps control of the eccentricity of elements, which is important for the FEM).
(a) Write an M-file that will perform the above refinement and with the following syntax:
`[newnodes, newtri] = meshrefine(nodes, tri)`
The input variable `nodes` is a two-column matrix of x- and y-coordinates of a given triangulation of a planar domain and `tri` is the corresponding three-column matrix of node numbers of the elements of the triangulation. (As usual, the node numbers are the rows of the nodes as they appear in the `nodes` matrix.) The output variables: `newnodes` and `newtri` are the corresponding matrices for the refined partition.
(b) With $\mathcal{T}_0$ being the triangulation of the hexagonal domain of Example 13.1 (see Figure 13.5), apply your M-file to construct and plot the next three successive triangulations: $\mathcal{T}_1$, $\mathcal{T}_2$ and $\mathcal{T}_3$.
(c) With $\mathcal{T}_0$ being the triangulation of the annular domain of Example 13.3 (see Figure 13.16(c)), apply your M-file to construct and plot the next three successive triangulations: $\mathcal{T}_1$, and $\mathcal{T}_2$.
(d) Comment on the performance of this refinement scheme for domains with polygonal boundaries (as in part (b)) versus domains with curved boundaries (as in part (c)). Can you suggest any modifications to help the above scheme better represent boundaries in cases of curved domains? Property (i) should still be maintained, and (ii) should be "essentially maintained" in that the minimum angle of any element of $\mathcal{T}_n$ should not be too much smaller than the minimum angle of all elements of $\mathcal{T}_0$. For any such ideas, build them into a modified M-file and experiment on some domains.

26. (*Examples of FEM with Mesh Refinement*) (a) For the BVP of Example 13.5, set up MATLAB code to perform the FEM starting with the triangulation of that example, and then after refining the triangulation (as in Exercise 25), re-solving the problem on the new triangulation and looking at the absolute value of the difference of the new FEM solution with the previous FEM solution (on the previous grid). Continue to iterate this process until the absolute value of the difference is less than 1e-4 or the FEM calculations take more than a few minutes, whichever happens first. Plot the successive FEM solutions as well as the difference graphs.
(b) Repeat the instructions of part (a) on the BVP of Exercise 2; compare the final FEM

solution with the exact solution given there.

(c) Repeat the instructions of part (a) on the BVP of Exercise 3 (using the initial triangulation given there).

27. *(An Adaptive Scheme for the FEM)* This exercise develops an example of an adaptive scheme for the FEM. General adaptive schemes recursively solve a BVP with the FEM (starting with any triangulation of the domain) and then attempt to locate those elements where the error of the FEM solution is greatest. The mesh is next refined in a way that puts more nodes near the elements that were identified in the error estimation. This process is then iterated until some stopping criterion (a sufficiently small estimated error or difference in successive FEM approximate solutions) allows an exit. Here is a basic outline of one such scheme:

(i) Start with any triangulation of a domain and solve the given boundary value problem with the finite element method.

(ii) For each element, note its *oscillation* (= max value − min value of computed solution on three vertices).

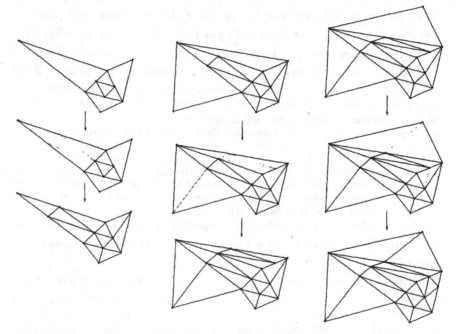

FIGURE 13.50: Illustration of adaptive mesh refinement scheme of Exercise 27. (a) (left) Step 1, (b) (middle) Step 2, and (c) (right) contingency plan for Step 3.

(iii) Flag those elements whose oscillations are "large" (with respect so some specified indicator, say more than double of the average).[17]

(iv) Refine the mesh accordingly so that each element flagged in (iii) gets split into three similar (triangular) elements as in Figure 13.49. Adjacent elements need to be refined accordingly so no hanging nodes remain. The two requirements are that the original node set is contained in the refined node set and no angle of any element gets too small (eccentricity requirement).

For definiteness, let us say that in (iii) the flagging criterion for elements is that the maximum

---

[17] We are using a rather basic error indicator. More sophisticated error indicators can be developed using advanced techniques of Sobolev spaces; see, for example, [CiLi-89], [Cia-02], or the classical reference [StFi-73] for details on such methods.

oscillation is more than double the average of all of the oscillations. In (iv) let us say that the eccentricity requirement stipulates that the minimum angle of any refinement cannot be less than 1/3 of the minimum angle, θ_{min}, of the original triangulation.

Balancing these two requirements makes the refinement scheme a delicate task. This sort of a scheme can be accomplished by iteratively applying a series of refinements that attempt (based on the two constraints) to isolate the "hanging nodes." We give an outline for such a scheme:

OUTLINE FOR ADAPTIVE MESH REFINEMENT SCHEME:
Step 1: After refining the flagged elements as in (iv), the new nodes introduced need to mesh into the next triangulation. Until they do become vertices of all adjacent elements, they will be referred to as "hanging nodes." Examine all neighboring elements of the flagged elements that were just refined; see Figure 13.50(a). If possible, we would like to contain the spread of green ("hanging nodes") but the problem is that we do not want any of the triangles to have very small angles. For each of the three neighbor triangles, if half the angle of the node opposite the hanging node is not too small ($< \theta_{min}/3$), then simply split it into two triangles by joining the hanging node of the first triangle to the opposite node of the neighbor triangle (Figure 13.50(a) has two such triangles[18]). Otherwise, we are forced to refine the neighbor triangle as in (iv), but this introduces two new hanging (green) nodes. (Figure 13.50(a) has one of these).
Step 2: If Step 1 introduced any new hanging (green) nodes (as it did in Figure 13.50(a)), look at the neighboring triangles and try to contain the hanging nodes as in Step 1. We may again introduce hanging nodes. (Figure 13.50(b) illustrates this). We continue to iterate this step until there are no longer any hanging nodes. There is one contingency we need to mention (if a neighboring triangle runs into another that was already refined), this is illustrated in Figure 13.50(c); below we explain what to do in such situations.
Contingency plan for Step 3: Figure 13.50(c) illustrates what to do if a neighbor triangle runs into one that was already refined. We do not refine any triangle twice (this will give some control on the convergence of the algorithm and prevent the possibility of an infinite loop). Instead, we revert to the original refinement (three subtriangles instead of two) to take care of the internal green node; see Figure 13.50(c).
(a) Write a MATLAB program that will perform the above adaptive scheme on the BVP and initial triangulation of Example 13.5. What happens when you run this program? Repeat, but now change the flagging criterion in (iii) to be that the oscillation of the FEM solution over an element exceeds 1/10. Repeat with 1/10 replaced by 1/100. Plot each refined mesh as well as the final FEM solution.
(b) Repeat the instructions of part (a) on the BVP of Exercise 2; compare the final FEM solution with the exact solution given there.
(c) Repeat the instructions of part (a) on the BVP of Exercise 3 (using the initial triangulation given there).
(d) Do you have any ideas for an alternative mesh refinement scheme (satisfying the two constraints mentioned above)?

28. (*Obtuse Angles in the Domain Are Sometimes Problematic for the FEM*) Engineers have known for some time, and mathematicians subsequently confirmed theoretically, that obtuse corners in the domain of a BVP can often slow down the convergence of the FEM near the boundary points with obtuse angles (see Section 8.1 of [StFi-73] or Section 5.6 of [AxBa-84]). Simple examples of domains with such obtuse angles are shown in Figure 13.51(b), (c). In general, the larger the obtuse angle, the greater the possible problems with the FEM. The extreme case is with an interior angle of 2π physically corresponding to a crack, fissure, or material interface in the domain; see Figure 13.51(c). This exercise will investigate such phenomena and explore stategies to mitigate problems that might arise. We will examine a certain Dirichlet problem for

---

[18] Note that at the first iteration, this could not occur with the stated eccentricity requirement since bisecting any of the original angles would result in angles at least as large as $\theta_{min}/3$; so this pathology in the figure could only occur in later iterations. In particular, for the first refinement, all hanging nodes could be isolated in Step 1.

the Laplace equation on such a domain where the exact solution is known. For any angle ω, where $0 < \omega \le 2\pi$, we let Ω_ω denote the subdomain of all points in the unit square $-1 \le x, y \le 1$ whose polar coordinates (r, θ) satisify $0 < \theta < \omega$. Thus the domains of Figure 13.51 are all examples of such domains. In particular, the domain in Figure 13.51(b) is $\Omega_{3\pi/2}$ and that of Figure 13.51(c) is $\Omega_{2\pi}$.

(a)   Show that on any such domain Ω_ω the function (given in polar coordinates) $u(r, \theta) = r^{\pi/\omega} \sin(\pi\theta/\omega)$ is harmonic (i.e., satisfies the Laplace equation $\Delta u = 0$) and vanishes on the *angular edges* (i.e., the rays of the angle emanating from the point O; see Figure 13.51.).

(b) For each of the domains in Figure 13.51 (for the one in Figure 13.51(a) use $\omega = 2\pi/3$), apply the FEM to solve BVP consisting of the Laplace equation with the boundary conditions $u = 0$ on the angular edges of the boundary and $u(r, \theta) = r^{\pi/\omega} \sin(\pi\theta/\omega)$ on the remaining portion of the boundary. Of course, you will need to convert the latter boundary conditions into cartesian (x, y) coordinates. Start off with the corresponding triangulations shown in Figure 13.52. Then apply the algorithm of Exercise 25 to successively refine the triangulations and re-solve with the FEM. For each triangulation, plot the exact error using the exact solution in part (a). Go through three refinements for each domain.

(c) Repeat part (b) for each of the three BVPs given there, but this time using the adaptive scheme of Exercise 27 in place of the refinement scheme of Exercise 25.

(d) Using the special form of the triangulations given, can you think of a more convenient refinement scheme for this problem? Make up a reasonable one and test it out for several iterations comparing with the exact solution at each step.

Suggestion: An elegant and illuminating way to do part (a) is to derive the Laplace operator in polar coordinates to be: $u_{xx} + u_{yy} = u_{rr} + (1/r)u_r + (1/r^2)u_{\theta\theta}$.

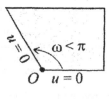 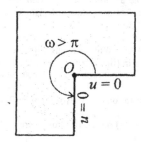 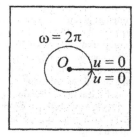

FIGURE 13.51: Simple examples of domains with different sorts of angles at a boundary point O. (a) (left) In general acute angles do not pose any problems for the FEM. (b) (middle) Obtuse angles can sometimes lead to slower convergence of the FEM. (c) (right) The larger the obtuse angle, the greater the potential difficulty. The extreme case is the slit domain. The indicated homogeneous Dirichlet boundary conditions on the angular edges is relevant for Exercise 28.

 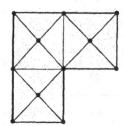 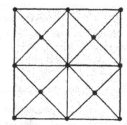

FIGURE 13.52: Initial triangulation for the domains of Figure 13.51 for Exercise 28.

29. Suppose that the FEM of this section is used to compute the solution of a BVP of form (10) whose exact solution is known to be a linear function $u(x, y) = ax + by + c$. Assume the integrals are all computed exactly and that the domain and triangulation are such that the boundary of the domain consists entirely of edges of the triangulation. Will the FEM solution always coincide with the exact solution? Either explain whether this is true or, if you are unable to do so, perform a series of numerical experiments to test this hypothesis.

    Note: Since the basis functions are piecewise linear, this seems to be the most general type of solutions that the FEM might be able to produce exactly. An example of such a BVP and triangulation is given in Exercise 1.

Appendix A: Introduction to MATLAB's Symbolic Toolbox

A.1: WHAT ARE SYMBOLIC COMPUTATIONS?

This appendix is meant as a quick reference for occasions in which exact mathematical calculations or manipulations are needed and are too arduous to expediently do by hand. Examples include the following:

1. Computing the (formula) for the derivative or antiderivative of a function
2. Simplifying or combining algebraic expressions
3. Computing a definite integral exactly and expressing the answer in terms of known functions and constants such as π, e, $\sqrt{7}$ (if possible)
4. Finding analytical solutions of differential equations (if possible)
5. Solving algebraic or matrix equations exactly (if possible)

Such exact arithmetic computations are known collectively as **symbolic computations**. MATLAB is unable to perform symbolic computations but the *Symbolic Math Toolbox* is available (or included with the Student Version), which uses the MATLAB interface to communicate with MAPLE®, a symbolic computing system. Thus, MATLAB has essentially subcontracted symbolic computations to MAPLE, and acts as a "middleman" so that it is not necessary to use two separate softwares while working on problems. Invoking such symbolic capabilities needs specific actions on the user's part, such as declaring certain variables to be symbolic variables. This is a safety device since symbolic calculations are usually much more expensive than the default floating point calculations and are usually not called for (see Chapter 5). It is important to point out that symbolic expressions are different data types than the other sorts of data types that MATLAB uses. Consequently, care needs to be taken when passing data from one type of data to the other. Moreover, most mathematical problems have answers that cannot be expressed in terms of well-known functions (e.g., $\ln(x)$, $\sqrt{x}$, $\arcsin(x)$) and/or constants (e.g., e, π, $\sqrt{2}$), and therefore cannot be solved symbolically.

There are also circumstances where the precision of MATLAB's floating point arithmetic is not good enough for a given computation and we might wish to work in more than the 15 (or so) significant digits that MATLAB uses as a default. As a middle ground between this and exact arithmetic, the Symbolic Toolbox also offers what is called **variable precision arithmetic**, where the user can specify how many significant digits to work with. We point out that there are a few special occasions where symbolic calculations have been used in the text.

The remainder of this appendix will present a brief survey of some of the functionality and features of the Symbolic Toolbox that will be useful for our needs. All of the MATLAB code and output given in a particular section results from a new MATLAB session having been started at the beginning of that section.

A.2: ANALYTICAL MANIPULATIONS AND CALCULATIONS

To begin a symbolic calculation, we need to declare the relevant variables as symbolic. To declare x, y as symbolic variables we enter:

```
>> syms x y
```

Let's now do a few algebraic manipulations. The basic algebra manipulation commands that MAPLE has are as follows: expand, factor, simplify; they work on algebraic expressions just as anyone who knows algebra would expect. The next examples will showcase their functionality. We point out that any new variable introduced whose formula depends on a symbolic variable will also be symbolic.

```
>> p2=(x+2*y)^2;,   p4=(x+2*y)^4;
>> expand(p2) %Multiplies out the binomial product.
→ans = x^2+4*x*y+4*y^2
>> expand(p4)
→ans =x^4+8*x^3*y+24*x^2*y^2+32*x*y^3+16*y^4
>> pretty(ans) %Puts the answer in a prettier form.
→       4     3      2 2       3      4
      x + 8 x y + 24 x y + 32 x y + 16 y
```

In general, for any sort of analytical expression exp, the command expand(exp) will use known analytical identities to try and rewrite exp in a form in which sums and products are expanded whenever possible.

```
>> pretty(expand(tan(x+2*y))) →
```

$$\cfrac{\tan(x) + 2\ \cfrac{\tan(y)}{1 - \tan(y)^2}}{1 - 2\ \cfrac{\tan(x)\,\tan(y)}{1 - \tan(y)^2}}$$

To clean up (simplify) any sort of analytical expression (involving powers, radicals, trig functions, exponential functions, logs, etc.), the simplify function is extremely useful.

```
>> simplify(log(2*sin(x)^2+cos(2*x)))    →ans =0
>> h=x^6-x^5-12*x^4-2*x^3+41*x^2+51*x+18;
```

```
>> pretty(factor(h))
```
→
$$(x + 2)(x - 3)^2(x + 1)^3$$

This function will also factor positive integers into primes. This brings up an important point. MATLAB also has a function `factor` that (only) does this latter task. Due to the limitations of floating point arithmetic, MATLAB's version is more restrictive than MAPLE's; it is programmed to give an error if the input exceeds $2^{32} \approx 4.2950e+009$.

```
>> factor(3^101-1)
??? Error using ==> factor
The maximum value of n allowed is 2^32.
>> factor(sym(3^101-1)) %declaring the integer input as symbolic
%brings forth the MAPLE version this command.
→ans = (2)^110*(43)*(47)*(89)*(6622026029)
```

Whereas the Student Version of MATLAB includes access to many of the Symbolic Toolbox commands that one might need to supplement MATLAB functionality, the complete Symbolic Toolbox (for MATLAB's professional version) includes unrestricted access to all of MAPLE's commands. All of the Symbolic Toolbox commands that we discuss in this Appendix are available with the Student Version. To learn more about additional Symbolic Toolbox commands available on the version of MATLAB that you are using, consult the Help menu.

The `factor` function is programmed to look only for real rational factors, so it will not perform factorizations such as $x^2 - 3 = (x + \sqrt{3})(x - \sqrt{3})$ or $x^2 + 1 = (x + i)(x - i)$. Recall (Chapter 6) that it is not always possible to find explicit expressions for all roots/factors of a polynomial, but nevertheless, by the fundamental theorem of algebra, any degree n polynomial always has n roots (counted according to multiplicity) that can be real or complex numbers. In cases where it is possible, the `solve` command can find them for us; otherwise, it produces decimal approximations.

`solve(exp,var)` →	If exp is a symbolic expression that involves the symbolic variable var, this command asks MAPLE to find all real and complex roots of the equation exp=0. In cases where they cannot be found exactly (symbolically), numerical (decimal) approximations are found. If there are additional symbolic variables, MAPLE solves for var in terms of them.

To solve the equation $x^5 - 5x^4 + 8x^3 - 40x^2 + 16x - 80 = 0$, we simply enter:

```
>>solve(x^5-5*x^4+8*x^3-40*x^2+16*x-80)
>> %shorter syntax if only one var
```

→ ans =   [2*i]   [2*i]   [5]
              [-2*i]   [-2*i]

The slightly perturbed polynomial equation $x^5 - 5x^4 + 8x^3 - 40x^2 + 16x - 78 = 0$, also has five different roots, but they cannot be expressed exactly, so MAPLE will give us numerical approximations, in its default 32 digits:

```
>> solve(x^5-5*x^4+8*x^3-40*x^2+16*x-78)
```

→ ans =   [-.2823772412563003180661278492544 9e-1 -
            2.1432362125064684675126753513414*i]
            [-.2823772412563003180661278492544 9e-1 +
            2.1432362125064684675126753513414*i]
            [.29428740076409528006464576345708e-1 -
            1.8429038593310837866143850920505*i]
            [.29428740076409528006464576345708e-1 +
            1.8429038593310837866143850920505*i]
            [4.9976179680984410076002964171595]

We can get the quadratic formula for the solutions of $ax^2 + bx + c = 0$ with the following commands:

```
>> syms a b c, solve(a*x^2+b*x+c,x)
```
→ ans =   [1/2/a*(-b+(b^2-4*a*c)^(1/2))]   [1/2/a*(-b-(b^2-4*a*c)^(1/2))]

Similarly, the Tartaglia formulas for the three solutions of the general cubic $ax^3 + bx^2 + cx + d = 0$, could be obtained.

A.3: CALCULUS

Table A.1 summarizes the Symbolic Toolbox commands needed to perform the most common "clerical" tasks in calculus: differentiation and integration.

TABLE A.1: Differentiation and integration using the Symbolic Toolbox.

Assume that f has been stored as a symbolic function of symbolic variables: $f(x)$ (or $f(x, y, ...)$), if we have a function of several variables.	
`diff(f,x)` →	Computes $f'(x) = \dfrac{df}{dx}$ $\left(\text{or} \ \dfrac{\partial f}{\partial x} \right)$.
`diff(f,x,2)` →	Computes $f''(x) = \dfrac{d^2 f}{dx^2}$ $\left(\text{or} \ \dfrac{\partial^2 f}{\partial x^2} \right)$.
`int(f,x)` →	Calculates (if possible) an antiderivative of $f(x)$: $\int f(x) dx$ (does not add on integration constant). If there are other variables, they are treated as constant parameters.
`int(f,x,a,b)` →	Calculates (exactly, if possible) the definite integral: $\int_a^b f(x) dx$ (does not add on integration constant). If there are other variables, they are treated as constant parameters.

EXAMPLE A.1: Use the Symbolic Toolbox to compute the following:

(a) $\dfrac{d}{dx}x^x$

(b) $\dfrac{\partial^3}{\partial x \partial y^2}\left(\dfrac{\cos(x+y^2+z^3)}{1+x^2+y^2}\right)$

(c) $\int \ln(x)\,dx$

(d) $\int \sin(x^2)\,dx$

(e) $\int_0^1 \sin(x^2)\,dx$

(f) $\int_{-\infty}^{-\infty} e^{-x^2}\,dx$

SOLUTION: Part (a):

```
>> syms x y z
>> diff(x^x)     →ans =x^x*(log(x)+1)
```

So the answer is $x^x(\ln x + 1)$.

Part (b):

```
>> f=cos(x+y^2+z^3)/(1+x^2+y^2);
>> pdf=diff(diff(f,y,2),x)     →pdf =
4*sin(x+y^2+z^3)*y^2/(x^2+1+y^2)*2/(x^2+1+y^2)^2*x-
2*cos(x+y^2+z^3)/(x^2+1+y^2)+4*sin(x+y^2+z^3)/(x^2+1+y^2)^2*x+8*cos(x+y^2+z^3)*y^2/(x^
2+1+y^2)^2-32*sin(x+y^2+z^3)*y^2/(x^2+1+y^2)^3*x-8*sin(x+y^2+z^3)/(x^2+1+y^2)^3*y^2-
48*cos(x+y^2+z^3)/(x^2+1+y^2)^4*y^2*x+2*sin(x+y^2+z^3)/(x^2+1+y^2)^2+8*cos(x+y^2+z^3)
/(x^2+1+y^2)^3*x
```

We shall refrain from putting this mess in usual mathematical notation, but we will do something else with it later (which is why we gave it a name).

Part (c): `>>int(log(x))`   → ans =x*log(x)-x

Part (d): >> `int(sin(x^2),x)` →ans =1/2*2^(1/2)*pi^(1/2)*FresnelS(2^(1/2)/pi^(1/2)*x)

This answer to part (d) needs a bit of explanation. Most indefinite integrals cannot be expressed in terms of the elementary functions. Using some additional **special functions** (e.g., Bessel functions, hypergeometric functions, the error function, and the above Fresnel sine function), additional integrals can be computed (but still only relatively few); thus MAPLE has found an antiderivative for us, but for most practical purposes this answer by itself is not so interesting. A similar result turns up (by the fundamental theorem of calculus) for the corresponding definite integral.

Part (e):>> `int(sin(x^2),x,0,1)` →ans =1/2*FresnelS(2^(1/2)/pi^(1/2))*2^(1/2)*pi^(1/2)

The following commands show how to get a more useful decimal answer out of this or any answer to a symbolic computation:

vpa(a,d) →	If a is a symbolic answer representing a number and d is a nonnegative number, this command will convert the number a to decimal form with d significant digits. vpa stands for variable precision arithmetic. The default value is d=32.[1]
digits(d) vpa(a) →	Has the same result as above, but now the default value of d=32 digits of MAPLE's arithmetic is reset to d in subsequent calculations.

```
>> vpa(ans)    →ans =.31026830172338110180815242316540
```

If we (for whatever reason) wanted to see the first 100 digits of π, we could simply enter:

```
>>vpa(pi,100) →ans=3.1415926535897932384626433832795028841971693993751 0582
                097494459230781640628620899862803482534211 7068
```

Part (f): Improper integrals are done with the same syntax as proper integrals.

```
>> int(exp(-x^2),x,-Inf, Inf)    →ans = pi^(1/2)
```

Thus we get that $\int_{-\infty}^{-\infty} e^{-x^2}\,dx = \sqrt{\pi}$.

Often, we need to evaluate a symbolic expression or substitute some of its variables with other variables or expressions. The following command subs is very useful in this respect:

subs(S,old,new) →	If S is a symbolic expression, old is a symbolic variable appearing in S (or a vector of variables), new is a symbolic number or symbolic expression (or a vector of such things having the same size as old), this command will produce the symbolic expression resulting from substituting in S each occurrence of old by the corresponding expression in new.

For example, suppose (in the setting of Example A.1) we wanted to compute

$$\frac{\partial^3}{\partial x \partial y^2}\left(\frac{\cos(x+y^2+z^3)}{1+x^2+y^2}\right)\Bigg|_{\substack{x=\pi \\ y=\pi/2 \\ z=0}}.$$

From what we have already computed, we could simply enter:

```
>> subs(pdf,[x y z], [pi pi/2 0])    → ans =-0.2016
```

---

[1] Thus, MAPLE uses approximately a 32-digit floating point arithmetic system in cases where exact answers are not possible. This is about double of what MATLAB uses and for many computations is overkill since large-scale calculations would proceed much more slowly. Thus, generally speaking, use of the Symbolic Toolbox should be limited to symbolic computations, except in the occasional instances where, say, the problem being solved is very ill-conditioned and roundoff errors run out of control with IEEE floating point arithmetic (see Chapter 5).

Since all symbolic variables were substituted with nonsymbolic (ordinary MATLAB floating point) numbers, the result is now a regular MATLAB floating point number. To retain the accuracy of symbolic computation in the substitution, we could instead enter:

```
>> exact=subs(pdf,[x y z], sym([pi pi/2 0])); %suppress messy output
>> vpa(exact) %could specify more or less digits here.
→ans = -.20163609585811087949860391144560
```

Note that the main difference is that in the latter we declared the numbers to be symbolic (exact):

fpn=double(sbn) →	If sbn is a (MAPLE) symbolic number, this command creates a (MATLAB) floating point number fpn from it essentially by rounding it off to about 16 digits of accuracy.
sbn=sym(fpn) →	If fpn is a (MATLAB) floating point number, this command creates a (MAPLE) symbolic number sbn from it by treating it as an exact number.

The Symbolic Toolbox has a simple way for computing Taylor series:

taylor(<fun>,n,a) →	If <fun> is a symbolic expression representing a function of a (previously declared) symbolic variable (say x), n is a positive integer, and a is a real number, this command will produce the Taylor polynomial of the function centered at $x = $ a of order (degree at most) n−1. The last input a is optional, the default value is a = 0.

EXAMPLE A.2: Obtain the 15th-order Taylor polynomial of $f(x) = x^2 \tan(x^3)$ centered at $x = 0$.

SOLUTION:

```
>> taylor(x^3*tan(x^2),16)    →ans =x^5+1/3*x^9+2/15*x^13
```

In the notation of Chapter 2, we can thus write $p_{15}(x) = x^5 + \dfrac{x^9}{3} + \dfrac{2x^{13}}{15}$.

A.4: ORDINARY DIFFERENTIAL EQUATIONS[2]

Analytic (symbolic) solutions of ordinary differential equations and systems of them, if they exist, can be found using the dsolve function from the Symbolic Toolbox.[3] Since the function has many available features, we roughly indicate the possible syntaxes for its use and give examples of each.

---

[2] Since this book does not assume that the reader has had any experience with differential equations, it is advised that those readers without such experience wait to read this subsection until they have started studying Part II of the book (ordinary differential equations).

[3] Although most ODEs (like indefinite integrals) do not have analytic solutions, this tool is occasionally useful when dealing with special well-known types of ODE which do have analytic solutions. The Symbolic Toolbox freely uses a collection of special functions when it looks for symbolic solutions.

dsolve('<diff_eq>') →	Looks for the analytic general solution of the differential equation: <diff_eq>, in which first, second, third, etc. derivatives are denoted by D, D2, D3, etc., using the default independent variable t.
dsolve('<diff_eq>','var') →	Works as above but specifies the independent variable to be var.

EXAMPLE A.3: Find, if possible, analytic general solutions of the following ODEs:

(a) $y' = y^2 - 2y$, $y = y(t)$

(b) $u'' + 5u' - 6u = \cos(x)$, $u = u(x)$

(c) $y'' = y^2 - 2y$, $y = y(t)$

SOLUTION: Part (a):

```
>> y= dsolve('Dy=y^2-2*y')
→y =2/(1+2*exp(2*t)*C1)
```

So we have the general solution, $y(t) = \dfrac{2}{1 + 2Ce^{2t}}$, where C is an arbitrary constant. Note that the dsolve did not even require us to declare any symbolic variables. The subs function, however, does require symbolic variables. Thus, if we try to set C1 equal to zero in y directly, we get an error message. But by first declaring C1 as a symbolic variable, we get the intended result:

```
>> subs(y,C1,0)
??? Undefined function or variable 'C1'.

>> syms C1
>> subs(y,C1,0)
→ans =2
```

Part (b): If we do not specifically declare x as the independent variable, x will be treated as a constant and we get an unintended solution of a more trivial differential equation.. The second MATLAB code below gives us what we want.

```
>> dsolve('D2u+5*Du-6*u=cos(x)')
→ans =-1/6*cos(x)+C1*exp(t)+C2*exp(-6*t)

>> dsolve('D2u+5*Du-6*u=cos(x)', 'x')
→ans = -7/74*cos(x)+5/74*sin(x)+C1*exp(x)+C2*exp(-6*x)
```

So we have the general solution:

$$u(x) = C_1 e^x + C_2 e^{-6x} - (7/74)\cos(x) + (5/74)\sin(x)$$

where C_1, C_2 are arbitrary constants.

Part (c):

```
>> dsolve('D2y=y^2-2*y', 'x')
→Warning: Explicit solution could not be found; implicit solution returned.
> In C:\MATLAB6p5\toolbox\symbolic\dsolve.m at line 292
→ans =[ 3*Int(1/(6*a^3-18*a^2+9*C1)^(1/2),a=``..y)-x-C2=0, -3*Int(1/(6*a^3-
18*a^2+9*C1)^(1/2),a=``..y)-x-C2=0]
```

Thus we see that, despite its simplicity (and similarity to the ODE in part (a)), the ODE of Part (c) does not have symbolic solutions.

The `dsolve` function can also solve initial and boundary value problems, the conditions need only be inserted as additional inputs after the DE:

dsolve('<diff_eq>','cond1', 'cond2', ..., 'var') →	Syntax is as above but with additional inputs corresponding to auxiliary conditions (boundary or initial) which we would like the solution to satisfy.

EXAMPLE A.4: Solve the following ODE problems.

(a) $\begin{cases} y'(t) = 2ty \\ y(1) = 1 \end{cases}$

(b) $\begin{cases} y'' + y = e^x \cos(x) \\ y(0) = 1, \ y'(\pi) = 0 \end{cases}$

SOLUTION:

Part (a):

```
>> y=dsolve('Dy=2*t*y','y(1)=1')       →y =1/exp(1)*exp(t^2)
```

Thus we get the exact solution $y(t) = e^{t^2 - 1}$

Part (b):

```
>> y=dsolve('D2y+y=exp(x)*cos(x)','y(0)=1', 'Dy(pi)=0','x')
→ans =-1/10*exp(x)*(sin(2*x)2*cos(2*x))*cos(x) +
(1/5*(cos(x)+2*sin(x))*exp(x)*cos(x)+2/5*exp(x))*sin(x)+4/5*cos(x)+(-3/5*cosh(pi)-
3/5*sinh(pi))*sin(x)
```

To plot a symbolic function, we could use the `subs` command to create vectors of *y*-coordinates and plot using MATLAB as shown in Chapter 1. Alternatively, the Symbolic Toolbox supplies a function `ezplot` that will directly and painlessly plot a symbolic function of a single symbolic variable.

ezplot(f, [a b]) →	If f represents a symbolic function of a single symbolic variable (say x), and a < b are real numbers, this command will produce a plot of f(x) over the interval [a, b].

With *y* still stored as the solution of the last boundary value problem, the following command will result in the plot shown in Figure A.1.

```
>> ezplot(y, [0 pi])
```

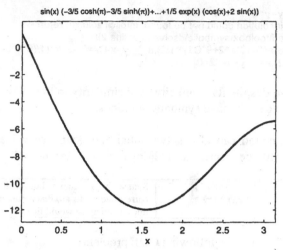

FIGURE A.1: Plot of the solution of the boundary value problem of Example A.4(b).

The final useful feature of `dsolve` is that it can solve systems of ODE. The syntax is a natural extension of the previous codes:

dsolve('<diff_eq1>','<diff_eq2>',... 'cond1','cond2', ...,'var') →	Syntax is as above but with additional differential equations with other unknown functions and a listing of all additional conditions to be satisfied by the unknown functions.

EXAMPLE A.5: Solve the following linear first order system of ODEs:

$$\begin{cases} x'(t) = 3x + 2y + z, & x(0) = 1 \\ y'(t) = x - y + z, & y(0) = 2 \\ z'(t) = 2x + 2y + 2z, & z(0) = 3 \end{cases}.$$

SOLUTION:

```
>> [x, y, z]=dsolve('Dx=3*x+2*y+z','Dy=x-
y+z','Dz=2*(x+y+z)','x(0)=1','y(0)=2','z(0)=3')
```

→x =4/41*(-328*exp(t)+369*exp(-1/2*(-3+41^(1/2))*t)-81*41^(1/2)*exp(-1/2*(-3+41^(1/2))*t)+81*41^(1/2)*exp(1/2*(3+41^(1/2))*t)+369*exp(1/2*(3+41^(1/2))*t))/(1+41^(1/2))/(-1+41^(1/2))

y =-2/41*(-738*exp(-1/2*(-3+41^(1/2))*t)-18*41^(1/2)*exp(-1/2*(-3+41^(1/2))*t)-164*exp(t)+18*41^(1/2)*exp(1/2*(3+41^(1/2))*t)-738*exp(1/2*(3+41^(1/2))*t))/(1+41^(1/2))/(-1+41^(1/2))

z =-4/41*(-369*exp(-1/2*(-3+41^(1/2))*t)+81*41^(1/2)*exp(-1/2*(-3+41^(1/2))*t)-492*exp(t)-81*41^(1/2)*exp(1/2*(3+41^(1/2))*t)-369*exp(1/2*(3+41^(1/2))*t))/(1+41^(1/2))/(-1+41^(1/2))

By themselves, these solutions do not appear to be very enlightening. But like any other symbolic functions, they can be manipulated and combined and vectors can be created from them using `subs`, so that much qualitative analysis, as is done in the text, can be performed.

Appendix B: Solutions to All Exercises for the Reader

NOTE: All of the M-files of this appendix (like the M-files of the text) are downloadable as text files from the ftp site for this text:

`ftp://ftp.wiley.com/public/sci_tech_med/numerical_differential/`

Occasionally, for space considerations, we may refer a particular M-file to this site. Also, in cases where a long MATLAB command does not fit on a single line (in this appendix), it will be continued on the next line. In an actual MATLAB session, (long) compound commands should either be put on a single line, or three periods (...) should be entered after a line to hold off MATLAB's execution until the rest of the command is entered on subsequent lines and the ENTER key is pressed. The text explains these and other related concepts in greater detail.

CHAPTER 1: MATLAB BASICS

EFR 1.1: `linspace(-2,3,11)`

EFR 1.2: `t = 0:.01:10*pi;   x = 5*cos(t/5)+cos(2*t);`
`y = 5*sin(t/5)+sin(3*t); plot(x,y), axis('equal')`

EFR 1.3: Simply run the code through MATLAB to see if you analyzed it correctly.

CHAPTER 2: BASIC CONCEPTS OF NUMERICAL ANALYSIS WITH TAYLOR'S THEOREM

EFR 2.1: `x=-10:.05:10; y=cos(x); p2=1-x.^2/2; p4=1-`
`x.^2/2+x.^4/gamma(5); p6=1-x.^2/2+x.^4/gamma(5)-x.^6/gamma(7); p8=1-`
`x.^2/2+x.^4/gamma(5)-x.^6/gamma(7)+x.^8/gamma(9); p10=p8-`
`x.^10/gamma(11); hold on, plot(x,p10,'k:'), axis([-2*pi 2*pi -1.5`
`1.5]), plot(x,p8,'c:'), plot(x,p6,'r-.'), plot(x,p4,'k--'),`
`plot(x,p2,'g'), plot(x,y,'+')`

EFR 2.2: Computing the first few derivatives of :

$$f(x) = x^{1/2}, \; f'(x) = \frac{1}{2}x^{-1/2}, \; f''(x) = -\frac{1 \cdot 1}{2 \cdot 2}x^{-3/2}, \; f'''(x) = \frac{1 \cdot 1 \cdot 3}{2 \cdot 2 \cdot 2}x^{-5/2},$$

$$f^{(4)}(x) = -\frac{1 \cdot 1 \cdot 3 \cdot 5}{2 \cdot 2 \cdot 2 \cdot 2}x^{-7/2} \; ..., \text{ leads us to discover the general pattern:}$$

$$f^{(n)}(x) = \; (-1)^{n+1}\frac{1 \cdot 3 \cdot 5 \cdots (2[n-1]-1)}{2^n}x^{-(2n-1)/2} \text{ (for } n \geq 2 \text{). Applying Taylor's theorem (with } a = $$

16, $x = 17$), we estimate the error of this approximation:

$$|R_n(17)| = \left| \frac{f^{(n+1)}(c)}{(n+1)!} 1^{n+1} \right| = \left| \frac{1 \cdot 3 \cdot 5 \cdots (2n-1)}{2^n (n+1)!} c^{-(2n+1)/2} \right| \le \frac{1 \cdot 3 \cdot 5 \cdots (2n-1)}{2^n (n+1)!} 16^{-(2n+1)/2} =$$

$$\frac{1 \cdot 3 \cdot 5 \cdots (2n-1)}{2^n (n+1)! \cdot 4 \cdot 16^n} = \frac{1 \cdot 3 \cdot 5 \cdots (2n-1)}{2^{5n+2} (n+1)!}. \text{ We use MATLAB to find the smallest } n \text{ for which this last}$$

expression is less than 10^{-10}; then Taylor's theorem will assure us that the Taylor polynomial of this order will provide us with the desired approximation.

```
>> n=2; ErrorEst=1*3/gamma(n+2)/2^(5*n+2);
>> while ErrorEst>1e-10, n=n+1; ErrorEst=ErrorEst*(2*n-1)/(n+1)/2^5;
end
>> n→n =7
>> ErrorEst→ErrorEst = 2.4386e-011 %this checks out.
```

So $p_7(17) = \sum_{k=0}^{7} \frac{1}{k!} f^{(k)}(16) \cdot 1^k$ will give the desired approximation. We use MATLAB to perform

and check it:
```
>> sum=16^(1/2)+16^(-1/2)/2; %first-order Taylor Polynomial
term = 16^(-1/2)/2; %first-order term
for k=2:7, term = -term*(2*(k-1)-1)/2/16/k; sum=sum+term; end, format
long
>> sum→sum = 4.12310562562925 (approximation)
>>abs(sum-sqrt(17))→ans =1.1590e-011 %actual error excels goal
```

EFR 2.3:   Using ordinary polynomial substitution, subtraction, and multiplication (and ignoring terms in the individual Maclaurin series that give rise to terms of order higher than 10), we use (9) and (10) to obtain: (a) $\sin(x^2) - \cos(x^3) =$

$$\left(x^2 - \frac{(x^2)^3}{3!} + \frac{(x^2)^5}{5!} - \cdots \right) - \left(1 - \frac{(x^3)^2}{2!} + \cdots \right) = -1 + x^2 - \left(\frac{1}{2!} - \frac{1}{3!} \right) x^6 + \frac{x^{10}}{5!} \cdots$$

(b) $\sin^2(x^2) = \left(x^2 - \frac{(x^2)^3}{3!} + \cdots \right) \cdot \left(x^2 - \frac{(x^2)^3}{3!} + \cdots \right) = x^4 - \frac{2}{3!} x^8 + \cdots$

In each case, $p_{10}(x)$ consists of all of the terms listed on the right-hand sides.

CHAPTER 3: INTRODUCTION TO M-FILES

EFR 3.1: In the left box we give the stored M-file; in the right we give the subsequent MATLAB session.

| ```
% script file for EFR 3.1:
listp2
power =2;
while power <= n
    power
    power=2*power;
end
``` | ```
>> n=5; listp2   → power = 2, power = 4
>> n=264;listp2
→ power = 2, power = 4, power = 8, power = 16,
power = 32, power = 64, power =128, power = 256,
>>n=2917;listp2
→ power = 2, power = 4, power = 8, power = 16,
power = 32, power = 64, power =128, power = 256,
power = 1024, power = 2048
``` |
|---|---|

Note: If we wanted the output to be just a single vector of the powers of 2, the following modified script would do the job:

```
% script file for EFR 3.1:  listp2ver2
power =2; vector = [ ]; %start off with empty vector
```

```
while power <= n
    vector = [vector power];
    power=2*power;
end, vector
```

For example, with this file stored, if we enter >> n=264; listp2ver2, we get the following vector output:

→vector = 2   4   8   16   32   64   128   256

EFR 3.2: With the boxed function M-file below saved, MATLAB will give the following outputs:

```
function f = fact(n)
% FACT f = fact(n) returns the factorial n! of a nonnegative integer
n
f=1;
for i=1:n
        f=f*i;
end
```

```
>> fact(4), fact(10), fact(0)
```
→ans = 24, 3628800, 1

EFR 3.3: At any (non-endpoint) maximum or minimum value $y(x_0)$, a differentiable function has its derivative equaling zero. This means that the tangent line is horizontal, so that for small values of $\Delta x \equiv x - x_0$, $\Delta y / \Delta x$ approaches zero. Thus, the y-variations are much smaller than the x-variations as x gets close to the critical point in question. This issue will be revisited in detail in Chapter 6.

EFR 3.4: We have only considered the values of y at a discrete set of (equally spaced) x-values. It is possible for a function to oscillate wildly in intervals between sets of discrete points (think trig functions with large amplitudes). More analysis can be done to preclude such pathologies (e.g., checking to see that there are no other critical points).

EFR 3.5: The M-file for the function is straightforward:

```
function y = wiggly(x)
%Function M-file for the mathematical function of EFR 3.5
y=sin(exp(1./(x.^2+0.5).^2)).*sin(x);
```

(a) >> x=-2:.001:2; plot(x,wiggly(x)) %plot is shown on left below
(b) >> quad(@wiggly,0,2,1e-5)   →ans = 1.03517910753379
(c) To better see what we are looking for, we create another plot of the function zoomed in near x = 0.
>> x=0:.001:.3; plot(x,wiggly(x)) %plot is shown (w/ other additions)
on right below.
We seek the x coordinates of the two points marked with "x's" in the figure below.

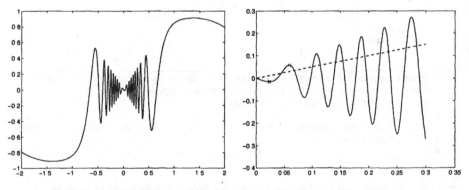

```
>> xmin=fminbnd(@wiggly,0,0.07,optimset('TolX',1e-5))
```
→xmin =0.02289435851906

```
>> xmax=fminbnd('-wiggly(x)',0,0.1,optimset('TolX',1e-5))
```
→xmax =0.05909071987402

Red and green x's can now be added to the graph as follows: >> hold on,
`plot(xmin,wiggly(xmin),'rx'), plot(xmin,wiggly(xmin),'gx')` (This also
gives us a visual check that we found what we were looking for.)

(d) To get a rough idea of the location of the x-value we are searching for, we now add the graph of the
line y = x/2 (as a black dotted line): >> `plot(x,x/2,'k--')`   From the graph, we see that the
intersection point we are looking for is the one closest to the midpoint of `xmin` and `xmax`.

```
>> xcross=fzero('wiggly(x)-x/2',(xmin+xmax)/2)
```
→ xcross =0.04479463640226

Let's do a quality check: >> `wiggly(xcross)-xcross/2`
→ans = 2.185751579730777e-016 (Very Good!)

CHAPTER 4: PROGRAMMING IN MATLAB

EFR 4.1: Simply run the code through MATLAB to see if you analyzed it correctly.

EFR 4.2: (a) The M-file is boxed below:

```
function [ ] = sum2sq(n)
%M-file for EFR 4.2
for a=1:sqrt(n)
    b=sqrt(n-a^2); %solve n=a^2+b^2 for b
    if b==floor(b); %checks to see if b is integer
        fprintf('the integer %d can be written as the sum of squares
of %d and %d', n,a,b)
        return
    end
end
fprintf('the integer %d cannot be written as the sum of squares', n)
```

(b) We now perform the indicated program runs:
>> `sum2sq(5)` →the integer 5 can be written as the sum of squares of 1 and 2
>> `sum2sq(25)` →the integer 25 can be written as the sum of squares of 3 and 4
>> `sum2sq(12233)` →the integer 12233 can be written as the sum of squares of 28 and 107

(c) The following modification of the above M-file will be more suitable to solving this problem:

```
function flag = sum2sqb(n)
%M-file for EFR 4.2b
flag=0; %will change to 1 if n can be written as a^2+b^2
for a=1:sqrt(n)
    b=sqrt(n-a^2); %solve n=a^2+b^2 for b
    if b==floor(b); %checks to see if b is integer
        flag=1;
        return
    end
end
```

The program has output 1 if and only if *n* is expressible as a sum of squares; otherwise the output is
zero. Now the following simple code will compute the desired integer *n*:
>> `for n=99999:-1:1, flag=sum2sqb(n);`
`if flag==0`
`fprintf('%d is the largest integer less than 100,000 not expressible`
`as a sum of squares',n)`
`break`
`end`
`end`
→99999 is the largest integer less than 100,000 not expressible as a sum of squares
(We did not have to go very far.)

(d) A minor modification to the above code will give us what we want; simply change the for loop to
`>> for n=1001:1:99999` (and the wording in the `fprintf` statement).
We then find the integer to be 1001.
(e) The following code will determine what we are looking for:
`>> for n=2:99999, flag=sum2sqb(n); if flag==0, count=count+1; end,`
`end`
`>> count`
`→count =75972`
Note: Part (e) took only a few seconds. If the programs were written less efficiently, for example, if we had run a nested loop by letting a and b run separately between all integers from 0 to $\sqrt{n}$ (or larger), some parts of this problem (notably, part (e)) could not be done in a reasonable amount of computer time.

EFR 4.3: (a) Before you run the indicated computations in a MATLAB session, try to figure out the output by hand. This will assure that you understand both the Collatz sequence generation process as well as the program. The reason for clearing the vector a at the end of the script is so that on subsequent runs, this vector will not start with old values from previous runs.
(b) The M-file is boxed below:

```
function n = collctr(an)
n=0;
while an ~= 1
    if ceil(an/2)==an/2   %tests if an is even
        an=an/2;
    else
        an=3*an+1;
    end
    n=n+1;
end
```

EFR 4.4: (a) The M-file is boxed below:

```
%raffledraw.m
%scriptfile for EFR 4.4

K = input('Enter number of players:   ');
N=zeros(K,26); %this allows up to 26 characters for each players
                %name.
n=input('Enter IN SINGLE QUOTES first player name:   ');
len(1)=length(n);
N(1,1:len(1))=n;
W(1)=input('Enter weight of first player:  ');
for i=2:K-1
    n=input('Enter IN SINGLE QUOTES next player name:  ');
    len(i)=length(n);
    N(i,1:len(i))=n;
    W(1)=input('Enter weight of this player:  ');
end
n=input('EnterIN SINGLE QUOTES last player name:  ');
len(K)=length(n);
N(K,1:len(K))=n;
W(K)=input('Enter weight of last player:  ');

totW = sum(W); %total weight of all players (=# of raffle tickets)

%the next four commands are optional, they only add suspense and
%drama to the raffle drawing which the computer can do in lightning
%time
fprintf('\r \r RANDOM SELECTION PROCESS INITIATED \r \r ...')
pause(1) %creates a 1 second pause
```

```
fprintf('\r \r ...SHUFFLING....\r \r')
pause(5) %creates a 5 second pause
%%%%%%%%%%%%%%%%%%%%%%%%%%%%

rand('state',sum(100*clock))
magic = floor(totW*rand); %this will be a random number between 0 and
                          %totW
count =W(1); %number of raffle tickets of player 1
if magic<=count
    fprintf('WINNER IS %s \r \r', char(N(1,1:len(1))))
    fprintf('CONGRATULATIONS %s!!!!!!!!!!!!!!!', char(N(1,1:len(1))))
    return
else count = count + W(2); k=2;
    while 1
        if magic <=count
          fprintf('WINNER IS %s \r \r', char(N(k,1:len(k))))
        fprintf('CONGRATULATIONS %s!!!!!!!!!!!!!!!', char(N(k,1:len(k))))
        return
        end
        k=k+1; count = count +W(k);
    end
end
```

(b) We now perform the indicated program runs:
>> raffledraw
Enter number of players: 4
Enter IN SINGLE QUOTES first player name: 'Alfredo'
Enter weight of first player: 4
Enter IN SINGLE QUOTES next player name: 'Denise'
Enter weight of this player: 2
Enter IN SINGLE QUOTES next player name: 'Sylvester'
Enter weight of this player: 2
Enter IN SINGLE QUOTES last player name: 'Laurie'
Enter weight of last player: 4

RANDOM SELECTION PROCESS INITIATED
... ...SHUFFLING....
→WINNER IS Laurie
→CONGRATULATIONS Laurie!!!!!!!!!!!!

On a second run the winner was Denise. If written correctly, and if this same raffledraw is run many times, it should turn out (from basic probability) that Alfredo and Laurie will each win roughly 4/12 or 33 1/3% of the time while Denise and Sylvester will win roughly 2/12 or 16 2/3% of the time.

CHAPTER 5: FLOATING POINT ARITHMETIC AND ERROR ANALYSIS

EFR 5.1: For shorthand we write: FPA to mean "the floating point answer," EA to mean "the exact answer," E to mean the "error" = |FAP-EA|, and RE to mean "the relative error" = E/|EA|.
(a) FPA = 0.023, EA = 0.0225, E = 0.0005, RE = 0.02222 ···
(b) FPA = 370,000 × .45 = 170,000, EA = 164990.2536, E = 5009.7464, RE = 0.030363...
(c) FPA = 8000 ÷ 120 = 67 , EA = 65.04878... , E = 1.9512195121 ···, RE = 0.029996...

EFR 5.2: (a) As in the solution of Example 5.3, since the terms are decreasing, we continue to compute partial sums (in 2-digit rounded floating point arithmetic) until the terms get sufficiently small so as to no longer have any effect on the accumulated sum.

$S_1 = 1$, $S_2 = S_1 + 1/2 = 1 + .5 = 1.5$, $S_3 = S_2 + 1/3 = 1.5 + .33 = 1.8$, $S_4 = S_3 + 1/4 = 1.8 + .25 = 2.1$,

$S_5 = S_4 + 1/5 = 2.1 + .2 = 2.3$, $S_6 = S_5 + 1/6 = 2.3 + .17 = 2.5$, $S_7 = S_6 + 1/7 = 2.5 + .14 = 2.6$,

$S_8 = S_7 + 1/8 = 2.6 + .13 = 2.7$, $S_9 = S_8 + 1/9 = 2.7 + .11 = 2.8$, $S_{10} = S_9 + 1/10 = 2.8 + .1 = 2.9$.

This pattern continues until we reach S_{20}: In each such partial sum S_k, $1/k$ contributes 0.1 to the cumulative sum. As soon as we reach S_{21}, the terms $(1/21 = 0.048)$ in floating point arithmetic become too small to have any effect on the cumulative sum so we have converged; thus the final answer is: $2.9 + 10 \times .1 = 3.9$.

(b) (i) $x^2 = 100$: Working in exact arithmetic, there are, of course, two solutions: $x = \pm 10$. These are also floating point solutions and any other floating point solutions will lie in some intervals about these two. Let's start with the floating point solution $x = 10$. In arithmetic of this problem, the next floating point number greater than 10 is 11 and (in floating point arithmetic) $11^2 = 120$, so there are no floating point solutions greater than 10. Similarly the floating point number immediately preceding 10 is 9.9 and (in floating point arithmetic) $9.9^2 = 98$, so there are no (positive) floating point solutions less than 10. Similarly, -10 is the only negative floating point solution. Thus there are exactly two floating point solutions (or more imprecisely: between 2 and 10 solutions).

(ii) $8x^2 = x^5$: In exact arithmetic, we would factor this $x^5 - 8x^2 = x^2(x^3 - 8) = 0$ to get the real solutions: $x = 0$ and $x = 2$. Because of underflow, near $x = 0$, we can get many (more than 10) floating point solutions. Indeed, since $e = 8$, if $|x| < 10^{-5}$, then both sides of the equation will underflow to zero so we will have a solution. Any number of form $\pm a.b \times 10^{-c}$, where a and b are any digits ($a \neq 0$) and $c = 6$, 7, or 8, will thus be a floating point solution, so certainly there are more than 10 solutions. (How many are there exactly?)

EFR 5.3: (a) As in the solution to Example 5.4, we may assume that $x \neq 0$ and write $x = .d_1 d_2 \cdots d_s d_{s+1} \cdots \times 10^e$. Now, since we are using s-digit rounded arithmetic, $fl(x)$ is the closer of the two numbers $.d_1 d_2 \cdots d_s \times 10^e$ and $.d_1 d_2 \cdots d_s \times 10^e + 10^{-s} \times 10^e$ to x. Since the gap between these two numbers has length $10^{-s} \times 10^e$, we may conclude that $|x - fl(x)| \leq \frac{1}{2} \cdot 10^{-s} \times 10^e$. On the other hand, $|x| > 100 \cdots 0 \times 10^e = 10^{e-1}$. Putting these two estimates together, we obtain the following estimate for the relative error: $\left| \frac{x - fl(x)}{x} \right| \leq \frac{\frac{1}{2} \cdot 10^{-s} \times 10^e}{10^{e-1}} = \frac{1}{2} \cdot 10^{1-s}$. Since equality is possible, we conclude that $u = \frac{1}{2} \cdot 10^{1-s}$, as asserted. The floating point numbers are the same whether we are using chopped or rounded arithmetic, so the gap from 1 to the next floating point number is still 10^{1-s}, as explained in the solution of Example 5.4.

(b) If $x = 0$, we can put $\delta = 0$; otherwise put $\delta = [fl(x) - x]/x$.

EFR 5.4: (a) Since $N - i \leq N$ when i is nonnegative, we obtain from (6) that

$$|fl(S_N) - S_N| \leq u[(N-1)a_1 + (N-1)a_2 + (N-2)a_3 + \cdots + 2a_{N-1} + a_N]$$
$$\leq u[Na_1 + Na_2 + Na_3 + \cdots + Na_{N-1} + Na_N] = Nu \sum_{n=1}^{N} a_n.$$

(b) Simply divide both sides of the inequality in (a) by $\sum_{n=1}^{N} a_n$ obtain the inequality in (b).

EFR 5.5: From $1 - \frac{1}{3} + \frac{1}{5} - \frac{1}{7} + \cdots = \frac{\pi}{4}$, we can write $\pi = 4 - \frac{4}{3} + \frac{4}{5} - \frac{4}{7} + \cdots = \sum_{n=0}^{\infty} (-1)^n a_n$, where $a_n = 4/(2n+1)$. Letting S_N denote the partial sum $\sum_{n=0}^{N}(-1)^n a_n$, Leibniz's theorem tells us that Error $\equiv |\pi - S_N| \leq a_{N+1} = 4/(2N+3)$. Since we want Error $< 10^{-7}$, we should take N large enough to satisfy $4/(2N+3) < 10^{-7} \Rightarrow 2N+3 > 4 \cdot 10^7 \Rightarrow N > (4 \cdot 10^7 - 3)/2 = 19{,}999{,}998.5$. Letting $N = 19{,}999{,}999$, we get MATLAB to perform the summation of the corresponding terms in order of increasing magnitude:

```
>> format long
>> Sum=0; N=19999999;
>> for n=N:-1:0
Sum=Sum+(-1)^n*4/(2*n+1);
end
>> Sum
→Sum = 3.14159260358979 (approximation to π )
>> abs(pi-Sum)
→ans = 4.999999969612645e-008 (exact error of approximation)
```

CHAPTER 6: ROOTFINDING

EFR 6.1: The accuracy of the approximation x7 is actually better than what was guaranteed from (1). The actual accuracy is less than 0.001 (this can be shown by continuing with the bisection method to produce an approximation xn with guaranteed accuracy less than 0.00001 (how large should n be?) and then estimating $|x7 - \text{root}| \leq |x7 - xn| + |xn - \text{root}| \leq 9 \times 10^{-4} + 1 \times 10^{-5} < 0.001$. So actually, $|f(x7)|$ is over 30 times as large as $|x7 - \text{root}|$. This can be explained by estimating $y'(\text{root}) \geq 30$ (do it graphically, for example). Thus, for small values of $\Delta x \equiv x - \text{root}$, $\Delta y / \Delta x$ gets larger than 30. This is why the y-variations turn out to be more than 30 times as large as the x-variations, when x gets close to the root.

EFR 6.2: (a) Since $f(0) = 1 - 0 > 0$, $f(\pi/2) = 0 - \pi/2 < 0$, and $f(x)$ is continuous, we know from the intermediate value theorem that $f(x)$ has a root in $[0, \pi/2]$. Since $f'(x) = \sin(x) - 1 < 0$ on $(0, \pi/2)$, $f(x)$ is strictly decreasing so it can have only one root on $[0, \pi/2]$.

(b) It is easy to check that the first value of n for which $\pi/(2 \cdot 2^n)$ $(= (b - a)/2^n)$ is less than 0.01 is $n = 8$. Thus by (1), using $x0 = 0$, it will be sufficient to run through $n = 8$ iterations of the bisection method to arrive at an approximation x8 of the root that has the desired accuracy. We do this with the following MATLAB loop:

```
>> xn=0; an=0; bn=pi/2; n=0;
>> while n<=8
xn=(an+bn)/2; n=n+1;
if f(x)==0, root = xn; return
elseif f(x)>0, an=xn; bn=bn;
else, an=an; bn=xn;
end
end
>> xn
→xn =0.73937873976088
```

(c) The following simple MATLAB loop will determine the smallest value of n for which $\pi/(2 \cdot 2^n)$ will be less than 10^{-12} (by (1) this would be the smallest number of iterations in the bisection method for which we could be guaranteed the indicated accuracy). (This could certainly also be done using logs.)

```
>> while pi/2/2^n>=1e-12, n=n+1; end
>> n
→n = 41
```

```
>> pi/2/2^41, pi/2/2^40   %we perform a check
→ans = 7.143154683921678e-013 (OK)   1.428630936784336e-012 (too big, so it checks!)
```

EFR 6.3: (a) The condition yn*ya > 0 mathematically translates to yn and ya having the same sign, so this is (mathematically) equivalent to our condition `sign(yn)==sign(ya)`.

(b) We are aiming for a root so at each iteration, yn and ya should be getting very small; thus their product yn*ya will be getting smaller much faster (e.g., if both are about 1e-175, then their product would be close to 1e-350 and this would underflow). Thus, with the modified loop we run the risk of a premature underflow destroying any further progress of the bisection method.

(c) Consider the function $f(x) = (x+.015)^{101}$, which certainly has a (unique) root $x = -0.015$ and satisfies the requirements for using the bisection method. As soon as the interval containing xn gets to within 1e-2 of the root, both y-values yn and ya would then be less than 1e-200; so their product would be less than 1e-400 and so would underflow to zero. This starts to occur already when $n = 2$ (xn = 0), and causes the modified if-branch to default to the else-if option—taking the left half subinterval as the new interval. From this point on, all approximations will be less than −0.5, making it impossible to reach the 0.001 accuracy goal.

EFR 6.4: The distance from x to e is less than MATLAB's unit roundoff and the minimum gap between floating point numbers (see Example 5.4 and Exercise for the Reader 5.3). Thus MATLAB cannot distinguish between the two numbers x and e, and (in the notation of Chapter 5) we have fl(x) = fl(e) = e (since important numbers like e are built in to MATLAB as floating point numbers). As a result, when MATLAB evaluates ln(x), it really computes ln(fl(x)) = ln(e) and so gets zero.

EFR 6.5: (a) If we try to work with quadratic polynomials (parabolas), cycling cannot occur unless the parabola did not touch the x-axis (this is easy to convince oneself of with a picture and not hard to show rigorously). If we allow polynomials that do not have roots, then an example with a quadratic polynomial is possible, as shown in the left-hand figure below. For a specific example, we could take $f(x) = x^2 + 1$. For cycling as in the picture, we would want $x_1 = x_0$. Putting this into Newton's formula and solving (the resulting quadratic) gives $x_0 = 1/\sqrt{3}$. One can easily run a MATLAB program to see that this indeed produces the asserted cycling. To get an example of polynomial cycling with a polynomial that actually has a root we need to use at least a third-degree polynomial. Working with $f(x) = x^3 - x = x(x-1)(x+1)$, which has the three (equally spaced) roots $x = 0, +1$ the graph suggests that we can have a period-two cycling, so we put $x_1 = x_0$ into Newton's formula. The resulting cubic equation is easily solved exactly (it factors) or with the Symbolic Toolbox (see Appendix A) or approximately using Newton's method. The solution $x_0 = 1/\sqrt{5}$ produces the period-two cycling shown in the right-hand figure below, as can be checked by running Newton's method.

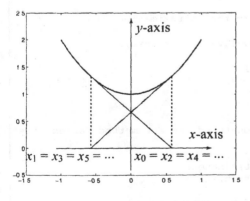

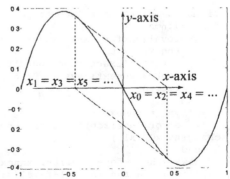

(b) On the right is an illustration of a period-four cycle in Newton's method. An explicit such example is furnished by $f(x) = x^3 - x - 3$. The calculations would be, of course, more elaborate than those of part (a); it turns out that x_0 should be taken to be a bit less than zero. (More precisely, about -0.007446; you may wish to run a couple of hundred iterations of Newton's method using this value for x_0 to observe the cycling.) By contemplating the picture, it becomes clear that this function has cycles of any order. Just move x_0 closer to the right toward the location where $f'(x)$ has a root.

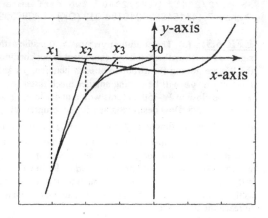

EFR 6.6: (a) The M-file is boxed below:

```
function [root, yval,niter] = secant(varfun,x0, x1, tol, nmax)
% input variables: varfun, x0, x1  tol, nmax
% output variables:  root, yval, niter
% varfun = the string representing a mathematical function (built-in,
% M-file, or inline) , x0  and x1 = the two (different) initial
% approx.
% The program will perform the Secant method to approximate a root of
% varfun near x=x0 until either successive approximations differ by
% less than tol or nmax iterations have been completed, whichever
% comes first.  If the tol and nmax variables are omitted default
% values of eps (approx. 10^(-16)) and 30 are used.
% We assign the default tolerance and maximum number of iterations if
% none are specified
if nargin < 4
   tol=eps; nmax=50;
end


%we now initialize the iteration
xn=x0;xnnext=x1;

%finally we set up a loop to perform the approximations
for n=1:nmax
   yn=feval(varfun, xn); ynnext=feval(varfun, xnnext);
   if ynnext == 0
      fprintf('Exact root found\r')
      root = xnnext; yval = 0; niter=n;
      return
   end
   if yn == ynnext
      error('horizontal secant encountered, Secant method failed, try
changing x0, x1')
   end
   newx=xnnext-feval(varfun, xnnext)*(xnnext-
xn)/(feval(varfun,xnnext)-feval(varfun,xn));
   if abs(newx-xnnext)<tol
         fprintf('The secant method has converged\r')
```

```
              root = newx; yval = feval(varfun, root); niter=n;
              return
       elseif n==nmax
              fprintf('Maximum number of iterations reached\r')
              root = newx; yval = feval(varfun, root); niter=nmax
              return
       end
       xn=xnnext; xnnext=newx;
end
```

(b) The syntax of this M-file is very close to that of newton:
>> f=inline('x^4-2'); [r y n] = secant(f, 2,1.5)
→The secant method has converged, r = 1.18920711500272,
y = -2.220446049250313e-016, n = 9
>> abs(r-2^(1/4))   →ans = 0
In conclusion, the secant method took nine iterations and the approximate root (r) had residual which
was essentially zero (in floating point arithmetic) and coincided with the exact answer $\sqrt[4]{2}$ (in floating
point arithmetic).

EFR 6.7: (a) For shorthand we write: HOC to mean "the highest order of convergence," and AEC
to mean "the asymptotic error constant." For each sequence, we determine these quantities if they
exist:
(i) HOC = 1; AEC = 1 (linear convergence), (ii) HOC = 1, AEC = 1/2 (linear convergence), (iii)
HOC = 3/2, AEC = 1, (iv) HOC = 2, AEC = 1 (quadratic convergence), (v) HOC does not exist.
There is hyperconvergence for every order $\alpha < 2$, but the sequence does not have quadratic
convergence.

(b) The sequence $e_n = e^{-3^n}$ has HOC = 3. In general, $e_n = e^{-k^n}$ has HOC = k whenever k is a positive
number.

EFR 6.8: Write $f(x) = (x-r)^M h(x)$, where M is the order of the root (and so $h(r) \neq 0$).
Differentiating, we see that the function $F(x) \equiv f(x)/f'(x)$ can be written as $F(x) = (x-r)H(x)$,
where $H(x) = h(x)/[Mh(x)+(x-r)h'(x)]$. Since $H(r) = 1/M \neq 0$, we see that $x = r$ is a simple root
of $F(x)$. Since $F'(x) = [(f'(x))^2 - f(x)f''(x)]/(f'(x))^2$, this method requires computing both
$f'(x)$ and $f''(x)$. The roundoff errors can also get quite serious. For example, if we are converging
to a simple root, then in the iterative computations of $F'(x_n) = [(f'(x_n))^2 - f(x_n)f''(x_n)]/(f'(x_n))^2$,
$(f'(x_n))^2$ will be converging to a positive number, while $f(x_n)f''(x_n)$ will be converging to zero.
Thus, when these two numbers are subtracted roundoff errors can be insidious. With higher-order roots
each of $(f'(x_n))^2$ and $f(x_n)f''(x_n)$ will be getting small very fast and can underflow to zero causing
Newton's method to stop. If the root is a multiple root and the order is known not to be too high then
this method performs reasonably well. If the order is known, however, the newtonmr method is a
better choice.

CHAPTER 7: MATRICES AND LINEAR SYSTEMS

EFR 7.1: Abbreviate the matrices in (1) by $DE = P$, and write $P = [p_{ij}]$. Now, by definition,
p_{ij} = (ith row of D) • (jth column of E) = $d_i \cdot e_{ij}$ (by diagonal form of D). But by the diagonal form of
E, e_{ij} (and hence also p_{ij}) is zero unless $i = j$, in which case $e_{ij} = e_i$. Thus
$p_{ij} = \{d_i e_i,$ if $i = j;$ 0, if $i \neq j$ and this is a restatement of (1).

EFR 7.2: (a) The M-file is boxed below:

```
function A=randint(n,m,k)
%generates an n by m matrix whose entries are random integers whose
%absolute values do not exceed k
A=zeros(n,m);
for i=1:n
    for j=1:m
        x=(2*k+1)*rand-k; %produces a random real number in (-k,k+1)
        A(i,j)=floor(x);
    end
end
```

(b) In the random experiments below, we print out the matrices only in the first trial.
```
>> A=randint(6,6,9); B=randint(6,6,9); det(A*B), det(A*B)-
det(A)*det(B)
```
→A =   9 -5 2 0 7 5   →B =   7 0 -6 3 6 -9

→A =	9	-5	2	0	7	5	→B =	7	0	-6	3	6	-9
	-1	-9	6	-1	2	6		3	-2	6	0	4	-1
	8	5	-6	-2	8	8		-4	-6	-6	3	-4	1
	-2	7	-8	-3	6	-9		-7	4	-2	7	7	2
	-7	-6	-6	2	-4	-6		0	8	6	3	6	3
	-9	5	-1	8	-1	-2		-3	-4	-3	1	4	-4

→ det(A*B) =   -1.9436e+010   →ans =   0

```
>> A=randint(6,6,9); B=randint(6,6,9); det(A*B), det(A*B)-
det(A)*det(B)
```
→ans = 6.8755e+009, 0
```
>> A=randint(6,6,9); B=randint(6,6,9); det(A*B), det(A*B)-
det(A)*det(B)
```
→ ans = 8.6378e+010, 0

The last output 0 in each of the three experiments indicates that formula (4) checks.
(c) Here, because of their size, we do not print out any of the matrices.
```
>> A=randint(16,16,9); B=randint(16,16,9); det(A*B), det(A*B)-
det(A)*det(B)
```
→ans = -1.2268e+035, 1.8816e+021
```
>> A=randint(16,16,9); B=randint(16,16,9); det(A*B), det(A*B)-
det(A)*det(B)
```
→ans =1.4841e+035, -6.9913e+021
```
>> A=randint(16,16,9); B=randint(16,16,9); det(A*B), det(A*B)-
det(A)*det(B)
```
→ans = 3.3287e+035, ans = 7.0835e+021

The results in these three experiments are deceptive. In each, it appears that the left and right sides of (4) differ by something of magnitude 10^{21}. This discrepancy is entirely due to roundoff errors!

Indeed, in each trial, the value of the determinant of AB was on the order of 10^{35}. Since MATLAB's (double precision IEEE) floating point arithmetic works with only about 15 significant digits, the much larger (35-digit) numbers appearing on the left and right sides of (4) have about the last 20 digits turned into unreliable "noise." This is why the discrepancies are so large (the extra digit lost came from roundoff errors in the internal computations of the determinants and the right side of (4)). Note that in part (b), the determinants of the smaller matrices in question had only about 10 significant digits, well within MATLAB's working precision.

EFR 7.3: Using the `fill` command as was done in the text to get the gray cat of Figure 7.3(b), you can get those other-colored cats by simply replacing the RGB vector for gray by the following: Orange → RGB = [1 .5 0], Brown → RGB = [.5 .25 0], Purple → RGB = [.5 0 .5]. Since each of these colors can have varying shades, your answers may vary. Also, the naked eye may not be able to distinguish between colors arising from small perturbations of these vectors (say by .001 or even .005). The RGB vector representing MATLAB's cyan is RGB = [0 1 1].

EFR 7.4: By property (10) (of linear transformations): $L(\alpha P_1) = \alpha L(P_1)$; if we put $\alpha = 0$, we get that $L(\vec{0}) = \vec{0}$ (where $\vec{0}$ is the zero vector). But a shift transformation $T_{V_0}(x,y) = (x,y) + V_0$ satisfies $T_{V_0}(\vec{0}) = \vec{0} + V_0 = V_0$. So the shift transformation T_{V_0} being linear would force $V_0 = \vec{0}$, which is not allowed in the definition of a shift transformation (since then T_{V_0} would then just be the identity transformation).

EFR 7.5: (a) As in the solution of Example 7.4, we individually multiply out the homogeneous coordinate transformation matrices (as per the instructions in the proof of Theorem 7.2) from right to left. The first transformation is the shift with vector (1,0) with matrix: $T_{(1,0)} \sim \begin{bmatrix} 1 & 0 & 1 \\ 0 & 1 & 0 \\ 0 & 0 & 1 \end{bmatrix} = H_1$. After

this we apply a scaling S whose matrix is given by $S \sim \begin{bmatrix} 2 & 0 & 0 \\ 0 & 1 & 0 \\ 0 & 0 & 1 \end{bmatrix} = H_2$. The homogeneous

cooordinate matrix for the composition of these two transformations is:
$M = H_2 H_1 = \begin{bmatrix} 2 & 0 & 0 \\ 0 & 1 & 0 \\ 0 & 0 & 1 \end{bmatrix}\begin{bmatrix} 1 & 0 & 1 \\ 0 & 1 & 0 \\ 0 & 0 & 1 \end{bmatrix} = \begin{bmatrix} 2 & 0 & 2 \\ 0 & 1 & 0 \\ 0 & 0 & 1 \end{bmatrix}$. We assume (as in the text) that we have left in the

graphics window the first (white) cat of Figure 7.3(a) and that the CAT matrix A is still in our workspace. The following commands will now produce the new "fat CAT":
```
>> H1=[1 0 1;0 1 0; 0 0 1]; H2=[2 0 0;0 1 0;0 0 1]; M=H2*H1
>> AH=A; AH(3,:)=ones(1,10);    %homogenize the CAT matrix
>> AH1=M*AH; % homogenized "fat CAT" matrix
>> hold on
>> plot(AH1(1,:), AH1(2,:), 'r')
>> axis([-2 10 -3 6]) % set wider axes to accommodate "fat CAT"
>> axis('equal')
```

The resulting plot is shown in the left-hand figure that follows.
(b) Each of the four cats needs to first get rotated by its specified angle about the same point (1.5, 1.5). As in the solution to Example 7.4, these rotations can be accomplished by first shifting this point to (0, 0) with the shift $T_{(-1.5,-1.5)}$, then performing the rotation, and finally shifting back with the inverse shift $T_{(1.5,1.5)}$. In homogeneous coordinates, the matrix representing this composition is (just like in the solution to Example 7.4).

$$M = \begin{bmatrix} 1 & 0 & 1.5 \\ 0 & 1 & 1.5 \\ 0 & 0 & 1 \end{bmatrix}\begin{bmatrix} \cos(\theta) & -\sin(\theta) & 0 \\ \sin(\theta) & \cos(\theta) & 0 \\ 0 & 0 & 1 \end{bmatrix}\begin{bmatrix} 1 & 0 & -1.5 \\ 0 & 1 & -1.5 \\ 0 & 0 & 1 \end{bmatrix}.$$

After this rotation, each cat gets shifted in the specified direction with $T_{(\pm1,\pm1)}$. For the colors of our cats let's use the following: black (rgb = [0 0 0]), light gray (rgb = [.7 .7 .7]), dark gray (rgb = [.3 .3. .3]), and brown (rgb = [.5 .25 0]). The following commands will then plot those cats:
```
>> clf, hold on  %prepare graphic window
>> %upper left cat,  theta = pi/6 (30 deg), shift vector = (-3, 3)
>> c = cos(pi/6);  s = sin(pi/6);
>> M=[1 0 1.5;0 1 1.5;0 0 1]*[c -s 0;s c 0;0 0 1]*[1 0 -1.5;0 1 -
1.5;0 0 1];
>> AUL=[1 0 -3;0 1 3;0 0 1]*M*AH;
>> fill(AUL(1,:), AUL(2,:), [0 0 0])
>> %upper right cat,  theta = -pi/6 (-30 deg), shift vector = (3, 1)
>> c = cos(-pi/6); s = sin(-pi/6);
>> M=[1 0 1.5;0 1 1.5;0 0 1]*[c -s 0;s c 0;0 0 1]*[1 0 -1.5;0 1 -
1.5;0 0 1];
>> AUR=[1 0 1;0 1 1;0 0 1]*M*AH;
```

```
>> fill(AUR(1,:), AUR(2,:), [.7 .7 .7])
>> %lower left cat,  theta = pi/4 (45 deg), shift vector = (-3, -3)
>> c = cos(pi/4); s = sin(pi/4);
>> M=[1 0 1.5;0 1 1.5;0 0 1]*[c -s 0;s c 0;0 0 1]*[1 0 -1.5;0 1 -
1.5;0 0 1];
>> ALL=[1 0 -3;0 1 -3;0 0 1]*M*AH;
>> fill(ALL(1,:), ALL(2,:), [.3 .3 .3])
>> %lower right cat,  theta = -pi/4 (-45 deg), shift vector = (3, -3)
>> c = cos(-pi/4); s = sin(-pi/4);
>> M=[1 0 1.5;0 1 1.5;0 0 1]*[c -s 0;s c 0;0 0 1]*[1 0 -1.5;0 1 -
1.5;0 0 1];
>> ALR=[1 0 3;0 1 -3;0 0 1]*M*AH;
>> fill(ALR(1,:), ALR(2,:), [.5 .25 0])
>> axis('equal'), axis off %see graphic w/out distraction of axes.
```

EFR 7.6: (a) This first M-file is quite straightforward and is boxed below.

```
function B=mkhom(A)
B=A;
[n m]=size(A);
B(3,:)=ones(1,m);
```

(b) This M-file is boxed below.

```
function Rh=rot(Ah,x0,y0,theta)
%viz. EFR 7.6; theta should be in radians
%inputs a 3 by n matrix of homogeneous vertex coordinates, xy
%coordinates of a point and an angle theta.  Output is corresponding
%matrix of vertices rotated by angle theta about (x0,y0).

%first construct homogeneous coordinate matrix for shifting (x0,y0)
to (0,0)
SZ=[1 0 -x0;0 1 -y0; 0 0 1];
%next the rotation matrix at (0,0)
R=[cos(theta) -sin(theta) 0; sin(theta) cos(theta) 0;0 0 1];
%finally the shift back to (x0,y0)
SB=[1 0 x0;0 1 y0;0 0 1];
%now we can obtain the desired rotated vertices:
Rh=SB*R*SZ*Ah;
```

EFR 7.7: (a) The main transformation that we need in this movie is vertical scaling. To help make the code for this exercise more modular, we first create, as in part (b) of the last EFR, a separate M-file for vertical scaling:

```
function Rh=vertscale(Ah,b,y0)
%inputs a 3 by n matrix of homogeneous vertex coordinates, a (pos.)
%numbers a for  y- scales, and an optional  arguments y0
```

```
%for center of scaling.  Output is homogeneous coor. matrix of scaled
%vertices.  default value of y0 is 0.

if nargin <3
    y0=0;
end
%first construct homogeneous coordinate matrix for shifting y=y0 to
%y=0
SZ=[1 0 0;0 1 -y0; 0 0 1];
%next the scaling matrix at (0,0)
S=[1 0 0; 0 b 0;0 0 1];
%finally the shift back to y=0
SB=[1 0 x0;0 1 y0;0 0 1];
%now we can obtain the desired scaled vertices:
Rh=SB*S*SZ*Ah;
```

Making use of the above M-file, the following script recreates the CAT movie of Example 7.4 using homogeneous coordinates:

```
%script for EFR 7.6(a):   catmovieNo1.m  cat movie creation
%Basic CAT movie, where cat closes and reopens its eyes.
clf, counter=1;

A=[0   0  .5  1   2  2.5  3  3  1.5  0; ...
     0  3   4  3   3   4   3  0  -1   0]; %Basic CAT matrix
Ah = mkhom(A); %use the M-file from EFR 7.6

t=0:.02:2*pi;  %creates time vector for parametric equations for eyes
xL=1+.4*cos(t); y=2+.4*sin(t); %creates circle for left eye
LE=mkhom([xL; y]); %homogeneous coordinates for left eye
xR=2+.4*cos(t); y=2+.4*sin(t); %creates circle for right eye
RE=mkhom([xR; y]); %homogeneous coordinates for right eye
xL=1+.15*cos(t); y=2+.15*sin(t); %creates circle for left pupil
LP=mkhom([xL; y]); %homogeneous coordinates for left pupil
xR=2+.15*cos(t); y=2+.15*sin(t); %creates circle for right pupil
RP=mkhom([xR; y]); %homogeneous coordinates for right pupil

for s=0:.2:2*pi
 factor = (cos(s)+1)/2;
 plot(A(1,:), A(2,:), 'k'), hold on
 axis([-2 5 -3 6]), axis('equal')
 LEtemp=vertscale(LE,factor,2); LPtemp=vertscale(LP,factor,2);
 REtemp=vertscale(RE,factor,2); RPtemp=vertscale(RP,factor,2);
 hold on
 fill(LEtemp(1,:), LEtemp(2,:),'y'), fill(REtemp(1,:),
REtemp(2,:),'y')
 fill(LPtemp(1,:), LPtemp(2,:),'k'), fill(RPtemp(1,:),
RPtemp(2,:),'k')
 M(:, counter) = getframe;
 hold off
 counter=counter+1;
end
```

(b) As in part (a), the following script M-file will make use of two supplementary M-files, AhR=reflx(Ah, x0) and, AhS=shift(Ah, x0, y0), that perform horizontal reflections and shifts in homogeneous coordinates, respectively. The syntaxes of these M-files are explained in Exercises 5 and 6 of this section.   Their codes can be written in a fashion similar to the code vertscale but for completeness are can be downloaded from the ftp site for this text (see the beginning of this appendix). They can be avoided by simply performing the homogeneous coordinate transformations directly, but at a cost of increasing the size of the M-file that we give:

```
%coolcatmovie.m: script for making coolcat movie matrix M of EFR 7.7
```

```
%act one:   eyes shifting left/right
t=0:.02:2*pi; counter=1;
A=[0   0   .5   1   2   2.5   3   3   1.5   0; ...
      0   3   4   3   3   4   3   0   -1   0];
x=1+.4*cos(t); y=2+.4*sin(t);xp=1+.15*cos(t); yp=2+.15*sin(t);
LE=[x;y]; LEh=mkhom(LE); LP=[xp;yp]; LPh=mkhom(LP);
REh=reflx(LEh, 1.5); RPh=reflx(LPh, 1.5);
LW=[.3 -1; .2 -.8]; LW2=[.25 -1.1;.25 -.6]; %left whiskers
LWh=mkhom(LW); LW2h=mkhom(LW2);
RWh=reflx(LWh, 1.5); RW2h=reflx(LW2h, 1.5); %reflect left whiskers
                                        %to get right ones
M=[1 1.5 2;.25 -.25 .25]; Mh=mkhom(M);   %matrix & homogenization of
                                      %cats mouth
Mhrefl=refly(Mh,-.25); %homogeneous coordinates for frown
for n=0:(2*pi)/20:2*pi
plot(A(1,:), A(2,:),'k')
axis([-2 5 -3 6]), axis('equal')
hold on
plot(LW(1,:), LW(2,:),'k'), plot(LW2(1,:), LW2(2,:),'k')
plot(RWh(1,:), RWh(2,:),'k')
plot(RW2h(1,:), RW2h(2,:),'k')
plot(Mhrefl(1,:), Mhrefl(2,:),'k')
fill(LE(1,:), LE(2,:),'y'), fill(REh(1,:), REh(2,:),'y')
LPshft=shift(LPh,-.25*sin(n),0); RPshft=shift(RPh,-.25*sin(n),0);
fill(LPshft(1,:), LPshft(2,:),'k'), fill(RPshft(1,:),
RPshft(2,:),'k')
Mov(:, counter)=getframe;
hold off
counter = counter +1;
end

%act two:   eyes shifting up/down
for n=0:(2*pi)/20:2*pi
plot(A(1,:), A(2,:),'k')
axis([-2 5 -3 6]), axis('equal')
hold on
plot(LW(1,:), LW(2,:),'k'), plot(LW2(1,:), LW2(2,:),'k')
plot(RWh(1,:), RWh(2,:),'k')
plot(RW2h(1,:), RW2h(2,:),'k')
plot(Mhrefl(1,:), Mhrefl(2,:),'k')
fill(LE(1,:), LE(2,:),'y'), fill(REh(1,:), REh(2,:),'y')
LPshft=shift(LPh,0,.25*sin(n)); RPshft=shift(RPh,0,.25*sin(n));
fill(LPshft(1,:), LPshft(2,:),'k'), fill(RPshft(1,:),
RPshft(2,:),'k')
Mov(:, counter)=getframe;
hold off
counter = counter +1;
end

%act three:   whisker rotating up/down then smiling
for n=0:(2*pi)/10:2*pi
plot(A(1,:), A(2,:),'k')
axis([-2 5 -3 6]), axis('equal')
hold on
fill(LE(1,:), LE(2,:),'y'),fill(LP(1,:), LP(2,:),'k')
fill(REh(1,:), REh(2,:),'y'),fill(RPh(1,:), RPh(2,:),'k')
LWrot=rot(LWh,.3,.2,-pi/6*sin(n)); LW2rot=rot(LW2h, .25,.25,-
pi/6*sin(n));
RWrot=reflx(LWrot, 1.5); RW2rot=reflx(LW2rot, 1.5);
```

```
plot(LWrot(1,:), LWrot(2,:),'k'), plot(LW2rot(1,:), LW2rot(2,:),'k')
plot(RWrot(1,:), RWrot(2,:),'k'),plot(RW2rot(1,:), RW2rot(2,:),'k')
if n == 2*pi
   plot(Mh(1,:), Mh(2,:),'k')
   for n=1:10, L(:,n)=getframe; end
   Mov(:, counter:(counter+9))=L;
   break
else
   plot(Mhrefl(1,:), Mhrefl(2,:),'k')

end
Mov(:, counter)=getframe;

hold off
counter = counter +1;
end

%THE END
```

EFR 7.8: (a) Certainly the zeroth generation consists of $1 = 3^0$ triangles. Since the sidelength is one, and the triangle has each of its angles being $\pi/3$, its altitude must be $\sin(\pi/3) = \sqrt{3}/2$. Thus, the area of the single zeroth generation triangle is $\sqrt{3}/4$. Now, each time we pass to a new generation, each triangle splits into three (equilateral) triangles of half the length of the triangles of the current generation. Thus, by induction, the nth generation will have 3^n equilateral triangles of sidelength $1/2^n$ and hence each of these has area $(1/2) \cdot 1/2^n \cdot [\sqrt{3}/2]/2^n = \sqrt{3}/4^{n+1}$.

(b) From part (a), the nth generation of the Sierpinski carpet consists of 3^n equilateral triangles each having area $\sqrt{3}/4^{n+1}$. Hence the total area of this nth generation is $\sqrt{3}(3/4)^n/4$. Since this expression goes to zero as $n \to \infty$, and since the Sierpinski carpet is contained in each of the generation sets, it follows that the area of the Sierpinski carpet must be zero.

EFR 7.9: (a) The 2×2 matrices representing dilations: $\begin{bmatrix} s & 0 \\ 0 & s \end{bmatrix}$ ($s > 0$), and reflections with respect to the x-axis: $\begin{bmatrix} -1 & 0 \\ 0 & 1 \end{bmatrix}$ or y-axis: $\begin{bmatrix} 1 & 0 \\ 0 & -1 \end{bmatrix}$ are both diagonal matrices and thus commute with any other 2×2 matrices; i.e., if D is any diagonal matrix and A is any other 2×2 matrix, then $AD = DA$. In particular, these matrices commute with each other and with the matrix representing a rotation through the angle θ: $\begin{bmatrix} \cos\theta & -\sin\theta \\ \sin\theta & \cos\theta \end{bmatrix}$. By composing rotations and reflections, we can obtain transformations that will reflect about any line passing through (0,0). Once we throw in translations, we can reflect about any line in the plane and (as we have already seen) rotate with any angle about any point in the plane. By the definition of similitudes, we now see that compositions of these general transformations can produce the most general similitudes. Translating into homogeneous coordinates (using the proof of Theorem 7.2) we see that the matrix for such a composition can be expressed as $\begin{bmatrix} s\cos\theta & -s\sin\theta & x_0 \\ \pm s\sin\theta & \pm s\cos\theta & y_0 \\ 0 & 0 & 1 \end{bmatrix}$ where s now is allowed to be any nonzero number. If the sign in the second row is negative, we have a reflection: If $s > 0$, it is a y-axis reflection; if $s < 0$, it is an x-axis reflection.

(b) Let T_1 and T_2 be two similar triangles in the plane. Apply a dilation, if necessary, to T_1 so that it has the same sidelengths as T_2. Next, apply a shift transformation to T_1 so that a vertex gets shifted to a corresponding vertex of T_2, and then apply a rotation to T_1 about this vertex so that a side of T_1 transforms into a corresponding side of T_2.

At this point, either T_1 and T_2 are now the same triangle, or they are reflections of one another across the common side. A final reflection about this line, if necessary, will thus complete the transformation of T_1 into T_2 by a similitude.

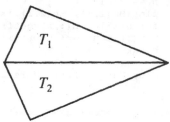

(c) It is clear that dilations, rotations, and shifts are essential. For an example to see why reflection is needed, simply take T_1 to be any triangle with three different angles and T_2 to be its reflection about one of the edges (see figure). It is clearly not possible to transform one of these two triangles into the other using any combination of dilations, rotations, and shifts.

EFR 7.10: (a) There will be only one generation; here are the outputs that were asked for (in format short):

```
A→       0    1.0000  2.0000      A1 →   0      0.5000  1.0000
         0    1.7321    0                0      0.8660    0
      1.0000  1.0000  1.0000           1.0000  1.0000  1.0000


A2 →  1.0000  1.5000  2.0000      A3→   0.5000  1.0000  1.5000
         0    0.8660    0                0.8660  1.7321  0.8660
      1.0000  1.0000  1.0000           1.0000  1.0000  1.0000


A1([1 2],2) → 0.5000            A3([1 2],2) → 1.5000
              0.8660                          0.8660
```

(b) Since the program calls on itself and does so more than once (as long as `niter` is greater than zero), placing a `hold off` anywhere in the program will cause graphics created on previous runs to be lost, so such a feature could not be incorporated into the program.

(c) Since we want the program to call on itself iteratively with different vertex sets, we really need to allow vertex sets to be inputted. Different vertex inputs are possible, but in order for the program to function effectively, they should be vertices of a triangle to which the similitudes in the program correspond. (e.g., any of the triangles in any generation of the Sierpinski gasket).

EFR 7.11: (a) S2, S1, S3, S2, S3, S2

(b) We list the sequence of float points in nonhomogeneous coordinates and in `format short`:
[0.5000 0.8660],   [0.2500 0.4330],   [1.1250 0.2165],   [1.0625 0.9743],   [1.5313 0.4871], [1.2656 1.1096].

(c) The program is designed to work for any triangle in the plane. The reader can check that the three similitudes are constructed in a way that uses midpoints of the triangle and the resulting diagram will look like that of Figure 7.15.

EFR 7.12: (a) As with `sgasket2`, the program `sgasket3` contructs future-generation triangles simply from the vertices and (computed) midpoints of the current-generation triangles. Thus, it can deal effectively with any triangle and produce Sierpinski-type fractal generations.

(b) For illustration purposes, the following trials were run on MATLAB's Version 5, so as to illustrate the flop count differences. The code is easily modified to work on newer versions of MATLAB by simply deleting the "flops" commands.

```
V1=[0  0];  V2=[1  sqrt(3)];  V3=[2  0];    %vertices of an equilateral
triangle
   test = [1  3  6  8  10];
```

`>> for i=1:5`	`>> for i=1:5`
`flops(0), tic,`	`flops(0), tic,`

`sgasket1(V1,V2,V3,test(i)), toc,`	`sgasket3(V1,V2,V3,test(i)), toc,`
`flops`	`flops`
`end`	`end`
→ (ngen =1) elapsed_time = 0.0600,	→ (ngen =1) elapsed_time = 0.1400,
ans =191	ans = 45
(ngen =3) elapsed_time = 0.2500,	(ngen =3) elapsed_time = 0.1310,
ans =2243	ans =369
(ngen =6) elapsed_time = 0.8510,	(ngen =6) elapsed_time = 0.7210,
ans =62264	ans =9846
(ngen =8) elapsed_time = 7.2310,	(ngen =8) elapsed_time = 6.2990,
ans =560900	ans =88578
(ngen =10) elapsed_time = 65.4640,	(ngen =10) elapsed_time = 46.7260,
ans =5048624	ans =797166

We remind the reader that the times will vary, depending on the machine being used and other processes being run. The above tests were run on a rather slow machine, so the resulting times are longer than typical.

EFR 7.13: The M-file is boxed below:

```
function []=snow(n)
S=[0 1 2 0;0 sqrt(3) 0 0];
index=1;
while index <=n
   len=length(S(1,:));
   for i=1:(len-1)
delta=S(:,i+1)-S(:,i);
perp=[0 -1;1 0]*delta;
T(:,4*(i-1)+1)=S(:,i);
T(:,4*(i-1)+2)=S(:,i)+(1/3)*delta;
T(:,4*(i-1)+3)=S(:,i)+(1/2)*delta+(1/3)*perp;
T(:,4*(i-1)+4)=S(:,i)+(2/3)*delta;
T(:,4*(i-1)+5)=S(:,i+1);
end
index=index+1;
S=T;
end
plot(S(1,:),S(2,:)), axis('equal')
```

The outputs of `snow(1)`, `snow(2)`, and `snow(6)` are illustrated in Figures 7.17 and 7.18.

EFR 7.14: For any pair of nonparallel lines represented by a two-dimensional linear system: $\begin{bmatrix} a & b \\ c & d \end{bmatrix}\begin{bmatrix} x \\ y \end{bmatrix} = \begin{bmatrix} e \\ f \end{bmatrix}$, the coefficient matrix will have nonzero determinant $\alpha = ad - bc$. The lines are

also represented by the equivalent system $\begin{bmatrix} a/\alpha & b/\alpha \\ c & d \end{bmatrix}\begin{bmatrix} x \\ y \end{bmatrix} = \begin{bmatrix} e/\alpha \\ f \end{bmatrix}$, where now the coefficient matrix

has determinant $(a/\alpha)d - (b/\alpha)c = 1$. This change simply amounts to dividing the first equation by α.

EFR 7.15: (a) As in the solution of Example 7.7, the interpolation equations $p(-2) = 4$, $p(1) = 3$, $p(2) = 5$, and $p(5) = -22$ (where $p(x) = ax^3 + bx^2 + cx + d$) translate into the linear system:

$\begin{bmatrix} -8 & 4 & -2 & 1 \\ 1 & 1 & 1 & 1 \\ 8 & 4 & 2 & 1 \\ 125 & 25 & 5 & 1 \end{bmatrix}\begin{bmatrix} a \\ b \\ c \\ d \end{bmatrix} = \begin{bmatrix} 4 \\ 3 \\ 5 \\ -22 \end{bmatrix}$. We solve this using left division, as in Method 1 of the solution of

Example 7.7:
```
>> format long
>> A=[-8 4 -2 1;1 1 1 1;8 4 2 1;125 25 5 1]; b=[4 3 5 -22]';
```

```
>> x=A\b
```
→x =   -0.47619047619048 (= a)
       1.05952380952381 (= b)
       2.15476190476190 (= c)
       0.26190476190476 (= d)

(b) As in part (a) and the solution of Example 7.7, we create the matrix A and vector b of the corresponding linear system: $Ax = b$. A loop will facilitate the construction of A:

```
>> xvals = -3:5; A = zeros(9) %initialize the 9 by 9 matrix A
>> for i =1:length(xvals)
A(i,:)=xvals(i).^(8:-1:0);
end
>> b = [-14.5  -12  15.5  2  -22.5  -112  -224.5  318  3729.5]'
```

We next go through each of the three methods of solving the linear system that were introduced in the solution of Example 7.7. We are working on an older and slower computer with MATLAB Version 5, so we will have flop counts, but the times will be slower than typical. The code is easily modified to work on the new version of MATLAB by simply deleting the flops commands. We show the output for x only for Method 1 (in format long) as the answers with the other two methods are essentially the same.

Method 1:

>> flops(0), tic, x=A\b, toc, flops	→x =	-0.00000000000000 0.00000000000000 0.50000000000000 -0.00000000000001 -6.00000000000000 -1.99999999999996 0.00000000000000 -17.00000000000003 2.00000000000000	→elapsed_time = 0.1300 →ans = 1125 (flops)

Method 2:

>> flops(0), tic, x=inv(A)*b, toc, flops	→elapsed_time = 0.3010 →ans = 1935 (flops)

Method 3:

>> Ab=A; Ab(:,10)=b; >> flops(0), tic, rref(Ab), toc, flops	→elapsed_time = 3.3150 →ans = 2175 (flops)

The size of this problem is small enough so that all three methods produce essentially the same vector x. The computation times and flop counts begin to demonstrate the relative efficiency of the three methods. Reading off the coefficients of the polynomial in order (from x), we get (after taking into account machine precision and rounding): $a = b = d = g = 0$, $c = 1/2$, $e = -6$, $f = -2$, $h = -17$, and $k = 2$, so that the interpolating polynomial is given by $p(x) = \frac{1}{2}x^6 - 6x^4 - 2x^3 - 17x + 2$. It is readily checked that this function satisfies all of the interpolation requirements.

EFR 7.16: As in Example 7.8, for a fixed n, if we let x denote the exact solution, we then have

$$b_n = H_n x = c(n)\left(1 \quad \frac{1}{2} \quad \frac{1}{3} \quad \cdots \quad \frac{1}{n-1} \quad \frac{1}{n}\right)'.$$ In order for b_n to have all integer coordinates, we need to

have $c(n)$ be a multiple of each of the integers 1, 2, 3, ..., n. The smallest such $c(n)$ is thus the least common multiple of these numbers. We can use MATLAB's lcm(a,b) to find the lcm of any set of integers with a loop. Here is how it would work to find $c(n) = \text{lcm}(1,2,...,n)$:

```
>> cn=1 %initialize
>> for k=1:n,  c(n)=lcm(cn, k), end
```

The remaining code for constructing the exact solution x, the numerical solution of Method 1, x_meth1, and the numerical solution of Method 2 x_meth2 are just as in Example 7.9. The flops commands in these codes should be omitted if you are using Version 6 or later. Also, since these computations were run on an older machine, the elapsed times will be larger than what is typical (but their ratios should be reasonably consistent). The loop below will give us the data we need for both parts (a) and (b):

```
>> for n=20:10:30
cn=1;  %initialize
for k=1:n,  c(n)=lcm(cn, k);  end
x = zeros(n,1); x(1)=cn;
bn = hilb(n)*x;
flops(0), tic, x_meth1=hilb(n)\bn; toc, flops
flops(0), tic, x_meth2=inv(hilb(n))*bn; toc, flops
Pct_err_meth1=100*max(abs(x-x_meth1))/cn,
Pct_err_meth2=100*max(abs(x-x_meth2))/cn
end
```

Along with the expected output, we also got some warnings from MATLAB that the matrix is either singular or poorly conditioned (to be expected). The output is summarized in the following table:

	Computer Time: $n=20/n=30$	Flop Count: $n=20/n=30$	Percentage of Maximum Error: $n=20/n=30$
Method 1:	0/0 seconds	10,339/27,481	0%/0%
Method 2:	0/0.01 seconds	20,312/63,509	512.5%/5400%

Note: The errors may vary depending on which version of MATLAB you are using.
(c) The errors with Method 1 turn out to be undetectable as n runs well over 1000. The computation times become more of a problem than the errors. MATLAB's "left divide" is based on Gaussian elimination with partial pivoting. After we study this algorithm in the next section, the effectiveness of this algorithm on the problem at hand will become clear.

EFR 7.17: (a) & (b): The first two are in reduced row echelon form. The corresponding general solutions are as follows: (for M_1): $x_1 = 3$, $x_2 = 2$; (for M_2): $x_1 = 2s - 3t - 2$, $x_2 = s$, $x_3 = 5t + 1$, $x_4 = t$, where s and t are any real numbers.

>> rref([1 3 2 0 3;2 6 2 -8 4])	→ans	1 3 0 -8 1
		0 0 1 4 1

(c) From the outputted reduced row echelon form, we obtain the following general solution of the first system: $x_1 = 1 - 3s + 8t$, $x_2 = s$, $x_3 = 1 - 4t$, $x_4 = t$, where s and t are any real numbers. Because of the arithmetic nature of the algorithm being used (as we will learn in the next section), it is often advantageous to work in format rat in cases where the linear system being solved is not too large and has integer or fraction coefficients. We do this for the second system:

>> format rat	→	1	-2	0	0	3/2	1
>> rref([1 -2 1 1 2 2;...	ans	0	0	1	0	1	3/2
-2 4 2 2 -2 0;...		0	0	0	1	-1/2	-1/2
3 -6 1 1 5 4;...		0	0	0	0	0	0
-1 2 3 1 1 3])							

From the output, we obtain the following general solution to the second system: $x_1 = 1 + 2s - 3t/2$, $x_2 = s$, $x_3 = 3/2 - t$, $x_4 = t/2 - 1/2$, $x_5 = t$, where s and t are any real numbers.

EFR 7.18: (a) The algorithm for forward substitution: $x_1 = b_1 / a_{11}$, $x_j = \left(b_j - \sum_{k=1}^{j-1} a_{jk} x_k\right)/a_{jj}$

(the first formula is redundant since the latter includes it as a special case) is easily translated into the following MATLAB code (cf. Program 7.4):

```
function x=fwdsubst(L,b)
%Solves the lower triangular system Lx=b by forward substitution
%Inputs:  L = lower triangular matrix,  b = column vector of same
%dimension
%Output:  x = column vector (solution)
[n m]=size(L);
x(1)=b(1)/L(1,1);
```

```
for j=2:n
    x(j)=(b(j)-L(j,1:j-1)*x(1:j-1)')/L(j,j);
end
x=x';
```

`>> L=[1 2 3 4;0 2 3 4;0 0 3 4;0 0 0 4]';`	→ans =	4
`>> b=[4 3 2 1]';`		-5/2
`>> format rat`		-5/6
`>> fwdsubst(L,b)`		-5/12

EFR 7.19: The two M-files are boxed below:

```
function B=rowmult(A,i,c)
% Inputs:   A = any matrix, i = any row index, c = any nonzero number
% Output:   B = matrix resulting from A by replacing row i by this row
% multiplied by c.
[m,n]=size(A);
if i<1|i>m
    error('Invalid index')
end
B=A;
B(i,:)=c*A(i,:);
```

```
function B=rowcomb(A,i,j,c)
% Inputs:   A = any matrix,  i, j = row indices,  c = a number
% Output:   B = matrix resulting from A by adding to row j the number
% c times row i.
[m,n]=size(A);
if i<1|i>m|j<1|j>m
    error('Invalid index')
end
if i==j
    error('Invalid row operation')
end
B=A;
B(j,:)=c*A(i,:)+A(j,:);
```

EFR 7.20: If we use gausselim to solve the system of Example 7.13, we get the correct answer (with lightning speed) with a flop count of 104 (if you have access to Version 5). In the table below, we give the corresponding data for the linear systems of parts (a) and (b) of EFR 7.16 (compare with the table in the solution of that exercise):

	Computer Time: $n = 20/n = 30$	Flop Count: $n = 20/n = 30$	Percentage of Maximum Error: $n = 20/n = 30$
Program 7.6	0.03/0.06 seconds	9,906/31,201	0%/0%

We observe that the time is detectable, although it was not when we used MATLAB's "left divide". Similarly, if we solve the larger systems of part (c) of EFR 7.16, we still get 0% errors for large values of n, but the times needed for gausselim to do the job are much greater than they were for "left divide".   MATLAB's "left divide" is perhaps its most important program. It is based on Gaussian elimination, but also relies on numerous other results and techniques from numerical linear algebra. A full description of "left divide" would be beyond the scope of this book; for the requisite mathematics, we refer to [GoVL-83].

EFR 7.21: Working just as in Example 7.14, but this time in rounded floating point arithmetic, the answers are as follows:

(a) $x_1 = 1$, $x_2 = .999$ and (b) $x_1 = .001$, $x_2 = .999$.

EFR 7.22: Looking at (28) we see that solving for x_j takes: 1 division + $(n-j)$ multiplications + $(n-j-1)$ additions (if $j < n$) + 1 subtraction (if $j < n$).
Summing from $j = n$ to $j = 1$, we deduce that:

Total multiplications/divisions = $\sum_{j=1}^{n} n - j + 1 = n^2 + n - n(n+1)/2 = (n^2 + n)/2$,

Total additions/subtractions = $\sum_{j=1}^{n-1}[n - j - 1 + 1] = \sum_{j=1}^{n-1}[n - j] = \sum_{j=1}^{n-1} j = (n^2 - n)/2$.

Adding gives the grand total of n^2 flops, as asserted.

EFR 7.23: Here we let $x = (x_1, x_2, \cdots, x_n)$ denote any n-dimensional vector and $\|x\|$ denote its max norm $\|x\|_\infty = \max\{|x_1|, |x_2|, \cdots, |x_n|\}$. The first norm axiom (36A) is clear from the definition of the max norm. The second axiom (36B) is also immediate: $\|cx\| = \max\{|cx_1|, |cx_2|, \cdots, |cx_n|\}$ $= |c| \max\{|x_1|, |x_2|, \cdots, |x_n|\} = |c|\|x\|$. Finally, the triangle inequality (36C) for the max norm readily follows from the ordinary triangle inequality for real numbers:

$$\|x + y\| = \max\{|x_1 + y_1|, |x_2 + y_2|, \cdots, |x_n + y_n|\}$$
$$= \max\{|x_1| + |y_1|, |x_2| + |y_2|, \cdots, |x_n| + |y_n|\} \le \|x\| + \|y\|.$$

EFR 7.24: (a) We may assume that $B \ne 0$, since otherwise both sides of the inequality are zero. Using definition (38), we compute:

$$\|AB\| = \max\left\{\frac{\|ABx\|}{\|x\|}, x \ne 0(\text{vector})\right\} = \max\left\{\frac{\|A(Bx)\|}{\|x\|} \cdot \frac{\|Bx\|}{\|Bx\|}, Bx \ne 0(\text{vector})\right\}$$

$$= \max\left\{\frac{\|A(Bx)\|}{\|Bx\|} \cdot \frac{\|Bx\|}{\|x\|}, Bx \ne 0(\text{vector})\right\} \le \|A\|\|B\|$$

(b) First note that for any vector $x \ne 0$, the vector $y = Ax$ is also nonzero (since A is nonsingular), and $A^{-1}y = x$. Using this notation along with definition (38), we obtain:

$$\|A^{-1}\| = \max\left\{\frac{\|A^{-1}y\|}{\|y\|}, y \ne 0(\text{vector})\right\} = \left(\min\left\{\frac{\|y\|}{\|A^{-1}y\|}, y \ne 0(\text{vector})\right\}\right)^{-1}$$

$$\underset{y=Ax}{=} \left(\min\left\{\frac{\|Ax\|}{\|x\|}, x \ne 0(\text{vector})\right\}\right)^{-1}.$$

EFR 7.25: (a) We first store the matrix A with the following loop, and then ask MATLAB for its condition number:
```
>>A=zeros(12);   for i=1:12,   A(i,:)=i.^(11:-1:0);   end,
c1=cond(A,inf)
```
`>> c1` →Warning: Matrix is close to singular or badly scaled. Results may be inaccurate. RCOND = 8.296438e-017.
→c1 = 1.1605e+016
(b) `>> c=norm(double(inv(sym(A))),inf)*norm(A,inf)` →c =1.1605e+016
`>> c-c1` →ans = 3864432
The approximation $c1$ to the condition number c is quite different, but relatively it at least has the same order of magnitude. If we choose a larger Vandermonde matrix here, we would begin to experience more serious problems as was the situation in Example 7.23.
(c) `>>   b = (-1).^(0:11).*(1:12); b=b';   % first create the vector b`
`>> z=A\b;   r=b-A*z;` (We get another warning as above.)
`>> errest=c1*norm(r,inf)/norm(A,inf)`   →errest = 0.0020

`>> norm(z,inf)→ans =8.7156e+004`

At first glance, the accuracy looks quite decent. The warnings, however, remove any guarantees that Theorem 7.7 would otherwise allow us to have.

(d) `>> z2=inv(A)*b; r2=b-A*z2;` -> Warning: Matrix is close to singular or badly scaled. Results may be inaccurate. RCOND = 8.296438e-017.

`>> errest2=c1*norm(r2,inf)/norm(A,inf)`   →errest2 = 2.3494

(e) As in Example 7.23, we solve the system symbolically and then get the norms that we asked for:

`>> S=sym(A); x=S\b;  x=double(x);`

`>> norm(x-z,inf)→ans =3.0347e-005`

`>> norm(x-z2,inf)→ans =3.0347e-005`

Thus, despite the warning we received, the numerical results are much more accurate than the estimates of Theorem 7.7 had indicated.

EFR 7.26: (a) Since $\lambda I - A$ is a triangular matrix, Proposition 7.3 tells us that the determinant $p_A(\lambda) = \det(\lambda I - A)$ is simply the product of the diagonal entries: $p_A(\lambda) = (\lambda - 2)^2 (\lambda - 1)^2$. Thus A has two eigenvalues: $\lambda = 1, 2$, each having algebraic multiplicity 2.

(b) `>> [V, D] = eig([2 1 0 0;0 2 0 0;0 0 1 0;0 0 0 1])`

→V =	1.0000	-1.0000	0	0	→D =	2	0	0	0
	0	0.0000	0	0		0	2	0	0
	0	0	1.0000	0		0	0	1	0
	0	0	0	1.0000		0	0	0	1

From the output of `eig`, we see that the eigenvalue $\lambda = 1$ has two linearly independent eigenvectors: [0 0 1 0]' and [0 0 0 1]', and so has geometric multiplicity 2, while the eigenvalue $\lambda = 2$ has only one independent eigenvector [2 0 0 0]', and so has geometric multiplicity 1.

(c) From the way in which part (a) was done, we see that the eigenvalues of any triangular matrix are simply the diagonal entries (with repetitions indicating algebraic multiplicities).

EFR 7.27: (a) The M-file is boxed below:

```
function [x, k, diff] = jacobi(A,b,x0,tol,kmax)
% performs the Jacobi iteration on the linear system Ax=b.
% Inputs:  the coefficient matrix 'A', the inhomogeneity (column)
% vector 'b', the seed (column) vector 'x0' for the iteration
% process, the tolerance 'tol' which will cause the iteration to stop
% if the 2-norms of differences of successive iterates becomes
% smaller than 'tol', and 'kmax' which is the maximum number of
% iterations to perform.
% Outputs:  the final iterate 'x', the number of iterations performed
% 'k', and a vector 'diff' which records the 2-norms of successive
% differences of iterates.
% If any of the last three input variables are not specified, default
% values of x0= zero column vector, tol=1e-10 and kmax=100 are used.

%assign default input variables, as necessary
if nargin<3, x0=zeros(size(b)); end
if nargin<4, tol=1e-10; end
if nargin<5, kmax=100; end
if min(abs(diag(A)))<eps
    error('Coefficient matrix has zero diagonal entries, iteration
cannot be performed.\r')
end

[n m]=size(A);
xold=x0;
k=1; diff=[];

while k<=kmax
```

```
        xnew=b;
        for i=1:n
            for j=1:n
                if j~=i
                    xnew(i)=xnew(i)-A(i,j)*xold(j);
                end
            end
            xnew(i)=xnew(i)/A(i,i);
        end
        diff(k)=norm(xnew-xold,2);
        if diff(k)<tol
          fprintf('Jacobi iteration has converged in %d iterations.\r', k)
                x=xnew;
                return
        end
        k=k+1; xold=xnew;
end
fprintf('Jacobi iteration failed to converge.\r')
x=xnew;
```

(b) >> A=[3 1 -1;4 -10 1;2 1 5]; b=[-3 28 20]';
>> [x, k, diff] = jacobi(A,b,[0 0 0]',1e-6);
→Jacobi iteration has converged in 26 iterations.
>> norm(x-[1 -2 4]',2)→ans = 3.9913e-007 (Error is in agreement with Example 7.26.)
>> diff(26)→ans = 8.9241e-007 (Last successive difference is in agreement with Example 7.26.)
>> [x, k, diff] = jacobi(A,b,[0 0 0]');
→Jacobi iteration has converged in 41 iterations. (With default error tolerance 1e-10)

EFR 7.28: (a) The M-file is boxed below:

```
function [x, k, diff] = sorit(A,b,omega, x0,tol,kmax)
% performs the SOR iteration on the linear system Ax=b.
% Inputs:  the coefficient matrix 'A', the inhomogeneity (column)
% vector 'b', the relaxation paramter 'omega', the seed (column)
% 'x0' for the iteration process, the tolerance 'tol' vector which
% will cause the iteration to stop if the 2-norms of successive
% iterates becomes smaller than 'tol', and 'kmax' which is the
% maximum number of iterations to perform.
% Outputs:  the final iterate 'x', the number of iterations performed
% 'k', and a vector 'diff' which records the 2-norms of successive
% differences of iterates.
% If any of the last three input variables are not specified, default
% values of x0= zero column vector, tol=1e-10 and kmax=100 are used.

%assign default input variables, as necessary
if nargin<4, x0=zeros(size(b)); end
if nargin<5, tol=1e-10; end
if nargin<6, kmax=100; end

if min(abs(diag(A)))<eps
    error('Coefficient matrix has zero diagonal entries, iteration
cannot be performed.\r')
end

[n m]=size(A);
xold=x0;
k=1; diff=[];

while k<=kmax
    xnew=b;
```

```
    for i=1:n
        for j=1:n
            if j<i
                xnew(i)=xnew(i)-A(i,j)*xnew(j);
            elseif j>i
                xnew(i)=xnew(i)-A(i,j)*xold(j);
            end
        end
        xnew(i)=xnew(i)/A(i,i);
        xnew(i)=omega*xnew(i)+(1-omega)*xold(i);
    end
    diff(k)=norm(xnew-xold,2);
    if diff(k)<tol
        fprintf('SOR iteration has converged in %d iterations\r', k)
        x=xnew;
        return
    end
    k=k+1; xold=xnew;
end
fprintf('SOR iteration failed to converge.\r')
x=xnew;
```

(b) We set the relaxation parameter equal to 1 for SOR to reduce to Gauss-Seidel:
```
>> A=[3 1 -1;4 -10 1;2 1 5]; b=[-3 28 20]';
>> [x, k, diff] = sorit(A,b,1,[0 0 0]',1e-6)
```
→ SOR iteration has converged in 17 iterations
```
>> norm(x-[1 -2 4]',2)
```
→ans =1.4177e-007 (This agrees exactly with the error estimate of Example 7.27.)
```
>> [x, k, diff] = sorit(A,b,.9,[0 0 0]',1e-6);
```
→SOR iteration has converged in 9 iterations

EFR 7.29: Below is the complete code needed to recreate Figure 7.41. After running this code, follow the instructions of the exercise to create the key.
```
>> jerr=1; n=1;
>> while jerr>=1e-6
x=jacobi(A,b,[0 0 0]',1e-7,n);
Jerr(n)=norm(x-[1 -2 4]',2); jerr=Jerr(n); n=n+1;
end
>> semilogy(1:n-1,Jerr,'bo-')
>> hold on

>> gserr=1; n=1;
>> while gserr>=1e-6
x=gaussseidel(A,b,[0 0 0]',1e-7,n);
GSerr(n)=norm(x-[1 -2 4]',2); gserr=GSerr(n); n=n+1;
end
>> semilogy(1:n-1,GSerr,'gp-')

>> sorerr=1; n=1;
>> while sorerr>=1e-6
x=sorit(A,b,0.9, [0 0 0]',1e-7,n);
SORerr(n)=norm(x-[1 -2 4]',2); sorerr=SORerr(n); n=n+1;
end
>> semilogy(1:n-1,SORerr,'rx-')
>> xlabel('Number of iterations'), ylabel('Error')
```

EFR 7.30: (a) By writing out the matrix multiplication and observing repeated patterns we arrive at the following formula for the vector $b \equiv Ax$ of size 2500×1. Introduce first the following two 1×50 vectors b', $\bar{b}$:

$$b' = [1\ 4\ -1\ 4\ -1\ \cdots\ 4\ -1\ 5],$$
$$\bar{b} = [0\ 2\ -2\ 2\ -2\ \cdots\ 2\ -2\ 3].$$

In terms of copies of these vectors, we can express b as the transpose of the following vector:

$$b = [b'\ \bar{b}\ \bar{b}\ \cdots\ \bar{b}\ \bar{b}\ b'].$$

(b)   We need first to store the matrix A.   Because of its special form, this can be expeditiously accomplished using some loops and the diag command as follows:

```
>> x=ones(2500,1); x(2:2:2500,1)=2;
>> tic, A=4*eye(2500); toc
→elapsed_time =0.6090
>> v1=-1*ones(49,1); v1=[v1;0]; %seed vector for sub/super diagonals
tic, secdiag=v1;
for i=1:49
if i<49
secdiag=[secdiag;v1];
else
secdiag=[secdiag;v1(1:49)];
end
end, toc
→elapsed_time =0.1250
>> tic, A=A+diag(secdiag,1)+diag(secdiag,-1)-diag(ones(2450,1),50)-
diag(ones(2450,1),-50); toc
→elapsed_time =12.7660
>> tic, bslow=A*x; toc
→elapsed_time = 0.2340
```

(c):   To see the general concepts behind the following code, read Lemma 7.16 (and the notes that precede it).

```
tic, bfast=4*x+[secdiag;0].*[x(2:2500);0]+...
[0;secdiag].*[0; x(1:2499)]-[x(51:2500); zeros(50,1)]-...
[zeros(50,1); x(1:2450)]; toc   →elapsed_time = 0.0310
```

(d)   If we take $N = 100$, the size of A will be $10,000 \times 10,000$, and this is too large for MATLAB to store directly, so Part (b) cannot be done.   Part (a) can be done in a similar fashion to how it was done when N was 50.   The method of part (c), however, still works in about 1/100th of a second.   Here is the corresponding code:

```
>>x=ones(10000,1); x(2:2:10000,1)=2;
>>v1=-1*ones(99,1); v1=[v1;0]; %seed vector for sub/super diagonals
>>tic, secdiag=v1; for i=1:99, if i<99, secdiag=[secdiag;v1];
else, secdiag=[secdiag;v1(1:99)]; end
end, toc
>> tic, bfast=4*x+[secdiag;0].*[x(2:10000);0]+...
[0;secdiag].*[0; x(1:9999)]-[x(101:10000); zeros(100,1)]-...
[zeros(100,1); x(1:9900)]; toc   →elapsed_time = 0.0100
```

EFR 7.31:   (a)   The M-file is boxed below:

```
function [x, k, diff] = sorsparsediag(diags, inds,b,omega,
x0,tol,kmax)
% performs the SOR iteration on the linear system Ax=b in cases where
% the n by n coefficient matrix A has entries only on a sparse set of
% diagonals.
% Inputs:   The input variables are 'diags', an n by J matrix where
% eachcolumn consists of the entries of one of A's diagonals.   The
% first column of diags is the main diagonal of A (even if all zeros)
% and 'inds' , a 1 by n vector of the corresponding set of indices
% for the diagonals (index zero corresponds to the main diagonal).
% the relaxation paramter 'omega', the seed (column) vector 'x0' for
% the iteration process, the tolerance 'tol' which will cause the
% iteration to stop if the infinity-norms of successive iterates
% become smaller than 'tol', and 'kmax' which is the maximum number
```

```
% of iterations to perform.
% Outputs:  the final iterate 'x', the number of iterations performed
% 'k', and a vector 'diff' which records the 2-norms of successive
% differences of iterates.
% If any of the last three input variables are not specified, default
% values of x0= zero column vector, tol=1e-10 and kmax=1000 are used.

%assign default input variables, as necessary
if nargin<5, x0=zeros(size(b)); end
if nargin<6, tol=1e-10; end
if nargin<7, kmax=1000; end

if min(abs(diags(:,1)))<eps
    error('Coefficient matrix has zero diagonal entries, iteration
cannot be performed.\r')
end

[n D]=size(diags);
xold=x0;
k=1; diff=[];

while k<=kmax
    xnew=b;
    for i=1:n
        for d=2:D %run thru non-main diagonals and scan for entries
that effect xnew(i)
            ind=inds(d);
            if ind<0&i>-ind %diagonal below main and j<i case
                aij=diags(i+ind,d);
                xnew(i)=xnew(i)-aij*xnew(i+ind);
            elseif ind>0&i<=n-ind %diagonal above main and j>i case
                aij=diags(i,d);
                xnew(i)=xnew(i)-aij*xold(i+ind);
            end
        end
        xnew(i)=xnew(i)/diags(i,1);
        xnew(i)=omega*xnew(i)+(1-omega)*xold(i);
    end
    diff(k)=norm(xnew-xold,inf);
    if diff(k)<tol
        fprintf('SOR iteration has converged in %d iterations\r', k)
        x=xnew;
        return
    end
    k=k+1; xold=xnew;
end
fprintf('SOR iteration failed to converge.\r')
x=xnew;
```

(b) In order to use this program, we must create the input matrix diags from the nontrivial diagonals of the matrix A. The needed vectors were constructed in the solution of EFR 7.30(b); we reproduce the relevant code:

```
>> v1=-1*ones(49,1); v1=[v1;0]; %seed vector for sub/super diagonals
secdiag=v1;
for i=1:49
if i<49
secdiag=[secdiag;v1];
else
secdiag=[secdiag;v1(1:49)];
end
end
```

We now construct the columns of diags to be the nontrivial diagonals of A taken in the order of the vector:

```
>> inds = [0   1   -1   50   -50]
>> diags = zeros(2500,5);
>> diags(:,1)=4; diags(1:2499,[2 3])=[secdiag  secdiag];
>> diags(1:2450,[4 5])= [-ones(2450,1) -ones(2450,1)];
```

We will also need the vectors x and b; we assume they have been obtained (and entered in the workspace) in one of the ways shown in the solution of EFR 7.30. We now apply our new SOR program on this problem using the default tolerance:

```
>> tic
>> [xsor, k, diff]=sorsparsediag(diags, inds,b,2/(1+sin(pi/51)),
zeros(size(b))); toc
→SOR iteration has converged in 222 iterations
→elapsed_time = 0.6510
>> max(abs(xsor-x))
→ans = 6.1213e-010
```

CHAPTER 8: INTRODUCTION TO DIFFERENTIAL EQUATIONS

EFR 8.1: In general, if a vector x is constructed with the MATLAB command x=a:h:b, where $a < b$ and $h > 0$ is the step size, then we can write: $x(n) = a + (n-1)h$. In the example on hand, $a = 0$, and $h = 0.01$, so $x(n) = (n-1)0.01$ which gives $n = 100x(n)+1$. Therefore, to use MATLAB to find y when $x = 3$, we should use the index $n = 100 \cdot 3 + 1 = 301$ and enter:
```
>> y(301)   →ans =1.7736
```

EFR 8.2: A calculus proof that the values where $P = k/2$ correspond to inflection points of solutions is given in the text (see the paragraph immediately following this EFR). There it is shown that $P''(t) = r(1 - 2P/k)f(P)$ where $f(P) = rP(1 - P/k)$. From this formula, we see that $P''(t)$ can vanish at no other (nonzero) values of P; there are no other inflection points. In fact, even if we allowed $P \le 0$, there would be no more.

EFR 8.3: (a) In the figure on the right, we have graphed the right side of the Gompertz equation $P'(t) = -sP\ln(P/k)$ and classified the unique equilibrium point.
(b) The plots can be accomplished using a loop analogous to that employed in the solution of Example 8.6.
```
>> f=inline('-0.024*P*
log(P/1)', 't', 'P');
>> hold on
>> for P0=0.1:.2:1.1
[t, y] = euler(f,0,200,
P0,0.1); plot(t,y)
end
```
The resulting plot is shown in the lower figure. Each of the six graphs maintains the same concavity; the

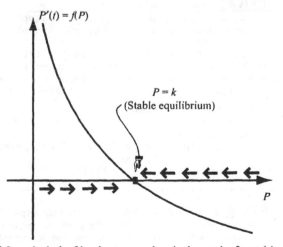

lack of inflection points can be deduced from the lack of local extreme values in the graph of part (a). Also, as expected from the stability graph of part (a), each of the solutions approaches the stable equilibrium solution $P \equiv 1(= k)$.

(c)   Following the suggestion, we introduce the new variable: $y = \ln(P/k)$.   Since $ke^y = P$,
different-tiation with respect to t gives:

$-ske^y y =$

$\dfrac{dP}{dt} = \dfrac{dP}{dy} y' = ke^y y'$ (we have taken into

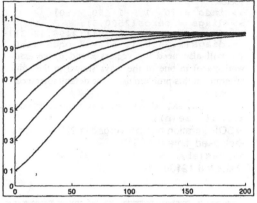

account the Gompertz equation).
Equating the first and last terms and
canceling common factors gives us
$y' = -sy$. Thus y satisfies the Malthus
growth equation, and we can write down
its general solution: $y = y_0 e^{-st}$.

Consequently, $P = ke^y = k \exp(y_0 e^{-st})$,

and so $P'(t) = Py_0 e^{-st}(-s) = ae^{-bt} P$,

where $a = -sy_0$ and $b = s$, as asserted.

EFR 8.4: Simply change h to 0.01 and let the loops for the improved Euler and Runge-Kutta
methods run from 1:200. (The correct size can again be seen by looking at size(t) after Euler
program is run with this step size). Here now are the commands to get the plot of Figure 8.12b:

```
subplot(2,2,1), s=1:.01:3; plot(s,yexact(s)), hold on,
plot(t,ye,'bo')
subplot(2,2,2), plot(t,abs(yexact(t)-ye), 'bo'), hold on,
plot(t,abs(yexact(t)-yie), 'gx')
subplot(2,2,3), plot(t,abs(yexact(t)-yie), 'gx'), hold on,
plot(t,abs(yexact(t)-yrk), 'r+')
subplot(2,2,4), plot(t,abs(yexact(t)-yrk), 'r+')
```

EFR 8.5: $\dfrac{dy}{dt} = 2ty \Rightarrow \int \dfrac{dy}{y} = \int 2t\,dt \Rightarrow \ln|y| = t^2 + C \Rightarrow y = \pm e^C \exp(t^2) = A\exp(t^2).$

EFR 8.6: The M-file is boxed below:

```
function [t,y]=impeuler(f,a,b,y0,hstep)
% input variables:  f, a, b, y0, hstep
% output variables:  t, y
% f is a function of two variables f(t,y).  The program will apply
% the improved Euler method to solve the IVP: (DE): y'=f(t,y), (IC)
% y(a)=y0 on the t-interval [a,b] with step size hstep.  The output
% will be a vector of t's and corresponding y's t(1)=a; y(1)=y0;

nmax=ceil((b-a)/hstep);
for n=1:nmax
   t(n+1)=t(n)+hstep;
   y(n+1)=y(n)+.5*hstep*(feval(f,t(n),y(n))...
      +feval(f,t(n+1),y(n)...
      +hstep*feval(f,t(n),y(n))));
end
```

EFR 8.7: The code is below and the resulting graph of the
error is shown on the right. From the graph, we see that the
maximum error is less than 1/10th of what was guaranteed by
Theorem 8.2.

```
>> f=inline('0.05*y','t','y');
>> [t, y]=euler(f,0,5,10,0.046 );
>> yexact=inline('10*exp(.05*t)','t');
>> plot(t,abs(y-yexact(t)))
```

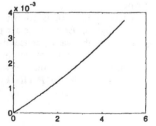

EFR 8.8: (a) The M-file is boxed below:

```
function [t, y] = rkf45(varf, a, b, y0, tol, hinit, hmin, hmax)
% input variables:  varf, a, b, y0, tol, hinit, hmin, hmax
% output variables: t, y, varf is a function of two variables
% varf(t,y).
% The program will apply the Runge-Kutta-Fehlberg (RKF45) method to
% solve the IVP:  (DE):  y'=varf(t,y), (IC)
% y(a)=y0 on the t-interval [a,b] with step size hstep.  The output
% will be a vector of t's and corresponding y's the last four input
% variables are optional and are as follows:
% tol = the target goal for the global error, default = 1e-5
% hinit = initial step size, default = 0.1
% hmin = minimum allowable step size, default = 1e-5
% hmax = maximum allowable step size, default = 1
% program will terminate with an error flag if it is necessary to
% use a step size smaller than hmin

%set default input variables as needed
if nargin<5, tol=1e-5; end
if nargin<6, hinit=0.1; end
if nargin<7, hmin=1e-5; end
if nargin<8, hmax=1; end

t(1)=a; y(1)=y0; n=1;
h=hinit;
flag =0; %this flag will keep track if maximum step size has been
reached.
flag2 =0; %this flag will keep track if minimum step size has been
reached.
while  t(n)<b               .
    k1=h*feval(varf,t(n),y(n));
    k2=h*feval(varf,t(n)+h/4,y(n)+k1/4);
    k3=h*feval(varf,t(n)+3*h/8,y(n)+3*k1/32+9*k2/32);
    k4=h*feval(varf,t(n)+12*h/13,y(n)+(1932*k1-7200*k2+7296*k3)/2197);
    k5=h*feval(varf,t(n)+h,y(n)+439*k1/216-8*k2+3680*k3/513-
845*k4/4104);
    k6=h*feval(varf,t(n)+h/2,y(n)-8*k1/27+2*k2-
3544*k3/2565+1859*k4/4104-11*k5/40);
    E=abs(k1/360-128*k3/4275-2197*k4/75240+k5/50+2*k6/55);

    if E > h*tol  %step size is too large, reduce to half and try
again
        hnew=h/2;
        if hnew<hmin %minimum step size has been reached, accept
approximation but set
% warning flag2
            flag2=1;
            t(n+1)=t(n)+h;
            y(n+1)=y(n)+16*k1/135+6656*k3/12825+28561*k4/56430-
9*k5/50+2*k6/55;
            n=n+1;
            h=hmin;
        else
            h=hnew;
        end
    elseif E < h*tol/4 %step size is too small, accept approximation
but double next step %size
        t(n+1)=t(n)+h;
        y(n+1)=y(n)+16*k1/135+6656*k3/12825+28561*k4/56430-
```

```
9*k5/50+2*k6/55;
    n=n+1;
    h=2*h;
    if h>=hmax
        flag=1;
        h=hmax;
    end
  else %accept approximation and proceed
    t(n+1)=t(n)+h;
    y(n+1)=y(n)+16*k1/135+6656*k3/12825+28561*k4/56430-
9*k5/50+2*k6/55;
    n=n+1;
 end
end
if flag ==1
    fprintf('In the course of the RKF45 program, the maximum step
size has been ... reached.')
end
if flag2 ==1
    fprintf('WARNING:  Minimum step size has been reached; it is
recommended to run ...
the \r')
    fprintf('program again with a smaller hmin and or a larger tol')
end
```

(b) The following commands will run the RKF45 program on the IVP of Example 8.7 with the default settings, and plot the error against the exact solution given in that example. The error plot is shown on the right.

```
>>
f=inline('2*t*y','t','y');
>> yexact =
inline('exp(t.^2-1)');
>> [t,
yrkf]=rkf45(f,1,3,1);
>> plot(t,abs(yrkf-
yexact(t)))
>> plot(t,abs(yrkf-
yexact(t)), 'rx')
>> xlabel('x-values'),
ylabel('y-values'),
>> title('Error for the RKF45 method')
>> size(t)→ans= 1  187
```

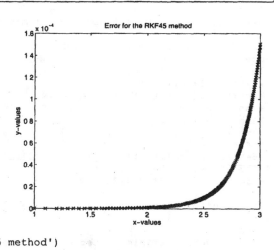

The last command shows us that RKF45 used 187 plotting points; and the figure shows that the density of them increases in the region on the right where the solution experiences its most rapid growth. Comparing with Figure 8.12b, we see that this error is about 10 times less than that of the Runge-Kutta method when the latter used 200 plotting points.

EFR 8.9: From the result of Example 8.19 (with $r = 2$), the region of numerical stability for Euler's method is $h < 1$, so any step size larger than one will eventually experience instability. The plot of Figure 18.21(a) resulted from using $h = 1.03$. With the same step size, the Runge-Kutta method gives a numerical solution that converges to zero, but (as is easily checked) is not very accurate. The plot of Figure 8.21(b) resulted from using a step size of $h = 1.43$ with the Runge-Kutta method.

EFR 8.10: Substituting $f(t, y) = ry$ (from the IVP (14)) and a constant step size $h_n = h$ into the recursion formula (17) $y_{n+1} = y_n + h_n[f(t_n, y_n) + f(t_{n+1}, y_{n+1})]/2$ produces: $y_{n+1} = y_n(1 + hr/2) + (hr/2)y_{n+1}$, or $y_{n+1} = \dfrac{(1 + hr/2)}{(1 - hr/2)} y_n$. Equivalently, $y_{n+1} = \mu^n y_0$, where $\mu \equiv \dfrac{1 + hr/2}{1 - hr/2}$. Since h is

positive and r is negative, we always have $\mu < 1$ and so $y_n \to 0$ as $n \to \infty$, regardless of the step size h. This proves the asserted unconditional numerical stability.

EFR 8.11: (a) The M-file is boxed below:

```
function [t, y] = adamsbash5(varf, a, b, y0, h)
% Performs the Adams-Bashforth fifth-order scheme to solve an IVP
% Calls on fifth-order Runge-Kutta scheme (rk5) to create the seed
% iterates.
% Input variables:  varf a function of two variables f(t,y)
% describing the ODE y' = f(t,y).  Can be an inline function or an M-
% file a, b = the left and right endpoints for the time interval of
% the IVP y0 the intial value y(a) given in the intial condition
% h = the step size to be used
% Output variables:  t = the vector of equally spaced time values for
% the numerical solution, y = the corresponding vector of y
% coordinates.

nmax=ceil((b-a)/h);
%first form the seed iterates using single step Runge-Kutta
[t,y]=rk5(varf,a,a+4*h,y0,h);

for n=5:nmax
 t(n+1)=t(n)+h;
 y(n+1)=y(n)+h/720*(1901*feval(varf,t(n),y(n))-2774*feval(varf,t(n-
1),y(n-1))...
    +2616*feval(varf,t(n-2),y(n-2))-1274*feval(varf,t(n-3),y(n-
3))+251*feval(varf,t(n-4),y(n-4)));
end
```

(b) The M-file is boxed below:

```
function [t, y] = adamspc5(varf, a, b, y0, h)
% Performs the Adams-Bashforth-Moulton fifth-order predictor-
% corrector scheme to solve an IVP.
% Calls on fifth-order Runge-Kutta scheme (rk5) to create the seed
% iterates
% Input variables:  varf a function of two variables f(t,y)
% describing the ODE y' = f(t,y).  Can be an inline function or an M-
% file a, b = the left and right endpoints for the time interval of
% the IVP y0 the intial value y(a) given in the intial condition
% h = the step size to be used
% Output variables:  t = the vector of equally spaced time values for
% the numerical solution, y = the corresponding vector of y
% coordinates.
nmax=ceil((b-a)/h);
%first form the seed iterates using single step Runge-Kutta
[t,y]=rk5(varf,a,a+4*h,y0,h);

for n=5:nmax
    t(n+1)=t(n)+h;
    %predictor
    y(n+1)=y(n)+h/720*(1901*feval(varf,t(n),y(n))-2774*feval(varf,...
    t(n-1),y(n-1))+2616*feval(varf,t(n-2),y(n-2))-1274*feval...
    (varf,t(n-3), y(n-3))+251*feval(varf,t(n-4),y(n-4)));
    %corrector
```

```
  y(n+1)=y(n)+h/720*(251*feval(varf, t(n+1),y(n+1))+646*feval..
  (varf, t(n),y(n)) -264*feval(varf, t(n-1),y(n-1))+106*feval...
  (varf, t(n-2),y(n-2))-19*feval(varf, t(n-3),y(n-3)));
end
```

EFR 8.12: (a) It is required to show that (assuming the differentiability assumptions of Taylor's theorem hold): $y(t+h) - y(t-h) - 2f(t,y) = O(h^2)$. Indeed using Taylor's theorem for the first two expressions on the left and the DE for the third, we obtain: $y(t+h) - y(t-h) - 2f(t,y)$
$= (y(t) + hy'(t) + O(h^2)) - (y(t) - hy'(t) + O(h^2)) - 2hy'(t) = O(h^2)$.

(b) In the notation of (18), the parameters for the midpoint method are: $K = 2$, $\alpha_1 = 0$, $\alpha_2 = 0$, $\beta_0 = 0$ (explicit), and $\beta_1 = 2$, and so the characteristic polynomial is given by: $P(\lambda) = \lambda^2 - (0 \cdot \lambda^1 + 1 \cdot \lambda^0) = \lambda^2 - 1$. Since the roots are $\lambda = \pm 1$, the stability theorem in the text implies that the midpoint method is weakly stable.

(c) The following code was used to produce Figure 8.25.

```
>> y(1)=1; t(1)=0; h=0.0001;
>> t(2)=t(1)+h; y(2)=y(1)+h*(20-4*y(1));
n=2;
while t(n)<=4
  t(n+1)=t(n)+h;
  y(n+1)=y(n-1)+h*(20-4*y(n));
  n=n+1;
end
plot(t,y), axis([0 4 0 7]), title('Weak Stability')
```

CHAPTER 9: SYSTEMS OF FIRST-ORDER DIFFERENTIAL EQUATIONS AND HIGHER-ORDER DIFFERENTIAL EQUATIONS

EFR 9.1: (a) Letting $y_1(t) = y'(t)$ and $y_2(t) = y''(t)$, the given IVP is equivalent to the following first-order system:

$$\begin{cases} y'(t) = y_1, & y(0) = 1 \\ y_1'(t) = y_2, & y_1(0) = 2 \; . \\ y_2'(t) = \sin(3t) - y_2 - e^t y, & y_2(0) = 3 \end{cases}$$

(b) Introducing $R_1(t) = R'(t)$ allows us to translate the second-order system into the following first-order system:

$$\begin{cases} R'(x) = R_1, & R(10) = 4 \\ R_1'(x) = RS + \sqrt{x^2 + 1}, & R_1(10) = -1 \; . \\ S'(t) = R_1 \cos(S), & S(10) = 1 \end{cases}$$

EFR 9.2: The M-file is boxed below:

```
function [t,x,y]=runkut2d(f,g,a,b,x0,y0,hstep)
% This M-file performs the Runge-Kutta method to solve a two-
% dimensional system of form:
% Dx(t)= f(t,x,y), x(a) = x0,  Dy(t) = g(t,x,y), y(a) = y0, on the
% interval a <= t <= b.
% Input variables: f and g inline functions (or M-files) for the
% derivatives of the unknown functions x(t) and y(t).  These must be
% specified as functions of the three variables: t, x, and y (in
% this order)a and b:  endpoints for the time interval on which the
% solution is sought x0, and y0, initial conditions for the unknown
% functions at t = a hstep, the step size (any positive number)
% output variables are three vectors of the same size: t, x and y,
```

```
% for the numerical solution.
t=a:hstep:b;x(1)=x0;  y(1)=y0;
[m nmax]=size(t);
for n=1:(nmax-1)
   k1x=feval(f,t(n),x(n),y(n));
   k1y=feval(g,t(n),x(n),y(n));
   k2x=feval(f,t(n)+.5*hstep,x(n)+.5*hstep*k1x,y(n)+.5*hstep*k1y);
   k2y=feval(g,t(n)+.5*hstep,x(n)+.5*hstep*k1x,y(n)+.5*hstep*k1y);
   k3x=feval(f,t(n)+.5*hstep,x(n)+.5*hstep*k2x,y(n)+.5*hstep*k2y);
   k3y=feval(g,t(n)+.5*hstep,x(n)+.5*hstep*k2x,y(n)+.5*hstep*k2y);
   k4x=feval(f,t(n)+hstep,x(n)+hstep*k3x,y(n)+hstep*k3y);
   k4y=feval(g,t(n)+hstep,x(n)+hstep*k3x,y(n)+hstep*k3y);
   x(n+1)=x(n)+1/6*hstep*(k1x+2*k2x+2*k3x+k4x);
   y(n+1)=y(n)+1/6*hstep*(k1y+2*k2y+2*k3y+k4y);
end
```

EFR 9.3: The MATLAB code that produced the plot on the right is given below. To see the flow direction along these solution curves, we fix one of them, and consider the (unique) point $(1, y)$ with $y < 1$ on the graph (directly below the equilibrium solution $(1, 1)$). At this point the first differential equation of the system, $x' = -x + xy$, tells us that x' is negative, so x is decreasing and this forces a clockwise flow orientation. A more general phase-plane analysis technique will be presented Section 9.2.

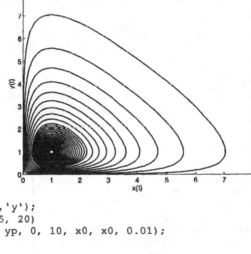

```
>> xp=inline('-x+x*y',
't','x','y');
>> yp=inline('y-x*y','t','x','y');
>> for x0 = linspace(.05, .95, 20)
     [t,xrk,yrk]=runkut2d(xp, yp, 0, 10, x0, x0, 0.01);
     plot(xrk,yrk), hold on
end
>> xlabel('x(t)'), ylabel('y(t)')
```

EFR 9.4: (a) Having $f > 1$ would correspond to removing more fish than are available, which is impossible.
(b) $x' = 0 \Rightarrow x(y - 1 - f) = 0$ and $y' = 0 \Rightarrow y(1 - x - f) = 0$ so if we also require $x, y \neq 0$ this gives $x = 1 - f$ and $y = 1 + f$ as the only (nontrivial) equilibrium point.

EFR 9.5: (a) From (6) we get that $\dfrac{dI}{dS} = \dfrac{dI/dt}{dS/dt} = \dfrac{I(rS - a)}{-rIS} = -1 + \dfrac{\rho}{S}$, provided $I \neq 0$. Viewed in the (I, S) plane, this DE is separable; integrating yields: $I(t) - I(0) = -S(t) + S(0) + \rho \ln(S(t)/S(0))$. Since $I' = I(rS - a) = Ir(S - \rho)$, we see that if $S(0) > \rho$, then $I'(0) > 0$ and I increases until $S = \rho$, after which I decreases (note by (6) that S is decreasing). Therefore, $\max I = I\big|_{S=\rho} = I(0) - \rho + S(0) + \rho \ln(\rho/S(0)) = N - \rho + \rho \ln(\rho/S(0))$, as asserted.
(b) As in part (a), we deduce that $dS/dR = -S/\rho$ so that S and R are related by Malthusian growth and thus $S(t) = S(0)e^{-R(t)/\rho}$. Since $R \leq N$, we get $S(t) \geq S(0)e^{-N/\rho} > 0$ so that $S(\infty) > 0$.

(c) We first observe that since eventually $S(t) < \rho$ the DE for I in (6) implies that eventually I will have exponential decay and so $I(t) \to 0$ as $t \to \infty$. Using (5), we can rewrite the equation $S(t) = S(0)e^{-R(t)/\rho}$ obtained in part (b) as $S(t) = S(0)e^{-[N-S(t)-I(t)]/\rho}$. If we take the limit of this equation as $t \to \infty$, it becomes: $S(\infty) = S(0)e^{-[N-S(\infty)]/\rho}$. We have so far shown that $S(\infty)$ is a (positive) root of the equation $x = S(0)\exp[-(N-x)/\rho]$. Consider the two sides of this equation as functions of $x > 0$: $f(x) \equiv x$ and $g(x) \equiv S(0)\exp[-(N-x)/\rho]$. Since $g'(x) = g(x)/\rho > 0$ and $g''(x) = g(x)/\rho^2 > 0$ we see that $g(x)$ is increasing and concave upward. Thus the equation $f(x) = g(x)$ can have at most two positive roots (draw a picture to see this). If there is only one root, we have nothing to prove, so assume there are two roots: $x_1 < x_2$ such that $g(x_i) = x_i$. For the larger root, we must have $g'(x_2) > f'(x_2)$, or $x_2/\rho > 1$, or $x_2 > \rho$ (draw a picture or use concavity to see this). But $S(\infty)$ cannot be greater than ρ since if it were then $I(t)$ would still be increasing. Therefore $S(\infty) = x_1$, as was to be proved.

To apply Newton's method (Program 6.2) we are seeking a root of $F(x) = S(0)\exp(-(N-x)/\rho) - x$ and in our example, $N = 763$, $S(0) = 762$, $a = .44036$, $r = 2.18e\text{-}3$, so $\rho = a/r = 202$:

```
>> f=inline('762*exp(-(763-x)/202)-x');
>> fp=inline('762*exp(-(763-x)/202)/202-1');
>> newton(f,fp,202)
→Exact root found
  →19.1758
```

Compare this with the approximately 22 susceptibles predicted from the PDE model after 14 days. Thus, theoretically, out of the 762 original susceptibles, all but about 3 who will contract the disease will have done so after 14 days.

EFR 9.6: (a) $x' = x(1-x-y/2) \Rightarrow$
x-nullclines: $x = 0$, $y = 2(1-x)$.
$y' = y(1-y-x/2) \Rightarrow$
y-nullclines: $y = 0$, $y = 1 - x/2$.
Equilibrium solutions: $(x,y) = (0,0)$, $(1,0)$, $(0,1)$, $(2/3, 2/3)$.
In the phase plane diagram on the right, the two x-nullclines (passing through $(0,2)$) are shown along with the y-nullclines (passing through $(2,0)$).
(b) The following code is similar to that employed in the solution of Example 9.5, and will produce a phase portrait similar to that of Figure 9.12.

```
>> dx=inline('x-x^2-x*y/2',
't', 'x', 'y');
>> dy=inline('y-y^2-x*y/2',
't', 'x', 'y');
>> x1=linspace(.5,1.8,11);
y1=2-x1;
>> x2=linspace(.2,.7,11); y2=.75-x2;
>> x0=[x1 x2]; y0 = [y1 y2];
>> size(x0)
→ans = 1      22
hold on
for k=1:22
[t,x,y]=runkut2d(dx,dy,0,20,x0(k),y0(k),0.01); plot(x,y)
end
```

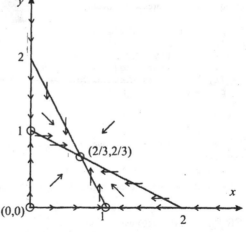

EFR 9.7: Equilibrium solutions of (14) are solutions of $X'(t) \equiv 0$ and thus are solutions of the

matrix equation $AX = \begin{bmatrix} 0 \\ 0 \end{bmatrix}$. From linear algebra (see Sections 7.1 and 7.2) the origin $\begin{bmatrix} 0 \\ 0 \end{bmatrix}$ will be a

unique solution if A is invertible. If A is not invertible, the solutions will consist of a line through the origin and hence the origin will not be isolated.

EFR 9.8: (a) In general, the SIRS system (7) yields the following S-I nullclines: $S' = -rIS + bR$ $= -rIS + b(N - I - S) \Rightarrow$ S-nullclines: $0 = -rIS - bI + bN - bS$, or $(rS + b)I = b(N - S)$, or

$I = b(N - S)/(rS + b) = (N - S)/(rS/b + 1)$,

$I' = rIS - aI \Rightarrow$ I-nullclines: $0 = rIS - aI$

$= rI(S - \rho)$ $(\rho = a/r)$ $\Rightarrow I = 0$ and

$S = \rho$. In the setting of Example 9.4, the parameters are as follows: $N = 10,000$, $r = 2$e-4, $a = 4$, $b = 0.25$, and so $\rho = a/r = 20,000$ and we get these specific nullclines: S-nullcline: $I = (10000 - S)/(8$e-4$*S+1)$, I-nullclines: $I = 0$, $S = 20000$. By testing the signs of the derivatives of S and I on the regions between nullclines, we obtain the phase-plane diagram shown on the right, where we have drawn the S-nullcline (curved) and the two I-nullclines (lines). The only equilibrium solution is (10000,0).

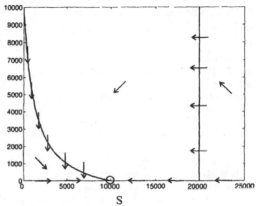

(b) and (c): To apply Theorem 9.3, we need to know that the equilibrium solution is isolated. This can be seen by extending the phase-plane diagram just drawn to include some values in the fourth quadrant (i.e., negative I-values and positive S-values), even though this quadrant bares no physical significance to the model at hand. Indeed, if we were to extend the phase-plane analysis to the whole fourth quadrant, the blue curve and vertical red line would intersect below to form only one new equilibrium solution leaving (10000,0) as isolated. We first use (7) to compute the form of the Jacobian matrix:

$$A = \begin{bmatrix} S_S & S_I \\ I_S & I_I \end{bmatrix} = \begin{bmatrix} -rI - b & -rS - b \\ rI & rS - a \end{bmatrix}.$$

From this we compute $tr(A) = r(S - I) - b - a$ and $\det(A) = (-rI - b)(rS - a) + rI(rS + b) = rb(I - S)$ $+ arI + ab$. We will be keeping the values of $b = 0.25$ and $r = 2$e-4 fixed in this part. From the analysis in part (a), (10,000, 0) will be the only (isolated) equilibrium solution as long as $\rho = a/r > 10,000$, or $a \geq 2$ (corresponding to the average infection lasting for 1/2 year, or 6 months rather than three months). In all of these cases, we may write: $tr(A) = 7/4 - a$, $\det(A) = a/4 - 1/2$. Thus, in the range $2 < a \leq 4$, $\det(A)$ is always positive and $tr(A)$ remains negative. We compute $tr(A)^2/4 - \det(A) = a^2/4 - 9a/8 + 81/64$ and see that this parabola has a minimum value of zero at $a = 2.25$ (this computation is done easily using MATLAB's Symbolic Toolbox). Thus, by part (b) of Theorem 9.3, the equilibrium solution (10000,0) will always be a stable node (this is corroborated with Figure 9.12 in the text in case $a = 4$), whenver $2 < a \leq 4$ and $a \neq 2.25$. In the special case $a = 2.25$, part (d) of that theorem tells us that we have either a stable node or spiral (a MATLAB plot would not be a reliable way to further determine the type of node due to the sensitivity of the problem to small changes in data). In case $a = 2$, $\det(A) = 0$, and Theorem 9.3 is inconclusive. Finally, in the range $0 < a < 2$ (corresponding to the average infection lasting more than 1/2 a year, and lasting indefinitely longer as the paramter a approaches zero), there will now be 2 equilibrium solutions: the original $P_1 = (10,000,0)$ and a new one $P_2 = (\rho, (10,000 - \rho)/(4a + 1))$, where $\rho = a/r = 5000a$. For P_1 we have $\det(A) < 0$ throughout the range $0 < a < 2$ so that by part (a) of Theorem 9.3, P_1 will be an unstable saddle point. For P_2 we will use the Symbolic Toolbox for the calculations.

```
>> b=1/4; r = 2e-4; syms a, rho = a/r; S=rho;
```

```
>> I = (10000-rho)/(4*a+1);
>> trA = r*(S-I)-b-a; detA = r*b*(I-S)+a*r*I+a*b;
>> ezplot(detA,[0,2]) %plot (not shown here) tells us the determinant
is always positive when 0<a<2.
>> ezplot(trA,[0,2]) %plot (not shown here) tells us the trace is
always negative when 0<a<2.
```

Thus we already know (from Theorem 9.3 or its corollary) that the equilibrium point P_2 is stable. To see the character of this stable equilibrium, we examine the graph of $\operatorname{tr}(A)^2/4 - \det(A)$ over our indicated range:

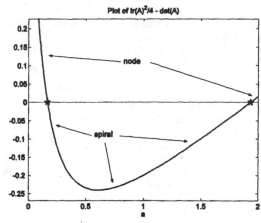

Plot of tr(A)²/4 - det(A)

```
>> ezplot(trA^2/4-
detA,[0,2]) %plot is shown
at right
```

From the plot we see that there are two places where a sign change occurs. We can solve for these numbers as follows:

```
>> double(solve(trA^2/4-
detA,a))
```

→ans =1.9336, -0.5989, 0.1653

Only the first and last of these are relevant for us; we add these special points on our graph:

```
>> hold on, plot(1.9336,0,'rp'),  plot(.1653,0,'rp')
>> xlabel('a'), title('Plot of tr(A)^2/4 - det(A)')
```

Theorem 9.3 tells us that when a is between these two points (marked with pentacles in the figure), the equilibrium point will be a stable spiral, and when it is to the left or right of them, it will be a stable node. When a coincides with one of these, the theorem tells us that P_2 will either be a spiral or a node.

We point out that it would not be feasible to solve the problem numerically at one of these borderline values to determine if there is a spiral or a node. This is because the problem is extremely unstable and sensitive to perturbations.

EFR 9.9: (a)

$$x' = x\left\{\tfrac{2}{3}(1-x/4) - y/(1+x)\right\}$$

$\Rightarrow$ x-nullclines: $y = \tfrac{2}{3}(1-x/4)(1+x)$,

$y' = sy(1-y/x)$

$\Rightarrow$ y-nullclines: $y = 0$, $y = x$.

Note that $x = 0$ is not an x-nullcline since y' is undefined at $x = 0$.   Equilibrium solutions: $(x,y) = (1,1)$, $(4,0)$. In the phase portrait diagram on the right, the x-nullclines is shown (curved) along with the the two y-nullclines (lines).

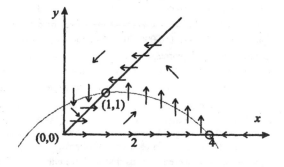

(b) To determine the character of the equilibrium solution (1,1), we will employ Theorem 9.3. Since the computations are long, we will use the Symbolic Toolbox.

```
>> syms x y s
>> xp = 2*x*(1-x/4)/3-x*y/(1+x); yp = s*y*(1-y/x);
>> A = [diff(xp,x)  diff(xp,y); diff(yp, x)  diff(yp, y)] %Jacobian
matrix
>> subs(det(A), [x y], [1 1])
```
→ans =5/12*s (This is always positive for s > 0.)
```
>> subs(Trace(A), [x y], [1 1])
```
→ans =1/12-s

Note that the latter (trace of the Jacobian) is positive whenever s < 1/12, and by Theorem 9.3, for such values of s, the equilibrium point (1,1) is an unstable node or spiral. In particular, it is repelling. Similarly, when s > 1/12, the theorem tells us that (1,1) is a stable node or spiral so is not repelling. When $s = 1/12$, Theorem 9.3 tells us that (1,1) is either a vortex or a spiral, but gives no additional information so we cannot use it to decide if (1,1) is repelling or not. Numerical computations of the solution would not be useful here because of the sensitivity of the problem to s being slightly less than 1/12 or slightly greater than 1/12.

(c) On all but the left side of the square R, the phase portrait of the solution of part (a) above shows that the direction fields never point outward, so it remains to deal with the left side. As (x,y) approaches the left side, we see from the system of DEs that $x' \to 0$ and $y' \to -\infty$. It follows that orbits that start in R can never reach the left side of R; they will first hit (from above) the green parabola, after which their horizontal velocity component will be positive. We have shown that no orbit that originates within the square R can ever exit R (i.e., R is a basin of attraction).

EFR 9.10: The code is below and the outputted plots appear on the right:

```
>> [t,X]=rksys('lorenz',0,50,[-8 8 27],0.1);
>> x=X(:,1); x=x';
>> for i=1:8
[ti,Xi]=rksys('lor
enz',0,50,[-8 8
27],0.1/2^i);
for j=1:501
xi(j)=Xi(2^i*j-
2^i+1,1);
end
subplot(3,2,i)
plot(t,x-xi)
x=xi;
end
```

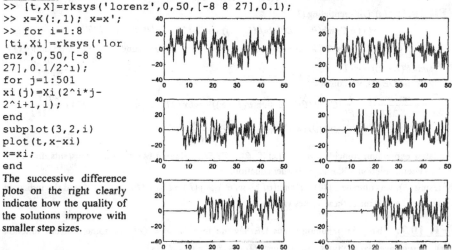

The successive difference plots on the right clearly indicate how the quality of the solutions improve with smaller step sizes.

EFR 9.11: (a) The code of part (a) of Example 9.8 needs only a very minor modification, namely the line defining dz should be modified to: `dz=inline('-32.174/1.5*sin(th)', 't','th','z');`

(b) The linear model can be explicitly solved using the Symbolic Toolbox:
```
>> syms th L g
>> dsolve('D2th +g*th = 0', 't')
→ans =C1*cos(g^(1/2)*t)+C2*sin(g^(1/2)*t)
```
The general solution, being a combination of cosines and sines with the same period, is certainly periodic.

For the nonlinear pendulum it is more difficult to prove periodicity. A phase-plane plot can be done with MABLAB (on increasingly longer long time intervals) to show that it is plausible that the solution is periodic, but this does not constitute a proof. The proof we give is motivated by considerations from physics, namely the conservation of energy in mechanics. The pendulum has two types of energies: kinetic and potential. The kinetic energy in physics is defined to be $K = \frac{1}{2}mv^2$ (where v is the velocity of the mass), so that for the pendulum, we have $K = \frac{1}{2}m(L\theta')^2 = \frac{L^2m}{2}(\theta')^2$. The potential energy in physics is defined (up to an additive constant) as $L = mgh$ where h is the height of the mass, so that for the pendulum, $L = mgL(1 - \cos\theta)$. The conservation of energy states that the total energy $E(t) = K + L = $ constant. It would suffice to prove this, since, when the pendulum comes back on the return trip, it will eventually have to stop (before bobbing back). At this time T, its kinetic energy will equal zero (as it was at time = 0), therefore, by the conservation of energy, the potential energy and hence θ would be the same value when time was zero. Thus, from time $t = T$ onward, the motion of the pendulum is identical (by the uniqueness theorem) to what it was from time $t = 0$ (since the IVPs are identical). This proves that T is the period of the pendulum. To make this rigorous, we need only prove the conservation of energy, i.e., that $E'(t) = 0$. Indeed, $E'(t) = K'(t) + L'(t) = L^2 m\theta'\theta'' +$ $mgL\sin\theta\theta' = Lm\theta'\left[L\theta'' + g\sin\theta\right]$. The bracketed expression is zero because of the DE: $L\theta'' + g\sin\theta = 0$. Related problems on periodicity of more general pendulum-like DEs have been the subject of much investigation; for some interesting surveys in this area we refer to the following two articles: [Maw-82] and [Maw-97].

CHAPTER 10: BOUNDARY VALUE PROBLEMS FOR ORDINARY DIFFERENTIAL EQUATIONS

EFR 10.1: (a) Differentiating (3)
$$y(x) = C\sinh(\theta x) + D\sinh(\theta(L-x)) + wLx/2T - w/\theta^2 T - wx^2/2T,$$
gives: $y'(x) = C\theta\cosh(\theta x) - D\theta\cosh(\theta(L-x)) + wL/2T - wx/T$ and $y''(x) = C\theta^2\sinh(\theta x) +$ $D\theta^2\sinh(\theta(L-x)) - w/T$.
To check the DE (2), we compute:
$$\frac{T}{EI}y + \frac{wx(x-L)}{2EI} = \frac{T}{EI}[C\sinh(\theta x) + D\sinh(\theta(L-x)) + wLx/2T - w/\theta^2 T - wx^2/2T] + \frac{wx(x-L)}{2EI}$$
$$= \frac{TC}{EI}\sinh(\theta x) + \frac{TD}{EI}\sinh(\theta(L-x)) - \frac{w}{\theta^2 EI}.$$
The latter expression coincides with $y''(x)$ if $\theta^2 = T/EI$. Having two arbitrary constants, this must be the general solution of the second-order equation.
(b) Using (3), we compute: $y(0) = D\sinh(\theta L) - w/\theta^2 T$, $y(L) = C\sinh(\theta L) - w/\theta^2 T$ and the indicated values for C and D make these values vanish.

EFR 10.2:   No. An inhomogeneous DE has the form $L[y] = r(x)$ for some nonzero function $r(x)$. If we have two solutions, y_1, y_2 then $L[cy_1 + dy_2] = (c+d)r(x)$.

EFR 10.3: (a) Nonlinear. (b) Linear, not homogeneous.

EFR 10.4: (a) Nonlinear. $f_y = 2y$ takes on negative values in $R = \{a \le t \le b, -\infty < y, y' < \infty\}$, so Theorem 15.1 does not apply.   (b)   Nonlinear. $f_y = t$ is not always positive in $R = \{a \le t \le b, -\infty < y, y' < \infty\}$ since $a = 0$, so Theorem 15.1 does not apply.

EFR 10.5:   The two associated IVPs of the given linear BVP are as follows:

$$(\text{IVP-1}) \begin{cases} y_1''(x) = \dfrac{1}{x}y_1' + x^4 & (\text{DE}) \\ y_1(1) = 1,\ y_1'(1) = 0 & (\text{IC-1}) \end{cases},$$

and

$$(\text{IVP-2}) \begin{cases} y_2''(x) = \dfrac{1}{x}y_2' & (\text{DE}) \\ y_2(1) = 0,\ y_2'(1) = 1 & (\text{IC-2}) \end{cases}.$$

Setting these up as two-dimensional linear systems and using the Runge-Kutta method

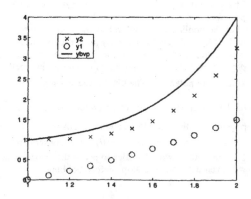

will be accomplished by the following MATLAB code:
```
>> f1=inline('u', 'x','y','u');
>> g1=inline('u/x+x^4','x','y','u');
>> f2=f1;
>> g2=inline('u/x','x','y','u');
>> [x,y1,u1]=runkut2d(f1,g1,1,2,1,0,.01);
>> [x,y2,u2]=runkut2d(f2,g2,1,2,0,1,.01);
```
To obtain the desired plots, we first verify the sizes of the solution vectors:
```
>> size(x)  →ans =1  101
>> plot(x(1:10:101),y1(1:10:101),'gx'), hold on,
plot(x(1:10:101),y2(1:10:101),'go')
>> ybvp=y1+(4-y1(101))/y2(101)*y2;  plot(x,ybvp)
>> ybvp(find(x==1.5))  →ans = 1.5892  (=value of solution when x = 1.5)
```

EFR 10.6:   (a) The M-file is boxed below. In order to facilitate the internal construction of the needed inline functions in terms of the inputted data for $p(t)$, $q(t)$, and $r(t)$, we have set up the program to input these functions as strings (so in single quotes) and with the independent variable being t. Inline functions cannot be constructed in terms of other inline functions, so if we had instead inputted p, q, and r as inline functions, we would have not been able to internally construct the needed inline functions to call on the Runge-Kutta program `runkut2d`. Thus, if we had gone this route, it would have been necessary to recode the Runge-Kutta program inside of this one.

```
function [t, y] = linearshooting(pstring, qstring, rstring, a, alpha,
b, beta, hstep)
%M-file for EFR 10.6
%This program will use the linear shooting method to solve a linear
%BVP of the following form: y''(t)=p(t)y'+q(t)y+r(t), y(a)=alpha,
%y(b)=beta
%Input variables: pstring = string for the function for p(t),
%qstring =string for the function for q(t), rstring = string for
%r(t), a, alpha, b, beta are numbers in the BVP, hstep is a
%positive number to be used in the Runge-Kutta method.
%Output variables: t and y, vectors of the same size that give the
%time values and associated numerical solution values.
%NOTE:  The first three input variables must be put in single quotes
%(so MATLAB will assign their data types to be strings).  Within the
%program, we will need to create inline functions in terms of the
%formulas for p(t), q(t),and r(t).  This would not be possible if
%instead we had these three functions inputted as inline functions.
%IMPORTANT:  the independent variable of the inputted strings for p,
%q and r must be t.

%Step 1:  Set up the functions for the linear systems corresponding
%two associated IVPs and solve each one.
%to the IVP-1:   y1''(t)=p(t)y1'+q(t)y1+r(t), y1(a)=alpha, y1'(a)=0
y1p = inline('u', 't', 'y', 'u');
u1p = inline(['(', pstring, ')*u+(', qstring, ')*y+'
rstring],'t','y','u');
[t,y1,u1]=runkut2d(y1p,u1p,a,b,alpha,0,hstep);
%IVP-2:  y2''(t)=p(t)y2'+q(t)y2, y2(a)=0, y2'(a)=1
y2p = inline('u', 't', 'y', 'u');
u2p = inline(['(', pstring, ')*u+', qstring, '*y'],'t','y','u');
[t,y2,u2]=runkut2d(y2p,u2p,a,b,0,1,hstep);

%Step 2:  Construct solution of BVP
y=y1+(beta-y1(find(t==b)))/y2(find(t==b))*y2;
```

(b) Looking at the BVP in Example 10.3, we see that the coefficient functions are as follows: $p(t) = 0$, $q(t) = 6.25\text{e-}6$, and $r(t) = 50t(t-50)/96000000$. Thus, we can solve and plot the solution of this problem using the program of part (a) as follows:

```
>> [t, y] = linearshooting('0','6.25e-6','50*t*(t-50)/96000000', 0,
0, 50, 0, .1);
>> plot(t,y)
```
The resulting plot is identical to that of Figure 10.2.

EFR 10.7: (a) In order to make this program more elegant, we would like to be able to call on Program 9.2, which has the following call format: `[t,X]= rksys(vectorf,a,b,vecx, hstep)`. In order for this to be feasible, we will need to internally construct an inline function for `vectorf` that consists of the right sides of the four DEs of the system that we will need to be (iteratively) solving. To make make the task more clear, we write down the system in terms of the variables that we will use in the program:

$$\begin{cases} y'(t) = yp, & y(a) = \text{alpha} \\ yp'(t) = f(t, y, yp), & yp(a) = mk \\ z'(t) = zp, & z(a) = 0 \\ zp'(t) = zf_y(t, y, yp) + zpf_{yp}(t, y, yp), & zp(a) = 1 \end{cases}$$

Thus the inline function `vectorf` should have inputs t (a number) and $\text{xvec} = [y, yp, z, zp]$ (a vector) and output the vector $[yp, f(t,y,yp), zp, z*fy(t,y,yp), zp*fy(t,y,yp)]$ (gotten from the right sides of the 4 DEs in the system). The problem is that, although we will be inputting the strings for f, fy and fyp with variables t, y, and yp, these will need to be internally be changed to t, xvec(1) and xvec(2), respectively, so that `vectorf`'s output will be expressed in terms of its input variables (the number t and the vector vecx). Thus, it will be convenient to make some string substitutions within the M-file; there is a useful command for this type of operation:

newstring= strrep(oldstring,'s1', 't1') →	If `oldstring` is any character string and `s1` and `t1` are string portions, this command will create another string `newstring` gotten by replacing all occurrences of `s1` by `t1`.

Here are some simple examples of the use of this command:
```
>> string = 'Jenny went out to dinner with Billy';strrep(string, 'to
dinner', 'dancing')
```
→ans =Jenny went out dancing with Billy
```
>> strrep('t*cos(yp)+(y+2)^2','yp', 'xvec(2)')
```
→ans =t*cos(xvec(2))+(y+2)^2
String manipulations can be a useful skill in writing certain types of programs; to see a brief synopsis of the numerous string related functions that MATLAB has, simply enter: `help strfun`. The annotated M-file is boxed below:

```
function [t, y, nshots] = nonlinshoot(a, alpha, b, beta, fstring,
fystring, fypstring,  tol, hstep, mk)
%M-file for EFR 10.7.
%This program will apply the non linear shooting method to
%solve the BVP:  y''(t)=f(t,y,y'),  y(a) = alpha, y(b) = beta
%Input variables:  a = left endpoint, alpha = left boundary value
%b = right endpoint, beta = right boundary value, fscript =
%inhomogeneity function, inputted as a script (in single quotes) with
%variables t, y and yp (y'),the next two input variables are the
%partial derivatives of f(t, y, y')with respect to y and y'
%respectively, also inputted as scripts in the same fashion.
%tol = tolerance, a positive number.  When successive approximations
%differ by less than tol at right endpoint, iterations stop.
%hstep = the step size to use in the Runge-Kutta method
%m0 = initial (shooting) slope; if this variable is not inputted
%the default value for m0 is (beta-alpha)/(b-a)
%Output variables:  t and y, two same sized vectors containing the
%time values and corresponding values of the numerical solution of
%the BVP, nshots, the number of iterations (shots) that were used in
%the nonlinear shooting method.
%This program internally will call on Program 9.2:  rksys

%set default if necessary
```

```
if nargin < 10
    mk = (beta-alpha)/(b-a);
end

%set up a vector-valued inline function for the 4 equation linear
%system that needs to be iteratively solved:
%Dy = yp,    y(a)=alpha,   Dyp = f(t,y,yp), yp(a) = mk
%Dz = zp,    z(a)=0,  Dzp = zfy(t,y,yp) + zpfyp(t,y,yp), zp(a) = 1
%fvec will have 4 components [Dy Dyp Dz Dzp] and will be an inline
%function of the 2 variables t (a number) and vec (a vector)
% representing the four numbers y, yp, z, and zp in this order:
%vec(1) = y, vec(2) = yp, vec(3) = z, and vec(4) = zp
%we first change the inputted strings to conform to these new
variables:
fstring=strrep(fstring,'yp','vec(2)');
fstring=strrep(fstring,'y','vec(1)');
fystring=strrep(fystring,'yp','vec(2)');
fystring=strrep(fystring,'y','vec(1)');
fypstring=strrep(fypstring,'y','vec(1)');
fypstring=strrep(fypstring,'yp','vec(2)');
fvec = inline(['[vec(2)   ', fstring, '  vec(4)  ', 'vec(3)*(',
fystring, ')+... vec(4)*(', fypstring, ')]'], 't', 'vec');
%Note: Some of the blank spaces left above were intentional and
%important to separate the four components of this vector valued
%function.

%start iterative loop
nshots =1;
while 1
    [t, X] =rksys(fvec,a, b, [alpha  mk   0   1], hstep);
    y = X(:,1); z=X(:,3);   %peel off the vectors we need
    Diff=y(length(y))-beta;
    if abs(Diff)<tol
    return
    end
    mk=mk-Diff/z(length(z)); nshots= nshots+1; %update slope and shot
counter
end
```

(b)   The BVP of Example 10.4 is now easily solved with this program; we run it with the two tolerances 0.01 and 1e-7 given in parts (a) and (b) of the example

```
>> [t, y, n] = nonlinshoot(1,0,2,-2,'-2*(y*yp+t*yp+y+t)', ...
        ' 2*yp-2','-2*y-2*t',0.01,0.01);   n  →n=3
>> [t, y, n] = nonlinshoot(1,0,2,-2,'-2*(y*yp+t*yp+y+t)', ...
        '-2*yp-2','-2*y-2*t',1e-7,0.01);
>> n   →n = 5
```

The number of shots agrees with what we had in that example, and the plots also agree (simply enter `plot(t,y)`).

(c)   In this BVP, we have $f(t,y,y') = te^y - \sin(\cos(t))y'$. We will need $f_y(t,y,y') = te^y$ and $f_{y'}(t,y,y') = -\sin(\cos(t))$. Since f_y is positive when t is and $f_{y'}$ is bounded, Theorem 10.1 tells us that the BVP has a unique solution. If we try to use the nonlinear shooting program of part (a) to solve the problem numerically with a tolerance of $h = 0.01$ (and the same Runge-Kutta step size), the program hangs, and actually enters into an infinite loop. To gain some insight on what has happened, we modify the program of part (a) so that it will display the variable Diff at each iteration (simply remove the semicolon at the end of the line that defines this variable). With this modification, here are the first several lines of output that we get:

```
>> [t, y, n] = nonlinshoot(1,0,3,-1,'t*exp(y) sin(cos(t))*yp', ...
't*exp(y)', '-sin(cos(t))',1,0.01,0);
```
→ Diff = Inf, Diff = NaN, Diff = NaN, Diff = NaN, ...

We briefly explain what has happened. We let MATLAB choose the initial value of the slope mk to be the default value $(y(3) - y(1))/(3-1) = -1/2$. What has happened is that the resulting IVP blew up to infinity too quickly due to the potentially very large te^y term present in the DE. Both y and z have become infinite at $t = 3$, and in computing mk MATLAB needed to divide an infinite quantity by another. This forced mk to be defined as NaN (not a number) and from that point on the iterations became meaningless and we entered into an infinite loop. It is not difficult to modify the program to force quit and give an appropriate error message in such an occurrence. Here, we simply experiment with different (more negative) initial values of mk so as to prevent such blowing up. We have quickly found that if we use $mk = -2$, things work fine:

```
>> [t, y, n] = nonlinshoot(1,0,3,-1,'t*exp(y)-sin(cos(t))*yp', ...
               't*exp(y)', '-sin(cos(t))',.01,0.01,-2);
>> n   →n = 3
>> [t, y, n] = nonlinshoot(1,0,3,-1,'t*exp(y)-sin(cos(t))*yp', ...
               't*exp(y)', '-sin(cos(t))',1e-7,0.01,-2);
>> n   →n = 5 (Only five shots needed.)
>> plot(t,y), grid on %plot is shown below
```

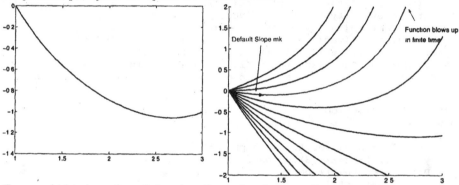

The second plot shows the pathology just discussed, and why nonlinearity made it necessary to "undershoot" the first shot, lest the solutions blow up in finite time. The code below constructs the second plot (without the embellishments):

```
>> f = inline('y',    't', 'x', 'y'); g = inline('t*exp(x)-...
            sin(cos(t))*y', 't','x','y');
>> clf, hold on
>> for mk=1:-.5:-4
   [t,x,y]=runkut2d(f,g,1,3,0,mk,0.01); plot(t,x)
end, axis([1 3 -2 2])
```

EFR 10.8: Using Taylor's theorem, we obtain:

$$f(a+h) = f(a)+hf'(a)+h^2 f''(a)/2+h^3 f'''(a)/6+O(h^4), \text{ and}$$

$$f(a-h) = f(a)-hf'(a)+h^2 f''(a)/2-h^3 f'''(a)/6+O(h^4).$$

From these we obtain that: $\dfrac{f(a+h)-2f(a)+f(a-h)}{h^2}$

$$= \frac{f(a)+hf'(a)+h^2 f''(a)/2+h^3 f'''(a)/6+O(h^4)-2f(a)}{h^2}$$

$$- \frac{-[f(a)-hf'(a)+h^2 f''(a)/2-h^3 f'''(a)/6+O(h^4)]}{h^2}$$

$$= \frac{2h^2 f''(a)/2+O(h^4)}{h^2} = f''(a)+O(h^2),$$

as asserted.

EFR 10.9: Since each p_i is a value of the function $p(t)$, we have $|p_i h/2| < M(2/M)/2 = 1$. Therefore, each of the nondiagonal entries of A is positive. Also, since each q_i is positive, each diagonal entry a_{kk} has absolute value greater than 2, so it suffices to show that the sum of the absolute values of the nondiagonal entries of any row is less than or equal to 2. Since all nondiagonal entries are positive they equal their absolute values. For the first and the last rows, there is only one nondiagonal entry which equals $1 \pm p_i h/2 < 2$. For all other rows, the sum of the nondiagonal entries equals $1 + p_i h/2 + 1 + p_i h/2 = 2$. This completes the proof.

EFR 10.10: If $v, w \in A$ then both are continuous and so must be their sum $v + w$, as well as any scalar multiple αv. Let $\mathscr{P}_1 : 0 = a_0 < a_1 < \cdots < a_{n+1} = 1$ and $\mathscr{P}_2 : 0 = b_0 < b_1 < \cdots < b_{m+1} = 1$ be two partitions of $[0, 1]$ over which $v'(x)$ and $w'(x)$ are continuous, respectively. It follows that $(v + w)'(x) = v'(x) + w'(x)$ is continuous with respect to the common refinement $\mathscr{P}_1 \cup \mathscr{P}_2$ of these partitions, and this function is bounded by the sum of the bounds for $v'(x)$ and $w'(x)$. More easily, the function αv has a piecewise continuous derivative with respect to $\mathscr{P}_1$ and is bounded by $|\alpha|$ times the corresponding bound for $v'(x)$. Finally since both of the functions v, w vanish at $x = 0$ and $x = 1$, so must the two functions: $v + w$ and αv, and we have thus proved that these latter two functions satisfy all of the requirements for membership in A.

EFR 10.11: The proof of the last EFR easily translates over to prove this result. The only change needed here is that the sum of two piecewise linear functions will also be piecewise linear (with respect to the common refinement partition).

EFR 10.12: (a) The proof we give works for any set of basis functions $\{\varphi_i(x)\}_{i=1}^n$ that are continuous, piecewise differentiable, and vanish at the endpoints $x = 0$ and $x = 1$. Since $a_{ij} = \langle \varphi_i', \varphi_j' \rangle = \int_0^1 \varphi_i'(x) \cdot \varphi_j'(x) dx$, we can use linearity of integration to write:

$$c'Ac = [c_1, c_2, \cdots, c_n] \cdot [\textstyle\sum_{j=1}^n a_{1j} c_j, \sum_{j=1}^n a_{2j} c_j, \cdots, \sum_{j=1}^n a_{nj} c_j]$$

$$= [c_1, c_2, \cdots, c_n] \cdot [\textstyle\sum_{j=1}^n \int_0^1 \varphi_1'(x)\varphi_j'(x)dx \, c_j, \sum_{j=1}^n \int_0^1 \varphi_2'(x)\varphi_j'(x)dx \, c_j,$$
$$\cdots, \textstyle\sum_{j=1}^n \int_0^1 \varphi_n'(x)\varphi_j'(x)dx \, c_j]$$

$$= [c_1, c_2, \cdots, c_n] \cdot [\textstyle\int_0^1 (\varphi_1'(x) \cdot \sum_{j=1}^n c_j \varphi_j'(x))dx, \int_0^1 (\varphi_2'(x) \cdot \sum_{j=1}^n c_j \varphi_j'(x))dx,$$
$$\cdots, \textstyle\int_0^1 (\varphi_n'(x) \cdot \sum_{j=1}^n c_j \varphi_j'(x))dx]$$

$$= \textstyle\sum_{i=1}^n c_i \cdot \int_0^1 (\varphi_i'(x) \cdot \sum_{j=1}^n c_j \varphi_j'(x))dx$$

$$= \textstyle\int_0^1 (\sum_{i=1}^n c_i \varphi_i'(x) \cdot \sum_{j=1}^n c_j \varphi_j'(x))dx$$

$$= \textstyle\int_0^1 (\sum_{i=1}^n c_i \varphi_i(x))' \cdot (\sum_{j=1}^n c_j \varphi_j(x))' dx.$$

We have shown that $c'Ac = \langle \Phi', \Phi' \rangle$, where $\Phi = \sum_{i=1}^n c_i \varphi_i(x)$ so that $c'Ac \geq 0$. Next assume that $c'Ac = 0$. By positive definiteness of the inner product, we may conclude that $\Phi'(x) = 0$ at all values of x except for the finite set of points where at least one of the $\varphi_i'(x)$ is not continuous. Since Φ, being admissible, vanishes at the endpoints, it follows that $\Phi(x) = 0$ for all x. But by linear independence of $\{\varphi_i(x)\}_{i=1}^n$, this forces the vector c to be zero. This completes the proof that the stiffness matrix is positive definite.

(b) In case of equal grid spacing $h_i = h$, by (38) and (39), the main diagonal entries of the stiffness matrix are all equal to 2/h, while the super and sub diagonal entries of this tridiagonal matrix equal $-1/h$. Note that when the DE of (24) is put into form (5), we have $p = q = 0$, and $r(x) = -f(x)$. The linear systems $Ax=b$, for the two methods are compared below:

Linear Rayleigh-Ritz:	Finite Difference:

$$A^{RR} = (1/h)\begin{bmatrix} 2 & -1 & & & \\ -1 & 2 & -1 & & \\ & -1 & 2 & 0 & \\ & 0 & & \ddots & -1 \\ & & & -1 & 2 \end{bmatrix},$$

$$A^{FD} = \begin{bmatrix} -2 & 1 & & & \\ 1 & -2 & 1 & & \\ & 1 & -2 & 0 & \\ & 0 & & \ddots & 1 \\ & & & 1 & -2 \end{bmatrix},$$

$$b^{RR} = \begin{bmatrix} \langle f, \phi_1 \rangle \\ \langle f, \phi_2 \rangle \\ \langle f, \phi_3 \rangle \\ \vdots \\ \langle f, \phi_n \rangle \end{bmatrix}$$

$$b^{FD} = h^2 \begin{bmatrix} -f(h) \\ -f(2h) \\ -f(3h) \\ \vdots \\ -f([n-1]h) \end{bmatrix}$$

If we multiply the left linear system by h and the right one by -1, then the matrices equate, and the ith term of the inhomogeneities on the right sides become $h\langle f, \phi_i \rangle = h\int_0^1 f(x)\phi_i(x)\,dx$ and $h^2 f(ih)$, respectively. By the mean value theorem for integrals, we can write $\int_0^1 f(x)\phi_i(x)\,dx$ $= f(\tilde{x}_i)\int_0^1 \phi_i(x)\,dx$, where $\tilde{x}_i$ is some number inside the interval $[ih-h, ih+h]$ (on which the hat function $\phi_i(x)$ lives). But (from Figure 10.11, for example), $\int_0^1 \phi_i(x)\,dx = h$, so we may conclude that $h\langle f, \phi_i \rangle = h^2 f(\tilde{x}_i)$, which now looks a lot like $h^2 f(ih)$. Thus the finite difference linear system looks very close to the linear Rayleigh-Ritz system. The latter uses an averaging process to measure $f(x)$ on each subinterval, whereas the former simply takes a point value.

EFR 10.13: With any linearly independent set of basis functions $\{\phi_k(x)\}_{k=1}^n$ the Galerkin method for the BVP gives the numerical solution $u = \sum_{k=1}^n c_k\phi_k(x)$ where the coefficients are determined by the matrix equation (36): $\sum_{j=1}^n \langle \phi_k', \phi_j' \rangle c_j = \langle f, \phi_k \rangle$ $(1 \le k \le n)$. Since $\phi_k(x) = \sin(k\pi x)$ we have $\phi_k'(x) = k\pi\cos(k\pi x)$, and we can use the angle addition formulas for sine and cosine from trig (or appeal to the Symbolic Toolbox) to verify the following orthogonality relations:

$$\langle \phi_k', \phi_j' \rangle = \begin{cases} k^2\pi^2/2, & \text{if } j = k \\ 0, & \text{if } j \ne k \end{cases}$$

Thus, the stiffness matrix is a diagonal matrix, and the system is very easy to solve: $c_k = 2\langle f, \phi_k \rangle/(\pi^2 k^2)$ so we need only actually compute $\langle f, \phi_k \rangle$. As a slightly different approach to what was used in the solution of Example 10.7, we solve these equations using a "for loop" with internally constructed inline functions. This approach may seem simpler to program but it uses more resources since a new (and complicated) function needs to be constructed at each iteration. The size of this problem is small enough so that computing time will not be a consideration.

To compare with the solution obtained in Example 10.7, we compute this Galerkin solution using the same vector x of 52 equally spaced points in [0,1]. The following code computes the corresponding y-coordinates, and plots the two graphs along with the error (we assume that the code of Example 10.7 has been run in this session and the solution was again plotted).

```
>> xG = linspace(0,1,52);
for k=1:50
  kst = num2str(k,2);
```

```
fphik = inline(['sin(sign(x-.5).*exp(1./(4*abs(x-.5).^1.05+.3))).*...
exp(1./(4*abs(x- .5).^1.2+.2)-100*(x-.5).^2).*sin(', kst, '*pi*x)']);
cG(k)=2*quadl(fphik,0,1)/pi^2/k^2;
end
>> y = zeros(size(x));
>> for k=1:50
    y = y + cG(k)*sin(k*pi*x);
end
>> hold on, plot(x,y),   plot(x,abs(c-y),'rx')
```

The graph on the left shows the three plots. Since for this example, the Rayleigh-Ritz method is exact, the red error graph is actually the error for the Galerkin method. To see it better, we plot it separately on the right, having used the following commands:

```
>> hold off, plot(x, abs(c-y), 'r-x'), title('Galerkin Error')
```

The error of the Galerkin method is relatively large in the middle portion. This is to be expected due to the highly oscillatory behavior of f in this region. We could attain better accuracy, of course, by using a larger value of n.

EFR 10.14: (a) First observe that the since w satisfies the BC of (42a), the function $u(x) \equiv w(x) - (1-x)\alpha - \beta x$ satisfies the BC of (42): $u(0) \equiv w(0) - \alpha = 0$ and $u(1) \equiv w(1) - \beta = 0$. Next we compute the left side of the DE of (42): $-(pu')' + qu = -(p(w' - (\alpha - \beta)))' + q(w - (1-x)u - \beta x)$
$= -(p'(w' - (\alpha - \beta))) - pw'' + qw - (1-x)\alpha q - \beta qx = -p'w' - p'w'' + qw + (\alpha - \beta)p' - (1-x)\alpha q - \beta qx$
$= -(pw')' + qw + F(x) = f(x) + F(x)$, where $F(x) \equiv p'(\alpha - \beta) - (1-x)\alpha q - \beta x q$. Thus $u(x)$ satisfies the BVP (42) with $f(x)$ being replaced by $f(x) + F(x)$.

(b) Define $\bar{w}(x) = w(t) = w(a + x(b-a))$, and similarly define $\bar{p}(x) = p(t)$, and in the same fashion the functions $\bar{q}(x)$, and $\bar{f}(x)$. By the chain rule, we can write $\bar{w}'(x) = (b-a)w'(t)$ and $\bar{w}''(x) = (b-a)^2 w''(t)$ where the derivatives of the barred function are with respect to x and those of the unbarred function are with respect to t. The same derivative relationships hold, of course, for the other matching pairs of barred and unbarred functions. If we take the DE of (42b), and change variables from t to x, we obtain:

$$-\left(p(t)w'(t)\right)' + q(t)w(t) = f(t) \Rightarrow -p'(t)w'(t) - p(t)w''(t) + q(t)w(t) = f(t)$$
$$\Rightarrow -\frac{\bar{p}'(x)}{b-a}\frac{\bar{w}'(x)}{b-a} - \bar{p}(x)\frac{\bar{w}''(x)}{(b-a)^2} + \bar{q}(x)\bar{w}(x) = \bar{f}(x)$$
$$\Rightarrow -\bar{p}'(x)\bar{w}'(x) - \bar{p}(x)\bar{w}''(x) + (b-a)^2\bar{q}(x)\bar{w}(x) = (b-a)^2\bar{f}(x)$$
$$\Rightarrow -\left(\bar{p}(x)\bar{w}'(x)\right)' + \bar{Q}(x)\bar{w}(x) = \bar{F}(x),$$

where $\bar{Q}(x) = (b-a)^2\bar{q}(x)$ and $\bar{F}(x) = (b-a)^2\bar{f}(x)$. Thus the function $\bar{w}(x)$ satisfies a BVP of the form (42a).

EFR 10.15: (a) The M-file is boxed below:

```
function [x,u] = rayritz(p, q, f, n)
% This program will implement the piecewise linear Rayleigh-Ritz
% method to solve the BVP:  -(p(x)u'(x))'+q(x)u(x)=f(x),  u(0)=0,
% u(1)=0.  The integral approximations (48) through (50) of Chapter
% 10 will be used. Input variables:  the first three:  p, q and f are
% inline functions representing the the the DE, n = the number of
% interior x-grid values to employ.  A uniform grid is used.
% NOTE:  The program is set up to require that the functions p, q,
% and f take vector arguments.
% Output variables:  x and u are same sized vectors representing the
% x grid and the numerical solution values respectively

x=linspace(0,1,n+2); h = x(2)-x(1);
% Use (48) and (49) of Chapter 10 to assemble diagonals of the
% symmetric tridiagonal stiffness matrix:
d = 1/(2*h)*(feval(p, x(1:n))+2*feval(p, x(2:n+1))+feval(p, ...
x(3:n+2)))+ h/12*(feval(q, x(1:n))+6*feval(q, x(2:n+1))+ ...
feval(q, x(3:n+2)));
% for off diagonals
offdiag= -1/(2*h)*(feval(p,x(2:n))+feval(p,x(3:n+1)))+...
      h/12*(feval(q,x(2:n))+feval(q,x(3:n+1)));
% by symmetry and to conform to syntax of 'thomas.m'
da = [offdiag 0]; %above diagonal
db = [0 offdiag]; %below diagonal

% Use (50) of Chapter 10 to construct vector b
b = h/6*(feval(f,x(1:n))+4*feval(f,x(2:n+1))+feval(f,x(3:n+2)));

% Use the Thomas method to solve the system Au=b and get solution
u = thomas(da,d,db,b);
u = [0 u 0]; %adjoin boundary values
```

(b) With the program above, the task is easily completed. We first store the coefficient functions for the DE (capable of taking vector inputs as stipulated in the notes of the code in part (a)), next we obtain the first three numerical solutions and then we look at successive differences on common x-grid values. Note that, for example, in passing from x1 to x2, since the step size is getting cut in half, the first internal grid value for x1 (in MATLAB notation x1(2)) will be the second internal grid value for x2, (in MATLAB notation x2(3)) and, hereafter, the indices of successive internal grid values for x1 will jump by 2's when looked at as indices of x2.

```
>> p = inline('ones(size(x))'); q = inline('6*ones(size(x))');
>> f = inline('exp(10*x).*cos(12*x)');
>> [x1, y1] = rayritz(p,q,f,99);
>> [x2, y2] = rayritz(p,q,f,199);
>> size(y1), size(y2) %Check the sizes of the vectors
→ans = 1  101, ans = 1  201
>> max(abs(y1(2:100)-y2(3:2:200))) →ans = 0.1019
>> [x3, y3] = rayritz(p,q,f,399);
>> size(y3) →ans = 1  401
>> max(abs(y2(2:200)-y3(3:2:400))) →ans = 0.0255
```

The following loop will now continue such iterations until the error of these successive differences gets less than 5e-5:

```
>> ynew = y3; count =3;
>> while Error > 5e-5
yold = ynew; count = count +1, numx = 2*numx;
[xnew, ynew] = rayritz(p,q,f,numx-1); Error = max(abs(yold(2:numx/2)-
ynew(3:2:numx)))
end
```

→ count =4, Error = 3.9833e-004

count =5, Error = 9.9578e-005
count =6, Error = 2.4896e-005

From the results we see that the difference of y6=ynew and y5 is less than 5e-5. We now get the exact error of y6 by invoking the Symbolic Toolbox to solve the DE exactly (as in the example):

```
>> yexact=dsolve('-D2y+6*y=exp(10*t)*cos(12*t)', 'y(0)=0', 'y(1)=0');
>> Yexact = double(subs(yexact,xnew));
>> max(abs(Yexact-ynew))  →ans= 8.3034e-006
```

Thus the successive differences turned out to give a good predictor of when we should stop the iteration process to meet the desired error goal. In the absence of exact solutions, this technique is used quite often in practice.

EFR 10.16: (a) Condition (i) states that we can write:

$$BS(x) = \begin{cases} a_1x^3 + b_1x^2 + c_1x + d_1, & \text{if } x \in [-2,-1], \\ a_2x^3 + b_2x^2 + c_2x + d_2, & \text{if } x \in [-1,0], \\ a_3x^3 + b_3x^2 + c_3x + d_3, & \text{if } x \in [0,1], \\ a_4x^3 + b_4x^2 + c_4x + d_4, & \text{if } x \in [1,2]. \end{cases}$$

The will show that the 16 parameters a_i, b_i, c_i, d_i will be uniquely determined by the other three conditions. Our strategy will be to first take all opportunities to either solve or eliminate any parameters, and then for the parameters that remain we will determine them by solving a linear system. Condition (ii) at the internal node $x = 0$, gives that $d_2 = d_3$, $c_2 = c_3$, and $b_2 = b_3$. The interpolation requirement that BS(0) = 1 tells us that $d_2 = d_3 = 1$. We now use the remaining conditions to obtain 12 linear equations for the remaining 12 parameters: $a_1, b_1, c_1, d_1, a_2, b_2, c_2, a_3, a_4, b_4, c_4, d_4$.

From (ii), we get the following six linear equations:
(continuity of BS at −1):

$$-a_1 + b_1 - c_1 + d_1 = -a_2 + b_2 - c_2 + 1, \text{ (and at 1) } a_3 + b_2 + c_2 + 1 = a_4 + b_4 + c_4 + d_4.$$

(continuity of BS' at −1):

$$3a_1 - 2b_1 + c_1 = 3a_2 - 2b_2 + c_2, \text{ (and at 1) } 3a_3 + 2b_2 + c_2 = 3a_4 + 2b_4 + c_4.$$

(continuity of BS'' at −1):

$$-6a_1 + 2b_1 = -6a_2 + 2b_2, \text{ (and at 1) } 6a_3 + 2b_2 = 6a_4 + 2b_4$$

From (iii) we get two equations:

$$(BS(-2) = 0): \ -8a_1 + 4b_1 - 2c_1 + d_1 = 0, \quad (BS(2) = 0) \ 8a_4 + 4b_4 + 2c_4 + d_4 = 0.$$

From (iv) we get the remaining 4 equations:

$$(BS'(-2) = 0) \ 12a_1 - 4b_1 + c_1 = 0, \quad (BS'(2) = 0) \ 12a_4 + 4b_4 + c_4 = 0,$$

$$(BS''(-2) = 0) \ 12a_1 + 2b_1 = 0, \text{ and } (BS''(2) = 0) \ 12a_4 + 2b_4 = 0.$$

Moving all variables to the left side and numbers to the right (and using the order of the 12 parameters listed above), leads to a matrix equation (for the 12 parameters represented by the vector x): $Ax = b$. We now use MATLAB to enter A and b and then to solve the system:

```
A = zeros(12); b=zeros(12,1);  A(1,[1 3 6])=-1; A(1,[2 4 5 7])=1;
b(1)=1; A(2,[8 6 7])=1; A(2,[9 10 11 12])=-1;  b(2)=-1; A(3,1)=3;
A(3,2)=-2; A(3,3)=1; A(3, 5)=-3; A(3,6)=2; A(3,7)=-1;  A(4,8)=3;
A(4,6)=2; A(4,7)=1; A(4, 9)=-3;  A(4,10)=-2;A(4,11)=-1;  A(5,1)=-6;
A(5,2)=2; A(5,5)=6; A(5,6)=-2;  A(6,8)=6;  A(6,6)=2; A(6,9)=-6;
A(6,10)=-2; A(7,1)=-8; A(7,2)=4; A(7,3)=-2; A(7,4)=1;  A(8,9)=8;
A(8,10)=4; A(8,11)=2; A(8,12)=1; A(9,1)=12; A(9,2)=-4; A(9,3)=1;
A(10,9)=12; A(10,10)=4; A(10,11)=1;  A(11,1)=-12; A(11,2)=2;
A(12,9)=12;  A(12,10)=2;
>> format long
>> x=A\b; x, format rat, x
→x =         0.25000000000000          1/4
            1.50000000000000          3/2
            3.00000000000000          3
```

2.00000000000000	2
-0.75000000000000	-3/4
-1.50000000000000	-3/2
-0.00000000000000	-1/4503599627370498
0.75000000000000	3/4
-0.25000000000000	-1/4
1.50000000000000	3/2
-3.00000000000000	-3
2.00000000000000	2

(The very small fraction in the second column is just roundoff error.) From these coefficients, we can express BS(x) as follows:

$$BS(x) = \begin{cases} \frac{1}{4}x^3 + \frac{3}{2}x^2 + 3x + 2, & \text{if } x \in [-2,-1], \quad -\frac{3}{4}x^3 - \frac{3}{2}x^2 + 1, & \text{if } x \in [-1,0], \\ \frac{3}{4}x^3 - \frac{3}{2}x^2 + 1, & \text{if } x \in [0,1], \quad -\frac{1}{4}x^3 + \frac{3}{2}x^2 - 3x + 2, & \text{if } x \in [1,2]. \end{cases}$$

It is readily checked (by hand or with the Symbolic Toolbox) that these formulas agree with the corresponding ones in the formula stated in the text.

(b) We create an M-file for the function BS(x); the following code produced the plot on the right.

```
function y = BSSpline(x)
%Basic cubic spline function
of %
%Chapter 10, (51),built to
accept %vector arguments.
for i=1:length(x)
    if x(i)>=0 & x(i)<=1
    y(i)=((2-x(i))^3-4*(1-
x(i))^3)/4;
    elseif x(i)>1 & x(i)<=2
        y(i)=(2-x(i))^3/4;
    elseif x(i)>2
        y(i)=0;
    else, y(i) = BSSpline(-
x(i));
    end
end
```

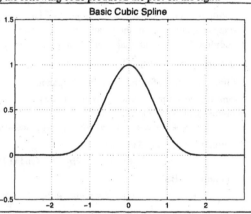

Basic Cubic Spline

```
>> x=-5:.01:5; plot(x,BSSpline(x)), grid on
>> axis([-3 3 -.5 1.5]), title('Basic Cubic Spline')
```

EFR 10.17: It is perhaps more helpful to do part (b) first, so we can get an idea of the functions that we need to answer some questions about. We proceed in this way. Using formula (52) in conjuction with the M-file for the basic cubic spline BS(x) constructed in the preceding EFR, the following code will produce the plots of the cubic spline basis functions when $n = 5$ (shown on the right):

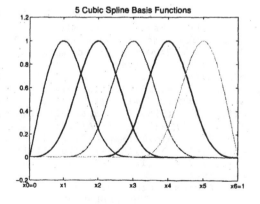

5 Cubic Spline Basis Functions

```
>> x=-5:.01:5;
>> n=5; h=1/(n+1);
>> x=linspace(0,1,n+2)
>> t=0:.01:1;
>> phi1=BSSpline((t-h)/h) - ...
BSSpline((t+h)/h);
>> phi2=BSSpline((t-2*h)/h);
>> phi3=BSSpline((t-3*h)/h);
>> phi4=BSSpline((t-4*h)/h);
```

```
>> phi5=BSSpline((t-n*h)/h)- ... BSSpline((t-(n+2)*h)/h);
>> plot(t,phi1,'r')
>> axis([0   1   -.2   1.2])
>> hold on, grid on
>> plot(t,phi2,'k', phi3, 'g', phi4, 'k', phi5, 'c' )
>> title('5 Cubic Spline Basis Functions')
```
Note: On MATLAB's graphics window, we used the "Axis Properties" subwindow from the "Edit" menu to change the x-axis tick marks to be at 0, 1/6, 2/6, ... and changed the corresponding labels to $x0 = 0, x1, x2, ...$

All of the properties stated about the cubic spline basis functions, except their linear independence, are directly inherited from those of $BS(x)$. Linear independence will take a bit more work to show than for the hat functions, but is not very difficult since the supports of the cubic spline basis functions (i.e., the intervals on which they are nonzero) do not have much overlap. Indeed, assume that for a fixed n, we have (*) $\sum_{i=1}^{n} c_i \phi_i(x) \equiv 0$ (for all x in [0, 1]) for some constants c_i. We must show that $c_i = 0$ for each i. Consider first x to be in the interval [0, x1]. Here, only the first two ϕ_i's are nonzero, so (*) becomes $c_1 \phi_1(x) + c_2 \phi_2(x) \equiv 0$. If either of c_1 or c_2 were nonzero, then we could solve for the corresponding ϕ_i in terms of being a constant multiple of the other one. This is clearly not possible, since (among many other reasons, just look at the picture) one of them has a zero derivative at $x = x1$ and the other does not. Consequently we may conclude that $c_1 = c_2 = 0$. The rest is now easy: On the next interval [x2, x3], only ϕ_1, ϕ_2 and ϕ_3 can be nonzero, but since we know already $c_1 = c_2 = 0$, (*) becomes $c_3 \phi_3(x) \equiv 0$ and this certainly forces $c_3 = 0$. We may continue this argument moving one new interval to the right at each step and concluding the successive c_i's must be zero. This completes the linear independence proof.

CHAPTER 11: INTRODUCTION TO PARTIAL DIFFERENTIAL EQUATIONS

<u>**EFR 11.1:**</u> (a) & (b): We do the plots only for surf; to create corresponding mesh plots simply replace all occurrences of surf below by mesh.
```
>> x=linspace(-5,5,30); y=x; [X,Y]=meshgrid(x,y);
>> Z=sin(X)*sin(Y)*exp(-sqrt(X.^2+Y.^2)/4);
>> surf(x,y,Z)
>> xlabel('x-values'), ylabel('y-values'), zlabel('z-values')
>> grid off %default view shown below left
>> view(90,0) %view from positive x-axis, shown below right
>> view(45, -30) %view shown from 30 degrees below xy-plane, shown on
>>                                %top of next page
```

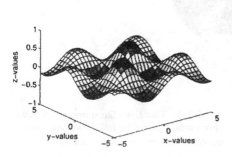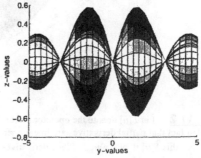

```
>> Z=sin(Y+cos(X));
>> clf
>> surf(x,y,Z)
>> xlabel('x-values'),
ylabel('y-values'), zlabel('z-
values')
>> grid off   %default view,
shown below left
>> view(80,20) %view from 10
degrees from pos. x-axis and
20 degrees below xy-plane,
%shown below right
>> view(45, 80) %View from 80
degrees above xy-plane, shown
at bottom.
```

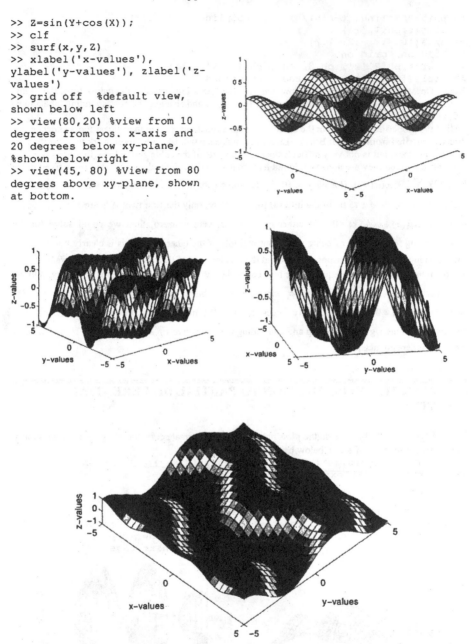

EFR 11.2: Let $L[u]$ denote the operator on the left side of (11). That L is a linear operator follows from the fact that partial derivatives are linear operators. For example, consider just the second term of $L[u]$, call this $S[u] = b(x,y)u_{xy}$. Since derivatives of sums equal the sums of the derivatives, we have,

$S[u+v] = b(x,y)[u+v]_{xy} = b(x,y)(u_{xy}+v_{xy}) = b(x,y)u_{xy} + b(x,y)v_{xy} = S[u] + S[v]$. Since the same is true for each term of $L[u]$, it follows that $L[u+v] = L[u] + L[v]$. Similarly, since constants can be pulled out of derivatives: for any constant c we have $S[cu] = b(x,y)[cu]_{xy} = cb(x,y)u_{xy} = cS[u]$, and because this is true for each term of $L[u]$, we get in the same fashion that $L[cu] = cL[u]$. Thus $L[u]$ is indeed a linear operator.

EFR 11.3: Computing the partial derivates directly (or using the Symbolic Toolbox), we get the following expressions for Δu:
(a) $\Delta u = 0 + 0 = 0$ so u is harmonic (same for any linear function).
(b) $\Delta u = 2 + 2 = 4 \neq 0$ so u is not harmonic.
(c) $\Delta u = 2 - 2 = 0$ so u is harmonic.
(d) $u(x,y)$ is only defined if $(x,y) \neq 0$ and in this region we have

$$\Delta u = \left[\frac{2}{x^2 + y^2} - \frac{4y^2}{(x^2 + y^2)} \right] + \left[\frac{2}{x^2 + y^2} - \frac{4x^2}{(x^2 + y^2)} \right] = 0,$$

so $u(x,y)$ is harmonic on its domain.

EFR 11.4: (a) Following the method and using the notation of Example 11.5, we skim through the details for the present example. Using (21)

$$4u_{i,j} - u_{i+1,j} - u_{i-1,j} - u_{i,j+1} - u_{i,j-1} = -h^2 q_{i,j},$$

we can write the 16 equations for the 16 unknown functional values; here are several of them (refer to the figure on the right):

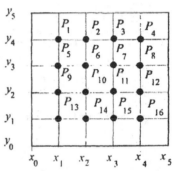

Examining this linear system shows that the coefficient matrix A of the linear system to be solved ($AU = C$) has exactly the same form as the one in the solution of Example 11.5, except now its size is 16×16. As in (19), we introduce vectors L, R, B, and T for the boundary data. It is also convenient to

Interior vertex	Linear equation [unknowns] = [knowns]
P_1	$4U_1 - U_2 - U_5 = -h^2 q_{1,4} + u_{0,4} + u_{1,5}$
P_2	$4U_2 - U_3 - U_1 - U_6 = -h^2 q_{2,4} + u_{2,5}$
P_3	$4U_3 - U_4 - U_2 - U_7 = -h^2 q_{3,4} + u_{3,5}$
P_4	$4U_4 - U_3 - U_8 = -h^2 q_{4,4} + u_{5,4} + u_{4,5}$
P_5	$4U_5 - U_6 - U_1 - U_9 = -h^2 q_{1,3} + u_{0,3}$
P_6	$4U_6 - U_7 - U_5 - U_2 - U_{10} = -h^2 q_{2,3}$
$\vdots$	$\vdots$
P_{13}	$4U_{13} - U_{14} - U_9 = -h^2 q_{1,1} + u_{0,1} + u_{1,0}$
P_{14}	$4U_{14} - U_{15} - U_{13} - U_{10} = -h^2 q_{2,1} + u_{2,0}$
P_{15}	$4U_{15} - U_{16} - U_{14} - U_{11} = -h^2 q_{3,1} + u_{3,0}$
P_{16}	$4U_{16} - U_{14} - U_{12} = -h^2 q_{4,1} + u_{5,1} + u_{4,0}$

introduce a length 16 vector Q whose values are the internal $q_{i,j}$-values given in the reading order (with the same relationship U has to $u_{i,j}$). Invoking MATLAB's index notation, the vector C of the system can be read off from the linear system on the left to take the following form:

$$C = -h^2 Q + [L(5) + T(2),$$
$$T(3), T(4), R(5) + T(5),$$
$$L(4), 0, 0, R(4), L(3),$$
$$0, 0, R(3), L(2) + B(2),...$$
$$B(3), B(4), R(2) + B(5)]'$$

Based on the above development, the following code will find the associated finite difference solution and create a surface plot of it.

```
%EFR11_4
%Script for solving the Poisson problem of EFR 11_4a
N=4; M=4; h = 1/(N+1); x=linspace(0,1,N+2); y=x; A=4*eye(N^2);
%form sub/super diagonals
```

```
a1rep=[0 -1 -1 -1]; a1=[-1 -1 -1];
for i=1:3, a1=[a1 a1rep]; end, a4=-1*ones(1,12);
%put these diagonal entries on A
A=A+diag(a1,-1)+ diag(a1,1)+diag(a4,-4)+diag(a4,4);
% key in vectors for boundary values:
L = zeros(size(y)); R = L; B = sin(pi*x); T = B/exp(2);
% Now (for the most complicated part), we construct the vector C
% First we construct a row vector for Q arising from the source term:
% We do this by creating an inline function for the inhomogeneity and
% then collecting the needed entries in the required reading order
% (using an appropriately designed loop).
q=inline('3*exp(-2*y).*sin(pi*x)', 'x', 'y');
row = 1;
for j=5:-1:2
    count=(row-1)*4+1;
    Q(:,count:count+3)=q(x(2:5),y(j)); row = row+1;
end
%By combining with the appropriate boundary values, we now construct
C:
C= -h^2*Q + [L(5)+T(2) T(3) T(4) R(5)+T(5) L(4) 0 0 R(4) L(3) 0 0...
    R(3) L(2)+B(2) B(3) B(4) R(2)+B(5)];  C=C';
%Now we are ready to solve the system, form the mesh, and plot
U=A\C; Z=zeros(4); Z(:)=U; Z=Z'; Z=[T(2:5); Z; B(2:5)];
for i=1:6, Lrev(i)=L(7-i); end
Z=[Lrev; Z'; R]';
for i=1:6, yrev(i)=y(7-i); end
surf(x, yrev ,Z), xlabel('x-values'), ylabel('y-values'), zlabel('u-
values')
```
The plot is the left-hand one shown below.

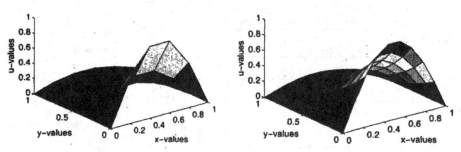

(b) The given function $u(x,y)$ is readily verified to satisfy all conditions of the BVP. By forming first the matrix of the values of the exact solution on the given grid, we can easily compute the maximum error and relative error:
```
[X, Y] = meshgrid(x,yrev);
Zexact=exp(-2*Y).*sin(pi*X);
Max_Error = max(max(abs(Z(2:N+1,2:N+1)-Zexact(2:N+1,2:N+1))))
Max_Relative_Error = max(max(abs(Z(2:N+1,2:N+1)-...
Zexact(2:N+1,2:N+1))./abs(Zexact(2:N+1,2:N+1))))
    0.2064
```
→Max_Error = 0.2064

→Max_Relative_Error = 0.5891

(c) The above code is easily modified to deal with the finer grid. We give it in a slightly more general context than was presented in part (a) and we omit the comments to save space. The resulting plot is the one on the above right.
```
N=8; M=8; h = 1/(N+1);
x=linspace(0,1,N+2); y=x;
```

```
A=4*eye(N^2);
alrep=[0 -1 -1 -1 -1 -1 -1 -1];
al=[-1 -1 -1 -1 -1 -1 -1];
for i=1:7, al=[al alrep]; end
aN=-1*ones(1,N^2-N);
A=A+diag(al,-1)+ diag(al,1)+diag(aN,-N)+diag(aN,N);
L = zeros(size(y)); R = L;
B = sin(pi*x); T = B/exp(2);
q=inline('(4-pi^2)*exp(-2*y).*sin(pi*x)', 'x', 'y');
row = 1;
for j=N+1:-1:2
    count=(row-1)*N+1;
    Q(:,count:count+N-1)=q(x(2:N+1),y(j)); row = row+1;
end
zer = zeros(1, N-2); %useful vector for constructing C
C= -h^2*Q + [L(9)+T(2) T(3) T(4) T(5) T(6) T(7) T(8) R(9)+T(9) ...
   L(8) zer R(8) L(7) zer R(7) L(6) zer R(6) L(5) zer R(5) L(4) ...
   zer R(4) L(3) zer R(3) L(2)+B(2) B(3) B(4) B(5) B(6) B(7) B(8)...
   R(2)+B(9)];
C=C';
U=A\C;
Z=zeros(N);
Z(:)=U;
Z=Z';
Z=[T(2:N+1); Z; B(2:N+1)];
for i=1:N+2, Lrev(i)=L(N+3-i); end
Z=[Lrev; Z'; R]';
for i=1:N+2, yrev(i)=y(N+3-i); end
surf(x, yrev ,Z), xlabel('x-values'), ylabel('y-values'),
zlabel('u-values')
```
The codes of part (b) for computing the errors will work here as well; the results, shown below, indicate significant improvements in the quality of the solution (both decrease nearly 100-fold!):
→Max_Error = 0.0030
→Max_Relative_Error = 0.0081

EFR 11.5: (a) In (21),

$$4u_{i,j} - u_{i+1,j} - u_{i-1,j} - u_{i,j+1} - u_{i,j-1} = -h^2 q_{i,j},$$

since all boundary terms are zero, whenever a boundary term is present on the left it can be deleted (rather than moved to the right side). Thus, the right sides of these equations will always take the same form, so we just have to describe how the left sides look. We use the grid-numbering scheme introduced in the section. For each node P_k, (21) gives an equation for the corresponding unknown function value $U_k = u(P_k)$. With the aid of the generic grid diagram on the right, we describe the various forms of the left sides of these equations:

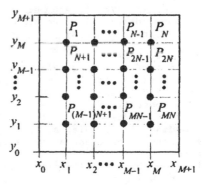

$k = 1$: $4U_1 - U_2 - U_{N+1}$

$k = 2:N-1$ $4U_k - U_{k+1} - U_{k-1} - U_{N+k}$

$k = N$: $4U_N - U_{N-1} - U_{2N}$

$k = N+1, 2N+1, ..., (M-2)N+1$: $4U_k - U_{k+1} - U_{k-N} - U_{k+N}$

$k = 2N, 3N, ..., (M-1)N$: $4U_k - U_{k-1} - U_{k-N} - U_{k+N}$

k between $N + 1$ and $(M - 1)N$,

but not of last two types: $4U_k - U_{k+1} - U_{k-1} - U_{k-N} - U_{k+N}$

$k = (M - 1)N + 1$: $4U_k - U_{k+1} - U_{k-N}$

$k = (M - 1)N + 2:MN - 1$: $4U_k - U_{k+1} - U_{k-1} - U_{k-N}$

$k = MN$: $4U_k - U_{k-1} - U_{k-N}$

Contemplating this linear system and putting it into matrix form: $AU = C$, we see that C is simply the column vector $-h^2 Q$ and A is the $NM \times NM$ banded matrix having the following 5 bands: 4's down the main diagonal, -1's down the diagonals at levels N (above main) and $-N$ (below main). At levels 1 and -1, the following vector appears: It begins with a sequence of $N - 1$, -1's, then the vector $[0\ -1\ -1 \ldots\ -1]$ with -1 repeated $N - 1$ times is tacked on $M - 1$ times. This being done, the following M-file is a straightforward modification of the code used in Example 11.5:

```
function [xgrid, ygrid, Zsol] = poissonsolver(q,a,b,c,d,h)
%M-file for EFR 11.5.  This program is designed to find the finite
%difference solution of the Poisson problem with zero Dirichlet
%boundary conditions on any rectangle R={a<=x<=b, c<=y<=d}:
%u_xx + u_yy = q(x,y) on R
%u(x,y) =   0 on bdy R
%Input variables:  q = inline function (or M-file) for inhomogeneity
%a,b,c,d = endpoints of the rectangle R (a<b and c<d), and h =
%uniform step size (x step = y step).  It is assumed that h divides
%into both b-a and d-c with an integer quotient.
%Output variables:  xgrid, ygrid, and Zsol,  the first two are
%vectors corresponding to the partition of [a, b] and [c, d],
%respectively determined by the step size h.  Zsol is the
%corresponding matrix of values over the rectangular grid, set up so
%that surf(xgrid, ygrid, Zsol) will result in a surface plot of the
%numerical solution.

%first check to see if h is a permissible step size:
if ((b-a)/h>floor((b-a)/h+eps))|((d-c)/h>floor((d-c)/h+eps))
    error('Inputted step size does not evenly divide into both side
lengths; try another step size')
end

N=floor((b-a)/h)-1; %number of internal x-grid points
M=floor((d-c)/h)-1; %number of internal y-grid points
xgrid=linspace(a,b,N+2); ygrid=linspace(c,d,M+2);

A=4*eye(N*M);
%form sub/super diagonals
a1=-ones(1,N-1); a1rep = [0 a1];
for i=1:M-1, a1=[a1 a1rep]; end
aN=-1*ones(1,N*M-N);
%put these diagonal entries on A
A=A+diag(a1,-1)+ diag(a1,1)+diag(aN,-N)+diag(aN,N);
% First we construct a row vector for Q, arising from the source
term:
% We do this by collecting the needed entries in the required reading
order (using an
% appropriately designed loop).
row = 1;
for j=M+1:-1:2
    count=(row-1)*N+1;
    Q(:,count:count+N-1)=q(xgrid(2:N+1),ygrid(j)); row = row+1;
end
C= -h^2*Q; ,  C=C';
%Now we are ready to solve the system.
```

```
U=A\C;
Z=zeros(N,M);
Z(:)=U;
Z=Z';
Z=[zeros(1,N); Z; zeros(1,N)];
Zsol=[zeros(1,M+2); Z'; zeros(1,M+2)]';
%rather than reverse the order of ygrid, we leave it in the usual
order,
%but change the ordering in Zsol to make it amenable to 3D plotting.
%Znew = zeros(size(Zsol));
for i=1:M+2, Znew(i,:)=Zsol(M+3-i,:); end, Zsol = Znew;
```

This M-file is very simple to implement. Both numerical solutions that we obtain are graphically indistinguishable from the exact solution, so we will show a plot of the latter (finer grid) numerical solution and also both error graphs.

```
>> q = inline('sin(2*pi*y).*(4*pi^2*(x-x.^3)+6*x)','x','y');
[x1,y1,Z1]=poissonsolver...
(q,0,1,0,1,.1);
[x,y,Z]=poissonsolver...
(q,0,1,0,1,.02);
>> surf(x,y,Z),
xlabel('x-values'),
ylabel('y-values'),
zlabel('u-values')
>> %Plot shown at right.

>> uExact=inline('(x.^3-...
x).*sin(2*pi*y)','x','y');
>> Zexact= (X.^3-...
X).*sin(2*pi*Y);
>> [X1,Y1]=meshgrid(x1,y1);

>> Z1exact= (X1.^3-X1).*sin(2*pi*Y1);
>> surf(x1,y1,abs(Z1-Z1exact))
>> xlabel('x-values'), ylabel('y-values'),
>> zlabel('Error')
>> %first error plot, below left
>> [X,Y]=meshgrid(x,y);
>> Zexact= (X.^3-X).*sin(2*pi*Y);
>> surf(x,y,abs(Z-Zexact))
>> xlabel('x-values'), ylabel('y-values'),
>> zlabel('Error')
>> %second error plot, below right
```

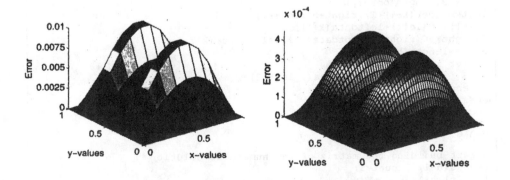

EFR 11.6:   (a) The M-file boxed below follows the strategies of the solution of Example 11.6:

```
function [Z, x, y] = triangledirichletsolver(n, leftdata, bottomdata,
slantdata)
% This program will solve the Dirichlet problem of Laplaces equation
% on the special isoceles triangle with vertices (0,0), (1,0), (0,1).
% The finite difference method will be used.
% The inputs are as follows: n = the number of interior grid points
% on both the x- and y-axis (so n+2= total # of x/y-grid values).
% leftdata = vector of boundary values on left side (size n+2, read
% top to bottom, bottomdata = vector of boundary values on bottom
% side (size n+2)slantdata = vector of boundary values on slant side
% (size n+2, read from top)
% The output variables are as follows:
% Z = the n+2 by n+2 matrix of the discrete solution's values
% x = vector of x grid values
% y = vector of y grid values (in reverse order to facilitate plots)
N=n*(n-1)/2; %number interior nodes (with unknown function values)
A=diag(4*ones(1,N)); border = [0]; count = 1;
for i=1:n-1
    border(i+1)=border(i)+count;
    count=count+1;
end

for k=2:length(border)
if k>2, pregap=right-left+1; end
left=border(k-1)+1; right=border(k);
if k<length(border)
postgap=right-left+1;
end
for i=left:right
if i<right %has right neighbor and top neighbor
A(i,[i+1    i-pregap])=-1;
end
if i>left %has left neighbor
A(i,i-1)=-1;
end
if k<length(border) %has bottom neighbor
A(i,i+postgap)=-1;
end
end
end

%Next we need to build the vector C
C=zeros(N,1);
for k=2:length(border)
left=border(k-1)+1; right=border(k);
C(left)=C(left)+leftdata(k+1);
C(right)=C(right)+slantdata(k+1)+slantdata(k);
if k==length(border)
    for i=left:right
        C(i)=C(i)+bottomdata(i-left+2);
    end
end
end

U=A\C;

%start building the matrix of the numerical solution
Z=ones(n-1); count=1;
for i=1:n-1
```

```
gap=border(i+1)-count+1; Z(i,1:gap)=U(count:(count+gap-1))';
count=count+gap;
end
Z=[ones(1,n-1);[slantdata(2) ones(1,n-2)];Z;bottomdata(2:n)];
Z=[leftdata', Z, [ones(1,n+1) bottomdata(n+1)]', slantdata'];
for i=1:n+2
    Z(i,i)=slantdata(i);
end
%We delete those values of the Z matrix which are not in the triangle
%except for those nodes adjacent to two diagonal nodes where we use
an
%average value
for i=1:n+2
    if i<n+2
    Z(i,i+1)=(Z(i,i)+Z(i+1,i+1))/2;
    end
    for j=i+2:n+2
        Z(i,j)=nan;
    end
end


h=1/(n+1);
x=0:h:1; y=x;
for i=1:length(y), yrev(i)=y(length(y)+1-i); end
y=yrev;
```

(b) The following commands will invoke the above M-file to solve the BVP of Example 11.6 with $n = 49$ internal grid values.

```
>> leftdata = [50 zeros(1, 50)]; bottomdata = [zeros(1,50) 50];
>> slantdata = [50 100*ones(1,49) 50];
>> [Z, x, y]= triangledirichletsolver(49,leftdata,bottomdata,
slantdata);
>> surf(x,y,Z)
```

A surface graph of the numerical solution will now appear. It can be rotated into looking exactly like the one in Figure 11.16b.

```
>> c=contour(x,y,Z,12), clabel(c,'manual')
```

The first of these two commands will create a contour (isotherm) plot with 12 contour lines. The second command has prompted users to click with their mouse at the locations (on the isotherms) where numerical values should be displayed. A plot as in Figure 11.17 can thus be constructed.

EFR 11.7: (a) The M-file boxed below follows the strategies of the solution of Example 11.7:

```
function [Z, x, y]=rectanglepoissonsolver(h, a, b, varf, leftdata,
rightdata, topdata, bottomdata)
% Program for solving the Dirichlet problem for the Poisson equation
% Laplace(u)=f on the rectangular domain: 0 <= x <= a, 0<= y <= b.
% Input variables:  h = common step size (assumed to divide evenly
% into both a and b), varf = inhomogeneity function (of x and y)for
% the Poisson equation, last four input variables give the Dirichlet
% boundary data on the various sides of the rectangle.  Horizontal
% data are assumed to be row vectors (reading from left to right) and
% vertical data are assumed to be column vectors (reading from top to
% bottom).  Output variables:  Z = matrix of values of the numerical
% solution at the grid values determined by the inputs.  x, y =
% corresponding x-grid vectors and y-grid vectors.  y-grid is assumed
% to read the values from top to bottom to facilitate plots.

%first check to see if h is a permissible step size:
if (a/h>floor(a/h)+eps) | (b/h>floor(b/h)+eps)
        warning('Inputted step size does not evenly divide into both side
```

```
lengths; unexpected results may occur')
end

N=floor(a/h)-1; %number of internal x-grid points
M=floor(b/h)-1; %number of internal y-grid points
xgrid=linspace(0,a,N+2); ygrid=linspace(0,b,M+2);

% Check to see data input vectors are correct size, if not exit
program
if ...
size(leftdata)~=size(ygrid')|size(rightdata)~=size(ygrid')|size(topda
ta)~=size(xgrid)|...
size(bottomdata)~=size(ygrid)
error('At least one of the boundary data vectors does not have
correct size corresponding to step size h, program will terminate')
end
% Creation of coefficient matrix A of the linear system AU=C:
% A=4*eye(N*M);
% form sub/super diagonals
a1=-ones(1,N-1); a1rep = [0 a1];
for i=1:M-1, a1=[a1 a1rep]; end
aN=-1*ones(1,N*M-N);
% put these diagonal entries on A
A=A+diag(a1,-1)+ diag(a1,1)+diag(aN,-N)+diag(aN,N);

% Creation of column vector C of the linear system AU=C:We use the
% decomposition (33), (34), (35)  of Chapter 11 to guide us.
% First we construct a row vector for F, arising from the source
% term:
% We do this by collecting the needed entries in the required reading
% order (using an appropriately designed loop).
row = 1; F=zeros(N*M,1);
for j=M+1:-1:2
    count=(row-1)*N+1;
    F(count:count+N-1)=feval(varf,xgrid(2:N+1),ygrid(j)); row =
row+1;
end
F=F';
% Since C = B-h^2*F, we also need to construct the vector B using the
% boundary data; we use (35) of Chapter 11.  We need to translate
% (35) into the notation of the inputted boundary data vectors.  For
% example, the vector 'leftdata', in the notation of g_i_j, has the
% following components for each MATLAB index (on left, so leftdata(i)
% is abbreviated as i):[1 2 3 ... M+1 M+2] = [g_0_M+1  g_0_M  g_0_M-1
% ... g_0_1  g_0_0] Similarly, the MATLAB components of rightdata
% are[:1 2 3 ... M+1 M+2] = [g_N+1_M+1  g_N+1_M  g_N+1_M-1 ...
% g_N+1_1  g_N+1_0] In the same fashion, here are the components of
% topdata and bottomdata [1 2 3 ... N+1 N+2] = [g_0_M+1  g_1_M+1
% g_2_M+1 ... g_N_M+1  g_N+1_M+1][1 2 3 ... N+1 N+2] = [g_0_0  g_1_0
% g_2_0 ... g_N_0  g_N+1_0]
% Here now is the construction of the vector B from (35):
Bcount=1;
for j=1:M
 if j==1
    for i=1:N
        if i==1, B(Bcount)=topdata(2)+leftdata(2); Bcount=Bcount+1;
        elseif i==N, B(Bcount)=topdata(N+1)+rightdata(2);
Bcount=Bcount+1;
        else, B(Bcount)=topdata(i+1); Bcount=Bcount+1;
        end
```

```
          end
   elseif j==M
       for i=1:N
           if i==1, B(Bcount)=bottomdata(2)+leftdata(M+1);
Bcount=Bcount+1;
           elseif i==N, B(Bcount)=bottomdata(N+1)+rightdata(M+1);
Bcount=Bcount+1;
           else, B(Bcount)=bottomdata(i+1); Bcount=Bcount+1;
           end
       end
   else
       for i=1:N
           if i==1, B(Bcount)=leftdata(1+j); Bcount=Bcount+1;
           elseif i==N, B(Bcount)=rightdata(1+j); Bcount=Bcount+1;
           else, B(Bcount)=0; Bcount=Bcount+1;
           end
       end
   end
end
end

%With F and B constructed (as row vectors), using (33) of Chapter
11, we
%can form the vector C:
C - B-h^2*F; C=C';

%Now we are ready to solve the system.
U=A\C;
Z=zeros(N,M);
Z(:)=U;
Z=Z';
% So far we have the numerical values in assembled in a matrix, we
% just need to attach the given boundary values appropriately.  For
% the corner interfaces between two boundary values (e.g., where the
% topdata vector meets the leftdata vector), we use the average
% value.

Z=[topdata(2:N+1); Z; bottomdata(2:N+1)];
NWavg=(leftdata(1)+topdata(1))/2;
NEavg=(rightdata(1)+topdata(N+2))/2;
SWavg=(leftdata(M+2)+bottomdata(1))/2;
SEavg=(rightdata(M+2)+bottomdata(N+2))/2;
Zsol=[NWavg leftdata(2:M+1)' SWavg; Z'; SWavg rightdata(2:M+1)'
SEavg]';
%rather than reverse the order of ygrid, we leave it in the usual
order,
%but change the ordering in Zsol to make it amenable to 3D plotting.
%Znew = zeros(size(Zsol));
for i=1:M+2, Znew(i,:)=Zsol(M+3-i,:); end, Zsol = Znew;
Z=Zsol; x=xgrid; y=ygrid;
```

(b) The program requires the inputted inhomogeneity function to accept vector arguments. Since the squareheatsource M-file constructed in Example 11.7 does not take vector inputs, we first need to modify the M-file accordingly (as shown in Example 4.4):

```
function z = squareheatsourcevec(x,y)
for i=1:length(x), for j=1:length(y)
if x(i)>=.25 & x(i)<=.5 & y(j)>=.65 & y(j)<=.9, z(i,j)=-800;
else, z(i,j)=0;
end, end, end
```

It is now straightforward to use the program of part (a) to solve our problem. We must contemplate what size vectors to use for the input boundary data vectors once we decide on the step size h. Since we will use $h = 0.02$, the x- and y-grids will both have $1/h + 1 = 51$ components.

```
>> leftdata=zeros(51,1); rightdata=5*ones(51,1); ...
xgrid = linspace(0,1,51); bottomdata=5*xgrid; topdata=bottomdata;
>>[Z, x,y]=rectanglepoissonsolver(.02,1,1, ...
@squareheatsourcevec,leftdata, rightdata,  topdata,bottomdata);
```

Surface graphs (and contour plots) of the numerical solution can be obtained just as was done in the solution of Example 11.7, and the resulting graphics will agree with those of Figure 11.19, as the reader should verify.

EFR 11.8: (a) The nodes, labeling scheme and ghost nodes are as illustrated below:

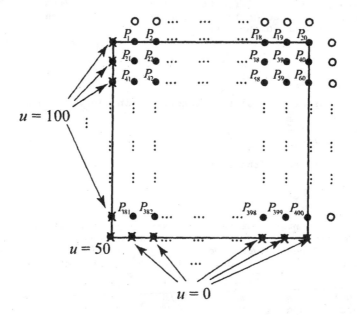

The nodes marked with X's have known values; as usual we took an average value for the corner point of a jump discontinuity. We will obtain a linear system for the 400 variables $U_k = u(P_k)$, $(k = 1, 2, ..., 400)$. Since the inhomogeneity of the PDE $f \equiv 0$, (26) becomes: $4u_{i,j} - u_{i+1,j} - u_{i-1,j} - u_{i,j+1} - u_{i,j-1} = 0$. To incorporate the boundary conditions, it is helpful to separate into cases:

CASE 1: Top-row nodes: $(P_1, \cdots, P_{20})$: (so $j = M$). In (26), $u_{i,M+1}$ corresponds to a ghost node. Using the Neumann boundary condition $u_y(x, 1) = 20$ with the central difference formula gives:

$$\frac{u_{i,M+1} - u_{i,M-1}}{2h} = 20 \underset{\text{Since } h=0.05}{\Longrightarrow} u_{i,M+1} = u_{i,M-1} + 2.$$

Similarly, if $i = 20$, (so we are at the right node), then we obtain (from the zero derivative conditions specified at the right side): $u_{21,M} = u_{19,M}$. If $i = 1$, the Dirichlet boundary conditions at the left side would tell us that $u_{0,M} = 100$. Summarizing and translating into k-indices gives the following:

$$4U_k - \underset{(=U_{k-1} \text{ if } k=20)}{U_{k+1}} - \underset{(=100 \text{ if } k=1)}{U_{k-1}} - 2U_{k+20} = 2 \quad (1 \le k \le 20).$$

CASE 2: Bottom-row nodes: ($P_{381}, \cdots, P_{400}$): (so $j = 1$). A similar argument (but this time we use the zero Dirichlet boundary condition for the values $u_{i,0}$) leads to the following:

$$4U_{380+i} - U_{380+i+1} - U_{380+i-1} - U_{360+i} = 0.$$
$$\text{\scriptsize}(=U_{399} \text{ if } i=20) \quad (=100 \text{ if } i=1)$$

CASE 3: Interior rows: ($P_{21}, \cdots, P_{380}$): (so $1 < j < M$). This case is easiest since the top and bottom edge boundary conditions never come into play; the equations are as follows:

$$4U_{20n+i} - U_{20n+i+1} - U_{20n+i-1} - U_{20(n+1)+i} - U_{20(n-1)+i} = 0.$$
$$\text{\scriptsize}(=U_{20n+i-1} \text{ if } i=20) \quad (=100 \text{ if } i=1)$$

The resulting linear system $AU = C$ is specified by the following $M \times M$ ($M = 20$) block matrices:

$$A = \begin{bmatrix} V_N & -2I_N & 0_N & \cdots & & \cdots & 0_N \\ -I_N & V_N & -I_N & 0_N & & & \vdots \\ 0_N & -I_N & V_N & -I_N & & & \\ \vdots & 0_N & -I_N & V_N & \ddots & & \vdots \\ & & & \ddots & \ddots & \ddots & 0_N \\ \vdots & & & & \ddots & \ddots & -I_N \\ 0_N & \cdots & & \cdots & 0_N & -I_N & V_N \end{bmatrix}, \quad U = \begin{bmatrix} w+v \\ w \\ w \\ w \\ \vdots \\ \vdots \\ w \end{bmatrix}$$

where V_N is the same matrix as W_N of (42), except that the (1,2) entry should be changed from -2 to -1, and $N = 20$. Also, $w = [100 \ 0 \ 0 \cdots 0]'$ and $v = [2 \ 2 \ 2 \cdots 2]'$. Once these matrices are entered into a MATLAB session, the system can be solved and one can obtain plots like those given in the text.

```
>> N=20;
>> A=diag(4*ones(1,N^2))-diag(ones(1,N^2-N), N)-...
          diag(ones(1,N^2-N),-N);
>> %next create vector for sub/super diagonals
>> v1=-ones(1,N-1); v=[v1 0];
>> for i=1:N-1
       if i<N-1
       v=[v v1 0];
       else
       v=[v v1]; vlow=v; vlow(N-1)=-2;
       end
end
>> A=A+diag(v,1)+diag(v,-1);

>> cblock = zeros(N,1); cblock(1)=100; C=cblock;
>> for i=1:N-1
       C=[C;cblock];
end
>> C(1:N)=C(1:N)+2*ones(N,1);

>> cblock = zeros(N,1); cblock(1)=100; C=cblock;
>> for i=1:N-1
       C=[C;cblock];
end
C(1:N)=C(1:N)+2*ones(N,1);
>> xgrid=0:.05:1;
>> for i=1:21
       ygrid(i)=xgrid(22-i);
end
>> U=A\C;
Z=zeros(N,N);
```

```
>> Z(:)=U; Z=Z';
>> Z=[100*ones(N,1) Z];
>> Z=[Z;[50 zeros(1,N)]];
>> mesh(xgrid,ygrid,Z) %produces a mesh plot of the solution
>> hidden off, xlabel('x-axis'), ylabel('y-axis')
```
A contour plot can now be produced in the usual fashion:
```
>> c=contour(xgrid,ygrid,Z,20)
>> clabel(c, 'manual')
```
We leave it to the reader to use their mouse to place the contour labels, and to repeat these computations with $N = 50$.

CHAPTER 12: HYPERBOLIC AND PARABOLIC PARTIAL DIFFERENTIAL EQUATIONS

EFR 12.1: The following commands will make the movie:
```
>> x=-5:.01:5; counter =1;
>> for t=0:.1:4; x1=x+t; x2=x-t;
      for i=1:1001
        u(i)=.5*(eg17_1(x1(i))+eg17_1(x2(i)));
      end
      plot(x,u), axis([-5 5 -1 3]) %We fix a good axis range.
      M(:, counter) = getframe; counter=counter+1;
end
```

Here is a possible playback mode:
```
>> movie(M,10,25)
```

EFR 12.2: (a) The M-file is boxed below:

```
function [] = dalembert(c,step, finaltime, phi, nu, range)
% This function M-file will produce a series of snapshots of the
% solution to the one-dimensional wave problem:  u_tt = c^2u_xx
% having initial displacement: u(x,0)=phi(x) and initial velocity
% u_t(x,0)=nu(x).
% The snapshots run from t=0 to t = finaltime in increments of step.
% Input variables:  c = wave speed (from PDE), step = positive number
% indicating time step for snapshot intervals, phi, nu = initial
% displacement and velocity functions for wave, respectively,  and
% range =4 by 1 vector of uniform axes range to use in plots.  The
% code is based on D'Alembert's solution of Theorem 12.1.
% Note:  Since 'quad' is used within the program on the function nu,
% it is necessary that nu be constructed to accept vector inputs.
x=range(1):.01:range(2);
sx = length(x);
%Set dimensions of subplot window
N = finaltime/step; %Number of shots
if N<=11
    N1=N+1; M=1;
elseif N>11&N<=21
 N1=ceil((N+1)/2); M=2;
else
    N1=ceil((N+1)/3); M=3;
end
counter =1;
for t=0:step:finaltime
x1=x+c*t; x2=x-c*t;
for i=1:sx
```

```
u(1)=.5*(feval(phi, x1(i))+feval(phi, x2(i)));
u(i)=u(i)+quad(nu, x2(i), x1(i));
end
subplot(N1,M,counter)
plot(x,u)
hold on
axis([range]) %We fix a good axis range.
counter=counter+1;
end
```

(b): Since the function quad gets used inside the above program (with the inputted function nu), we must ensure that the nu is constructed in a way so that it can take vector inputs.

`>> nu = inline('zeros(size(x))')`

The following command will reproduce the results of Figure 12.5:

`>> dalembert(1, .5, 5, @EX12_1, nu, [-5 5 0 2])`

(c) We first create an M-file for the function $v(x)$

```
function y = EFR12_3nu(x)
for i = 1:length(x)
if abs(x(i))<1, y(i)=1;
else y(i)=0;
end
end
```

The function phi, need not take vector inputs:

`>> phi = inline('0', 'x')`

A bit of experimenting will show that a y-range of 0 to 2.5 serves well for this problem.

`>> dalembert(1, .5, 5, phi, @EFR12_3nu, [-5 5 0 2.5])`

(d) If we change the finaltime input in both parts (a) and (b) to 10, rerun the program, and examine the output, we will see that in both parts, the wavefront will reach $x = 10$ at (approximately) $t = 9$. This agrees with the theoretical fact that the wavefronts (here there are two moving in opposite directions) are traveling at speed c. The snapshots show that the starts of the disturbances are propagating to the left and to the right with speed $c = 1$ unit of space per one unit of time.

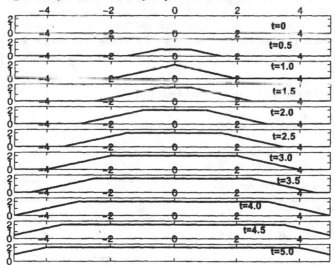

EFR 12.3: (a) In order to prevent the M-file from running into logical dilemmas, we define it for all values of x (note that formula (13) excludes definitions at enpoint values: $x = 0$, L, $-L$, etc.). We use formula (13) together with an if-branch for the structure of the M-file below (cf. Example 4.4).

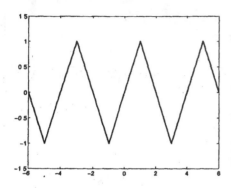

```
function y = phihat(x)
for i = 1:length(x)
    if (x(i)<=2)&(x(i)>=0),
y(i)=1-abs(1-x(i));
    elseif (x(i)<2)&(x(i)>=-2),
y(i)=-phihat(-x(i));
    else n = floor((x(i)+2)/4);
    r = x(i)-4*n; y(i) =
phihat(r);
    end
end
```

(b) The following commands were used to produce the plot shown above:
```
>> x=-6:.05:6; plot(x,phihat(x))
>> axis([-6 6 -1.5 1.5]), grid on
```

EFR 12.4: Letting $\hat{\varphi}(x)$ and $\hat{v}(x)$ denote the periodic extensions (specified by (13)) of the functions $\varphi(x)$ and $v(x)$, respectively, the method of reflections and d'Alembert's theorem tell us that the function

$$\hat{u}(x,t) = \frac{1}{2}[\hat{\varphi}(x+ct) + \hat{\varphi}(x-ct)] + \frac{1}{2c}\int_{x-ct}^{x+ct} \hat{v}(s)ds,$$

provides a solution to the finite string problem (11) (if we restrict x to the domain $[0, L]$). For such x and for any positive integer n, if we substitute $t = t + nL/c$ into this equation, we see that the integral extends from $x - ct - c(nL/c) = x - ct - nL$ to $x + ct + cnL/c = x + ct + nL$. But since the integrand is a period L extension of an odd function, it follows that the portions of the integral from $x - ct - nL$ to $x - ct$ and from $x + ct$ to $x + ct + nL$ must be zero. Similarly, since $\hat{\varphi}(x)$ is periodic of period L, we may conclude that

$$\frac{1}{2}[\hat{\varphi}(x + c(t + nL/c) + \hat{\varphi}(x - c(t + nL/c))]$$

$$= \frac{1}{2}[\hat{\varphi}(x + ct + nL) + \hat{\varphi}(x - ct - nL)] = \frac{1}{2}[\hat{\varphi}(x + ct) + \hat{\varphi}(x - ct)].$$

It follows that $\hat{u}(x, t + nL/c) = \hat{u}(x,t)$ and this is the asserted periodicity statement.

EFR 12.5: (a) To contruct the initial profile "pulse" function, we make use of the basic cubic spline function BS(x) given by formula (51) of Chapter 10. In EFR 10.16, we constructed an M-file BSSpline for this function. We will also need an M-file for its derivative—the following M-file gives such a construction based on the formula (53) of Chapter 10. Note that because we will be using a variant of this function for the nu input of the dalembert M-file, we need to

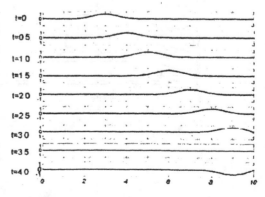

construct it in a way that it will accept vector inputs (cf. Example 4.4):

```
function y = BSprime(x)
%Derivative of basic cubic spline
%function of Chapter 10, (52),
%function is built to accept
%vector arguments.
for i=1:length(x)
    if x(i)>=0 & x(i)<=1
y(i)=3/4*(4*(1-x(i))^2-(2-x(i))^2);
    elseif x(i)>1 & x(i)<=2
        y(i)=-3/4*(2-x(i))^2;
    elseif x(i)>2
        y(i)=0;
    else, y(i) = -BSprime(-x(i));
    end
end
```

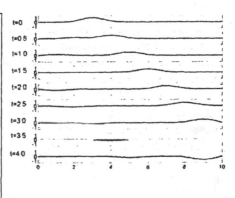

We represent the pulse in Figure 12.10 with $(u(x,0)\equiv)\varphi(x)=BS(x-3)$. Since we are given that the initial velocity of the pulse is 2 (units to the right per unit time), we can compute $(u_t(x,0)\equiv)v(x)=\dfrac{d}{dt}\varphi(x-2t)\Big|_{t=0}=-2\varphi'(x)=-2BS'(x-3)$ and thus $(u_t(x,0)\equiv)v(x)=-2BS'(x-3)$.

Since we will be using the method of reflections we need to create M-files for the odd periodic extensions of these functions. The resulting M-files are as follows:

```
function y = EFR12_5phihat(x)
if (x>=0)&(x<=10)
  y=BSSpline(x-3);
elseif (x<0)&(x>=-10)
y = -EFR12_5phihat(-x);
else q=floor((x+10)/20);
y=EFR12_5phihat(x-20*q);
end
```

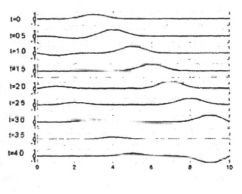

Again, we need to construct the second one so it will accept vector inputs (as required by dalembert).

```
function y = EFR12_5nuhat(x)
for i=1:length(x)
if (x(i)>=0)&(x(i)<=10)
  y(i)=-2*BSprime(x(i)-3);
elseif (x(i)<0)&(x(i)>=-10)
y(i) = -EFR12_5nuhat(-x(i));
else q=floor((x(i)+10)/20);
y(i)=EFR12_5nuhat(x(i)-20*q);
end
end
```

The series of snapshots can now be accomplished with the following single command:
```
>> [x,ua]=dalembert(2,.5,4,@EFR12_5phihat,@EFR12_5nuhat,[0 10 -1.5
1.5]);
```
The output is shown as the first figure in the series of three for this EFR (appearing on the last page).
Digression: After completing all three parts of this exercise, it will be useful to make some comparisons. For such a purpose, it would be helpful to have additional output data for the snapshot profiles that our dalembert program computes. It is a simple matter to modify our program accordingly to produce output data (with or without the snapshots).

(b) Here, the only change will be in the constant appearing in the formula for $v(x)$, arguing as in Part (a), we find that $v(x)=-BS'(x-3)$. If we modify the M-file 'EFR12_5nuhat' accordingly (let's call the modified M-file as 'EFR12_5nuhatb'), we can then get our snapshots with the correspondingly modifed 'dalembert' call:

```
>> [x,ub]=dalembert(2,.5,4,@EFR12_5phihat,@EFR12_5nuhatb,[0 10 -1.5
1.5]);
```

The resulting graphic is the middle one appearing in the series.

(c) Similarly, to get the final series of snapshots, we just need to change the M-file for $v(x)$ to correspond to the formula $v(x) = -4BS'(x-3)$. This being done (and the M-file stored as 'EFR12_5nuhatc'), we obtain the third series of snapshots with the corresponding call on 'dalembert'.

We point out some observations from the snapshots. In Part (a), we simply get an undistorted pulse moving at speed two (and after it reflects on the right end, it switches direction and moves upside down to the left at speed two). In part (b) wave's initial velocity is slower than the speed of the PDE (natural speed for the string), the pulse is slightly weaker but still moves to the right at speed two, but we also get a secondary smaller pulse moving to the left at the same speed (so after it reflects and moves to the right but will be upside down). In part (c) when the initial velocity is faster than the speed of the PDE, we get a stronger main pulse moving to the right at speed two as well as a secondary pulse that moves in the opposite direction but with upside down orientation.

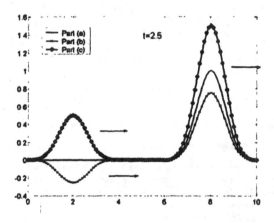

The relative strengths of these pulses are difficult to detect from the subplots shown above. To get a clearer picture, we plot in a single axis window the three solution snapshots when t = 2.5 (assuming the solution matrix for Part (c) was stored as a matrix 'uc'). The plot shown at the right was created with the following commands:

```
>> size(ua) → ans = 9 101
>> plot(x,ua(6,101))
>> hold on, plot(x,ub(6,:),'r-x')
>> plot(x,uc(6,:), 'bo-')
```

EFR 12.6: (a) Taylor's theorem tells us that we may write:

$$f(x+h) = f(x)+f'(x)h+f''(x)h^2/2+O(h^3) \text{ and } f(x-h) = f(x)-f'(x)h+f''(x)h^2/2+O(h^3).$$

Since $O(h^3)/h = O(h^2)$, subtracting the second of these equations from the first and then dividing by $2h$, results in (30).

(b) Under the assumptions on $u(x,t)$, we may apply the centered difference approximation (30) in the time variable to obtain the $O(k^2)$ estimate: $u_{i,1}-u_{i,-1} \approx 2kv(x_i)$ or $u_{i,-1} \approx u_{i,1}-2kv(x_i)$. If we substitute this latter approximation into (23) with j = 0:

$$u_{i,1} = 2(1-\mu^2)u_{i,0}+\mu^2\left[u_{i+1,0}+u_{i-1,0}\right]-u_{i,-1} = 2(1-\mu^2)\varphi(x_i)+\mu^2\left[\varphi(x_{i+1})+\varphi(x_{i-1})\right]-u_{i,-1} \text{ and then}$$

solve for $u_{i,1}$, we arrive at (31) with local truncation error: $O(h^2+k^2)+O(k^2) = O(h^2+k^2)$.

EFR 12.7: (a) The program will be identical to onedimwave, except that the single line defining 'U(2,i)' should be changed to: U(2,i)=U(1,i)+k*feval(nu,x(i)); so as to correspond to (29).

(b) The following loop will create a window with the 10 asked for plots using onedimwavebasic:

```
>>for n=0:9
```

```
d=2^n; %doubling factor
[xbas, tbas, Ubas] = onedimwavebasic(phi, nu, pi, A, B, 8, 10, d*30,
c);
subplot(2,5,n+1)
plot(xbas,abs(Ubas(d*30,:)-uExact(xbas,8)))
end
>> title('Error plots for ''onedimwavebasic'', N = 10, M = 30, 60,
120, ...')
```

The resulting graphic is shown in the upper left portion below. To get the corresponding graphic when
$N = 40$, simply use the same code but change the seventh input of 'onedimwavebasic' from 10 to
40. Also, the same two codes will produce the corresponding plots for 'onedimwave', simply replace
this as the M-file name, and keep all else the same.

By examining the plots below, we see that the results of both methods are quite similar, with the
accuracy of 'onedimwave' being slightly better than that of 'onedimwavebasic'. The instability
is only evident in the first two plots (for each method) when $N = 40$. This corroborates well with the
CFL stability condition, which states (since the wave speed is one) that we should have $k \leq h$; indeed,
this happens exactly on the third plots. Notice also that in all cases (after instability), the maximum
errors seem to decrease by about 50% each time that we double the time step until the gains begin to
taper off (e.g., the last three frames when $N = 10$). No matter how large we take M, it seems
reasonable that there will be a certain threshold of accuracy that we will be limited to if we keep the x-
grid fixed.

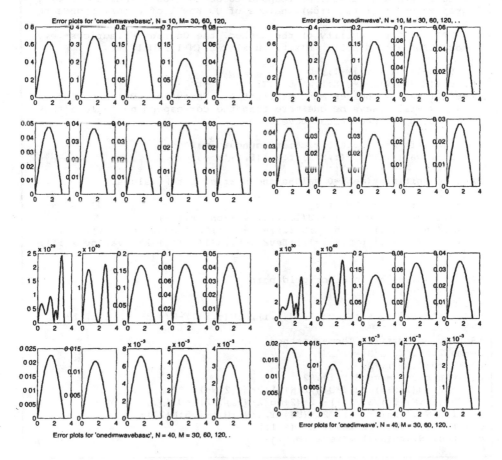

EFR 12.8: We leave it to the reader to rerun the code of Example 12.6 by replacing onedimwave with onedimwavebasic and to compare the graphical results.

EFR 12.9: (a) The M-file is boxed below:

```
function [x, t, U] = onedimwaveimpl_4(phi, nu, L, A, B, T, N, M, c)
% solves the one-dimensional wave problem u_tt = c^2*u_xx
% using implicit method with parameter omega=1/4.  The Thomas method
% is used.
% Input variables:  phi=phi(x) = initial wave profile function
% nu=nu(x) = initial wave velocity function, L = length of string, A
% =A(t) height function of left end of string u(0,t)=A(t), B=B(t) =
% height function for right end of string u(L,t)=B(t), T= final time
% for which solution will be computed, N = number of internal x-grid
% values,  M = number of internal t-grid values, c = c(x,t,u,u_x)
% speed of wave.  Functions of the indicated variables must be stored
% as(either inline or M-file) functions with the same variables, in
% the same order.
% Output variables: t = time grid row vector (starts at t=0, ends at
% t=T, has M+2 equally spaced values),  x = space grid row vector,  U
% =(N+2) by  (M+2) matrix of solution approximations at corresponding
% grid points x grid will correspond to second (col) indices of U, y
% grid values to first (row) indices of U. Row 1 of U corresponds to
% t = 0.
% CAUTION:  For stability of the method, the Courant-Friedrichs-Levy
% condition should hold:  c(x,t,u,u_x)(T/L)(N+1)/(M+1) <1.

h = L/(N+1); k = T/(M+1);
U=zeros(M+2,N+2); x=0:h:L; t=0:k:T;
% Recall matrix indices must start at 1.  Thus the indices of the
% matrix will always be one more than the corresponding indices that
% were used in theoretical development.

%Assign left and right Dirichlet boundary values.
U(:,1)=feval(A,t)'; U(:,N+2)=feval(B,t)';

%Assign initial time t=0 values and next step t=k values.
for i=2:(N+1)
   U(1,i)=feval(phi,x(i));
   mu(i)=k*feval(c,0,x(i),U(1,i),(feval(phi,x(i+1)))...
-feval(phi,x(i-1)))/2/h)/h;U(2,i)=(1-mu(i)^2)*feval(phi,x(i)) ...
+mu(i)^2/2*(feval(phi,x(i-1))+feval(phi,x(i+1))) + k*feval(nu,x(i));
end


%Assign values at interior grid points
for j=2:(M+1)
for i=2:(N+1)
mu(i)=k*feval(c, t(j), x(i), U(j,i), (U(j,i+1)-U(j,i-1))/2/h)/h;
end
%Set up vectors for Thomas method
a=-mu(2:N).^2; a(N)=0;
d=4+2*mu(2:N+1).^2;
b(1)=0; b(2:N)=-mu(3:N+1).^2;
cT=4*(2-mu(2:N+1).^2).*U(j,2:N+1)+2*mu(2:N+1).^2.* ...
   (U(j,1:N)+U(j,3:N+2))- 2*(2+mu(2:N+1).^2).*U(j-...
   1,2:N+1)+mu(2:N+1).^2.*(U(j-1,1:N)+U(j-1,3:N+2));
cT(1)=cT(1)+mu(2)^2*feval(A,t(j+1));
cT(N)=cT(N)+mu(N+1)^2*feval(B,t(j+1));
```

```
     U(j+1,2:(N+1))=thomas(a,d,b,cT);
end
```

(b) We leave such experiments to the reader. In particular, we suggest trying to choose the paramters N and M in such a way as to reduce the artificial "noise" (cf., Figure 12.16). Does going outside the stability ranges for the explicit method seem to help much?

EFR 12.10: In an analogous fashion to how (28) was derived, Taylor's theorem gives us that:

$$u(x_i, y_j, k) = u(x_i, y_j, 0) + ku_t(x_i, y_j, 0) + (k^2/2)u_{tt}(x_i, y_j, 0) + O(k^3).$$

Now, if we assume that the PDE (wave equation) continues to hold for $u(x,y,t)$ on the initial plane $t = 0$, we can write: $u_{tt}(x, y, 0) = c^2 \Delta u(x, y, 0) = c^2[\varphi_{xx}(x, y) + \varphi_{yy}(x, y)]$.   Using the central difference approximations (Lemma 10.3) on these second derivatives of φ and substituting into the first estimate produces the following:

$$u(x_i, y_j, k) = u(x_i, y_j, 0) + ku_t(x_i, y_j, 0) + (c^2k^2/2h^2)\{\varphi(x_{i+1}, y_j) + \varphi(x_{i-1}, y_j) + \varphi(x_i, y_{j+1}) + \varphi(x_i, y_{j-1})$$
$$- 4\varphi(x_i, y_j)\} + O(k^3) + O(h^2 + k^2).$$

Since $O(k^3) + O(h^2 + k^2) = O(h^2 + k^2)$, $\varphi(x, y) = u(x, y, 0)$, $\mu^2 = c^2 k^2/h^2$, and $v(x, y) = u_t(x, y, 0)$, if we use multi-index notation, this approximation tranlates to the following $O(h^2 + k^2)$ estimate:

$$u_{i,j}^1 \approx (1 - 2\mu^2)\varphi(x_i, y_j) + kv(x_i, y_j) + \mu^2/2\{\varphi(x_{i+1}, y_j) + \varphi(x_{i-1}, y_j) + \varphi(x_i, y_{j+1}) + \varphi(x_i, y_{j-1})\},$$

as desired.

EFR 12.11: We will show that the local truncation error of the Crank-Nicolson method is $O(h^2 + k^2)$ when viewed as a discretization of the PDE at (x_i, t_j). A similar argument will show the same estimate is valid if we were to discretize at (x_i, y_{j+1}). To clean up the notation a bit we will write (x, t) in place of (x_i, t_j) for the remainder of this proof. This proof will be more delicate than others since we really need to carefully use both Taylor's theorem and the PDE to estimate the error. We need to estimate the left side of (49) minus the right side, by expanding all terms using Taylor's theorem based at (x_i, y_j):

$$\frac{u_{i,j+1} - u_{i,j}}{k} - \frac{\alpha}{2}\left[\frac{u_{i+1,j} - 2u_{i,j} + u_{i-1,j}}{h^2} + \frac{u_{i+1,j+1} - 2u_{i,j+1} + u_{i-1,j+1}}{h^2}\right] - \frac{1}{2}\left[q(x_i, t_j) + q(x_i, t_{j+1})\right]. \qquad (*)$$

We invoke Taylor's theorem to first separately estimate each term:

$$\frac{u_{i,j+1} - u_{i,j}}{k} = \frac{u(x, t+k) - u(x, t)}{k} = \frac{[u(x, t) + ku_t(x, t) + k^2/2u_{tt}(x, t) + O(k^3)] - u(x, t)}{k}$$
$$= u_t(x, t) + (k/2)u_{tt}(x, t) + O(k^2)$$

$$\frac{1}{2}\left[q(x_i, t_j) + q(x_i, t_{j+1})\right] = \frac{1}{2}\left[q(x, t) + q(x, t+k)\right] = \frac{1}{2}\left[q(x, t) + [q(x, t) + kq_t(x, t)] + O(k^2)\right]$$
$$= q(x, t) + (k/2)q_t(x, t) + O(k^2)$$

$$\frac{u_{i+1,j} - 2u_{i,j} + u_{i-1,j}}{h^2} = \frac{u(x+h, t) - 2u(x, t) + u(x-h, t)}{h^2}$$
$$= \frac{[u(x, t) + hu_x(x, t) + (h^2/2)u_{xx}(x, t) + (h^3/6)u_{xxx}(x, t) + O(h^4)] - 2u(x, t) +}{h^2}$$
$$+ \frac{[u(x, t) - hu_x(x, t) + (h^2/2)u_{xx}(x, t) - (h^3/6)u_{xxx}(x, t) + O(h^4)]}{h^2} = u_{xx}(x, t) + O(h^2).$$

Similarly, $\dfrac{u_{i+1,j+1} - 2u_{i,j+1} + u_{i-1,j+1}}{h^2} = u_{xx}(x, t+k) + O(h^2)..$ We use Taylor's theorem once again to

obtain: $u_{xx}(x, t+k) = u_{xx}(x, t) + k u_{xxt}(x, t) + O(k^2).$ If we invoke all of these estimates into (*), the expression can be rewritten in the following form (for further notational convenience, we omit functional arguments since they are all (x, t)):

$$u_t + (k/2)u_{tt} + O(k^2) - (\alpha/2)\Big[u_{xx} + O(h^2) + u_{xx} + k u_{xxt} + O(h^2 + k^2) \Big] - q - (k/2)q_t + O(k^2) =$$

$$\big(u_t - \alpha u_{xx} - q \big) + (k/2)\big\{ u_{tt} - \alpha u_{xxt} - q_t \big\} + O(h^2 + k^2).$$

Now, the expression in parentheses is zero by the PDE. The expression in braces is the time derivative of the first expression. Thus, if the solution and q are sufficiently differentiable, this expression will also be zero and we are left with the desired $O(h^2 + k^2)$ estimate.

EFR 12.12:   (a) The M-file is boxed below:

```
function [x, t, U] = fwdtimecentspace(phi, L, A, B, T, N, M, alpha,q)
% solves the one-dimensional heat problem
% u_t = alpha(t,x,u)*u_xx+q(x,t)
% using the explicit forward time centered-space method.
% Input variables:  phi=phi(x) = initial wave profile function
% L = length of rod, A =A(t)= temperature of left end of rod
% u(0,t)=A(t), B=B(t) = temperature of right end of rod u(L,t)=B(t),
% T= final time for which solution will be
% computed, N = number of internal x-grid values,  M = number
% of internal t-grid values, alpha =alpha(t,x,u,u_x)= diffusivity of
rod.
% q = q(x,t) = internal heat source function
% Output variables: t = time grid row vector (starts at t=0, ends at
% t=T, has M+2 equally spaced values),  x = space grid row vector,
% U = (M+2) by (N+2) matrix of solution approximations at
corresponding
% grid points. x grid values will correspond to second
(column)entries of U, y
% grid values to first (row) entries of U. Row 1 of U corresponds to
% t = 0.

h = L/(N+1);  k = T/(M+1);
U=zeros(M+2,N+2); x=0:h:L; t=0:k:T;
% Recall matrix indices must start at 1.  Thus the indices of the
% matrix will always be one more than the corresponding indices that
% were used in theoretical development.

%Assign left and right Dirichlet boundary values.
U(:,1)=feval(A,t)'; U(:,N+2)=feval(B,t)';

%Assign initial time t=0 values and next step t=k values.
for i=2:(N+1)
    U(1,i)=feval(phi,x(i));
end

%Assign values at interior grid points
for j=2:(M+2)
for i=2:(N+1)
mu(i)=k*feval(alpha,t(j-1),x(i),U(j-1,i))/h^2;
qvec(i)=feval(q,x(i),t(j-1));
end
% First form needed vectors and matrices, because we will be using
the
```

```
% thomas M-file, we do not need to construct the coefficient matrix
T.
V = zeros(N,1); V(1)=mu(2)*U(j-1,1); V(N)=mu(N+1)*U(j-1,N+2);
Q = k*qvec(2:N+1)';

%We now form the next time level approximation.  Notice we have
avoided
%matrix multiplication.
U(j,2:N+1)=(1-2*mu(2:N+1)).*U(j-1,2:N+1)+mu(2:N+1).*(U(j-1,1:N)+U(j-
1,3:N+2)));
end
```

(b) The M-file is boxed below:

```
function [x, t, U] = backwdtimecentspace(phi, L, A, B, T, N, M,
alpha,q)
% solves the one-dimensional heat problem
% u_t = alpha(t,x,u)*u_xx+q(x,t)
% using the backward-time central-space method.
% Input variables:  phi=phi(x) = initial wave profile function
% L = length of rod, A =A(t)= temperature of left end of rod
% u(0,t)=A(t), B=B(t) = temperature of right end of rod u(L,t)=B(t),
% T= final time for which solution will be
% computed, N = number of internal x-grid values,  M = number
% of internal t-grid values, alpha =alpha(t,x,u)= diffusivity of rod.
% q = q(x,t) = internal heat source function
% Output variables: t = time grid row vector (starts at t=0, ends at
% t=T, has M+2 equally spaced values),  x = space grid row vector,
% U = (M+2) by (N+2) matrix of solution approximations at
corresponding
% grid points. x grid values will correspond to second
(column)entries of U, y
% grid values to first (row) entries of U. Row 1 of U corresponds to
% t = 0.

h = L/(N+1);  k = T/(M+1);
U=zeros(M+2,N+2); x=0:h:L; t=0:k:T;
% Recall matrix indices must start at 1.  Thus the indices of the
% matrix will always be one more than the corresponding indices that
% were used in theoretical development.

%Assign left and right Dirichlet boundary values.
U(:,1)=feval(A,t)'; U(:,N+2)=feval(B,t)';

%Assign initial time t=0 values and next step t=k values.
for i=2:(N+1)
    U(1,1)=feval(phi,x(i));
end

%Assign values at interior grid points
for j=2:(M+2)
for i=2:(N+1)
mu(i)=k*feval(alpha,t(j),x(i),U(j-1,i))/h^2;
qvec(i)=feval(q,x(i),t(j));
end
% First form needed vectors and matrices, because we will be using
the
% thomas M-file, we do not need to construct the coefficient matrix
T.
Q = k*qvec(2:N+1)';
V = zeros(N,1); V(1)=mu(2)*U(j,1); V(N)=mu(N+1)*U(j,N+2);
```

```
%Now perform the matrix multiplications to iteratively obtain
solution
% values for increasing time levels.
c=U(j-1,2:(N+1))'+V+Q;
a=-mu(2:N+1); b=a; a(N)=0; b(1)=0;
U(j,2:N+1)=thomas(a,1+2*mu(2:N+1),b,c);
end
```

EFR 12.13: (a) We first need to construct an M-file for the inhomogeneity function, since it involves cases:

```
function y = phiEFR12_13(x)
for i = 1:length(x)
    if x(i)<=3 & x(i)>=1, y(i) =
100;
    else, y(i)=0;
    end
end
```

The remaining input functions can be stored as inline functions:
`alpha=inline('3','x','t','u') ; q = inline('0','x','y');`
`A=inline('0'); B=inline('100*(1-exp(-t))', 't');`
It is now a simple matter to run the two programs and obtain the desired numerical graphs. We will use $N = 80$ internal x-grid values and $M = 20$ internal time-grid values. This gives equal spacing of the time and space grids.
`>>[x, t, UCN] = cranknicolson(@phiEFR12_13, 4, A, B, 1, 80, 20, alpha,q);`
`>> plot(x,UCN(22,:))  %plot of the CR solution profile at time t = 1.`
`>> %compare w/ Figure 12.27b`
`>> [x, t, UBT] = backwdtimecentspace(@phiEFR12_13, 4, A, B, 1, 80, 20, alpha,q);`
`>> plot(x,UBT(22,:))  %plot of the BT solution profile at time t = 1.`
`>> %compare w/ Figure 12.27a`
(b) With the data from part (a), the desired surface plots are readily obtained by the following commands. The results are shown below.
`>> surf(x,t,UCN), xlabel('x-values'), ylabel('t-values'), title('Crank-Nicolson')`
`>> surf(x,t,UBT), xlabel('x-values'), ylabel('t-values'), title('BTCS')`

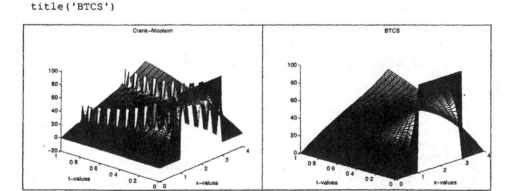

EFR 12.14: (a) The code of the Example 12.9 just needs a minor modification (in the line defining U(j,1)). Since it is a short code, we give it here and provide details on how to obtain Figure 12.28b:
`h=1/20; k=1/1850; mu=k/h^2;`
`N=21; M=1851;`
`U=zeros(M+1,N); x=0:h:1; t=0:k:1;`

```
%Assign initial time t=0 values and next step t=k values.
for i=1:N, U(1,i)=100; end
%Assign values at interior grid points
for j=2:M+1
U(j,2:N-1)=(1-2*mu)*U(j-1,2:N-1)+mu*(U(j-1,3:N)+U(j-1,1:N-2));
U(j,1)= (1-2*mu)*U(j-1,1)+2*mu*(U(j-1,2)-h* U(j-1,1)^1.5);
U(j,N)= (1-2*mu)*U(j-1,N)+2*mu*(-h*U(j-1,N) +U(j-1,N-1));
end
```

```
>> plot(x,U(1,:)) %initial temperature
>> hold on, plot(x,U(11,:)), plot(x,U(21,:)), plot(x,U(81,:)),
plot(x,U(121,:))
>> plot(x,U(241,:)), plot(x,U(441,:)), plot(x,U(661,:)),
plot(x,U(801,:))
>> plot(x,U(1201,:)), plot(x,U(1600,:))
>> axis([0 1 0 111]), xlabel('space'), ylabel('temperature')
>> gtext('Initial temperature (t=0)') %Use the mouse to put in the
first label.
>> [10 20 80 120 240 440 660 800 1200 1600]/1850 %time data for other
labels
→ans = 0.0054  0.0108  0.0432  0.0649  0.1297  0.2378  0.3568  0.4324  0.6486  0.8649
>> gtext('t = 0.005')   %Use the mouse to place this label, repeat for
rest of labels.
```
(b) Physically, the heat in the rod will continue to be lost forever as the temperature distribution decays (but never reaches zero). The fact that it will never be totally lost follows from the fact there is an exponential decay of heat from the right BC and (eventually) less than exponential decay from the left BC. Since $T^{1.5} > T$ only when $T > 1$, in fact as T approaches zero the ratio $T/T^{1.5} = 1/\sqrt{T} \rightarrow \infty$, it follows that eventually the heat will be lost much, much faster on the right end than on the left. Thus, the temperature at the right end should eventually fall below that of the left. To see this with MATLAB, we will have to let the solver code of part (a) run for more time. It is not immediately clear how long we should run it (until the temperatures at the ends fall to be less than one), but after some experimentation, we see that letting it run until $t \sim 2$ will be sufficient. The code of part (a) is easily modified to obtain the two plots we give below. The first one is for $t = 2$ and the second is for $t = 4$.

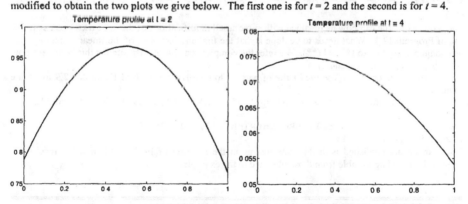

(c) Physically, the BC conditions mean that heat is being absorbed at the left end at a rate proportional to the temperature there and is being lost at the right end at a rate proportional to the temperature there. It follows that the left end will always be hotter than the right, and there will be a net exponential gain of heat absorption of the rod, the rate being equal to the difference of the left temperature less the right temperature. Since the diffusion takes time for the heat from the left side to make it to the right, the difference in temperatures (between left and right end) will continue to increase. The temperture in the heat in the rod will thus increase without bound. To confirm this phenomenon on MATLAB, the solver code in Part (a) needs only to have changed the line defining U(j,1) to read as follows.

```
U(j,1)= (1-2*mu)*U(j-1,1)+2*mu*(U(j-1,2)+h* U(j-1,1));
```

Below we include two plots.

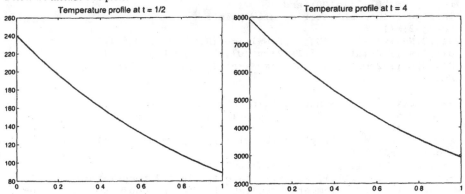

EFR 12.15: (a) The x-grid will have two ghost nodes, just as in Example 12.9:
$$-h = x_0 < 0 = x_1 < \cdots < x_N = L < x_{N+1} = L.$$

The central difference approximation for the left BC gives $[u_{2,j} - u_{0,j}]/2h \approx au_{1,j} + b$. Solving this for the ghost node value produceds $u_{0,j} \approx u_{2,j} - 2h(au_{1,j} + b)$. Assuming the PDE is valid on the left boundary, we substitute this approximation into the discretization (50):
$$-\mu u_{i-1,j+1} + 2(1+\mu)u_{i,j+1} - \mu u_{i+1,j+1} = \mu u_{i-1,j} + 2(1-\mu)u_{i,j} + \mu u_{i+1,j} + k[q_{i,j} + q_{i,j+1}]$$
(when $i = 1$), to obtain
$$2(1 + \mu + \mu ha)u_{1,j+1} - 2\mu u_{2,j+1} + 2hb\mu = 2(1 - \mu - \mu ha)u_{1,j} + 2\mu u_{2,j} + k[q_{i,j} + q_{i,j+1}] - 2hb\mu \quad (*).$$

Similarly, the central difference approximation on the right BC gives $u_{N+1,j} \approx u_{N-1,j} + 2h(cu_{N,j} + d)$, and when this is incorporated into (50), we obtain:
$$-2\mu u_{N-1,j+1} + 2(1 + \mu - \mu hc)u_{N,j+1} - 2hd\mu = 2\mu u_{N-1,j} + 2(1 - \mu - \mu hc)u_{N,j} + k[q_{N,j} + q_{N,j+1}] + 2hd\mu$$
(**).

The required M-file can now be obtained with some small modifications of the cranknicolson M-file of Program 12.3. What needs to be done is that the first and last rows of the linear system need to be changed according to (*) and (**). We refer the complete code to the ftp site for this book (see note at the beginning of this appendix).

(b) The following code will use the M-file of part (a) to re-solve the BVP of Example 12.9 using the same grid.

```
>> phi = inline('0'); q = inline('0', 'x', 't'), alpha = inline('1',
't', 'x', 'u')
>> [x, t, U]=cranknicolsonRobinLR(phi, 1, [1 0], [-1 0],
1,21,1851,alpha, q)
```

Plots can be accomplished with this data just as in the solution of EFR 12.14, and the results are graphically indistinguishable from those obtained in the example.

CHAPTER 13: THE FINITE ELEMENT METHOD

EFR 13.1: For Φ_3, the code given in Example 13.1 just needs a small modification to accommodate the change of node. Indeed, in the for loop, the three modified lines are: if ismember(3,T(L,:))==1,index=find(T(L,:)==3);nv=[T(L,1:2)3]; nv(index) =T(L,3);. The new output of the modified loop is then the following matrix A:

$$\rightarrow A = \begin{matrix} 1 & -1 & 0 & 1 \\ 5 & -1 & 0 & 1 \end{matrix}$$

From this we can write: $\Phi_3(x,y) = \begin{cases} -x & +1, & \text{if } (x,y)\in T_1, \\ -x & +1, & \text{if } (x,y)\in T_5, \\ & 0, & \text{otherwise.} \end{cases}$

In a similar fashion, we find that:

$$\Phi_4(x,y) = \begin{cases} \frac{2}{3}x - y - \frac{2}{3}, & \text{if } (x,y)\in T_3, & -x - y + \frac{7}{2}, & \text{if } (x,y)\in T_4, \\ \frac{2}{3}x \quad -\frac{2}{3}, & \text{if } (x,y)\in T_6, & y+1, & \text{if } (x,y)\in T_7, \\ -x + y + \frac{7}{2}, & \text{if } (x,y)\in T_8, & 0, & \text{otherwise.} \end{cases}$$

EFR 13.2: (a) As a quadrilateral has four vertices and a linear function in x and y has only three parameters (see equation (2)), linear functions are not versatile enough to accommodate arbitrarily specifying four numerical values at the vertices of a quadrilateral.
(b) A generic function of x and y having four parameters is a so-called bilinear function: $axy + bx + cy + d$, these are often used for quadrilateral elements. In the special case where the elements are rectangles parallel to the axis, the matter is further discussed in some of the exercises at the end of Section 13.2.

EFR 13.3: (a) The M-file is boxed below:

```
function voronoiall(x,y)
% M-file for EFR 13.3
% inputs:  two vectors x and y of the same size giving, respectively,
% the x- and y-coordinates of a set of distinct points in the plane;
% outputs:  none, but a graphic will be produced of the Voronoi
% regions corresponding to the point set in the plane,
% including the unbounded regions
n=length(x);
xbar = sum(x)/n; ybar = sum(y)/n; %centroid of points
md = max(sqrt(x-xbar).^2 + sqrt(y-ybar).^2);
%maximum distance of points to centroid
mdx = max(abs(x-xbar)); mdy = max(abs(y-ybar));
%max x- and y- distances to averages
% We create additional points that lie in a circle of radius 3md
% about (xbar, ybar).  We  deploy them with angular gaps of 1 degree,
% this will be suitable for all practical purposes.
xnew=x; ynew=y;
for k = 1:360
    xnew(n+k)=xbar+3*md*cos(k*pi/180);
    ynew(n+k)=ybar+3*md*sin(k*pi/180);
end
voronoi(xnew,ynew)
axis([min(x)-mdx/2 max(x)+mdx/2  min(y)-mdy/2  max(y)+mdy/2])
```

(b) With this program, we can easily re-create Figure 13.9(b):
```
>> N=[1 1;5/2 1;0 0;1 0;5/2 0;7/2 0;1 -1;2.5 -1];x=N(:,1); y=N(:,2);
>> voronoiall(x,y)
```

EFR 13.4: (a) Although the scheme we used in the solution of part (c) of Example 13.2 can be adapted for this triangulation, we will introduce a slightly different approach. Specifically, we will take advantage of the fact that intersections of circles (centered at (0,0)) all have the same boundary angles. At each iteration, we will deploy nodes on circles of equally spaced radii in the annular sector domains: $\Omega_n = \{(x,y)\in\Omega : 1/2^n < \text{dist}((x,y),(0,0)) < 2\cdot(1/2^n)\}$ for $n = 1, 2, \ldots$ By the special shape of Ω, we get the following exact formula for the area of Ω_n: $\text{Area}(\Omega_n) = \frac{1}{2}\cdot\frac{5\pi}{3}[(2\cdot 2^{-n})^2 - (2^{-n})^2] = \frac{5\pi}{2}2^{-2n}$.

Thus, if we were to deploy (approximately) 100 nodes in Ω_n with a uniform grid, the gap size s should (approximately) satisfy: $100s^2 = \frac{5\pi}{2}2^{-2n}$ or $s = \sqrt{\pi/40}\cdot 2^{-n}$. We use this for the gaps between radii, and, on average, arrange for a similar gap size between adjacent nodes on a given circle of deployment. Since the domain is not convex, we will use additional ghost nodes (as in Example 13.3) to help us detect and remove unwanted elements.

```
%Script for EFR 13.4a
count=1;
for n=1:7
s=sqrt(pi/40)/2^n;
len = 5*pi/2/2^n; %avg. arclength of node circular arc in Omega_n
nnodes= ceil(len/s); %number of nodes to put on each circular arc
ncirc = ceil(1/2^n/s);
%number of circlular arcs w/ to put nodes on Omega_n
rads = linspace(2/2^n, 1/2^n+s/2, ncirc);
%radii of circular arcs with nodes
angles = linspace(pi/6, 11*pi/6, nnodes); %angles for node deployment

%deploy nodes:
for r=rads
    for theta = angles
        x(count)=r*cos(theta); y(count)=r*sin(theta); count=count+1;
    end
end
end
%the final portion takes a slightly different approach since we want
%to deploy nodes throughout the whole sector (not just the annulus).
%We will thus want the circles of deployment to have radii all the
%way down to s(gap size), but on the smaller circles we should deploy
%less nodes
n=8; s=sqrt(pi/40)/2^n;
len = 5*pi/2/2^n;
%avg. arclength of node circular arc in Omega_n-outer circles
nnodes= ceil(len/s);
%number of nodes to put on each outer circular arc
rads = linspace(2/2^n, 0, ceil(2/2^n/s));
%radii of circular arcs with nodes
angles = linspace(pi/6, 11*pi/6, nnodes); %angles for node deployment

%delploy nodes
for r=rads
    for theta = linspace(pi/6, 11*pi/6, ceil(len/s*r/(2/2^n)))
        x(count)=r*cos(theta); y(count)=r*sin(theta); count=count+1;
    end
end

% Put in extra ghost nodes to detect bad elements
% There are several ways to do this, we will deploy them in a
% sufficient pattern on the positive x-axis.
nnodes=count-1; %number of nodes (=932)
for k=0:7
  x(count)=1/2^k; y(count)=0; count=count+1;
  x(count)=.75/2^k; y(count)=0; count=count+1;
end
for k=rads
    if k>0
      x(count)=k; y(count)=0; count=count+1;
  end
end
tri = delaunay(x,y);
%The following two commands will plot the triangulation containing
%the ghost nodes, the latter indicated by pentacles.  This plot and a
```

```
%magnification are shown in the %figure below.
```

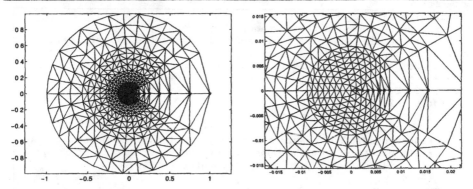

By the way that the ghost nodes were deployed, the unwanted elements are precisely those that have a ghost node as one of their vertices. The remaining code will search and destroy these elements, it is modeled after that of Example 13.3. The final triangulation and a zoomed view are shown in the two figures below.

```
plot(x(nnodes+1:count-1), y(nnodes+1:count-1), 'rp')
hold on, trimesh(tri,x,y), axis('equal')
%>> size(tri)
%ans =
%       1876              3
badelcount=1;
for ell=1:1876
    if max(ismember(nnodes+1:count-1, tri(ell,:)))
        badel(badelcount)=ell;
        badelcount=badelcount+1;
    end
end
clf
tri=tri(setdiff(1:1076,badel),:);
x=x(1:nnodes); y=y(1:nnodes);
trimesh(tri,x(1:nnodes),y(1:nnodes)), axis('equal')
```

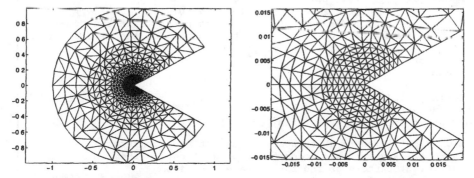

(b) We take the vertices of the domain to be: (0, 0), (2,0), (2,1), (−1, 1), (−1, −2), and (0, −2). The code below presents yet another variation of node deployment schemes. The crucial part (Stage 2 in the code below) is the deployment of nodes inside the circle with center (0, 0) and radius 0.8. We put an equal number of nodes (13) on each such circle. Because of the exponential decay of the radii, the gaps between radii remain close to the arclength gaps on the corresponding circles. The code below uses ghost nodes and they can be viewed by executing the code up to the line with the first `trimesh`

command (as in part (a)). We show only a figure of the final triangulation along with a zoomed view
(without axes).

```
%Script for EFR 13.4b
%We deploy the nodes in three stages
%Stage 1:  Outside the circle of radius 1, center (0,0), squarelike
%grid with gap size s = 0.2
%We can do boundary and interior nodes together:
count=1;
for xt=-1:.2:2
    for yt=-2:.2:1
        pt=[xt yt]; %test point
        if norm(pt,2)>.8+.1 & ~(xt>0 & yt<0)
        %these conditions ensure the test point is in the domain and a
        %safe distance from the boundary of the outer circle of Stage 2
            x(count)=xt; y(count)=yt; count=count+1;
        end
    end
end
%Stage 2:  Put nodes on concentric circles with exponential decay of
%radii
angles=0:pi/16:3*pi/2;
%this vector of angles will not change in the loop
for k=1:40
    r=.8^k;
    for theta = angles
      x(count)=r*cos(theta); y(count)=r*sin(theta); count=count+1;
      if k==0 & (x(count-1)<-.95|y(count-1)>.95)
        count=count-1; end %discard points too close to domain boundary
      end
end
%Stage 3:  Put nodes on the inside of the last circle of Stage 2
gap=3*pi/4*r/13;
%approx. gap size gotton by dividing arclength of last circle
%by number of nodes that were put on it
xvec=linspace(-r,r,2*ceil(r/gap)+1); yvec=xvec;
for xt=xvec
    for yt=yvec
      pt=[xt yt]; %test point
        if norm(pt,2)<=r-gap/2 & ~(xt>0 & yt<0)
            %these conditions ensure the test point is in the domain
            %and a safe distance from the boundary of the circle
            x(count)=xt; y(count)=yt; count=count+1;
        end
    end
end
%plot(x,y,'rp')
tri = delaunay(x,y);
hold on, trimesh(tri,x,y), axis('equal')

% Now we put in extra ghost nodes to detect bad elements
% There are several ways to do this, we will deploy them in a
% sufficient pattern on the ray theta = - pi/4
nnodes=count-1; %number of nodes
x(count)=1; y(count)=-1; count=count+1;
for k=0:40
x(count)=.8^k*cos(-pi/4); y(count)=.8^k*sin(-pi/4); count=count+1;
end
for k=r:-gap:gap
x(count)=k*cos(-pi/4); y(count)=k*sin(-pi/4); count=count+1;
```

```
end
x(count)=gap/2*cos(-pi/4); y(count)=gap/2*sin(-pi/4); count=count+1;
tri = delaunay(x,y);
clf, plot(x(nnodes+1:count-1), y(nnodes+1:count-1), 'rp')
hold on, trimesh(tri,x,y), axis('equal')
size(tri)
%ans =
%         2406              3
badelcount=1;
for ell=1:2406
    if max(ismember(nnodes+1:count-1, tri(ell,:)))
        badel(badelcount)=ell;
        badelcount=badelcount+1;
    end
end
clf
tri=tri(setdiff(1:2406,badel),:);
x=x(1:nnodes); y=y(1:nnodes);
trimesh(tri,x,y), axis('equal'),  axis off
```

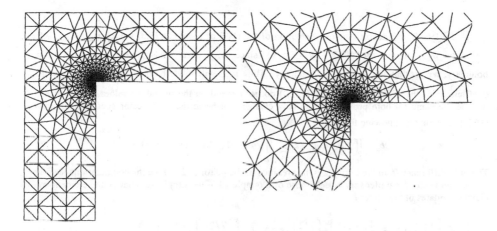

EFR 13.5: (a) In multivariable calculus, it is proved that the gradient vector of a function of two variables points in the direction in which the partial derivative is maximum and has magnitude equal to this maximum partial derivative. Also, the gradient is perpendicular to the direction in which the partial derivatives are zero. Since ϕ is a linear function, its gradient is a constant vector. Since the function ϕ is zero on the line joining v_1 and v_2, it follows that the gradient must be perpendicular to this side of T and therefore it must be parallel to $\vec{a}$ (the opposite direction would have negative partial derivative). The magnitude of the partial derivative, since the function is linear, can be gotten by taking the difference quotient of the values of ϕ at the tip and tail of $\vec{a}$ over the length of the vector $\vec{a}$ and this completes the proof of (a).

(b) The integral will be unchanged if we perform a rotation change of variables (which has Jacobian = 1), so we may assume that the line joining v_1 and v_2 is the x-axis. Write $v_1 = (a, 0)$, $v_2 = (b, 0)$ and let c denote the x-coordinate of v_3. Thus $a \le c \le b$ and $\phi(x, y) = y / \|\vec{a}\|$. Assume first that $a < c < b$. The height of the triangle at any value of $x \in [a, b]$ is given by:

$$h(x) = \begin{cases} \dfrac{\|\vec{a}\|}{c} \cdot \left(\dfrac{x-a}{c-a}\right), & \text{if } x \le c, \\[3mm] \|\vec{a}\| \cdot \left(1 - \dfrac{x-c}{b-c}\right), & \text{if } x > c. \end{cases}$$

Thus we may compute:

$$\iint_T \phi(x,y)\,dxdy = \int_a^b \int_0^{h(x)} \frac{y}{\|\vec{a}\|}\,dydx = \int_a^b \frac{h(x)^2}{2}\,dx = \frac{\|\vec{a}\|}{2}\int_a^c \left(\frac{x-a}{c-a}\right)^2 dx + \frac{\|\vec{a}\|}{2}\int_c^b \left(1-\frac{x-c}{b-c}\right)^2 dx.$$

These two integrals are easily done by u-substitution. In the first one, we let $u = \dfrac{x-a}{c-a}$, so $du = \dfrac{dx}{c-a}$

and the integral becomes: $\displaystyle \int_a^c \left(\frac{x-a}{c-a}\right)^2 dx = (c-a)\int_0^1 u^2 du = \frac{1}{3}(c-a)$. Similarly, the second integral is

$\dfrac{1}{3}(b-c)$; combining these gives $\displaystyle \iint_T \phi(x,y)\,dxdy = \frac{1}{3}\frac{\|\vec{a}\|}{2}(b-a) = \frac{1}{3}\text{Area}(T)$, as asserted. In the

remaining case that $c = a$ or $c = b$ (so the triangle is a right triangle), the function $h(x)$ can be written as a single formula and the above proof simplifies.

EFR 13.6: If u_{old} denotes the exact solution of the BVP of Example 13.5, we let $u_{new} \equiv u_{old} + 1$. Certainly u_{new} satisfies the PDE $-\Delta u = f(x,y)$ since u_{old} does. Also, since $u_{old} \equiv 1$ on the boundary, we get that $u_{new} \equiv 2$ on the boundary. Thus u_{new} will solve the modified BVP. Since the coefficients of the stiffness matrix (see (15 $^\ell$)) do not depend on the boundary values, this matrix A will be the same for both problems. The only change will be in the load vector coefficients; now (16 $^\ell$) takes on the following form:

$$b_\alpha^\ell = \iint_{T_\ell} f\Phi_{i_\alpha}\,dxdy - 2\sum_{s \ne 4,5}\iint_{T_\ell} \nabla\Phi_s \cdot \nabla\Phi_{i_\alpha}\,dxdy \quad (1 \le \alpha \le 3).$$

The only difference from the example is the presence of the factor of 2. Since the computations leading to the load vector b parallel very closely those of Example 13.5, we simply summarize the element-by-element updates of the vector b:

$$\ell = 1:\ b = \begin{bmatrix} 2 \\ 0 \end{bmatrix},\ \ell = 2:\ b = \begin{bmatrix} 2+3/2 \\ 0 \end{bmatrix} = \begin{bmatrix} 7/2 \\ 0 \end{bmatrix},\ \ell = 3:\ b = \begin{bmatrix} 7/2 \\ 0+3/2 \end{bmatrix} = \begin{bmatrix} 7/2 \\ 3/2 \end{bmatrix},$$

$$\ell = 4:\ b = \begin{bmatrix} 7/2 \\ 3/2+11/6 \end{bmatrix} = \begin{bmatrix} 7/2 \\ 10/3 \end{bmatrix},\ \ell = 5:\ b = \begin{bmatrix} 7/2+2 \\ 10/3 \end{bmatrix} = \begin{bmatrix} 11/2 \\ 10/3 \end{bmatrix},\ \ell = 6:\ b = \begin{bmatrix} 11/2+3/2 \\ 10/3 \end{bmatrix} = \begin{bmatrix} 7 \\ 10/3 \end{bmatrix},$$

$$\ell = 7:\ b = \begin{bmatrix} 7 \\ 10/3+3/2 \end{bmatrix} = \begin{bmatrix} 7 \\ 29/6 \end{bmatrix},\ \ell = 8:\ b = \begin{bmatrix} 7 \\ 29/6+11/6 \end{bmatrix} = \begin{bmatrix} 7 \\ 20/3 \end{bmatrix}.$$

If we solve the resulting matrix equation $Ax = b$; we get (to 4 decimals) $x(1) = 1.9869$, and $x(2) = 1.9179$. Comparing with the solutions of the original system: $x(1) = 0.9278$ and $x(2) = 1.0484$, we see that the numerical solutions are somewhat close, but definitely do not differ by one. With finer triangulations, the exact relationship would be made more apparent.

EFR 13.7: (a) We first draw a picture of the region S on which the integration is to take place, and realize it lying between two functions of x; see the left figure below. It is convenient to break the integral up into two pieces since the top function experiences a formula change at $x = \cos(\pi/4)$.

```
>> syms x y
>> Int_A = quad2d(x*y^2,0, cos(pi/4), 0, x)+quad2d(x*y^2,cos(pi/4),
1, 0, sqrt(1-x^2))
```

→ Int_A = 0.0236 (This is the numerical approximation to the first integral in "format short.")
(b) The picture, shown on the right below, shows that the curves intersect when *x* is negative. We first find this intersection point:

```
>> xmin = fzero(inline('x^2-1-exp(x)'), -1) → xmin =-1.1478
>> Int_B = quad2d(exp(1-x^2-2*y^2),xmin, 0, x^2-1, exp(x)) →Int_B =
2.0661
```

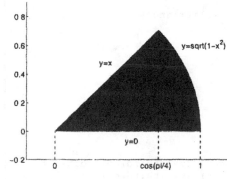
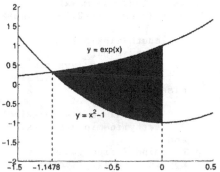

Note: Both plots were created in MATLAB using the built-in function `patch`. The syntax is as follows:

`patch([x xrev],`
`[flow ftop_rev], [r g b])`
→
	This command will produce a graphic of a shaded region between two functions. Here x is a vector of x-coordinates for an interval on which the corresponding vectors `flow` and `ftop_rev` represent two functions. The function represented by `flow` has its graph lying below the one represented by `ftop_rev`. The syntax requires also as input the vector `xrev` that is the vector x taken in reverse order. The vector `ftop_rev` (representing the top function) correspondingly needs to be inputted in reverse order. The final input is a 3×1 rgb vector of numbers between 0 and 1 that will determine the color of the patch.

As an example, we give the code used to create the second plot:

```
>> x=xmin:.01:0;
>> for i=1:length(x), xrev(i)=x(length(x)+1-i); end
>> patch([ x xrev], [x.^2-1  exp(xrev)], [.5 .5 .5]), hold on
>> t = -1.5:.01:.5;  plot(t,t.^2-1,'b'), plot(t,exp(t),'b'),
>> axis([-1.5 .5 -2 2])
>> gtext('y = exp(x)') %use mouse to place text on graphic window
>> gtext('y = x^2-1') %use mouse to place text on graphic window
```
Some embellishments were done to the graph using menu options on the graphics window.

EFR 13.8: (a) The M-file is boxed below:

```
function integ=triangquad2d(fun,v1,v2,v3)
% M-file for EFR 13.8.  This function will integrate a function
% of two variables x and y over a triangle T in the plane.
% It uses the M-file 'quad2d' of Program 13.1
% Input variables:  fun = a symbolic expression (using one or both of
% the symbolic variables x and y,  v1, v2, and v3:  three length 2
% vectors giving the vertices of the triangle T in the plane.
% NOTE: Before this program is used, x and y should be declared as
% symbolic variables.syms x y u
```

```
vys = [v1(2) v2(2) v3(2)];
vxs = [v1(1) v2(1) v3(1)];
minx=min(vxs); maxx=max(vxs);
miny=min(vys);
minyind =find(vys==miny);
minxind =find(vxs==minx);
maxxind =find(vxs==maxx);
if length(minxind)==2|length(maxxind)==2 %triangle has a vertical
side
    if length(minxind)==2
        vertx=minx;
        vertymax=max(vys(minxind));
        vertymin=min(vys(minxind));
    else
        vertx=maxx;
        vertymax=max(vys(maxxind));
        vertymin=min(vys(maxxind));
    end
    thirdind = find(vxs~= vertx);
    topslope=sym((vys(thirdind)-vertymax)/(vxs(thirdind)-vertx));
    botslope=sym((vys(thirdind)-vertymin)/(vxs(thirdind)-vertx));
    ytop=topslope*(x-vxs(thirdind))+vys(thirdind);
    ylow=botslope*(x-vxs(thirdind))+vys(thirdind);
    integ = quad2d(fun,minx,maxx,ylow,ytop);
else %no vertical sides so vertices have 3 different x coordinates
    midind = find(vxs>minx& vxs<maxx); %index of middle vertex
    longslope=sym((vys(maxxind)-vys(minxind))/(maxx-minx));
    ylong = longslope*(x-vxs(minxind))+vys(minxind);
    if vys(midind)>subs(ylong,x,vxs(midind));
    %long edge lies below mid vertex
    topleftslope = sym((vys(midind)-vys(minxind))/(vxs(midind)-...
                    vxs(minxind)));
    toprgtslope = sym((vys(midind)-vys(maxxind))/(vxs(midind)- ...
                    vxs(maxxind)));
    ytopleft = topleftslope*(x-vxs(midind))+vys(midind);
    ytoprgt = toprgtslope*(x-vxs(midind))+vys(midind);
    integ = quad2d(fun,minx, vxs(midind),ylong,ytopleft)+...
        quad2d(fun,vxs(midind),maxx,ylong,ytoprgt);
    else %long edge lies above mid vertex
    botleftslope = sym((vys(midind)-vys(minxind))/(vxs(midind)- ...
                    vxs(minxind)));
    botrgtslope = sym((vys(midind)-vys(maxxind))/(vxs(midind)- ...
                    vxs(maxxind)));
    ybotleft = botleftslope*(x-vxs(midind))+vys(midind);
    ybotrgt = botrgtslope*(x-vxs(midind))+vys(midind);
    integ = quad2d(fun,minx,vxs(midind),ybotleft,ylong)+...
        quad2d(fun,vxs(midind),maxx,ybotrgt,ylong);
    end
end
```

(b) To recalculate the integrals of Example 13.5, the following commands will suffice and result in the
same outputs that were obtained in the example:

```
>> syms x y
>> v1 = [1 3]; v2 = [5 1]; v3 = [4 6]; %Triangle of Example 13.5
>> triangquad2d(2*x*y^2,v1,v2,v3)  →ans = 724.8000
>> triangquad2d(sin(x*y*sqrt(y)),v1,v2,v3)  →ans = 0.1397
```

The remaining integrals can be done in the same swift fashion. We store separately the vertices of the
two triangles *T1* and *T2*:

```
>> v1 = [0 0]; v2 = [6 0]; v3 = [12 2]; %Triangle T1 of EFR 13.8
>> V1 = [1 3]; V2 = [3 2]; V3 = [2 5]; %Triangle T2 of EFR 13.8
```

```
>> int_1 = triangquad2d(1,v1,v2,v3) →int_1 = 6
>> int_2 = triangquad2d(1,V1,V2,V3) →  int_2 = 2.5000
>> int_3 = triangquad2d(2*x^2,v1,v2,v3) →  int_3 = 504
>> int_4 = triangquad2d(sin(x^2),V1,V2,V3) →  int_4 = -0.2998
```
These numerical calculations are all in agreement with the exact answers that were provided.

EFR 13.9: We will use modification of the method used in part (c) of Example 13.2. In that example, a similar node deployment was required on the same domain, except that there we wanted more nodes to be focused near the boundary point (1,0) and here we want the focus area to be the boundary point (cos(3), sin(3)). What we will do is very slightly modify the node deployment code of the example (since here we want less nodes) and then simply rotate the node set by an angle of $\theta = 3$ (using the rotation transformation of Section 7.2). The rotation idea is quite a natural one; it could be circumvented, but then we would need a more serious modification of the code of the example. Since the codes are long, we indicate only the changes needed for the present problem.

Referring to the notations of the solution of part (c) of Example 13.2, in the determination of the gap size s to use in the region Ω_n, we will use roughly 10 nodes (rather than 100) per such region, so s should satisfy $10 \cdot s^2 \le \text{Area}(\Omega_n) \le \frac{3\pi}{2} 2^{-2n}$ or $s \le \sqrt{3\pi/20} \cdot 2^{-n}$. If we then run through 8 iterations (n runs from 0 to 7) of deploying nodes just as in the example, we see that the nodes are a bit sparse in the first two regions. To mitigate this, we make s a bit smaller in the first two iterations. If we replace the first three lines of the node deployment code of the example with the following four lines (and run the rest of the code), we will arrive at a triangulation that looks quite appropriate.

```
>> n=0; nodecount=1;
>> while n<8
s=sqrt(3*pi/20)/2^n;
if  n==0, s = s/3; elseif n ==1, s=s/2; end
```
This node set should now be rotated by an angle of $\theta = 3$, and this is done using the rotation matrix of Section 7.2:
```
>> Rot=[cos(3)   -sin(3); sin(3)  cos(3)]*[x; y];
>> xn = Rot(1,:); yn = Rot(2,:);  %newly rotated nodes for desired
triangulation.
```
Now, in order to be able to more easily use the assembly code of Example 13.7, we should reorder the nodes so that the boundary nodes appear last. This is accomplished with the following commands:
```
bdyind = find(xn.^2 + yn.^2 > 1 - 10*eps);
size(xn), size(bdyind)  →   1    123,    1     38
intind = setdiff(1:123, bdyind); %these are the indices of interior
nodes
xn = [xn(intind) xn(bdyind)]; %reordered x-coordinates of nodes
yn = [yn(intind) yn(bdyind)]; %reordered x-coordinates of nodes
tri=delaunay(xn,yn);
trimesh(tri,xn,yn), axis('equal') %triangulation is shown below left
```

We now store the boundary values:
```
for i=86:123
th=cart2pol(x(i),y(i));
if th<0, th=th+2*pi; end
%need to ensure th is in domain of boundary data function
c(i)=ex_13_7_bdydata(th);
end
```

The assembly code of the example will now work very well in this situation; we need only change the first three lines as follows:
```
N=[xn' yn'];
E=tri;
n=85; m=123; syms x y
```
When this and the rest of the code is run, we will have created the numerical solution's values stored as a vector c. The exact solution's values can be created just as in the example (the code is verbatim) and

stored as a vector cp. This being done, the following command will plot the error of the numerical solution; the plot is shown on the right.

```
>> trimesh(E,xn,yn,abs(c-cp))
```

Notice that the maximum error is seen to be smaller than that obtained in part (b) of the example (cf. Figure 13.38b), using a lot less nodes but a more appropriate node deployment strategy.

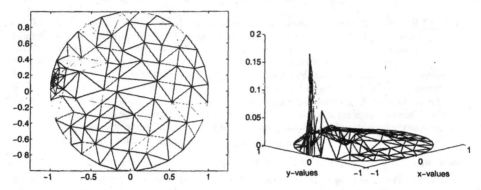

EFR 13.10: (a) The M-file is boxed below:

```
function int = gaussianintapprox(f,V1,V2,V3)
% M-file for numerically approximating integral of a function f(x,y)
% over a triangle in the plane with vertices V1, V2, V3
% Approximation is done using the Gaussian quadrature formula (24)
% of Chapter 13.
% Input Variables:  f = an inline function or an M-file of the
% integrand specified as a function of two variables: x and y
% V1, V2, V3 length 2 row vectors containing coordinates of the
% vertices of the triangle.  Output variable: int = approximation
A=feval(f,(V1(1)+V2(1))/2,(V1(2)+V2(2))/2);
B=feval(f,(V1(1)+V3(1))/2,(V1(2)+V3(2))/2);
C=feval(f,(V2(1)+V3(1))/2,(V2(2)+V3(2))/2);
M=[V1 1;V2 1; V3 1];
area=abs(det(M))/2;   %See formula (5) of Chapter 13
int=area*(A+B+C)/3;
```

(b) After creating and storing the triangulation for part (c) of Example 13.7, and then the boundary values (just as was done in the example), the first three lines of the assembly code should read as follows:

```
>> N=[x' y'];
>> E=tri;
>> A=zeros(n); b=zeros(n,1);
```

(Same as before, except now we do not need symbolic variables.) The rest of the assembly code only needs changing in the two places where the numerical integrator triangquad2d was used. To save space, we include the relevant modified passages here; the ftp site for this book includes a file for the complete code.

```
%update stiffness matrix
for i1=1:length(intnodes)
for i2=1:length(intnodes)
fun1 = num2str(intgrad(i1,:)*intgrad(i2,:)',10); %integrand for
(15ell)
fun=inline(fun1,'x', 'y');
integ=gaussianintapprox(fun,xyt,xyr,xys);
A(intnodes(i1),intnodes(i2))=A(intnodes(i1),intnodes(i2))+integ;
end
end
```

```
%update load vector
for i=1:length(intnodes)
for j=1:length(bdynodes)
fun1 = num2str(intgrad(i,:)*bdygrad(j,:)',10); %integrand for (16ell)
fun=inline(fun1,'x', 'y');
integ=gaussianintapprox(fun,xyt,xyr,xys);
b(intnodes(i))=b(intnodes(i))-c(bdynodes(j))*integ;
end
end
```

Whereas the original code took about an hour to run (on the author's computer), the modified assembly code took only a few seconds. Moreover, an examination of the error plot (against the exact Poisson solution) shows the errors of the two methods to be about the same.

EFR 13.11: For completeness, we include a full code for the FEM. Assume that the node set (vectors x and y) have been constructed as in Example 13.3. Although from the construction it is clear that the nodes on the inner circle came first and those on the outer circle came last, before we triangulate, we give a code that will automatically reindex so that the interior nodes precede the boundary nodes:

```
m = length(x); %m = total number of nodes.
cnt1=1; cnt2=1;
for i=1:m
if norm([x(i) y(i)],2)<1+4*eps %tests if node is on inner circle
bdy1(cnt1)=i; cnt1=cnt1+1;
elseif norm([x(i) y(i)],2)>2-4*eps %tests if node is on outer circle
bdy2(cnt2)=i; cnt2=cnt2+1;
end
end
n=m-length(bdy1)-length(bdy2); % n = total number of interior nodes
xnew=[x(setdiff(1:m, union(bdy1, bdy2))) x(bdy1) x(bdy2)]; x=xnew;
ynew=[y(setdiff(1:m, union(bdy1, bdy2))) y(bdy1) y(bdy2)]; y=ynew;
```

Next, we form the Delaunay triangulation using the code of Example 13.3. This being done, we assign the boundary values:

```
c(n+1:n+length(bdy1))=2;
for i=n+length(bdy1)+1:m
th=cart2pol(x(i),y(i));
c(i)=cos(2*th);
end
```

The remaining code below will perform the FEM and create the plot of the numerical solution (Figure 13.39). We use Method 2 (Gaussian quadrature for the integrals). We include the code for completeness only, but because of the way we have prepared things, the remaining code is identical to that of the preceding EFR (if we had written it out there):

```
N=[x' y'];
E=tri;
A=zeros(n); b=zeros(n,1);
[L cL]=size(E);
for ell=1:L
nodes=E(ell,:);
bdynodes=nodes(find(nodes>n));
intnodes=setdiff(nodes,bdynodes);

%find gradients [a b] of local basis functions
% ax + by +c; distinguish between int node
%local basis functions and bdy node local basis
%functions

for i=1:length(intnodes)
xyt=N(intnodes(i),:); %main node for local basis function
onodes=setdiff(nodes,intnodes(i));
```

```
%two other nodes (w/ zero values) for local basis function
xyr=N(onodes(1),:);
xys=N(onodes(2),:);
M=[xyr 1;xys 1;xyt 1]; %matrix M of (4)
abccoeff=[xyr(2)-xys(2); xys(1)-xyr(1); xyr(1)*xys(2)-...
xys(1)*xyr(2)]/det(M); %coefficients of basis function on triangle#L,
see formula (6a)
intgrad(i,:)=abccoeff(1:2)';
end


for j=1:length(bdynodes)
xyt=N(bdynodes(j),:); %main node for local basis function
onodes=setdiff(nodes,bdynodes(j));%two other nodes (w/ zero values)
for local basis function
xyr=N(onodes(1),:);
xys=N(onodes(2),:);
M=[xyr 1;xys 1;xyt 1]; %matrix M of (4)
abccoeff=[xyr(2)-xys(2); xys(1)-xyr(1); xyr(1)*xys(2)-...
xys(1)*xyr(2)]/det(M); %coefficents of basis function on triangle#L,
see formula (6a)
bdygrad(j,:)=abccoeff(1:2)';
end
%update stiffness matrix
for i1=1:length(intnodes)
for i2=1:length(intnodes)
fun1 = num2str(intgrad(i1,:)*intgrad(i2,:)',10); %integrand for
(15ell)
fun=inline(fun1,'x', 'y');
integ=gaussianintapprox(fun,xyt,xyr,xys);
A(intnodes(i1),intnodes(i2))=A(intnodes(i1),intnodes(i2))+integ;
end
end


%update load vector
for i=1:length(intnodes)
for j=1:length(bdynodes)
fun1 = num2str(intgrad(i,:)*bdygrad(j,:)',10); %integrand for (16ell)
fun=inline(fun1,'x', 'y');
integ=gaussianintapprox(fun,xyt,xyr,xys);
b(intnodes(i))=b(intnodes(i))-c(bdynodes(j))*integ;
end
end


end
sol=A\b;
c(1:n)=sol';
>> trimesh(tri,x,y,c)
>> xlabel('x-axis'), ylabel('y-axis')
```

EFR 13.12: (a) Rerun the triangulation code of the solution of Example 13.2(a). The construction was done in a way that the boundary nodes came last. So we will be able to adapt the assembly code of EFR 13.11 quite simply. (The only change will be in dealing with the load vector, because of the presence of the inhomogeneity function.)

```
>> m=length(x); %number of nodes
>> n = min(find(x.^2+y.^2>1-10*eps))-1;  %number of interior nodes
```

Now we easily modify the (boxed) code of the preceding EFR to work for the present situation. There are some changes here due to the fact that the boundary data is now all zero, but we do have a nonzero inhomogeneity fucntion $f(x,y)$, and thus (16^ℓ) takes on the following more simple form: $b_\alpha^\ell = \iint\limits_{T_\ell} f\Phi_{i_\alpha}\,dxdy$ $(1 \le \alpha \le 3)$. Thus, the "updating the load vector" portion should be replaced by:

```
%update load vector
for il=1:length(intnodes)
xyt=N(intnodes(i),:); %main node for local basis function
onodes=setdiff(nodes,intnodes(i));
%two other nodes (w/ zero values) for local basis function
xyr=N(onodes(1),:);
xys=N(onodes(2),:);
M=[xyr 1;xys 1;xyt 1]; %matrix M of (4)
abccoeff=[xyr(2)-xys(2); xys(1)-xyr(1); xyr(1)*xys(2)-...
xys(1)*xyr(2)]/det(M); %coefficients of basis function on triangle#L,
                   %see formula (6a)
%since we cannot mix M-file and inline functions to input into
%another M-file, we recode the gaussianintapprox M-file
atemp=num2str(abccoeff(1),10); btemp=num2str(abccoeff(2),10);
ctemp=num2str(abccoeff(3),10);
phixy=inline([atemp, '*x+', btemp, '*y+',ctemp],'x','y');
Atemp=feval(@EFR13_12f,(xyt(1)+xyr(1))/2,(xyt(2)+xyr(2))/2)*...
feval(phixy,(xyt(1)+xyr(1))/2,(xyt(2)+xyr(2))/2);
Btemp=feval(@EFR13_12f,(xyt(1)+xys(1))/2,(xyt(2)+xys(2))/2)*...
feval(phixy,(xyt(1)+xys(1))/2,(xyt(2)+xys(2))/2);
Ctemp=feval(@EFR13_12f,(xyr(1)+xys(1))/2,(xyr(2)+xys(2))/2)*...
feval(phixy,(xyr(1)+xys(1))/2,(xyr(2)+xys(2))/2);
M=[xyr(1) xyr(2) 1;xys(1) xys(2) 1; xyt(1) xyt(2) 1];
area=abs(det(M))/2;
integ=area*(Atemp+Btemp+Ctemp)/3;
b(intnodes(il))=b(intnodes(il))+integ;
end
```

Also, the loop portion of the assembly code commencing with "for j=1:length(bdynodes)" can be deleted since the boundary node gradients that it creates will not be needed in (16^ℓ). With these modifications, the code will produce the numerical solution of Figure 13.39(b), once an M-file for the inhomogeneity function is created (due to the cases in its definition, an inline construction is not feasible):

```
function n = EFR13_12f(x,y)
if norm([x y]-[0 .5],2)<.25
z=20;
else
z=0;
end
```

(b) The assembly instructions are exactly as in part (a), after we have created the node set and triangulation according to the specifications. The following code will create such a triangulation:

```
% node deployment, use concentric circles centered at (0, 1/2)
% except for on the boundary
% Step 1 inside Omega1 (small circle) has 50% of nodes
% d1=common gap size
% avg radius = 1/8, avg. circumf= pi/4,
% avg no. of nodes on circ = pi/4/d1
% number of circles = 1/4/d1
% setting 50% of 800 = [pi/4/d1][1/4/d1] gives
d1=sqrt(pi/16/400);
x(1)=0; y(1)=.5;
nodecount=1; ncirc=floor(1/4/d1); minrad=1/4/ncirc;
for i=1:ncirc, rad=i*minrad; nnodes=floor(2*pi*rad/d1);
anglegap=2*pi/nnodes;
```

```
for k=1:nnodes
  x(nodecount+1)=rad*cos(k*anglegap);
  y(nodecount+1)=rad*sin(k*anglegap)+.5;
  nodecount = nodecount+1;
end
end

% step 2:  inside annulus Omega2 has 25% of nodes
% d2=common gap size
% avg radius = 3/8, avg circumf = 3pi/4,
% avg no of nodes on circ = 3pi/4/d2
% number of circles = 1/4/d2
d2=sqrt(3*pi/16/200); ncirc=floor(1/4/d2);minrad=1/4+(d1+d2)/2;
%blend interface
for i=1:ncirc
  rad=minrad + (i-1)*d2; nnodes=floor(2*pi*rad/d2);
  anglegap=2*pi/nnodes;
  for k=1:nnodes
    x(nodecount+1)=rad*cos(k*anglegap);
    y(nodecount+1)=rad*sin(k*anglegap)+.5;
    nodecount = nodecount+1;
end
end

% step 3:  inside region Omega3 has 15% of nodes
% d3 = common gap size
% avg radius = 3/4, avg arclength (approx)= (2pi +pi)/2*3/4=9pi/8
% number of circles = 1/2/d3
d3=sqrt(9*pi/16/120); ncirc=floor(1/2/d3); minrad=1/2+(d2+d3)/2;
%blend interface
for i=1:ncirc
  rad=minrad + (i-1)*d3; nnodes=floor(2*pi*rad/d3);
  anglegap=2*pi/nnodes;
   for k=1:nnodes
    xtest=rad*cos(k*anglegap); ytest=rad*sin(k*anglegap)+.5;
    if norm([xtest ytest],2)<1-d3/2 %don't put nodes too close to bdy
 x(nodecount+1)=xtest; y(nodecount+1)=ytest; nodecount = nodecount+1;
    end
   end
end

% step 4: inside region Omega4 has 10% of nodes
% d4 = common gap size
% avg radius = 5/4, avg (approx) arclength =  5pi/4
% number of circles = 1/2/d4
d4=sqrt(5*pi/8/80); ncirc=floor(1/2/d4); minrad=1+(d3+d4)/2; %blend
interface
for i=1:ncirc
  rad=minrad + (i-1)*d4; nnodes=floor(2*pi*rad/d4);
  anglegap=2*pi/nnodes;
  for k=1:nnodes
    xtest=rad*cos(k*anglegap); ytest=rad*sin(k*anglegap)+.5;
    if norm([xtest ytest],2)<1-d4/2 %don't put nodes too close to bdy
 x(nodecount+1)=xtest; y(nodecount+1)=ytest; nodecount = nodecount+1;
    end
  end
end

% step 5: put nodes on boundary
% if bdy point is touches Omega3 use d3 spacing
```

```
% otherwise use d4 spacing
theta=0;
while theta<2*pi-d4
  x(nodecount+1)=cos(theta); y(nodecount+1)=sin(theta);
  nodecount = nodecount+1;
  if norm([cos(theta) sin(theta)]-[0 .5], 2)<.5
    theta=theta+d3;
  else
    theta=theta+d4;
  end
end
```

EFR 13.13: (a) The M-file is boxed below:

```
function lineint = bdyintapprox(fun, tri, redges)
% function M-file for EFR 13.13
% inputs will be 'fun', an inline function (or M-file) of vars x, y;
% a matrix 'tri' of nodes of a triangle in the plane, and a 2-column
% matrix 'redges', possibly empty ([ ]), containing, as rows,   the
% corresponding node indices (from 1 to 3 indicating nodes
% by their row in 'tri') of nodes which are endpoints of segments of
% the triangle which are part of the 'Robin' boundary (for an
% underlying FEM problem).  Thus the rows of 'redges' can include
% only the following three vectors:  [1 2], [1 3], and [2 3].  (Or
% permutations of these.) The output, 'lineint' will be the the
% Newton-Coates approx. ((31) of Chapter 13) line integral of 'fun'
% over the Robin segments of the triangle.
lineint=0;
[rn cn] = size(redges); %rn = number of Robin edges
if rn == 0
    return
end
for i=1:rn
    nodes = redges(i,:);
    N1=tri(nodes(1),:); N2=tri(nodes(2),:);
    N1x=N1(1); N1y=N1(2); N2x=N2(1); N2y=N2(2);
    vec = N2-N1;

approx=norm(vec,2)/6*(feval(fun,N1x,N1y)+4*feval(fun, (N1x+N2x)/2, (N1y
+N2y)/2)+feval(fun,N2x,N2y));
    lineint=lineint+approx;
end
```

(b)
```
>> tri1 = [0 0;2 0;0 3]; tri2=tri1/10;
>> f1 = inline('4','x','y'); f2=inline('cos(pi*x/4+pi*y/2)','x','y');
>> redges1 = [1 2;2 3]; redges2 = [1 2; 1 3];
>> Int1=bdyintapprox(f1,tri1,redges1)    →Int1 = 22.4222
>> abs(Int1-8-4*sqrt(13))    →ans = 1.7764e-015 (Error for first approximation)
>> bdyintapprox(f1,tri2,redges1)    →ans = 2.2422
>> Int2=bdyintapprox(f2,tri1,redges2)    →Int2 = 0.3619 (Error for first
approximation)
>> abs(Int2-2/pi)    →ans = 0.4882
```
It is not surprising that the error for the first integration was as small as machine precision, since the method is exact for polynomials of degree up to three and we are integrating a constant function. A similar accuracy would hold for the integral over the smaller triangle. The second integration had a very large error and this was due to the fact that the integrand experiences a lot of variation on the edges. A similarly large error (although a bit smaller relatively) would occur if we looked at the integral of the second function over the smaller triangle (the error would be 0.1263). When we utilize

this integrator in our FEM codes, we can use a fine enough partition (in the portions of the boundary where the data has more variation) to prevent such problems.

EFR 13.14: The PDE and the Dirichlet portion of the BCs are plainly satisfied, so we have only to check the Neumann BC on the parabolic portion of the boundary. A tangent vector to a point on the parabola $y = \varphi(x) \equiv x(10 - x)$ is given by $\vec{\tau}(x) = (d/dx)(x, \varphi(x)) = (1, 10 - 2x)$. Since this tangent vector has positive x-component, an outward-pointing normal vector can be obtained from it by rotating $\vec{\tau}(x)$ by an angle of $\pi/2$ (see Section 7.2). Dividing this vector by its Euclidean norm (see Section 7.6) gives the outward pointing unit normal vector: $\vec{n} = \vec{n}(x) = (2x - 10, 1) / \|(2x - 10, 1)\|_2 =$

$\dfrac{(2x - 10, 1)}{\sqrt{4x^2 - 40x + 101}}$. Taking the dot product with the gradient of u $\nabla u(x, y) = (0, y/25)$ of the exact

solution given produces the stated Neumann BC.

EFR 13.15: The triangulations for this problem have been already done and can simply be imported. The main task is to set up the assembly process. In the notation of (10), we have: $p \equiv 1, q \equiv 0, g \equiv 0, r = 2$ (on Γ_2), $h = 40$ (on Γ_2), $f(x)$ is as specified. Thus, $c_s = 0$ $(s > n)$ and the element matrix analogues of (28) and (29) (cf., (15^ℓ) and (16^ℓ) become:

$$a^\ell_{\alpha\beta} = \iint\limits_{T_\ell} [\nabla\Phi_{i_\alpha} \cdot \nabla\Phi_{i_\beta}]\,dxdy + 2\!\!\int\limits_{\Gamma_2 \cap T_\ell}\!\! \Phi_{i_\alpha}\Phi_{i_\beta}\,ds \quad (1 \le \alpha, \beta \le 3)\text{, and}$$

$$b^\ell_\alpha = \iint\limits_{T_\ell} f(x, y)\Phi_{i_\alpha}\,dxdy + 40\!\!\int\limits_{\Gamma_2 \cap T_\ell}\!\! \Phi_{i_\alpha}\,ds \quad (1 \le \alpha \le 3).$$

This is just a bit more involved than the assembly equations for Example 13.8, since in the former there were no line integrals in the first (stiffness matrix coefficient) equations. Nonetheless, the assembly code of the example can be easily adapted to fill our present needs. We first need to store an M-file for the inhomogeneity function $f(x,y)$:

```
function z = EFR13_15f(x,y)
if x>=4 & x<=6 & y>=10 & y<=15
z=200;
else
z=0;
end
```

Before running the assembly code below, we assume that the triangulation code of Example 13.8(a) has been run. In particular, the following variables have been created: `nint` = the number of interior nodes, n = the number of interior/Robin nodes, m = the number of nodes, `dir1` = m = node index for (0,0), and `dir2` = `nint` + 1 = the node index for (10,0). As in the example, in the first part of the code, we need not compute gradients of basis functions corresponding to Dirichlet nodes, since the Dirichlet boundary values are all zero. The only new technical issue here is that in the computation of the load coefficients (in the first integral), since it is awkward to mix inline functions and M-files into a single function, we choose to simply recode the `gaussianintapprox` M-file (which is a rather short code).

```
N=[x' y']; E=tri; A=zeros(n); b=zeros(n,1); [L cL]=size(E);
for ell=1:L
nodes=E(ell,:); %global node indices of element
percent=100*ell/L %optional percent meter will show progression.
intnodes=nodes(find(nodes<=n)); %global interior/Robin node indices
%find coefficients [a b c] of local basis functions
% ax + by +c; for  int/robin nodes
for i=1:length(intnodes)
xyt=N(intnodes(i),:); %main node for local basis function
onodes=setdiff(nodes,intnodes(i));
%global indices for two other nodes (w/ zero values) for local basis
function
xyr=N(onodes(1),:); xys=N(onodes(2),:);
```

```
M=[xyr 1;xys 1;xyt 1]; %matrix M of (4)
%local basis function coefficients using (6B)
abccoeff=[xyr(2)-xys(2); xys(1)-xyr(1); xyr(1)*xys(2)-...
          xys(1)*xyr(2)]/det(M);
intgrad(i,:)=abccoeff(1:2)'; abc(i,:)=abccoeff';
end

% determine if there are any Robin edges
marker=0; %will change to 1 if there are Robin edges.
roblocind=find(nodes==dir1|nodes==dir2|(nodes<=n & ...
          nodes >=(nint+1)));
%local indices of nodes for possible robin edges
if length(roblocind>1
elemnodes = N(nodes,:);

%now find robin edges and make a 2 column matrix out of their local
%indices.
rnodes=nodes(roblocind); %global indices of robin nodes
count=1;
for k=(nint+1):(n-1)
if ismember(k,rnodes) & ismember(k+1,rnodes)
robedges(count,:)=[find(nodes==k)  find(nodes==k+1)]; count=count+1;
marker =1;
end
end
end

%update stiffness matrix
for i1=1:length(intnodes)
for i2=1:length(intnodes)
if intnodes(i1)>=intnodes(i2)
%to save some computation, we use symmetry of the stiffness matrix.
fun1 = num2str(intgrad(i1,:)*intgrad(i2,:)',10);
          %integrand for (15ell)
fun=inline(fun1,'x', 'y'); integ=gaussianintapprox(fun,xyt,xyr,xys);
A(intnodes(i1),intnodes(i2))=A(intnodes(i1),intnodes(i2))+integ;

%now add Robin portion, if applicable
%robin edges were computed above
if marker==1
ai1 = num2str(abc(i1,1),10); ai2 = num2str(abc(i2,1),10);
bi1 = num2str(abc(i1,2),10); bi2 = num2str(abc(i2,2),10);
ci1 = num2str(abc(i1,3),10); ci2 = num2str(abc(i2,3),10);
prod=inline(['2*(',ai1,'*x+',bi1, '*y+', ci1,')* ...
      (',ai2,'*x+',bi2, '*y+',ci2,')'],'x','y');
A(intnodes(i1),intnodes(i2))=A(intnodes(i1),intnodes(i2)) ...
          +bdyintapprox(prod,elemnodes, robedges);
end
end
end
end

%update load vector
for i1=1:length(intnodes)
ai1 = num2str(abc(i1,1),10); bi1 = num2str(abc(i1,2),10);
ci1 = num2str(abc(i1,3),10);
phi=inline([ai1,'*x+',bi1, '*y+', ci1],'x','y');
```

```
%since we cannot mix M-file and inline functions to input into
%another M-file, we basically must recode the gaussianintapprox M-
%file
Atemp=feval(@EFR13_15f,(xyt(1)+xyr(1))/2,(xyt(2)+xyr(2))/2)*...
      feval(phi,(xyt(1)+xyr(1))/2,(xyt(2)+xyr(2))/2);
Btemp=feval(@EFR13_15f,(xyt(1)+xys(1))/2,(xyt(2)+xys(2))/2)*...
      feval(phi,(xyt(1)+xys(1))/2,(xyt(2)+xys(2))/2);
Ctemp=feval(@EFR13_15f,(xyr(1)+xys(1))/2,(xyr(2)+xys(2))/2)*...
      feval(phi,(xyr(1)+xys(1))/2,(xyr(2)+xys(2))/2);
M=[xyr(1) xyr(2) 1;xys(1) xys(2) 1; xyt(1) xyt(2) 1];
area=abs(det(M))/2;
integ=area*(Atemp+Btemp+Ctemp)/3;
b(intnodes(il))=b(intnodes(il))+integ;

%now add Robin portion, if applicable
%robin edges were computed above
if marker==1
prod=inline(['40*(',ail,'*x+',bil, '*y+', cil,')'],'x','y');
b(intnodes(il))=b(intnodes(il))+ ...
        bdyintapprox(prod, elemnodes, robedges);
end
end
clear roblocind rnodes robedges
end
A=A+A'-A.*eye(n); %Use symmetry to fill in remaining entries of A.

sol=A\b;
c(1:n)=sol';
c(n+1:m)=0;

%The result is now easily plotted using the 'trimesh' function of the
%last section:

x=N(:,1); y=N(:,2);
trimesh(E,x,y,c)
hidden off
xlabel('x-axis'), ylabel('y-axis')
```
The above code will produce a plot of the FEM solution.

EFR 13.16:   In the notation of (10), we have: $p \equiv 1$, $q \equiv 0$, $f \equiv 0$, $g \equiv 100$(on Γ_1), $r = 0,1$, or 2 (on Γ_2), and $h = 0, 20$, or 30(on Γ_2). The element matrix analogues of (28) and (29) (cf., (15$^\ell$) and (16$^\ell$)) thus become:

$$a_{\alpha\beta}^\ell = \iint_{T_\ell} [\nabla \Phi_{i_\alpha} \cdot \nabla \Phi_{i_\beta}] dxdy + r \int_{\Gamma_2 \cap T_\ell} \Phi_{i_\alpha} \Phi_{i_\beta}\, ds \quad (1 \le \alpha, \beta \le 3), \text{ and}$$

$$b_\alpha^\ell = \iint_{T_\ell} f(x,y)\Phi_{i_\alpha}\, dxdy + 40 \int_{\Gamma_2 \cap T_\ell} \Phi_{i_\alpha}\, ds \ - $$
$$\sum_{s=n+1}^{m} 100 \left[\iint_{T_\ell} [\nabla \Phi_s \cdot \nabla \Phi_{i_\alpha}] dxdy + r \int_{\Gamma_2 \cap T_\ell} \Phi_s \Phi_{i_\alpha}\, ds \right] \quad (1 \le \alpha \le 3).$$

The triangulation is new but can be accomplished with the various techniques that we have developed so far. Here is the complete annotated code for our construction. The code also introduces some special variables used to store important node numbers corresponding to the eight corner nodes on the boundary.

```
%Mesh Generation
A =36-pi*(4+1); %area of region
```

```
delta = sqrt(A/2500); count = 1;


%place interior nodes first
for i=1:ceil(6/delta), for j=1:ceil(6/delta)
xt=i*delta; yt=j*delta; xy=[xt yt];
if norm(xy,2)>2+delta/2 & norm(xy-[6 0],2)>2+delta/2 & ...
   norm(xy-[6 6],2)>2+delta/2 &norm(xy-[0 6],2)>2+delta/2 ...
   & norm(xy-[3 3],2)>1+delta/2 & xt<6-delta/2 & yt<6-delta/2
x(count)=xt; y(count)=yt; count=count+1;
end, end, end
nint=count-1; %number of interior nodes

%now deploy boundary nodes; we will group them according to their
%boundary conditions; as usual, the Robin nodes precede the Dirichlet
%nodes.  At the corners there is some ambiguity since the normal
%vector is undefined.  We make some conventions that Robin
%conditions take precedence over Neumann conditions, and for Neumann
%conditions at an interface, we simply average the values of the
%normal derivative values.

%Helpful Auxilliary Vectors:
v1=linspace(2,4,2/delta); lenv1=length(v1);
thetaout=linspace(0,pi/2,pi/delta); %node angular gaps for big
quarter circles
lenthout=length(thetaout);
thetain=linspace(0,2*pi,2*pi/delta); %node angular gaps for smaller
interior cirlce
lenthin=length(thetain);

%Neumann conditions with zero boundary values:
for i=2:lenv1 %east
x(count)=6; y(count)=v1(i); count=count+1;
end
for i=2:lenthout %northeast
x(count)=6+2*cos(-pi/2-thetaout(i)); y(count)=6+2*sin(-pi/2-...
   thetaout(i)); count=count+1;
end
toprightindex=count-1

for i=2:lenv1 %north
x(count)=6-v1(i); y(count)=6; count=count+1;
end
topleftindex=count-1

for i=2:lenthout %northwest
x(count)=2*cos(-thetaout(i)); y(count)=6+2*sin(-thetaout(i));
count=count+1;
end

for i=2:lenv1-1 %west
x(count)=0; y(count)=6-v1(i); count=count+1;
end
lastwestind=count-1
firstsouthind=count

for i=2:lenv1-1 %south
x(count)=v1(i); y(count)=0; count=count+1;
end
```

```
lastsouthind=count-1

%Now we move on to the two Robin portions
firstswind=count
for i=1:lenthout %southwest
x(count)=2*cos(thetaout(i)); y(count)=2*sin(thetaout(i));
count=count+1;
end
lastswind=count-1
firstseind=count

for i=1:lenthout %southeast
x(count)=6+2*cos(pi/2+thetaout(i)); y(count)=2*sin(pi/2+thetaout(i));
count=count+1;
end
n=count-1 %number of interior and Robin nodes
lastseind=n;

% finally put in the Dirichlet nodes
for i=1:lenthin
x(count)=3+cos(thetain(i)); y(count)=3+sin(thetain(i));
count=count+1;
end
m=count-1    %number of nodes

%ASIDE:  Enter these commands to plot the nodes
%plot(x(1:nint),y(1:nint),'b.'), axis('equal')
%hold on
%plot(x(nint:m),y(nint:m),'rp'), axis('equal')

%Since the domain is not convex (in 5 spots) we will use the
%technique of Example 13.3 of introducing 5 ghost nodes that will
%yield a triangulation from which it will be easier to delete the
%unwanted triangles

x(m+1)=3; y(m+1)=3;
x(m+2)=5; y(m+2)=1;
x(m+3)=5; y(m+3)=5;
x(m+4)=1; y(m+4)=5;
x(m+5)=1; y(m+5)=1;
tri=delaunay(x,y);
trimesh(tri,x,y,'LineWidth', 1.2), axis('equal')
%Plots the triangulation
axis('equal')
%Now we need to delete all elements which have a node with index in
%the range m+1 to m+5.
size(tri)  %ans =5224  3,  so there are 5224 elements
badelcount=1;
for ell=1:5224
if sum(ismember(m+1:m+5, tri(ell,:)))>0
badel(badelcount)=ell;
badelcount=badelcount+1;
end
end

tri=tri(setdiff(1:5224,badel),:);
x=x(1:m); y=y(1:m);
trimesh(tri,x,y), axis('equal')
```

To facilitate writing the assembly code, we store the following M-files for the functions *r* and *h*:

```function z=h_EFR13_16(x,y)```
```%Inhomogeneity function for```
```Neumann/Robin BC of %EFR13.16.```
```if y>2+eps & y<6-eps```
`    z=0;`
```elseif y>=6-eps, z=20;```
```elseif y<eps, z=-20;```
```elseif (y>=eps & y<=2+eps)|[x y]==[2```
```0]|[x y]==[4 0]```
`    z=30;`
`end`
```if y==0 & (x==2|x==4), z=5; end```
```if y==2, z=15; end```
```if y==6 % (x==2|x==4), z=5; end```
	```function r=r_EFR13_16(x,y)```
	```%u-coefficient function```
	```for %Neumann/Robin BC of```
	```%EFR13.16.```
	```if (y>=0&y<=2) & x<=2```
	`    r=1;`
	```elseif (y>=0&y<=2) & x>=4```
	`    r=2;`
	`else`
	`    r=0;`
	`end`

The assembly code is long, but it can be done by combining elements of the others we have developed so far. For space considerations we will refer the complete assembly code to the FTP site for the book (see beginning of this appendix for the URL.)

References

[Abb-66] Abbott, Michael B., *An Introduction to the Method of Characteristics*, American Elsevier, New York (1966)

[Aga-00] Agarwal, Ravi P., *Difference equations and inequalities. Theory, methods, and applications*, Second edition, Marcel Dekker, Inc., New York, (2000)

[Ahl-79] Ahlfors, Lars Valerian, *Complex Analysis, Third Edition*, McGraw-Hill, New York (1979)

[Ame-77] Ames, William F., *Numerical Methods for Partial Differential Equations*, Barnes and Noble, New York (1977)

[Ant-00] Anton, Howard, *Elementary Linear Algebra*, Eighth Edition, John Wiley & Sons, New York (2000)

[Apo-74] Apostol, Thomas. M., *Mathematical Analysis: A Modern Approach to Advanced Calculus (2nd Edition)*, Addison-Wesley, Reading, MA (1974)

[Arn-78], Arnold, Vladimir. I., *Ordinary Differential Equations*, MIT Press, Cambridge, MA (1978)

[Asm-00], Asmar, Nakhlé, *Partial Differential Equations and Boundary Value Problems*, Prentice-Hall, Upper Saddle River, NJ (2000)

[Atk-89] Atkinson, Kendall E., *An Introduction to Numerical Analysis, Second Edition*, John Wiley & Sons, New York (1989).

[AxBa-84] Axelsson, Owe, and Vincent A. Barker, *Finite Element Solution of Boundary Value Problems*, Academic Press, Orlando, FL (1984)

[Bar-93] Barnsley, Michael F., *Fractals Everywhere. Second Edition*, Academic Press, Boston, MA (1993)

[BaZiBy-02] Barnett, Raymond A., Michael R. Ziegler, and Karl E. Byleen, *Finite Mathematics, For Business, Economics, Life Sciences and Social Sciences, Ninth Edition*, Prentice-Hall, Upper Saddle River, NJ (2002)

[Bec-71] Beckmann, Petr, *A History of π, Second Edition*, The Golem Press, Boulder, CO (1971)

[BeEp-92] Bern, Marshall and David Eppstein, *Mesh Generation and Optimal Triangulation*, In F.K. Hwang and D.-Z. Du, editors, *Computing in Euclidean Geometry*. World Scientific Publishing, River Edge, NJ (1992)

[Br-93], Braun, Martin, *Differential Equations and Their Applications*, Springer-Verlag, New York (1993)

[BrCh-93] Brown, James W., and Ruel V. Churchill, *Fourier Series and Boundary Value Problems, Fifth Edition*, McGraw-Hill Inc., New York (1993)

[BrChSi-03] Brown, James W., Ruel V. Churchill, and H. Jay Siskin, *Complex Variables and Applications, Seventh Edition*, McGraw-Hill Inc., New York (2003)

799

[BuFa-01] Burden, Richard, L., and J. Douglas Faires, *Numerical Analysis, Seventh Edition*, Brooks/Cole, Pacific Grove, CA (2001)

[But-87] Butcher, John C., *The Numerical Analysis of Ordinary Differential Equations: Runge-Kutta and General Linear Methods*, John Wiley & Sons, New York, 1987.

[Cia-02] Ciarlet, Phillipe G., *The Finite Element Method for Elliptic Problems*, Soc. of Industrial and Applied Math.(SIAM), Philadelphia, PA (2002)

[CiLi-89] Ciarlet, Phillipe G. and Jacques Louis Lions, *Handbook of Numerical Analysis, Volume II: Finite Element Methods (Part 1)*, North Holland, Amsterdam (1989)

[Col-42] Collatz, Lothar, *Fehlerabschätzung für das Iterationsverfahren zur Auflösung linearer Gleichungssysteme* (German), Zeitschrift für Angewandte Mathematik und Mechanik. Ingenieurwissenschaftliche Forschungsarbeiten, vol. **22**, pp. 357-361 (1942)

[Con-72] Conway, John H., *Unpredictable iterations*, Proceedings of the 1972 Number Theory Conference, University of Colorado, Boulder, Colorado, pp. 49-52, (1972)

[Cou-43] Courant, Richard, *Variational methods for the solution of problems of equilibrium and vibrations*, Bull. Amer. Math. Soc., vol **49**, pp. 1-23 (1943)

[Cow-73] Cowper, E. R., *Gaussian quadrature formulae for triangles*, International Journal of Numerical Methods in Engineering, vol **3**, 405-408 (1973)

[CrNi-47] Crank, John, and Phyllis Nicolson, *A practical method for the numerical evaluation of solutions of partial differential equations of the heat conduction type*, Proceedings of the Cambridge Philosophical Society, vol. **43**, pp. 50-67 (1947)

[Del-34] Delaunay, Boris, *Sur la sphere vide*, Izv. Akad. Nauk SSSR, Otdelenie Matematicheskii i Estestvennyka Nauk, vol. **7**, pp. 793-800 (1934)

[DuCZa-89] DuChateau, Paul, and David Zachmann, *Applied Partial Differential Equations*, Harper & Row, New York (1989)

[Dur-99] Duran, Dale, R., *Numerical Methods for Wave Equations in Geophysical Fluid Dynamics*, Springer-Verlag, New York (1999)

[Ede-01] Edelsbrunner, Herbert, *Geometry and Topology for Mesh Generation*, Cambridge University Press, Cambridge, UK (2001)

[EdK-87] Edelstein-Keshet, Leah. *Mathematical Models in Biology*, McGraw-Hill. New York (1987).

[Ela-99] Elaydi, Saber N., *An Introduction to Difference Equations, Second edition.* Springer-Verlag, New York (1999)

[Eng-69] England, Roland, *Error Estimates for Runge-Kutta type solutions to systems of ordinary differential equations*, Computer Journal, vol. **12**, pp. 166-170 (1969)

[Epp-02] Epperson, James F., *An Introduction to Numerical Methods and Analysis*, John Wiley & Sons, New York (2002)

[Fal-86] Falconner, Kenneth J., *The Geometry of Fractal Sets*, Cambridge University Press, Cambridge, UK (1986)

[Feh-70] Fehlberg, Erwin, *Klassische Runge-Kutta Formeln vierter und niedrigerer Ordnung mit Schrittweiten-Kontrolle und ihre Anwendung auf Wärmeleitungsprobleme*, Computing, vol. **6**, pp. 61-71, (1970)

[Gea-71] Gear, C. William, *Numerical Initial Value Problems in Ordinary Differential Equations*, Prentice-Hall, Englewood Cliffs, NJ, 1971

[GiTr-83] Gilbarg, David, and Neil S. Trudinger, *Elliptic Partial Differential Equations of Second Order*, Springer-Verlag, Berlin (1983)

[GoVL-83] Golub, Gene, H., and Charles F. Van Loan, *Matrix Computations*, The Johns Hopkins University Press, Baltimore (1983)

[GoWeWo-92] Gordon, Carolyn, David L. Webb, and Scott Wolpert, *One cannot hear the shape of a drum*, Bulletin of the American Mathematical Society *(N.S.)* **27** (1992), no. 1, pp. 134-138.

[Gre-97], Greenbaum, Anne, *Iterative Methods for Solving Linear Systems*, SIAM, Philadelphia, PA (1997)

[HaLi-00] Hanselman, Duane, and Bruce Littlefield, *Mastering MATLAB 6: A Comprehensive Tutorial and Reference*, Prentice Hall, Upper Saddle River, NJ (2001)

[Hea-00] Heathcote, Herbert, *The mathematics of infectious diseases*, SIAM Review 42 (2000), pp. 599-653

[HiHi-00] Higham, Desmond J., and Nicholas J. Higham, *MATLAB Guide*, SIAM, Philadelphia, PA (2000)

[HiSm-97], Hirsch, Morris and Stephen Smale, *Differential Equations, Dynamical Systems and Linear Algebra*, Academic Press, New York (1997)

[HoKu-71] Hoffman, Kenneth, and Ray Kunze, *Linear Algebra*, Prentice-Hall, Englewood Cliffs, NJ (1971)

[HuLiRo-01] Hunt, Brian R., Ronald L. Lipsman, and Jonathan M. Rosenberg, *A Guide to MATLAB: for Beginners and Experienced Users*, Cambridge University Press, Cambridge, UK (2001)

[Hur-90] Hurewicz, Witold, *Ordinary Differential Equations*, Dover Publications, New York (1990)

[IsKe-66] Isaacson, Eugene, and Keller, Herbert B., *Analysis of Numerical Methods*, John Wiley and Sons, New York (1966)

[John-82] John, Fritz, *Partial Differential Equations, Fourth Edition*, Springer-Verlag, New York (1982)

[Joh-87] Johnson, Claes, *Numerical Solutions of Partial Differential Equations by the Finite Element Method*, Cambridge University Press, Cambridge, UK (1987)

[Kac-66] Kac, Marc, *Can one hear the shape of a drum?*, American Mathematical Monthly, vol. **73**, pp. 1-23 (1966)

[Kah-66] Kahan, William M., *Numerical linear algebra*, Canadian Mathematical Bulletin, vol. **9**, pp. 757-801 (1966)

[Kel-68] Keller, Herbert B., *Numerical Methods for Two-Point Boundary Value Problems*, Blaisdell Publishing, Waltham, MA (1968)

[KeMcK-27] Kermack, W. O., and A. G. McKendrick, *A contribution to the mathematical theory of epidemics*, Proceedings of the Royal Society of London, Series A, vol. **115**(772), pp. 700-721 (1927)

[KaKr-58] Kantorovich, Leonid V., and Vladimir I. Krylov, *Approximate Methods of Higher Analysis*, P. Noordhoff Ltd., Amsterdam (1958).

[Kev-00] Kevorkian, Jerry, *Partial Differential Equations, Analytical Solution Techniques*, Springer-Verlag, New York (2000)

[Kol-99] Kolman, Bernard, and David R. Hill, *Elementary Linear Algebra, Seventh Edition*, Prentice-Hall, Upper Saddle River, NJ (1999)

[Kry-62] Krylov, Vladimir I., *Approximate Calculation of Integrals*, MacMillan, New York (1962)

[Lag-85] Lagarias, Jeffrey C., *The 3x+1 Problem and Its Generalizations*, American Mathematical Monthly vol. **92**, pp. 3-23, (1985)

[Lam-91] Lambert, John D., *Numerical Methods for Ordinary Differential Systems, The Initial Value Problem*, John Wiley & Sons, New York, 1991

[Lan-84] Langford, William F., *Numerical studies of torus bifurcations*, Internationale Schriftenreihe zur Numerischen Mathematik, vol. **70**, pp. 285-295 (1984)

[Lan-99] Langtangen, Hans P., *Computational Partial Differential Equations*, Springer Verlag, Berlin (1999)

[Lau-91] Lautwerier, Hans A., *Fractals: Endlessly Repeated Geometrical Figures*, Princeton University Press, Princeton, NJ (1991)

[Log-94] Logan, J. David, *An Introduction to Nonlinear Partial Differential Equations*, John Wiley & Sons, New York (1994)

[Mat-99] Matilla, Pertti, *Geometry of Sets and Measures in Euclidean Spaces. Fractals and Rectifiability*, Cambridge University Press, Cambridge, UK (1999)

[Maw-82] Mawhin, Jean, *Periodic oscillations of forced pendulum-like equations*, Lecture Notes in Mathematics No. 964, pp. 458-476, Springer-Verlag, New York (1982)

[Maw-97] Mawhin, Jean, *Seventy five years of global analysis around the forced pendulum equation*, Proceedings of the Equadiff Conference at Brno in 1997, pp. 115-145 (1997)

[Mor-85] Morley, Tom, *A simple proof that the world is three-dimensional*, SIAM review, vol. 27, no. 1, pp. 69-71 (1985)

[Mor-85b] Morley, Tom, Errata: *A simple proof that the world is three-dimensional*, SIAM review, vol. 28, no. 2, p. 229 (1986)

[Mur-03] Murray, James D., *Mathematical Biology, Volume I: An Introduction*, Springer-Verlag, New York (2003)

[Neu-98] Neumaier, Arnold, *Solving ill-conditioned and singular linear systems: a tutorial on regularization*, SIAM Review, vol. **40**(no. 3), pp. 626-666 (1998)

[OkBoSu-92] Okabe, Atsuyuki, Barry Boots, Kokichi Sugihara, and Sung-Nok Chiu, *Spatial Tessellation: Concepts and Applications of Voronoi Diagrams*, John Wiley & Sons, Chichester UK (1992)

[OeS-99] Oliveira e Silva, Tomás, *Maximum excursion and stopping time record-holders for the 3x+1 problem: computational results*, Math. Comput. vol. **68**, pp. 371-384, (1999).

[PSMI-98] Pärt-Enander, Eva, Anders Sjöberg, Bo Melin, and Pernilla Isaksson, *The MATLAB Handbook*, Addison-Wesley, Harlow UK (1998)

[PSJY-92] Peitgen, Heinz-Otto, Dietmar Saupe, H. Jurgens, and L. Yunker, *Chaos and Fractals: New Frontiers of Science*, Springer-Verlag, New York (1992)

[ReSh-97], Reichelt, Mark W. and Lawrence F. Shampine, *The MATLAB ODE suite*, SIAM Journal of Scientific Computing, vol. **18** no. 1, pp. 1-22 (1997)

[ReRo-92] Renardy, Michael, and Robert C. Rogers, *An Introduction to Partial Differential Equations*, Springer-Verlag, New York (1992)

[RiMo-67] Richtmyer, Robert D., Morton, K. W., *Difference Methods for Initial Value Problems*, Second Edition, John Wiley & Sons, New York (1967)

[Ros-00] Rosen, Kenneth H., *Handbook of Discrete and Combinatorial Mathematics*, CRC Press, Boca Raton, FL (2000)

[Ros-96] Ross, Kenneth A., *Elementary Analysis: The Theory of Calculus, Eighth Edition*, Springer-Verlag, New York (1996)

[Rud-64] Rudin, Walter. *Principles of Mathematical Analysis, Second Edition*, McGraw-Hill, New York (1964)

[Sch-95] Schreiber, Peter, *The Cauchy-Bunyakovsky-Schwarz inequality*, in Hermann Grassmann, (Lieschow, 1994) pp. 64-70, Ernst-Moritz-Arndt Univ., Greifswald, Germany (1995)

[Sch-73] Schultz, Martin. H., *Spline Analysis*, Prentice-Hall, Englewood Cliffs, NJ (1973)

[ShAlPr-97] Shampine, Lawrence F. , Richard Allen and Steve Pruess, *Fundamentals of Numerical Computing*, John Wiley & Sons, New York, (1997)

[Smi-85] Smith, Gordon D., *Numerical Solution of Partial Differential Equations*, Oxford University Press, New York (1985)

[Smo-83] Smoller, Joel, *Shock Waves and Reaction-Diffusion Equations*, Springer-Verlag, New York (1983)

[Sni-99] Snider, Arthur David, *Partial Differential Equations, Sources and Solutions*, Prentice-Hall, Upper Saddle River, NJ (1999)

[StBu-92] Stoer, Josef, and Roland Bulirsch, *Introduction to Numerical Analysis*, Springer-Verlag: TAM Series #12, New York (1992)

[Sta-79] Stakgold, Ivar, Green's Function and Boundary Value Problems, John Wiley & Sons, New York (1979)

[Str-88] Strang, Gilbert, *Linear Algebra and Its Applications, Third Edition*, Prentice-Hall, Englewood Cliffs, NJ (1988)

[StFi-73] Strang, Gilbert., and George J. Fix, *An Analysis of the Finite Element Method*, Prentice-Hall, Englewood Cliffs, NJ (1973)

[Str-92] Strauss, Walter A., *Partial Differential Equations, An Introduction*, John Wiley & Sons, New York (1992)

[StVa-78] Strauss, Walter. A. and Luis Vazquez *Numerical solution of a nonlinear Klein-Gordon equation*, Journal of Computational Physics, vol. 28, pp. 271-278 (1978)

[SuDr-95] Su, Peter and Robert L. Drysdale, *A comparison of sequential Delaunay triangulation algorithms*, in Proceeding of the ACM 11th Annual Symposium on Computational Geometry, pp. 61-70, ACM, Vancouver, CANADA (1995)

[Tho-95a] Thomas, James W., *Numerical Partial Differential Equations, Finite Difference Methods*, Springer-Verlag, New York (1995)

[Tho-95b] Thomas, James W., *Numerical Partial Differential Equations, Conservation Laws and Elliptic Equations*, Springer-Verlag, New York (1995)

[TrBa-97] Trefethen, Lloyd N. and David Bau, III, *Numerical Linear Algebra*, SIAM, Philadelphia, PA (1997)

[Vor-08] Voronoi, Georges, *Nouvelles applications des paramèmetres continues à la théorie des formes quadratiques*, J. Reine Angew. Math., vol. 133, pp. 97-178 (1907) and vol. 134, pp. 198-287 (1908)

[Wei-65] Weinberger, Hans F., *A First Course in Partial Differential Equations*, John Wiley & Sons, New York (1965)

[Wil-88] Wilkinson, James H., *The Algebraic Eigenvalue Problem*, Clarendon Press, Oxford, UK (1988)

[You-71], Young, David M., *Iterative Solution of Large Linear Systems*, Academic Press, New York, (1997)

[Zau-89] Zauderer, Erich, *Partial Differential Equations of Applied Mathematics*, Second Edition, John Wiley & Sons, New York (1989)

[ZiMo-83] Zienkiewicz, Olgierd C., and Kenneth Morgan, *Finite Elements and Approximation*, John Wiley & Sons, New York (1983)

MATLAB Command Index

FAMILIAR MATHEMATICAL FUNCTIONS OF ONE VARIABLE

Algebraic	$\texttt{sqrt(x)}$ $(= \sqrt{x}$), $\texttt{abs(x)}$ $(= |x|)$
Exponential/ Logarithmic	$\texttt{exp(x)}$ $(= e^x$), $\texttt{log(x)}$ $(= \ln(x)$), $\texttt{log10}$
Trigonometric	sin, cos, tan, sec, etc, asin, etc, asin, etc, sinh, cosh, etc., asinh, etc.,

MATLAB COMMANDS AND M-FILES: NOTE: Textbook-constructed M-files are indicated with an asterisk after page reference. Optional input variables are underlined. Most of the auxiliary M-files representing specific functional data needed to solve examples in Part III have been omitted.

SYMBOLIC TOOLBOX COMMANDS:

General Index

PURE AND APPLIED MATHEMATICS

A Wiley-Interscience Series of Texts, Monographs, and Tracts

Founded by RICHARD COURANT

Editors Emeriti: MYRON B. ALLEN III, DAVID A. COX, PETER HILTON, HARRY HOCHSTADT, PETER LAX, JOHN TOLAND

*Now available in a lower priced paperback edition in the Wiley Classics Library
†Now available in paperback.

*Now available in a lower priced paperback edition in the Wiley Classics Library.
†Now available in paperback.

Printed in the USA/Agawam, MA
March 14, 2018

671506.002